THE

MECHANICS' MAGAZINE,

MUSEUM,

Register, Journal,

AND

GAZETTE,

OCTOBER 6th, 1838—MARCH 30th, 1839.

VOL. XXX.

" Our needful knowledge, like our needful food,
Unhedged, lies open in life's common field,
And bids all welcome to the vital feast."

YOUNG.

LONDON:

PUBLISHED FOR THE PROPRIETOR BY W. A. ROBERTSON,
MECHANICS' MAGAZINE OFFICE, PETERBOROUGH-COURT.

1839.

PRINTED BY W. A. ROBERTSON, 6, PETERBOROUGH-COURT, FLEET STREET.

INDEX

TO THE THIRTIETH VOLUME.

NEW PATENTS.

ENGLISH.

SCOTCH.

IRISH.

Mechanics' Magazine,
MUSEUM, REGISTER, JOURNAL, AND GAZETTE.

No. 791.]   SATURDAY, OCTOBER 6, 1838.   [Price 3*d*.

COCKS'S MACHINE FOR MAKING LONG AND CIRCULAR BLOCKS OF INDIA RUBBER.

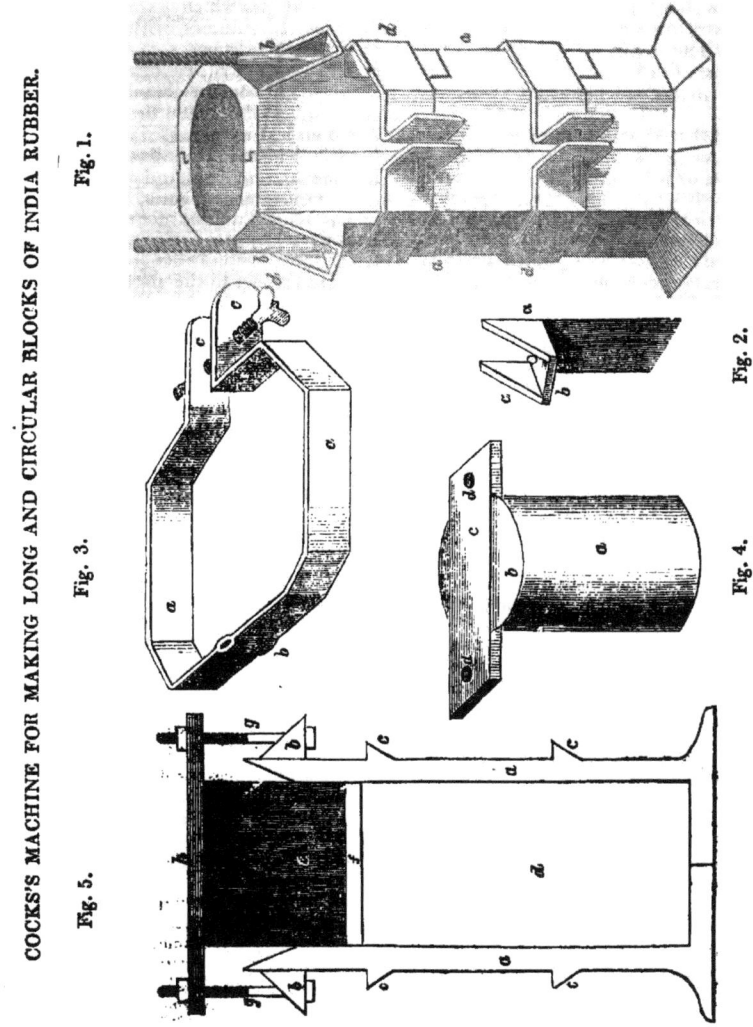

Fig. 1.

Fig. 2.

Fig. 3.

Fig. 4.

Fig. 5.

COCKS'S MACHINE FOR MAKING LONG AND CIRCULAR BLOCKS OF INDIA RUBBER.

Sir,—I take the liberty to transmit for your publication a sketch of an iron box for making *long,* and *circular blocks of india rubber.* The rubber, when prepared, is pressed into the box, and kept there until porperly hardened. It is then removed and cut into circular slices three quarters of an inch thick by a common saw, without teeth; a sharp edge is given by means of a file; these pieces are then cut into tapes and finally into threads.

Since the completion of this box, I am told that it is an infringement of a patent granted to a Mr. Keen; if this be the case, or if a patent has been granted to any other person for the same invention, or for a similar purpose, the fact was altogether unknown to me when I designed the before-described machine, which is certainly original as far as I am concerned. If I have been preceded by Mr. Keen or any one else, I shall be obliged by being so informed in your valuable Journal.

I am, Sir, your obedient servant,
W. P. COCKS.

Tottenham, Sept. 6, 1838.

Fig. 1. The box, composed of two halves, *a a. b b,* ears on each side through the bottom of which bolts are put to secure the iron cross piece, fig. 5, *h.   c,* interior of the box; *d d,* iron hoops to keep the halves firmly together.

Fig. 2, *a a,* sides of the ear; *b,* bottom of the ear with a bolt hole ½ an inch in diameter.

Fig. 3, *a a,* iron hoops 1½ inch broad, ½ inch thick : *b,* an hinge joint; *c c,* ears; *d,* a thumb screw passing through the ears, to lighten the box.

Fig. 4, a circular piece of hard wood made accurately to the bore of the box, *a,* with a stop *b* on its top for the cross bar of iron 2½ inches broad, and ¾ of an inch thick ; *c,* the holes in the bar *d* correspond with the holes in the ears. When properly pressed the top is secured by means of the screws and nuts. Vide Fig. 5.

Fig. 5; *a a,* section of the box ; *b b,* ears; *c c,* knobs ¾ of an inch thick, to prevent the slipping down of the hoops ; *d,* space occupied by the rubber ; *e,* the circular wooden top ; *f,* the bottom of the plunger shod with copper ; *g g,* screws ; *h,* bar of iron for securing the top.

ON FIRE-ESCAPES—REPLY TO MR. J. D. PARRY.

Sir,—With every disposition to give Mr. J. D. Parry's description of "an excellent fire-escape," kind and judicious consideration, I must be permitted to observe, that his ideas upon the subject generally, are somewhat crude, and that he might peruse the numerous articles on *fire-escapes* which have appeared in your last 20 volumes, with manifest advantage.

The parish of St. Pancras has certainly attained most unenviable distinction for the number of its fatal fires. One life has been lost in Mr. Parry's immediate neighbourhood in the interim of his writing the communication dated Sept. 4th, and its appearance in your 787th Number, and it is only a few months since two lives were lost at a fire in the same district.

With respect to the particular invention alluded to by Mr. Parry, I beg to inform him that these escapes were made (and I believe invented) by a rope and twine merchant, living, up to within a recent period, in King-street, Snow-hill; who had a print for many years exhibited in his window, showing the application of the *canvass trough escape.*

Although this escape might have passed for "an excellent one" 20 years ago, it is at this time superseded by many better contrivances, most of which, have at different times been described in your pages. The canvass trough is somewhat expensive, it is bulky and cumbersome; it also demands considerable dexterity and presence of mind in the parties employing it, and after all, its real efficiency depends very much upon external assistance—for Mr. Parry's plan of a steadying-weight is altogether impracticable. This escape is wholly inapplicable to all houses having deep shops run out over what was originally a garden plot in front, or such as are surrounded by iron spiked railings—objections which are equally fatal to the new mode of affixing the canvass trough beneath a ladder.

As an *internal* or domestic fire-escape, the most conveniently deposited, and most simple and efficient in use, is the plain belt and rope running through a pulley, the hook of which is attached to an eye-bolt previously provided, in a few seconds. A communication is thus instantly formed, the parties in danger may descend, or assistants may ascend with the greatest ease and rapidity. If apparatus for this purpose is not simple in its construction, and easily comprehended by the party in danger, as well as by those whom chance may bring to their assistance, it is very likely to increase the danger by increasing the confusion of ideas. The simplicity of this apparatus is its chief characteristic, and it has been pronounced by all who have used it, the best extant. As an *external* escape the same apparatus affixed to the portable ladder-top, surpasses all other inventions in the several points which really constitute "an excellent fire-escape." As escapes of this description (by Merryweather) are rather numerous in Mr. Parry's neighbourhood, I hope the interest he feels in this subject will induce him to take an early opportunity of inspecting them, and then judge for himself between this, and any other inventions for the same excellent purpose.

The fishing-rod elevator, after being tried in 20 or 30 different ways, by as many would-be inventors, seems at length to be entirely abandoned, having given way to more effectual methods of communication not depending upon the efforts of the persons in jeopardy. Mr. Parry is very much out in his estimate of the weight of a rod or pole capable of bearing a person descending; there are two or three of Ford's spar-fire-escapes in the south-western district of St. Pancras, capable of reaching from 30 to 34 feet high, an inspection of which would enlighten Mr. Parry considerably on this branch of the subject.

It is a very frequent, though a very absurd practice, to attempt a sort of comparison between deaths by burning and those by drowning; the fact is, the two cases present no parallel, and do not admit of comparison. Man can, if he pleases, become familiar with water, he may by practice attain the art of floating, swimming, diving, and also habituate himself to remain for a considerable period of time submerged beneath the surface. Again, should animation be suspended by accidental immersion, there is every chance, within certain limits, by prop r treatment, of restoring the vital action. But with the other element, fire, it is impossible to be on similar terms; it is not possible to become familarized to the effects of fire—nor can the suffocat-

ing influence of its pernicious attendant, smoke, be supported many minutes. The contact of the dreadful element itself rapidly destroys the organization of the human frame, and this effect carried but to a comparatively short extent, death is perhaps lingering, but inevitable. and the torture excessive. No course of treatment, however skilful or judicious, can subsequently restore the vital spark.

A writer in the *Morning Herald* a few months since, describing the last anniversary dinner of the Royal Humane Society, took occasion to observe that " so far as the preservation of that precious gift of the Creator, 'human life,' can be rescued from destruction by *water*, we have made some progress; but, in the means of escaping from the opposite devouring element, *fire*, we seem not at all to keep pace with the advancement of the age in general knowledge. We have, it is true, many machines called ' fire-escapes,' as we suppose *par excellence*, but we have no list of persons rescued by them to lay before our readers."

Now it happened, singularly enough, that on the very day Grandmamma put forth this cynical lament, a meeting was held at the Freemason's Hall, Great Queen street, Lincolns-inn-fields, Lord Teignmouth in the chair, at which upwards of 20 medals were given to policemen, firemen and others, who in the most gallant manner, at great personal risk, had saved the lives of several persons during the present year. In several instances their preservation had been accomplished by means of the portable fire-ladders (including one whole family in Goodman's Fields, rescued by Loader* of the fire-brigade), "the sway and spring of which," says Grandmamma, "are so great, that there can be no safety for human beings attempting to escape by them"! As these proceedings, if reported, would have entirely confuted the old lady's logic of the preceding day, she took especial good care to *burke* the account of the meeting, so that she might continue to exclaim, " we have no list of persons rescued to lay before our readers!"

The subject itself is fraught with diffi-

---

* Loader received two medals at this meeting, for his successful and heroic conduct upon two occasions, in preserving lives from fire. This young man is a credit to his country, and one of the brightest ornaments in the intrepid corps to which he belongs—may he have many equals!

culties, and sufficiently discouraging without such pitiful taints as the foregoing, being periodically put forward by the public press. Do what you will, make fire-escapes the most perfect things in existence, be the police as numerous and as watchful as possible, still in spite of all provisions, loss of life by fire will occur. It will sometimes happen, as it has more than once of late, that life will be extinct before discovery of the fire is made, what ground is there for reproach in such cases,—what power but that of the Omnipotent could avert the calamity?

The Royal Humane Society have hitherto most pertinaciously resisted all applications for extending their efforts to counteract the baneful effects of fire, and from the numberless instances in which they must necessarily have failed, and thereby, to a certain extent, have incurred the odium of the ignorant, they have wisely, if not humanely, declined interfering.

Notwithstanding these disheartening circumstances no expedient should be left untried that seems likely to abridge the number of these tragedies; the attention of the ingenious should still be directed to improve the mechanism, and the best energies of the humane, to ensure the application of suitable escapes. It is gratifying to know that great improvements in these matters have, within the last few years, obtained throughout the metropolis; and we do not often now read such horrifying tales of assistance implored in vain, as the daily press of seven years since presented too often to our gaze.

I remain, Sir, yours respectfully,
WM. BADDELEY.

London, Sept. 27, 1838.

FIRE ENGINEERING—RESCUE SERVICE

Sir,—Your indefatigable and valuable correspondent, Mr. Baddeley, in his unceasing efforts to perfect a system for extinguishing fires, is evidently actuated by the ardent desire of conferring a benefit on the public; but however sincere and disinterested his views may be, and however extensive his experience and knowledge, he is not infallible; and, as at the present time several honorary fire-brigades are either on the point of being formed, or but yet in their infancy, I take the liberty of soliciting the insertion of a few remarks on the rules laid down by that gentleman in No. 786 of your Magazine.

The cause of the errors into which Mr. Baddeley has fallen, appears to be either his not knowing the regulations observed in properly organised corps, or the inaccuracy of the description given by Mr. Ober of the continental "bags." It is extremely easy for any one in the constant habit of addressing the public to lay down principles and rules, which may, to the generality of readers, appear extremely plausible, and " dictated by sound sense and past experience ;" but those who have made fire engineering either their study or profession, are well aware that these theories do not at all times accord with practical exigencies. Before I make any direct observation on the contents of the letter in question, allow me to offer a short *exposé* of the principles which corps of this description on the *Continent* observe, as respects the rescue of persons and property from the flames.

1st. A number of posts are established in the town, according to its size and locality; besides which, small detachments are stationed at proper intervals and positions. The men on guard are always ready; regularly exercised and instructed; well officered, and in perfect discipline.

2nd. The first post, or detachment, that arrives at a fire (which arrival ought to take place in five minutes at farthest after the alarm) will reconnoitre the position of the fire, and ascertain the contents and principal localities of the building. The next step must entirely depend on the circumstances of the case; if there be any life in danger, the rescue of the inmates should supersede every pecuniary consideration; if not, the air should be excluded as much as possible by shutting the doors, windows, &c. if possible, and water directed so as to preserve the partitions and other communications. Upon the arrival of other detachments or posts, the service is divided into three divisions, each under the superintendence of an experienced officer. 1st division, rescue of property, &c.; 2nd division, attack of the fire; 3rd division, supply of water and other necessaries,—the whole under the command of a superior officer.

3rd. The rescue division observes the following orders (varying them according to circumstances as ordered by the commander of that division): The preservation of human life to be considered as the most important object; papers, jewels, plate, pictures, works of art, &c., according to localities. If the fire appears likely to spread and become general throughout the house, the furniture and remaining contents of the rooms most threatened, to be removed at once, and if their size prevent their removal, they should be broken up and thrown out rather than suffered to remain to increase the flames; this step must be determined by the superior officer. As regards the "bags" mentioned by Mr. Ober, and which are the means by which valuables, glass, furniture, &c. are rescued without danger of injury, and which are at the same time one of the most simple and efficacious fire-escapes yet known, the accompanying sketches will perhaps enable those who are interested in fires of adding something to their already extensive knowledge of these matters.

Fig. 1.

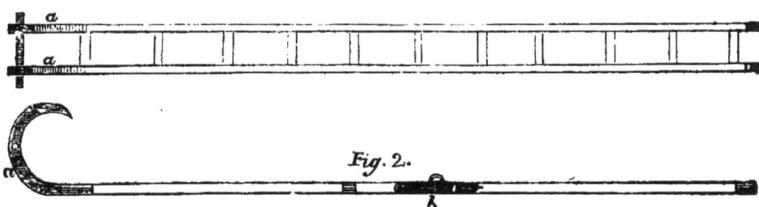

Figs. 1 and 2 represent a ladder, the upper extremities of which are curved and strengthened by iron straps *a a*, terminating in solid iron points; in the middle of the ladder are joints *b*, which enable it to be folded up for convenience of transport (when opened two bolts retain it in its proper position). The length of the ladder when open is about 8 feet; it is light and portable, and costs in Paris about forty francs, (1*l*.12*s*). The manner of its application is as follows: One man places his back against the wall of the house under the window of the first floor, and taking the ladder by the lower extremity hooks it on to the sill of the said window, another man ascends and enters *into* the room, the other follows him and stands on the sill of the window (being held by the belt by the man inside), he then lifts the ladder and hooks it on the sill of the next upper window; this is repeated until the required story is attained; they then throw down the end of a small line to which the persons below fasten the end of the *sac de sauvetage*, or "bag." Fig. 3 represents this latter apparatus, which consists of a long bag of canvass fastened

Fig. 3.

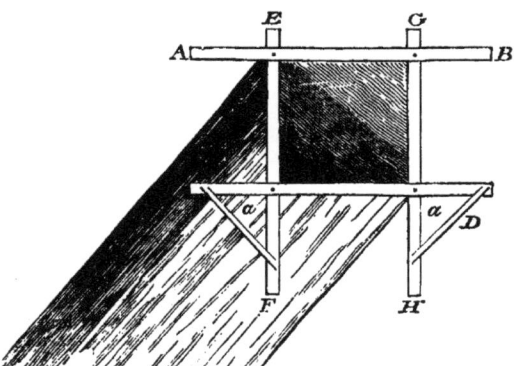

to folding frame A B C D; when drawn

up and opened inside the window, the cross bars A B and C D, prevent it from falling out, and the upright ones E F and G H, rest on the floor and keep it in a proper position; small bars or bolts *a a*, retain it square.   When thus fixed, the lower end of the long " bag" is held out so as to produce a proper angle of declivity, and any person or thing placed in the mouth proceeds at an easy rate to the bottom, without the least danger or inconvenience. It may also be proper to remark, that the usual length of these *sacs de sauvetage* is, in Paris, about 50 feet; and that when it is necessary to increase the angle of declivity beyond the limits of the length, a cord is attached to the end, by which it may held out in the proper position.   The price of this fire-escape, which, perhaps, has been the means of saving more lives than any other extant, is not five pounds.

Mr. Baddeley, in adverting to the course pursued on the continent, is entirely in error, when he says (in speaking of the rescue of property) that " the consequences of this mode of proceeding have proved most disastrous," and that " many thousand pounds worth of property have been consumed that would most assuredly have been preserved had all the assistance present been concentrated upon extinguishing the flames." I beg to refer him to the fires of the Theatre Italia; to that of the Palace of the Emperor of Russia, and to the one of the Royal Exchange ; these are but a few of the many that might be instanced to prove his assertion unfounded.

Mr. Baddeley also assures us, that " wherever a skilful fire police is formed, their attention is now exclusively directed to the suppression of the fire, and all the energies of the men being directed to this one point, their efforts are nine times out of ten crowned with extraordinary success." The fire police of Paris is certainly as skilful in saving people from the flames and extinguishing the latter, as either of those very skilful ones mentioned by Mr. B. If we take into consideration the immense size of the houses; their construction; the difficulty of excluding the air, &c., we shall find the parellel between them and those of London, Manchester or Edinburgh, much in favour of the first. I will not, however, be quite so bold as to declare, that nine times out of ten they succeed, nor can I think, that unless it be the object

of the London brigade to let the houses and people burn, that this very extraordinary success is always the result of their skilful efforts; I am the more disposed to think this from having seen in your valuable Magazine, an account given, I suppose by a namesake of Mr. Baddeley, in which the ratio of the extraordinary success before mentioned was very different.

As to the fire at Southampton, I should rather doubt that the exertions and time *wasted* in saving property would have extinguished the flames, and the inference drawn by Mr. Baddeley is therefore incorrect on that ground.

In a word, although the remarks of that gentleman may be dictated by sound sense, I very much question their being founded on the same, and think that upon referring to " past experience," their correctness would be rather *disproved* than otherwise by the reference.

I remain, Sir,

Your obedient servant,

FELIX M. SIMEON.

Unity-street, Bristol, Sept. 8, 1838.

ANCIENT FIRE-ENGINE.

Sir,—I send you a rough sketch of a fire-engine which was undoubtedly the first machine of this kind ever used. I observed it some time ago spoken of in a little pamphlet, containing a " Life of Sir Samuel Morland," some extracts from which were printed in your Magazine.   The drawing is taken out of an old book on surveying, by a person named Cyprian Lucar, entitled " *A Treatise named* Lucarsolace, *devided into fovor bookes which in part are collected out of diverse authors, in diverse languages, and in part devised by Cyprian Lucar, gentleman: imprinted at London, by Richard Field for John Harrison, and are to be sold at his shop in Paule's Churchyard at the signe of the greyhound—A.D.* 1590." It is dedicated to William Roe, Esq , Alderman of the City of London, and brother-in-law of the author.   In the eighth chapter of the fourth book, after speaking of a method for making fresh water from sea water, Lucar says :—

" I may recite an infinite number of other examples which will prove the necessary use of water. and the diverse qualities of diverse waters, but to avoide tediousnesse I will of purpose passe them over in silence, and here, at the end of this chapter, set before your eyes a type of a "squirt" which hath beene devised to cast much water upon a burning house, wishing a like squirt and plenty of water to be alwaies in a readinesse where fire may do harme, for this kind of squirt may be made to holde an hoggeshed of water, or if you will, a greater quantity thereof, and may so be placed on his frame that with ease and a smal strength, it sahl be mounted, imbased or turned to any one side right against any fired marke, and made to squirt out the water upon the fire that is to be quenched."

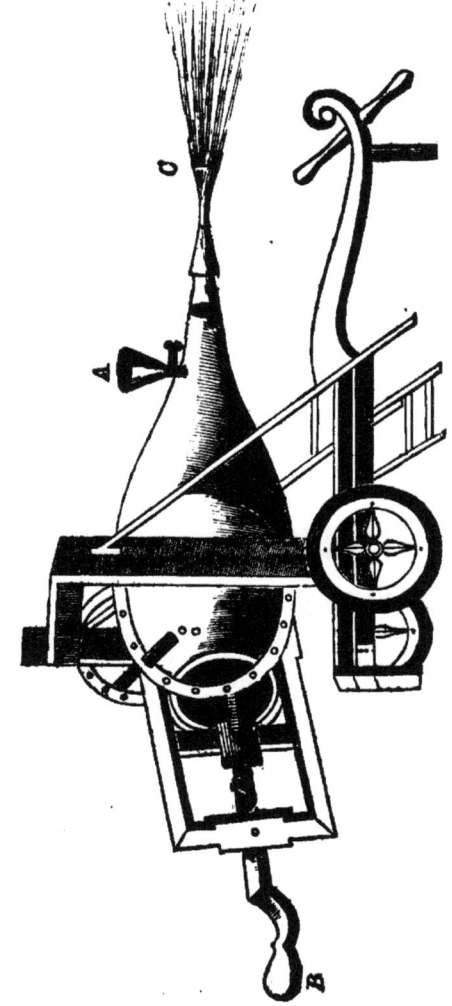

ANCIENT FIRE ENGINE.

The author does not give an account of the operations of the different parts of the machine, it appears, however, (as mentioned in the above pamphlet) to consist of a kind of hollow cone, moveable on a wooden frame, and open at the vertex C, into which might be inserted a long pipe, or even a hose, to convey the water to any part of the building. The water is put into the machine through the funnel A, and is ejected by means of the piston B, at the other extremity. The principle is precisely similar to that of the squirt now in use.

The existence of this machine at so early a period is perhaps not generally known, and is curious, inasmuch as it does away with the claims of Sir Samuel Morland and others to the invention of the "fire-engine," since the description and figure in the *Lucarsolace* was printed above 100 years prior to his time, and to that of Greatorix, whom Evelyn mentions as having invented one in the year 1656. It is not of so rude and imperfect a construction as might be expected, and when well managed, and built on a large scale, was no doubt tolerably effective in its operations. The inventor of it is not mentioned, which is a very singular omission on the part of the author, as he speaks of it without claiming it himself or attributing it to any other person.

I remain, Sir

Yours obediently,

J. C. W.

———

HANCOCK'S STEAM-CARRIAGE INVENTIONS.
—TOOTHED GEAR FOR LOCOMOTIVES.—
COLES' FRICTION-WHEEL CARRIAGES.—
WHITELAW'S BOILER FEEDING APPARATUS.

Sir,—Sir J. Anderson's steam-carriage boiler, described in No. 775 of your Journal, must be the invention of some one who is ignorant of the improvements that have taken place in steam-carriage boilers; the inventor of such a boiler, instead of advancing, is retrograding. If the intended company will not employ Walter Hancock's simple and ingenious boiler, they had better keep the money intended for shares in their pockets, as no other boiler yet invented, but Mr. Hancock's, will answer for locomotive purposes on common roads, and they cannot make any modified plan of his boilers, without infringing his patent right.

Walter Hancock's improvements in the locomotive steam-engine are ingenious, beautiful, and simple, and are nearly all applicable in those engines working on railroads; his boiler is more simple than the tubular boiler, and equally, if not more, efficient in generating steam in abundance, and rapidly, and is easier made, and kept in repair; locomotive steam-engines with his improvements, would be much less likely to need repairs so often, as the present complicated railway locomotives do; and by adopting his improvements, the expense of such engines would be much reduced, both in prime cost, and in the outlay for repairs.

I think as James Watt deserves the credit of being the great improver of the low-pressure steam-engine, and Trevethick of being the great improver of the high-pressure steam-engine, so does Walter Hancock deserve to rank with them, as the great improver of the locomotive steam-engine, and the high-pressure steam-boiler. In Luke Hebert's *Mechanics' Encyclopædia* your readers will find, in the article, "Railroads and Locomotive engines," a very interesting account of Walter Hancock's improvements in those engines. The editor does full justice to them, he praises the improvements, and particularly points out the originality and value of his boiler; and generously advocates his cause, as the great improver of the locomotive steam-engine.

In Nos. 784 and 786 of your Journal, I see some account of toothed wheel and pinion gear, proposed to be used in locomotive steam-engines, instead of the crank. In the years 1835—6, the locomotive steam-engines, (of excellent workmanship), on the Baltimore and Washington railroad, had vertical tubular boilers, and vertical cylinders; the wheels were low, as far as I can remember, not more than two feet in diameter; the motive power was communicated by toothed wheels and pinions, to the front wheels; the speed was not great.

Coles's patent railway carriages, described in No. 773 of your valuable Journal, is no improvement; I have little doubt but the axles of the friction wheels, and the rubbing of the wheels' edges against the other axles, will increase the amount of friction, and make it greater than the friction would be with fewer large axles; besides greater expense and complexity, and extra trouble required to keep them in order.

Mr. Whitelaw's plan for feeding steam-engine boilers, described in No. 769 of your Journal, I dare say would answer very well, but it is far from simple as described in your Magazine, and will add much to the cost of an engine; besides requiring additional attention from the engineer who attends the engine to which his plan is applied.

I remain, Sir,
Yours faithfully,
ARTHUR TREVELYAN.
Anglesey, Sept. 28, 1838.

———

APPARATUS FOR REGISTERING SPEED OF STEAM VESSELS.

Sir,—I take the liberty of sending you a plan for an apparatus for continuously registering the rate a vessel passes through the water, considering that its importance will claim a place in your columns. The principle is the regulating the escapement of a motive force, which may be acquired by increasing the tension of a spiral spring, similar to that of a common clock. The escapement of a common watch or clock is regulated by the constant action and reaction of the balance-wheel or pendulum. The escapement may be quickened, retarded, or entirely stoped, and the whole movement regulated accordingly. My plan is to effect this regulation by means of a rod which ascends and descends according to the speed of the vessel. This action is transmitted to the rod by the force of the current passing the vessel, or by the velocity of the vessel passing through the water, the results being the same. The method I propose for ascertaining the velocity of the vessel is by placing a metal blade or plate to oppose the current, and measuring the pressure against it by means of a spiral spring, placed so that the contraction shall measure the pressure. The faster the vessel passes through the water the stronger the current, and consequently, the more powerful the pressure the spring is contracted, and the rod descends. When the current, or the speed of the vessel decreases, the pressure also decreases, and the spring being relieved the rod ascends.

The following plan may enable the principle to be better understood. The regulation of the escapement may be otherwise managed than here shown, it is merely drawn to show the principle.

A, the side view of the metal blade or plate; B, the joint; C,C, the crank; D a chain fixed to the crank and the rod E; E, the rod; F, the spiral spring; G, a common balance-wheel, similar to that

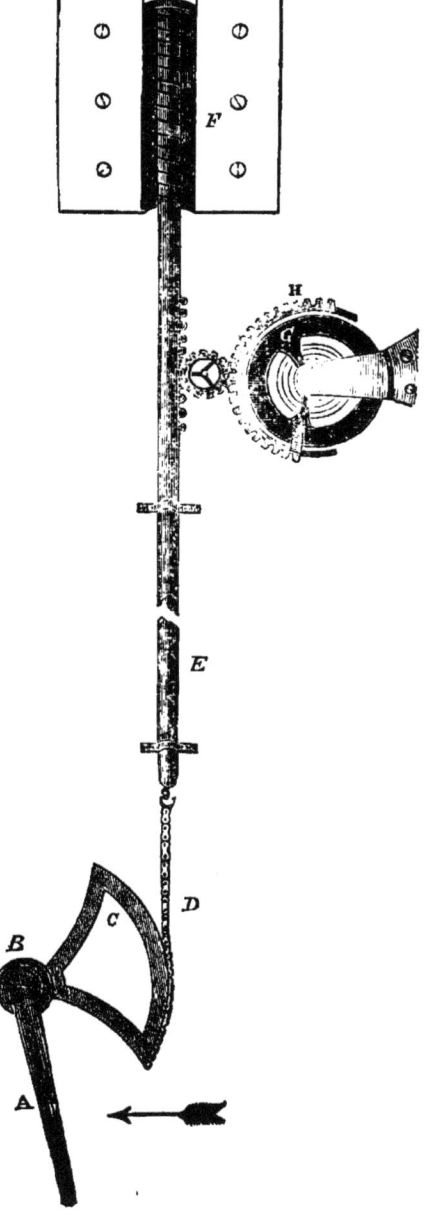

of a common watch with a pendulum spring; H, a segment of a wheel with a projecting leaf with a notch in it, I, to receive the pendulum spring, so that by the rising and falling of the rod the notch is moved, and so lengthens and shortens the pendulum spring, and regulates the escapement.

I am, Sir,
Your obedient servant,
A. J. MURRAY.

17, Camden Cottages, Camden New Town,
September 15, 1838.

INTELLECTUAL TASTES IN ARTISANS.

The labourers in the great cause of popular enlightenment have been steadily pursuing their course for some years. Their object is not indeed one to be suddenly gained, but of the seed sown it is reasonable to look for some fruits arising as time passes on. The following instances of intellectual tastes and pursuits in artisans will, it is hoped, be acceptable to those who rejoice to recognise in all, without regard to station in the world, the possession and the use of understanding. Two artisans in working trim, though on a holiday, were intently conversing together the other day, as they walked near London Bridge. A passer-by heard from one of them the words, "I always think of Julius Cæsar." Whether the reference was made to point the moral of the vanity of ambition;— whether Cæsar's bridge-building achievements were contrasted with those of the constructor of the grand pile over which the conversing parties had just passed; or whether the view of the Greenwich railroad suggested the idea of the Romans and their road-making, is immaterial to our purpose. The fact remains; in these our days, the mechanic conversing with his friend, is not confined in his topics to his shop, his home, or his ale-house, but thinks and speaks of Julius Cæsar.

On a certain green-painted door in the neighbourhood of Bethnal Green was lately to be seen, written roughly in chalk, the following problem:—"What sum at the same per cent, will amount to £78 8s. in two years? No guessing—show a theorem!" Virgil's first known couplet (and that by no means a very poetical one) is said to have been posted on the gate of Augustus. And here we have the enthusiasm of a young mathematician

delighted, as all young mathematicians are with their first introduction to arithmetical and algebrical problems. May he, like the poet, rise by degrees to higher themes! We liked the mathematical exactness of the question—"no guessing—show a theorem!" and should certainly have been tempted to answer the question on the spot but for want of chalk.*

A few years since the people of Hull determined to erect a column to the memory of their townsman, Wilberforce. Among the lookers-on who were one day witnessing the process of pile-driving for the foundations, there arose a discussion as to the utility of such erections. One observed, that they were at any rate good for trade, as affording employment. His neighbour suggested, not in the exact words, but to their effect, that the reproductive employment of capital was the best. He did not go on to discuss the utility of sinking capital thus as an incentive to public and social virtue: but it was a good deal to have advanced thus far. The political economy sounded not the worse for its coming, in homely phrase, from a man in a fustian jacket. We hope he may be as clear-headed and well informed, on the subject of wages and commercial intercourse, and his knowledge will be useful to his neighbours as well as improving to his own mind.

These instances of interest taken by labouring men in intellectual pursuits have occurred within a short time to one observer. How widely then is it reasonable to suppose these tastes are spread! Nor least is it pleasing to observe the simple and unostentatious manner in which they are manifested. Here at least there is nothing of the pride of knowledge and the pedantic love of display, to which, perhaps, is mainly owing the appearance of absurdity in the attempts which are occasionally made to throw ridicule on the propagators of knowledge, by putting scientific language

---

* A solution of the problem is here added—
Let a = the sum required

$$x \cdot \overline{1 + \frac{x}{100}}^2 = 78 \quad 8$$

$$x + \frac{2x^2}{100} + \frac{x^3}{10,000} =$$

$$x^3 + 200x^2 + 10,000x = 784,000.$$

into the mouths of the people. Hear them talk in their own plain and expressive way, of subjects which you are used to see treated only in the language of books, and ridicule is the last idea which will arise in your mind. Knowledge, evidently, digested and fairly applied, commands respect in all stations; and whether it appears among our labouring brethren in the shape of history, mathematics, or political economy, in all shapes, at all times, it is welcome! May all manifestations of it encourage those to whose exertions it is in any measure to be attributed, to new and greater labours in so high and holy a cause! These are hardly the beginning; the end—who shall imagine?

W. REES.

2, Holles-street, Clare Market.

TREVELYAN'S PRINTING PRESS.

[From the Report of the Cornwall Polytechnic Society.]

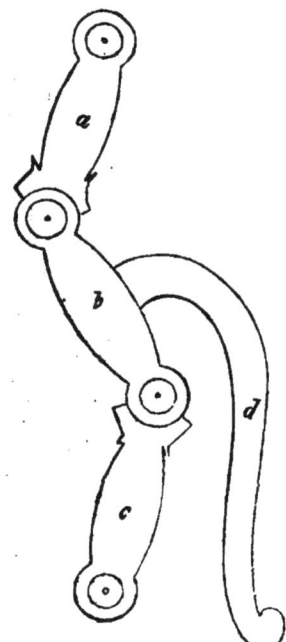

The framing of this press was of the ordinary construction. The improvement consisted in the arrangement of its levers, for giving and regulating the pressure. These levers are contained in the hollow of the piston, and are shown in the above figure: a, b, c, are the principal levers. That marked a, is jointed to the head of the press, the lower part of c, is jointed to the back of the platin, and b is placed between them, and works in sockets on their opposite ends. The crooked arm, d, is merely intended to change the direction of the force, and to increase the leverage; any power applied at its lower end causes the lever, b, to turn towards a vertical position, and at the same time, brings the others nearer to the same line. This, of course, forces the platin down on the type, with a pressure which increases rapidly as it descends, a quality which renders this arrangement very efficient for printing purposes, particularly as it admits of the nicest adjustment of power requisite for the different kinds of work.

HOUSE-MOVING IN AMERICA.

House-moving is curious, but it does not appear that much is gained by the process. A flooring of beams is introduced below the foundation of the house, and rests on three or more beams; these beams resting on others, on which they are slid along, impelled by powerful screw-jacks, and by greasing the surfaces of the beams that come in contact.

"In consequence of the great value of labour, the Americans adopt, with a view to economy, many mechanical expedients, which, in the eyes of British engineers, seem very extraordinary.

"Perhaps the most curious of these is the operation of moving houses, which is often practised in New York. Most of the old streets in that town are very narrow and tortuous; and, in the course of improving them, many of the old houses were found to interfere with the new lines of street; but instead of taking down and rebuilding those tenements, the ingenious inhabitants have recourse to the more simple method of moving the whole, *en masse*, to a new site. This was, at first, only attempted with houses formed of wooden frame-work, but now the same liberty is taken with those built of brick. I saw the operation put in practice on a brick house, at No. 130, Chatham-street, New York, and was so much interested in the success of this hazardous process, that I delayed my departure from New York for three days, in order to see it completed. The house measured 50 feet in depth, by 25 feet in breadth of front, and consisted of four stories, two above the ground floor, and a garret story at the top the whole being surmounted by large chimney stacks. This house, in order to make room for a new line of street, was moved back 14 feet 6 inches from the line which the front wall of the house originally occupied; and as the operation was curious, and exceedingly interesting in an engineering point of view, I shall endeavour to describe the manner in which it was accomplished."

After describing the operation, Mr. Stevenson proceeds:—

"The operation is attended with very great risk, and much caution is necessary to prevent accidents. Its success depends chiefly upon getting a solid and unyielding base for supporting the screw-jacks, and for the prolongation of the beam to the new site which the house is to occupy. It is further of the utmost importance that, in working the screws, their motion should be simultaneous, which, in a range of 40 or 50 screw-jacks, is not very easily attained. The operation of drifting the holes through the walls also requires caution, as well as that of removing the intermediate pieces between those of the beams, which pass through both walls. The space between the beams is only 2 feet, and the place of the materials removed is, if necessary, supplied, while the house is in the act of moving, by a block of wood which rests on the beams. The screw-jacks, by which the motion is produced, require also to be worked with the greatest caution, as the cracking of the walls would be the inevitable consequence of their advancing unequally.

Notwithstanding the great difficulty attending the successful performance of this operation, it is practised in New York without creating the least alarm in the inhabitants of the houses, who, in some cases, do not even remove their furniture while the process is going forward. The lower part of the house which I saw moved was occupied as a carver and gilder's shop; and, on Mr. Brown, under whose directions the operation was proceeding, conducted me to the upper story, that he might convince me that there were no rents in the walls or ceilings of the rooms, I was astonished to find one of them filled with picture frames and plates of mirror glass, which had never been removed from the house. The value of the mirror glass, according to Mr. Brown, was not less than 1500 dollars, which is equal to about 300*l.* sterling; and so much confidence did the owner of the house place in the success and safety of the operation, that he did not take the trouble of removing his fragile property. I understood from Mr. Brown, that the whole operation of removing this house, from the time of its commencement till its completion, would occupy

about five weeks; but the time employed in actually moving the house 14½ feet, was seven hours. The sum for which he had contracted to complete the operation was 1000 dollars, which is equal to about 200*l.* sterling. Mr. Brown mentioned, that he and his father, who was the first person who attempted to perform the operation, had followed the business of 'house-movers' for fourteen years, and had removed upwards of 100 houses, without any accident, many of which, as in the case of the one I saw, were made entirely of brick. I also visited a church in 'Sixth' Street, capable, I should think, of holding from 600 to 1000 persons, with galleries and a spire, which was moved 1100 feet; but this building was composed entirely of wood, which rendered the operation much less hazardous.—*Stevenson's Engineering of America.*

ROAD-MAKING IN AMERICA.

The roots of the felled trees are often not removed; and in marshes, where the ground is wet and soft, the trees themselves are cut in lengths of about 10 or 12 feet, and laid close to each other across the road, to prevent vehicles from sinking; forming what is called in America a "Corduroy road," over which the coach advances by a series of leaps and starts, particularly trying to those accustomed to the comforts of European travelling.

Some interesting experiments have lately been set on foot at New York, for the purpose of obtaining a permanent and durable City road, for streets over which there is a great thoroughfare. The place chosen for the trial was the Broadway, in which the traffic is constant and extensive.

The specimen of road-making first put to the test was a species of causewaying or pitching; but the materials employed are round water-worn stones, of small size; and their only recommendation for such a work appears to be their great abundance in the neighbourhood of the town. The most of the streets in New York, and, indeed, in all the American towns, are paved with stones of this description; but owing to their small size and round form, they easily yield to the pressure of carriages passing over them, and produce the large ruts and holes for which American thoroughfares are famed. To form a smooth and durable pavement, the pitching-stones should have a considerable depth, and their opposite sides ought to be as nearly parallel as possible, or in other words, the stones should have very little taper. The footpaths, in most of the towns, are paved with bricks set on edge, and bedded in sand, similar to the "clinkers," or small, hard-burned bricks, so generally used for road-making in Holland.

The second specimen was formed with broken stones; but the materials, owing chiefly, no doubt, to the high rate of wages, are not broken sufficiently small to entitle it to the name of a "Macadamised road." It is, however, a wonderful improvement on the ordinary pitched pavement of the country; and the only objections to its general introduction, as already noticed, are the prejudicial effects produced on it by the very intense frost with which the country is visited, and the expense of keeping it in repair.

The third specimen is rather of an original description. It consists of a species of tessellated pavement, formed of hexagonal billets of pine wood, measuring 6 inches on each side, and 12 inches in depth. From the manner in which the timber is arranged, the pressure falls on it parallel to the direction in which its fibres lie; so that the tendency to wear is very small. The blocks are coated with pitch or tar, and are set in sand, forming a smooth surface for carriages, which pass easily and noiselessly over it. There can be no doubt of the suitableness of wood for forming a roadway; and such an improvement is certainly much wanted in all American towns, and in none of them more than in New York. Some, however, have expressed a fear that great difficulty would be experienced in keeping pavements constructed in this manner in a clean state, and that, during damp weather, a vapour might arise from the timber, which, if it were brought into general use, would prove hurtful to the salubrity of large towns.

In the northern parts of Germany, and also in Russia, wooden pavements are a good deal used. My friend, Dr. D. B. Reid, informs me that at St. Petersburg, a wooden causeway has been tried with considerable success. The billets of wood are hexagonal, and are arranged in the manner of the American pavement. At first, they were simply embedded in the ground; but a great improvement has been introduced, by placing them on a flooring of planks laid horizontally, so as to prevent them from sinking unequally. This has not, so far as I know, been done in America.—*Stevenson's Engineering of America.*

AMERICAN STEAM-BOILER EXPLOSIONS.—AN ACT FOR INSTITUTING EXPERIMENTS UPON SAFETY APPARATUS.

An Act authorising the appointment of persons to test the usefulness of invention to improve and render safe the boilers of steam-engines against explosions.

Be it enacted by the Senate and House of Representatives of the United States of America in Congress assembled, That the President of the United States be, and he hereby is, authorised to appoint three persons, one of whom at least shall be a man of experience and practical knowledge in the construction and use of the steam-engine, and the others, by reason of their attainments and science, shall be competent judges of the usefulness of any invention designed to detect the causes of explosion in the boilers; which said persons shall jointly examine any inventions made for the purpose of detecting the cause and preventing the explosion of boilers, that shall be presented for their consideration; and if any one or more of such inventions or discoveries justify, in their judgment, the experiment, and the inventor desires that his invention shall be subjected to the test, then the said persons may proceed and order such preparations to be made, and such experiments to be tried as in their judgments may be necessary to determine the character and usefulness of any such invention.

Sec. 2. The said Board shall give notice of the time and place of their meeting to examine such inventions, and shall direct the preparations to be tried, at such place as they shall deem most suitable and convenient for the purpose; and shall make full report of their doings to Congress at their next session.

Sec. 3. To carry into effect the foregoing objects, there be, and hereby is, appropriated out of any money in the Treasury not otherwise appropriated, the sum of six thousand dollars; and so much thereof as shall be necessary for the above purposes shall be subject to the order of the said board, and to defray such expenses as shall be incurred by their direction, including three hundred dollars to each, for his personal services and expenses: *Provided however,* And their accounts shall be settled at the Treasury, in the same manner as those of other public agents.

———

THE CANAL LIFTS ON THE GRAND WESTERN CANAL: BY JAMES GREEN, M. INST. C. E.

[From the Proceedings of the Inst. of Civ. Eng.]

The lift which is the subject of the following paper was erected by Mr. Green in the year 1835, on the Grand Western Canal, and has been in operation ever since. Lifts are not intended to supersede the use of canals in all cases, but in those in which a considerable ascent is to be overcome in a short distance, and in which the water is inadequate to the consumption of a common lock, or in which the funds are inadequate to the execution of the work on a scale adapted to such locks.

This lift is 46 feet in height, and consists of two chambers, similar to those of a common lock, with a pier of masonry between them; each chamber being of sufficient dimensions to admit of a wooded cradle, in which the boat is to ascend or descend floats. The cradle being on a level with the pond of the canal, a water-tight gate, at the end of the cradle, and of the pond of the canal, is raised up, and leaves the communication betwixt the water in the canal and in the cradle free, and the boat swims into or out of the cradle.

The cradles are balanced over three cast-iron wheels of 16 feet in diameter, to the centre of one of which is fitted spur and bevil gear, so that the motion may be given by machinery worked by the hand, without any preponderating weight of water in the cradle, when scarcity of water renders this necessary. To this hand gear in also attached brake-wheels and a brake-lever for regulating the motion.

It is obvious that the weights of the additional length of the suspending chains on the side of the cradle which is the lowest, must be counterbalanced; for this purpose there is attached, to the under-side of each cradle, a chain of equal weight per foot with the suspending chain, and this elongates under the ascending, and is shortened under the descending cradle; thus the disparity in the weights due to the suspending chain is obviated.

It is so arranged that the water in the upper cradle is about two inches below the level of the water in the pond; the consequence of which is, that the upper cradle has a slight preponderance first, sufficient to set the machinery in motion; the weight of this water is generally about one ton; it may however be regulated at pleasure.

The strength of materials is the great desideratum in machinery of this nature, and though the lift here described is but 46 feet, and the boats about eight tons, the same method is applicable to much greater heights and larger tonnage. The advantages of these lifts over common locks are, great economy of construction, and a great saving of time and water.

The time occupied in passing one boat up and another down this lift of 46 feet is three minutes, whereas thirty minutes would be required to attain the rise of 46 feet by locks; thus the saving in time amounts to 9-10ths for boats of eight tons.

The quantity of water consumed is about two tons for eight tons of cargo, whereas in common locks it is about three tons of wa-

ter per ton of cargo; the saving is, therefore, 22 parts out of 24, or very nearly 92 per cent. If the trade were all downward, there would, by the use of the lifts, be carried from the lowest to the highest level of the canal, a quantity of water equal to the loads passed down.

Mr. Green stated that in some parts of the canal it had been found impracticable to get a sufficient drain to empty the chamber—they were compelled, therefore, to use a half-lock of eighteen inches fall; that there were seven lifts and one inclined plane on the canal, effecting a rise of 262 feet in eleven miles. That he should not recommend them as applicable to boats of more than 20 or 30 tons. The width of larger boats was an obstacle. They were extremely advantageous for narrow canals; for boats of 50 or 60 feet in length, and about 30 tons.

Mr. Parkes remarked, that he considered the question of narrow canals as a most important one—the advantages to be derived from narrow canals was a subject to which sufficient attention had not been paid.

The President called attention to the remark in Mr. Green's paper respecting the quantity of water carried up from one level to another in a downward trade wherever these lifts are used; then a coal country on a high level may supply itself with as much water as it sends down coal. The subject of inclined planes being alluded to, especially those of the Morristown Canal of 200 feet each, where a rise of 1600 feet is effected by eight inclined planes, Mr. G. remarked, that more water and time must be expended, the friction and length being much greater. In the lifts there was only as much water consumed as was equal to the load, but that he should not consider them as practically applicable to more that 60 or 70 feet. Favourable levels with ascents of more than 60 or 70 feet could seldom be found; could he have had the choice of the line in this particular instance, he should have effected by four lifts the rise for which seven are now employed.

RENDEL'S FLOATING BRIDGE ACROSS THE HAMOAZE FROM DEVONPORT TO TORPOINT.

[From the Minutes of Proceedings of the Institution of Civil Engineers.]

The floating bridge now described is used as a system of communication betwixt the opposite shores of the Tamer, a little to the north of Devonport. The width of the river at this site is 2550 feet, at high water, and its greatest depth at spring tides 96 feet. The ordinary velocity of the stream is 3¼ knots an hour, but under heavy land floods, it is increased to 5 knots. The line of passage is directly at right angles to the current; this, combined with the exposure of the site, and the rapidity of the current, rendered an attempt to apply a twin boat, similar to those at Dundee, a total failure.

The floating bridge is a large flat-bottomed vessel, of a breadth or width nearly equal to its length, namely, 60 feet long, and 50 feet wide, divided in the direction of its length into three divisions, the middle being appropriated to the machinery, and each of the side divisions to carriages and traffic of all kinds. These side divisions have decks, raised from 2 feet to 2 feet 6 above the line of floatation, and carriages, horses, &c. pass on and off the deck by strong commodious platforms or draw-bridges, communicating with the landing-places, and over which carriages of all kinds drive on and off the bridge without difficulty or inconvenience. The bridge is guided in its passage by two chains, which, passing through it, over cast-iron wheels, are laid across the river, and secured to the opposite shore ; thus forming, as it were, a road, along which the bridge is made to travel forward and back from shore to shore. The peripheries of the wheels are cast with sockets, fitted to the links of the chain, so that when the wheels are stationary the bridge is moored by the chains ; when the wheels revolve, the bridge moves in the opposite direction. Two steam engines, of 11 horse power each, are employed to turn these wheels. The author then describes the details of the wood-work, and the dimensions of the several parts ; the draw-bridge, and the landing-places, or inclined planes, formed on each shore ; the galleries ; the engine-house and machinery ; the chains and balance weights ; the accommodation, and regulations of the bridge.

The peculiar feature in these works are the balance weights. There would have been great difficulties in fixed moorings ; the ends of the chains are attached to weights, suspended in shafts 16 feet square, and 20 feet deep, sunk in the landing-place above high water mark. The weights are cast-iron boxes, loaded with about five tons each. Thus the additional length, requisite when the vessel is in the middle of the river, is obtained. Were the chains fixed to the shores, they would be too short, and consequently unnecessarily strained at this time, or so long as to allow the vessel to make lee way in her approach to the landing-place. This is altogether avoided by the balance weights ; for as the vessel leaves the shore, the weights rise, and the chains lengthen, so as to adjust themselves to an easy curve ; and as it approaches the other shore, the balance weights on that side fall, the chains are shortened,

and the draw-bridges or platforms are brought straight and steadily to the landing-places.

The economy, both as regards first cost and annual expenses of these floating-bridges, no less than their superior accommodation to every other mode of crossing estuaries, has already given Mr. Rendel the opportunity of establishing several: the latest was that at Southampton, across the Itchin ferry, over which there are 12 coaches daily, and great carriage traffic, although the public have the option of crossing a fixed bridge over the same river, and only a short distance farther round.

The Lords Commissioners of the Admiralty having sanctioned the establishment of a similar bridge across Portsmouth Harbour from Portsmouth Point to Gosport Beach, and a bill is now before Parliament to incorporate a company for carrying the work into effect. The great national importance of this harbour, and the well-known jealousy of the Board of Admiralty in all matters connected with its economy, furnish the best proof that these bridges, though requiring chains to be laid across the river, do not occasion the slightest impediment to the navigation or tidal currents.

Mr. Rendel stated that the chains were kept bright by the rubbing which they received on the bed of the river. The bed, consisting of mud and not of gravel, the chains only scoured, and did not perceptibly wear. The chains, which have been in use four years, have not been sensibly diminished. They had tried chilled segments—these wore the chains ; they consequently returned to good grey iron. Three sets of segments are worn out in the course of a year. In reply to a question respecting the deviation of the bridge under the action of the wind and current, Mr. R. stated, that he had never known it diverge more than by its breadth, or, 50 feet ; owing to the particular form of the bridge, and the small draft of water, the current had but little effect.

The usual weight of the balance boxes is five tons ; but in hard weather, it is usual to add a ton more. He conceived that no comparison could possibly be instituted betwixt the relative advantages of the floating bridge and the twin boat ; the latter requiring very expensive wharfs—those at Dundee, for instance, having cost upwards of 25,000*l*., and still there is much attention and care required in bringing the boats to their piers. But the floating bridge requires no such expensive appendages ; the chains on which it works, when the wheels are in motion, becoming the most secure fastening when the engines are stopped. The chains also act as a pilot and crew, two persons only being required in a vessel of this kind, viz. an engine-man and one on the decks to attend to the drawbridges.

Mr. Vignoles remarked, that the plan now proposed would obviate many difficulties which occurred in the case of railroads; there were many situations in which the floating bridge might be adopted with great service, and he could not refrain from expressing his admiration of the great forethought, skill, and design, which were here exhibited; at the minuteness with which the details had been studied; and, not the least, the adaptation of the balance weights for the chains; the chains not having elongated, proves of itself how completely they answered their intended purpose.

On a subsequent occasion Mr. Rendel remarked, that as the same velocity could not be acquired in the manner proposed as by a paddle-boat, the question to be considered was, as to the advantage of employing the floating bridge in preference, in particular cases. They had to consider to what width of ferry the floating bridge is applicable; what the maximum velocity; what the expense of piers for paddle-boats. The great disadvantage of paddle-boats results from the difficulty of making fast, and of getting the cargo on and off. If we take as the measure of advantage the facility afforded by the floating bridge, then its superiority is very great. But the question is one of time, as well as of accommodation. He was of opinion, that a velocity of more than eight miles could not be attained by these boats. Now if a paddle-boat could be impelled ten miles an hour, the time of landing, which would amount to ten minutes, would compensate for the increased speed.

He conceived that the chain might be applied to a distance of three miles; the time of crossing and the expense of the chain were the only limits. Mr. Parkes remarked, that Mr. Rendel undervalued the advantages of his plan. In crossing the Mersey, for instance, excepting at high and low water, they had to run up or down, whereas the floating bridge would go straight across. There was great loss of time and uncertainty with the paddle-boats. They were frequently only a quarter of an hour in crossing the Mersey, but he had himself been three quarters of an hour in crossing from Liverpool to Birkenhead. A simple beach being sufficient for landing was a great gain, whereas to get paddle-boats alongside, extensive and expensive piers are required.

NOTES AND NOTICES.

Steam Navigation to India.—Captain Sir John Ross passed through this town the other day on his return from Greenock, whither he had gone to inspect a splendid steamer, built for the company by Messrs. Scotland and Sinclair, of that town. This beautiful ship will be launched in about a month from the present date, and will be ready to sail for Calcutta direct or shortly after Christmas next, calling only at Hout's Bay, Cape of Good Hope. In extreme length the said vessel is 201 ft., in breadth 40, in depth 25, measuring 1,200, and will carry 1,500 tons, with accommodation for 150 passengers. Her name, we believe, has not yet been fixed; but, be this as it may, she will be the first of a most superior line of steamers direct to Calcutta—itself a very high achievement, which the uncertainty of the Red Sea route is daily rendering of increased moment.—*Dumfries Courier.*

Discovery of a New Continent.—M. Doubonzel, an officer on board the *Zelle* (the expedition to the South Pole), has written a letter, dated Valparaiso, March 30, confirming the details given by M. D. D'Urville. A new circumstance mentioned by him is the discovery of a great continent to the south of South Shetland. "We carefully explored and determined," says that officer, "forty leagues of coast, notwithstanding the surrounding ice. This discovery is a real service to nautical and geographical science."—*Le Constitutionnel,* Sept. 30.

The Royal William, which left Liverpool on her second voyage for New York on the 21st ultimo, had on board 67 cabin passengers. The amount of fares for passengers, freight, cargo, parcels and letters, considerably exceeded 3,000*l.* She carries also with her 40 tons of the new scientific and useful invention, for which Mr. C. W. Williams, the able and enterprising manager of the City of Dublin Steam Packet Company, has lately obtained a patent. From the circumstance of each ton of this discovery being able to do the work of three of the common coal, the proprietors of the vessel have been enabled this voyage to take 50 tons of cargo, with the full complement of passengers.

Description of Fuel.

	Tons.	Cwt.	Qr.	lb.
Coal	310	14	1	0
Peat stone fuel	52	1	0	7
Total	362	15	1	7

Draught of water at starting.	Feet.	Inches.
Lighter than on previous voyage	0	6
Forward	13	11
Aft	14	4

The *Railway Map* of England and Wales continues on sale, in a neat wrapper, price 6d.; and on fine paper, coloured, price 1s.

LONDON: Printed and Published for the Proprietor, by W. A. Robertson, at the Mechanics' Magazine Office, No. 6, Peterborough-court, between 135 and 136, Fleet-street.—Sold by A. & W. Galignani, Rue Vivienne, Paris.

Mechanics' Magazine,

MUSEUM, REGISTER, JOURNAL, AND GAZETTE.

No. 792.]　　　　SATURDAY, OCTOBER 13, 1838.　　　　[Price 3*d.*

MACHINE FOR PROVING TENACITY OF IRON AND COPPER FOR THE CONSTUCTION OF STEAM BOILERS.

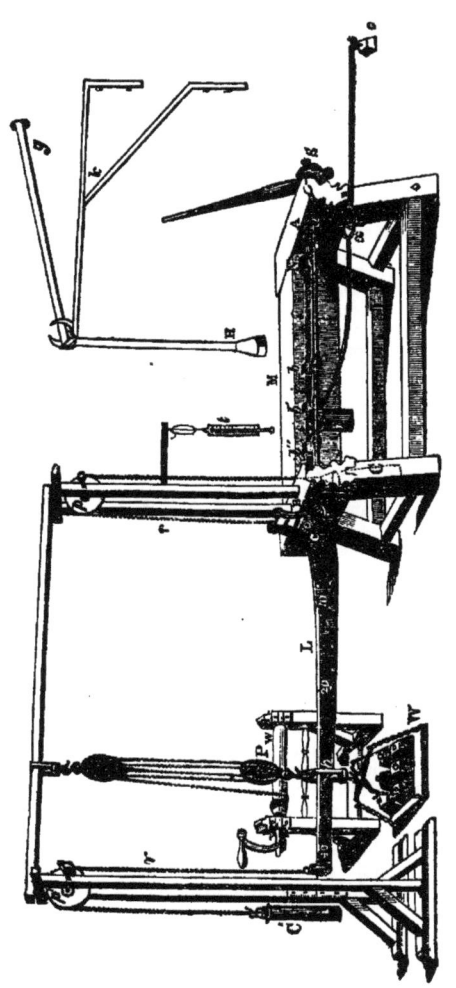

REPORT OF THE COMMITTEE OF THE FRANK-
LIN INSTITUTE ON THE EXPLOSIONS OF
STEAM BOILERS, OF EXPERIMENTS MADE
AT THE REQUEST OF THE TREASURY DE-
PARTMENT OF THE UNITED STATES.—
PART II. CONTAINING THE REPORT OF
THE SUB-COMMITTEE TO WHOM WAS RE-
FERRED THE EXAMINATION OF THE
STRENGTH OF THE MATERIALS EMPLOYED
IN THE CONSTRUCTION OF STEAM BOILERS.
[From the *Franklin Journal*.]

The sub-committee, to whom was referred the examination of the strength of materials employed in the construction of Steam
Boilers, beg leave to submit the following
report :—

While it is important to know the causes
which may produce a dangerous developement of elastic forces in the interior of steam
boilers, it is obviously not less so, to understand aright the efficacy of those means on
which we rely for confining or controlling
their energies. Hence, in investigating the
causes of explosions, it is both natural and
expedient to examine separately those facts
and principles which concern the *divellent*
and the *quiescent* forces respectively. The
number and variety of circumstances, which
affect the character and durability of materials of which steam boilers are formed, are
probably not less than of those which tend
to modify the action of the fluids which
they contain. In this view of the importance to be attached to the subject of the
strength of materials, it may be considered
remarkable, that while numerous investigations have been made as to the causes of
danger, so little should have been attempted
in regard to the most direct and obvious
means of security. Before the series of experiments here detailed had been commenced,
the necessity for such an investigation had
been repeatedly pointed out, in public and
private lectures on the steam engine ; the
reasons assigned for instituting the inquiry,
being the very general and unsatisfactory
nature of those results, which are given in
practical treatises, respecting the strength
of metals, as dependant on the mode of manufacture, and on the different temperatures
and other circumstances to which they are
exposed. We had, it is true, a considerable
number of results, obtained at different periods, by experiments on the direct cohesion
of wrought iron,* they were, however, in

* The following brief Table contains some of the
general results, obtained by different authors, as the
strength of wrought iron.

Name of the Experimenter.	Strength in lbs. per sq. inch.
Muschenbroek	73.100
Perronet—on square bars	61.083
Perronet—on round bars.................	60.086
Buffon	84.730

general, undertaken for purposes very different from those which prompted the present
investigation.

Few of the experimenters had in view the
influence of temperatures on tenacity ; and
even those data which they furnish for calculating the proper thickness of metal to be
employed at ordinary temperatures, in constructing steam boilers, are liable to much
uncertainty, owing to the diversity in the
results themselves. Laborious and protracted as has been this investigation,
still the practical importance of the subject
has appeared to warrant a careful survey,
and a diligent comparison of the various
facts which might influence the practice of
those who desire to attain a secure action in
the steam boiler.

Without entering, therefore, into all the
delicate questions, which, had a mere scientific view been indulged, we might have
been prompted to examine, it has been the
aim of the committee to obtain and present
such classes of facts as both scientific and
practical men may make subservient to their
respective purposes.

The questions, which in the course of this
inquiry, it has been found necessary to investigate, may be classed under three general divisions.

1. *Principal.*
2. *Incidental.*
3. *Subsidiary.*

I. *Principal.*—1. What is the absolute
tenacity per square-inch bar of rolled boiler
iron, at ordinary temperatures, and to what
irregularities is it liable ?

2. The same for rolled copper ?

3. What is the effect of increased temperature on the tenacity of iron and copper?

4. What in the tenacity of wrought iron,
manufactured by other means than rolling
into plates ;—as by rolling it into bars or
rods, by hammering and wire-drawing ?

5. What are the relative advantages of
iron made by refining from different sorts
of pig metal and their mixtures ?

Name of the Experimenter.	Strength in lbs. per sq. inch.
Poleni	63.390
Rennie—on " English iron"	55.843
Brunel—" Best English"	68.544
Brunel—" Best-best English"	72.352
Telford—Welsh iron...................	65.520
Telford—Staffordshire iron............	60.928
Telford—Swedish iron.................	64.960
Brown—Welsh iron	57.075
Brown—Swedish	49.796
Brown—Russian iron	59.472
Martin—(Fr.) St. Chambaud iron.......	49.000
Martin—Fourchambauldt iron	47.964
Martin—Superior English	52.823
Martin—English best cable	49.251
Rennie—Copper......................	33.792

6. What is the comparative value of sheet iron manufactured by the processes of puddling, blooming, and piling respectively; and in the last case, what influence have repetitions of the process?

7. What is the effect of piling into the same slab, iron of different degrees of fineness?

8. What is the comparative tenacity of rolled iron, in the longitudinal, diagonal, and transverse directions of the rolling respectively?

9. What influence may be produced, by long and repeated use, towards modifying the character of boiler iron?

II. *Incidental.*—1. What is the specific gravity of the specimens submitted to examination?

2. What elasticity is found in the metals under different circumstances of the trial?

3. What relation exists between the force which will produce a permanent elongation in a bar, and that which will entirely overcome its tenacity?

4. What amount of elongation may the several kinds of metal undergo before fracture?

5. Does the amount of *constriction* or diminution of area, at the section of fracture, bear any relation to the absolute strength of the metals, to the direction in which the strips are cut from the plate, to the breadth and thickness of the strips themselves, or to the temperature under which the trial is made?

6. What is the effect of the rivets on the total strength of a boiler?

III. *Subsidiary.*—1. What is the friction of the apparatus employed to determine tenacities?

2. What is the amount of its elasticity?

3. What is the latent heat of the vapour of water?

4. What is the specific heat of iron, copper and glass, respectively?

5. What is the rate of heating of a given mass of liquid, when subjected to the direct action of a solid of higher temperature?

6. At what rate will the same mass of liquid change its temperature by the action of air alone?

From the foregoing statement, it will be seen that more than 20 distinct topics have demanded the attention of the committee. They have felt strongly inclined to embrace some other points of great practical and scientific importance, but the time already unavoidably consumed, and the very limited means which the other branches of inquiry and experiment on explosion left to be appropriated to the purposes of this sub-committee, compelled the relinquishment, for the present, of those objects which do not immediately concern the construction and use of steam boilers.

The discussion of the questions above enumerated, will necessarily follow an order somewhat different from that in which they are here stated. A view of the apparatus, employed by the committee, claims the first notice. The origin and preparation of the materials to be tested, will also precede the detail of experiments.

Machine for proving the strength of materials.—The apparatus used, by the committee, for the direct determination of the principal questions regarding the strength of the specimens submitted to examination, is represented in plate I. (See front page). M is a strong frame of oak timber, the two longer sides five feet in length, 14 inches deep, and 6 inches thick.

The two shorter, or end pieces, which project beyond the *sides* to the distance of 3 inches, are each two feet 8 inches long, seven and a half inches thick, and 14 inches deep.

Between the two side pieces, (one of which is in the figure removed, to exhibit the interior or working parts,) is a space 14½ inches wide, affording room for a screw, cross-head, guide-rods, connecting blocks and wedges, to hold the specimens under trial; and also for the heating apparatus in experiments at high temperatures.

These four massive blocks or beams of timber are held together by strong screw bolts, passing through mortises in the end pieces, along tenons into screw nuts imbedded in the timber of the longitudinal beams. The frame is supported, as represented in the figure, by four firm trussel legs, 6 inches square, tied together near the bottom, and fastened as well to the ties as to the frame above, by mortising and bolting. The top of the frame is 3 feet 8 inches above the floor on which the machine rests.

Through one end A, of the frame M, about 6 inches below the top, and centrally between the two side beams of the frame, passes the screw S, 2¾ inches in diameter, and 3 feet long, cut into threads $\frac{6}{10}$ths of an inch apart. Near the head of the screw is a neck turned rather deeper than the threads, to allow a clamp collar to embrace it; which, together with a strong cast iron plate, against which the head of the screw works, prevents any longitudinal motion of the screw itself.

N is the box or nut of this screw, which by the revolution of S, either approaches to or recedes from the end A of the frame; *s s*, are two guide-rods, one on each side of the screw, level with its axis and near the inner faces of the longitudinal beams of the frame, serving to support a cross head that contains in its central ring the nut N, and

B 2

embraces by loops at its extremities the two guide-rods. The purpose of these loops is to prevent the nut from turning by the revolution of the screw.

The cross head thus secured is united by two strong straps or bars of iron *i i*, 2 inches wide by half an inch thick, to a block of iron *b*, which is also furnished with two projecting arms that rest on the guide-rods already described. This block as well as the two others *b'* and *b''*, is 4 inches long, 4 inches deep, and 1¼ inches thick, being perforated centrally in the direction of its thickness with a hole in the form of the frustum of a square pyramid, the purpose of which is to admit of wedges placed within them to hold the bars of metal under trial. A more detailed description of these will be given hereafter.

The block *b'* is connected to *b* by a separate pair of straps *i i*, and has arms reposing on the guide-rods, or when necessary, admitting a vertical semi-revolution, so as to be laid over backward between the straps *i i*. This latter disposition of the block *b'* was made whenever specimens of 20 or 30 inches in length were to be tried ; but when those of only a few inches in length were under trial, *b'* was used in the position represented in the figure.

The block *b''* is connected by the strong iron straps *i'' i''*, which pass freely through a suitable opening in the head B of the frame, to the lever L. One of these straps is seen at *e*, the other being on the posterior side of the lever, with which they are united by means of a steel bolt turned with care and well polished. The straps are kept in place by a head, screw nut and washers, on the bolt. This lever is of the *rectangular* kind, the longer arm being horizontal, the shorter vertical, and the angular point being in the axis of a second or lower bolt which serves as a fulcrum.

At the end next the frame, the lever has a breadth or depth of 7 inches and a thickness of 1 inch. Towards the opposite extremity or that on which the weights are placed, it diminishes to a breadth of 4 inches, and a thickness of ⅝ths of an inch. The upper edge of the beam is straight to within 24 inches of the broader end, where it curves upwards, affording a massive support for the upper bolt already described. In a vertical direction beneath the bolt, and in the prolongation of the upper straight edge of the lever, is the position already indicated, of the second steel bolt, serving for a gudgeon, on which the lever turns. The distance betwen the axes of the two bolts, is 2.914 inches, which is therefore the length of the shorter arm of the lever. The bolts are very nearly of the same diameter, being

each about 1.086 inches. The lower bolt rests against a plate of cast iron, having suitable projecting cheeks, with bearings adapted for its reception.

A strap from the top of each cheek comes down over the bolt, and is fastened with a thumb screw, to prevent the lever being thrown out of place by the recoil of the machine. The two guide-rods *s s*, pass through this cast iron plate, as well as through that which serves as a collar to the screw head S, on the opposite end of the frame. The lever is formed of the best wrought iron, and weighs 164$\frac{3}{16}$ pounds, the matter being so distributed that if not neutralised by counter weights, its effect in straining any bar attached horizontally to the upper bolt would have been equal to 2495¼ lbs. To obviate this, and to prevent the weight of the lever from adding anything to the friction, it is accurately counterpoised by means of weights C and C,' corresponding to the parts of its mass which they are respectively required to sustain. Thus the weight of C, the larger counterpoise, is 103 pounds 12 ounces, that of C' 60 pounds 7 ounces. The former is, however, increased to counterpoise likewise, one-half the weight of the two straps *i'' i''*, the other half resting, as will be seen, on the horizontal guide-rods *s s'*

The axes of the pulleys *p p'*, over which the cords *r r'* pass, are furnished with cavities to receive steel pivot-points, in order to reduce, as far as practicable, the friction of these parts. The diameter of these pulleys is 12 inches.

The iron stirrup, to which the cord *r* is attached, is applied to the lower bolt or fulcrum of the lever, the projecting ends of which roll on straight, horizontal edges, forming the bottom of two loops with which the stirrup is furnished.

By means of the suspending apparatus above described, the lever is enabled to obey any force acting vertically on its longer arm, with the advantage of ample strength and stiffness, combined with the condition of a theoretical lever, in respect to the *gravity of parts*.

There are two modes of operating by which a bar of metal, placed in the machine between *b'* and *b''*, might be broken, so as to ascertain the tenacity.

The first is to apply the force of the screw S to strain the bar in raising a weight W suspended at any convenient point on the arm *h* of the lever ; the second is to employ the screw only to regulate the height of that arm, and to restore it when relieved of the weights, to the horizontal position, whenever the extension of the bar had allowed it to fall below that position.

The latter method was, with very few exceptions, adopted by the committee,—both because it allowed of a more exact determination of the breaking weights, by a small addition at a time, and because it rendered the effect of the friction constant in its kind, being always in opposition to the gravitating force of the weight W, and *subtractive*, in the calculation.—In order to apply this mode of action without requiring correction for the stiffness of the cord *r'* and the friction of the pulley *p'*, it was only necessary, after adjusting the weight C', to remove so much as would allow the arm *h* of the lever to descend upon the slightest jarring of the machine. The tenacity of the bar, and the friction at the fulcrum, were then the only resistances to the motion of the weight W.

The purpose of the tackle of pulleys P, is to elevate the scale pan and weight after they have descended to the floor, in order, by turning the screw S, to counteract the elongation of the bar under trial, and again to commence operations with the descending motion of the arm *h*. The power of the operator is applied to the tackle by means of the windlass *w*, furnished with crank, ratched wheel and click. The upper edge of the lever was graduated into parts distant from each other just ten times the length of the shorter arm.

By the aid of these several appendages, the machine allows the most gradual additions to be made to the divellent force applied to the specimens, breaking each with a descending movement, and consequently, rendering the *friction* definite in the direction of its influence, being, as before stated, always subtractive.

The very few cases in which the mode of operation rendered it *additive*, are particularly mentioned in the tables.

At the outer extremity of the lever, and in the prolongation of its upper edge, is placed a style *z*, serving as an index to the graduated arc *a*, which is divided into minutes of a degree. The point of the style is 10 feet 3 inches from the axis of motion in the lever, and the length of the entire circumference which it would describe 772.8276 inches. Hence each degree is 2.14674 inches, and each minute, as measured on the arc, is .035779 of an inch. The whole extent of the arc *a* is about 5°, the zero, or point of horizontal position being placed 3 degrees from the upper extremity. The chief use of this arc was to determine approximately the *elasticity of the bars*, and of the machine itself, as preliminary to that inquiry. The weights W, in the scale-pan, (which, with its suspending chains, cross-bars, &c. weighs 56 pounds,) were, in every case, applied on the lever, at the third mark, a distance from the axis of motion 30 times as great as that between the axes of the two bolts, or 30 × 2.914 = 87.42 inches.

Friction of the Machine.—The amount of friction of the machine already described for testing the tenacity of metals, was an object requiring particular investigation before any thing more than a comparative value could be assigned to the results which were afforded by the experiments.

To determine this point, it was deemed advisable to ascertain under various loads what proportion the weight which was sustained by the machine, after it had been raised by the screw S till the index stood at zero on the arc *a*, bore to that which, after the lever was relieved and then loaded again with the same weight, would cause it once more to descend to zero.

Between the heads *b'* and *b''* was placed a strong bar of iron 1 inch wide by ⅜ths of an inch thick. Two methods were then pursued for the purposes of mutual verification.

1. A certain weight was placed in the pan suspended at *h*, and the screw S turned until, as before mentioned, it was raised to a level so that *z* stood at zero on the graduated arc. The windlass *w* was then employed to raise the scale-pan and entirely relieve the lever. On again restoring the weights, the index remained some minutes of a degree above 0, and an additional quantity of weight was necessary to bring it once more down to that level. As, in this case, the weight added served to increase the friction, it is manifest that the comparison of it with the whole weight, itself included, must be necessary in order to show the relation between the weight at first raised and that part of it which represented the friction of the machine. When the weight was raised by the screw, the bar which connected the heads *b' b''*, must have sustained a strain composed of the weight raised *added to* the friction of the machine ; whereas, when the weight was let on by the windlass while the index was at some distance above 0, the bar sustained a strain represented by the weight borne, *diminished by* the friction.

II. The lever was caused to rest on a solid support near the extremity, the index being opposite to zero on the arc, and in that position the scale-pan was loaded with the weight under which it was intended to try the friction. The screw S was then carefully turned to strain the bar and bring the loaded arm of the lever barely off of the support. The weights were next raised by the windlass, and the recoil of the machine raised the lever to a certain elevation ; from which it was once more depressed by replacing the weights upon it, and adding such an amount as would just depress the arm to the level of its original support.

The first of the above methods gives the *double*, and the second the *single*, friction of the machine. The following table exhibits the weights in the scale-pan, the weight representing the friction when the first method was employed; the same for the second method; and the ratio of the friction to the total weight sustained by the lever. A correction is required particularly at the higher pressures, on account of the increased elasticity in the machine under the added weights, which actually brought the index down to zero sooner than it would have arrived at it, by the simple effect of a strain upon the lever regarded as inflexible.

The machine was kept constantly well oiled, still a trifling difference may possibly have existed in regard to its condition at different times; but no influence of this sort was ever found sufficient to determine the rupture of a bar, after the weight had been taken up, the gudgeons newly oiled, and the same weight replaced which it had borne previous to that operation.

TABLE I.

No. of the experiment.	Weight applied to the lever.	Double Friction.			Single Friction.		
		Weight required to counterpoise the double friction.	Ratio of friction to weight by the method of double friction.	No. of repetitions furnishing the mean result on double friction.	Weight required to counterpoise single friction.	Ratio of friction to weight by the method of single friction.	No. of repetitions furnishing the mean result on single friction.
1	56	6.00	.050 +	3	3.00	.050 +	1
2	112	10.82	.051 +	7	5.79	.051 +	6
3	168	17.62	.052 +	4	9.06	.051 +	4
4	224	24.85	.055 +	10	12.75	.053 —	5
5	280	28.16	.050 +	6	14.71	.050 —	7
6	336	35.30	.053 +	5	17.25	.049 —	6
7	392	38.33	.047 +	4	20.42	.050 —	4
8	448	42.50	.048 +	4	22.64	.048 +	7
9	504	45.50	.045 +	1	26.80	.050 +	7
		Mean .050		Total 44	Mean .050 5		Total 47

From the above Table it appears that the second method gave results more nearly in accordance with each other than the first, but it will also be noticed that forty-four observations with the first gives a mean value sensibly identical with that obtained from forty-seven experiments with the second method of trial above described. We were hence led to adopt 5 per cent. on the weight as the effect of the friction of the machine. The bolts are of well polished steel and the lower bearing of cast iron, and the upper one or the eyes of the straps i″ i″, of wrought iron.

This relation of friction to pressure between these substances, as deduced from the experiments of the committee, will be found to correspond very nearly with that obtained by Mr. Wood when operating on railway carriages.*

Elasticity of the Machine.—In order to determine, in particular cases, the amount of elasticity exhibited by the bars under trial, it became necessary to ascertain the elasticity of the machine, when loaded with different weights. Several series of trials were accordingly made, expressly with a view to this object.

Putting into the machine, in place of a bar to be broken, one which was intended not to yield sensibly to the strains applied, and not in any case to be permanently elongated by them, different weights were appended to the lever, and allowed to remain until the latter had become stationary. They were then carefully raised by the windlass, and the lever allowed to rise by the recoil until it became entirely free from strain. The number of degrees and minutes on the arc a, traversed by the index, was then noted, the weight replaced, and the trial repeated until it was ascertained that no error of observation had occurred.

Three series of operations were performed,

---

* See Wood on Railways, Smith's edition, Philadelphia. 1832, p. 202. The mean of nine out of twelve experiments here detailed is exactly 5 per cent. for the friction between steel and cast iron.

: each beginning with the lever 3° 30' above zero, when entirely unloaded, but fully in contact with its bearings. Weights were then added by 56 pounds at a time, and the depression below 3° 30', produced by each addition, was noted. This was continued until weights had been added sufficient to bring the index to 0, which was effected with 11 weights of 56 pounds each.

If the *lever* had been entirely inflexible, the natural sine of the angle of elevation after being relieved might have been considered the measure of the compression of parts sustained by the frame, links, &c. under each weight; for as the shorter arm of the lever is only 2.914 inches in length,

while the bar and connecting straps are more than 5 feet, the direction of the horizontal bar may be considered sensibly constant.

The following Table contains the results of the experiments just described, together with those of another set in which the operations were in every respect similar, except that the weights were applied by 37½ pounds at each time instead of 56 pounds. The natural sines are added; by comparing which with the respective *compressing forces*, it will be found that the law which governs the *elasticity* of the machine is, that *the latter is proportionate to the fifth root of the cube of the compressing force.*

TABLE II.

No. of the trial.	Weight in pounds producing compression.	Recoil of the lever in minutes of a degree.	Natural sine of the angle of elevation of the lever after the recoil.	REMARKS.
1	37.5	40.'	.0116353	Comparing the first with the last experiment by the formula $\left(\dfrac{616}{37.5}\right)x = \dfrac{\text{nat. sin. } 211.5'}{\text{nat. sin. } 40'}$ we get $x = 0.594$.
2	56.	47.2	.0136713	
3	75.	58.	.0168707	The 3rd and 16th, by a similar comparison, give $x =$ 0.608
4	112.	74.7	.0218149	The 4th and the 19th give $x =$ 0.606
5	150.	86.	.0250138	
6	168.	95.	.0276309	The 6th and 8th give $x =$ 0.600
7	187.5	100.	.0290847	The 7th and 17th give $x =$ 0.627
8	224.	112.7	.0328644	The 8th and 10th give $x =$ 0.595
9	262.5	120.	.0348995	
10	280.	128.7	.0375158	The 10th and 12th give $x =$ 0.560
11	300.	134.	.0389692	The 11th and 18th give $x =$ 0.641
12	336.	143.4	.0415850	
13	375.	154.5	.0447818	The 13th and 15th give $x =$ 0.583
14	392.	157.4	.0456536	
15	412.5	166.3	.0482687	
16	448.	171.9	.0500119	
17	504.	186.	.0540788	The 2nd and 17th give $x =$ 0.625
18	560.	199.7	.0581448	Mean 0.603
19	616.	211.5	.0613389	Hence the mean of the above 10 comparisons gives $x = .603$, which, by rejecting the last figure, furnishes the law above stated.

Another set of trials was made, loading the lever with weights by 7 pounds at a time from 0 to 609 pounds; and from the results of this and the preceding series a table was constructed, furnishing the column of elasticities of the machine for every observed depression under given weights when testing the elasticity of bars of iron. By deducting

the elasticity due to the machine alone from that obtained by observation, we get the measure, in minutes of a degree, of the elasticity of the bar.

In the table of elasticities actually observed will be found various numbers between 5' and 73'. To facilitate the comparison of each observed elasticity with the length of the bar on which the trial was made, the following table is annexed, in which the natural sine belonging to each number of minutes has been multiplied by 2.914, the length in inches of the shorter arm of the lever.

TABLE III.

Observed elasticity in parts of the arc.	Corresponding extension and recoil of the bar in inches.	Observed elasticity.	Extension and recoil in inches.	Observed elasticity.	Extension and recoil in inches.	Observed elasticity.	Extension and recoil in inches.	Observed elasticity.	Extension and recoil in inches.
5'	.0042323	19'	.0160829	33'	.027965	47'	.039835	61'	.051695
6	.0050788	20	.0169295	34	.028821	48	.040680	62	.052539
7	.0059253	21	.0177760	35	.029665	49	.041529	63	.053385
8	.0067718	22	.0186525	36	.030508	50	.042370	64	.054230
9	.0076183	23	.0194690	37	.031360	51	.043223	65	.055076
10	.0084648	24	.0203455	38	.032200	52	.044060	66	.055920
11	.0093114	25	.0211618	39	.033050	53	.044903	67	.056792
12	.0101579	26	.0220083	40	.033839	54	.045750	68	.057639
13	.0110041	27	.0228548	41	.034769	55	.046596	69	.058484
14	.0118506	28	.0237013	42	.035580	56	.047439	70	.059330
15	.0126972	29	.0245478	43	.036425	57	.048288	71	060173
16	.0135437	30	.0253941	44	.037270	58	.049158	72	.061019
17	.0143902	31	.0262406	45	.038115	59	.050009	73	.061865
18	.0152367	32	.0270871	46	.038989	60	.050848	74	·062710

Instead of the numbers in the table, a tolerably near approximation to the true temporary elongation corresponding to each observed elasticity of the bar in minutes, might have been obtained, by multiplying the number of minutes by .000847 inches, the length of one minute on the arc of a circle, the radius of which is 2.914 inches. This would give a result sensibly correct, especially for all numbers of minutes under 60.

Sources from which the materials were obtained.—The materials on which the committee have performed the experiments detailed in this report, were procured from various sources, a considerable quantity having been collected previous to their appointment by one of its members, then making arrangements for a private course of investigations on several scientific and practical points, relating to tenacity. Other specimens were voluntarily offered, or kindly supplied at the request of the committee by the different manufacturers, or other persons to whom application was made for that purpose. In several instances, more specimens of the same iron were furnished than will be found mentioned in the tables as derived from the same quarter;—the whole number obtained being about 250, and the number tried about 150. As the aim of the experiments was the establishment of such practical truths as might be found generally useful in regard to the manufacture and employment of materials

for steam boilers, it was not deemed necessary to enter into a minute comparison of the merit of different manufacturers from whom the materials were received, nor to limit the inquiry to any given number of specimens or of trials on those derived from each source. The reader will, however, be able to institute such comparisons as his curiosity may dictate,—the tables furnishing all the facts, (as well as the names of the manufacturers when known,) which have been obtained by the committee, in regard to the origin and manufacture of the different specimens.

Among the names of those from whose manufactories specimens have been received, are Messrs. Mason and Miltenberger, H. S. Spang and Son, Barnet Shorb, H. Blake and Co., and Shoenberger and Son, of Pittsburg; S. E. H. and P. Ellicott, and E. T. Ellicott and Co., of Baltimore; the Salisbury Iron Company, of Salisbury, Connecticut; Messrs. Yeatman and Woods, of Nashville, Tennessee; Mr. Massey, of Maramec, Missouri; R. Lukens, of Coatesville, Chester county, Pennsylvania; George Pennock, McWilliamstown, in the same county; Messrs. Grubbs, Lancaster county, Pa.; Hardman Phillips, Esq., Clearfield county; and Messrs. Valentine and Thomas, Centre county, Pa. To Messrs. A. and G. Ralston the committee were indebted for specimens of boiler, bolt, and railroad iron, of English manufacture, which served as means of comparison between the foreign and the domestic material; and from other importers they procured those of Russian and Swedish manufacture, for the same purpose. All the samples of American iron thus far mentioned, were manufactured by the aid of charcoal. A single specimen furnished by Mr. P. Ritner, of Carthouse's Place, on the West Branch of Susquehanna river, was made by smelting with coke.

The specimens of boiler copper, tried by the committee, were obtained from the establishment of John M'Kim, jr. and Son, of Baltimore.—To the above, and several other gentlemen who were active in procuring the materials, and otherwise forwarding the objects of this inquiry, the committee are bound to offer their grateful acknowledgments.

Preparation and gauging of the specimens.—The experiments were made on materials in several different forms; and as the results are in some measure dependent on the circumstances now referred to, it seems proper to describe those several conditions, together with the method of obtaining the areas of transverse sections at the points of fracture.

As the greater number of experiments was, of course, made on materials manufactured expressly for steam boilers, the mode of preparing these is of most importance. The strips were in general cut, by shears, from the plates. about 2 or 2½ feet long, and 1 inch wide ; and with a view to determine the tenacity in different directions, they were cut either lengthwise, crosswise, or diagonally of the direction in which the plate had been rolled.

The tables will be found to indicate, in all cases where rolled iron is under consideration, the direction of the slitting.

On specimens of this kind, trials were made in three ways. *First,* by finding and measuring the area of the smallest section, as the strip came from the shears, and placing it in the machine, applying force till that or some other section gave way. When not broken at the smallest section, the actual area of the point of fracture was ascertained approximately by measuring, after fracture, at a short distance on each side of the broken part, taking care to keep just outside of the constriction or part sensibly diminished by the strain. After thus determining the area previous to trial, a portion of the bar was replaced, and other fractures made, until the specimen had been used up. Fractures on bars, tried in this manner, are referred to in the tables as made at *original sections.* But as the slitting of bars in the manner described necessarily caused· some diminution of strength along the edges, and as from accidental causes this diminution was often very unequal, it was apparent that the irregularity in the *strength* might frequently be greater than that in the *breadth* of a strip. To ascertain the mean effect of the shears on bars of this breadth, the second method of trial was adopted.

This consisted in filing away a section of the metal on each side of the strip, in the form of the segment of a circle. At different points, these sections were filed to different depths, with a view of ascertaining how far beneath the surface the metal had been affected. The scale of oxide was also in some cases filed from the surface, but in most instances where the rolls had left the iron tolerably smooth, it was thought best to take the measurements of thickness, as they must be taken in practice with the surface in its natural state. In some instances, it will be found that the fractures did not occur in the filed section, even when a considerable portion of the whole material had been filed away. In general, however, about one-eighth of the breadth ot the bar being removed by the two opposite sections, the sound part of the metal was attained, and gave results nearly proportionate to the areas of the remaining sections.

But as neither the rolling nor the hammering of iron can give a perfect uniformity of structure, and as consequently the results on very deeply filed sections would not always prove uniform in their indications of strength, it became necessary, in order at once to remove the irregularities proceeding from the slitting, and to compare the advantage of different modes of manufacture, and different kinds of metal employed, as well as to ascertain the maximum and the minimum strength of the same bar at various temperatures, to employ the third method of preparation, that of filing away the edges of the inch bars till they were reduced to three-quarters of an inch in width throughout their whole length, and also removing completely the scale from both faces, and rendering the thickness as nearly as possible uniform throughout. The bars treated in this manner were next divided through their whole length into spaces of one inch each, marked across with a steel point, numbered at every inch, and subsequently gauged at every mark, both in breadth and thickness. In these measurements, as well as those applied in the two other methods of preparation, the gauging was carried to thousandths of a lineal inch in both directions, giving the areas, true to millionths of a superficial inch.

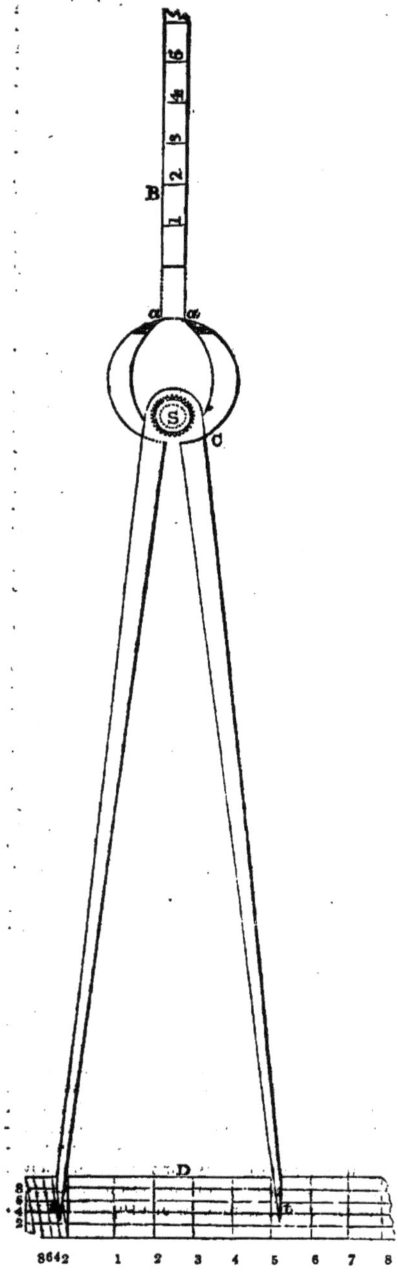

Plate II. (opposite column) represents the apparatus used for this purpose, and a portion of a bar prepared for gauging. C is a pair of proportional callipers of brass, pointed at *a a* with steel. S is a screw head projecting half an inch above the face of the instrument, and is one-third of an inch in diameter, being a trifle less in length than the thickness of the two arms of the callipers. The distance S *t* is 10 times that of S *a*, so that the space between the points *a a* is read into tenths, hundredths, and thousandths, when that between *t t*, on the diagonal scale D, is found in inches, tenths, and hundredths.

Specimens of hammered iron and of iron formed into bars by rolling and slitting, were tried with a view to certain comparisons; and in these cases all the three modes of preparation applied to specimens of boiler iron were likewise employed. In a few instances specimens were received from the manufacturers in a form which required no alteration before trial, but in the majority of cases they were to be either slit or hammered and filed to adapt them to the purpose of these experiments. In the treatment of boiler-iron, *heating* before trial was, with few exceptions, avoided. The tables will be found to contain a few experiments on upsetting, annealing, and hammer-hardening. They will also exhibit a very limited number of trials on cast iron and steel; but as these materials enter sparingly into the composition of steam boilers, and as their tenacity has been formerly much more extensively examined than that of boiler-plate, it was not considered within the purpose of the present investigation to do more than present a few verifications of the correctness of those results on which practical men commonly rely.

The bars of cast iron were tried as they came from the mould, or with very little filing to remove the irregularities of the surface. The specimens of copper were all reduced by filing to a good degree of uniformity, and gauged as already described.

At the foot of each column of original areas in each bar gauged throughout its length, will be found the mean area, and under areas of "sections of fracture" are the mean areas of the points broken.

(To be continued.)

MARINE STEAM BOILERS AND ENGINES.—SAFETY STEAM-BOAT COMPANY.

Sir,—The steam-boat boilers that burst in America are generally of one construction, and that is cylindrical, and of small diameter, with a flue running through the centre to the funnel. The bursting of steam boilers, both on land and at sea, is also getting more frequent in this country, and it appears to be owing to cylindrical boilers of small diameter, or too small water space, becoming more common. It appears that from the frequent bursting of boilers of the above construction, and the accidents arising therefrom, that such boilers are totally unfit to be used in steam-boats; and are not, even on land, to be much relied on when used to generate high-pressure steam, whatever situation they may be placed in. The public should be warned against exposing their lives and property to such a risk; a public demonstration of dislike to cylindrical boilers would soon cause others of a safer construction to be substituted in their place.

In boats fitted with cylindrical boilers, the engines can be used either as high or low-pressure; on enquiring on board such a boat, they are always called low-pressure engines.

High-pressure boilers, whether on land or in steam-boats, whether stationary or locomotive, are only safe when divided into numerous small compartments, such as Perkins's steam generators, or James's tubular boiler, now in general use in locomotive engines on railroads; or, best of all, Walter Hancock's boiler; a boat fitted with such boilers, and Hall's or some other system of surface condensation, would be even safer than a boat fitted with low-pressure boilers. When I was at New York, in 1835, I saw a boat, called the *Water Witch*, fitted with tubular boilers of similar construction to those used in railroad locomotives, but on a much larger scale; toothed wheels and pinions are employed, instead of cranks, to communicate the motive power to the paddle-wheels in the towing steam-boats on the river Hudson, N. A.

The water space in all steam engine boilers ought never to be less than 7 inches in depth, excepting in boilers divided into small compartments.

Raub's safety apparatus for steam boilers, described in No. 734 of your Journal, is an excellent invention, ingenious and simple; all steam boilers ought to be fitted with it.

A real Safety Steam-boat Company is wanted: if such a company was formed there is little doubt of its being encourged by the public, and being a good speculation. The boats on the different stations

to be known by the name of the Safety Boat; the boats to be built of iron, the hull to be divided into its several compartments with water-tight bulk-heads of iron; the boilers to be low-pressure, or safety high-pressure, such as Hancock's, the tubular boiler, or others; the boilers to be supplied with fresh water, by condensing the waste steam, by means of Hall's, or some other, surface-condensing apparatus; the vessel to carry boats according to the average number of passengers she is likely to have, and these boats to be made life-boats, which can be done at a small expense; the vessel also to carry as many of that very simple, cheap, and efficient apparatus, Mackintosh's water-proof-cloth life-preservers, as they may be likely to want, in case of being wrecked.

I am surprised that Witty's smoke-consuming furnace, described in No. 523 of your Journal, has not got into general use: the apparatus is simple, of little cost, and answers perfectly the end intended; it is applicable to steam boiler furnaces either on land or at sea; there must be some prejudice amongst engineers against it, or it would ere this time have got into general use. A smoke-consuming furnace is much wanted, and nothing could be more simple and efficient than it is; none of the inventions for that purpose that have come out are at all equal to it.

Hall's patent smoke-consuming furnace, described in No. 752 of your Journal, is exactly similar to the principle described in No. 734 as Coad's smoke consumer; it is ingenious, but is more expensive and complex than Witty's.

Chanter and Co.'s smoke-consuming furnace, described in No. 771 of your Journal, appears to me to be a modification of Witty's, but is certainly no improvement on it.

In some manufactories the steam boiler fires are fed with fuel by means of hoppers, which scatter the coal lightly over the fire; but these hoppers require an upright shaft with an apparatus attached to it to shake the coal out, which shafts are driven by the engine, thus consuming a part of its power: they are much more expensive, and not nearly so simple as Witty's smoke consumer.

I remain, Sir,
Yours faithfully,
ARTHUR TREVELYAN.

Anglesey, Oct. 3, 1838.

INTERNAL CIRCULAR BOILER FLUES.

Sir,—I have been hoping that the liability of circular flues having the pressure outside to collapse, and the serious consequences likely to ensue, would have called the attention of competent persons to the subject, and that they would have given a rule for estimating the pressure such flues would bear, sufficiently near the truth for practical purposes; but I have been disappointed.

If the remarks I made in your 781st Number are correct, the extent of the deviation from the true form must be an important element of the calculation; as the deviation which would be unimportant in a small flue, would be of great importance in a large one.

Tredgold, who is, perhaps, as good an authority in matters of calculation of this kind as any one who has written on the subject in your Magazine, says, in the first edition of his work on the steam engine, page 259, "it is indifferent whether the curve be concave or convex to the pressure, provided it have either abutments as an arch, or forms a complete circle;" and, in the next page, he gives a rule which estimates the strength to be in direct proportion to the thickness.

Yours, &c.

C. G. JARVIS.

3rd Oct. 1838.

NEW MATERIAL FOR PENDULUMS.

Sir,—When I read the article on fireproof houses, in No. 778, (p. 232,) I was much gratified with the result of the experiments, and it struck me at the same time that the composition would answer another useful purpose. If the composition has the properties assigned to it by our acute friend Mr. Baddeley, (and I sincerely hope he is not deceived,) —that it remains of the same bulk under all temperatures, that it will neither expand nor contract—if a rod of it is sufficiently cohesive to bear a few pounds weight, it would form an excellent substitute for a compensating pendulum. I hope some of your scientific correspondents will take the hint, try the experiment, and make the result known, through your valuable publication.

I am, Sir, yours, respectfully,

T. W.

Birmingham, Oct. 3rd, 1838.

ON FIRE-ESCAPES.
"Discordia Concors."

Sir,—The preservation of human life is a subject so deeply involving all the best charities, that all controversy upon it should be conducted, as I am happy to see it *is* conducted in your Magazine, with an entire absence of those acrimonious feelings which would be a great slur and stain. I am quite aware of the worth and talent of Mr. Baddeley; and could I answer satisfactorily his objections, his fame would bear so trifling a diminution. My ideas may be somewhat "crude," for I am not a scientific person; though an attentive observer, who is not so, might make a happy hit. I am also a "Tyro;" but, notwithstanding the oversight as to Mr. Wivell's "Escape," (which yet did not affect the paragraph,) I *have* seen most of the modes of "fire escape" in London.

My object was to bring, or revive, into notice, the "*canvas* (flexible) *trough escape:*" this done, it must stand on its own merits. Mr. Baddeley feels it necessary to condemn it; including, also, Mr. Wivell's, so much lauded by many, which only proves how "doctors disagree." By a coincidence, however, which has, doubtless, struck him as curious, in a letter immediately following, by another correspondent, the Parisan "*salvage bag,*" which is precisely the same thing as regards the style of descent, (whilst the management is more complex,) is pronounced to be "one of the most simple and efficacious fire-escapes yet known," and to have "perhaps been the means of saving more lives than any other extant." The price, also, is about what I stated.

Mr. Baddeley's scientific knowledge and experience being so much greater than my own, I will suppose the "steadying-weight" impracticable; whilst I regret that it is so. But is the objection to the difficulty of lifting the weight and ejecting it eight or ten feet,—to the inefficiency of its resisting power when on the ground,—or the danger of breaking the trough by the shock, when fixed? Certainly, without an expressive cry of "*garde!*" there might be danger to the "brains" of persons below; and, however small each one's quantity may be, he has, generally, a bigotted inclination to retain it,

The question as to which mode of escape will least tax the confidence and presence of mind of the party to be saved, must be one of private opinion, and probabilities; but one of such *intense* interest that it loudly calls for the most anxious and sympathetic attention. Let any one but fancy himself in such an agonising situation, and he will know how to feel for the victim; or let him see the " poor remains" on the morning after a fire! I confess that, to me, the idea of a person's slinging himself, and descending by a rope, (even were it done at noon-day, and for experiment,) is infinitely more formidable than that of entering a protector trough or bag, where the danger would be scarcely seen,— which is " half the battle." I allow that the former would be *cheaper*, and very much better than nothing; but in the case of delicate health, or age— particularly in females—is there a strong chance of success? Nothing, however, can be better than mooting these points, for the public mind is informed, and the opinion of the majority, and of different classes, is attainable.

You will observe, that the Paris "bag" is *fifty* feet long, with means of rescuing the parties beyond its length. Every London " escape" which I have seen or read of, is decidedly *too short* for contingencies: the existence of a house a little higher than the rest, or even of an attic story, seems to have been thrown out of account;* besides, the inventors seem to have taken in for granted that it can be held nearly perpendicularly; whereas, in the case of flames bursting forth from the lower story, it might be necessary to have the base at least 15 feet off; and though no failure has yet been reported from the insufficient length of the escape, how very melancholy and cruelly tantalizing it would be thought if it did!

Let the humble suggestion I threw out of having the " escapes" fortified by some anti-combustible solution not fall to the ground. Ignition has, also, not yet occurred, but there must always be a chance. A " portable staircase" has, perhaps, been considered too wild an

idea for notice; but from some opinions I have heard, I think a practicable plan will be produced. Should Mr. Baldeley notice these observations, I trust he will consider them unpresuming and respecful.

I am, Sir, yours, respectfully,

J. D. PARRY.

Oct. 6, 1838.

SUBURBAN GAS-WORKS.

Sir,—Would it not be practicable to have gas, for lighting and heating, brought to London from the coal countries, by pipes along the railroad walls, or some other part thereof, and, by that means, save the metropolis the vast expense of freight on coals, which is, on an average, one hundred pounds a-ship, besides lighterage, metage, City-dues, carting, and lumbering of every description; and, likewise, do away with the nuisance of gas-houses about London? I am informed that seven or eight shillings, at the pit's mouth, will purchase a ton of coals, which will yield from 800 to 1200 cubical feet of gas, making the price per thousand, exclusive of the cost of pipes, &c., nine-pence, for which we now pay the companies nine shillings; but this is an outside estimation, for there is in the pits a great quantity of small coal, at present useless, which would be as convertible as the other, and would, if brought into use, reduce the cost, we might fairly calculate, to one-half, or about four-pence per thousand feet. Should it be objected to on the ground of the great pressure required to send the gas to London, I would propose, that stations at certain distances, perhaps fifteen miles apart, be established, with gasometers at each, keeping always a good supply, so that the pressure would be divided to any required degree. Perhaps some of your intelligent and practical readers will say something on the practicability of the above scheme. My object in troubling on the subject is, that if there be a chance of its becoming useful to the community, it may not be lost for want of discussion amongst men of science.

I am, Sir,

Your obedient servant,

G. N.

Vauxhall Bridge Road, Oct. 5, 1838.

---

* I know that houses in Paris are, generally, higher than in London; but how many are there here of from 40 to 50 feet, and upwards, perpendicular height. Of what use would an escape of 34 feet be, in the highest?

TOOTHED GEAR FOR LOCOMOTIVE ENGINES.

Sir,—I am sorry to perceive, by Mr. Trevelyan's letter, (page 9 of your last Number,) that he has altogether misunderstood the purport of the writer's observations, in the article "Speed on Railways," relative to toothed gear for locomotive engines. The proposition was not to employ wheel and pinion gear *instead of the crank*, but in conjunction with it, so as to moderate the velocity of the piston without detracting from the speed of travelling.

" Some engineers contend, that the cranks by which the continuity of the *working axle* is broken in two places, is a monstrosity in engineering, and a complete violation of every principle of sound mechanical science; that the *working axle* on which the chief part of the weight of the engine must be thrown, in order to give the driving-wheels sufficient adhesion, is thereby rendered the weakest part of the machine." It was, therefore, proposed to connect the piston-rods to a separate crank, and *convey* the power of the engine to the working axle, by the intervention of wheel and pinion gearing. Some practical difficulty having been stated to impede the adoption of this plan, I took occasion, in your 786th Number, to explain, that Messrs. Heatons, of Birmingham, had perfected and patented a peculiar method of conveying the power from the crank to the working axle of the driving-wheels, by means of tooth and pinion gearing, capable of considerable variation as to the relative proportions of time and power.

This letter Mr. Trevelyan, also, seems to have mistaken; but I trust this explanation will put him right. When Watt was precluded from using the crank, he invented his ingenious contrivance, well known as the sun-and-planet-motion; but neither this, or any other system of toothed gearing, will ever, successfully, rival the crank. Such methods may be advantageously employed for transmitting, and for varying of motion, but it is only in a very few instances that toothed gearing will be judiciously employed for the purpose of converting right-lined into circular motion.

I remain, Sir, yours, respectfully,
WM. BADDELEY.

London, Oct. 9, 1838.

PROPOSAL TO ERECT PUBLIC BAROMETERS AND THERMOMETERS.

Sir,—Having had occasion recently to visit Geneva, my attention while at that place was arrested by a stone pillar, which I observed on the quay by the side of the lake. On approaching it I discovered a barometer and thermometer fixed into it for the public use. It struck me forcibly that this idea might be advantageously borrowed by ourselves, 1. These instruments, of infinite utility to sea-going people (particularly the first), might be conspicuously placed in all the ports, towns, and villages on our coasts, on the quays, bridges, or near the customhouses. There is no question that by habitual attention to the state of the barometer by those who go out to sea in small craft, such as fishermen, many valuable lives might be saved, and many large families preserved from destitution, since it is well known that this instrument foretels gales even more accurately than rain. 2. Their utility to the practical agriculturist is well known. For his use they might be placed in the churchyard or porch, or market-place, of every town and village in the kingdom, so as to be universally accessible. 3. The thermometer would be useful to all that numerous class of retail shopkeepers who deal in articles susceptible of injury from temperature, as meat, fish, beer, tallow, &c. 4. The general knowledge of the properties of these instruments would draw the attention of the public to the science of met-orology, and thus become a source both of amusement and instruction. 5. The pillars or other erections used for their establishment and security might occasionally be made otherwise useful and ornamental : useful, for example, in breaking the thoroughfare of broad streets at the crossings and thus becoming a source of security to foot-passengers while crossing, or for supporting lamps; ornamental inasmuch as there is no reason why good taste should not be exhibited in these little structures. That above-mentioned is a circular pillar of white stone, divided into compartments in the form of Gothic arches in bas-relief ; the instruments are let into the stone, in two of them at opposite sides, and the whole protected by a neat iron railing. It is decidedly ornamental to the quay. It is possible that in the beginning they might be wantonly injured, but in time the evidence of their general usefulness would cause them to be respected by all classes.

F. R. S.

NOTES AND NOTICES.

Substitute for Emery.—Topaz, the discovery of which in the United States was first announced in this journal *(Silliman's American Journal)*, many years since has continued to occur in such abundance, (although not *in general* beautiful,) that the owner of the locality has been induced to crush it to powder as a substitute for emery. The hardness of the topaz is such (8) as to place it next to corundum, (9) with the exception of spinella, automolite and chrysoberyl, which approach nearer to corundum than topaz ; but they have never been found in the quantity that the latter occurs at Monroe. And we understand that those who have made use of this substitute find, that for all common purposes it answers very well.

Portable Iron Steamers.—We have observed a very large iron steam vessel appear on the north bank of our river, immediately under the Broomielaw, as if by magic. On inquiry, we find it was by those successful engineers, Messrs. Tod and M'Gregor, and it was taken down on carriages, in three pieces, to where she was to have her carpenter work completed, by those scientific gentlemen, Messrs. Hedderwick and Rankine. This vessel is, we understand, 185 feet long upon decks, and her model is much admired. She is destined to ply between Glasgow and Liverpool, being the first of a line of iron steam vessels building for that trade, by a company lately established. The successful operations of the large iron steam vessel, the *Rainbow*, now plying between London and Antwerp, will, we have no doubt, secure for these vessels, when started, a favourable reception.—*Glasgow Chronicle.*

Cleansing Casks.—Mr. Peter Walker, brewer, Liverpool, late of the Fort Brewery, Ayr, and son of the late Andrew Walker, Esq., of Bonville, and coal-master, Gairbraid colliery, Glasgow, has obtained her Majesty's letters patent for an apparatus for cleansing beer and other fermented liquors. The advantages, and saving to the proprietor of a brewery of thirty barrels length brewing four times a-week, will be at least 300l. sterling per annum, exclusive of men's wages. It is peculiarly calculated to assist the private brewer, being applicable to vessels of any size and in any situation.

Percussion Cannon Locks.—The percussion principle has been adapted to the locks of great guns by Commander Henderson, R.N., by a method at once so simple and effective, as promises at no distant period to supersede all others. The apparatus consists of two square pieces of iron, a common fowling-piece nipple, and an iron cap to cover the nipple. The two pieces of iron are made just large enough to cover the groove about the touch-hole, and are connected with each other in the form of a hinge. One of these is fastened to the gun, by means of a screw, to the left of the touch-hole, and has the iron cap fastened to it in the same manner. The other piece of iron has the nipple screwed on to its centre, and of course communicates with the touch-hole of the gun when folded down. The gun being loaded, the cartridge pricked, and tube introduced, a common copper cap (such as is used for fowling-pieces) is put upon it ; the iron cap is then brought over the copper one, when a tap with a wooden mallet never fails to ignite it, and discharge the gun. When fired, the plate of iron with the nipple is thrown back upon the other, thereby exposing the touch-hole, and giving room for the vent to be closed by the thumb in the usual manner. It is not the least of its advantages that the ship's armourer can fit the guns of a first-rate in this manner in a few days, and it does not interfere with the present equipment, as percussion, or other locks, may be used at the option of the commanding officer, according to the circumstances of the case, as the match or salamander may be also.—*Naval and Military Gazette.*

Trevelyan's Printing Press.—Sir,—Observing in your last Number an account of Trevelyan's printing press, I beg to inform you that I have a press at the Polytechnic Institution, Regent-street, that I constructed 3 years ago, in the working of which the impression is produced in the same manner as described of Trevelyan's press. If Mr. Trevelyan can show priority of invention I have no desire to strip him of the credit or advantages attending it ; if not, I hope you will do me the justice you have often awarded to others. I am, Sir, &c.

J. W. WAYTE,

Pressman and Engineer to the Morning Advertiser.

Eastern Counties' Railway.—Within these few days we have been along the line which is under contract as far as Romford, and on the entire distance the utmost activity prevails. From Dog-row, near the Mile-end-road, where the temporary terminus is to be, the ground is cleared of the houses and buildings which have hitherto impeded the operations of the company ; but we understand that this part will be commenced and finished forthwith. For a considerable distance on the London side of the Regent's canal, the embankment to carry the railway is in course of formation. A brick and stone, [A mistake. The superstructure is of iron; the piers, only, of brick and stone.] bridge, of a substantial and handsome design is nearly completed over the canal; between which and the river Lea the ground is fenced off, and the road bridges, four in number, in course of execution. From the river Lea to Ilford, the *whole* of the brick-work and masonry is finished, with the exception of (in a few instances) the parapets and coping. The embankments over the Stratford marshes will be shortly completed, the permanent rails being already laid on a portion of it, and for a considerable distance in the cutting by Maryland-point. The company have employed, on this part of the line, two powerful locomotive engines, in place of horses. for drawing the earth-waggons. In the Stratford marshes they have a contrivance to facilitate the tipping of the earth-waggons, which consists of a moveable stage of about 40 feet in length, on which are laid two lines of rails, corresponding to those on the embankment ; one end of the stage rests on the embankment, the other end is supported by frame-work on wheels ; on this stage the waggons are run, and the contents tipped with great rapidity. The brick-work and masonry is generally to be praised. For a considerable distance on the London side of Ilford (where a station is to be made) the permanent rails are also laid. At Ilford the company are proceeding with great spirit, all the houses and buildings being cleared away, and the excavations proceeding rapidly. The works, as far as Romford, are being carried on with the same activity, and by the time the earth-work between Stratford and within about a mile of Ilford is finished (which alone remains to be done), the line from Ilford to Romford will be completed, so that within a very few months from the present time, the public will be able to avail themselves of this line as far as Romford ; and from the immense traffic on the great eastern road. an immediate and profitable return will be made to the shareholders.—*Civil Engineers' Journal.*

Promenading in Railroad Coaches.—In papers on railroads published in the *Scotsman* in 1824, [See *Mech. Mag.* vol. iii, pp. 211, 237, 245] we anticipated that the locomotive coaches might be made of such a breadth as to permit the passengers to walk in them ; and that, by joining one coach to another, and having open communications between them, the walk might be of such a length as to serve for recreation, as on the deck of a steam-boat. Everybody knows the irksomeness of sitting long in a stage coach without the power of stretching the limbs, or getting away from a disagreeable companion. The idea we threw out is already in part realized in America, as we learn from the subjoined statement from *Stevenson's*

American Engineering. The Great Western Railway may admit of this improvement in consequence of its increased breadth :—"The *passenger carriages* of the American railways are extremely large and commodious. They are seated for 60 passengers, and are made so high in the roof, that the tallest person may stand upright in them without inconvenience. There is a passage between the seats, extending from end to end, with a door at both extremities; and the coupling of the carriages is so arranged, that *the passengers may walk from end to end of the whole train without obstruction.* In winter they are heated by stoves. The body of each of these carriages measures *from 50 to 60 feet in length,* and is supported on two four-wheeled trucks, furnished with friction-rollers, and moving on a vertical pivot, in the manner formerly alluded to in describing the construction of the locomotive engines. The flooring of the carriages is laid on longitudinal beams of wood, strengthened with suspension rods of iron." —*Scotsman.*

Scientific Instructions.—It is very instructive to throw even a rapid glance over the directions drawn up by scientific bodies for the use of travellers; in them we see the positive degree to which various sciences have attained, and the desiderata still to be sought for; we can form beforehand a general idea of the countries about to be explored, and we see what has been done from interval to interval. Many travellers obtain glimpses of phenomena which are too readily adopted as facts, and they are inserted in our elementary books of science as such. It would be well if the compilers of these could procure a sight of the instructions above alluded to, and thus ascertain whether the problem be really solved or no. We have been particularly led to these reflections by a perusal of those supplied by the academy, for the savans who are to attend the army in Algiers; not only are they extremely interesting in themselves, but they so ably and clearly show what has been done, and what is still to do in that part of the world, that we should like to see them printed in the form of a pamphlet for general distribution. The geological part, drawn up by M. Elie de Beaumont, in particular, gives an excellent picture of the nature of the soil, and its connexion with the great desert of the Sahara makes a further knowledge highly desirable. Some curious inquiries are recommended concerning the plague, blindness, and hydrophobia, and M. de Freycinet has particularly desired some observations to be made either to deny or confirm the received opinion, that there are no tides in the Mediterranean. M. Biot has contrived a new apparatus in order to procure water from a great depth in the sea, and it is to be tried at Algiers, as well as in the *Bonite.* Most important phenomena concerning the under-currents of the Mediterranean, still require to be ascertained; such as, whether the cold water which flows from the pole, do, or do not, enter the straits of Gibraltar, &c.—*Athenæum.*

Chinese method of preparing Eggs.—Eggs of certain ducks are prepared in China so as to keep for one or even two years; for ten eggs they take half a pint of ashes of cypress wood, or bean stalks (some use potash), ⅔ths of powered chalk, and two ounces of pulverised coarse salt. This is wetted with a strong infusion of tea, so as to form a paste, with which the eggs are entirely covered, they are then put into an earthen vessel and hermetically sealed.

Stone for the new Houses of Parliament.—The inquiry which was committed to Mr. Barry, M. de la Beche, and others, as to the stone to be employed in the erection of the new houses of parliament, is at length closed, and specimens collected, received from quarries in nearly every part of the kingdom. Some time, however, must elapse before the particular kind is fixed on, as the specimens are to form the subject of very careful examination and analysis in London.

Foul Air of the Thames Tunnel.—In a communication on the progress of the works at the Thames Tunnel, made to the Institution of Civil Engineers, at one of its last meetings, Mr. Brunel stated that the excavators were much more inconvenienced by fire than water. Some of the gases which issue forth ignite very rapidly, and the reports from Guy's Hospital stated some of the men to be so much injured by breathing them that but small hopes are entertained of their recovery. The explosions are frequent, and put out the candles of the workmen, but the largeness of the space prevents their being dangerous. These deleterious gases issue from the mud of the river, and enter from a crevice at the top. Chloride of lime had been used, but without success. Inhalation of the gas produces sickness and other disagreeable sensations.

New Botanic Gardens.—The interior portion of the Regent's-park, late the nursery of Mr. Jenkins, will shortly be laid out as botanical gardens, the Commissioners of Woods and Forests having granted a lease to a society newly formed, under the title of the Royal Botanic Society of London, at the head of which is Duke of Richmond as president. The object of the this society is the formation of an extensive botanic garden, with a library, museum, and conservatories, so that medical and scientific, as well as merely ornamental botany will respectively receive the attention commensurate with their importance. Public exhibitions and lectures will also be given periodically during the season. On the council are the names of the Duke of Devonshire, Lord Teignmouth, Sir George Staunton, Sir Astley Cooper, Professor Don, &c.

Eastern Counties' Railway.—We understand that Mr. Braithwaite is acting in a most extraordinary manner in the execution of this great work. When a bridge is to be built, or an excavation made, he absolutely refers to the estimate, and limits the expenditure to the sum therein laid down! By following this rule rigidly, instead of a million or two of excess, it is probable that the difference on the entire work, between the estimate and the cost, will be a mere trifle. In that case, the Eastern Counties' Railway will be indeed, a prodigy.—*Spectator.*

Greenwich Railway.—There is every prospect, as we are informed, of the line being opened by the end of the present month. Scarcely any thing now remains to be done except the iron work of the bridge over the Ravensbourne Creek, and this is now in active progress. The permanent way between Deptford High-street and the Greenwich terminus, is in a very forward state. This portion of the line is laid on longitudinal wooden sleepers, with three-feet bearing. The receipts from passengers continue good: the increase on the last nine months is nearly 1,200*l.*—*Railway Times.*

The Railway Map of England and Wales continues on sale, in a neat wrapper, price 6d.; and on fine paper, coloured, price 1s.

☞ *British and Foreign Patents taken out with economy and despatch; Specifications, Disclaimers, and Amendments, prepared or revised; Caveats entered; and generally every Branch of Patent Business promptly transacted. A complete list of Patents from the earliest period* (15 Car. II. 1675,) *to the present time may be examined. Fee* 2s. 6d.; *Clients, gratis.*

LONDON: Printed and Published for the Proprietor, by W. A. Robertson, at the Mechanics' Magazine Office, No. 6, Peterborough-court, between 135 and 136, Fleet-street.—Sold by A. & W. Galignani, Rue Vivienne, Paris.

𝔐𝔢𝔠𝔥𝔞𝔫𝔦𝔠𝔰' 𝔐𝔞𝔤𝔞𝔷𝔦𝔫𝔢,

MUSEUM, REGISTER, JOURNAL, AND GAZETTE.

| No. 793.] | SATURDAY, OCTOBER 20, 1838. | [Price 3*d*. |

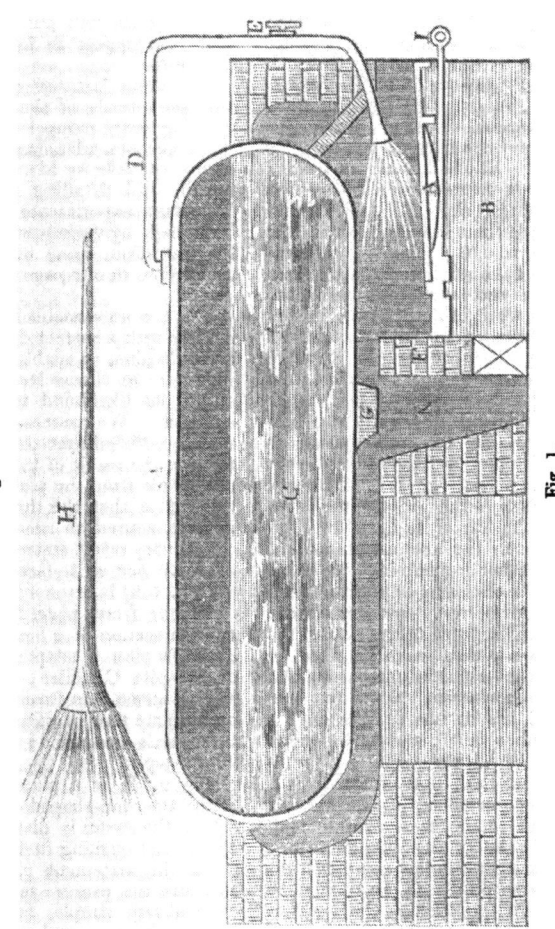

IVESON'S PATENT MODE OF EFFECTING THE COMBUSTION OF SMOKE.

Fig. 2.

Fig. 1.

IVESON'S PATENT MODE OF EFFECTING THE COMBUSTION OF SMOKE.

Smoke has always been regarded as a nuisance, and for a long time as a waste of fuel. As the age of steam advanced, and as its benefits were extended over, and enriched the face of the earth, so did the concomitant evil, smoke, throw a cloud abroad which bade fair to obscure the face of the sky. So great became the evil, that in process of time every one not directly interested in its cause complained aloud, and seemed willing to forego the grand effects of the steam-engine rather than put up with its smoke. Petition upon petition poured into the Houses of Parliament praying the legislature to compel the manufactories to put out their fires or swallow their smoke. Learned societies offered premiums to incite its members to search for an antidote; chemists and mechanicians racked their brains to obtain the desired secret. A parliamentary committee was appointed to examine into the subject; and numerous were the plans which were brought under its notice, and described in the reports of its proceedings. An Act was the result of this inquiry, rendering manufacturers indictable for a nuisance if the smoke of their chimneys annoyed their neighbours, and giving the judge power to compel the culprit to adopt any efficient plan for remedying the evil. This law has, however, never strictly been enforced, but the fear of its infliction compelled many to try and adopt various plans for either consuming or doing away with the offending vapour. Chimneys hundreds of feet high were built; hoppers, reverberatory furnaces, and numerous other devices were patented, or suggested, (not a few in our pages), and carried into operation. Some few succeeded, and are in use to this day; a greater number failed, and have gone to the "tomb of all the Capulets." Another incentive than that of fear of the law has, however, been of late stirring up inventors to devise a plan to consume the smoke of the manufacturer's furnace. Competition has compelled the users of steam-power to economize in the means of producing that power. The capability of the fuel to produce heat has been pushed to the utmost. Now the dregs—the smoke—only remains, and that must, and will, be turned to profitable account, and love of gain will effect what fear of loss could not accomplish.

If the expectations which have been raised in the present instance prove to be well founded, the bane has been found to carry with it its antidote—a small jet of steam to be all that is necessary to disperse a vast column of smoke. *Auld Reikie* is the birth-place of the discovery now under notice; that city will now, mayhap, deserve this title no more; she will purify herself of her soot, and appear as fair as any city of the south.

For these last few weeks the papers and periodicals of the day have been teeming with paragraphs setting forth the enormous advantages of the grand discovery made by Mr. Iveson, of Edinburgh, and detailing some strikingly important experiments stated to have been made by various eminent men upon the invention, some of which we have transferred to our pages. (See vol. xxix, p. 408.)

We are now enabled to present our readers with a correct description of Mr. Iveson's plan, to enable them to form a judgment for themselves upon its merits, and of its likelihood to answer the desired end. We confess that we are by no means satisfied that the advantages set forth as the result of the experiments are obtainable from the plan described. We think also that the theory by which it is endeavoured to account for the extraordinary result stated, that of a saving of forty per cent. (see *Mech. Mag.* last vol., p. 408) is erroneous.

In our front page, fig. 1, is a longitudinal section of a boiler to which Mr. Iveson's plan is adapted. A, fire-bars; B, ash-pit; C, boiler; D, tube for conveying steam into furnace; E, stop-cock to regulate flow of steam; F G, bridges afterwards described; I, dander plate; K, dust-pit.

Fig. 2, H is a plan of the tube D, with the fan-shaped termination, by which the steam is distributed over the top of the burning fuel in the furnace.

In the statement circulated by the patentee the process is described to be "peculiarly simple, and of cheap and easy application, and may be adapted to any existing furnace in a short time. It consists in the admission of steam into the furnace, and discharged over the fuel at any expedient place. The best and easiest method is by the introduction of a pipe from the boiler above the door of the furnace, with a fan-shaped termination suitable to the size of the furnace, reaching beyond the dumb plate, and perforated with minute apertures, as shown in the drawing, so as to throw the steam in small jets down upon and over the whole breadth of the fire. The quantity required does not exceed one-tenth of the steam generated.

"The effect of the process is the prevention of smoke, and, *under proper management*, the creation of a great additional amount of heat. As the quantity of fuel is much diminished, it is necessary to contract the fire space considerably in height and width, but leaving the same surface of the boiler exposed to the fire; to raise the back bridge so as to contract the throat; and also, a little way beyond the bridge, to construct an inverted arch as shown in the drawing, which will propel the flame downwards, and precipitate the ashes. The lower part of the arch must be on a line with the upper surface of the bridge; and. the space between them may be equal to that between the bridge and the boiler.

"As the introduction of steam into the furnace greatly increases the draught, it must, in other respects, be checked by every expedient means; it is necessary to work with the damper much lower than usual, and the great height of the chimney-stalk will be unnecessary. The fresh fuel should be placed in the first instance on the dumb or charring plate, so as to cause the flame and smoke to be exposed to the action of the steam."

We observed by the Edinburgh papers, that Mr. Iveson's plan had been applied to the *Royal Adelaide* steam-ship, trading between Leith and London, and we looked forward with considerable interest for her arrival in the Thames to witness the working of the plan. We are sorry to have to state that the application has in this instance not succeeded. Perhaps the failure was owing to the difficulties and disadvantages always incident to first experiments and early workings of new operations; or it might be for want of the "proper management" stated in the preceding quo-

tation to be necessary to obtain the great saving in fuel set forth as the result of the previous experiments; and if such be the case, we shall be happy to be able to publish the causes of the failure, if the parties interested think fit to supply us with them.

The boilers of the *Royal Adelaide* were fitted, as nearly as possible, in the manner shown in the engraving on our front page, having one of the pipes with fan-shaped terminations to each furnace During the whole of the voyage the engineers experienced a great scarcity of steam for the engine, and about six hours more than the usual time were occupied in performing the voyage. The fire bars also were completely destroyed. The smoke, however, was almost entirely consumed. The furnaces were consequently restored to their original condition, and the *Adelaide* returned to Leith belching forth dark clouds of smoke as was her wont; nay, the smoke seemed even to rise higher, and spread abroad its murky shadow with renewed vigour, as if in triumph over the defeat of its would-be vanquishers.

Still, therefore, will the steamer's vapoury pennant roll in the breeze, and still will "the smoke that so gracefully curls," indicate the approach from the broad Atlantic of the steam-ship that, more truly than its rival, of which the words were written,—"Walks the water like a thing of life."

HINTS ON THE CONSTRUCTION AND MANAGEMENT OF IRON AND OTHER STEAM-BOATS: BY JUNIUS REDIVIVUS.

Sir,—Years have passed away since, in your pages, I expressed my conviction that the time would come when iron would be the material in common use for the purpose of constructing the hulls of vessels intended for navigation. I was led to this conviction, not by the consideration of workshop details, but by a general impression that nature has provided stores of various kinds for the uses of man, suited to the various conditions of his constantly progressing intellect. Some of your readers may, perchance, deem me a visionary in my ideas; but this is of little consequence, provided their publication tends to elicit thought.

In the earliest condition of man—savage man I mean—the food most nearly assimilating to him was the flesh of the lower animals. But for this provision of nature—this natural elimination of food fitted to his wants—he must, in common with many of the lower animals, have disappeared from the earth almost as soon as created. He could scarcely have subsisted on roots, at least in cold climates ; and a long time intervened ere his inventive powers converted peculiar grasses into wheat, oats, rye, barley, and other grains; and though the lower animals are still used as food, invention has been constantly at work to vary their physical qualities, making them, by art, still more fit for man's purposes. As intellect shall continue to make progress, new discoveries in chemistry will continue to devise new means of procuring food, and at length enable us to put an end to the coarse —and, to the delicate imagination—the humiliating process of entombing the dead bodies of the lower animals in our own persons. So far from advocating abstinence from animal food, I consider it the best we have in our present limited state of knowledge ; and as an essential ingredient in human progress, yet, still only a temporary ingredient. Knowledge will eventually be ours to prepare food from inorganic matter, as stimulating and as nutritious as that now furnished by the lower animals, who are made to die, that we may live.

In man's earlier time, his fuel was wood ; but, as population thickened, wood lessened in quantity, and nature's next provision was laid bare to his view. Coal was the fuel suited to his improved condition, when his knowledge was sufficient to devise means of raising it from its subterranean store-houses. He first consumed the upper strata, and his knowledge became enlarged to penetrate to still deeper stores. Through the instrumentality of coal, have the modern wonders of locomotion been achieved. But coal is a constantly-decreasing quantity not reproduced by nature ; already may we calculate on a not very distant period when it will cease to exist. But, ere that period shall have arrived, man's constantly progressing knowledge will have devised means of extracting heat by chemical agencies, as superior to coal as coal is to wood.

In like manner, nature seems to have provided trees for us, for the purpose of ship-building—fitted for the slow progress hitherto attained, but of too slow growth for our increased activity of locomotion. The thickening of the population in England cannot afford forest space ; and the ample stores of iron, useless, in former times, from want of the knowledge necessary to render them available, are now more valuable than the organic matter of the forests.

Several iron steam-boats have been sent to India for river use, built in London. Many are in use on the Clyde, plying between Glasgow and Greenock ; and the first* trial of a sea-going iron vessel is now about to be made between Glasgow and Liverpool. It is also the largest yet built, being of five hundred tons measurement ; and the form is well proportioned. The builders expect to get from 12 to 15 knots speed out of her.

As befals most novel plans, prognostics are rife, that it is impossible to bring iron steam-ships into general use, especially for the sea. The reasons alleged are as follows :—

First, The liability of iron to rust, and the consequent quick destruction of the vessel.

Secondly, The danger of breaking large holes in them when striking against obstacles.

Thirdly, Their liability to leak, from the vibration of the engine.

Fourthly, The alleged impracticability of navigating them across the ocean, owing to their effect on the comp as .

An answer to the first objection, viz. rust, may be found in the galvanic process by means of zinc—thus rendering the iron indestructible by ordinary oxydation.

The second objection is an unsound one ; for an iron vessel striking against an obstacle would merely dinge in, and rebound ; while the wooden one would be cut through, and the planks started with an equivalent shock.

The third objection—the liability to leak, from the vibration of the engine— is of more importance. But this objection may also be made to wooden vessels. It is true that wood may swell, and diminish the effect of a leak, which

---

* Our esteemed correspondent appears to have overlooked the *Rainbow*, now trading between Antwerp and London, a notice of the performances of which appears in another part of our Number. ED. M. M.

is not the case with iron. I have made passages in vessels belonging to "respectable" companies, and watched the boilers leaking fearfully, and the bearings of the working beams lifting and falling, and the combings of the hatches working loose against the butt-ends of the deck planks; while the water was running down, and the grim skeleton of the *Forfarshire*, mocking on the distant rocks, seemed to say—" ere long, I may have a companion to share my solitude, leaving a wild sea again strewn with the fragments of humanity." The fact is, there ought to be no mischievous vibration from an engine taking place on the hull of a steam-vessel. It is an evidence of imperfect construction—wasted power, used for destructive purposes, and serving to impair speed.

The last objection—impairing the efficiency of the compass—does not seem well founded. All steam-vessels have a large proportion of iron used in them, yet still the compass works. The method of obviating this difficulty—if difficulty it be—is simply to make the attraction equal on all sides. I would indicate an easy experiment to your numerous readers, some of whom will, perhaps, furnish the results to your pages. Take a common cast-iron pitch kettle, and suspend a ship's compass in the hollow; then varying its position to and from the centre, and nearer to, and further from, the bottom: note the effect of the attraction, and whether it varies. Theory would seem to indicate that, if the iron be disposed in equal quantities at equal distances, the effect should be *nil*. It would then be well to try whether the application of pieces of zinc to the iron would produce any change of effect.

Three positive advantages are to be found in the use of iron vessels.

First, Their incombustibility. This quality in an ocean steam-ship is a *sine qua non*—for how a wooden vessel, roasted to dryness by continuous heat, is to be extinguished when once fairly on fire, I do not well comprehend.

Secondly, Their great saving in expense of outlay as to first cost, being only about one-half the expense of wooden vessels.

Thirdly, Their greater buoyancy in the water—enabling them to carry a greater cargo, with the same displacement of water, as a wooden vessel.

The first requisite in a steam-vessel intended to cross the Atlantic, is safety. Against fire, the safety is found in the incombustible nature of the material. Against water, safety may be found by a peculiar construction of the deck—realizing the long-sought object of Mr. Ballingall, making the vessel a life-boat. In the first place, the hold, throughout its length, should be divided into compartments, by athwart positions or bulkheads; each compartment being watertight, independent of the rest. The deck-beams should be air-tight hollow cylinders of sheet-iron, and the spaces between them sheet-iron boxes, also airtight. The angles between the square sides of the boxes and the curved sides of the beams, should contain wooden thwarts to which to nail the deck planks. A double caulking might thus be used —first to the iron casings, and secondly to the planks. The rivetting should be performed, not by the uncertain and unsafe process of the workman's hammer, which hardens the iron, and disposes it to fly with a trifling strain, but by the process of pressure lately *patented* by Mr. Fairbairn, which makes every rivet equally certain, without impairing its tenacity. The best mode of fixing the deck planks to the timber thwarts would be by means of Drake's wooden screw trenails, the most efficient fastening I have seen.

The next requisite is speed. In this consideration, after attaining the very best proportion and form of the parts, the most important matter is size. The smaller the vessel, the smaller is the rate of speed, even in smooth water—just as a long-legged animal gets on faster than a short-legged one. But, on rough water, the retarding friction increases with compound proportion; while the mere distance is increased to the small vessel, which has to pass up and down the opposite slopes of the wave. The large vessel, on the contrary, goes direct through it. If a vessel were constructed so large, that the largest waves of the ocean bore the same proportion to her that the ripple of a calm does to a small vessel, a speed might be attained on the water possibly nearly equal to that at present usual on our railways; and the evil of sea-sickness, so fearful an infliction to passengers, entirely removed. Supposing harbours to be formed fit for such vessels, it would scarcely be possible to construct them of wood. The specific strength of the material limits the size of construction; but not so with iron. The strength of iron may be increased to any amount, *i. e.* the size of the component parts.

The third requisite is economy. The economy of the iron construction has been already alluded to; but increased size and increased speed are still more fruitful sources of economy—being an economy of expenditure, far more important than economy of outlay. A large number of passengers can be carried at a far cheaper rate than a small number; and steam can also be used with a less proportionate waste in large engines, than it can in small ones. Beyond all this, there is the saving of time, involving expensive maintenance and much wear and tear. A passenger would rather pay his thirty guineas for a five days' trip, than for one of twelve days; and the actual maximum of speed for ocean steamers being yet far from ascertained, owing to the expense of the experiments on the large scale, we cannot yet prescribe a limit. One thing, however, has been satisfactorily ascertained—that the largest vessels have proved the swiftest;—and this seems to indicate the true principle, which has yet to be worked out.

In the science of water locomotion by by steam, one most important principle seems to have been wholly overlooked, the necessity of getting rid of all undue vibration, in order to attain the maximum of speed. It is well known that a vessel which is too " crank" or rigid, will not sail well. She will not yield to the motion of the water and thus produces a partial concussion. The American schooners belonging to Baltimore carry the heavy sails in their long span by its own elastic strength, with scarce any aid from shrouds, and the heavy yards of the Mediteranean feluccas have the same elasticity. If a Baltimore schooner be tightly stayed up with rigging its wondrous speed will depart from it, and it will be as a common vessel. This fact was once proved by our dock-yard wise-acres at Portsmouth, who thought to make the vessel more seaworthy by trummelling her up in a multifarious tracery of ropes. The French privateers during the war understood the advantage of elastic yielding, when

by knocking out the wedges from their beams they were enabled to gain extra speed and escape their swift-heeled adversaries.

I once in your pages advocated the plan of applying springs between the axes and circumferences of paddle-wheels; my reason for this was, that I had noticed the mischievous vibration caused by the strokes of the paddle-blades on the surface of the water, causing annoyance to the passengers and retardation to the vessel.

I have since thought more deeply on the subject, and am satisfied, that the principle of making the framings of the engine a fixture of the vessel's hull is altogether wrong. The vessel itself should resemble a fish—not a dead fish but a living one—not a loose mass falling inertly from one form to another—but a well-proportioned body, with an elastic power of yielding to pressure within certain limits. and of acquiring its true form by virtue of its elasticity the moment the pressure is removed. This is the reason why boats without decks sail better than boats with decks. This is the reason why Thames wherries and Deal gigs are the fastest rowing boats in the world. Large vessels made of sheet iron, if properly constructed, with the rivet hands flush and smooth, are better adapted for this elastic yielding and smooth gliding motion than any other construction.

When a Thames waterman rows a wherry he does not keep his body rigid, but sways it to and fro, according to the motion, and he is careful to enter his oar in the water with as little shock as possible. Were the waterman tied fast to the thwarts, with his limbs rigid and prevented from feathering his oars, the result would be a great decrease of speed.

In a steam-vessel the engine is the rower, improperly fastened to the hull, and transmitting to it every shock and concussion, annoying the passengers and impairing the speed. The true method of construction would be to frame the engine separately, and then to attach it to the hull of the vessel by the intervention of springs or elastic substances. By this means the motion of the vessel would not be unduly retarded, and the power of the engine would be confined to the axis and wheels.

It is, I believe, a known fact, that a boat can be "sculled" with a single oar astern faster than she can be rowed. The principle of sculling is that of forcing a wedge down an inclined plane alternating from side to side. On this principle is constructed the stern sculler patented by Mr. Taylor. Instead of the shaft athwart ship to which paddle-wheels are attached, he uses a longitudinal shaft passing through a stuffing-box into an opening formed between the stern post and the dead wood. At right angles with this shaft is placed a kind of oar blade working in the opening with a continuous revolution. Being set at an angle with the plane of its revolution, this blade is continually cutting against an inclined plane, and thus the vessel is forced onwards.

The experiments exhibited with this simple instrument were as follows :—

A model boat proportioned to one of the best government steamers was set moving, by a clock-spring power in a trough of water about 30 feet long.

The spring being wound up to its full power, the boat made the distance in fifty seconds. This was repeated more than once with a very trifling variation.

The paddle-wheels being removed, the sculler was applied, and the distance was accomplished in thirty seconds. A repetition of the experiment gave the same result, and there was no apparent reason to question the perfect fairness of the experiment.

There are two reasons to account for the superiority of the sculler over the paddles. First, that the sculler always works in unbroken water, whereas, the paddles frequently act against mere froth, churned by themselves. Secondly, the sculler acts in deeper, and therefore denser water, always immersed, without any load to lift at the return stroke; whereas, the paddles work in surface water varying their immersion with the roll of the vessel, and lifting a load as they emerge.

The objects I advocate for the purpose of challenging discussion in your pages may be thus summed up:

First, The use of iron steamers for ocean navigation.

Secondly, An increase of size till the maximum of speed be attained.

Thirdly, Hollow air-tight decks, and a divided hold, to ensure safety.

Fourthly, The separation of the engines and paddles from the hull by the intervention of springs.

Fifthly, Galvanization of the iron to prevent decay.

Sixthly, Experiments on a large scale to compare the sculling and paddling processes.

I remain, Sir, yours, &c.

JUNIUS REDIVIVUS.

October, 1838.

CIRCULAR BOILERS AND CYLINDRICAL FLUES.

Sir,—The remarks of A. Trevelyan, in your Number of to-day, seem calculated to prejudice the public against all circular *boilers,* which are, perhaps, the safest as to *form* that can be made. The accident they are most liable to is, the *collapse* of the circular flue by pressure on the *outside.* The danger is in the *flue*, not in the *boiler ;* and many have *no internal flue.*

" A public demonstration of dislike to cylindrical boilers," would be a public demonstration of gross and most unwarrantable prejudice.

Cylindrical boilers are not more likely to be used to generate steam of a greater pressure than they were intended to bear, than other boilers are.

I am, Sir, yours, &c.

C. G. JARVIS.

Oct. 13, 1838.

DEANE'S IMPROVED DOUBLE-ACTION LEVERED LOCK.

Sir—To put a thing under lock and key is, unhappily, not at all times to make it secure. For all people know, sufficiently well, that it is not love alone which laughs at locksmiths. To find out any contrivance which may render assurance doubly, or even singly sure, is what has long been desired by those who have treasures to take care of, or secrets to conceal.

Intricacy in the structure of the lock has been considered the grand secret of success ; and makers have multiplied grooves and wards to a great extent, but not to the complete satisfaction of minds only moderately suspicious.

We have here, however, a lock which, with great simplicity of construction unites the most entire security. A few moments' inspection would do much more than any description of ours to satisfy any competent judge as to the correctness of these assertions.

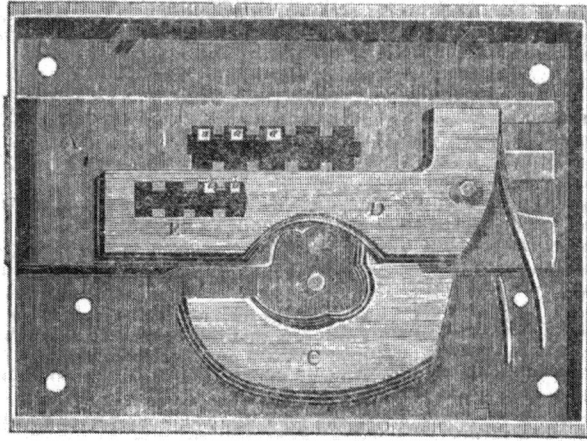 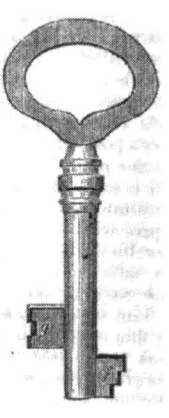

In the figure it will be seen, that the bolt A has racks or indentations, fitted to the pins which we will call *a a a*. The bolt cannot pass until these pins are severally moved down by the *lower* levers at C. But the bolt has also two pins *b b*, held by the racks in the upper lever D, which must be raised at the same moment with the lower levers, or the bolt cannot pass. If any one of these five remains unreleased, there can be no movement to lock, or unlock.

The key has two bits F and G, acting on the levers C and D. F, with its three notches, moves the three levers C, which act severally on the pins *a a a*, whilst G, with its two notches, raises the two levers D, releases the pins *b b*, and at the same time throws the bolt.

The superiority of this lock consists, 1st. In its simplicity of parts and arrangement, greatly excelling in this respect the celebrated patent locks; 2nd, In the extra bit *g*, which constitutes, in fact, a double key, and adds exceedingly to the security of the lock; 3rd, That it is impossible to pick such a lock as this; the five stays, and the bolt itself, must all move together, or not at all, a thing which one might defy the most skilful pick-lock, with whatever tools, to effect, without the assistance of the key. Or should not this be thought sufficient, the number of upper and under levers may be increased, each additional one adding, of course, to the security of the lock; 4th, An impression in wax, or any other composition, to imitate the key, would be of no avail; for so great is the exactness and accuracy of construction required, that even under the most favourable circumstance of having the key itself for a pattern, it would scarcely be possible to avoid leaving some one of the racks or pins untouched, which would render all the rest of the labour unavailing.

Deane's double-action levered lock has been submitted to the inspection of several practical and scientific men, who have given very high and laudatory opinions of its excellence and efficiency.

B.

MANUFACTURE OF SILK-WORM GUT.

Sir,—You would greatly oblige a numerous class of your subscribers, who, like myself, are fond of angling, if you would request some one of your correspondents competent to the task, to furnish you for publication in your Magazine, with a detailed account of the process for making silk-worm gut. This article, so essential to our gentle art, comes chiefly, I am told, from Spain and Turkey; but of late years nothing but gut of a very inferior quality has been imported, and the London fishing-tackle makers all declare that really good, strong, and sound salmon gut is not to be procured for love or money. Owing to this scarcity, a most extravagant price is demanded, and it is surprising that, considering the great profit which the wholesale importers would derive from taking a little pains to induce the manufacturer to make a better article, nothing of this sort has ever been attempted. As far as I can learn, the process is exceedingly simple, and is generally performed by women or children, who take the silk-worm as it is about to spin its cocoon, and draw it out into a glutinous thread between their finger and thumb, fifteen or twenty inches in length, when it hardens, by exposure to the air, and is made up into hanks or skeins containing one hundred such lengths or threads. This operation would appear to be performed in the most careless manner, as in such hanks there are seldom more than a dozen or two good round pieces of gut, the rest being all "*stale, flat,* and *unprofitable.*" The retail price of a hank varies from *sixpence* to two or three guineas, according to quality! Now, if there be no mystery in the art, and any silk-worm in a proper state be capable of being converted into a length of gut, we may calculate the profit to be made on one hundred silk-worms when converted into a three-guinea hank. What occurs to me, in my ignorance of the real process, is, that the glutinous matter of the worm might, at some time or other before it hardens into the beautiful transparent thread which it ultimately becomes, be rolled, or be drawn through a gauge, so as to ensure its being of an uniform thickness and roundness for a certain length, such as we sometimes find it. One-third of the length is always quite useless; but why it is so I do not know. Now, as we have in India every variety of silk-worm, and some of them of the wild species, yielding a particularly strong silk, which is often spun into fishing-lines, all that is wanting is to instruct our countrymen who have extensive filatures in Bengal, in the process of making silk-worm gut

with such improvements (if practicable) as I have suggested, and to make them aware of the extent of the demand for it as a marketable article, and it cannot be doubted that, ere long, we should draw all our supplies from our own territories in the east, instead of being indebted, as hitherto, to foreign countries. May I hope that you will bring this subject into notice by calling for information regarding it, and perhaps some ingenious person may be induced to make a few experiments with the silk-worms reared at home, and ascertain whether it be possible by some such means as are above alluded to, to manufacture a superior article to what is generally to be found in the shops. Amateur anglers will tell you that they frequently pay sixpence or a shilling for single threads or hairs of salmon gut, and would gladly purchase whole hanks if they were to be had. I know not what quantity of silk-worm gut is annually imported into England, but this could be ascertained at the Custom-house, and would prove to the proprietor of a silk-worm establishment that it would be well worth his while to turn his attention to the manufacture of it on a large scale.

I remain, Sir, &c.

PISCATOR.

Edinburgh, October 9, 1838.

FRENCH AND ENGLISH FIREMEN.— RESCUE SERVICE, &c.

Sir,—I am much obliged to Mr. Felix M. Simeon, for the complimentary remarks with which he introduces his communication, at page 94 of your 791st Number; but I must be permitted to show, that while charging me with error, he has fallen into considerable error himself.

An attentive reader would hardly have supposed that when I was speaking of, and offering advice to, *honorary* fire-brigades in England, or the *voluntary* fire-associations of the Continent, I could have included in either of these classes, the numerous and well-trained force of the French Sapeur Pompiers. No *volunteer force,* I apprehend, can ever hope to rival, either in point of numbers or discipline, the military firemen of Paris. In our own case, it would be absurd to expect that any "posts" can be established—"detachments stationed at proper intervals," or, "men kept constantly on guard and always ready." Nor is such a service at all likely to be extensive enough, "to be formed into three divisions, each under the superintendence of an experienced officer," so as to undertake simultaneously and independently, the rescue of property, attack of the fire, supply of water, &c. With this explanation, I proceed to notice the comparison instituted by Mr. F. M. Simeon, between the Sapeur Pompiers of Paris and the fire-engine establishment of London, as well as the fire-police of Edinburgh and Manchester. Mr. Simeon states, that " if we take into consideration the immense size of the houses, their construction, the difficulty of excluding air, &c., we shall find the parallel between them much in favour of the first." But how this comes about is by no means clear, at least to my limited comprehension. If the immense size of the buildings is to be considered, neither London, Edinburgh, nor Manchester, are much behind the Parisian capital, the size of the Edinburgh houses is notorious.

With respect to " their construction," " difficulty of excluding air," &c., London buildings stand pre-eminent. The circumstance of the existence of stone stairs throughout the Parisian houses, is one of the greatest safeguards against the rapid spread of fire that can possibly be devised, and I have more than once or twice asserted in your pages, that the general introduction of incombustible stairs would tend more than any other provision that could be made, to limit the extent and diminish the danger of London fires; in fact, it would completely change their character.

There is considerable disparity in point of numbers between the corps alluded to; the Sapeur Pompiers of Paris are, I believe, *one thousand* strong; the fire-engine establishment of London musters *not quite one hundred;* while Edinburgh is very efficiently protected by *eighty* firemen, Manchester by *forty!* The extent of ground, number of buildings, and value of the property effectually protected by the London fire establishment, may be set down as being ten times greater than in Paris; and yet, after the burning of the "Theatre Italia" in January last, M. Paulin, the Col. Commandant of the Sapeur Pompiers, required an augmentation of his force, as being *too few.*

Besides, in Paris all classes of persons turn out to assist in extinguishing fires; and in addition to the firemen, there may usually be seen private citizens, National Guards, troops of the line, Municipal Guards, Sergens de Ville, and police officers, all actively employed. The French laws are exceedingly strict in compelling all passers-by—whether the fashionable beau in his ball dress, the lowest of the canaile, or the monarch himself—to hand the water or pump the engines as he may be directed. At the conflagration of the "Theatre Italia," a number of well-dressed persons who had just issued from Musard's ball, and also from the concert of M. Valentino, submitted with the best possible grace, to join the chain, and pass the water-buckets, notwithstanding the intense cold which prevailed at the time.

The following table will show at a glance the relative proportions which the fires in the two capitals bear to each other; viz.—

	In London.	In Paris.*
In the year 1833 the number of fires was	458	151
1834 ,,	482	190
1835 ,,	471	213
1836 ,,	564	191

The following fires of very serious magnitude have occurred in London within the last five years; viz.—

1834, October 16. Houses of Lords and Commons.
1835, March 2. Silver-street, Golden-square.
1836, March 26. Western Exchange and Burlington Arcade.
1836, August 30. Fenning's Wharf.

1837, December 28. Davis's Wharf.
1838, January 10. Royal Exchange. Against which, I believe, the Parisians can set nothing beyond the " Theatre Italia" in January last.

---

* The numbers of Parisian fires are copied from a report published in your 721st Number, an authentic copy of which was forwarded to Mr. F. M. Simeon by M. Paulin.

Whenever a comparison has been made, in my hearing, between the promptness, skill, and intrepidity of London and Parisian firemen, it has always been to the advantage of the former; a glance at the foregoing elements will show the fairness of this conclusion.

I am well aware that Mr. Simeon is a great admirer of the Sapeur Pompiers, and I know that he made a proposition to Government some time since to establish a similar corps in London; but his project was not entertained.

When alluding to the evil consequences of certain practices on the continent, I certainly did not include the City of Paris, and as to other places, if the accounts transmitted to us by the public journals are to be believed, it is an absolute fact that in America, and elsewhere, many thousand pounds worth of property have been sacrificed to the flames *solely by the diversion of efforts from the extinction of the fire, to the attempted rescue of property.* That such has too often been the case in this country, both in provincial towns and in London, I have the evidence of my own senses. Constant attendance at the earliest stages of most of the large, and very many of the smaller fires in London, for the last eighteen years, qualify me to speak with some degree of confidence on these subjects; at the same time I do not presume to consider myself "infallible."

I confess I do not quite comprehend the paradoxical remark of Mr. Simeon, that my arguments may be dictated by, but are not founded upon, sound sense; but let that pass, I am quite content that all my opinions should be tested by "past experience," and if this is but fairly and impartially done, I shall be quite satisfied with the result. This communication has extended to an inconvenient length, I must, therefore, for the present postpone any further remarks, but I will take an early opportunity of adding some further particulars of the Parisian fire-apparatus, to that which Mr. Simeon has forwarded. In the meantime,

I remain,
Yours respectfully,
WM. BADDELEY.

London, October 11, 1838.

REVIEW OF FIRE-ESCAPES, &c.

Sir,—I have much pleasure in affording such further information on the subject of certain fire-escapes, as Mr. Parry's last letter seems to require. When opinions are combated, or questions proposed in the style adopted by Mr. Parry, it is really a pleasure to answer them; at the same time, I beg to state that the deference paid me by Mr. Parry is far beyond what I have any pretensions to. His former letter would have been noticed somewhat more in detail, but that I feared to trespass on your space, or your readers' patience, by a recapitulation of matters already expatiated upon at some length in your previous volumes. In briefly pronouncing sentence of impracticability against Mr. Parry's proposed steadying-weight for the canvass trough, I imagine a little reflection would supply the reasons, as being obvious; but as Mr. Parry enquires what is the objection? I would beg to direct attention to the following points.

Every person must at once admit, that a *heavy* weight must necessarily be employed; now the lifting and throwing out of a heavy weight to a proper distance, during the excitement produced by impending danger, requires great skill, and is a performance that few persons could achieve—females, aged persons, or invalids, are quite out of the question. But supposing the weight to be ejected by sufficient manual dexterity, still there is no certainty that it will fall and continue at the required spot.

These and some other difficulties are inseparable from the plan; but the fact is, that no weight capable of being wielded in the manner supposed, would do what is required, *i. e.* maintain the angular extension of the trough from the burning building. The weight of the parties descending through the trough, as they approached the ground, would drag the weight, unless *very heavy*, and so render the descent perilous in the extreme.

There is one objection to all these troughs not hitherto broached in your pages; I allude to its indelicacy of operation. Such of your readers as witnessed a certain notable exhibition at the west-end of the town, some time back, when a young female was persuaded to slide down through the trough to demonstrate its safety, will remember the result, and perfectly well understand to what I refer.*

The hurried manner in which escape is always effected from midnight conflagrations, with little save the sleeping garments about the person, renders the *exit* from the canvas-spout anything but pleasant. This inconvenience could be avoided, it is true, by descending *head foremost*, but the bigoted inclination of parties to retain their stock of brains, renders this course quite as objectionable as the other—the remedy's as bad as the disease. I dislike unnecessary levity, as I despise the mock modesty that would stand upon trifles in a case of life or death; at the same time there are certain common decencies of life that should never be needlessly outraged.

With respect to Mr. Simeon's remark, that the canvas spout has "*perhaps* been the means of saving more lives than any other extant;" I would only say, *perhaps not*. At any rate, its operation may probably be considered less objectionable in Paris than it would be here.

The "portable staircase" has long been a favourite scheme with many persons, and considerable ingenuity has been directed to this point, but hitherto without any success. Nothing sufficiently portable has yet been produced, and I fancy, never will. Most of the attempts in this way have been described in your pages,† and some have been rewarded by the Society of Arts; but all have the insurmountable defect of being too cumbersome.

With much respect,

I remain, Sir, yours truly,

WM. BADDELEY.

London, October 15, 1838.

IMPORTANT IMPROVEMENTS IN STOCKING WEAVING.

Our townsman, Mr. Robert Scott, has invented some ingenious machinery for working the stocking frame, which (though since the time of its original inventor it has remained *in statu quo*), is now placed in the first rank of improvements, inasmuch as, prior to Mr. Scott's machine, a man could not manufacture more than 24 pairs per week, while this invention will enable a man and boy to produce 20 pairs per hour, or 100 dozens per week! The perfecting of this machine has been a work of immense labour, time, and capital, and reflects the highest credit on the ingenuity and perseverance of its talented inventor, and we confess we regret exceedingly that anything should deprive him of his due reward. "It may be questioned, however," says a correspondent, "whether the advantages resulting from the present form of the machine — distributing the labour to all branches of the family of the poor, in their cottages, are not greater than those obtained by attaching it to steam power, and congregating the weavers in large factories. The time may come, indeed, when this may be the best form of employing labour, but it is at present accompanied with inconveniences of no trifling amount. The progress of mechanical invention, however, cannot be stayed by these advantages. We must avail ourselves of improvements as they arise, rather than suffer other districts by adopting them, to rob us of that ancient inheritance, our stocking manufacture. The worsted branch of the hosiery trade has been so long established in this county that we should look with suspicion upon any improvements of machinery adopted by other counties, before they are brought into actual competition. Several patents have, we understand, been obtained at Nottingham, for power stocking-frames, of which the Nottingham papers have spoken very highly, and which, most likely, will result in something being done, though we are informed that all of them are much inferior in speed to the invention of Mr. Scott, whose improvements in the mode of working the common frame render it capable of being attached to power whenever the wants of the trade or the competition of our neighbours, may render it necessary. The chief advantages of his plan are the rendering the labour of working a frame much easier and pleasanter when applied to hand labour, and the attainment of a speed greater than any other invention, the expense when attached to steam power, being small, and the wear and tear being less than by hand." Mr. Scott's engagement in another county prevents his further carrying out his intentions with respect to them; and he therefore purposes to sell his machines by auction, which will, no doubt, excite very considerable interest and competition.—*Leicestershire Mercury.*

---

* This circumstance was not stated in a certain lying puff, which appeared soon afterwards in the *Ladies Newspaper.*

† I would take the liberty of mentioning that the *Engineers' and Mechanics' Cyclopædia*, contains a very tolerable account of the principal fire-escapes, up to the time of its publication.

THATCHED HOUSE NUISANCES.

(From Loudon's *Suburban Gardener*.)

A thatched cottage is an object of admiration with many persons who have not had much experience of country life; and, accordingly, we find several in the neighbourhood of London. Such cottages have, perhaps, the gable end covered with ivy, the chimney tops entwined with Virginian creepers, and the windows overshaded by roses and jasmines. The ivy forms an excellent harbour for sparrows and other small birds, which build there in quantities in spring, and early in summer, and roost there during winter. In June, as soon as the young birds are fledged, all the cats in the neighbourhood are attracted by them, and take up their abode in the roof of the house every night, for several weeks —the noise and other annoyances occasioned by which we need only allude to. We say nothing of the damp produced by the deciduous creepers and the roses, as we have already mentioned that; but we must here notice another evil, which is not so obvious, though quite as serious, and this is, the numerous insects generated in the decaying thatch, and more especially that loathsome creature the earwig, which in autumn, whenever the windows are open, comes into the house in quantities, and finds its way into every closet, chink, piece of furniture, and even books and papers. All cottages of this kind harbour snails and slugs in the ivy, and spiders under the eaves of the thatched roof; and wherever there are spiders, there are also abundance of flies. As there is always a garden attached to such cottages, it is almost certain, if on clayey soil, to abound in snails, slugs, worms, and, if the situation is low, perhaps newts. Some of these from the doors, or, at all events, the back door, being generally kept open, are quite sure to find their way, not only into the kitchen, but even into the pantry and cellars. Slugs, when very small, will enter a house through a crevice in the window or a crack in the door, find their way to the moist floor of the pantry or the cellar, and remain there for weeks, till they are of such a size that they cannot retreat. There are few persons, indeed, who do not experience a feeling of disgust at seeing the slimy traces of a slug in any part of their house, not to speak of finding them in dishes in which food is kept, or even on bread; or at discovering an earwig in their bed or on their linen. The kitchen, in low damp cottages of every kind, almost always swarm with beetles and cockroaches, and the pantry with flies; while, from the closeness and want of ventilation in the rooms, it is almost impossible to keep fleas, &c., from the beds.

If a large dog be kept in or near the house, as it frequently is, or if a stable or cowhouse be near, the fleas from the dog, the horses, or the cows, which are larger than the common kind, will overspread the carpets, and find their way to the sofas and beds. Having lived in cottages of this kind in the neighbourhood of London, we have not stated a single annoyance that we have not ourselves experienced; and we have purposely omitted some. Two of these, offensive smells and rats, are the infallible results of the want of proper water-closets and drainage; but these evils, great as they may seem to be, are much easier to remedy then the others already mentioned—which are, in a great measure, inseparable from the kind of house. Two others, the danger ot setting fire to a thatched roof, and its liability to be injured by high winds, are sufficiently obvious; but it would hardly occur to any one who had not lived in a house of this description in the neighbourhood of London, that a thatched roof is of all roofs the most expensive, both when first formed, and afterwards to keep in repair. A plumber or a slater, to repair a lead or a slate roof, may be found everywhere in the suburbs of large towns; but a professional thatcher must be sent for from the interior of the country. For example, the nearest cottage-thatchers to London are in the hundreds of Essex on the east, and in Buckinghamshire on the west.

ROYAL CORNWALL POLYTECHNIC SOCIETY.

SIXTH MEETING.

The sixth annual exhibition of works of art and science, under the patronage of this society, was opened on Tuesday, at the Hall in Falmouth. There was an unusually large display of original works of art and science, and the general appearance of the room was enriched by the exhibition of some excellent paintings, and articles of *vertu* from the galleries and cabinets of several ladies and gentlemen of Falmouth and its vicinity. Davies Gilbert, Esq., took the chair, and in the course of his speech observed that there were several models of steam-engines, which, although they contained nothing new, yet showed that Cornwall possessed a great many young men who applied themselves to the advancement of mechanics. Great attention had been bestowed on the important object of saving the excessive labour of ascending and descending the mines. To accomplish this, three or four plans were proposed, the models of which were before them. One displayed a complete novelty. It was a mode of lowering and lifting persons by means of a counter-balance of water. It was well known that some of our largest steam engines lifted 72,000,000 lbs. a foot high, by the consumption of a bushel of coal; and it was found that to lift 100 men, each weighing 150 lbs., 200 fathoms, a quarter of a bushel would be sufficient. The cost of this was two-pence; and if an additional penny were allowed for leakage, it would be possible for the small cost of three-pence, applied according to the proposed plan, to purchase a saving of human labour and human health, such as could hardly be conceived. He had no doubt they should shortly have miners lifted out of mines in this way; and a very liberal subscription had been entered into by gentlemen concerned in mines, to carry the plan into operation. (Cheers.)

Mr. Jordan, the Secretary, then read the awards of premiums, amongst which were the following for mechanical subjects:—

Ten Guineas, by Sir Charles Lemon, Bart. and R. W. Fox, Esq., for the best reports of a series of experiments made with the wedge for blasting rocks, invented by R. W. Fox Esq. To Capt. R. Dunstan, of Wheal Vyvyan mine.

Ten Pounds, by John Hearle Tremayne, Esq., for the best available method or improvement on the plans already suggested, for facilitating the ascent and descent of miners. To John Phillips, of Halsetown.

Three Guineas, by Charles Fox, Esq., for the best model (either original or copy) not less than 18 inches in length, of a life boat, which shall be judged most manageable in a storm. To F. J. Adams, of St. Ives. An extra prize of Two Pounds was also awarded to John Phillips, of Halsetown, for his model.

Three premiums, by Charles Fox, Esq., the first of Three Guineas, for the best description and drawings of the least inconvenient and inexpensive, and, at the same time, most efficient means of securing a fortnight's supply of bread and water, within reach of a ship's crew, in the event of their not being able to go below deck, owing to the vessel being water-logged, or to other causes; the number of the crew may be estimated in the proportion of 15 tons register to each man. First Premium of Three Guineas awarded to John Payne, of Falmouth; second Premium of Two Guineas, awarded to F. J. Adams, of St. Ives.

The total amount of premiums awarded was 51l. 8s.

In delivering a premium to Mr. Phillips, the President stated that one great cause of disease in miners was climbing ladders to the surface. He himself, had, in his youth, experienced the injurious effects of this mode of ascent; and, he was convinced that to devise a better method was one of the greatest objects which this society could have in view. He had great pleasure in stating that this had been actually achieved in Germany, in the deep mines of the Hartz forest; and he believed, upon the principle of the model now on the table. This plan, was, to hold by projecting handles when passing from place to place. Mr. Alfred Fox had applied to the Hanoverian Ambassador for a description of it; and he hoped at the next meeting to lay it before the society. He had said already that this machine would lift 100 men 200 fathoms for about three-pence; but the machine itself would be very costly: and, as mines were exceedingly precarious, mine-owners had been unwilling to incur that expense. He was therefore happy to say, that besides donations from other gentlemen, 100l. would be given by Mr. Charles Fox, through this society, to that mine that should first adopt the principle. (Cheers.)

Davies Gilbert, Esq., in allusion to the comparison made with other societies, stated four or five had been established in imitation of this; and, it should never be forgotten that they owed this society, useful and splendid as it was, to the exertions of two young ladies, under 20 years of age, seconded by the judgment and liberality of their relations, a family long known through Cornwall, but now distinguished for their acquirements from one end of Europe to the other.

R. W. Fox, Esq., made some observations on the ascent and descent of miners, in consequence of his having conversed with a gentleman who had ascended and descended by a machine in the King William Mine, in the Hartz. That mine was 360 fathoms deep; and the principle there adopted was similar to that to which the premium was this day awarded. This plan differed from that which obtained the first prize last year, inasmuch as there was now the additional security afforded by taking hold of the stays of the ladder. This single circumstance warranted the judges in deciding that the present plan should have a premium. The apparatus descended about 200 fathoms out of 360, partly on an inclined shaft, and the other part perpendicularly. Some parts of the mine were three-quarters of a mile from the place of ascent, yet the miners who work in those parts, walk that distance through the levels besides coming down 40 fathoms, in order to be raised to the surface by the apparatus. They had no idea of danger; and he stated this publicly, because a working miner had lately in a letter to him, expressed an opinion that the plan was a dangerous one, and also that the prizes offered by this society tended to the injury of the miner.

NOTES AND NOTICES.

Discouragement of Science by the Austrian Government.—It is a remarkable fact, that Vienna is the only European capital in which there is no academy or association for the cultivation of science organized under the sanction and the encouragement of the State. It can hardly be supposed that the Austrian Government should dread the effects of an increased activity of the human intellect. According to D'Alembert, princes encourage learning for the sake of diverting the minds of their subjects from the consideration of their practical interests and political rights; and one would suppose that the examples of Peter the Great and Frederick the Great, who both did their utmost to give science a permanent abode in their respective capitals, would be sufficient to inspire the Austrian statesman with confidence, if he were at all disposed to favour the progress of science. The Academy of Sciences of St. Petersburgh has experienced to such an extent the munificence of its imperial patrons that its fixed revenue is now tenfold that assigned to it by its founder, Peter the Great. Among the philosophers who lent a hand to the organisation of it, was the celebrated Leibnitz, who also made great exertions, and for some time with every prospect of success, to bring about the establishment of a similar institution in Vienna. The Court seemed favourable to the design, which yet, unaccountably, was never carried into execution. It was revived about 60 years later, under Maria Theresa (in 1773), and then the foundation of an Austrian Academy of Sciences seemed quite certain; but, unfortunately, the produce of the sale of the National Almanac formed a large item in the estimate of the contemplated funds, and when all the arrangements were complete, a petition of the almanac-mongers to the Empress, setting forth the injury with which they were threatened by the institution of the academy, was sufficient to upset the philosophical fabric. Nothing further was dreamt of the special cultivation of science in Austria till last year, when 12 men, well known for their learning and abilities, presented, by the hands of the Arch-Duke Lewis, a petition for the establishment of an Academy of Sciences at Vienna. No notice has, we believe, been taken of this petition; and we presume that Prince Metternich does not deem it becoming in a fond and paternal Government to give its subjects the pains of thinking.—*Athenæum.*

Encke's Comet.—(From the *Cambridge Chronicle.*)—The peculiar clearness of the atmosphere and heavens at the time of the comet's passing the meridian, October 4, 13h. 30m., induced me to look for it, but with little hope of success, the moon having passed but half an hour before and was shining very brightly; the comet was 30 degrees above her; I saw some very faint object which did not appear to be a star, and took an observation of it, though very roughly, there being no light in the room, and found that it agreed with the tabular place very nearly. I then looked with the 20-feet telescope (presented by the Duke of Northumberland), but could not find it, owing, most likely, to the adjustments of the instruments not yet being sufficiently accurate to insure an object being in the field. I afterwards looked with a five-feet equatorial, the telescope of which is very good, and found some stars which I knew were very near the comet, and, after excluding all possible light from the room, again saw the comet, and watched it till 4 o'clock, during which time I got three pretty good observations of it, from which it appears that it moved through a minute and a half of declination in one hour—proving, beyond a doubt, that it was the comet; it is a nebuloid, filling a space of about three quarters of a minute, and at the time of observation followed three stars of the seventh or eighth magnitude, in the form of a triangle, about two minutes. It had been looked for very attentively with the 20-feet telescope on several previous evenings, but at each time the atmosphere was very moist, which after a short time dimmed the object glass.

Cambridge Observatory.   J. GLAISHER.

Tribute of Respect.—The engineers, firemen, &c. on board of the *Countess of Galloway* steam-packet, having resolved to offer a token of respect to P. V. Fitzgerald, fireman, in testimony of their gratitude for his attention while acting as coal-trimmer, and of congratulation on his advancement to fireman, presented that gentleman, on the 5th inst., a hand-some silver-mounted tobacco pipe. Peter acknowledged the compliment in a manly style, and declared that he was sure, with such an instrument, whether on land or sea, in *getting up steam*, he would always be sure to think of his kind friends on board the *Countess.*—*Dumfries Times.*

The *Railway Times* contained, in one paper, a list of no fewer than twenty accidents (some of them attended with fatal results) that have occurred to stage-coaches (taking no note of accidents to private vehicles) within the last few weeks. This is intended as a set off against railway accidents.

Nova Scotia Coal Mines.—The fuel used in the few last voyages of the *Great Western* steam-boat was obtained from the coal mines of Picton, in Nova Scotia, and it answered every purpose required. This fact will, no doubt, greatly contribute to establish the competition necessary to keep up a regular line of steam-vessels throughout the year between England and the United States, which is so much desired by all parties connected by business or otherwise with America. However this may be, the very fact that the coals of Nova Scotia have been successfully used by the *Great Western* steam-vessel is a matter of much importance.

British Alum Manufactories.—The only alum manufactories now worked in Great Britain are those of Whitby, in England, and of Herlett and Campsie, near Glasgow, in Scotland; and these derive the acid and earthly constituents of the salt from a mineral called alum slate. This mineral has a bluish or greenish black colour, emits sulphurous fumes when heated, and acquires thereby an aluminous taste. The alum manufactured in Great Britain contains potash as its alkaline constituent; that made in France commonly contains ammonia, either alone or with variable quantities of potash. Alum may in general be examined by water of ammonia, which separates from its watery solutions its earthly basis in the form of a light floculent precipitate. If the solution be dilute, this precipitate will float along as an opalescent fluid.—*Dr. Ure's Dictionary of Manufactures and Mines.*

To preserve Wall-nails from Rusting.—I beg to communicate a little valuable information to those who use many nails for fastening wall trees. I use cast nails about one inch and a quarter long and heat them pretty hot, in the fire-shovel, over the fire, but not red, and then drop them into a glazed flower-pot saucer, half filled with train oil. They absorb a great deal of oil, and thus prepared never become rusty, and will last many years. The effluvia of the oil also, for a long time, I fancy, keeps insects from the trees.—*Magazine of Domestic Economy.*

Hydrogen Gas.—A scientific chemist, of great celebrity in France, has lately visited this country, for the purpose of taking out a patent for an economical process, by which he obtains from the decomposition of water, hydrogen gas, for the purpose of lighting houses and streets. His process has for some time been in very successful operation in France, but the method has been kept secret. He has now, however, undertaken to light the Royal Printing Office in Paris, with gas procured in the manner above-mentioned.—*Birmingham Gaz.*

Extraordinary Steam-boat Speed.—Extract of a letter from a distinguished and experienced naval officer, describing his passage to Antwerp and back in the *Rainbow* steam-ship—" Underneath I give you the times we were abreast of the places marked against each, premising we started in the *Rainbow* from Blackwall, Sunday, the 16th instant, at two minutes past one, p. m. Passed Gravesend, 2.18, being 22 miles, in one hour 16 minutes; the Nore, 3.44; Margate, 5.44; Ostend Light, bearing E. by S., 10.25; Flushing, 1.25. Took a pilot on board, and anchored for nearly three hours; at five went fairly a-head again, and got up to Antwerp between nine and ten in the morning. On our return, we started Wednesday, 19th, from Antwerp, at 12 minutes past noon, passed the different places as under:—Ostend Light, bearing S. S. W., 6.12; North Foreland, 11.10; Margate, 11.15; Nore, 1.12; Gravesend, 3.3; stopped 11 minutes, and arrived alongside Brunswick Wharf, Blackwall, 5.13; making the passage, upwards of 210 miles, in 16 hours and 50 minutes. She is perfectly easy, and has the least tremulous motion of any steamer I was ever in, and it is a curious fact, that both in going and returning there was not one case of sea-sickness on board, although there was sufficient motion to have caused it to some in any other vessel. But there is not the slightest smell either from the bilge-water or from the engines."—*True Sun.*

Electricity of Wood.—M. Ratt, a cabinet-maker, when planing wood, remarked that several chips manifested electric phenomena. By means of an electrometer he ascertained, that in certain species, especially those of America, the electricity was positive, while in others it was negative, and this particularly characterized the French woods.

Wooden Paper.—MM. Montgolfier, paper-makers, have, it is said, substituted wooden chips for rag in their manufacture; and besides this, they expect soon to have on sale a wooden pasteboard, which shall be impervious to the wet, and prove an economical substitute for slate, in the covering of the roofs of buildings.

M. Dulong.—M. Dulong is no more, and in him the scientific world has sustained a serious loss. He was well known for his important labours with respect to caloric, and the progress of modern chemistry. His health had long been declining, for he was obliged to resign his perpetual secretary-ship to the Academy of Sciences on that account; on office to which he was appointed on the death of Baron Cuvier. M. Coriolis succeeds him at the Polytechnic School.—*Athenæum.*

New Discovery.—A foreigner states that he has discovered a cheap and easy process for hardening iron through its whole substance, so as to defy the efforts of the file or drill. In all cases of wear and tear this invention will offer immense savings, and in many cases where brass is now used to resist friction, that expensive metal will be rendered unnecessary, many other advantages will accrue from the introduction of this process, particularly in the increased protection, capable of being given to property in various ways.

Asphalte Experiments.—By the proceedings of the Prussian Association for promoting the Arts, we see that they have laid down a pavement of nine different compositions, of asphalte, coal tar, oil, sand, lime, chalk, &c. Observations will be made to ascertain the durability and relative properties of the several compositions.

Galvanic Telegraph.—The highly scientific mode of making instantaneous telegraphic communications by galvanic power, which has so long been considered attainable, has already been put to the most decided test on the London and Birmingham Railway, under the direction of Professor Winston, and Mr. Stephenson, the engineer to the company.

Four copper wires, acted upon at each end of the line at pleasure, by the agency of very simple galvanic communicators, have been laid down on the line of the London and Birmingham Railroad, to the extent of twenty-five miles. They are inclosed in a strong covering of hemp, and each terminus is attached to a diagram, on which the twenty-four letters of the alphabet are engraved, in relative positions with which the wires communicate, by means of moveable keys, and indicate the terms of the communication. The gentlemen are fully satisfied that communications to almost any extent may be made instantaneously by the agency of galvanism.—*Midland Counties' Herald.*

Steamers between New York and the West India Islands.—It is stated, that a Steam Company is about to be formed in New York, and that two superb steamers will be fitted up for the purpose of establishing a line of communication between that port and Kingstown, in the island of Jamaica. The steamers are to touch at St. Thomas and Havannah, The article adds, "it is not generally known, that Jamaica is destined to be connected by *three* lines of steam communication with the isthmus of Panama, opening a direct line of steam navigation to Canton, by Owybec, and to New Holland, by Otaheite. When this great event is established, Jamaica will become the most important commercial entrepot in the *world!*"

Discouragement of Inventions by the Prussian Government.—The Prussian authorities, we know not for what reason, have lately completely set their faces against the introduction of improvements in machinery and manufactures—refusing to grant patents to inventors on the most frivolous pretexts—such as, that the invention had been alluded to in some newspaper or publication, that it was not of sufficient importance, and such like reasons. We could mention three improvements in steam-engines, and several new processes of manufacture, one of a most important nature, and which is being extensively adopted in this country, patents for which, have, within the last six months, been refused. We fear that there is "something rotten in the State of" Prussia; that the inventions for which patents are applied for, instead of being *not* of sufficient importance, are *too* important, and *too* valuable to be allowed to slip through a functionary's hands who may chance to have some friend to whom the disclosure of their communication would be useful.

Map of the Southampton Railway.—The cheapest and best Railway Map (next, of course, to our own, of all the railways,) is that executed by Mr. Jobbins, the lithographer, and lately published by Mr. Tyas, of Cheapside. We noticed the publication of the first division, of the Southampton line to Woking some time since (No. 774); the map is now carried on to Basingstoke, and includes a very considerable portion of the adjacent country. It is an important point to observe, that every building, path, and lane, laid down in the ordnance survey, is contained in this map. To travellers by the train it will prove most useful, as there is attached a list of fares, times of starting, distances of towns in the neighbourhood of the line from the various stations, and other necessary information. The plan of the station will also tend to elucidate to the curious the movements of the trains, and the mode of shifting from one line to another.

The Railway Map of England and Wales continues on sale, in a neat wrapper, price 6d.; and on fine paper, coloured, price 1s.

☞ British and Foreign Patents taken out with economy and despatch; Specifications, Disclaimers, and Amendments, prepared or revised; Caveats entered; and generally every Branch of Patent Business promptly transacted. A complete list of Patents from the earliest period (15 Car. II. 1675,) to the present time may be examined. Fee 2s. 6d.; Clients, gratis.

LONDON: Printed and Published for the Proprietor, by W. A. Robertson, at the Mechanics' Magazine Office, No. 6, Peterborough-court, between 135 and 136, Fleet-street.—Sold by A. & W. Galignani, Rue Vivienne, Paris.

𝔐𝔢𝔠𝔥𝔞𝔫𝔦𝔠𝔰' 𝔐𝔞𝔤𝔞𝔷𝔦𝔫𝔢,

MUSEUM, REGISTER, JOURNAL, AND GAZETTE.

| No. 794.] | SATURDAY, OCTOBER 27, 1838. | [Price 3d. |

UPTON AND CO.'S PATENT ROTARY STEAM-ENGINE.

Fig. 2.

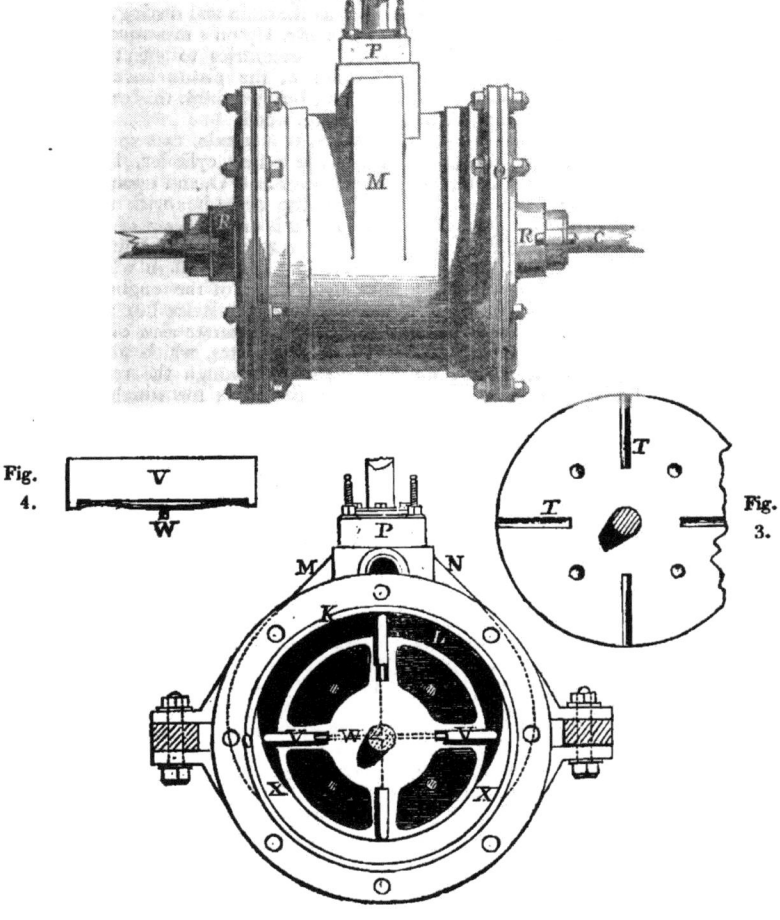

Fig. 4.

Fig. 3.

Fig. 1 .

UPTON AND CO.'S PATENT ROTARY STEAM-ENGINE.

So numerous have been the failures —equal in number, we may say, to the attempts—to construct a rotary engine which will compete with the ordinary reciprocating machine; so deceptive to the most skilful examiner have been the results of experiments upon a small scale, or for a brief space of time, that it becomes a matter of no small hazard for any one who would stand well with the engineering and mechanical world to express an opinion favourable to an invention asserted to effect the object so much desired, but so generally despaired of in attainment. So many also have been the explorers in the field—so completely has every mechanical movement been examined and applied, that it would be a matter of no less hazard to assert the originality of such an invention : so nice, mayhap, will be the distinction between the contrivance of the successful inventor of the rotary steam-engine, and a variety of his precursors, that the best evidence of its novelty will be its successful operation.

Amongst the many rotary engines which we have seen, and the multitude which we have read or heard of, we think we shall be borne out by future experience in saying, that the engine now about to be described (invented and patented by Mr. John Upton of Battersea), comes nearest to the accomplishment of the desired end ; under any other circumstances than those which accompany the search after the rotary engine, we should have said that the end was attained.

On our front page, figure 1, is an end or side view of the engine, with its exterior plate removed, to show the several parts in the interior. The engine consists of an outer cylinder, K, within which turns the revolving cylinder L, which contains the four pistons, V V V V. One piston is connected to that on the opposite diameter of the revolving cylinder by a rod W; the two others are similarly connected : these rods work freely through C, the shaft of the engine, and upon which the revolving cylinder is fixed. X X is what Mr. Upton calls a " triple excentric," being formed of three segments of circles. The middle-segment works close against the circumference of the revolving cylinder ; and the side segments are so formed, that as the piston on one side of the revolving cylinder is thrust into its socket or aperture therein, by pressing against the segment on that side, the opposite one is allowed to protrude in exactly the same proportion—still keeping close against its segment on the other side. Upon the correct formation of this excentric—the exact coincidence of operation of the one side segment to the other—does the smooth working of this engine depend : in fact, we regard this point as the main and distinguishing feature of Mr. Upton's invention ; applications of eccentrics to effect a certain position of the piston have been numerous, but we think this one not only novel, but good.

M, N, is a nozzle, cast with, or fixed upon the outer cylinder, having two steam-ways at O O, and upon the nozzle is fixed the steam box-with slide-valves P. Fig. 2 is an end view of the engine with the side plates Q Q bolted on. R, the stuffing box, through which passes the main shaft of the engine. M the nozzle, and P the steam-box and valves.

Fig. 3 is a separate view of one of the interior side plates, which are fixed by bolts passing through the revolving cylinder ; the holes for which bolts are seen in this figure, as well as in fig. 1. T T T T are grooves, in which the ends of the pistons V V V V work steam-tight. Fig. 4 is a separate view of one of the pistons ; at its juncture with the connecting rod W, there is a spring to keep it with a gentle and regular pressure against the inner circumference of the outside cylinder K and the eccentric ; and also to compensate for the effects of wear. According as steam is admitted by one or other of the steam-ways O O, so may the engine be made to revolve one way or other. By connecting the eduction-hole with a condensing apparatus of any kind, a condensing engine may be formed.

To the railway locomotive, and to the steam-boat, would the rotary action of this engine be its most important application, and we hope that the expectations raised by inspection of its working, and examination of its details, will not be disappointed.

Z.

SUBSTITUTE FOR MOUNTAIN BARO-METERS.

Sir,—In the selections that have been published from the proceedings of the British Association, at Newcastle, there is an account of a substitute for Mountain Barometers proposed by Sir John Robison, Secretary of the Royal Society of Edinburgh. As considerable errors appears to have crept into the reports regarding this instrument, I deem it worth while to describe more fully its real character, premising that it is much simpler and more easily managed by unskilled travellers, than the description given at page 469 of your last volume would lead persons to suppose.

The instrument exhibited at the meeting consisted of a glass tube, about 1.25 inches diameter, and about 14 inches long, with a small bulb at the end. The capacity of the bulb appeared to be three or four times that of the inside of the tube. The stem of the tube was graduated by divisions, which had been experimentally formed by the instrument-maker, in the following way:—At a time when the mercurial barometer was at 30 inches, and Farenheit's thermometer at 62°, the instrument was suspended within the receiver of an air-pump, over a cup containing water; the air in the receiver being exhausted to a degree of rarefaction corresponding to 29 inches of the barometer, the instrument was then lowered until its lower end was immersed in the cup of water; air being admitted into the receiver, the water rose in the tube of the instrument, and its height was carefully marked. The instrument was again suspended in the receiver, and the exhaustion repeated until the barometer-gauge indicated 28 inches; the immersion in the cup was then made, and a second mark put upon the stem. By continuing this process, the graduation of the stem was carried on as far as was thought requisite, when the instrument became ready for use.

It will be evident on reflection, that with a number of tubes graduated in this manner, a traveller ariving at a station in the midst of mountains may ascertain the tension of the air on the summits of all of them, by sending a messenger to each, with one or more of these tubes, and a tin case containing a little water. The messenger taking up the tubes with their stems open, the air

within them partakes of the density of the atmosphere at the station visited, and if when at the summit the mouth of the tube is put into the water, and left in it while the messenger descends, the water will rise in the stem with the increasing density of the atmosphere, and will indicate, by its height, the degree of rarefaction of the air at the upper station, if the barometer at the lower one stands at 30 inches; if it be more, or less, a corresponding correction must be made for the difference.

If the temperature of the air at the upper and lower stations was the same, nothing further would require to be done; but as this will seldom or never be the case, unless the instrument at both stations can be put into water containing melting snow, it is necessary, where accuracy is required, to send up a thermometer with the messenger, that the temperature of the instrument at the upper station may be noted, and a correction made for any difference that is observed between that and the one at the lower station.

This instrument is equally adapted for the use of aeronauts, and will, I have no doubt, be very advantageously employed for many other purposes; ingenious in design, beautifully simple in its construction, accurate in its results, extremely portable, and easily managed, it strongly recommends itself to scientific tourists, &c. as a valuable addenda to their more costly apparatus.

I remain, Sir,
Yours respectfully,
WM. BADDELEY.

London, October 23, 1838.

FIRE-ESCAPES.—BANKER'S FIRE-PROOF HOUSE

Sir,—Fire-escapes are either too cumberous or too complicated ever to be of much service at a fire, particularly when the reason of few persons is settled enough at those times to use them; even those specially appointed for their management frequently appear to want the necessary presence of mind; the escapes from their failure sometimes cause accidents, instead of preventing them. The best life-preserver in cases of fire is such as I saw at New York, N. A., and that was a volunteer band of humane and active individuals, who, on such occasions, were seen running to the scene of destruction, drawing after them, by means of ropes attached to it, a light two-wheeled spring carriage, which carried ladders of various lengths; but the ladders in general use in the United States, both for building and other purposes, are of a much better construction than those used in this country, which are too narrow to stand steady, the measurement of two Yankee ladders which I took, are as follows:—long, 35 feet; wide at the bottom, 3 feet; wide at the top, 2 feet. Another,—long, 25 feet; wide at the bottom, 2 feet 2 inches; wide at the top, 1 foot 14 inches.

More than 20 years ago I remember seeing in an upper room in a house in Lincoln's Inn Fields, occupied by a relative, a canvass tube fastened by hooks to eyes fixed in the joist of the floor; in case of fire this tube was to be thrown out of the window, the end of which reached the ground, to be held by some of the assembled people; those in the house were to get in and slide through. It would not be a bad plan if every house was furnished with a similar apparatus.

The public should not put faith in advertisements of fire-proof boxes and closets, unless a safe or box is invented that will stand the heat of a glass-house furnace, as I believe the heat in a house thoroughly gutted by fire is not much less; but in partial fires, fire-proof boxes and closets may assist in preserving papers, books, &c.

In the year 1835 I was at New York, during the great fire, and saw plenty of boxes and closets, which every means had been used to make fire-proof; but after the fire, however sound they might appear on the outside, everything within was charred to a cinder; in some instances the closets were in the cellars, but the heat was so great from the burning ruins, that the papers even in those did not escape.

The only way to preserve papers, books, &c., in case of a fire in a city, is to keep them in a dry well, as is done at one of the banking houses in London; or what is more convenient, to erect a building out of contact with all other houses, built with walls sufficiently thick, not only to prevent outward heat from penetrating, but to support a solid roof of either stone, cement, brick, or iron; there must be no windows on the outside, but the light is to be obtained from within,

by having a court in the centre of the building, the interior of the house to be heated by warm air, proceeding from a hot water apparatus situated in the cellar. Such a building, even on a small scale, might be erected near, but detached from, the main building of banking and other houses, where valuable papers are now kept.

I remain, Sir, yours faithfully,
ARTHUR TREVELYAN.
Anglesey, Oct. 19, 1838.

PUBLIC BAROMETERS AND THERMOMETERS.

Sir,—On perusal of the interesting communication of "F. R. S.," in No. 792, suggesting the erection of pillars on the quays in seaport towns, &c., with barometers and thermometers affixed for the public use, the accomplishment of so desirable an object appeared to me to be far distant, unless, indeed, in isolated instances, through the bounty of philanthropic individuals.

That the instruments *would* be consulted to advantage by a large class of persons on our coasts, as pilots, fishermen, watermen, &c., there can be no question; and it is not a little recommendatory to the suggestion thrown out by "F. R. S." that parties of all the above classes *do* make a constant practice of inspecting the barometers in the shops of opticians, usually situated near the water side in ports, where their avocations lead them to a considerable distance from the shore; the description of boats' gear for the day being generally decided by the indications with which they have become familiar.

I have been lead to bear testimony to the above from the practice having often struck me as very judicious, and which indeed one would not have been prepared to expect from men whose habitual usages dispose them rather to place reliance on the ordinary prognostications of the weather, which would be qualified or confirmed by constant observance of the variation in the instruments.

I am, Sir, your obedient servant,
NAUTICUS.
Woolwich, Oct. 16, 1838.

STEAM NAVIGATION TO INDIA.

Sir,—Your correspondent " H.'s " remarks upon my communications have forcibly impressed me with the truth of the sentiment advanced by an eminent writer, when speaking of some quotations which had been made from his own writings on a particular subject :—

" It is obvious that he who wishes to judge of them fairly, must view them in their proper place, accompanied with their respective proofs and illustrations ; and that to tear them from their connection and exhibit them in their naked form, as though they had been expressed in the author's own terms, is a direct appeal to prejudice. The obvious design is to deter the reader at the outset, and to dispose him to prejudge the cause before it is heard. To mingle in the course of controversy, insinuations and inuendoes which have no other tendency than to impair the impartiality of the reader is too common an artifice ; but such an open barefaced appeal to popular prejudice is of rare occurrence. It is an expedient to which no man will condescend who is conscious of possessing superior recourses."

I have only to ask the readers of your Magazine to refer to my letters on the subject of Indian Steam Navigation, where they will find the *gravamen* of my complaint was, that steam navigation in India had not been introduced to the extent that it ought to have been; that there was strong grounds for attributing this to the *direct* or *indirect* influence of the Indian government, and showed that the steam ships now in India, were, with two exceptions, of a very inferior description. In his last letter " H." states that a private steamer was employed there—and reminded me that some iron steam-vessels were sent out in 1834. Will he state the size and places where these iron vessels are employed?

I regret having fallen into the error of speaking of Colonel Chesney's expedition to the *Euphrates* as having been to the *Tigris*, but I presume your more candid readers at once perceived and pardoned my mistake, although " H." would not.

My conviction remains unaltered, that the steam flotilla in the East Indies is absolutely contemptible, as compared with the resources and wants of that part of the British empire.

I am, Sir, your obedient servant,
GEORGE BAYLEY.

September 14, 1838.

P.S.—I cannot forbear observing, that the iron steam-vessels sent out in 1834, were intended for river navigation only, and therefore cannot fairly be deemed a part of the *large steam flotilla* in the East Indies of which we have heard so much from " H."

G. B.

DREDGE'S MATHEMATICAL SUSPENSION BRIDGE.

Sir, — In your Magazine, Number 789, page 468, I find you have misreported the subject I submitted to the mechanical section of the British Scientific Association at Newcastle. Having always perceived a readiness on your part to correct mistakes of this sort, I ask you the favour of inserting in your valuable journal my paper " On Suspension-bridges and Chain-piers" referred to. In the *Athenæum* it was also the same as you have reported it.

The paper I have alluded to was as follows :—

I beg to offer to your notice a mathematical suspension chain, and some experiments showing the difference from, and the superiority to, the ordinary chain used in the construction of bridges and piers.

In every bridge there are two forces or actions—vertical and horizontal.

1st, In the plan I propose, the vertical force, or gravity, is borne by the arch, commencing at the centre of the bridge, and progressively increasing to the abutments.

2nd, The horizontal force is sustained by the roadway, which is rendered a rigid line.

3rd, The roadway is attached to the chains by a series of rods diagonally suspended, so that the entire weight of the structure concentrates by its nearest direction on the bases or towers respectively.

4th, The chains diminish from each base progressively, and gravity diminishes in the same ratio and in the same direction, by which the centre of the bridge is relieved of all vertical pressure.

5th, The chains being but lightly affected at their extremities, the moorings required are but trivial.

On the other hand, by the ordinary method of construction :—

1st, The vertical and horizontal forces are borne by the arch, and its centre is a *fulcrum* instead of the commencement of gravity.

2nd, The roadway being destitute of horizontal force, it is subject to great undulation and lateral motion.

3rd, The roadway being attached to the chains by rods suspended vertically, therefore half the weight of structure concentrates on its centre, and is oppressive inversely, according to its versed sine.

4th, The chains are mostly parallel throughout, and half of the whole weight of the structure is direct vertical pressure on the centre of the bridge.

5th, The chains are loaded at their extremities with all their horizontal force; the moorings required must be strong accordingly.

The advantages of a suspension bridge, or pier, on the mathematical principle, are, in fact, stability and economy, combined with lightness, strength, and simplicity of construction; these advantages are derived by passing the weight through a compound series of inclined planes to each respective base; there the horizontal force is transferred from the chains to the roadway, by which the roadway is changed from neutrality to power; and by the subtraction of this force from the chains, more power, together with a considerable reduction of material, is obtained.

The following experiments will prove these points, in four separate trials by model, 4 feet 6 inches span, and 6 inches deflection, of equal portions of material.

	Parallel Chain Model.				Mathematical Chain Model.
1st Trial. Bath, Jan. 2, 1838.	— { Broke down when loaded with six sacks of horse beans.			} — {	Bore six sacks of horse beans, 7 sacks malt, 2 cwt. cast-iron, and 11 men, all this did not break it down.
	cwt.	qr.	lbs.	cwt. qr. lbs.	
2nd Trial. Bristol, Jan. 6, 1838.	13	3	25	} — { 34 1 25 } — {	And each broke on adding more weight
3rd Trial. Bristol, Jan. 13, 1838.	13	0	0	} — { 33 0 0 } — {	Ditto.
4th Trial. Bristol, Jan. 13, 1838.	23	0	0	} — { 61 0 17 } — {	Ditto.

Of the mathematical design given for crossing the Avon at the St. Vincent's Rocks, Clifton; the following are its dimensions, weight, ultimate power, and estimate:—

The distance between the points of suspension is 700 feet, the deflection of the chains $\frac{1}{10}$th the chord line; the weight of the chains and diagonals 60 tons; the weight of the roadway 130 tons; and the transverse section of the chains altogether, at the base, 132 square inches.

M. O'Nally (engineer of the Fribourg Suspension-bridge,) found, that the best iron when drawn into wire, is equal to sustain 52 tons per square inch; and Telford ascertained that a bar of 5 inches is equal to sustain 27 tons per square inch; therefore, it is in fair ratio to presume, that an inch bar is equal to sustain 30 tons, and having proved it by experiment, I have made my calculations accordingly.

	Inches. Power per inch.	Ultimate power.
The transverse section of the chains at the base 132 square inches.	132×30 tons	Tons. = 3960 tons.
	Tons.	Tons.
Weight of chains and diagonals............	$60 \times 1,5^{*}$ =	90 ....
Ditto of platform, &c....................	$130 \times 2,$† =	260 ....
Surface 17,400 feet, two feet room for one person, it would contain 8,700 persons at 140lbs. each, at 543 tons..	$543 \times 2 = 1086$	
	Total.................. 1436	
"	Surplus power, in Tons 2524	
	3960 = 3960	

* The deflection being one-tenth the chord line, and the centre of gravity of the mathematical chains being one-third their length from each base, their weight multiply by 1, 5 (1,66 is the real factor, but the curve yields some advantage.)

† The roadway centres of gravity being half between the extremes and centre of the bridge, the weight would multiply by 2.5 were it on the common bracket principle, but inasmuch as it is sustained by the segment of a circle, and a considerable portion being sustained without affecting the chains, I therefore multiply the entire weight by 2, which amply shows its natural tension.

The chains would be composed of inch bars of iron 12 feet long, with tranverse bolts.

The platform of 2 inch planking, with tranverse beams, altogether requiring about 6,000 cubic feet of timber.

	£
Weight of chains and diagonals 90 tons at 25l. per ton ..	2,250
Palisades and other iron, 20 tons at 20l. per ton	400
600 cubic feet of timber	900
	3550
Cost of fixing, say..........	2450
Total estimate	6000

The chains of the Menai Bridge are 260 square inches transverse, and 1,710 feet long, their weight upwards of 1,900 tons; on the mathematical principle an average of 30 square inches transverse, and 1,200 feet long would have been sufficient, and the weight of the chains only 70 tons.

The chains of the Clifton Bridge, as proposed on the old plan, averages 477 square inches, transverse; the one I have proposed averages only 37 square inches transverse, and either equal to sustain three times more load than can be fairly put on them.

The above paper excited a discussion, in which several members opposed my plan. I produced a drawing of a bridge copied from the *Saturday Magazine*, March 1834, at Wandipore, which clearly proved that the principle in the construction of bridges I now advocate was acknowledged 195 years ago; whereupon my plan received the unanimous approbation of the Mechanical Section.

Your Report alluded to the Victoria Bridge, Bath, therefore I will give you a short account of that bridge. The span is 150 feet; deflection of the chains one-sixth the chord line; the transverse section of the entire chains at the base 48 square inches, and their weight between the points of suspension only five tons; the width of the roadway is 18 feet, it is composed of oak transverse beams, and longitudinal planking 2 inches thick, which is covered with a composition of coal tar and lime, with about 50 tons of gravel, and afterwards Macadamized. The chains were commenced putting together in November 1836, and notwithstanding that rough weather hindered the workmen upwards

of a fortnight, and the days being short, yet the bridge was opened to the public the following month; and from December 1836 all description of weights have been allowed to pass over it. The cost of the masonry was upwards of 500l., and the entire cost of the bridge, including the masonry, was under 1760l. Its total length from one mooring to the other extremity is upwards of 330 feet, and the total weight of iron in the bridge, inclusive of its fence, which is wholly iron, is about 21 tons.

I have only to add, that being an admirer of fair play, I fully depend on receiving the same at your hands.

I remain, Sir,
Your humble servant,
JAMES DREDGE.

Bath, October 19, 1838.

MANUFACTURE OF SILK-WORM GUT.

Sir,—In your Number of this day, I see a letter signed *Piscator*, on the subject of "silk-worm gut" used by anglers, and I venture to assure you, that his remarks and suggestions are extremely cogent. I have seen the article made in Italy, and have a hundred times thought of making it in England. But such have been the varied turmoils of my life, that procrastination, forced or casual, has suspended my project. However, I have this very autumn procured a packet of silk-worm eggs from Italy, which I intend (if in a situation to do so) to hatch and rear, and convert the produce into silk-worm gut, by a manipulation somewhat superior, I hope, to the slovenly performance of the Spanish, Italian, and Hindoo silk spinners, who waste one-half of the material, and make a bad article of nine-tenths of the other half.

The cylindrical mass of silk gum contained in the worm cannot be drawn into "gut," without a certain previous preparation, performed upon the living worm itself, and which most silk producers keep, or affect to keep, a secret.

I should be glad to join my knowledge and labour to any one desirous of employing the trifling capital for supplying all the anglers of Great Britain and America with an article essential to their craft, as far superior to that in present use, as a chain cable is to a rope of rags. I can make the "gut" of several

yards long without a join, and if needs
be, as thick as a pack-thread. But,
apropos of thickness, three lengths twisted together *with the fingers* (not the machine) is far better than a single gut of
equal thickness to the three, as much
less liable to crack when handled dry.

I have hit upon some great improvements in the materials and construction
of fishing-rods, which I shall soon give
to the public in a new "Art of Angling."
I have the honour to be, Sir, &c.
F. MACERONI.
3, St. James-square, Oct. 20, 1838.

BERNDSON'S IMPROVED PUMP-BUCKET

Sir,—I beg leave to send you a sketch
(from memory) of a pump bucket, adapted for being used in mines, where the
quality of the water may be such as to
preclude the use of metal, or of leather
packing; this was shown lately at Newcastle, and was said to be the invention
of a Mr. Berndson, a Swedish engineer.

The perspective view, fig. 1, and section,
fig. 2, will, perhaps, sufficiently explain
this simple construction to most of your
readers; but lest any misapprehension
should arise from the imperfection of the
sketch I add a short description of the
parts :—

Fig. 1.   Fig. 2.   Fig. 3.

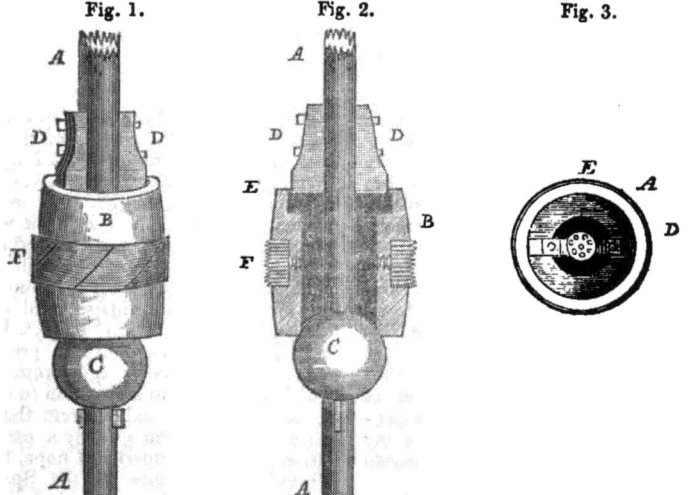

A is the lower end of the spear, or
pump-rod, having attached to it the
shoulders or pushing pieces D D, and
the valve or ball C.

B is the box, or bucket, having a
cylindric passage (of about half its exterior diameter) bored through it; this
passage is enlarged at its upper orifice,
as shown at E, to receive and to admit
of the alternate upward and downward
movement of the pieces D D; the lower
orifice is slightly champered, to make a
water-tight seat for the ball C, when the
upward stroke of the pump-rod brings
it against the box.

F is a deep right-angled gutter, cut
into the box B, in which the packing is

lodged. In the specimen shown at
Newcastle, this packing was made of
birch bark, and consisted of diagonally
cut parts, so as to form an appending
piston when pressed outwards, *during
the upward stroke,* by the column of
water in the pump, which acts on the
inner face of the packing through small
openings, communicating through the
sides of the cylindric passage above-
mentioned.

The sketch represents the relative situations of the parts during the upward
stroke of the pump-rod, when C acts as
a valve, closing the aperture. At the
commencement of the downward stroke,
the bucket B remains for a moment at

rest, until the shoulder pieces D D have reached the bottom of the enlargement at E, when the pump-rod begings, and continues, to push the bucket B before it, until it has reached the bottom of its course. During this period the ball C has been pushed away from the lower orifice, and a free passage has been left for the water to get through the bucket.

As this bucket admits of being made of green wood, and by any country workman who has access to a common turning lathe, it may be made available on many occasions, where other materials or more elaborate workmanship may not be easily attainable.

K. H.

AN AVAILABLE POWER NEGLECTED.

Sir,—If coal-gas is manufactured under a pressure of two atmospheres (or as much higher as consists with perfect safety) and be received into a reservoir, provided with a safety-valve, weighted at the required pressure, and from thence made to act on a piston as in a steam-engine, and so pass off into the gasometer,— I calculate (as Brother Jonathan says), the gas would be much purer, by being suddenly released from a state of great density to that of corresponding rarity, and the gas-*factor* would be enabled to sell his commodity at a lower rate to the *consumer*, if it nearly paid itself in *labour*, previous to combustion?

So many are the uses to which this power might be applied,—especially in a large town, where a greater quantity of gas is needed, and where its services as a working power are in more request, that it would be quite superfluous to point them out, for the locality of the works would vary its application, I shall therefore merely show how the above may be effected.

The pipe where the coal-tar comes over must be furnished with a stop-cock, which I shall call No. 1; there must be another on the retort, and a third in the pipe which leads to the purifiers. The first is to let out the coal-tar and ammoniacal water as they accumulate, (or it might be better done by making that reservoir gas-tight and strong enough to bear the working pressure, and so no cock would be required,

except one to let it out at the bottom when too full, which might be ascertained by using a float, as in the steam-engine boiler). When a retort wants changing, stop off No. 3, and open No. 2, which will let out the high-pressure gas contained in the retort without mischief, and then the lid can be removed with perfect safety. The purifying process is the same as at present used. When the gas is pressing on one side of the piston at two atmospheres, the other side, in connexion with the gasometer, is only sustaining a pressure of about six ounces per square inch.

If the above be of no other use, it can at least be applied to fill vessels with portable gas?

I am, Sir,
Yours very respectfully,
WM. PEARSON.
Bishop Aukland, Oct. 10, 1838.

CIVIL ENGINEERING IN AMERICA.[*]

This " Sketch," as it is modestly entitled, is a work of much more value than several which make higher pretensions to finish. It is a book on an interesting subject, by a man who understands that subject well; and, therefore, in our opinion, likely to be popular. It is singular that authors and publishers have not yet discovered that the public is, in general, partial to works of this kind, in which a professional man, writing mainly, it may be supposed, for professional readers, yet takes a view of his subject sufficiently popular to be intelligible to those who, without much previous intimacy with the matter in hand, feel a wish to become acquainted with its principal bearings. General readers of this class are getting to be weary of general writers. They feel, like Göethe, that " they have ignorance enough of their own." They find likewise that they not only derive more instruction, but more amusement, from books of travels, in which something more is aimed at than merely to furnish a fund of " light reading," which in fact has, in our days, become the heaviest of all

[*] Sketch of the Civil Engineering of North America, comprising remarks on the harbours, river and lake navigation, lighthouses, steam navigation, water-works, canals, roads, railways, bridges, and other works in that country. By David Stevenson, Civil Engineer. London, 1838. 8vo. pp. 320.

reading, from being furnished in such overpowering quantities as altogether to glut and destroy the appetite.

To readers of this class, and, as we have said, we are confident the class is a larger one than authors and publishers generally imagine, Mr. Stevenson's book will be highly acceptable. It furnishes, in a brief and agreeable shape, a mass of well-digested information on a subject to which it would not be easy to find many rivals in interest and importance. Civil Engineering and its products form a prominent feature of the general system of life and society in all countries at the present day; but more so, perhaps, in the United States of North America than in any other, not even England excepted. The stranger who surveys the face of our country will find many things to attract his attention in the labours of man besides, and perhaps beyond, our canals and our railways; but, in America, the civil engineer reigns triumphant. It is he who has created almost every thing remarkable that the land can show beyond the beauties it received from the hand of nature. A person may feel a warm interest in Italy, and yet care nothing about the embankments of the Po; in France, and yet be ignorant of the canal of Languedoc; but all who take an interest in North America must take an interest in its Civil Engineering.

Mr. Stevenson's book is the product of a tour of only three months in Upper and Lower Canada, and the most interesting parts of the United States. In so short a space of time it would have been impossible for him to collect the information contained in his work, but for the kindness and liberal spirit with which he was received by several distinguished Americans, to whom he very properly returns his thanks in the preface. We are gratified to find that he speaks so highly of the attentions shown him by all classes of persons in America, and their readiness to communicate freely every kind of information. The English have repeatedly received the praises of foreigners for the same liberality, which is not, we believe, so common on the continent of Europe, and we are glad to find that their posterity do not degenerate.

The work is divided into twelve chapters, each of which is devoted to a parti-cular branch of the subject. Mr. Stevenson usually commences each chapter with a general view of the matter to be treated of, which he then proceeds to illustrate with his own observations, and in conclusion, with a summary of those of others, in such cases as he has not an opportunity of observing himself. These observations are illustrated, wherever necessary, by cuts and engravings, executed with great neatness, which are at once a useful and agreeable accompaniment to the work.

The subjects of the different chapters are nearly all enumerated in the title-page, and follow in much the same order they there appear in. The two most prominent are, of course, Steam Navigation and Railways. On the former subject the following remarks are interesting, and may serve as a specimen of the style of the work :—

" It would be improper to compare the present state of steam navigation in America with that of this country, for the nature of things has established a very important distinction between them. By far the greater number of the American steam-boats ply on the smooth surfaces of rivers, sheltered bays, or arms of the sea, exposed neither to waves nor to wind; whereas most of the steam-boats in this country go out to sea, where they encounter as bad weather and as heavy waves as ordinary sailing vessels. The consequence is, that in America a much more slender build and a more delicate mould give the requisite strength to their vessels, and thus a much greater speed, which essentially depends upon these two qualities, is generally obtained. In America the position of the machinery and the cabins, which are raised above the deck of the vessels, admits of powerful engines, with an enormous length of stroke being employed to propel them; but this arrangement would be wholly inapplicable to the vessels navigating our coasts, at least to the extent to which it has been carried in America.

" But, perhaps, the strongest proof that the American vessels are very differently circumstanced from those of Europe, and, therefore, admit of a construction more favourable for the attainment of great speed, is the fact that they are not generally, as in Europe, navigated by persons possessed of a knowledge of seamanship. In this country steam navigation produces hardy seamen; and British steamers being exposed to the open sea in all weathers, are furnished with masts and sails, and must be worked by persons who, in the event of any accident happening to the machinery, are capable of sailing the vessel, and who must, therefore, be experienced seamen. The case is very different in America, where, with the exception of the vessels navigating the lakes, and one or two of those which ply on the eastern coast, there is not a steamer in the country which has either masts or sails, or is commanded by a professional seamen. These facts forcibly show the different state of steam navigation in America, a state very favourable for the attainment of great speed and a high degree of perfection in the locomotive art." pp. 117-8.

These observations may serve as an answer to those who have lately been expressing their wonder at the comparative apathy with which the Americans seem to be witnessing the transfer of the business of conveying passengers to and from Europe from the hands of their own " New York liners" to those of European steamers. It is altogether " out of their line." The appearance of the *Great Western,* with its below-deck cabins and its invisible machinery (the bull is inevitable) and its masts and sails, must have been as " something new and strange" to the eyes of the dwellers on the Hudson, as to those of the dwellers on the Thames would be that of an American steam-boat from the island of Manhattan, with its roofed-over lower-deck, and its up-stairs promenade-deck, and its machinery all turned inside out, for the investigation of the curious. But most probably the thoughts of the American steam-boat builders will soon " suffer a sea-change." Be it so—we have the start, and the honour of having had the start, which are both something.

The conveniences of steam-boat travelling in America have been so often and so justly celebrated, that Mr. Stevenson's account of canal travelling will excite some surprise.

" The canal travelling in many parts of America is conducted with so little regard to the comfort of passengers as to render it a very objectionable conveyance. The Americans place themselves entirely in the power and at the command of the captains of the canal-boats, who often use little discretion or civility in giving their orders, and strangers, who are not accustomed to such usage, and would willingly rebel against their tyranny, are in such cases compelled to be guided by the majority of voices, and quietly submit to all that takes place, however disagreeable it may be. About eight o'clock in the evening every one is turned out of the cabin by the captain and his crew, who are occupied for some time after the cabin is cleared in suspending two rows of cots or hammocks from the ceiling, arranged in three tiers, one above another. At nine the whole company is ordered below, when the captain calls the names of the passengers from the way-bill, and at the same time assigns to each his bed, which must immediately be taken possession of by its rightful owner, on pain of his being obliged to occupy a place on the floor, should the number of passengers exceed the number of beds, a circumstance of very common occurrence in that locomotive land. I have spent several successive nights in this way, in a cabin only 40 feet long by 11 feet broad, with no less than forty passengers; while the deafening chorus produced by the croaking of the numberless bull-frogs that frequent the American swamps, was so great as to render it often difficult to make one's-self heard in conversation, and, of course, nearly impossible to sleep. The distribution of the beds appears to be generally regulated by the size of the passengers, those that are heaviest being placed in the berths next the floor. The object of this arrangement is partly to ballast the boat properly, and partly in the event of a breakdown, to render the consequence less disagreeable and dangerous to the unhappy beings in the lower pens. At five o'clock in the morning all hands are turned out in the same abrupt and discourteous style, and forced to remain on deck, in the cold morning air, while the hammocks are removed and breakfast is in preparation. This interval is occupied in the duties of the toilette, which is not the least amusing part of the arrangement. A tin vessel is placed at the stern of the boat, in which every one washes, and fills for his own use, from the water of the canal, with a gigantic spoon formed of the same metal ; a towel, a brush, and a comb, intended for the general service, hang at the cabin door, the use of which, however, is fortunately quite optional. The breakfast is served between six and seven o'clock, dinner at eleven, and tea at five. The American canal travelling certainly forms a great contrast to that of Holland and Belgium. The boat in which I was conveyed on the canal between Ghent and Bruges, for example, was commodiously fitted up with separate state-rooms, containing one berth in each, and was in other respects a most comfortable and agreeable conveyance. But I trust the reader will not form an estimate of American travelling from what has just been said, nor take this single specimen of it as a criterion of the whole. In the eastern, and earlier settled districts of the country, no such grievances have to be suffered, and there are many hundreds of persons in that part of the United States who hardly believe in their existence. So long as the traveller keeps on the east of the Alleghany mountains all goes on smoothly, but if he attempts to cross their summits, and penetrate into the "far west," he must look for treatment such as I have described. There is, indeed, as great a difference in this respect between the seaward and interior states of North America as there is between the counties of Kent and Caithness."—pp. 196-199.

Our extract has extended to such a length that we have only left ourselves room to add, that even Mr. Stevenson's book does not explain the subject, and that we trust some brother of the profession will take the hint thrown out in the preface, and publish the results of a longer and more extended tour. The improvements in America are so rapid that an account of them might well form the subject of a perennial annual.

THE ARCHITECT, ENGINEER, AND OPERATIVE BUILDER'S CONSTRUCTIVE MANUAL. BY CHRISTOPHER DAVY, ARCHITECT.*

Indebted as the pages of this journal have been to the pen of Mr. Davy, through nearly the whole period of its existence, for much valuable support as a contributor on architectural and engineering subjects, it is with more than ordinary pleasure we welcome his appearance as the author of a book of his own, having for its united and legitimate objects to benefit the distinguished profession to which he belongs, and to obtain for himself a higher standing among his cotemporaries than is to be achieved by any occasional magazine contributions, however ingenious or useful. The work is entitled "The Architect, Engineer, and Operative Builder's

---

* Being a practical and scientific treatise on the construction of artificial foundations for buildings, railways, &c.; with a comparative view of the application of piling and concreting to such purposes; also an investigation of the nature and properties of the materials employed in securing the stability of buildings. Illustrated by examples selected from the most important architectural and engineering works of this country. To which is added, an analysis of the principal legal enactments affecting the operations of the practical builder, with notes of cases occurring in actual practice. Part I. Illustrated by 13 Plates and Woodcuts.

Manual;" but the scope of it is more correctly indicated by the alternative title given in our foot note, which describes it as being limited in its application to " foundations," " materials," and " legal enactments effecting the operations of practical builders." A "Manual" is commonly understood to include *all* that relates to any business *in hand*, and which should, when wanted, be *at hand*; the Latin synonyme of *Vademecum* describes the thing to the life. A *Manual*, too, should be something *handy*; fitter for the waistcoat than the greatcoat-pocket. Now, our friend Mr. Davy's Manual, is, more properly speaking, and in truth, not a Manual, but a book of reference—in goodly octavo—on the particular subjects of which it treats; a sort of chamber counsel, to be consulted by all who would wish to build lasting fabrics on secure foundations, and steer clear of the law and its manifold meshes. But to the matter of the work;—what, after all, is *the matter* to be most looked to; the Part (I.) before us contains—1st, A chapter of excellent observations on strata and subsoil, with descriptions of the best sort of boring apparatus and modes of using them.— 2nd, Another chiefly on timber and metallic piling.—3rd, A full account of the composition and application of *Béton* (translated from the French of M. Belidor.—4th, An analysis of British limestones.—5th, Modes of calcining limestones, including Dr. Higgins's experiments and descriptions of the most approved English, Scotch, and Irish kilns. And, 6th, Professor Fuch's experiments on mortar—Constituents of ordinary mortar—Examples from public works— Varieties of sand—Examples from the disaggregation of British sandstones— Action of acids on sands, &c. &c.

The information collected on each and all of these subjects is of great interest and of the first authenticity. Here and there we have felt disposed to quarrel with Mr. Davy, for adopting too implicitly the dicta of those who have gone before him—as for instance, when he assumes it to be matter of history that Mr. Kyan has " *ascertained* that albumen is the primary cause of putrefactive fermentation, and subsequently of the decomposition of vegetable matter,"—whereas, all that Mr. Kyan has *really* ascertained is, that a certain effect is produced, or supposed to be produced, by his process, for which neither himself, nor any one for him has yet been able to adduce the *sufficient reason ;* but it is only due to him to observe, that he always gives his authorities, and acknowledges almost always the extent to which he has drawn upon them; so that the reader is in every case enabled to judge for himself, whether the author's conclusions are right or wrong. We hope—indeed we make no doubt—the work will be liberally patronised by the profession. If it has only as many readers as there are persons to whom Mr. Davy's contributions to the *Mechanics' Magazine* have supplied pleasure and profit, he will have no lack of success to complain of; neither will he have more than his talent and industry really merit.

NEW FRENCH MACHINERY FOR MAKING BRICKS, TILES, &c. INVENTED BY M. TERRASSON.

[Translated from the *Receuil Industriell.*]

After remarking upon the properties necessary in a machine that shall completely fulfil the objects of an artificial brick maker, the difficulties to be surmounted in such an invention, especially that which arises from the adhesion of the clay to the mould, and the failure, wholly or in part, of all former contrivances for this purpose, the inventor thus describes his plan :—

This machine is divided into two parts.

1. Machine for kneading the clay.

2. Machine for moulding and transporting the bricks to the area, or yard.

The same motive power may accomplish both objects.

Machine for Kneading.—Horse or water power, &c., may be used at pleasure. That which I use at Teil is moved by two horses. The yard is on the side of an argillaceous hill. Two large sheds are erected, one of which is two stories high ; the upper serving for the horse floor, the large wheel, and the several connecting parts. The ground floor below contains the moulding and transporting machine.

The other shed is a drying room and store house ; it is situated so as to serve as a place of discharge by the shortest way possible.

The hill forms a slightly inclined plane ; it furnishes the earth, and from its position facilitates the transportation of the clay.

Machine for Kneading and Drawing Out the Clay.—Two horses attached to the two arms of a lever, turn a large wheel connected with a trundle, which impels a vertical shaft working in a cylinder or field vat. This shaft is furnished with spiral cutters ; the moistened earth is put into the vat, the cutters in turning, knead it, divide it, and force it, by its own weight and the impulsion of the screw, which operates like the screw of Archimedes, excepting that it reverses the motion. The clay arrives at the bottom of the vat exceedingly well prepared, and then passes through a hole, round or square, which may be enlarged at pleasure by means of a screw which presses against a slide. We thus obtain a roll or fillet of clay of any desired dimensions.

Machine for Moulding and Transporting the Bricks.—This machine, which gears in to the one above, is placed below the vat in such a manner that the clay, already formed of a due calibre, enters the mould, which is nothing more than a chaplet or hollow frame, moving horizontally, with an adjusted velocity, and turning upon and around cylinders. A large cylinder is placed over the mould, and another below, forming a rolling press which flattens and compresses the clay.

After having passed with the mould under this large cylinder, the clay is perfectly level or even ; a boy having previously placed boards in the mould, and upon these boards the bricks are made. The frame returns empty, the boards continue their horizontal course, bringing a fillet of clay well pressed and levelled ; two wires, stretched by a weight, and placed on each side of the board, free the clay, and remove all the roughness which, in being stripped, the mould might have left on the fillet.

The boards continue their horizontal course, being urged forward by those which the child continues to place in the mould ; they arrive at a point immediately under one end of a balance weight (bascule) formed of a frame, across which wires are stretched, and which, as the weight falls off, in the twinkling of an eye, 10 or 12 bricks. The machine stops an instant to give time for this important operation, which may be repeated five times in a minute. A clock bell gives notice when the stroke is given. When the bricks are cut off, and the balance weight rises, the chaplet continues its motion, and two boys take hold of the board.

By means of a long file of cylinders upon a slightly inclined plane, the boards are carried to any desired distance without labour, being hurried down like waggons on an inclined railroad.

I have stated that five strokes of the balance weight may be given per minute, making 50 bricks of large size. By dividing them, 100 of less size would be made ;—indeed, if the children could take them off, 150 might be made, a result which, however incredible, is nevertheless true, as I have proved before the authorities of the country, who were quite incredulous before they saw it. I have even gained bets upon it.

I have hitherto employed only horse power, but there would be great advantage in using steam, wind, and especially water ; the velocity of the shaft being more regular and rapid with a water fall, and with a greater force than two horses, the kneading and other operations would be better performed.

By a similar arrangement tiles are made, earthen water pipes may be moulded, concave bricks for vaults, and in fact all sorts of bricks, refractory and others.

LIST OF ENGLISH PATENTS GRANTED BETWEEN THE 27th OF SEPTEMBER AND THE 25th OF OCTOBER, 1838.

John White, of Haddington, North Britain, iron monger, for certain improvements in the construction of ovens and heated air-stoves. September 27 ; six months to specify.

John Bourne, of Dublin, engineer, for certain improvements in steam-engines, and in the construction of boilers, furnaces and stoves. October 8 ; six months.

Jehiel Forbes Norton, of Manchester, merchant, for certain improvements in stoves or furnaces, and in instruments or apparatus for making the same. October 8 ; six months.

Henry Dunington, of Nottingham, lace manufacturer, for improvements in warp machinery, and in fabrics produced by warp machinery. October 8 ; six months.

George Haden, of Trowbridge, Wilts, engineer, for improvements in the manufacture of a soap or composition applicable to the felting and other purposes employed in the manufacture of woollen cloth, and other purposes to which soap is usually employed. October 8 ; six months.

Charles Sanderson, of Sheffield, steel manufacturer, for a certain improvement in the art or process of smelting iron ores. October 11 ; six months.

Matthew Heath, of Furnival's-inn, London, Esq., for improvements in clarifying and filtering water, beer, wine, and other liquids. October 11 ; six months.

John Woolrich, of Birmingham, professor of chemistry, for an improved process for manufacturing carbonate of lead, commonly called white lead. October 11 ; six months.

John Fowler, of Birmingham, gent., for certain improvements in preparing or manufacturing sulphuric acid. October 16 ; six months.

William Brockedon, of Queen-square, Middlesex, Esq., for a combination of known materials, forming a substitute for corks and bungs. October 17 ; six months.

Henry Meyer, of Piccadilly, wax-chandler and oil merchant, for improvements in the manufacture of lamps. October 17 ; six months.

Elias Robison Handcock, of Dublin, for improvements in castors for furniture and other purposes. October 17 ; six months.

George Harrison, of Carlton-house-terrace, Middlesex, surveyor, for improvements for supplying air for promoting and supporting the combustion of fire in close stoves and furnaces, and for economising fuel therein. October 17 ; six months.

William Edward Newton, of 66, Chancery-lane, Middlesex, for improvements in the construction of bridges, viaducts, piers, roofs, truss-girders, and stays for architectural purposes. October 17 ; six months.

John George Bodmer, of Manchester, in the county of Lancaster, engineer, for his invention of certain improvements in the machinery or apparatus for carding, drawing, roving, and spinning cotton, flax, wool, silk, and other fibrous substances. October 22 ; six months.

William Jeakes, of Great Russell-street, Bloomsbury, for a mode of applying ventilating apparatus to stoves constructed on Dr. Arnott's principle. October 22 ; six months.

William Edward Newton, of Chancery-lane, mechanical draftsman, for an improved method or methods of preparing certain substances for the preservation of wood and other materials used in the construction and fitting-up of houses, ships, and other works, which improvements are also applicable to other useful purposes. October 22 ; six months.

John Henfrey, of Weymouth-terrace, Shoreditch, engineer and machinist, for certain improvements in the manufacture of hinges or joints, and in the machinery employed therein. October 25 ; six months.

LIST OF SCOTCH PATENTS GRANTED BETWEEN THE 22nd OF SEPTEMBER AND THE 22nd OF OCTOBER, 1838.

Robert William Sievier, of Henrietta-street, Cavendish-square, Middlesex, gent., for certain improvements in looms for weaving, and in the mode or method of producing figured goods or fabrics. Sealed 1st of October, 1838; four months to specify.

John Robb, of No. 13, Commercial-road, Hutchesontown, Glasgow, mechanic, for a machine for preparing wood for joiners, carpenters and others. October 2.

Edmond Henzé, of Fenton's Hotel, Saint James-street, Middlesex, merchant, in consequence of a communication made to him by a certain foreigner residing abroad, for improvements in the manufacture of Dextrine. October 8.

Robert William Sievier, of Henrietta-street, Cavendish-square, gent., for certain improvements in rigger and pully bands for driving machinery, and ropes and lines for other purposes. October 11.

James Nasmyth, of Patricroft, near Manchester, engineer, for certain improvements in machinery, tools, or apparatus for cutting or planing metals and other substances, and in securing or fastening the keys or cottars used in such machinery, and other machinery where keys or cottars are commonly applied. October 11.

Thomas Ridgway Bridson, of Great Bolton, Lancaster, bleacher, for certain improvements in the construction and arrangement of machinery or apparatus for stretching, mangling, drying, and finishing woven goods or fabrics, and part or parts of which improvements are applicable to other useful purposes. October 12.

William Angus Robertson, of the Patent Agency Office, Peterborough-court, Fleet-street, London, for certain improvements in the manufacture of hosiery, shawls, carpets, rugs, blankets, and other fabrics, being a communication from a certain foreigner residing abroad. October 12.

John Seaward, of the Canal Iron Works, Poplar,

Middlesex, engineer, for an improvement in condensing steam-engines. October 12.

Joshua Wordsworth, of Leeds, machine-maker, for certain improvements in machinery for heckling and dressing flax, hemp and other fibrous matter. October 18.

John Melling, of Liverpool, for certain improvements in locomotive steam-carriages to be used upon railways, or other roads, part or parts of which improvements are also applicable to stationary steam-engines, and to machinery in general. October 18.

Horace Cary, of Marrow-street, Limehouse, Middlesex, Bachelor of Medicine, for improvements in the manufacture of white lead. October 18.

Henry Huntley Mohun, of Regent's-park, Middlesex, M.D., for improvements in the composition and manufacture of fuel, and in furnaces for the consumption of such, and other kinds of fuel. October 18.

LIST OF IRISH PATENTS GRANTED IN SEPTEMBER, 1838.

Pierrie Armand Lecomte de Fontainemoreau, for improvements in wool combing. September 14.

Edward Davy, of Crediton, for improvements in saddles and harness. September 22.

John Hanson, of Huddersfield, for improvements in machinery, or apparatus for making or manufacturing pipes or tubes from metallic substances. September 22.

Robert William Sevier, of Henrietta-street, Cavendish-square, for improvements in looms for weaving, and in the mode or method of producing figured goods or fabrics. September 25.

Arthur Dunn, of Stamford Hill, for improvements in the manufacture of soap. September 28.

NOTES AND NOTICES.

Launch of the Archimedean Steam-Vessel.—On Thursday a steam-vessel, named the *Archimedean*, was launched from the yard of Mr. Wynn, of Millwall, nearly opposite Deptford. The principle upon which the vessel is proposed to be propelled (Ericsson's) is one which has long been in agitation, and which has already been experimentally tried with considerable success upon a vessel of eight tons, and of four-and-a-half horse power, and the objects which it is desired should be attained are at once speed, and the ready application either of steam or sailing power. The engine will be placed amid-ships, as in the steam-vessels now in use, and the propeller or paddle, which is under the stern, will be worked by a communicating shaft, acting upon a screw, called the Archimedean screw, in the application and use of which the invention is grounded. The propeller being placed under the stern, the inconvenience arising from the paddle now in use, which act themselves as a backwater, is avoided, and great benefit will be derived in seas when the wind is on the beam, when, instead of a great portion of the power being lost as now, the revolutions of the paddles will continue with as good effect as in calm weather. Should any circumstance render it necessary to remove the steam power, the wheel may be immediately unshipped, or its action upon the water may be prevented, and sailing power may then be applied. The vessel has been built at the yard of Mr. Wynn, under the direction of Mr. Smith, and is of exceedingly elegant construction. Its dimensions are as follow:—Extreme length fore and aft, 155 feet; length between perpendiculars, 197 feet; breadth of beam, 22 feet 6 inches; depth of hold, 13 feet; diameter of screw, 7 feet; length

of screw 8 feet; and it is intended to apply engines of 45-horse power.—*Chronicle.*

New American Steam Ship.—The steam-ship *Liverpool* made an experimental trip on Saturday from Liverpool to Dublin. The weather was very boisterous, but the ship made her way at the rate of about 10 miles an hour and reached Dublin in 12 hours and 21 minutes. In returning, the wind was not quite so strong, and blew from the west, and the distance from Kingstown harbour to Liverpool was accomplished in 10 hours and 42 minutes. The *Liverpool Albion* says—"The result fairly justifies the conclusion that, when all on board is completed, the stiffness of the engines worn off, and the improvements made which this first voyage has suggested, she will be one of the most ocean-worthy and swiftest vessels hitherto built, and become a credit to the owners and to the port." She started on her first trip to New York last Saturday.

Foreign Railroads.—Mr. Stephenson, the celebrated railroad engineer, has been engaged by the Florence and Leghorn Railroad Company to make the requisite surveys and plans for that line. Two English engineers have already arrived at Florence, to commence the preliminary works.

Iveson's Smoke-consuming Plan.—Sir,—I perceive an article making the round of the newspapers, upon the saving of fuel, and the consumption of smoke, said to be the discovery of a Mr. Iveson, manager of the Castle Silk Mills, Edinburgh, on which, with your leave, through the medium of your valuable columns, I wish to offer to the public the following remarks:—It is part of a plan which I mentioned to several gentlemen so long ago as 1835, for the consumption of smoke, the economising and condensation of fuel in bulk, applicable to steamers in long voyages. I may here observe, that this fuel is not more expensive than the best Newcastle coal, and somewhat less than one-half the bulk. The use of the fuel will involve a slight alteration in the frame-work of the furnace, and render the smoke-funnel or chimney no longer necessary. Among the gentlemen to whom I mentioned my discovery in 1835, was the Rev. George Jacque, of Auchterarder, N.B., and again in conversation with him upon the same subject, in July 1837, he suggested an improvement in the mode of applying the jet of steam, which was to take it from the exhauston-pipe after the steam had done its work; and early last winter it was mentioned to Matthias Dunn, Esq., of this town. Having some specimens of the fuel prepared, I shall be happy to show them to any person interested in this important subject.

I am, Sir, your most obedient servant,
J. M. ARNOTT.

33, Blackett-street, Newcastle-upon-Tyne, October 15, 1838.

Statue of James Watt in Greenock.—The statue of James Watt, by Sir Francis Chantrey, is now placed in the building erected for it in Union-street. It is an 8 feet figure, of statuary marble, and weighs upwards of 2 tons, and the pedestal, which is of Sicilian marble, weighs about 3 tons. On the front of the pedestal is the following inscription, from the classic pen of Lord Jeffrey:—"The inhabitants of Greenock have erected this statue of James Watt, not to extend a fame already identified with the miracles of steam, but to testify the pride and reverence with which he is remembered in the place of his nativity, and their deep sense of the great benefits his genius has conferred on mankind. Born XIX January, MDCCXXXVI. Died at Heathfield, in Staffordshire, August XXV., MDCCCXIX." On the right of the pedestal is a shield containing the Arms of Greenock, and on the left strength and speed. On the back is an elephant, in obvious allusion to the beautiful parallel drawn by Mr. Jeffrey between the steam-engine and the trunk of that animal, which was equally qualified to lift a pin or to rend an oak.

Bridge over the Danube.—The new suspension bridge over the Danube, between Buda and Pesth, which will be begun the next spring, is a colossal undertaking. Two piers of granite and the red marble of Neudorf, 35 feet thick, and 150 feer above the level of the foundation, will support the whole structure. There will consequently be three openings for the water to pass through, the middle passage being 640 feet in width, and each of those at the sides 270 feet, making in all 1,180 feet. The entire length of the bridge will be 1,600 feet. Cast-iron beams will support the platform, which is to be 37 feet wide, viz., 25 feet for the carriage-way, and six feet for each foot-path. The whole will be suspended by 12 chains, weighing together upwards of 2,000 tons.

The last of the First Steam-boat.—A curious fate has attended the first actually employed steam-vessel, which was constructed for the late Mr. Miller of Dalswinton, by Mr. William Symington, on the Forth and Clyde Canal, near Falkirk, and from seeing which Fulton took his idea of introducing steam-vessels on the rivers of the United States. In the formation of some new works at the eastern extremity of this canal, it lately became necessary to fill up the course of a stream, which formerly ran into the harbour of Grangemouth; among the rubbish and materials used for this purpose, the venerable remains of Mr. Miller's vessel have been buried—strongly reminding us of the lines of the poet—

" Imperial Cæsar, dead and turned to clay,
May stop a hole and keep the wind away."

Sir James Anderson's Steam-coach is finished, and will start from Buttevant for this city in a fortnight, travelling at 15 miles an hour.—*Limerick Paper.*

The Suspension Bridge at Freyburg, the longest in the world, was completed and thrown open in 1834. The engineer who constructed it is M. Chaley, of Lyons. Its dimensions, compared with those of the Menai bridge, are as follows:—

	Length.	Elevation.	Breadth.
Freyburg	905ft.	.. 174ft	.. 28ft.
Menai	580	.. 130	.. 25

It is supported on four cables of iron wire, each containing 1056 wires, the united strength of which is capable of supporting three times the weight which the bridge will ever be likely to bear, or three times the weight of two rows of waggons, extending entirely across it. The cables enter the ground on each side obliquely for a considerable distance, and are then carried down vertical shafts cut in the rock, and filled with masonry, through which they pass, being attached at the extremity to enormous blocks of stone. The materials of which it is composed are almost exclusively Swiss; the iron came from Berne, the limestone masonry from the quarries of the Jura, the woodwork from the forest of Freyburg : the workmen were, with the exception of one man, natives,who had never seen such a bridge before. It was completed in three years, at an expense of about 600,000fr. (25,000*l.* sterling.)—*Hand-book for Switzerland.*

Sweden.—The Swedish government has recently published a military map, in which are figured the works of public utility, commenced, continued, or finished under the reign of the present king, Charles-Jean Bernadotte. There are fifteen canals, eight ports, eight roads, nine lines of defence, the expense of which has amounted to 77,177,095fr., all furnished, (without borrowing) by the royal treasury.

Fermentation an act of Vegetation.—M. Turpin has lately published his observations upon certain phenomena, which he considers sufficient to show, that the act of fermentation, concerning which chemists have been so much embarrassed, is owing to the rapid development of infusorial plants. He states, that all yeast, of whatever description, derives its origin from the separation from organic tissue, whether animal or vegetable, of spherical particles of extreme minuteness, which particles, after a certain time, rise to the surface of the fluids in which they are immersed, and there germinate. Their germination is said to be caused by a certain amount of heat, and by contact with atmospheric air. The carbonic acid obtained by fermentation is ascribed to the infusorial plants. M. Turpin considers the act of adding yeast to liquids, when fermentation is languid, as equivalent to sowing millions of seeds in a favourable soil. He calls the yeast plant of beer *Torula cerevisiæ* : he considers each infusion to have its peculiar plant, and he names the whole race of such beings *Levurians.* No doubt the yeast of beer consists of minute molecular matter, the particles of which are globular; and that those particles produce, from their sides, other particles like themselves, which eventually separate from the parent, but we do not know that they are *therefore* plants.—*Athenæum.*

Life-saving Buoys.—Sir,—I think it would be an excellent plan, if the Trinity-house Commissioners were to order a set of iron handles, to be fixed round all their buoys that are anchored in the sea, and in mouths of rivers, for marking rocks, shoally bottoms, or channels, in case of a vessel being wrecked, or a boat upsetting, near such a buoy. Some of the immersed crew might gain it, and by means of the handles hold on until other assistance came. In making new buoys, or repairing old ones, these handles might be easily attached, as one of the hoops might be forged with them on it, or they might be rivetted to the hoop, the increased cost and trouble of making would be very trifling.—*Arthur Trevelyan.*

The Compass in Iron Vessels.—Sir,—In reading your Magazine of this day, I observe among the hints upon iron and other steam boats by Junius Redivivus, his alluding to the alleged impracticability of navigating iron-boats across the ocean, owing to their effect upon the. compass. I beg to say, that two or three years back I tried some experiments by placing the compass *within an iron rim, and in an iron desk,* in order to make attraction equal, and to neutralize the partial attraction of the ship's iron, and I enclosed the account to the Admiralty, and I think I enclosed it to you, but in neither case was it thought worthy notice ; the idea seems now revived by Redivivus, and I certainly believe it much more likely to be successful than the experiments that I read of lately, of turning the compass upside down. I had thought of it for years, particularly to counteract the *partial* attraction of the ship's iron. Your constant reader. C. P. ASTON.

26, Little Windmill-street, Golden-square,
October 20, 1838.

The Railway Map of England and Wales continues on sale, in a neat wrapper, price 6d. ; and on fine paper, coloured, price 1s.

☞ *British and Foreign Patents taken out with economy and despatch; Specifications, Disclaimers, and Amendments, prepared or revised; Caveats entered; and generally every Branch of Patent Business promptly transacted. A complete list of Patents from the earliest period (15 Car. II. 1675,) to the present time may be examined. Fee 2s. 6d.; Clients, gratis.*

LONDON: Printed and Published for the Proprietor, by W. A. Robertson, at the Mechanics' Magazine Office, No. 6, Peterborough-court, between 135 and 136, Fleet-street.—Sold by A. & W. Galignani, Rue Vivienne, Paris.

𝔐echanics' 𝔐agazine,

MUSEUM, REGISTER, JOURNAL, AND GAZETTE.

| No. 795.] | SATURDAY, NOVEMBER 3, 1838. | [Price 3d. |

SAVORY'S MAGNETIC TIMEPIECE.

Fig. 1.

SAVORY'S MAGNETIC TIMEPIECE.

Great curiosity has been excited in the neighbourhood of the Exchange for some days past, in consequence of the exhibition in the window of Mr. T. Cox Savory, the well known silversmith, of a time piece, apparently consisting of only a dial of glass, on the centre of which an index hand turns, and points correct time, and without any visible mechanism. Our ingenious correspondent, Mr. E. Whitley Baker, has handed us a communication, suggesting that the motion of the hand is obtained by means of a magnet. The following is the description of a method, by which a magnet may be made to act in a timepiece similar to that exhibited by Mr Savory—and if the plan be different from that which this gentleman has adopted, Mr. Baker has given to the world certainly a most ingenious and valuable contrivance.

On our front page, fig. 1, is a perspective view of the (supposed) magnetic clock as made by Mr. Savory. The dial is of clear glass, with the hours painted upon it, so that it is evident, the dial does not move. This dial is bordered by a ring or rim of brass, or other metal, supported by an elegant pedestal.

Figure 2, shows the *modus operandi* of the timepiece. A A is the frame, or rim containing the dial ; within this frame is placed the ring B B, which has a certain number of teeth on its outer edge, and which, on its inner edge, works freely on small friction-wheels placed in such a manner as to support the toothed ring correctly in the frame ring. C is a connecting shaft, hidden in the inside of one of the scrolls connecting the frame with the pedestal, having on its upper end the endless screw D, and on its lower end the be-

Fig. 2.

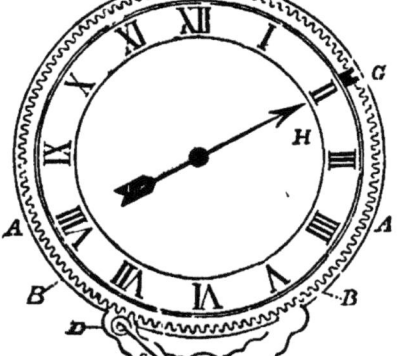

velled wheel E, the teeth of which take into another bevelled wheel F, the shaft of which leads to the the minute hand pivot of works similar to those of an ordinary watch, or timepiece. These works may be placed in the pedestal of the clock. G is a magnet fixed on any part of the toothed ring B B. H is an arrow or index hand having a steel point, and revolving freely on its pivot in the centre of the dial. The re-

lative numbers of teeth necessary in the wheels may vary according to the will of the maker—as, a pinion of 10 working in a ring of 120; or a pinion of 30 in a ring of 360—and so on, according to the size or other circumstances attending the making of each particular timepiece.

Motion being given to the bevel led wheel F, by the mainspring and works of a watch, as before mentioned, will be communicated by the shaft C, with its bevelled pinion and endless screw, to the toothed ring B B, and cause it to perform one revolution in twelve hours. The magnet G being fixed on the ring B B will revolve with it, and the steel point of the arrow H will follow its course, and indicate the time by the hours on the dial.

There is a large timepiece in the Adelaide Gallery, working in a manner apparently the same; but the mechanism is on so large a scale that it can easily be detected. A watch is concealed in the tail end of the hour hand, and moves it by means of a shaft connected to a fixed point on the axis.

SYMINGTON'S STEAM VESSELS.

Sir.—Concerning the remains of Mr. Symington's boat lately discovered at Grangemouth, and noticed in your last Number, they are not, as imagined, those of the one used in the experiments of Messrs. Millar and Symington in 1789, but those of the steam-boat constructed by Mr. Symington for Lord Dundas in 1800 and 1801.

The first boat used in the canal in 1789 was dismantled at Grangemouth and removed to Bruce-Haven. The second boat, the one now found, not only towed vessels on the Forth and Clyde canal in 1801, but also from the Forth up the Carron into Grangemouth, and carried Mr. Fulton, the American engineer, a distance of eight miles in an hour and twenty minutes, on this very canal.

I deem it an act of justice to the memory of Lord Dundas, who, with the exception of Mr. Symington, did more for steam navigation than any other individual, to furnish this explanation.

In order to prevent future mistakes, I may add, that Mr. Symington fitted and propelled four boats with the steamengine:—The first, in 1788, on Dalswinton Lake; the second in 1789; the third in 1801; and the fourth, in 1803, on the Forth and Clyde canal. The first and second were under the patronage of Mr. Millar; the third and fourth under the patronage of Lord Dundas.

I remain, Sir,
Your most obedient humble servant,
ROBERT BOWIE.

44, Burr-street, Oct. 31, 1838.

REMARKS ON THE PROPER SOURCE OF FOOD FOR MAN. — REPLY TO JUNIUS REDIVIVUS.

Sir,—Without entertaining the most distant desire of being classed with those —if such indeed there be—who deem Junius Redivivus a "visionary in his ideas," I take the liberty of adverting to one little circumstance in his last communication, in which he has carried his speculations, not only beyond the bounds of probability, but has really exceeded the limits of possibility.

I allude to the remark made by Junius Redivivus, at page 36 of your last Number, that "knowledge will eventually be our's to prepare food from *inorganic* matter, as *stimulating* and as *nutritious* as that now furnished by the lower animals, who are made to die that we may live." Now, it is held by physiologists to be a well-ascertained and unalterable law of nature, that life can only be sustained by that which has itself possessed life. It is undoubtedly true, that metals, minerals, and other inorganic substances are continually employed in medicine as stimulating and disturbing agents, or as sedatives; but no nutritious powers were ever supposed to exist in any of these substances. Indeed, our present chemical knowledge of the constituents, general powers, and properties of most inorganic matters, though possibly far from perfect, is quite sufficient to enable us positively to deny the existence in them of any nutritive characters.

Under these circumstances, we are fully justified in taking it for granted, that no further advancements in knowledge can ever find, amid the vast range of inorganic matter, any efficient substitute for the food at present obtained from organic nature, in all the numerous varieties of animal and vegetable life.

I remain, Sir,
Yours respectfully,
WM. BADDELEY.

London, October 22, 1838.

IMPROVEMENT IN SERAPHINES.

Sir,—I fancy that the effect of the trumpet-stop in organs, might be given to seraphines, by merely using *one* bell-shaped vent, made of fine brass, similar to

E 2

that of a French-horn. This stop might be used or taken off as follows :—suppose the diminuendo pedal down (which will close in the instrument and so soften its tone) lift a valve by drawing out a stop, or touching a pedal, which shall cause the only vent for the sound to be through the orifice of the brazen bell. This, *I know*, will give a rich, brassy sort of tone to the instrument, because I saw a bassoon once, which had a brass trumpet-'ike end fixed to it, and was wonderfully improved thereby. *Query*, therefore, If this will succeed for seraphines, might it not also for organs ? Let there be a pipe with a trumpet vent, laid horizontally or slanting over the range of pipes intended for trumpet music, and let each of them be butted into it, which, I suppose, will produce the desired effect, and so make the cost of the trumpet stop much less, as the present method requires each pipe to have a separate bell-end.

Can any of your correspondents inform me what is the cost of a set of seraphine tones ready for fixing in the frame, and where they may be had; also the best mode of tuning them ?

I am, &c.
WM. PEARSON.
Bishop Aukland, Oct. 10, 1838.

IVISON'S PATENT METHOD OF CONSUMING SMOKE.

Sir,—A copy of the *Mechanics' Magazine* of the 20th having been sent me, my attention was called to the article on Ivison's patent with which I have made a conjunction of interests with my own, noticed in your Magazine long ago.

I never pay any attention to anonymous communications on such subjects; but yours is editorial, and from so highly respected a quarter, as well as so special in its allegations, as to induce me to wish to afford you the most authentic information on the subject. I therefore transmitted the copy of your Magazine to the officers of the company to whom the ship *Adelaide* belongs, requesting them to state authoritively the facts of the case; and I have received the following letter, to which, I have only to add, that the captain of the vessel himself further told me, that the captain of a steam-ship, which sailed in company with him, ignorant that any alteration had been made, mentioned,

when he met him in London, that the *Adelaide* appeared like a schooner, from the entire absence of smoke, and inquired what he had done.

Copy. Messrs. Ogilvie and Crichton, to Mr. Bell.
Dated Leith, 24th October, 1838.

"Dear Sir,—We are sorry hat the particular circumstances under which Ivison's patent was tried in the *Royal Adelaide*, should have led to such erroneous reports as have appeared in the *Mechanics' Magazine*. When we agreed to try it, we mentioned to you, that owing to one of this company's steam-ships being about to undergo extensive alterations, the *Royal Adelaide* would for several weeks be much hurried in performing her voyages. In fact, until last week, when she was thrown out of her turn by a heavy gale of wind, her stay here has seldom exceeded twelve hours.

"Hence, it has never been in our power to replace the three flanges which were deranged during the vessel's voyage to London on the 8th and 9th ultimo. From the temporary manner in which they were fitted, they were very liable to accident in feeding the furnaces, which made it desirable to discontinue the use of them, until there was time to fit them more substantially.

"There is not the least foundation for the report that the use of them was discontinued from the loss of speed, or injury to the furnace bars, as so far from having made a long passage, Captain Allen remarks, that 'though a deep loaded ship, with a very heavy E. N. E. sea to contend with, we made a 48 hours passage, being under our average passage up.'

"Captain Allen states in answer to the allegation, that the fire-bars were completely destroyed, that 'On his arrival in London a few of the old fire-bars were bent, which were straightened there, and are still in use; *this* we are obliged to do in *every* voyage with *old bars*.'

"We are, &c.
(signed)   "Ogilvie and Crichton."

While addressing you, however, on the point, you would deem it extraordinary were I to say nothing on the general merits of the discovery. In addition, therefore, to the statement made by the editor of the *Mining Journal* on the 31st of August, on the result of an actual operation conducted by himself, I may mention, that for a long time the original engine has been worked to the full extent of her power, one-half the day with, and one-half the day without, Iveson's patent, and the proportion of coals used was as 56 to 82; the work without the patent being so good as to give 6.25 lbs. of water evaporated to the pound of very ordinary Scotch coal. Allegations having been made, that the admission of steam into the furnace might perhaps be injurious to the boiler, the opportunity was taken the other day, when a storm of wind had blown down a great part of the chimney stalk (and when, by-the-bye, the engine within half an hour did her duty as well as previously—proving, that high stalks are unnecessary) to execute some repairs, and to inspect the boiler; when several engineers of eminence, separately, but unanimously, reported, that although it had been used with steam for many months, it was altogether uninjured.

I rely on your giving the earliest information possible to the public, on this matter. As it is very necessary that unintentional errors should immediately be put right.

I am, Sir, yours, truly,
WILLIAM BELL.

Office for Conjoined Patents,
Edinburgh, Oct. 25, 1838.

[We readily give insertion to Mr. Bell's letter, in explanation of the statements made in our 793rd Number, but we do not see that it shows any material difference as to facts. The information upon which our remarks were founded, we received from a party in whom we place implicit confidence, and for whose integrity we can vouch, and lest we might have misunderstood his report, we requested him to explain or confirm our remarks, which he has done as follows:—"My authority is derived from statements made by those connected with the vessel, and what I witnessed myself. They said the six hour's lost on the passage up, was owing to the inadequate supply of steam to the engines occasioned by the use of the apparatus. About sixty-fire bars were sent on shore rendered completely useless on the voyage, until they were repaired. The whole of the fan-shaped pipes were removed, with the exception of one, and they were restoring the furnaces to their original condition, as opportunities occured, without detaining the vessel." That the apparatus had the effect of consuming the smoke, we distinctly acknowledged, as Mr. Bell may perceive, by a re-perusal of the article in question.—ED. M. M.]

PLANS FOR THE DESTRUCTION OF SMOKE FROM STEAM-ENGINE FURNACES, ETC.

Sir,—Should Mr. Ivison's patent mode of affecting the combustion of smoke, (described at page 34 of your 793rd Number) contrary to my expectations, prove practically useful, I fancy his patent will avail him but little. In the description as given above, *Auld Reikie* is said to be the birth-place of this discovery; which it so happens is quite true, for a precisely similar process was invented and repeatedly exhibited in Edinburgh many years ago, by Messrs. Nasmyths, now machine-makers in the neighbourhood of Manchester, and it was more than once before the Society of Arts.

Mr. Ivison's plan embraces two points—a saving of fuel, and the prevention of smoke. With respect to the former of these objects he appears to have been long since anticipated; very early in the year 1824, Mr. Evans, of Bread-street, Cheapside, promulgated his discovery of a method of economising fuel, by the introduction of steam at the lower parts of the furnace; a description of which appeared in Number 49 of the *Register of Arts*, and in Number 45 of the *London Mechanics' Register*, which plan Messrs. Gilman and Sowerby subsequently contended was a direct infringement of a patent granted to them for a similar purpose, in April, 1825. There was considerable economy in Mr. Evans's plan, inasmuch as he only employed the *waste steam* after it had performed its office in giving motion to a high-pressure steam-engine. Mr. Evans seems to have missed the smoke-consuming property of steam thus applied, he having employed coke for the purpose of getting rid of the smoky nuisance.

It appears to me, that if any thing like the *tenth-part* of the steam generated, is required to ensure the combustion of smoke—unless *waste steam*, as in Mr. Evans's plan, can be employed—economy of fuel is altogether out of the question, and the result of the experiment on board the *Royal Adelaide*, Edinburgh steamer, seems to warrant this conclusion.

Mr. Evans noticed a singular phenomenon, attending this application of steam as a supporter of combustion, "whatever the measure of water decomposed may be," says Mr. Evans, "I am led to suppose that a greater quantity is re-formed, having been obliged, since using steam for this purpose, to have a cistern made at the bottom of the chimney, with a pipe to conduct the condensed water to a vessel, in which there appears to be collected more water than the steam used would account for; to what this may be attributed experience must determine." Has Mr. Ivison noticed this effect in any of his experiments?

In the year 1825, a correspondent of the *Birmingham Chronicle*, communicated an account of a successful experiment in destroying smoke by steam; at the close of his letter he observes, "I have sent the above for insertion in your paper, that the world may not be pestered with *patents*, in a matter where every man who has an engine may be his own chimney-doctor."

The real fact of the matter is, that the steam process for destroying smoke, falls far short of several other plans, which depend for their efficiency upon the smoke being generated in small quantities in the front of the furnace, and burned in its passage to the chimney, by passing over the mass of intensely ignited fuel. This is the true theory, and its perfect efficacy will be greater or less, in proportion to the skill exercised in reducing it to practice.

I remain, Sir,

Yours respectfully,

WM. BADDELEY.

London, October 24, 1838.

ON THE DEGREE OF SECURITY TO PROPERTY ATTAINABLE BY LOCKS.

Sir,—There are few subjects, if we except the "healing art," about which there has been so much quackery—or with respect to which, the public have been so completely gulled—as the best method of securing property by means of *locks*. The ordinary vendors of locks have, for the most part, been intent upon selling those locks by which they could obtain the greatest profit, upon the smallest possible outlay, without the slightest regard being had to *security*. In other instances, different shopkeepers have become interested in, or prejudiced in favour of some one particular invention, which has in each case been put forward as the "only secure lock," to the exclusion of all others; and this, very often without understanding—frequently without caring to know—any thing about the real practical merits or defects of the article they so strongly recommend. The immense number of cheap (?) locks that are made to "sell," many without either tumblers or wards, gives rise to the truism, at the commencement of an article by "B.," at page 39, of your 793rd Number, that "to put a thing under lock and key is unhappily, not at all times to make it secure."

Incalculable injury frequently arises from the employment of insecure locks, under which is very often deposited property of the greatest importance, such as cash, deeds, jewellery, private ledgers, &c., the insecurity of which, holds out a temptation to the *over-curious* and *dishonest*, too great to be resisted. Of this fact, our police reports furnish us with daily proofs, though it is but fair to suppose that by far the greater number of frauds never meet the public eye: such as purloining of valuable documents, making additions or erasures, &c., which are not discovered till too late for remedy.

The Commissioners of the Metropolitan Police, have lately directed public attention to the number of robberies that are continually effected by persons entering houses with false or skeleton keys, and the various superintendents have issued circulars, cautioning housekeepers against the use of locks opened with such facility—observing, "that the thief eludes the vigilance of the police, by appearing to belong to the premises he is entering."

It is quite untrue, as stated by "B.," in the communication referred to, that a *secure lock has long been a desideratum;* no such deficiency really exists, and if it did, it certainly would not be supplied by Messrs. Deanes lock. Locks founded upon a similar principle, and affording equal security, are proved to have been used in Egypt upwards of 4,000 years ago, and we have abundant evidence to show that very curious and secure locks have been employed in this and other countries for many centuries.

As a brief history and description—

nay, a bare enumeration of the many *secure locks* that have been made, would form an article far too voluminous for your pages; I shall content myself with adverting to the two grand principles of construction, upon which have been founded nearly all the secure locks of the last sixty years.

The first of these is the lock patented by Mr. Barron, in 1774, which has been designated as " the first scientific attempt to improve this very important instrument." The security of Barron's lock, consisted in the employment of two or more tumblers, so arranged, that they required to be raised to unequal heights, before the bolt could be thrown; the key, therefore, had a number of steps upon the edge of the bit, of unequal heights, adapted to elevate each tumbler through the required space."

Barron's lock, when well made, with two or t ree tumblers and *a good box of wards*, is extremely difficult, if not impossible to pick; for the wards, which are useless as preventives to the operation of throwing up a single tumbler, greatly impede so delicate a proceeding as elevating two or more to a precise height. This arrangement admits of almost endless variations, and is extremely simple in itself. Very beautiful and secure locks have been made upon this principle ever since its introduction by Barron, each manufacturer priding himself upon some little variation, by which he supposed increased security was obtained. Of all the varieties of Barron's locks that have of late years been introduced to the public, Chubb's has perhaps obtained the greatest celebrity, and the persevering and business-like manner in which the ingenuous inventor has contrived to keep it before the public eye, has contributed in no small degree to the " run" it has had. The chief characteristic in this lock, and that which marks it as Chubb's, is the employment of a spring lever, called a detector, which locks the bolt fast upon any of the tumblers being elevated beyond its assigned range, and shows that an attempt has been made to open it by a false instrument. In Barron's, and also in Bramah's locks, the picker has no means of knowing whether the tumblers are lifted too high or not; but in Chubb's he has only to put the detector *hors de*

combat in the first instance, by a correct thrust from the outside of the door, so as to fix it fast in its place; the detector then becomes a stopper to the undue ascent of the tumblers, and the extent of their range is thereby correctly ascertained; thus it appears the *detector* might be be converted into a *director* of the means of opening the lock.

The second secure principle of lockmaking—second *only in date*—is that patented by Mr. Bramah in 1784, just ten years after Barron's. Bramah's lock presented an entirely new feature; it depended for its security upon a revolving barrel, the rotation of which was prevented by four or more steel sliders, which required to be depressed in unequal proportions, until a notch on the edge of each was brought in a right line with a fixed steel plate called the locking-plate—this being done, the barrel is free to turn round, and a bolt is thrown by a pin on the barrel working in a suitable scroll on the bolt-tail. Notches of varying depths are cut in the end of the key-pipe, for the purpose of giving the required depression of the sliders; on the key being withdrawn, the sliders are thrown up again, by an internal spiral spring. These locks as now made, with *seven guards*, and the additional securities for which the late Mr. Russell received a reward from the Society of Arts, are, unquestionably, the most secure ever produced.

In the manufacture of these locks, the key is first made, and the lock fitted to it. The machinery employed by Mr. Riddle, of Blackfriars-aoad, for the production of these keys, is unrivalled for its accuracy and beauty of its construction, and for the extraordinary power of its changes—which are so extensive, that *duplicates* are not likely to occur for a long series of years. In Barron's locks, the security depends greatly (and in those of his improvers (?) wholly) upon the number of tumblers. In Bramah s locks the security consists in the number of sliders, and as the number of these are increased, so are the number of changes that can be introduced to prevent the occurrence of duplicates. The following table shows at once, the increased number of changes that can be obtained by increasing the number of the guards or tumblers, from one up to seven.

Number of Guards.	Number of Changes.	Increased number of changes by twice altering the *situation* of each notch.	Further increased number of changes by twice altering the *depth* of each notch.
1	1	2	4
2	2	8	32
3	6	36	216
4	24	192	1536
5	120	1200	12000
6	720	8640	103680
7	5040	70560	987840

In Messrs. Deane's lock, five tumblers only are employed, and the only novelty is in the mode of placing them, some above and some below the axis of the key, by which no advantage whatever is obtained. Granting the correctness of " B.'s" third proposition, (page 40) " That it is impossible to pick such a lock as this ;" I must deny the truth of all the rest. In his concluding proposition, " B." says, " An impression in wax, or any other composition, to imitate the key, would be of no avail; for so great is the exactness and accuracy of construction required, that even under the most favourable circumstance of having the key itself for a pattern, it would scarcely be possible to avoid leaving some one of the racks or pins untouched, which would render all the rest of the labour unavailing." This is entirely false, and the publication of such a statement is a gross imposition on the public; Deane's keys have been imitated, from impressions taken in sealing-wax, and also in yellow soap, and their locks opened, by persons who are not locksmiths ! The real truth is, that the Barron's key, as made by Chubb, Deane, and some others, without wards, is much easier imitated than a common warded key. A glance at page 40, would suffice to show any person at all conversant with the arts of copying, that to make a fac-simile of such a key as is there represented, is easy in the extreme; if the key is in possession, a plaster of Paris mould, and a pewter or type-metal casting, gives the thing at once—if not, an impression in any ductile material is quite sufficient. I have by me two of Chubb's best locks each furnished with "*false keys*," one made of tin plate, the other of thin brass, both from sealing-wax impressions, which open and shut the respective locks fully as well as the real keys. So long as the key is kept secret, any of these locks may be safely relied on; but the key once obtained for a few moments, and there is an end to all security.

Messrs. Deane's double-bitted key is an awkward-looking instrument, possessing neither security nor convenience; it has been the object with skilful lockmakers to make their *keys* as small and portable as possible, for the sake of convenience, and also to reduce the size of the key-hole as much as possible, so as to check any fraudulent attempt upon the tumblers at the outset. In this respect Mr. Riddle seems to have surpassed all his competitors, having succeeded in introducing a portable seven-guarded key into his patent ever-pointed pencils, the seal-top of which forms the handle of the key, which affords a ready means of securing the writing-desk, the jewel, cash, or deed-box, &c. and of carrying the key concealed from observation, protected from accident, and no incumberance to the wearer. When made in gold, these keys form a neat appendage to a watch-chain, or placed within a signet-ring they can be worn on the finger. I mention these facts, to show how essentially portability has been considered, and to what an extent ingenuity has been exercised to meet the desideratum.

Without trespassing further upon your valuable space, I would beg to refer such of your readers as are curious in these matters, to the articles on locks in Hebert's *Mechanics' Cyclopedia,* and to the new edition of the *Encyclopedia Brittanica,* of both of which I have availed in the foregoing statement.

Yours obliged,

B. B.

London, Oct. 27, 1838.

ON THE CAUSES OF STEAM-BOILER EXPLOSIONS.

Sir,—I have read, with much interest, the several contributions to your valuable pages, relative to the circumstances connected with the bursting of steam boilers; and the hypothetical and practical opinions as to the several causes of such disastrous occurrences, which have been furnished by many eminent and talented individuals; and have, upon every occasion, risen from the perusal with disappointed expectations, resulting from the conflicting, contradictory, and unsatisfactory testimony which they have supplied.

In the investigation of this truly important subject, much greater attention appears to have been paid to practice than to theory; and to its analysis, rather than to its synthesis; or, in other words, to an endeavour to deduce the several causes from the several effects, rather than endeavour to deduce the several effects from the several causes.

The inquiry appears to me to have been made from the effect—the explosion, backward to the cause—the generation of heat, and its impartation to water to create steam; rather than progressively from the cause toward the effect; gradually and satisfactorily elucidating the subject from the base to the pinnacle, instead of from the pinnacle to the base.

The foundation of the subject appears to me to be the "Theory of Heat." What are we to understand by the terms latent and active heat? Is heat a fluid composed of material atoms, solid, spherical, indestructible, and smaller than the material and ponderable atoms of which all substances are composed; and sufficiently small to pass through the interstices are presented by the union of the ponderable atoms of which all substances are compounded; and it is by virtue of this atomic form, and this difference in their magnitude, that heat is able to permeate the metallic medium, and combine with the water in the boiler, and eventually convert it into steam? Does the combustion of fuel in the furnace liberate heat from its constituent state of latency, and thus render it free to yield obedience to the laws to which it is subject; and is it in obedience to such laws, that thereby its transition from the fuel to the water is effected? And, if so, what are those

laws? What are the covenants, conditions, restrictions, and limitations in appropriating heat, subject to such laws, to the generation of steam to be used as a motive power? These, and an innumerable series of inquiries, it appears to me, should be consecutively made by those who are deeply interested in the subject, and who have time and talent to devote to the pursuit.

In suggesting to others the mode by which, I think, the subject should be investigated, I have to apologise for recommending others to perform a task, from which I shrink from an endeavour to accomplish myself, from a consciousness of not possessing any of the requisite qualifications to effect it; but I forward you a copy of the work which I have just published, and beg to refer you to the articles under the heads of "Heat," "Explosion," and "Steam;" and should you think their transference to your pages will furnish an incentive to a more competent exposition of the subject, by an abler pen, I shall feel pleasure in having proved the inciting cause, should it lead to so desirable a result.

I am, Sir, your obedient servant,
G. H. WIGNEY.

Brighton, Oct. 6, 1838.

[The following are extracts from the articles referred to by Mr. Wigney as pubished in his *Cyclopædia*. We shall take a future opportunity of noticing his work.]

Heat.—To abstract the free or active heat possessed by a body, it is sufficient to place in contact or proximity with it, another body possessing less heat than itself; and by the law of equal diffusion to which heat is subject, the transmission from the major to the minor possessing body will ensue; but to abstract a part or the whole of the latent heat of a body, it is necessary to place such body in close contact or proximity with another body which possesses not only as much less heat as is equal to the abstraction of the whole of the free or active heat of such body, but a part also or the whole of its latent heat, the latent heat in such a case becoming active, in obedience to the law of equal diffusion.

Latent and active heat are arbitrary terms, used to denominate heat in a twofold state; —latent heat when it exists in bodies in a quiescent state, and active when in a state of motion or liable thereto. But, strictly speaking, heat cannot be permanently latent, being ever liable to detection, liberation, freedom and activity, as the result of abstraction, induced by the law of equal diffusion, or the decomposition of the containing body by a variety of methods; and therefore it is necessary not only to conclude, that there are not two species of heat, termed latent and active, but that the single species is occasionally latent or active as induced by circumstances.

The impartation of heat to a body will not decrease its weight, provided a sufficient quantity of heat is not imparted to that body, as will cause one single ponderable atom to be separated from the rest, beyond the limits of aggregate attraction; but as soon as such is produced then is there an actual diminution in the weight of that body, and in amount equal to the number of ponderable atoms which are placed beyond the sphere of such attraction.

When the impartation of heat to a body is sufficient to separate the ponderable atoms one from another, and thereby such body is increased in magnitude; then is its specific gravity decreased, but not its actual weight. The decrease in the specific gravity of a body, implies that an equal bulk of that body in an expanded state by heat, will not weigh so much as an equal weight of that body previous to its expansion.

The atoms of heat being subject to the primary law of equal diffusion, and the secondary law of recession; the transition of heat from a body as induced by the law of equal diffusion, is in the direction of the abstracting body; but if induced by the law of recession, unopposed by the law of equal diffusion, its transition is in the line of perpendicular ascension.

When the operation of the primary law is unopposed by the secondary law, heat will invariably and uninterruptedly ascend from the body from which it emanates; and whenever the primary law is in operation, it is always strongly opposed by the secondary.

Place a body possessing the most active heat, above a body possessing the least, and there will be a transmission of a portion of heat from the upper to the lower; but reverse the position of the two bodies, and the transmission of the greater to the lesser possessing body will be much more abundant.

From these facts may be inferred, the mechanical advantages to be derived, as relates to the economical and prompt transmission of heat, in placing the recipient over the imparting body, arranging for its perpendicular transmission, and extending as far as is mechanically convenient, the receiving surface.

By the impartation of heat solid bodies are converted into fluids and vapours; and by the abstraction of heat, vapours and fluids are converted into solids.

A solid consists of ponderable atoms united together in close contact by attraction of cohesion, and the interstices presented as the result of such union, partially or totally filled with imponderable atoms.

A fluid consists of ponderable atoms, not in a state of cohesive contact, commixed with imponderable atoms, in greater quantity than is required to fill up the interstices presented by a solid; yet are they held together by attraction of aggregation, and in this state of semi freedom are subject to diffusive motion within the sphere of such attraction.

A vapour consists of ponderable atoms blended with a much greater quantity of imponderable atoms than is the case with a fluid; consequently its ponderable atoms are separated to a greater extent than are those of a fluid; and as in the case of a fluid, the whole of its constituent atoms are confined within the limits of the sphere of attraction of aggregation, and are susceptible of diffusive motion within that sphere.

By the impartation of the imponderable atoms of heat to a solid, the first effect is to completely fill the interstices presented by the union of the ponderable atoms of which the solid is composed; continue the impartation, and the attraction of cohesion which kept the ponderable atoms in close contact is overpowered, and they are separated one from the other, and the solid is thereby converted into a fluid: continue yet further the imparation and the fluid is converted into a vapour, by a separation of the ponderable atoms yet further from each other: proceed yet further with the impartation, and you separate the ponderable atoms beyond the precincts of the sphere of aggregate attraction, and its annihilation as an integral body is effected.

Reverse the operation and cause ponderable atoms having an affinity to each other, to come within the sphere of mutual attraction, intermixed with the imponderable atoms of heat to a considerable amount, and you constitute a vapour: next abstract a sufficient quantity of heat, as will enable the ponderable atoms in obedience to the law of attraction, to approximate close to each other but not cohere, and you then constitute a fluid; continue the abstraction of heat until the ponderable atoms cohere, and then you constitute a solid.

These analytical and synthetical operations may be illustratively performed on ice as the basis of the first process; and by an admixture of the right proportions of oxygen and hydrogen gases as the preliminary of the second.

Both vapours, fluids, and solids are conductors of heat in variable proportion; and as a general rule solids are better conductors than fluids, and fluids than vapours.

The conductive power of heat of solids, fluids and vapours, is in the ratio of the proximity of their ponderable atoms to each other, and to their constituent amount of latent heat.

Bodies which are incombustible at a moderate temperature, receive and part with caloric without experiencing a partial or total change; but combustible bodies are partially or totally decomposed in proportion to the quantity of heat received.

The impartation of heat alone causing the decomposition of a body, may be said to effect it mechanically; but the impartation of heat as an agent, to aid and facilitate the fulfilment of the laws of affinity and attraction subsisting between the atoms of which the combustible body is composed, and a body in contact or proximity, may be said to effect the decomposition chemically.

Latent heat becomes active in every case of the condensation of solids, fluids, or vapours; and whether the condensation is effected by mechanical pressure, causing the ponderable atoms of which the body is composed to approach nearer to each other; or by chemical attraction causing the ponderable to approach nearer to each other, as frequently occurs in the blending of two or more bodies together, the liberation of latent heat and its evolution from the body is a resulting consequence. —pp. 179-185.

Steam.—The general mode of producing steam to serve as a motive power, or as a medium of the impartation of heat to some distant body or substance, consists in the charging of a metal vessel from one-half to three-fourths full with water, which vessel termed the boiler, is nearly imbedded in brick-work over a furnace, the flue of which extends in height around the sides of the boiler, nearly to the water line, or that point to which it is intended the boiler should be kept constantly charged, during the whole period of the generation of steam for any specific purpose, the space above the water line being reserved as a reservoir for the steam generated. The purport of limiting the height of the flue to a short distance below the water line, is that no more heat may be imparted to the boiler, than the water is capable of receiving; and the purport of keeping the boiler constantly charged up to the water line is, that the whole of the heat imparted to the boiler, may find a recipient within in the water with which it is kept constantly supplied: and the purport of constructing the flue round the sides of the boiler, is to present as large a surface of metal as is possible, consistent with the several other arrangements for the reception of heat from the burning fuel, and the impartation of it to the water within the boiler; and all the mechanical arrange-

ments in the formation and fixing the boiler, and the construction of the furnace, flues, &c. should be effected upon the best principles with a view to the generation of heat and its subsequent impartation to the water in the boiler, upon the most economical and advantageous terms. And supposing this to be done, we will next examine the principle of generating heat in the furnace; and we find that the fuel being ignited, its combustion is supported and continued by the access of atmospheric air to it, which passes through the ash pit and between the furnace bars, and coming in contact with the in candescent fuel, as much heat is imparted to it as is sufficient to separate the ponderable atoms of which it is composed beyond the sphere of mutual attraction, and consequently the constituent atoms of oxygen of such air unite with the constituent atoms of the carbon of the fuel, between which there is a powerful affinity and they together form carbonic acid; and an impartation of heat to such carbonic acid occurring from the decomposed fuel and air, it is converted into carbonic acid gas of less specific gravity than the atmosphere, and therefore evolates and passes off by the flue and chimney with the nitrogen, which previously formed a constituent portion of such air. In this process of combustion, the decomposition of the fuel and a large amount of atmospheric air is effected, and as both the fuel and air are compounded of a large proportion of heat, which heat is latent while forming a constituent portion of either body, but free the moment of the dissolution of either as integral bodies, so such constituent heat being liberated, is rendered ready for its impartation to other bodies; and we have next to review the principle of impartation by which such heat is transmitted from the furnace to the boiler, and from the boiler to the water within it. Heat in a state of freedom is subject to two laws, a primary and secondary—the primary termed equal diffusion, and the secondary recession from the centre of the earth, or gravitation towards the centre of the sun. The heat liberated from the decomposed fuel and air, traverses the bottom and sides of the boiler in its passage from the furnace through the flue to the chimney, and if not traversing the surface of such portion of the boiler to which it has access too rapidly, the whole of the heat may be imparted to the boiler in obedience to the law of equal diffusion; but if passing more rapidly or in greater abundance than can pass by transition to the boiler in obedience to such law, then a portion of such heat will find its way to the chimney in obedience to the secondary law, the law of recession, and thereby be lost to the purpose for which it was rendered free. The heat which is thus transferred from the burning fuel to the boiler, induced by the law of equal diffusion, the gases emanating therefrom being of a much higher temperature than the boiler, is transmitted from the boiler to the water within, in obedience to the primary law; the temperature of the boiler being higher than that of the water, and the heat thus transferred to the water, ascends to its upper service in obedience to the secondary law, the law of recession; and thus the first portion of heat imparted to the water, although received at its lowest surface, ascends to the upper, and the temperature of the upper surface becomes greater than the lower.   The decomposition of fuel and air, and the impartation of the resulting liberated heat continuing, the temperature of the upper surface of the water soon amounts to 212 degrees, when the formation of steam commences: and the safety and eduction valves of the boiler being closed, the steam which is generated on the upper surface of the water evolates, and the generation and evolation continuing, the space above the surface of the water would soon be filled with steam, was it not already filled with air; but as such air is elastic, the volume is compressed into a smaller space by the expulsion of a portion of its constituent heat through the metal composing the top of the boiler, and the boiler being so charged with heated water, steam, and condensed air, we will suppose that either the safety or the eduction valve is opened and the air rushes out, leaving the boiler charged with water and steam only, the valve being again shut. We will next suppose that the temperature of the water from the upper to the lower surface, has attained to about 212 degrees or boiling heat, and the boiler being completely full of water and steam, the question is, can any more heat be imparted to either water or steam ? we think not, because the receptive capacity of either water or steam for heat is not to a greater amount than about 212 degrees, and because no addition can be made to a plenum. But it is said that steam is compressible and expansive, and that in these two properties consists the accumulation and the exercise of the motive power with which it is endued, and that the amount of the power generated is proportionate to the extent of the compression; and hence we hear of the generation of steam, and its compression within the boiler to such an extent as to cause its pressure upon the internal surface of the boiler to be equal to the pressure of so many atmospheres ; but we are decidedly of opinion that steam is neither compressible nor expansive, (unless in a vaccuum) and although a volume of steam may be

evolved from a boiler, and occupy a tenfold space subsequent to its liberation than it possibly could in the boiler, yet that is by no means a confirmation to us that steam is either compressible or expansive ; and we deem that the proofs offered in favour of the doctrine of its compressibility or expansibility are diametrically opposed to the possibility of the occurrence of either, and to all the well known results, which ensue from the impartation and abstraction of heat. If we ask what is the effect of the impartation of heat to water, we are told to cause its expansion or increase in bulk, until its diminution commences by the generation of steam, and its evolation from the surface of the water ; and we are told that if the impartation is continued, that the boiler will eventually be filled with water and steam, and up to the point of repletion expansion has been the result of the impartation of heat ; and then it is endeavoured to convince us, that by a yet further impartation of heat, that a compression or diminution of the steam which first occupied the whole space within the boiler above the water is effected by the production of more steam by heat subsequently imparted ; but to suppose that heat can cause both the expansion and the contraction of a fluid appears to us to be impossible, and not supported or countenanced by any analogous proofs ; for if we examine the resulting effects of the condensation of atmospheric air, we find that heat is evolved and not imparted in the process, and that it is only by the expulsion of a portion of the constituent heat, that the ponderable atoms can be brought closer together, and thereby cause its volume to be reduced in bulk. It will then perhaps be asked, to what do we attribute the increase of power furnished to steam by the impartation of heat in the ratio of the amount imparted, and to the production of a greater amount of steam than the space within the boiler above the water could possibly contain without compression, or a diminution of its bulk, and its subsequent expansion or increase of its bulk, which it experiences on its liberation from the boiler, and also the cause of the bursting of the b er, in case the steam is not liberated p.evious to the generation of such an amount, and its compression to such an extent, as to cause an internal pressure upon the boiler which it is unable to bear ? To which our reply is, that we conceive that the formation of more steam than the space above the water within the boiler will contain is impossible pre ious to its exit ; that the compression and expansion is apparent and not real ; that an accumulation of heat in the metal of the boiler and its appendages, and in the surrounding brick-work and atmosphere, occurs in consequence of the inability of either water or steam to receive more heat ; that the heat accumulated in the boiler and the surrounding media, presses upon the water and steam within the boiler in virtue of the law of equal diffusion, with a power and force proportionate to the extent of endowment conferred upon it by such law, and the difference in the amount of the heat accumulated and communicable, and the amount possessed by the water and steam within the boiler ; that the pressure upon the internal surface of the boiler is a reverberating, reflective or recoiling pressure, resulting from the pressure of heat upon the water and steam, in its endeavour to force its admission into the water and steam in obedience to the law of equal diffusion, and not an expansive force exerted by an increase in the bulk of either water or steam, beyond their receptive capacities for heat ; that between the impulsive power of heat, which may be compared to a power operating upon a lever, and the resisting power of the water and steam, which may be compared to a weight to be raised by such imaginary lever, the sustaining boiler may be compared to the *fulcrum* of such lever, and which if not of sufficient strength to sustain the force of pressure exerted upon it by the impulsive and resisting powers, its disruption necessarily ensues. We conceive also that the amount of the motive power of steam which is generated within the boiler, previous to its bursting, or its exit by the safety or eduction valves, is no more than the difference between its specific gravity and that of the surrounding atmosphere—a difference so slight as will be found very far inferior to account for the rapidity and force of its exit ; but we atribute all the acquired force to the formation of an additional quantity of steam subsequent to the bursting of the boiler, or the opening of the safety or eduction valves, by the very rapid transition of the accumulated heat in the boiler and surrounding media to the water within the boiler, as soon as the first amount generated has made its exit, and that the apparent expansion of a large amount of steam compressed into a small space previous to its exit from the boiler, is but the rapid formation of a large additional quantity, subsequent to the liberation of the first amount created.

We admit of the expansibility of steam within the cylinder of the steam engine, resulting from the creation of a vacuum therein by the ascent or descent of the piston, and the consequent absence of atmospheric air, which if present would keep the constituent atoms of steam within their ordinary limits; but such expansion is not the effect of an innate power impelling its constituent atoms to occupy a greater space, but of the removal or absence of a restraining power which under customary circumstances, circumscribes the limits of the space which steam can occupy ; and hence such expansion is unproductive of impulsive power.

If then the theorem which we have advanced is correct, it appears to us that the liability to burst, to which a steam boiler is subject, is due to several causes. We conceive that no created substance is able to resist the diffusive power of heat, and that such diffusive power is proportionate in amount to the difference between the temperature of the impartive and the receptive medium. That the accumulation of heat in the metal of the boiler beyond the receptive capacity of the water and steam within the boiler, causes a recoiling force or pressure upon the surface of the metal of the boiler, proportionate in power to the amount of heat accumulated. That the heat which is thus imparted to the metal composing the boiler and retained by it, separates the ponderable atoms of which it is composed further from each other, and thereby diminishes the force of attraction of cohesion and aggregation which holds those atoms united, to an extent proportionate to the amount of heat imparted and retained, and the consequent separation to which its ponderable atoms are subject ; and that as what is termed the strength or tenuity of the metal is dependant upon the force of the power of attraction of cohesion and aggregation by which those ponderable atoms are held united together, so in proportion to the amount of heat imparted and retained must be the distance of removal of those ponderable atoms from each other, and the consequent diminution of the force and power of the attraction of cohesion and aggregation subsisting between those atoms, and the resulting diminution in the strength or tenuity of the metal.

In the generation of steam as a motive power, the purport is not merely to render a portion of the water of less specific gravity than the atmosphere, by the impartation of as much heat as will separate its ponderable atoms so far apart, as to occupy such a space as will render the whole of less weight than the bulk of atmospheric air which they together have displaced, and thereby enable the new compound called steam, merely to evolate from the surface of the water which supplied the ponderable atoms as a principle, and the imponderable as an agent ; but the purport, in addition, is to generate an impulsive as well as an evolating power, and to invest steam with it, in order that it may accomplish the several purposes for which it is created. If steam is generated to an amount just sufficient to occupy the space between the water within the boiler and its upper portion, and the safety and eduction valves are closed, and the temperature of the water and steam within the boiler, and the metal composing the boiler, are at 212 degrees of heat, and no more heat is imparted, and the safety or eduction valves are then opened, it will be found that the steam which has been generated will evolate from the boiler with as much force as will be induced by the amount of the difference in the specific gravity of such steam, and the superincumbent atmosphere; but such force will not furnish a motive power adequate to the need of any mechanical operation; and if the safety or eduction valves were allowed to continue open, and the impartation of heat was continued, so as to generate steam with a rapidity no greater than such steam could be evolved from the valve without impediment or restraint, we might cause the whole of the water in the boiler to evaporate, without furnishing a motive power to an adequate amount, and during the process up to its termination, we should find that the temperature of the water, steam, or boiler, never exceeded about 212 degrees; but if we generate steam faster than it can pass such safety or eduction valves free from restraint, we still find that the water and steam continue at the same temperature, but that in the same given period of time, a much larger amount of steam passes by such valve, and consequently with a much greater degree of force than it had previously passed, and thereby an available motive power is furnished. Such being the case, we therefore conclude, that in order to invest steam with a motive power, it must be generated faster in the steam compartment of the boiler, than it can pass free and unrestricted by the safety or eduction valves; that the smaller the orifices are which such valves cover, the greater will be the force with which the steam will pass through them; and the force with which the steam generated will pass, will be proportionate to the difference in the amount created, which would pass in virtue of its lesser specific gravity than the atmosphere, and which will be able to pass impulsively. And what is the impulsive power? The heat imparted from the fuel to the boiler, and from the boiler to the water. But it is said, that neither water nor steam can receive any more heat after it has attained to the thermometic temperature of about 212 degrees; and if neither can receive any more heat, how then can the heat imparted to the boiler, and which cannot be received by the water within the boiler, or the steam already generated, impart an impulsive power to such steam as issues from

the orifice of the safety or eduction valves? The reply to which is, that if heat is imparted to the boiler faster than the water can receive it, the surplus quantity will be accumulated and retained by the metal composing the boiler, and such heat will press upon the steam already generated, with a force proportionate to the amount accumulated, multiplied by the power of equal diffusion to which such heat is subject; and should the amount thus accumulated so far diminish the tenuity of the metal, and increase the recoiling pressure upon the internal surface of the boiler as to exceed its ability to bear it, its disruption must then necessarily occur.

We have already observed, that it is usual to charge a steam boiler in the first instance with water to a definite height, and that it is usual to endeavour to regulate the supply of water to the boiler during the process of the generation of steam to an amount, and with a rapidity proportionate to its consumption in such creation, so as that its depth shall not exceed or fall short of such point which is called the water line; and that the height of the furnace flue around the sides of the boiler, is but a little short of such water line, and that the purport of such arrangement is to endeavour to cause the impartation of as much heat as is furnished by the fuel to the boiler as can possibly be effected, and to impart no more than the water within can receive; and hence it necessarily follows that if the supply of water to the boiler is not sufficient to cover any portion of that part which is exposed to the action of the fire within the flue, and that the accumulation of heat will occur in the metal which is thus exposed, and the impartation being above the water, such heat will in obedience to the laws of equal diffusion and recession, rapidly ascend and become diffused throughout the entire surface of the boiler, and gradually accumulate to the amount of disruption of the boiler, provided the impartation of heat is continued, and its subsequent abstraction is not effected. And it should be well understood, that a subsequent abstraction may become a dangerous remedy if injudiciously administered, for if instead of diminishing the supply of heat, and allowing the escape of the surplus heat accumulated in the boiler by gradual radiation, a large amount of water is injected into the boiler to raise the supply to the water line, and such water is cold and too rapidly injected, the probable result will be the disruption of the boiler, by the too rapid increase of the amount of difference between the temperature of the boiler and the water, thereby increasing the amount of the power of equal diffusion to which the heat in the boiler is subject, causing a consequent greater recoiling pressure upon the boiler.

It is the opinion of some, that one of the causes of the bursting of steam boilers, is the decomposition of the water within by its primary conversion into steam, and a subsequent decomposition of such steam, a combination of the oxygen of the water with the metal composing the boiler, leaving its hydrogen subject to ignition by the highly heated metal, and that the disruption of the boiler is to be attributed to the explosive decomposition of such hydrogen ; but without questioning the possibility of such an occurrence in a wrought iron boiler, we very much doubt the probability of it.

The following summary list of causes or means by which the disruption of a steam boiler may be effected, we conceive may be useful.

To charge a boiler full with cold water ; to load the safety valve with a weight greater than the cohesive strength or tenuity of the metal of which such boiler is composed ; to shut the eduction valve, and then to impart heat to such water by medium of the metal of the boiler, to a sufficient amount to cause disruption, and probably to a much less amount than is sufficient to raise the temperature of such water to 212 degrees.

To charge a boiler to the usual water line, load the safety valve as before, shut the eduction valve, and impart heat, and disruption will probably occur, before the lower surface of the water in the boiler has attained to the temperature of 212 degrees.

To properly charge the boiler to the water line, and continue the supply to such point, neither above nor below, during the whole process of the generation of steam, and impart heat to the boiler more rapidly than the water within can receive it, and to an amount in accumulation in the metal composing the boiler sufficient to diminish its tenuity to a disruptive extent, in conjunction with the recoiling pressure of the heat accumulated upon the internal surface of the boiler ; the safety valve being at the same time properly loaded, and the eduction valve opened and shut at an unrestricted speed of the engine.

To properly charge the boiler to the water line, to reduce the charge below the water line during the process of the generation of steam by insufficient injection, and thereby increase the area of the steam compartment ; to impart heat to the boiler with a rapidity, and to an amount perfectly safe and proper, with the boiler charged and kept charged to the water line, but owing to the increased area of the steam compartment, and the diminished area of the water compartment, the accumulation of heat in the boiler beyond the receptive capacity for heat of the

water and steam may be such as to cause the disruption of the boiler, although the safety valve may be properly loaded and the eduction valve op ned sufficiently frequent.

To overload the safety valve ; the accidental adherence of the safety valve ; the irregular or inefficient opening of the eduction valve ; the irregular or insufficient reception of steam by the cylinder of the engine ; an accidental or designed overloading of the engine, or an endeavour to cause it to perform more work than it is able to accomplish, thereby causing a diminished consumption of steam and a consequent reactive pressure of the steam upon the boiler; an accretion of calcareous earth by deposition from the water to the internal surface of the boiler, the consequent additional impediment presented to the rapid transition of heat from the boiler to the water and the resulting accumulation of heat in the metal and accreting substance ; the liability of such calcareous coating being separated in portions from the surface of the boiler by the effect termed blistering, and the transmission of heat from such part of the boiler to the water being nearly suspended, an accumulation of heat in the metal may occur to the extent of its disruption. These, and most probably other causes which do not at present offer themselves to our memory, may severally or conjunctively effect the bursting of steam boilers.

We have stated that the capacity of both water and steam for thermometric heat does not exceed about 212 degrees, and it may be here necessary to explain, that although the thermometric temperature of steam does not exceed that of water, yet that its constituent amount of heat far exceeds that of water, and that in the conversion of water into steam, it is not only necessary first to raise the temperature of the water to about 212 degrees of thermometric heat, but to impart a large additional quantity of heat to such water to convert it into steam, and that the cause which prevents such additional quantity from passing by transition to the mercury of the thermometer is because such heat becomes latent in virtue of its constituency, and is therefore not subject to the law of equal diffusion to any body which may have attained to its own thermometric temperature. In the generation of steam for the purpose of imparting heat to the other bodies or substances, should they be situated at some distance from the steam boiler, it is necessary that such steam should be furnished with a motive power to enable it to reach its destination, and it is necessary that the conveying pipes should-be covered with some non-conducting heat material, in order to prevent an unnecessary diminution of that motive power, and a loss of valuable heat by abstraction in its passage, and it will be found that an abstraction of the constituent heat of such steam by any body or substance will be accomplished, while the abstracting substance is of a lower thermometric temperature than about 212 degrees. pp. 327-336.

NOTES AND NOTICES.

Origin of Railroads.—The German newspapers on mentioning the approaching completion of the railroad from Brunswick to the Harz, take occasion to remark, that the Harz was the " cradle of railroads" altogether. It was, they say, to miners from that district, whom Queen Elizabeth imported into England for the purpose of improving the English process of mining, that the first railroad ever seen in England owed its origin. So early as 1676, they remark, coals were conveyed from the mines in the vicinity of Newcastle-upon-Tyne to the neighbouring river by means of an imperfect railroad, which enabled as much work to be done, with the assistance of one horse, as formerly with that of four—and a similar contrivance was, they assert, long before in use in Germany. If it be true that the invention was practised in Germany previous to its application at Newcastle, it is not very easy to see why the phrase, " so early as 1676," should be made use of ; and, if it is to miners invited into England by Queen Elizabeth that its introduction is due, it is rather singular that it should be " so late as 1676" before it was introduced, no less than seventy years after the death of that illustrious princess. The whole story bears about it marks of incorrectness, if not of fabrication.

Steam-carriages on common roads.—Mr. Hancock and Sir James Anderson are, it appears, to have a foreign rival. On the 14th of October, M. Dietz performed a journey on the high road, from Brussels to Ghent, in a locomotive steam-carriage, to which several others were attached. The road was crowded with spectators ; but, to the amazement of all, M. Dietz succeeded in guiding his train not only without any accident or inconvenience to the public, but without ever being compelled to stop his movements or to abate his speed. A performance which must, indeed, have required a remarkable degree of dexterity anywhere, and which, we think, Mr. Hancock might safely challenge his foreign opponent to repeat on his scenes of action, the City-road and Cheapside. There seems to be an unusual degree of activity prevalent just now in this highly interesting and important branch of inventions, and we believe the time cannot now be far distant, when it must either " come to something," or come to a stand-still. M. Dietz, however, does not confine his exertions to steam-carriages only. It is stated in the foreign journals that the King of the Netherlands has granted him a privilege, and made pecuniary advances for the execution of another plan of his—a patent mechanical carriage without steam, in which two horses are to draw six carriages, containing eighty passengers, on common roads. Great things are to be done " if the scheme succeeds." We suppose so.

Railway Masks.—Half masks of gauze are now sold at Leipzig, for the protection of the eyes during railway journies. The price is two groschen each.

Chelsea Water-works.—At a recent meeting of the Directors of the Chelsea water-works, Mr. Lyon the Secretary laid before the Board a statement of the progress of the new filtering receivers which are now nearly completed. They occupy a space of 48,000 square yards superfices. The filtering bottoms consist of 12,000 cubic yards of pebble, siliceous gravel, and shelly sand ; and they have been constructed at an expense of 57,000*l*. The pipes are laid eight feet below the surface of this bottom, through which the water is filtered and cleansed before it passes into the wells from which the mains are supplied. It was further stated, that the brewers in the district consider the water as good for their purposes as that derived from the Artesian springs, and the custom of the Company had been thereby vastly increased since the adoption of the new filters. In the hearing of the reporter, however, it was remarked by a proprietor that the project of Artesian springs would compel the other water companies of the metropolis to follow the example of the Chelsea Company, and to adopt this cheaper and readier means of supplying filtered water without the intervention of Mr. George Robins.

Mousseline-de-laine Manufacture.—This appears to be a new species of manufacture, and as such is likely to become very valuable. The *Glasgow Constitutional* says :—The Mousseline-de-laines were first introduced into this country about three years ago, in a fabric composed wholly of wool, within the reach only of the wealthier classes. To meet, however, the pretensions of all ranks in society, a mixed fabric constituting of cotton and wool, was substituted, coarser wools being employed ; and the trade having got into a number of hands, the Mousseline-de-laine may now adorn the person of any one who can command the price of a common chintz. Formerly immense quantities were imported from France, on payment of a heavy duty ; and when first introduced sold at most extravagant prices, but now the foreign goods have been met with such active opposition from the British manufacturer, that they are nearly driven out of the market. France has always had the reputation of producing the finest printed goods in the world ; and had our Gallic neighbours not met with such powerful competitors in the British, they might, for a long time, and for the higher description of goods, have commanded a complete monopoly of the British market. At this moment the number of hands engaged in this trade is enormous. Besides those who are occupied at the printing tables, a great proportion, nearly one-half, of the hand-loom weavers in Scotland are in full operation in manufacturing the cloth. In almost all the small manufacturing villages in the west there is scarcely a loom idle ; and we are pretty safe in saying that there is not a respectable weaver who will find any difficulty in obtaining employment from the Mousseline-de-laine manufacturers. This trade has given a mighty impulse to the wool trade, but at the same time it has considerably weakened the hands of the cotton-spinners, who are complaining of the small demand for particular sorts of their yarns. Besides other important attributes, they possess one intrinsic advantage—they are not at all ignite on coming in contact with flame, like muslins or calicoes.

The Railway Map of England and Wales continues on sale, in a neat wrapper, price 6d. ; and on fine paper, coloured, price 1s.

☞ *British and Foreign Patents taken out with economy and despatch ; Specifications, Disclaimers, and Amendments, prepared or revised ; Caveats entered ; and generally every Branch of Patent Business promptly transacted. A complete list of Patents from the earliest period (15 Car. II. 1675,) to the present time may be examined. Fee 2s. 6d. ; Clients, gratis.*

LONDON: Printed and Published for the Proprietor, by W. A. Robertson, at the Mechanics' Magazine Office, No. 6, Peterborough-court, between 135 and 136, Fleet-street.—Sold by A. & W. Galignani, Rue Vivienne, Paris.

Mechanics' Magazine,

MUSEUM, REGISTER, JOURNAL, AND GAZETTE.

No. 796.] SATURDAY, NOVEMBER 10, 1838. [Price 3*d*.

WEATHERDON'S ROTARY SAW-CRANK.

WEATHERDON'S ROTARY SAW-CRANK.

Sir,—On perusing some of the earlier numbers of your instructive Magazine, I observe in vol. vi. a plan for a new rotary motion by a Mr. De Tir. Should you consider the drawing and description of the one I have sent you, which I completed some time since and called a "rotary saw-crank," likely to interest your readers, I without further apology give it.

The following is a brief description of the drawing, the model of which was made of wood, and worked extremely well. A A is part of the framework C, which passes through the blocks B B, and works in grooves to keep the frame in place. L F is a half-toothed wheel (or rather less), strongly fixed to the shaft or axle D; the dotted lines, showing the hold which the teeth has on crossing its centre at right angles. S S are the side frames or saws, the teeth of which are on either side, directly opposite to the spaces on the other; so that the impulse given to the last teeth L S, at the conclusion of the upstroke, will press forward the first tooth F, on the half wheel, into the first opposite space F S, as shown by the drawing; when the frame will descend on the tooth F, in gear with the half wheel, and transfer it by the impulse of the teeth L H, at the conclusion of the downstroke, into the space G at the top of the frame, completing with the up and down-stroke, one revolution of the shaft D, and so on, thereby producing a uniform rotary motion.

I am, Sir,
Your obedient servant,
B. F. WEATHERDON.
London, Sept. 20, 1838.

PROPELLING MACHINERY FOR STEAM-VESSELS.

Sir,—At page 400 of your last volume, I observe an article, descriptive of a new apparatus for propelling steam-ships, invented by Mr. J. Jephson O. Taylor, of No. 51, Gracechurch-street. In a letter recently received from Sir John Robinson, Secretary of the Royal Society, Edinburgh, he informs me that this method was carefully tried at Leith several years ago, by the late Mr. Waddell, who, having made a fortune as a ship-builder in India, and having re-turned to his native country, spent his time and a considerable proportion of his income in scientific and in practical researches. To a high degree of talent and great practical skill, Mr. Waddell joined sagacity and perseverance, and as he grudged no outlay in his experiments, he seldom took up a subject without pursuing it until he was satisfied that he had achieved what he had attempted—or that success was unattainable. Few persons have made more numerous or better devised experiments on the different modes of propelling steam-vessels; among other plans, he tried that now proposed by Mr. Taylor, which he varied in many different ways, and ultimately thought he had succeeded in putting it into a useful working form.

On this he prepared a small vessel, in which he made a voyage from Leith to the Western coast of Africa; this voyage satisfied him that he was mistaken, and it is understood, that latterly he had come to the conclusion, that no form of propeller was equal in effect to a well made paddle-wheel: provided the diameter was large, and the immersion of the floats not too great.

It might not have been worth while to advert to the above circumstances, had it not been that the last modification which Mr. Waddell gave to his stern propeller, included a great improvement over the form it had previously, which was that now employed by Mr. Taylor. Instead of having the vanes at the extremity of a rigid shaft, Mr. W. applied an universal joint immediately outside the stern of the vessel. The part of the shaft beyond the joint was supported by a collar hanging from an outrigger, and from this collar, chains passed to the quarters of the vessel; by this arrangement he dispensed with a rudder, as by applying the quarter-chains as tiller-ropes, the shaft and its vanes were made to traverse sufficiently to alter the direction of the ship's course as required. By means of the pendulous support, the shaft was capable of adjustment to the most favourable angle of depression, and might also be elevated entirely out of the water when in harbour, or exposed to injury from other vessels.

I regret that it is not in my power to supply a detailed account of Mr. Waddell's labours in this matter, because they would furnish the desiderata required by Junius Redivivus at page 39, viz.: a series of "Experiments on a large scale to compare the sculling and paddling processes."

I remain, Sir, yours, respectfully,
WM. BADDELEY.
London, October, 25, 1838.

ROE'S PATENT WATER-CLOSET.

The intricacy of water-closet machinery in the plans upon which they have hitherto been fitted, has been a great drawback upon the comfort of these conveniences. Upon any part getting out of order, no one but a practised plumber understands the various connections, and the apparatus remains for a time useless; mayhap, from something breaking or getting out of place, a whole house is flooded with water, and furniture damaged or destroyed. A water-closet has recently been brought into use, invented by Mr. F. Roe, of Chamberwell, in which nearly all the old and intricate apparatus of wires, cranks, box-valves, spring boards, springs, and such like, are dispensed with.

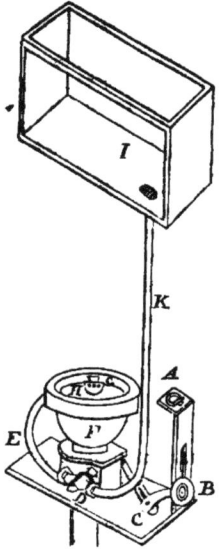

In the above figure of Roe's closet,

A is the handle; B, weight to bring down handle; C, lever, to open D, the ship cock or valve; E, pipe to arm of basin; F. basin; G, recess in basin to to supply the basin; H, fan; I, cistern. This simple arrangement, may, of course be modified in a variety of ways to suit different situations. One particular advantage of the plan is, that the apparatus being all in the closet, and none in the supply cistern, any number of closets may be supplied from one cistern, situated at any part of the house, so that it is at a level high enough to produce a strong current of water.

Z.

MR. UTTING'S ASTRONOMICAL TABLES.

Sir,—Severe indisposition has prevented me from replying sooner to *Nautilus's* defence of some of Mr. Utting's Astronomical Theorems.

Nautilus, in his last letter (No. 773) on this subject, remarks "The *Scotch Dominie* and his followers have pinned their faith, that in no case in the solar system is $\frac{P\,p}{P-p} = \frac{T\,t}{T-t}$; where P and p are the siderial, and T and t the tropical periods of any two planets." As one of the supporters of the *Dominie*, I have said so; and I repeat, that there is not a single known case in the solar system where an equality exists between the astronomical expressions $\frac{P\,p}{P-p}$ and $\frac{T\,t}{T-t}$. *Nautilus* asserts, that the "*Dominie's*" proposition has been mere assertion unsupported by any arguments better than an arithmetical comparison, founded upon false principles, &c." He gives us (what he no doubt considers) a demonstration of the universal truth of the equality of the expressions $\frac{P\,p}{P-p}$ and $\frac{T\,t}{T-t}$; but I will candidly inform him, that any assertion the *Scotch Dominie* has made, is worth a waggon load of such demonstrations as he has thought proper to give, and if I fail in proving this, then let it be said, that *Kinclaven is one of the most ignorant pretenders* that ever attempted to impose upon the readers of the *Mechanics' Magazine.*

Nautilus seems to have a horror at arithmetical calculations (which the *Cambridge Student* designated, "the drudgery of science,") in this I perfectly agree with him, and to prevent one having occasion (in the following investigations) to allude to that vulgar branch of science, I shall here premise, that the symbol a, in what follows, is understood to be the measure of four right angles, or three hundred and sixty degrees.

Let P and p represent the siderial, and T and t the tropical periods of any two planets; x and y the measure of the precession of their equinoxial points (*Nautilus* designates it recession, and I agree with him) in terms of a, for a tropical year. The $t : y :: T : \dfrac{T\,y}{t}$. That is, $\dfrac{T\,y}{t}$ is the recession of the equinoxial points of the second planet for the time T. Also $a : a-x :: P : T$, or $P\,x = a$ (P—T) $\therefore a = \dfrac{P\,x}{P-T}$. Similarly $a = \dfrac{p\,y}{p-t}$; hence, $\dfrac{P\,x}{P-T} = \dfrac{p\,y}{p-t}$. Now we shall assume it as a truth, that $\dfrac{P\,p}{P-p} = \dfrac{T\,t}{T-t}$; and if, from this supposition, we deduce results that are perfectly inconsistent with well-known astronomical facts, then it follows that there cannot be an equality between the expressions $\dfrac{P\,p}{P-p}$ and $\dfrac{T\,t}{T-t}$. Well, on the faith of their equality, we obtain from this last equation $P = \dfrac{p\,T\,t}{T\,t+p\,t-p\,T}$; and substituting the value of P in the equation $\dfrac{P\,x}{P-T} = \dfrac{p\,y}{p-t}$, we obtain—

$\dfrac{p\,T\,t\,x}{T\,t+p\,t-p\,T} \times \dfrac{T\,t+p\,t-p\,T}{T\,z\,(p-t)} = \dfrac{p\,y}{p-t}$, and this equation, by reduction, becomes to $x = T\,y \therefore x = \dfrac{T\,y}{t}$; that is, the recession of the equinoxial points of the first planet is $\dfrac{T\,y}{t}$. But the recession of the equinoxial points of the second planet was proved to be $\dfrac{T\,y}{t}$.

Hence, we arrive at this startling result:

that the recession of the equinoxial points of any two planets for the time T are equal, or, finally, if the recession of the equinoxial points of the earth for any time whatever be x, the recession of the equinoxial points for all the planets in the system for the same time will also be x. They have now nothing more to do but to consult their astronomical books, to cure them of this blunder.

Nautilus has made a most unwary attack upon the *Cambridge Student*. In "thumbing over" (this expression I borrow from N. himself) the pages of "Maddy's Principles of Plane Astronomy," he finds that $t = \dfrac{T\,S}{T+S}$. Very likely he may also find $p = \dfrac{P\,S}{P+S}$; from which we may easily deduce, that in the case of the earth and moon, there is an equality between the expressions $\dfrac{P\,p}{P-p}$ and $\dfrac{T\,t}{T-t}$; and all this I allow to be true, of the earth and moon; but the *Cambridge Student* in none of his letters has made the slightest allusion to these bodies. The truth is, *Nautilus* having found, that in the case of the earth and moon, that $\dfrac{P\,p}{P-p} = \dfrac{T\,t}{T-t}$, he very unwisely concludes, that the same must be true in regard of any two of the planets. He might have with equal propriety concluded—that, "because the three angles of a plane triangle are equal to two right angles;" the three angles of a spherical triangle are so too. Perhaps in my next communication I may point out to Nautilus the false steps he has used in his attempted demonstration of the equality of the expressions $\dfrac{P\,p}{P-p}$ and $\dfrac{T\,t}{T-t}$ as they regard the planets.

I am, Sir, yours, &c.
KINCLAVEN.

Oct. 31, 1838.

———

THE MECHANICAL CLOCK-PUZZLE— CORNHILL.

Sir,—This clock, placed in the window of Mr. Savory, seems to have caused some little noise and wonderment; some of the papers state that even clock-makers themselves are puzzled to account for the motion of the hand in front of the glass dial-plate.

I think the thing can be managed simply enough in the following way :—The stand contains, no doubt, all the works necessary for the movements of an ordinary clock, and these are set in motion, either by a pendulum or spring. The circular dial-plate of glass is firmly fixed upon this stand, upon which the figures are written; behind this is another circular plate of glass; this plate has the hour-hand fastened to it by a small arbor projecting through the front plate.—Now it is evident, that this second, or back plate, may rotate with the hour hand without the motion being seen; there is a small train of wheels leading up the branch of the stand terminating in a small drum; upon this drum the periphery, or edge of the back plate of glass (which carries the hand) rests; it is therefore evident, that any motion may be given to this invisible glass-wheel, or roller, by the motion of the drum, in exactly the same manner as the motion is now given in ordinary clocks by a small pinion driving the large wheel, which carries the hour hand. Indeed, the clock might be made more complete by introducing another plate of glass behind the second one; this could carry the minute-hand upon a hollow spindle, which would pass through the other two to the front, and which third plate of glass might receive its motion by another hidden train of wheels and drum like the former. I have no doubt, whatever, that this is the way the movement is effected, and not by any galvanic or magnetic agency.

I am, Sir,

Your obedient servant.

[Two other correspondents suggest somewhat similar plans to the above.]

It appears to me that there are *two* glasses (of which I annex a sketch *in section*) and the following is my explanation of the mechanism :—

Let A A represent the section through the centre of the front plate, on which the figures for the hours are painted. This is fixed in the outer frame, or ring, C C; B B is another plate of glass cemented into a ring of brass D D, which ring has teeth cut on its outer edge; E

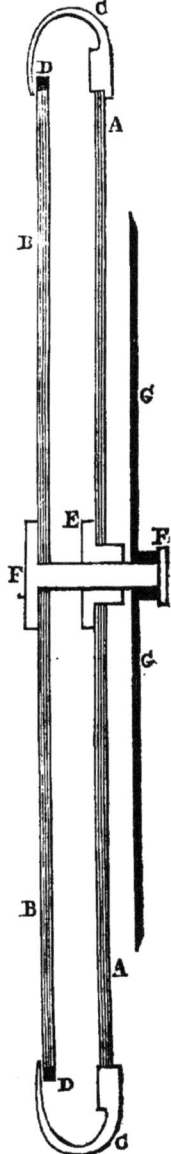

is a collar of brass cemented into a hole in the centre of the front plate; F, a pivot fixed in a similar manner through the back plate. F passes through E, and the index G is attached to it. The motion is communicated to the plate B B by means of a small wheel or endless screw, which passes up one of the supports of the dial, and works in the teeth of the rim D. By having G placed on the pivot F, without being too tight, we can alter the position of the hand if required, without deranging the inner works. On the magnetic principle, there must be some other index, or hand, at the back of the stand, to turn the wheel carrying the magnet; and this does not appear to be the case in Mr Savory's clock.

I must remark, that this explanation suggested itself to me, by observing the peculiar appearance of the clock face; which reflected the light exactly as *two* or more plates of a transparent medium usually do; but did not look like a single plate.

I have purposely drawn the two glass plates at a greater distance from each other than is required; for the purpose of better showing the principle.

Your obedient servant,

J. P.

Monday, November 5, 1868.

Sir,—I have just read Mr. Baker's explanation of Savory's clock, in your last Number. It is certainly very ingenious, but there are many difficulties in the way; one, not the least, that the magnet must be so small, and therefore weak, as to be liable to constant disturbance. The following I propose as the true solution: The dial is composed of two close, thin, plates of talc (not glass): one fixed, on which the figures are painted; the other carrying the hand, and having a toothed frame round its edge, and made to revolve in the manner explained by Mr. Baker. Sixteen guineas is the price of the clock.

Your obedient servant,

W. SCARFIELD GREY.

1, Cloisters, Temple, Nov. 7, 1838.

THE LOWCA COAL PIT EXPLOSION.

Having devoted no small portion of our columns to those inventors who have endeavoured to render less dangerous the occupation of the miner, we have thought that it would not be uninteresting to record the extensively fatal explosion which occurred on Wednesday, the 31st ult., in which forty lives were lost. This explosion is certainly a most striking instance of the callousness to danger which results from a continual exposure thereto; and a convincing proof, (applicable as also to the case of steam boiler explosions,) that the certainty of those who have the danger, in a great measure, under their control, being the first sufferers in case of neglect or carelessness, is no sufficient hostage for the security of those who trust their lives in their hands. Here, it appears, the two overmen entered to examine the workings with *naked candles*. What avails the lamp of a Davy, an Upton, or any other of the numerous philanthropists who have turned their talents to the subject, if the miners will not, or their masters cannot compel them to, use the means of safety offered to them by science?

The following account is abridged from the *Whitehaven Herald* and *Cumberland Packet*—the plan of the mine is from the former.

On Wednesday morning, the 31st October, a tremendous explosion of carburetted hydrogen gas took place in one of the coal pits belonging to Mr. H. Curwen, of Workington Hall, situate at Lowca, in the parish of Harrington, between three and four miles from Whitehaven. The regulation adopted in "John Pit," we understand, was for the work people to wait about 200 yards from the bottom of the shaft, at what is called "the steer," until the overman and deputy-overman had gone into the workings and seen that all was free from danger. At this spot thirty-four human beings were congregated when the foul air took fire, and they were swept, with six others, to instant destruction.

The shaft of "John Pit" is ninety-five fathoms in depth. From the bottom of the shaft a rolley-way of 200 yards in length conducts to "the steer;" here a drift of 300 yards in length branches off to the south, and another to the north of about 200 yards in length. These main drifts, and the workings which branched from them, it was the duty of the overman and deputy-overman to examine every morning, and ascertain that the pit was free from foul air and danger. The body of the overman, Harrison Kay, was found in the south drift, and that of William Hetherington in the north drift. Now, as the force of the explosion drove the poor men and boys who were at "the steer" into the north arm of the workings, it follows, as a reasonable inference, that the torrent rushed from the south drift, where an immense quantity of carburetted hydrogen must have accumulated and been ignited. From a light carried by Harrison Kay, therefore, it is probable that the explosion arose, as the usual practice of Kay and Hetherington was for them to separate at "the steer," the former surveying the south side of the mine and the latter the north. As a further proof of the accuracy of this conjecture, we may also observe, that Hetherington had not been touched by the fire, but had died from suffocation by the styth. It is much to be feared that Kay was so imprudent as to take a naked light in his hand, and enter the fatal magazine, notwithstanding the admoni-

tion of Darling, the fireman, only a few minutes before the explosion took place.

"John Pit" had until very lately been ventilated by "Hodgson Pit," as the upcast-shaft; but latterly the ventilation was carried on by a new air course into "Jane Pit;" and it is supposed that owing to a fall from the roof, the new air-course had been choked up, and hence the accumulation of foul air which led to the present awful and melancholy loss of life. Indeed, this interruption to the circulation may fully account for the accumulation of the inflammable air, which, when intermixed with atmospheric air in certain proportions, explodes on the application of a naked flame.

The annexed plan of the pit, from the *Whitehaven Herald,* will by reference to the account, better explain the circumstances attending the explosion.

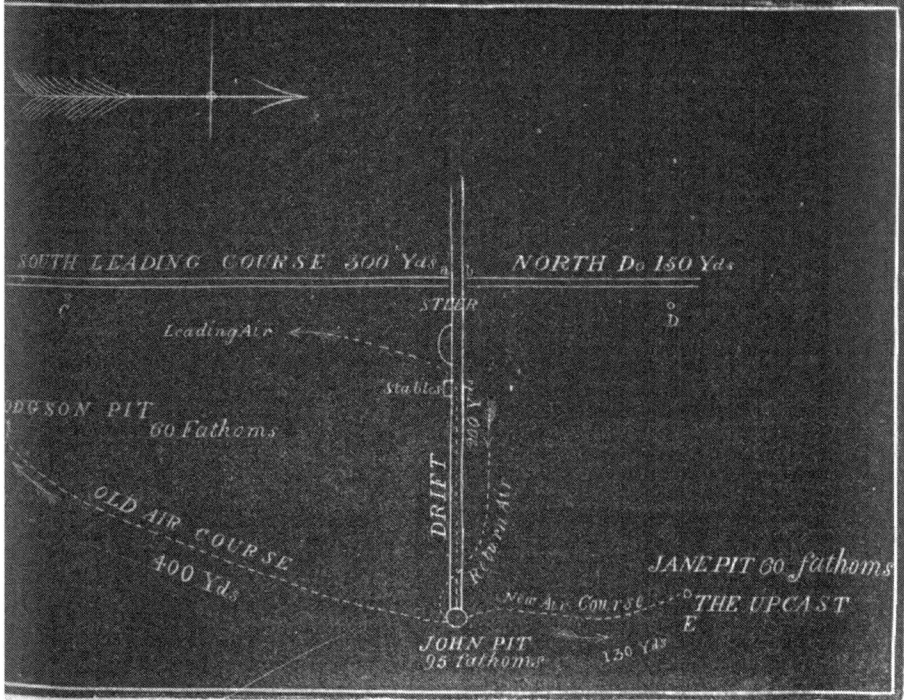

A The "steer," in which part of the men and boys were awaiting return of overmen.

B Working in which they were driven by force of blast.

C Place in which Kay the overman was found.

D Place where deputy-overman Hetherington was found.

E Place of the supposed fall in air-course.

THE BRITISH ASSOCIATION.[*]

The German Association, of which the British is the offspring, goes on its way rejoicing—and not only so, but sees rising around it a number of similar institutions as healthy as their parent. The "meeting of schoolmen" has just taken place, a meeting of Jurisconsults is said to be in contemplation. The British Association may also be considered to have met with no inconsiderable share of success. It has been doubted, and more than doubted, whether it be useful, and even whether it be not pernicious; but it cannot be denied, that it has drawn into its magic circle almost all the genius and talent, as well as much of the quackery, that swell the train of British science; that it has filled the inns of country towns and the columns of London newspapers—and when parliament was up, and no very interesting murder afloat, really attracted a good share of public attention.

It is, however, we believe, in success alone that, after all, any great resemblance will be found between the two associations of Germany and England. The elements of success have, we apprehend, been very different. It was, indeed, justly urged, at the time when the suggestion for a British Association of the kind was first made public, that the circumstances which had given birth to the idea in Germany were the very circumstances in which that country was most contrasted with our own. Germany is split into a variety of different governments; England, Ireland, and Scotland are united in one. Germany has no capital; England possesses that enormous "city of the world" which is the capital of half the globe. In Germany it may be requisite to provide a place, and specify a time for "Jurisconsults" to meet and talk over subjects of jurisprudence; but who in England would ever dream of supplanting by an assemblage on some selected day, in some obscure country town, the real and breathing interest of the first day of term in Westminster Hall?

Yet the British Association has, after all, succeeded; and why? We are inclined to think for the very opposite reason to that which promoted the success of the parent institution. The British Association has a tendency to uncentralize—to call away attention for a short time from the great city and the common objects of attention, and to bring under observation a new field, which, if less interesting, is also less hackneyed. It is not that men meet there who never met before—how few are there of any celebrity who encountered one another lately at Newcastle, who had not previously seen each others faces in London—but that they meet in a new place. It is not that scattered rays of light are collected into a new focus, but that the lens already constructed is brought to force its blaze on a new material. And it is obvious, that an "experiment" of this kind may be just as interesting; and as useful as an experiment of the other. That it is interesting has led to its successful repetition—and useful it certainly is in more ways than one.

That some decentralizing power is not only useful, but necessary in England, is, indeed, more than one opinion; it is a fact, which will hardly admit of dispute. The often-repeated aphorism, that "Paris is France," is, after all, perhaps less true than one that has never been hackneyed, it may be, never promulgated—that "London is England." How pitiably provincial our provincial towns—how still more pitiably our provincial cities? Where in the rest of the world can we point to two such masses of population as these contained in Liverpool and Manchester, professing to belong to an intellectual and cultivated community, and giving so little evidence of the fact? We have seen books from the presses of Nismes, of Montauban, of towns in France so obscure, that at this moment we cannot recall their names, to which we never saw an equal from the presses of those two flourishing and wealthy cities of England—for cities they are. This very day we saw a numerous series of volumes by a private society for the cultivation of science at Mulhausen, in Alsace, which made us blush as we thought of the slow progress and shabby appearance of those of the society at Manchester. A book has lately been published in America, called the "Boston Book," containing the productions of more than eighty authors of some pretensions, con-

* Report of the Seventh Meeting of the British Association for the Advancement of Science, held at Liverpool, in September 1837. Vol. vi, London, 1838, 8vo., pp. 710.

nected either by birth or residence with Boston. When shall we have such a book from any provincial town in England? In the face of the disgraceful apathy on literary and scientific subjects which so notoriously pervades the "provinces," it is useless to exclaim that we are not "centralized," because the men of the counties can mend their roads, without applying, as in France, to the capital for an order. This helpless reliance on the capital for every kind of information—for every shade of opinion—this almost total passiveness of the intellect—is the sign and symptom that centralization is doing its worst; and it is, we think, in its tendency to counteract this, to promote more circulation in the extremities, that the British Association will find an element of success and usefulness, equal to that which its parent society enjoys in a course of action almost directly the reverse.

We have, however, detained our readers too long from the volume in hand, we shall not detain them long with it. With a general resemblance to those which have preceded it, it is favourably distinguished from them by the greater share of attention it gives to practical science, a branch of science which, it might not be difficult to prove, possesses theoretical advantages over theoretical. There are no less than three treatises in this volume, on the different effects produced in the strength and composition of iron by the use of the hot and cold blast They occupy ninety pages of the more lengthy class of articles—the "Reports on the State of Science," which occupy more than five hundred pages of the seven hundred comprised in the volume. In the "Notices and Abstracts of Communications to the Association," only eight pages out of one hundred and fifty are devoted to mechanical science, and the notices are therefore even less complete than those which appeared in the *Athenæum* at the time of the meeting, and were transferred from its pages to those of the *Mechanics' Magazine.*

Under these circumstances we have only to state, that we have no further observations to make.

———

THE BOOMARANG.—NOTES ON COCHRAN'S RIFLES, IMPROVED COFFEE POT, AND NAPTHA LAMPS.

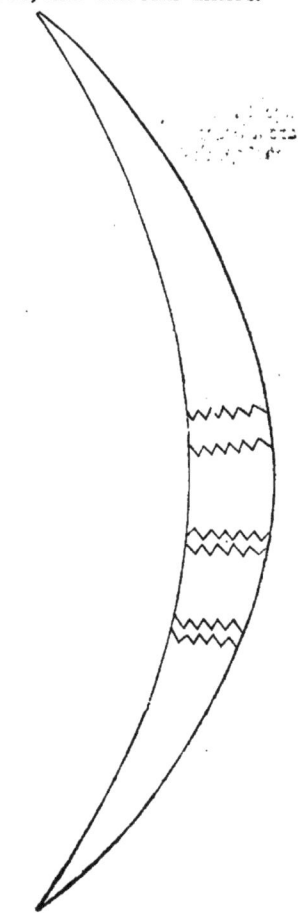

Sir,—In No. 780 of your Journal, I see an account of experiments with the Boomarang, by Captain John Norton. The above is a sketch of one, in my father's museum; it was in the possession of Lady Wilson, Charlton House, Kent, and at her death, in the year 1819, was left with her museum to my father. The sketch is drawn to a scale of one-fifth the size of the instrument, it is made of a hard wood, the zigzag lines are merely for ornament, and only on one side; in the widest part it is 2 in. ⅝ths.

and following the curve it is 2 ft. 3½ in. long; it is 1½ in. thick in the centre, and is brought gradually to a sharp edge, both on the concave and convex edges.

Cochran's many-chambered pistol, described in No. 723 of your Journal, is not a new invention; at my father's residence at Wallington, Northumberland, are a brace of pistols of excellent workmanship, the maker's name on them is Gorgo, London—the owner of them was Sir W. Blackett, who died in 1777; each of the pistols has three chambers, which carry each a charge; after one chamber has been discharged the next touch-hole is turned round to the pan. There is also an ingenious contrivance in the lock, by which the pistol primes itself. When the pistol is discharged it always turns round in the hand, owing to the intensity of the recoil, caused by the muzzle not being in a direct line with the chamber containing the charge.

The improved coffee-pot, described in No. 775 of your Journal, was patented by some one in London about five years back; soon after they came out I bought one, and used it for the twelve months preceding my visit to America. I found that it made most excellent coffee; it is, I think, the most ingenious and the best method of making coffee yet invented, and it makes it better and sooner than any other; I at first used the lamp, using naptha as the burning fluid, but afterwards found the most convenient and best method was to raise the steam over a clear fire.

In the Notices to No. 777 of your Journal, a light obtained from a mixture of spirits of turpentine and alcohol, is described as a new discovery in Paris. When I was at New York, in the year 1835, spirit gas lamps, of various shapes and sizes, made of tin, pewter, glass, and china, with wicks or without wicks, were used in many shops in that city, and generally to a certain extent in all the cities in the United States. The spirit was as pellucid and had the smell of spirits of turpentine, but had not the oily feel; the light obtained by it was brilliant, equal to the best gas made at gas works. The spirit is so inflammable, that the lamp with careless persons would be highly dangerous; I do not think there is the least chance of it ever superseding gas, made at gas works, it is neither so cheap, convenient, or safe, but in country shops, in towns where there is no gas work, it would be found very useful. In London naptha is cheaper than oil; besides the light from one naptha lamp is so great that it would save the use of many oil lamps in the same shop. The inventors name in America is Jennings. The high price of spirits of wine would prevent that spirit from being used in this country, here the retail price of spirits of wine is 24s. a gallon. In America, in the year 1835, alcohol sixty over proof was one dollar (4s. 2d. sterling) a gallon, and such a price was thought high at that period. I find on trial, that spirits of turpentine and alcohol, when mixed together, will not enter into combination; perhaps, if they were distilled together they might unite.

I remain, Sir,
    Yours faithfully,
        ARTHUR TREVELYAN.

Anglesea, Sept. 26, 1838.

SUBSTITUTES FOR THE COMMON YEAST USED IN MAKING BREAD.

Sir,—Since the members of the Temperance Society of this town formed the resolution to abstain from all fermented liquors (which is now four years), a strong desire has existed to procure some kind of yeast or barm equal to that in common use, so that the latter could be dispensed with, as well as the intoxicating drinks from which it is made. With this view, we have spared neither trouble or expense in trying the different substitutes which were to be found in receipt books, and the numerous temperance publications now issued, and not any of them have given us satisfaction equal to the two receipts which I now venture to transmit for insertion in your useful work.

The subject is not only worthy the attention of the friends of the temperance cause, but those also who do not practice or approve of the total abstinence principle, may find it very much to their advantage to have these simple, cheap, wholesome, and useful substitutes at hand, more particularly when the common yeast cannot be procured; at such times they are invaluable; inasmuch as it must be of importance that the "staff of life," which we are using from three to four times each day, should be light, palatable, digestable, nutritive, and as free from deliterious qualities as possible; which cannot be always said of the bread when the common yeast is used

The first receipt is excellent; and as buttermilk contains much real nutriment, it is to be preferred in country places to the common yeast, more particularly during summer, when it can be conveniently procured at a cheap rate, and is sufficiently acid. It gives the bread a rich flavour, unequalled by any other substitute. Several ladies of Barnsley have used it for more than two years with great satisfaction, except during the winter, when, from the buttermilk not being sufficiently acid, the bread was not so light as when the barm made from the second receipt was used.

In order to prevent any unpleasant taste to the bread, which will arise if too great a portion of the carbonate of soda be used, I would suggest, that the proper quantities be carefully weighed and folded up separately, to suit the weight of the flour taken for each baking. The druggist can easily do this, and should not charge more than 1d. or 1½d. per ounce, according to the quality of the article. If this precaution be not adopted, probably the persons by whom the bread is made will err either in putting in too much or too little, which will cause it to be sad, have an unpleasant flavour, and be unwholesome as food.

The ingredients of the latter receipt are all nutritive except the hops; and when one pint of the yeast can be purchased for 1d., it is certainly cheaper than the common barm obtained from ale or any kind of intoxicating liquor, which contains no nutriment whatever, but frequently the bread made by it is stained and flavoured with a portion of that deleterious quality which causes so much disease among ale and porter drinkers.

This yeast should be kept constantly on sale at the temperance coffee-houses in each town; the friends of the temperance cause, after giving it a fair trial, using means to recommend it to the attention of the public.

RICHARD BAYLDON.

Barnsley, Oct. 1838.

First Receipt.—(Sour buttermilk and carbonate of soda.)

Mix with 12 lbs. of flour 1 oz. of carbonate of soda, along with the usual quantity of salt. Knead the whole up with *sour* buttermilk; if very sour, half water and half buttermilk will do; but all buttermilk is preferable, which will be no worse if kept one, two, or three weeks before used; *the more acid the better*. The dough will be ready for baking in a quarter of an hour, as the fermentation goes on while kneading; but it will take no harm by standing one, two, or three hours.

The buttermilk must be *acid*, the soda pounded *small*, and *well mixed* with the flour, and the oven brisk, or the bread will probably be sad, and taste of the soda.— *Altered from Rothwell's receipt.*

Second Receipt.—(Potatoes, flour, hops, and sugar.)

Put 1 oz. of hops into a coarse bag, and boil them in two quarts of water; pare, boil, and mash 1 lb. of potatoes very well, and press them through a cullender into the hop water. Place the mixture on the fire until it begins to boil, then empty it into an earthen vessel with a narrow bottom, in which there has been previously mixed half a pound of flour with a gill of cold water, in the form of a paste; stir it well while pouring in, and when it is about the warmth of new milk, put in 4 oz. of dry flour, and 1 lb. of tee-total barm; or, if that cannot be had, half a pound of common yeast, which, though not so good, may answer the purpose; let it stand in the vessel covered up, in a situation where it will keep its temperature. It takes from four to twenty-four hours to ferment, according to the state of the weather. When it begins to lower in the vessel, it is fit for immediate use, or may be preserved when put in a bottle and corked up for several weeks. Should it be frozen, it will be no worse after being thawed.

In case you have no barm wherewith to begin; make about a pint or quart in the manner above directed, except in one particular; instead of putting any barm with the dry flour into the mixture, put two or three spoonsful of sugar with the flour; bottle it immediately, and having tied down the cork, set it where it will keep warm, and in twenty-four or thirty hours this will answer to ferment with, instead of the common barm. But it is always better to preserve some of the old for this purpose.

Directions for use.—Take 12 or 14 lbs. of flour; when you have mixed the salt with it in your kneading vessel, as is usual, make a hole in the middle, and pour in 1 lb. of the barm; let the water for kneading be two parts of boiling water to one of cold, in winter; and in summer, an equal quantity of each; the water should be soft. When the dough is of proper consistence, cover it up, and *keep it warm whilst it rises*; which will probably be from five to ten hours. If kneaded at night, it will be fit for baking in the morning; but, if it should not then be ready (which may be the case if kept too cool in the night), by applying a hot iron plate under the vessel containing the dough, it will in a short time be fit for the oven.— *Altered from Wild's receipt.*

FAILURE OF THE "LIVERPOOL" AMERICAN STEAM-SHIP.

A passenger on board the *Liverpool* during her late unfortunate expedition, westward and back again, in a letter which appeared in the *Athenæum* of last Saturday, gives the following particulars of the attempt, and opinions of the cause of failure:—

"We left port on Saturday the 20th—more than fifty passengers on board—in high spirits. The weather was then fair, but did not long continue so. The sea had run high for some days before, in consequence of long-prevailing violent west winds: it soon became a serious obstacle to our progress. Bad weather came on—rains and squalls. Still the boat went on bravely. At times the sea, which grew worse and worse, broke over her, fore and aft, sweeping all before it, and giving her not unfrequently tremendous dead digs, which, as we lay in our berths at midnight—or tried to lie—seemed absolutely to *take up* the ship and give her a shaking, as a dog does a rat. During this time it appears some damage was done. Some small leakages were sprung about the upper part of the vessel, such as might be expected in a new one under such circumstances, causing a little transient alarm, but probably without much reason. The fore cabin suffered severely: at one time the water, as I now hear, was some inches deep there. I also understand that the cargo, to the amount of 150 tons, appears to be damaged throughout. An accident at one time happened to the machinery, which occasioned a suspension of operations for some hours. Still we pushed on, not much exhilarated by such a beginning, but yet more and more convinced of the staunch qualities of the *Liverpool* as a sea-boat, and moreover satisfied with the behaviour and management of the captain and all the officers on board. Thus matters stood when we were suddenly notified of the captain's resolution to turn back—a great sensation arose of course—a council was called—every cabin and berth turned out their cadaverous-looking tenants, sea-sick, sleepy and all. It seemed that the engineer had sent in a written report of the state of the fuel, from which it appeared, on a comparison of quantities and distances, that there was not enough on board to carry us through the voyage; and that consequently we must seek absolute safety in retreat. To this nothing could be said; we acquiesced with the best grace we could. At the end of between 900 and 1,000 miles, on the expiration of the *sixth* day, we turned round and went before the gale—the ship dashing through the surge with an eagerness which seemed to say that no time was now to be lost.

"And now, you will ask, what was the cause of this difficulty? Want of coals, and nothing else. The ship is a fine sea-craft, —nothing can be said against her;—she is as staunch as wood and iron could make her. The commander, and all his subordinates, did their duty like old sailors;— nothing that skill or science could do was omitted. Our progress, in point of fact, was satisfactory. In the worst weather, with raging seas, the wind against us, all but a few hours, and generally amounting to little short of a gale, we yet made at the rate of more than 150 miles a day—something like 6½ miles the hour. Even at this rate we anticipated completing the voyage at most in about twenty-one days, more probably in eighteen. But this was not to be done without coals; and the calculation seemed to be, that, having started with about 564 tons, including 100 of William's resined and condensed peats, called "patent fuel," we had already consumed something like *half of our stock;* which proved that, instead of 564 tons, 800 would be the minimum of the quantity required to carry us through. This extraordinary consumption will excite surprise. The explanation of the ship's going to sea, provided as she was, with such a consuming power, will be called for. This question we have looked into as well as we could, having examined the papers and all the officers from whom information was to be had, and that information being freely given. It would appear that the ship was not sufficiently tried before starting. She went to Dublin, but that was no trial at all. More than this, it comes out that a very material alteration was made in a part of the machinery, *after the Dublin trip*, and *without super adding the least pretence of an experiment thereon*, by which the consumption of coal was increased nearly 700lb. the hour. Other disclosures I might add, but I have said sufficient till an answer appears to explain this. The return voyage to Cork was made in three days. The vessel showed great powers of speed as well as strength."

EXPERIMENTS ON SPONTANEOUS EVAPO-
RATION. BY JAMES P. ESPY.

(From the *Franklin Journal* for August.)

On the 2nd April 1831, I hung up two porous earthen pots, which I kept constantly filled with water, one in the shade, and the other in the sun. The superficies of each was thirty square inches. I supplied these, from day to day, from two vials each containing 12 ounces of water, avoirdupois.

The pot in the sun evaporated,

April 2nd to April	8th	12 oz.
8	17	12 "
17	27	12 "
27	May 5	12 "
May 5	1	12 "
12	18	12 "
18	24	12 "
24	29	12 "
29	June 4	12 "
June 4	10	12 "
10	19	12 "
19	26	12 "

The pot in the shade evaporated,

April 2nd to April	10th	12 oz.
10	20	12 "
20	30	12 "
30	May 10	12 "
May 10	20	12 "
20	27	12 "
27	June 3	12 "
June 3	12	12 "
12	26	12 "
26	July 6	12 "
July 6	15	12 "
15	24	12 "
24	Aug. 3	12 "

On the same day, April 2nd 1831, I also placed three tumblers of glass in the sun, one of them, in the open air, kept filled with water, and two sunk in the ground up to the rim, one of them was kept filled with water, the other with wet earth. From the 2nd of April to the 19th of May the tumbler in the open air had evaporated 21½ ounces, and each of those sunk in the earth 11 ounces, avoirdupois. On the 12th of June the sunk tumbler with water evaporated 21½ ounces, and on the 13th of June, that is, one day longer, the tumbler with wet earth had evaporated 21½ ounces from the 2nd of April. The experiments with the two sunk tumblers were soon after discontinued in consequence of an accident; but the tumbler in the open air had evaporated 21½ ounces more on the 18th of June, and 21½ ounces more on the 24th of July. The area of the interior of the rim of each of these tumblers was 12 square inches.

It will be seen from these experiments, that about 2½ times as much evaporated from a square inch of surface of the porous pot in the sun as from the sunk tumblers in the sun, which can be accounted for from the readiness with which the vapour, as soon as formed, would be removed from the surface of the porous pot; for I have demonstrated since, by experiment, that if the film of vapour is not removed from the surface of a humid body, by the motion of the air, evaporation ceases; as air, I find, is not pervious to the vapour of water to any considerable extent.

From these experiments it may be calculated how much is evaporated from a humid surface of earth in a given time, at the season of the year in which the experiments were made, and unless I have made a mistake in a rough calculation, the reader will find that about 2.70 inches, in perpendicular depth were evaporated from each square inch of moist earth from the 2nd of April to the 4th June, and from 2nd April to the 12th of June, 3.04 inches.

Wishing to know how much more rapidly evaporation goes on when the vapour is rapidly removed from the humid surface, I took two towels of 8,000 square inches area each, or counting both surfaces 16,000 square inches. I made these towels wet, and hung one of them up in a close room by two of its corners; and in the same room I swung the other towel about, continuing the experiment for 8 minutes, for two successive experiments.

Experiment 1st, evaporation from one at rest	119 grains.
" " one in motion	1,153 "
Experiment 2nd, " one at rest	104 "
" " one in motion	1,172 "

The temperature of the air at the beginning of 1st experiment was 74°, and dew point 53.5°; at the end of 2nd experiment, temperature 74, and dew point 58.4.

A third experiment was made by blowing upon one of the towels with a fan, instead of agitating it in my hands, and the following is the result of the operation continued for 8 minutes:

Experiment 3rd, towel at rest, lost 107 grs.
" towel fanned " 669 "

Temperature at beginning of experiment 75°, dew point 58°. Temperature at the end of experiment 75.6, dew point 61°.

(Copied from my original minutes, this 7th July 1838, Philadelphia.)

DEAR'S PROCESS OF SEPARATING AND
REMOVING THE BITTERINGS FROM THE
KETTLES OR BOILERS USED IN THE
MANUFACTURE OF SALT.

A patent was lately granted to David
Dear, of the town of Salina, county of
Onondaga, and state of New York, for an
improved mode of separating the bitterings
from, and cleaning the same from salt kettles
or boilers of any description used in the
manufacture of salt. The nature of the invention consists in using any of the properties of ashes, such as ley, kelp, or potash,
in such quantity as shall be necessary to
slack, soften, or remove the bitterings from
the kettles or boilers of any description
used in the manufacture of salt. without
cooling down the fire underneath the kettles
or boilers of any description, in which salt
is, or may be, manufactured.

In the first place (says the patentee), as
the kettles or boilers of the salt block are in
full operation in making salt, I commence
with any two or more of the kettles or
boilers in the block, and dip the brine out
of them. I then fill the kettles or boilers
so emptied with ley, or with fresh water,
and dissolve kelp or potash therein, or such
other alkali as shall have the same effect, in
sufficient quantity to make a strong ley
thereof, which ley when heated to a boiling
state has such an effect upon the bitterings
adhering to the kettles or boilers, as to
either slack or soften to such a degree, that
they may be removed from, and taken out
of the kettle or boiler with a ladle made for
the purpose; after removing the bitterings
from the kettle or boiler, proceed to dip
and clean the ley from the same, and pour
it into the next empty kettle or boiler, and
then fill up the first kettle or boiler, so
cleaned of its bitterings and ley, with brine
from the third kettle or boiler; proceed in
the same manner until the kettles or boilers
in the salt block are all cleaned of their
bitterings. If the ley should become too
weak by being too much used strengthen it
by dissolving more alkali therein.—*Franklin
Journal.*

VARIOUS METHODS OF BRONZING CASTS,
ETC.

Bronzing is the art of giving to objects of
wood, plaster, &c. such a surface as makes
them appear as if made of bronze, The
term is sometimes extended to signify the
production of a metallic appearance of any
kind upon such objects. They ought first
to be smeared over smoothly with a coat of
size or oil varnish, and when nearly dry,
the metallic powder made from Dutch foil,
gold leaf, mosaic gold, or precipitated copper, is to be applied with a dusting bag,
and then rubbed over the surface with a
linen pad; or the metallic powders may be
mixed with the drying oil beforehand, and
then applied with a brush. Sometimes fine
copper, or brass filings, or mosaic gold, are
mixed previously with some pulverized bone
ash, and then applied in either way. A mixture of these powders with mucilage of gum
arabic is used to give paper or wood a bronze
appearance. The surface must be afterwards burnished. Copper powder precipitated by clean plates of iron, from a solution of nitrate of copper, after being well
washed and dried, has been employed in
this way, either alone or mixed with pulverized bone-ash. A finish is given to works
of this nature by a coat of spirit varnish.

A white metallic appearance is given to
plaster figures by rubbing over them an
amalgam of equal parts of mercury, bismuth,
and tin, and applying a coat of varnish over
it. The iron coloured bronzing is given by
black lead or plumbago, finely pulverized
and washed. Busts and other objects made
of cast iron acquire a bronze aspect by being well cleaned and plunged in solution of
sulphate of copper, whereby a thin film of
this metal is left upon the iron.

Copper acquires by a certain treatment a
reddish or yellowish hue, in consequence of
a little oxide being formed upon its surface.
Coins and medals may be handsomely
bronzed as follows: 2 parts of virdigris and
1 part of sal ammoniac are to be dissolved
in vinegar; the solution is to be boiled,
skimmed, and diluted with water until it
has only a weak metallic taste, and upon
further dilution lets fall no white precipitate.
This solution is made to boil briskly, and is
poured upon the objects to be bronzed, which
are previously made quite clean, particularly
free from grease, and set in another copper
pan. This pan is to be put upon the fire that
the boiling may be renewed. The pieces under operation must be so laid that the solution has free access to every point of their
surface. The copper hereby acquires an
agreeable reddish brown hue, without losing
its lustre. But if the process be too long
continued, the coat of oxide becomes thick,
and makes the objects appear scaly and dull.
Hence they must be inspected every 5 minutes, and be taken out of the solution the
moment their colour arrives at the desired
shade. If the solution be too strong, the
bronzing comes off with friction, or the
copper gets covered with a white powder,

which becomes green by exposure to air, and the labour is consequently lost. The bronzed pieces are to be washed with many repeated waters, and carefully dried, otherwise they would infalibly turn green. To give fresh-made bronze objects an antique appearance, three quarters of an ounce of sal ammoniac, and a dram and a half of binoxalate of potash (salt of sorrel) are to be dissolved in a quart of vinegar, and a soft rag or brush moistened with this solution is to be rubbed over the clean bright metal till its surface becomes entirely dry by the friction. This process must be repeated several times to produce the full effect ; and the object should be kept a little warm. Copper acquires very readily a brown colour by rubbing it with a solution of the common liver of sulphur, or sulphuret of potash.

The Chinese are said to bronze their copper vessels by taking 2 ounces of verdigris 3 ounces of cinnabar, 5 ounces of sal ammoniac and 5 ounces of alum, all in powder, making them into a paste with vinegar, and spreading this pretty thick like a pigment on the surfaces previously brightened. The piece is then to be held a little while over a fire, till it becomes uniformly heated. It is next cooled, washed, and dried ; after which it is treated in the same way once and again till the washed-for colour is obtained. An addition of sulphate of copper makes the colour incline more to chesnut brown, and of borax more to yellow. It is obvious that the cinnabar produces a thin coat of sulphuret of copper upon the surface of the vessel, and might probably be used with advantage by itself.—*Dr. Ure's Dictionary of Arts, &c.*

NOTES AND NOTICES.

Railways.—The people of the North (writes a valued correspondent) have all along been sneered at for their great partiality for railway investments ; with what justice will be best perceived on an inspection of the present prices of the great railways now in active operation :—The present price of the London and Birmingham Railway, including the quarter shares is 205l., whilst the amount paid up is only 95l., showing a profit to the original subscriber of 110l., per share ! ! The present price of the Grand Junction, including the half shares, is 250l., whilst the amount paid up is only 110l., showing a profit to the original subscriber of 140l., per share ! The present price of the Liverpool and Manchester Railway, including the half and quarter shares, is 320l., whilst the amount paid up is only 155l., showing a profit to the original subscriber of 165l., per share ! Nor are these prices fanciful or fictitious, the dividends already made by the Grand Junction and Liverpool and Manchester Railways, and the present brilliant prospects of the London and Birmingham Railway, more than justifying them. Here, then, we at once see the source of the two to three millions of money gained by the town of Liverpool, and a clear proof that in northern sagacity there is not one whit of deterioration. *Allons donc*, the prospect for all the railways now in the course of formation is encouraging beyond our utmost expectations, and forms a spur and pleasing hope for such as are yet in embryo.—*Whitehaven Herald.*

Parallax of the fixed Star.—This important and valuable problem, which has for so many centuries been an object of inquiry amongst astronomers, has, it appears, by letters received in this country, been solved by Professor Bessel, of Konigsburg. His observations were made on the double star, No. 61, in the constellation Cygnus, whose distance he has ascertained to be 660,000 times the radius of the Earth's orbit, or 62 trillions and 700 billions of miles in round numbers. The details of this discovery will be communicated at an early meeting of the Royal Astronomical Society.

New Dye-plant.—In the South of Russia, numerous tufts of Hurmala or rue of the steppes, has been remarked. It is called Inserlik by the Tartars, and its botanical name is *Peganum harmala.* It sometimes covers extensive plains in the Tartar country: its root is strong and coriaceous, resists the plough, and is an invincible obstacle to cultivation. It is not useful for cattle. its odour being so disagreeable that they will not touch it, but it is likely to prove of immense service to the Russians in their manufactures. Attempts were formerly made to dye cloth of a red colour with the seeds, but it was a complicated process, and has been since abandoned : M. Goebel, professor of Chemistry at the University of Dorpat, having analysed these seeds, has ascertained the nature of their colouring matter, and invented a much simpler method of extraction. It is superior to most of the ordinary substances which produce seed, serves equally well for silk, wool, cotton, and linen, presenting every shade from rose to crimson, and not being subject to fade. Half an ounce of the extract is sufficient for dying six square archines, or more than three yards, of a deep crimson.

Railroads simplified.—The Polish General Dembinski has, it is said, discovered a method by which it will be rendered unnecessary to level hills and bridge over valleys in future railroad operations. He has taken out patents for the invention both in England and France. It is rather singular that such an invention should originate with a native of country emphatically described as the " Land of Plains," a country which in its surface presents fewer obstacles to the introduction of a railway system than any other in Europe, with the single exception of Russia, but in which political circumstances seem destined entirely to crush and extinguish even the few advantages with which nature has endowed it.

Arsenic.—A manufacturer of painted papers has been innocently poisoning his neighbours with the arsenic which he used in his colours. After a great deal of research, and many experiments, it was ascertained that this substance was filtered through the soil of the court-yard in which it was used, and thus reached the well from which the victims procured water for culinary and drinking purposes.

Lithographic Stone.—A vein of the lithographic stone has been found on the banks of the Rhone, which is said to be superior to that of Munich, although it has some unsightly marks occasionally, which, it is supposed, will disappear as the workmen penetrate farther into the mountain.

Australian Museum.—It would not be easy to imagine a more gratifying evidence of a young colony's progress in civilisation than that which is given by a handsome volume, a few copies of which have lately reached this country. The book is " A Catalogue of the Specimens of Natural History and Miscellaneous Curiosities deposited in the Australian Museum." It is very handsomely printed "by James Tegg and Co., at the Atlas Office, George-street, Sydney," an offshoot from the well-known "Thomas Tegg, at the Old Mansion House, Cheapside, London." The collection of natural history appears to be very considerable, as the list of it occupies nearly all the seventy-one rather closely printed pages of which the pamphlet consists. The "miscellaneous curiosities" are chiefly, as might be expected, specimens of the dresses, weapons, and utensils of the natives, presented by Major Mitchel, the able and intelligent colonial surveyor. We are glad to observe, that "the museum is open for the inspection of the public, every Tuesday and Friday, from 11 to 4." In fact, everything about the establishment does honour to all parties concerned, with one slight exception, which we hope to see amended in the next edition of the catalogue. The short "Advertisement" which is prefixed, is drawn up in a style which may, perhaps, pass for grammatical at Sydney, but, most certainly will not in London, though backed by the authority of the "Secretary to the Museum, George Bennet, F.L.S."

Berlin and Potsdam Railway.—The opening of the Berlin and Potsdam Railway took place on the 29th of October. The buildings and carriages were adorned with flags and flowers, and a couple of bands of music added to the gaiety of the scene, one in the train and the other at the "terminus." At twelve o'clock 280 persons started in eleven carriages, drawn by two locomotives. Among the passengers were the Crown Prince and numerous other members of the royal family, with officers of all the government establishments. The train arrived in forty-one minutes at Potsdam ; halted to allow the party to take refreshments, and accomplished the return journey in thirty-eight minutes and a half. Everything appears to have gone off well.

Asphaltum Pavement.—The promoters of asphaltic paving on the Continent mention with triumph, that some which has lately been laid down in the "Electorial Street," at Warsaw, has borne with success a trial which must remove all doubt of its stability. A train of carriages lately passed through Warsaw, carrying machinery from the manufactory of Cockerill, at Seraing in Belgium, for the use of a great cotton spinnery, belonging to a Mr. Geier, at Loda. The load carried by some of these carriages was nine tons, their own weight was a ton and a half, and the combined weight of 10½ tons, passed over the new asphaltic pavement without occasioning the slightest fracture or other injury. The circumstance is noticeable in itself but its accessories give it additional interest. The "great cotton factory at Loda," beyond Warsaw, and the machinery supplied to it from the workshops of Englishmen settled in Belgium, are striking facts in the history of the "march of manufactures."

March of Illumination.—On the 19th of October the members of the Parisian Academy of Sciences were present at some experiments in a new method of illumination proposed by M. Gaudin, which, it is said, were so completely successful as not only to satisfy but to enrapture the scientific spectators. M. Gaudin's illumination is of three degrees :—The first is calculated to supplant the use of common gas, supplying a brighter and whiter light ;—the second, which is called "star-light," is brighter still, and purposed to be introduced into lighthouses ; a focus of the size of a nut gives out a blaze which it requires the aid of green spectacles to survey without injury. The third, which is called "sun-light," possesses all the brilliancy of the rays of that luminary, and has the same effect in dazzling the eyes. Such is the first account of a discovery which, if what is stated be true, will soon dazzle all Europe.

Substitute for Steam. — The Corfu newspaper mentions that on the 23rd of September an important experiment was tried in the harbour of Corfu on a new invention, by a Greek of the name of Mauras. It consists of a machine capable of moving ships without the aid of fire or wind, and without the slightest danger—in fact, a complete substitute for the steam-engine in navigation. Nothing is said of expense, and no further particulars are given—And the curious must therefore ttrust to their old companion, Time, to ascertain the real merit of the invention.

March of Science.—The correspondent of a weekly *Scientific* Periodical, requests to be made acquainted with the *best* method of softening steel ; upon which the editor volunteers a reply, kindly directing him to put it in the fire, make it red-hot, and then lay it on a *dry stone to cool !* The same work lately contained a description of an improved oxy-hydrogen blow-pipe, the principal novelty of which consisted in the employment of a condensing syringe with a *bell-shaped working barrel ! Oh tempora, oh mores.* P. P.

Kréosote.—This interesting substance was recently discovered by M. Reichenlach, in impure pyroligneous acid ; it has been applied with great success, both internally and externally. In tooth-ache, a single drop introduced into the cavity, previously dried with a bit of cotton, will in most cases give immediate relief. The employment of this substance in ring-worm, and similar diseases of the skin has been attended with the most advantageous results, some practitioners considering it a specific. In diabetes, spitting of blood, and catarrhal affections, kréosote has been administered in doses of from two to six drops, three or four times a day, either in solution (two drops being soluble in one ounce of distilled water) or in the form of pills, mixed with liquorice root powder, and mucilage. In affections of the bronchia and lungs, it may be most advantageously applied in the form of vapour ; for this purpose pour into an inhaler (capable of containing a quart) a pint and a half of water, at a temperature of 150 degrees, adding to it thirty or forty drops of kréosote : mix by agitation and inhale the vapour through the tube. It may also be usefully employed undiluted, as an application to corns, warts, &c. ; but it is highly important that this substance be employed in a pure state, entirely free from certain deleterious principles naturally combined with it in the compounds from which it is obtained. W. B.

The Railway Map of England and Wales continues on sale, in a neat wrapper, price 6d. ; and on fine paper, coloured, price 1s.

☞ *British and Foreign Patents taken out with economy and despatch ; Specifications, Disclaimers, and Amendments, prepared or revised ; Caveats entered ; and generally every Branch of Patent Business promptly transacted. A complete list of Patents from the earliest period (15 Car. II. 1675,) to the present time may be examined. Fee 2s. 6d. ; Clients, gratis.*

LONDON : Printed and Published for the Proprietor, by W. A. Robertson, at the Mechanics' Magazine Office, No. 6, Peterborough-court, between 135 and 136, Fleet-street.—Sold by A. & W. Galignani, Rue Vivienne, Paris.

𝔐echanics' 𝔐agazine,

MUSEUM, REGISTER, JOURNAL, AND GAZETTE.

| No. 797.] | SATURDAY, NOVEMBER 17, 1838. | [Price 3d. |

HEINEKEN'S SIMPLE TRANSIT INSTRUMENT.

Fig. 1

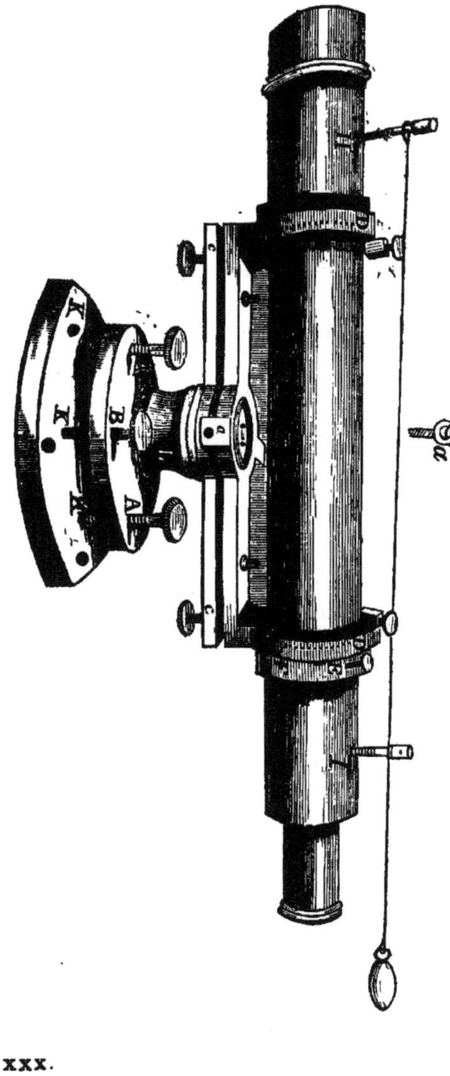

Fig. 2.

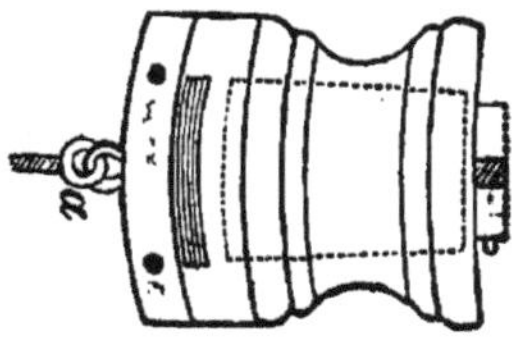

DESCRIPTIONS OF VARIOUS INVEN-
TIONS BY THE REV. N. S. HEINE-
KEN.

Sir,—I forward you a description and
drawings of some contrivances of mine,
which are at your service should you
think fit to give them a place in the *Mechanics' Magazine.*

Believe me, Sir,

Respectfully yours,

N. S. HEINEKEN.

Sidmouth, Sept. 20, 1838.

No. 1.—*Simple Transit Instrument.*

The object of the drawing Fig. 1, is to
show the application of the common
level, slightly altered, to the purposes of
the transit instrument. My aim has
been in this contrivance to enable any
clock-maker to construct for *himself* an
instrument by which he may ascertain
the time for rating his clocks; not indeed with the accuracy attained by the
usual transit instrument, but still with
far greater accuracy than by the ordinary means of the dial and the meridian line. The instrument may be constructed at a trifling cost, and, as before
stated, may be made to serve the purpose
of a level (the spirit-level being added)
when not in use as a transit.

The instrument is attached to a wall
or side of a window, &c., situated as near
as may be in the meridian of the place,
by means of four screws, K K K, fig. 1,
passing through the larger plate. A second plate, having four adjusting screws
and a stem, and containing the axis to
which the frame of the telescope is fixed,
is connected with the first plate by means
of a simple loop of metal, a, fig 2, or a
ball and socket, or Hook's joint. When
the instrument is attached to the wall,
the two adjusting screws, A A, are placed
vertically by means of a fine plumb-line,
and dots marked in their centres. The
axis of the telescope (represented by
dotted lines in fig. 2) is levelled by these
two screws, the adjustment being made
by means of a star or other object seen
by direct vision and by reflection as
usual. The two other screws, B B, serve
to place the telescope in the meridian.
The telescope is placed at right angles to

the axis by means of a fine plumb-line,
suspended from the groove in the screw
H, and passing over a corresponding
groove in I, and this adjustment is made
by the screws $c\,c$, the telescope being
of course placed vertically, and the indications of the plumb-line observed during
a revolution in the conical collars of the
telescope frame. If required, of course
the usual plumb-line apparatus could be
attached. The revolution of the telescope
in the collars furnishes the means of
making the collimation adjustment, the
wire plate being moveable for this purpose; it also allows of reversion during
an observation; one of the conical rings
turning in the collars is fixed to the tube,
the other is moveable, and can be clamped
by a screw, E. This moveable ring has
upon it two marks at exactly opposite
points, F F; a mark, f, is also made
upon the frame in which the ring turns.
When, therefore, the hairs have been
made vertical, the mark F upon the ring
is brought to coincide with f on the
frame, and the ring then fixed by its
screw E. The hairs will, therefore, be
vertical when the second F is brought
into the same position by a revolution
of the telescope. The screws D D serve
to clamp the telescope. The axis being
conical (see fig. 2) can be tightened at
any time by the insertion of a turn-key
under the telescope, without altering the
previous adjustments, and a divided
circle might be attached to the stem Q,
if required, for placing the telescope at
any angle of elevation; or a common
quadrant applied to the telescope for ordinary purposes. The construction of
the telescope is of course the same as in
other instruments in which cross hairs
are employed. And in conclusion I may
observe that every direction for placing
a transit instrument in the meridian, will
be found in the 2nd edition of Mr. Sims's
admirable treatise on surveying instruments. I may state, also, that I have
for some time used a telescope merely
fixed to the frame $c\,c$ (without the contrivance of the collars, &c.), for ascertaining the *rate* of watches, by the passage of the fixed stars across the wires,
when I did not require the *true* time.

No. 2.—*Description of an addition to the common Parallel Rule for the purpose of ruling equidistant lines.*

Fig. 3.

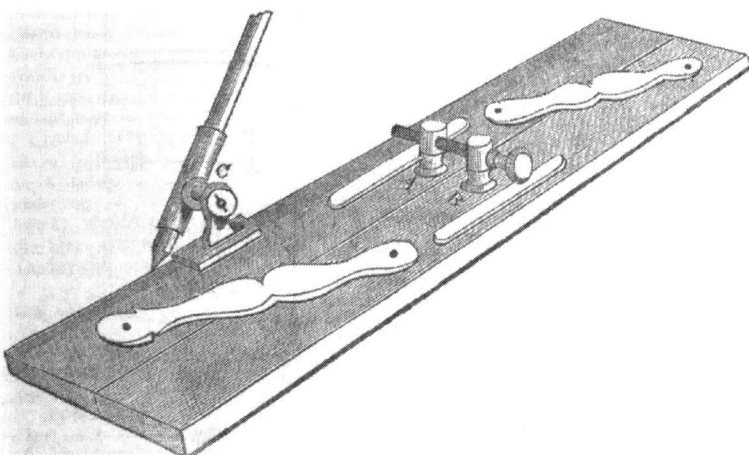

The sketch fig. 3 is the representation of a contrivance to enable a draughtsman to rule any number of equi-distant parallel lines, which are frequently required in mechanical and other drawings. It consists merely in the addition to the common parallel rule of an adjusting screw for limiting the distance between the lines, and a socket for holding the pen or pencil so as to enable the draughtsman to rule with greater accuracy than by hand alone. The rule, however, can be used with or without either of the additions, by withdrawing the adjusting screw, &c. The addition is made to the most common parallel rule, and also in the simplest way, so that the artist or engraver may be enabled to make it himself. For more accurate purposes the treble parallel rule would be preferable, and the socket containing the pencil, &c. should slide upon a raised plate of metal fixed along the edge of the rule, and perpendicular to its face; or the usual dividing point and frame might be adopted for *short* lines. It is evident that by having the edge of the rule indented, waving lines might be drawn.

Mode of using the Rule.—The method of using the instrument is to withdraw the adjusting screw, so that the rule shall open the required distance, press the lower half of the rule firmly to the paper with two of the fingers, and the upper with the other fingers, apply the socket containing the pencil to the edge of the rule, elevate, or depress the pencil by means of the point at C till it touches the paper, and draw a line. Now, with the two fore-fingers slide the upper half of the rule on the paper till stopped by the screw, the other half being held down firmly by the other fingers, and repeat the process for the lower half till this is stopped by the upper, then draw a second line, and so on. If it be required to draw lines upon a copper plate, as in etching, it will be necessary to allow the rule to rest upon borders of card-board, or the like, elevated a little above the plate, both to prevent the rule from slipping and from injuring the plate.

The adjusting screw is tapped only into the pillar A, the other half of the screw is left plain and slides through B. The pillars are fixed into the rule by nuts counter-sunk in the under side of the rule, but not so tightly as to prevent the pillars from turning a little, in order that they may accommodate themselves to the different distances to which the blades of the rule may be opened.

No. 3.—*Adaptation of Cavallo's Pearl Slip Micrometer to the Reflecting Telescope.*

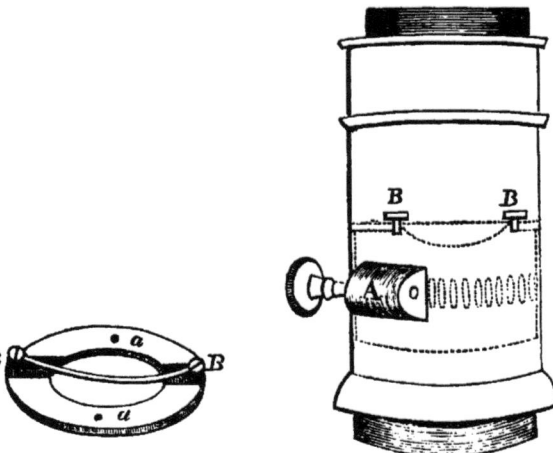

Fig. 5.   Fig. 4.

The objections to this simple micrometer have been that it has not been easily applicable to the reflecting telescope, as in consequence of the eye-piece being made in one tube, there has been no means of changing the position of the micrometer; and also that the divisions of the micrometer, in its ordinary form, are distorted by the eye-glass. I have endeavoured to remove these objections thus;—I have racked the tube which carries the diaphragm of the eye-piece, and applied an endless screw. By this means I can place the micrometer in any required position. The second defect I have sought to lessen by giving to the pearl slip a curvature, whose radius is the focal distance of the eye-glass. The plate carrying the pearl slip is attached to the diaphragm tube by two long screws, and by removing this plate, a circular pearl micrometer, or a plain diaphragm, may be substituted. Thus the same eye-piece becomes available for several purposes.

Fig. 4 represents the eye-piece. The endless screw is contained in the box A, and the dotted lines show the racking of the diaphragm tube, and the micrometer secured to its plate and curved by the two screws B B.

Fig. 5 represents the pearl slip, and its plate, removed from the eye-piece. B B are the two flat-headed screws which confine and give the requisite curvature to the micrometer; *a a* the two holes for the insertion of screws for the purpose of attaching it to the diaphragm tube.

———

No. 4.—*Improvement on Davy's Electrical Telegraph.*

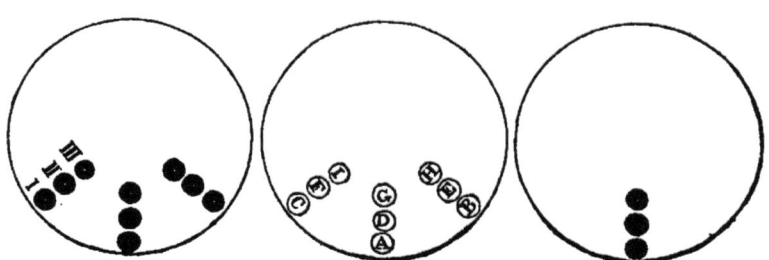

Your correspondent, Mr. Davy, sometime since, described what he considered to be the principle of the electrical telegraph exhibiting at Exeter Hall (see No. 758). I was induced to make a small model according to his plan, which I found fully to answer. Since then I have made the following alterations, by which nine letters may be shown with only one needle and one coil. The top of the box, in which the needle, &c. are enclosed, is pierced with three sets of three holes each, and numbered 1, 2, 3, as in figure 6. Below this is the needle carrying a disc pierced with three holes, which correspond with *one* row of the aperture above, viz. 1, 2, 3. Under the needle is another disc, having the letters $\left\{\begin{array}{l} \text{I. G. H.} \\ \text{F. D. E.} \\ \text{C. A. B.} \end{array}\right\}$ in position exactly corresponding with the apertures in the upper disc. Below this again is the coil.

Now, when the needle is in one position, say its ordinary one, the letters A. D. G. will appear,—with polarity in the coil, C. F. I., &c. By this construction, therefore, a saving may be effected in coils, needles, &c., though a coil and needle will be required for intimating to the person who is to read the words, what compartment he is to look at for the several letters; some loss of time will also be occasioned. My object, however, in contriving this, has been to make it an amusing experiment for a class. It may, perhaps, afford a hint which may be applied to more important purposes. Other sets of letters could be added at different parts of the discs if required.

Fig. 6 represents the top of the model of the electrical telegraph, with the sets of apertures I. II. III.

Fig. 7, the disc, which is fixed to the needle with its apertures.

Fig. 8, the disc beneath the needle, &c. upon which are the letters.

WALKER'S CHIMNEY-SWEEPING MACHINE.

Sir,—Having some time ago observed a paragraph in the newspapers relating to a legacy of 200*l.*, stated to have been left by a lady to any person who should invent a machine, which would, to the satisfaction of the legislature, supersede the use of climbing-boys in chimney-sweeping, I set to work, (not that I had much expectation that I should obtain the 200*l.*, but for a love of mechanics, blended with the hope of being useful to the public) and devised a plan of which the following is a description:—

The machine consists of a rod composed of a number of lengths, formed of strips of whalebone or cane bound together. The brushes and scrapers are fixed at right angles to the top of this rod, and the handles of them are elastic, in order that they may contract or extend, according to the size of the chimney they are used in. Beyond these brushes and scrapers, and forming a continuation of the principal rod, I place what I call a guide-rod, which is about 10 feet long, made tapering to the end, like, and of the same materials as, a fishing-rod. On the top of this guide-rod there is a ball, to prevent any rough places or defects in the chimney stopping its passage upwards. The purpose of this guide-rod is, that when it comes to a sharp angle in a chimney, it may bend up, and guide the brushes and principal rod round the corner. Should it bend the wrong way, if the sweeper twists the machine, it will right itself.

I leave you herewith a model, for your own inspection and satisfaction; and perhaps you will allow any of your readers who may desire so to do, to examine it.

I am, Sir, your obedient servant,
JOSEPH WALKER.

May 22, 1838.

THE CORNHILL CLOCK—ANOTHER METHOD OF PRODUCING MOTION WITHOUT APPARENT MECHANISM.

Sir,—There have been two or three attempts described in the *Mechanics' Magazine* to account for the manner in which motion is communicated to the index of the clock now exhibiting in Savory's shop at Cornhill. I have not, myself, seen the clock, but I can hardly imagine that any of the expedients described by your correspondents, can be that which is resorted to; setting aside their clumsiness, they all require the revolution of a large circle, unsupported by a central axis, which, in practice, would necessarily involve immense friction and require great power. In my opinion, a much more feasible plan might be modified from the acknowledged contrivance of the dial exhibited in the Polytechnic Institution, although its particular application, the displayed, is clumsy and objectionable.

Where the works of a watch are concealed in the *feather* part of an arrow, not only is a great counter-balancing weight rendered necessary at the pointed end, but also, a connecting bar, or shaft, to transfer the motion to the centre. The following plan appears to me to be an obvious improvement, and quite capable of producing all that is attributed to Savory's clock, at least so far as the drawing in front of No. 795, *Mechanics' Magazine*, represents it:—

Take a small Geneva watch, (of which there are several scarcely bigger than a shilling,) divest it of its outer casing, dial, pendant, and hands, and let the place of the pendant be supplied by the shaft and point of an arrow, proportioned to the size of the dial to be used, and let the feather part be affixed diametrically opposite—it will then present the annexed appearance in which the watch is at *a*.

Next, let a brass pipe which tightly fits on the hour-hand arbor of the watch, be cemented into the centre of the glass dial, so as to be flush with its surface. If then the hour-hand arbor be inserted into the cemented pipe, the index will move round the dial once in twelve hours, correctly indicating the time. To wind up, the index may be unfixed, if, as is usually the case, the winding arbor be on the face. There need then be no appearance of works or machinery to betray the contrivance, and the only weight necessary being in the centre, and equally distributed, the index might be as light and delicate as could be desired.

It is evident that there would be no difficulty in affixing a light index at the *back* of the transparent dial, by means of a central pipe, squared internally, to fit the minute-hand arbor of the watch, and long enough to pass through the cemented pipe and dial; reaching to the *minute-hand* arbor, and affixing thereon. There would then be *two* indices, as in an ordinary clock, one for hours, and the other for minutes; thus rendering exhibition much more wonderful and useful than Savory's.

NAUTILUS.

Nov. 12, 1838.

I perceive by your last Number that *Kinclaven* has shaken off his "*long indisposition*" (to attempt a reply, *ne c'est pas?*). He labours hard again to invest the question in the mist, which it was the object of my communication of last June (No. 773) to dispel; he does not make even an attempt to disprove either of the two simple positions on which the truth or falsehood of my proposition entirely hinges, (all the rest of my letter being merely a plain statement of undisputed facts) mamely—

Firstly, $a : b : : T : t$; and, secondly, if so, $\left(\dfrac{360-a}{T} - \dfrac{360-b}{t}\right) = \left(\dfrac{360}{T} - \dfrac{360}{t}\right)$.

In the first page of his letter he writes thus:—" I repeat that there is not a *single* known case in the solar system where an equality exists between the astronomical expressions $\dfrac{P\,p}{P-p}$ and $\dfrac{T\,t}{T-t}$;" but, on the succeeding page he says " there *is* an equality between the expressions $\dfrac{P\,p}{P-p}$ and $\dfrac{T\,t}{T-t}$," " *I allow it to be true.*"—(vide pp. 83-84, No. 796.)

This is only a small spice of the mess into which he is floundering, but as he promises a further communication, it would be a pity to spoil sport by interrupting him, until he has got over head and ears. I hope, however, that when he is convinced of his error, he will not *again* (as at page 245, vol. 27) limp off by saying it is " *not necessary*" that two persons should agree about a simple matter of fact.

NAUTILUS.

METHOD OF FINDING THE CURVES OF THE TEETH OF WHEELS FOR MILL-WORK.

Sir,—For the purpose of giving a general explanation of the method, a particular case of which was described in No. 752, let it be required to find the form of the curved part of the tooth of a spur-wheel, A, of *any* diameter, which is to work with another wheel, B, of any diameter, either greater, equal to, or less than the diameter of A; the part of the tooth that is within the pitch line of each wheel being formed by radial lines in the usual manner.

Fig. 1.

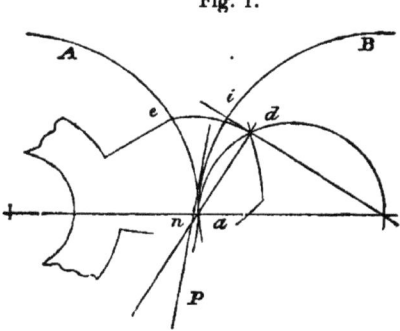

From three centres, on a straight line (see fig. 1), draw the pitch lines of A and B touching each other, and a semi-circle half the diameter of the pitch line of B, touching the pitch line of A and the centre of B.

From the centre of A, and with a radius equal to that of the ends of the teeth of A, intersect the semi-circle at the point d.

Through the point d, and the centre of B, draw a radial line cutting the pitch line of B in the point i.

From the point a, where the pitch lines of A and B intersect the straight line, set off an arc, a e, upon the pitch line of A, equal in length to the arc a i. (This may be done by stepping, or by a method I shall describe hereafter.)

Draw an indefinite straight line through the points d a.

Raise a perpendicular P to the points d and e, midway between them.

The point n, in which P cuts d a, is the centre from which to strike the curved part of the tooth, with a radius n d or n e, whether n is within or without, or upon the pitch line.

I have said enough of "dropping the line" in No. 752.

If several pinions of various diameters are required to work correctly with the same wheel, the object may be effected by making the teeth of the wheel, or those of some or all of the pinions, curved within the pitch lines as well as without; for it is demonstrated by writers on this subject that, if the epi-

cycloid without the pitch line of one wheel, is formed by the same generating circle as the epicycloid within the pitch line of the other, the wheels will work correctly together. But it is not desirable that any of the teeth should be thinner at the root, than they would be if formed by radial lines: and therefore it is best to form the teeth of the wheel, and smallest pinion, by radial lines within the pitch lines, and by such curves without the pitch lines as if they alone were to work together; and to form that part of the tooth that is without the pitch line of each of the other pinions, as if it alone were to work with the wheel; and then to find the curve of that part of the tooth that is within the pitch line of each of the pinions except the smallest, in the following manner:—

Fig. 2.

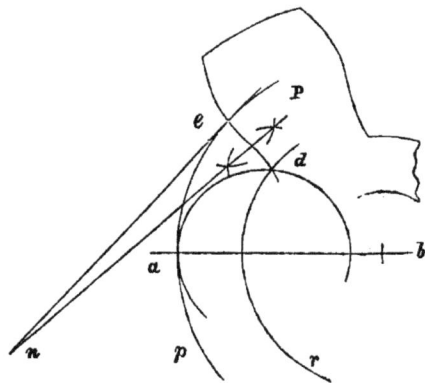

Upon a straight line a b, fig. 2, draw the pitch line of the pinion p, and another circle r from the same centre, as far within the pitch line as the ends of the teeth of the wheel it works with are without their own pitch line.

On the line a b, draw a circle half the diameter of the pitch line of the smallest pinion, touching p in a, and cutting r in d.

Set off an arc a e, equal in length to the arc a d, and draw a tangent e n to the pitch line p at the point e.

Raise a perpendicular P to the points d and e midway between them.

The point n, in which P cuts e n, is the centre from which to strike the part of the tooth within the pitch line, with a radius n d or n e.

In practice, the radial line will, in most cases, be sufficiently accurate, and will be preferred on account of its easy execution; but I have given this method because some writers attribute great importance to pinions of various diameters working correctly with the same wheel; and this is a close approximation to correctness with circular curves.

Yours, &c.

C. G. JARVIS.

Nov. 12, 1838.

In my paper in No. 752, line 21, for *n m*, read *p m*.

IMPROVEMENTS IN WATER-CLOSETS.

Sir,—From the sweeping remark of " Z." (at page 83 of your last Number) that " the intricacy of water-closet machinery in the plans upon which they have hitherto been fitted, has been a great drawback upon the comfort of these conveniences," it would seem as if he was desirous of leading your readers to suppose that there are no exceptions —that all plans are " intricate, &c." save and except the patent water-closet, erroneously stated to have been *invented* by Mr. Roe.

It is somewhat singular, that " Z." has so very briefly noticed the gist of Mr. Roe's plan, which is the rim or recess G (vide engraving, p. 83,) at the upper part of the basin, which performs the office of a box-valve by allowing a proper quantity of water to enter and remain in the basin, after the supply has been shut off by turning the stop-cock D. It must be admitted that there is considerable simplicity in the plan of the water-closet *brought out* by Mr. Roe, and described at page 83; but there is also much fallacy in ascribing certain defects to " all the old and intricate apparatus," and making an empty parade of " spring-boards, springs and such like," while the *new* plan is equally open to objections. It is true that " wires, cranks, and box-valves are dispensed with"—but we have, instead, a cock, which, however well made and perfect at first, is sure to become worn in a comparatively short space of time, and leak in proportion to the pressure of the superincumbent column of water. By means of box-valves the water is shut off within the cistern, but in Mr. Roe's

plan, the whole length of pipe from the cistern to the closet, which in some situations will be considerable, is exposed to a continual pressure, and is liable to burst by the sudden closing of the cock, or from the effects of frost in winter, when " a whole house may be flooded with water, and furniture damaged or destroyed." In the event of a valve leaking within the cistern, the only inconvenience is the loss of a little water, which escapes without annoyance through the usual channel; but a leak in any part of " the apparatus in the closet" is likely to be attended with serious additional inconvenience, so that what is considered by " Z." as one particular advantage of this plan, strikes me as being a particular disadvantage. I am at a loss to understand how Mr. Roe's water-closet is free from the objection made by " Z." to all others, viz. that " upon any part getting out of order, no one but a practical plumber understands the various connections," and is capable of restoring the apparatus to its pristine perfection !

For my own part, I am free to confess that I entertain a strong objection to all water-closets supplied from " cisterns high enough to produce a strong current of water," as I well know they are all liable at times to cause a waste of water, and in this respect, by the bye, Mr. Roe's plan surpasses all others ; let any person, either from ignorance or with a mischievous intention, keep the handle elevated, the consequence will be the running to waste of the water with very great rapidity.

I have seen but one plan that comes up to my ideas of propriety on this subject, and that is the water-closet of Messrs. Lambert and Son, which draws a supply of water from a cistern *beneath the closet*, by means of one of their double-acting pumps. The raising of the handle injects a stream of water to wash away the contents of the basin. On depressing it, a second stream of water enters and remains in the basin. In this plan there can be no great waste of water, except wilfully, and then only by dint of continued personal exertion. Messrs. Lambert's contrivance was noticed by Mr. Wm. Baddeley (at page 83 of your 27th volume) who very justly observed that, " the water being raised by the party using the closet, is used in the *exact quantity*, and at the *precise time*, it is required, thereby most effectually preventing both loss of labour and waste of water."

I am, Sir,
Your obliged humble servant,
Z. Z.

London, Nov. 11, 1838.

SMOKE-CONSUMING FURNACES.—PHILOSOPHY OF HUMAN FOOD, ETC.

Sir,—In my note to you respecting iron steam ships I omitted adverting to one point of the subject which is at present exciting some attention, I allude to the perfecting the process of combustion, and thus getting rid of smoke by converting it into fuel. I do not calculate much on the amount of economical saving of fuel in this matter, for the principal loss may be roughly estimated by the volume of smoke, and it would take a large number of cubic feet of smoke to compress into the bulk and weight of a ton of coals. The amount of combustible matter escaping in the form of invisible gas must be very small. The great loss sustained is in the want of proper economy in using the heat evolved from the coals; roasting the stokers alive instead of producing steam with it. But there are sufficient reasons of another kind to make it desirable to consume the smoke—cleanliness and beauty.

If the combustion of the smoke could be accomplished by means of the *waste* steam of the engine the matter would be simple enough, but to rob the engine of its working power is a manifest mischief. There is one great principle to be attended to in all projects for smoke consuming; it is to make all the smoke pass through a stratum of burning fuel sufficiently hot to flash it into flame. This principle was adopted several years back by Mr. Cutler, in what he denominated a gas stove. The bottom or fire-bars frame of a common register stove was made to rise and fall by means of a cog-wheel and endless chain, in an iron box beneath. The grating being sunk to the extreme depth, the whole box was filled with small coal to the ordinary level; on this the fire was made, and the coals gradually heating below, all the smoke and gas evolved was converted into flame as fast as produced. Chimneys used with this stove did not require sweeping above once in two or three years. As the fire

burned downwards the grate was gradually raised by means of a winch handle, till the whole of the coals were burned. The disadvantages of this stove were, that if the coals happened to be exhausted in the day time or before bed time, it was necessary to fill the box with coals again, and to re-light the fire. There were no simple means of re-charging. The stove went out of use, probably on this account. A simple mode of retaining all these advantages without the disadvantages would have been to place the magazine of coals in the two side jambs of the register,—a large space at present quite useless. The feeding place for the coals would be a close-fitting door on each side the mouths of the hoppers, if I may so call them, opening into the fire. By this mode the gas would be distilled and burn in passing through the fire. The application of the poker from time to time would remove the coke from the mouths of the hoppers to the middle of the grate.

I do not conceive that it would be difficult to feed the fires of marine engines by means of hoppers placed in a similar manner. The coals might fall down pipes in the form of a segmental curve, to arrest their too rapid descent, and a stoking rod passed through them occasionally would aid them when required, without any current of cold air getting in to chill the furnace and lower the steam; or a piston might be applied to force them on. These coals would have their gas distilled and burned previous to reaching the furnace bars, and a stoking rod introduced through a small hole in the furnace door would be sufficient to thrust the coke forward or draw it back. By this process the stokers would be saved from the effects of the radiant heat. The idea is a crude one, but if it incites any one to work out the principle in practice, it may become of value. The hoppers might be surrounded by water to prevent their burning.

In your last week's Number Mr. Baddeley states that it is a recognised philosophic principle that life cannot be supported by any nutriment but living matter, or that which has been living matter. To this I must still rejoin, that philosophy is not yet a perfect science. We do not, I believe, convert actual life to the purposes of food, except in the two cases of school-boys swallowing live frogs, and grown people swallowing live oysters, if indeed, in the latter case, the life of the oyster be not extinguished by forcing open. Living beings, whether animals or vegetables, are but chemical compositions with life superadded, and if the life be taken away, they still remain chemical compositions, till decomposition takes place. The marble rock is but a congeries of living beings or their exuviæ, and the milk which nourishes the human infant carries in its current portions of the marble rock, without which bones would not be. The coal which feeds our fires is but the exuviæ of forests, and they again are but the representatives of the sun's rays. It is a known fact that the gases contribute largely to the increase as well as diminution of human bodies, according as they be wholesome or unwholesome. Witness butchers, and butchers' wives, and sons and daughters. It is said that the gases of meat promote their fulness and health. Yet, surely, these gases are not living matter. But to return to the opinion of the philosophers; there is a living philosopher, Humboldt, who has recorded that certain tribes of Indians, dwelling on the banks of the Onnow, do, during certain periods of the year, subsist on a white earth which they dig up. It is not to the purpose to say that they live poorly on it; they do live, and the fact is substantiated that man may subsist on inorganic matter. To say that organized matter is better for him is merely to assert that man's skill is yet imperfect. The truth is, that all nature is a series of gradations. We scarcely know where the stone ends and the vegetable begins, or where the vegetable ends and the animal begins. The mushroom furnishes a food strongly resembling animal matter, and so does the olive. There is nutriment in heat, or why does warm food please us and satisfy us better than cold? Why do we eat less in warm weather than in cold? I may be told that heat is merely an assimilater. Well, then, the heat assimilates and becomes part and parcel of the nutriment. All the ingredients of which the human body is composed are to be found in inorganic matter. Our chemistry is at fault if we cannot use the ingredients, whatever source they may be derived from.

I remain, Sir, yours, &c.

JUNIUS REDIVIVUS.

Nov. 6, 1838.

REMARKS ON HALL'S, BELL'S, IVISON'S, AND OTHER PLANS FOR CONSUMING SMOKE AND SAVING FUEL IN STEAM-ENGINE FURNACES.

Sir,—The many contradictory statements that have been circulated respecting several recent projects for economising fuel in steam-engines, induces me to forward to you a few observations, which you may perhaps deem worthy of insertion in your Magazine.

Bell's patent has for its object the consumption of smoke and economy of fuel, by supplying heated air to the furnace, instead of air at the ordinary temperature of the atmosphere. An iron pipe is placed in the furnace, through which the air for maintaining the combustion of the fuel passes. The air is thus raised to a high temperature, say 600° or 800°, and whilst at this high temperature it is passed through flues in the boiler, by which its temperature is diminished to nearly the boiling point of water, when it enters the ash-pit, and passing up through the bars maintains the combustion of the fuel. Now, smoke, it may be mentioned, is the product of imperfect combustion, caused either by the want of oxygen or the want of heat. If air be admitted to a flame in deficient quantity, smoke is produced, because the hydrogen of the burning body will monopolize the oxygen, and the carbon, having no oxygen wherewith to combine, will be thrown off unconsumed, in the form of soot. If the temperature of the flame be artificially lowered, then there will cease to be a sufficiency of heat for the ignition of the carbon, which will therefore be again thrown off in the form of soot. It is from this cause that a long wick causes a candle to smoke; the radiation of heat from the wick reduces so much the temperature of the flame that after the volatilization of the wax, or tallow, there is too little heat left to ignite the carbon.

The combustion of smoke, by the admission of an extra supply of air into the furnace, holds out specious promises of success; but, although numerous modifications of the principle have been tried, none that I am aware of has been entirely successful. The most common arrangement is to have air-valves at the bridge, through which a sufficiency of air is admitted to effect the combustion of the smoke; but the difficulty of adjusting these valves to suit all the varying conditions of the fire is such, that it has been found altogether unmanageable in practice. Mr. Bell supports the combustion of the fuel in his furnace by heated instead of cold air, but many of his predecessors have done the same thing, and some of them have done it in a much better way. The admission of heated air below the furnace bars *will not* effect the combustion of the smoke, because, before it reaches the flame by which the smoke is produced, all its oxygen has been appropriated by the lower or coaky stratum of the fire. To remedy this evil Mr. Bell admits air *above* the bars as well as beneath them, by which means hot air is introduced directly to the flame, the spot where its presence is required. This is an ingenious and judicious arrangement, but I suspect it is one to which Mr. Bell has no legitimate title. In a patent, sealed by Mr. Hall, of Basford, in 1836, this very same principle is clearly developed, and forms, indeed, the main feature of the patent. If smoke be caused either by the want of oxygen, or the want of heat, as this is the only principle that can supply both deficiencies—it is the only principle by which smoke can be expected to be efficiently and advantageously consumed—it has the advantage, too, of extreme simplicity. Indeed, it appears to possess every desirable quality, and it can hardly be doubted, that with fair play, and with the advantage of the improvements in the details of the plan which experience always suggests, it will become eminently successful.[*]

Ivison's patent is for the admission of a jet of steam into the furnace above

* It is extraordinary that this patent of Mr. Hall's has not been more generally adopted, but the probability is that it is but little known. It certainly is deserving of greater publicity and more vigorous prosecution, else it will be caught up and appropriated by others, who will then reap the advantage. And in all matters where the judgment of the public is appealed to, it is a great point gained to have the start of all competitors; it is important that the attention of the public should be arrested, when it is unoccupied with the contemplation of any similar project. Bell's and Ivison's plans may already be said to have died a natural death; the public curiosity is sated—the public interest is extinguished. Now, therefore, is Mr. Hall's time for pushing his inventions into notice, before some new aspirant for public attention steps in before him, throws his plans into the shade, and perhaps by the production of some passable, though less perfect, plan, renders Mr. Hall's inventions unnecessary, or even unimportant.

the bars. Steam is composed of oxygen and hydrogen, and it is supposed that the steam is decomposed, the oxygen combining with the uncombined carbon of the smoke, and forming carbonic oxide, setting hydrogen free. It is not improbable that such a decomposition really takes place; but, even although it does, what is the advantage? The hydrogen and carbonic oxide immediately on their formation will mix with the carbonic acid, resulting from the combustion of the coal, and form with it an incombustible mixture. The smoke, therefore, is not in reality burned, but is merely rendered invisible, and therefore no superior heating efficacy can be expected to result from the process. But even supposing that the gases produced by the combination of the steam with the smoke were burned, in what way could the supply of steam be made always proportionate to the ever-varying production of smoke, so that there might neither be excess nor deficiency? There will generally be either too much or too little steam admitted, which will occasion either a production of smoke or a waste of fuel.

Taken in the gross Ivison's plan does not occasion a saving of fuel, but rather an increased consumption. A tenth of the steam produced by the boiler is injected into the furnace, which of course occasions a direct loss of one-tenth of the fuel, without economy or conpensation. The fierce draught, too, caused by the steam, is very severe upon the furnace bars. The plan has been now applied to many furnaces, but in no single instance could I learn that there had been a saving in fuel, and almost in every instance the furnace bars were speedily destroyed.

But whatever be the merits or defects of Mr. Ivison's patent, he can hardly be held deserving of praise for the one, or of censure for the other.

The plan of admitting steam into a furnace as a remedy for smoke was not originated by Ivison; it has been long known and often practised, but it has as often been thrown aside, as effecting no economy of fuel, but rather an increased consumption, was found to result from the process. In 1827 the Messrs. Nasmyth, two ingenious young engineers of Edinburgh, applied steam as a preventive for smoke, in the Edinburgh gas works, and their object, which was merely the prevention of smoke, they effectually attained, but there was no saving of fuel. In 1834 the effect of a jet of steam, admitted above the bars in the manner practised by Mr. Ivison, was tried in the furnaces of a steam vessel, on the Clyde, but no benefit was found to result from its employment.

Mr. Williams, of Liverpool, has recently taken out a patent for using compressed peat, or peat saturated with coal tar, instead of coal as fuel. But wherein resides the novelty of the advantage of this process? Was the compression of peat originated by Mr. Williams, or does he imagine that his manufactured fuel can be procured as cheaply as peat, which is already compressed and saturated with bitumen in the great laboratory of nature? Mr. Kingston, of Woolwich, has also proposed the manufacture of an artificial fuel, by mixing coal dust and coal tar. I am at a loss to divine in what respect either this or the preceding kind of fuel can be superior to coal. They are both, indeed, artificial coal, produced with more trouble and greater expense than common coal, and they do not possess a single advantage that I can discover to entitle them to a preference.

I may sum up what has been advanced in a few words. Of the various schemes for saving fuel that have of late been so much agitated, there is not one that is free from many serious objections, nor is there one (if we except Hall's) in which the endeavours of the inventor seem to have been pointed in the right direction. The prevention of smoke is the necessary precursor of the economization of fuel, but economy of fuel is not the necessary sequence of the prevention of smoke. The want of an efficient means of achieving both these objects is one that has yet to be supplied, the public expectation has yet to be fulfilled; and he who appears first in the field, and secures the precedency of his compeers; he who avails himself of the flood-tide of expectation, which is now at its highest, will assuredly be led on to fame and fortune.

I am, Sir,
Your obedient servant,
WILLIAM FIELD HENDERSON.
Dundee, Nov. 12, 1838.

VARIOUS MODES OF BOOKBINDING.

Bookbinding is the art of sewing together the sheets of a book, and securing them with a back and side boards. Binding is distinguished from stitching, which is merely sewing the leaves without bands or backs; and from half-binding, which consists in securing the back only with leather, the pasteboard sides being covered with blue or marble paper; whereas in binding, both the back and the sides are covered with leather.

Bookbinding, according to the present mode, is performed in the following manner:—The sheets are first folded into a certain number of leaves, according to the form in which the book is to appear; viz. two leaves for folios, four for quartos, eight for octavos, twelve for duodecimos, &c. This is done with a slip of ivory or boxwood, called a folding stick; and in the arrangement of the sheets the workmen are directed by the catch-words and signatures at the bottom of the pages. When the leaves are thus folded and arranged in proper order, they are usually beaten upon a stone with a heavy hammer, to make them solid and smooth, and are then condensed in a press. After this preparation they are sewed in a sewing press, upon cords or packthreads called bands, which are kept at a proper distance from each other, by drawing a thread through the middle of each sheet, and turning it round each band, beginning with the first and proceeding to the last. The number of bands is generally six for folios, and five for quartos, or any smaller size. The backs are now glued, and the ends of the bands are opened, and scraped with a knife, that they may be more conveniently fixed to the pasteboard sides; after which the back is turned with a hammer, the book bing fixed in a press between boards, called backing-boards, in order to make a groove for admitting the pasteboard sides. When these sides are applied holes are made in them for drawing the bands through, the superfluous ends are cut off, and the parts are hammered smooth. The book is next pressed for cutting; which is done by a particular machine called the plough, to which is attached a knife. See the figures and descriptions *infra*. It is then put into a press called the cutting press, betwixt two boards, one of which lies even with the press, for the knife to run upon; and the other above for the knife to cut against. After this the pasteboards are cut square with a pair of iron shears; and last of all, the colours are sprinkled on the edges of the leaves, with a brush made of hog's bristles; the brush being held in the one hand, and the hair moved with the other.

Different kinds of binding are distinguished

by different names, such as law binding, marble binding, French binding, Dutch binding, &c. In Dutch binding, the backs are vellum. In French binding a slip of parchment is applied over the back between each band, and the ends are pasted upon the inside of each pasteboard. This indorsing, as it is called, is peculiar to the French binders; who are enjoined, by special *ordonnance*, to back their books with parchment. The parchment is applied in the press, after the back has been grated to make the paste hold. The Italians still bind in a coarse thick paper, and this they call binding *alla rustica*. It is extremely inconvenient, as it is liable to wear without particular care.

A patent was obtained in 1799 by Messrs. John and Joseph Williams, stationers in London, for an improved method of binding books of every description. The improvement consists of a back, in any curved form, turned a little at the edges; and made of iron, steel, copper, brass, tin, or of ivory, bone, wood, vellum, or, in short, any material of sufficient firmness. This back is put on the book before it is bound, so as just to cover without pressing the edges; and the advantage of it is that it prevents the book, when opened, from spreading on either side, and causes it to rise in any part to nearly a level surface. In this method of binding the sheets are prepared in the usual manner, then sewed on vellum slips, glued, cut, clothed, and boarded, or half boarded; the firm back is then fastened to the sides by vellum drawn through holes, or secured by inclosing it in vellum or ferret wrappers, or other materials pasted down upon the boards, or drawn through them.

A patent was likewise obtained in 1800 by Mr. Ebenezer Palmer, a London stationer, for an improved way of binding books, particularly merchants' account-books. This improvement has been described as follows :—Let several small bars of metal be provided about the thickness of a shilling or more, according to the size and thickness of the book; the length of each bar being from half an inch to several inches, in proportion to the strength required in the back of the book. At each end of every bar let a pivot be made of different lengths, to correspond to the thickness of two links which they are to receive. Each link must be made in an oval form, and contain two holes proportioned to the size of the pivots, these links to be the same metal as the hinge, and each of them nearly equal in length to the width of two bars. The links are then to be riveted on the pivots, each pivot receiving two of them, and thus holding the hinge together, on the principle of a link-chain or hinge. There must be two holes or more of different sizes, as may be required, on each bar of the hinge or chain; by means of these holes each section of the book is strongly fastened to the hinge which operates with the back of the book, when bound, in such a manner as to make the different sections parallel with each other, and thus admit writing without inconvenience on the ruled lines, close to the back.

The leather used in covering books is prepared and applied as follows :—Being first moistened in water, it is cut to the size of the book, and the thickness of the edge is paired off on a marble stone. It is next smeared over with paste made of wheat flour, stretched over the pasteboard on the outside, and doubled over the edges within. The book is then corded, that is, bound firmly betwixt two boards, to make the cover stick strongly to the pasteboard and the back; on the exact performance of which the neatness of the book in a great measure depends. The back is then warmed at the fire to soften the glue, and the leather is rubbed down with a bodkin or folding stick, to set and fix it close to the back of the book. It is now set to dry, and when dry the boards are removed; the book is then washed or sprinkled over with a little paste and water, the edges and squares blacked with ink, and then sprinkled fine with a brush, by striking it against the hand or a stick; or with large spots, by being mixed with solution of green vitrol, which is called marbling. Two blank leaves are then pasted down to the cover, and the leaves, when dry, are burnished in the press, and the cover rolled on the edges. The cover is now glazed twice with the white of an egg, filleted, and last of all polished, by passing a hot iron over the glazed colour.

The employment in bookbinding of a rolling press for smoothing and condensing the leaves, instead of the hammering which books have usually received, is an improvement introduced several years ago into the trade by Mr. W. Burn. His press consists of two iron cylinders about a foot in diameter, adjustable in the usual way, by means of a screw, and put in motion by the power of one man, or of two if need be, applied to one or two winch-handles. In front of the press sits a boy who gathers the sheets into packets, by placing two, three, or four upon a piece of tin plate of the same size, and covering them with another piece of tin plate, and thus proceeding by alternating tin plates and bundles of sheets till a sufficient quantity have been put together, which will depend on the stiffness and thickness of the paper. The packet is then passed between the rollers, and received by the man who turns the winch, and who has time to lay the sheets on one side, and to hand over the tin plates by the time that the boy has prepared a second packet. A minion bible may be passed through the press in one minute, whereas the time necessary to beat it would be twenty minutes. It is not, however, merely a saving of time that is gained by the use of the rolling press; the paper is made smoother than it would have been by beating, and the compression is so much greater, that a rolled book will be reduced to about five-sixths of the thickness of the same book if beaten. A shelf, therefore, that will hold fifty books bound in the usual way would hold nearly sixty of those bound in this manner, a circumstance of no small importance, when it is considered how large a space even a moderate library occupies, and that book cases are an expensive article of furniture, the rolling press is now substituted for the hammer by several considerable bookbinders.

One of the greatest improvements ever made in bookbinding is, apparently, that for which Mr. William Hancock has very recently obtained a patent. After folding the sheets in double leaves, he places them vertically, with the edges forming the back of the book downwards in a concave mould, of such rounded or semi-cylindrical shape as the back of the book is intended to have. The mould for this purpose consists of two parallel upright boards, set apart upon a cradle frame, each having a portion or portions cut out vertically, somewhat deeper than the breadth of the book, but of a width nearly equal to its thickness before it is pressed. One of these upright boards may be slidden nearer to or farther from its fellow, by means of a guide bar, attached to the sole of the cradle. Thus the distance between the concave bed of the two vertical slots in which the book rests, may be varied according to the length of the leaves. In all cases about one-fourth of the length of the book at each end projects beyond the board, so that one half rests between the two boards. Two or three packthreads are now bound round the leaves thus arranged, from top to bottom of the page in different lines, in order to preserve the form given to the back of the mould in which it lay. The book is next subjected to the action of the press. The back, which is left projecting *very slightly* in front, is then smeared carefully by the fingers with a solution of caoutchouc, whereby each paper-edge receives a small portion of the cement. In a few hours it is sufficiently dry to take another coat of a somewhat stronger caout-

chouc solution. In forty-eight hours, four applications of the caoutchouc may be made and dried. The back and the adjoining part of the sides are next covered with the usual band or fillet of cloth, glued on with caoutchouc; after which the book is ready to have the boards attached, and to be covered with leather or parchment as may be desired.

We thus see that Mr. Hancock dispenses entirely with the operations of stitching, sewing, sawing-in, hammering the back, or the use of paste and glue. Instead of leaves attached by thread stitches at two or three points, we have them agglutinated securely along their whole length. Books bound in this way open so perfectly flat upon a table without strain or resilience, that they are equally comfortable to the student, the musician, and ther merchant. The caoutchouc cement moreover being repulsive to insects, and not effected by humidity, gives this mode of binding a great superiority over the old method with paste or glue, which attracted the ravages of the moth, and in damp situations allowed the book to fall to pieces. For engravings, atlasses, and ledgers, this binding is admirably adapted, because it allows the pages to be displayed most freely, without the risk of dislocating the volume; but for security, three or four stitches should be made. The leaves of music-books bound with caoutchouc, when turned over, lie flat at their whole extent, as if in loose sheets, and do not torment the musician like the leaves of the ordinary books, which are so ready to spring back again. Manuscripts and collections of letters which happen to have little or no margin left at the back for stitching them by, may be bound by Mr. Hancock's plan without the least eucroachment upon the writing. The thickest ledgers thus bound, open as easily as paper in quire, and may be written on up to the innermost margin of the book without the least inconvenience.

Having inspected various specimens of Mr. Hancock's workmanship, I willingly bear testimony to the truth of the preceding statement.

Cloth-binding.—Nothing places in so striking a point of view the superior taste, judgment, and resources of London tradesmen over those of the rest of the world, than the extensive substitution which they have recently made of embosed silks and calicoes for leather in the binding of books. In old libraries, cloth-covered boards indeed may occasionally be seen, but they have the meanest aspect, and are no more to be compared with our modern cloth-binding, than the *jupon* of a trull, with the ballet dress of Taglioni. The silk or calico may be dyed of any shade which use or fancy may require, impressed with gold or silver foil in every form, and variegated by ornaments in relief, copied from the most beautiful productions in nature. This new style of binding is distinguished not more for its durability, elegance, and variety, than for the economy and dispatch with which it ushers the offspring of intellect into the world. For example, should a house eminent in this line, such as that of Westleys, Friar-street, Doctors'-Commons, receive 5000 volumes from Messrs. Longman & Co. upon Monday morning, they can have them all ready for publication, within the incredibly short period of two days; being far sooner than they could have rudely boarded them upon the former plan. The reduction of price is not the least advantage incident to the new method, amounting to fully 50 per cent. upon that with leather.

The dyed cloth being cut by a pattern to the size suited to the volume, is passed rapidly through a roller press, between engraved cylinders of hard steel, whereby it receives at once the impress characteristic of the back, and the sides, along with embossed designs over the surface in sharp relief. The cover thus rapidly fashioned, is as rapidly applied by paste to the stitched and pressed volume ; no time being lost in mutual adjustments ; since the steel rollers turn of the former, of a shape precisely adapted to the latter. Hard glazed and varnished calico is moreover much less an object of depredation to moths, and other insects, than ordinary leather has been found to be.—*Dr. Ure's Dict. of Arts, &c.*

NOTES AND NOTICES.

The *"Liverpool" American Steamer.* — Cork, Nov. 5.—As to the cause of the great consumption of fuel, I have but to confirm what was said before. The fault is in some of the flues or bridges,—I am not engineer enough to describe it technically ; but no one denies that, *after* the little pretence of an experiment to Dublin and back, and before starting for New York, an alteration was made in some of these avenues, by the removal of bricks or otherwise, to which, at least, an additional consumption of 700lb. the hour is immediately to be traced. This was unknown, it is said, to the company ; perhaps even to the agent. No matter. It is not unknown now ; at least it is not here denied. But here lies the fault ; the ship was got off in too great a hurry. It was inevitable, in such a flurry, to be absurdly *punctual* to a day, (which is the only apology I hear of,) that some deficiencies should occur. I hope it may prove a lesson to all candidates for Transatlantic navigation in future. At all events, the cause itself must not suffer on account of such a proceeding as this. You may feel some interest in knowing, what I hear from the best authority, that this company are having a depot of coals established at Fayal, for

the greater security or comfort of their winter navigation. This, no doubt, may sometimes be a convenience, though not one, I hope, necessary to be relied on, as that island, I believe, is at least 300 miles out of the regular course, to the south.— P.S. *Monday afternoon.*—Just as I expected. The *Liverpool* has come into Cove from an "experimental trip,"—experimental on the re-alterations made here, which, of course, should have been made and tried at Liverpool. The result is "highly satisfactory;" that is, the boat has made 185 miles in twenty-four hours, with a high wind all the way, and a head-wind part of it; and this she has done with a consumption of *thirty tons and a fraction.* On the strength of this proceeding, such as it is, we shall leave port again early to-morrow. Meanwhile, it is announced nearly, if not quite 700 tons of fuel will be on board, with which we have reason to be satisfied, especially as we are already a day or two on our way. Under these circumstances, and with a good boat,—being only seven inches deeper than before,—it will be strange if we cannot accomplish the voyage. We hope to be in New York in eighteen days at the farthest.—*Correspondent of Athenæum.*

The Diagraphe.—The great invention of M. Gavard, the Diagraphe, perhaps the most ingenious instrument of the day, and the effect of which, by a well-combined system of pullies, levers, and powerful lenses, is to make the point of the draftsman's pencil follow the motion of the pupil of his eye, has been applied to the illustration of this volume (*Notices sur l'Hotel de Cluny, et sur le Palais des Thermes.* By M. Du Sommerard) with admirable success. The most delicate chasings of the works of Benvenuto Cellini, the finest carvings of a series of ebony or oaken cabinets, such as we believe are not to be met with in any other collection, could evidently have been only imperfectly represented without the aid of an instrument like this. As it is, however, they are given with the most mathematical exactness, and, *therefore,* are exceedingly bold and beautiful.—*Foreign Quarterly Review.*

Polar Expeditions.—The favourite objection to Polar discovery has been, that, if we were to round the northern limit of America to-morrow, it might be impassable the day after, and for ever ; and that under no circumstances can it form a regular track for the merchant ships of England. Yet, how can we be assured that such a conjecture is true ? Who is to limit the inventions of man, or the faculties of Nature? Steam has already defied time and tide in the temperate zone ; why may not some invention, as applicable to the ice as steam is to the wave, sweep vessels across the Pole, with as much ease as boilers and paddle-wheels now carry them across the Equator? Why may not the time come when the minerology of the Pole will both attract commerce and furnish the means of reaching itself by Thames wherries? Why may not some mineral be discovered as attractive as the magnet, or as explosive as pulvis fulminans, and manageable as steam, with ten times its power? If any chemist will supply potassium for the experiment, we might go to the Pole burning or boiling the ice every league of the way before our keel. The chemists have already given us concentrated cold, why may we not expect them to discover concentrated fire? If they can give us at this moment as much of the essence of frost in a wine-glass as would congeal a hogshead of claret, a fact which is every day displayed at every itinerant lecture, is fire to be so much more unmanageable, that we can never condense enough of it to thaw an iceberg? The thing will be done yet, and we shall have parties of pleasure to the Pole ; Hudson's Bay itself but a larger and more crooked Serpentine river, and the Arctic Circle but a half-way house. With this chance before us, who is to despise discoveries, because they have not turnpikes and toll-houses already at every half mile?—*New Monthly Magazine.*

Indelible Writing.—The French Consul in Dublin, (says an Irish Paper,) has offered 36,000 francs (£1440) for paper on which writing cannot be obliterated.

Temperance Yeast.—Sir,—In your last Number I observe two receipts which may be very useful. Some time ago I heard of *dry yeast*, which I understood was imported from Germany or Holland. I should be glad to know the manner of preparing it, or where it can be purchased.
Nov. 10, 1838.   C. C. C. C.

Steam-Boats.—The following letter has been received by the Lord Provost of Glasgow from the Honourable Fox Maule, Under Secretary of State:—" Whitehall, Nov. 2, 1838. My Lord—I am directed by Lord John Russell to acknowledge the receipt of your lordship's letter of the 26th of October, forwarding a memorial from the parliamentary trustees for improving the navigation of the river Clyde, and enlarging the harbour of Glasgow; and I am to inform your lordship that the recent accidents to steam-boats have been viewed with much concern by Lord John Russell, and he will carefully consider what measures are best calculated to prevent a recurrence of such calamities. I have the honour to be, my lord, your lordship's obedient servant, F. Maule.—The Lord Provost of Glasgow.'

Huskisson's Monument.—During the frost of last winter the marble tablet erected at Parkside, to the memory of the late Mr. Huskisson, was destroyed by the bank against which it was placed pressing too heavily against it. A new monument has recently been substituted for the old one, at the expense of the shareholders of the Liverpool and Manchester Railway Company.

Railways in Denmark.—Denmark is of all countries the most unfortunately situated with regard to railways. In the mainland of Jutland there is not sufficient population and capital to justify the establishment of one—in the rich island of Seland the distances are so small, and water carriage so easily procured, that there again the new discovery is of no use. To crown all, there is one line in the Danish dominions the execution of which would be advantageous to commerce—that on the road from Hamburgh to Lubeck, which would connect the German Ocean and the Baltic Sea; but this line, if executed, would destroy the commerce of the Sound, and place Copenhagen up a sort of "no thoroughfare," instead of in the highway of nations. Denmark, therefore, is one of the few countries in which, amid the universal uproar around, a dead silence, interrupted only by an occasional whisper, prevails on the engrossing subject.

The Austrian "Lloyds."—England is not the only country in which speculation is sometimes carried a little too far. Among others the famous establishment called, in imitation of our own, the Austrian "Lloyds" appears to have launched out too freely. At the last settlement of accounts a deficit of 200,000 florins was found to exist, the shares which were lately quoted at 110, are in consequence now at 75, and the scale of the establishment is about to be reduced, and some of its undertakings given up altogether. Throughout Germany in general the railroad, and other schemes, appear to be at present under a cloud, and in France the case is almost the same. We are more sorry than surprised at this.

Many of the railroads proposed in Germany were, from towns of little consequence to towns of none, and where the capital was to come from in the first instance, or how it was to be repaid in the last, was equally beyond the imagination of everybody but the projectors. The present check will only have the effect of killing off a few schemes which ought never to have been born; those that are sound at core will weather the blight.

Substitute for the Sun.—The newly-invented light of M. Gaudin, on which experiments were recently made at Paris, is an improved modification of the well-known invention of Lieutenant Drummond. While Drummond pours a stream of oxygen gas, through spirits of wine, upon unslaked lime, Gaudin makes use of a more etherial kind of oxygen, which he conducts through burning essence of turpentine. The Drummond light is fifteen hundred times stronger than that of burning gas; the Gaudin-light is, we are assured by the inventor, as strong as that of the sun, or thirty thousand times stronger than gas, and of course ten times more so than the Drummond. The method by which M. Gaudin proposes to turn the new invention to use is singularly striking. He proposes to erect in the island of the Pont Neuf, in the middle of the Seine and centre of Paris, a light-house, five hundred feet high, in which is to be placed a light from a hundred thousand to a million gas pipes strong, the power to be varied as the nights are light or dark. Paris will thus enjoy a sort of perpetual day; and as soon as the sun of the heavens has set, the sun of the Pont Neuf will rise.

Russia and America.—Mr. Gerstner, the great German railroad engineer, is now on his way to Bristol, to embark in the Great Western for the great western world, to study the railway system in the United States. He declares his intention to be not only to investigate the mechanical improvements introduced in America, but to study the method of management adopted, by which more satisfactory financial results have been arrived at than in similar concerns in Europe. Mr. Gerstner is the engineer of the Ysarskoe-selo railway. Had he performed this journey before he executed that work, which is said to present financial results the reverse of satisfactory, he would probably have learned that one of the first elements of success in a railway is to choose a tolerable terminus.

Bridge at Pesth.—The Austrian government is said to be unwilling to ratify the engagement of the deputation of the Hungarian diet with the great banking-house of Sina, respecting the permanent bridge to be erected over the Danube, at Pesth, considering the conditions as favourable beyond what the deputation was authorised to grant. A strong party against the new bridge is said to exist at Pesth. We hope the rumours of these obstacles may turn out unfounded, not only on account of the cause of general improvement, but because the employment of English engineers and architects, for the erection of a bridge in that distant land, would be an honourable, and we think a merited, tribute to the talents of our countrymen.
The Railway Map of England and Wales continues on sale, in a neat wrapper, price 6d.; and on fine paper, coloured, price 1s.

☞ *British and Foreign Patents taken out with economy and despatch; Specifications, Disclaimers, and Amendments, prepared or revised: Caveats entered; and generally every Branch of Patent Business promptly transacted. A complete list of patents from the earliest period (15 Car. II. 1675,) to the present time may be examined. Fee 2s. 6d.; Clients, gratis.*

LONDON: Printed and Published for the Proprietor, by W. A. Robertson, at the Mechanics' Magazine Office, No. 6, Peterborough-court, between 135 and 136, Fleet-street.—Sold by A. & W. Galignani, Rue Vivienne, Paris.

Mechanics' Magazine,

MUSEUM, REGISTER, JOURNAL, AND GAZETTE.

| No. 798.] | SATURDAY, NOVEMBER 24, 1838. | [Price 3d. |

PENNY'S IMPROVED CANNON.

Fig. 1.

Fig. 2.

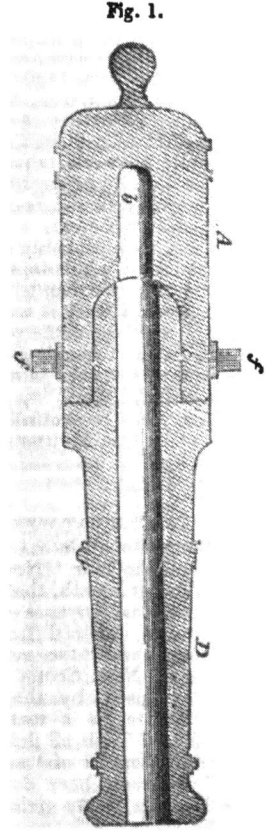
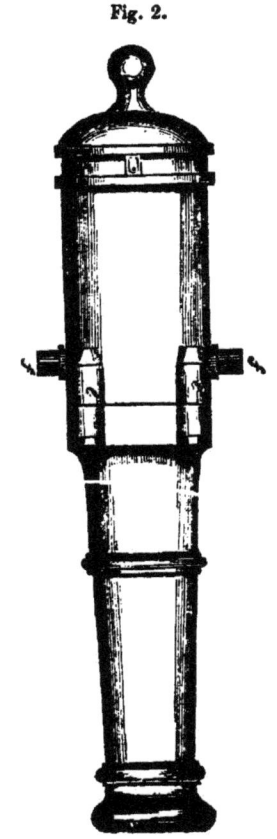

PENNY'S IMPROVED CANNON.

Sir,—It has been asserted by a benevolent writer, with much apparent reason, that if subjects were wise, war is a game that kings could never play at. Now, although knowledge is rapidly extending its civilizing influences throughout the known world, but few of us, I fear, have yet attained this beatific state of wisdom : we may therefore at least be pardoned, if we occasionally turn our inventive talents towards the improvement of warlike engines—and the more so, when it is a generally-received opinion that the attainment of universal excellence in the art of war is more likely than any other circumstance to be productive of universal peace.

Numberless notices of the progressive march of improvement in this direction have been recorded in your Journal—several by Col. Maceroni, himself a host in such matters—to these I would now beg leave to add a brief description of an ingenious contrivance by Mr. Penny, for increasing the portability, and thereby facilitating the transport of heavy artillery, and also for rendering them more generally useful in the twofold capacity of heavy battering guns and mortars.

Those among your readers who are aware of the extreme jealousy usually evinced by military men towards all suggestions relative to martial matters emanating from civilians, will feel no surprise that our Board of Ordnance have declined availing themselves of the advantages which Mr. Penny's method of construction seems likely to afford : neither have they shown any disposition to prosecute an inquiry into the practical merits of the plan.

Mr. Penny's improvement consists in casting the gun in two separate lengths of nearly equal weight, as shown in the accompanying drawings; fig. 1 being a section, and fig. 2 a plan of one of these cannons. The breach end A used by itself, forms a *mortar*, b is the chamber for the powder, the large cylindrical cavity c, the barrel for shells. When the barrel D is added, a long heavy battering gun is formed. It will be seen in this fig. how the end of the muzzle fits into the barrel of the mortar, making a close joint; the method of securing the joint is by four strong screw-bolts and nuts passing through suitable eyes, cast on either piece for this purpose, as shown at e, fig. 2. f f are the trunnions.

The advantages claimed for this gun by its inventor, are, greater facility in the casting, &c., and a reduced liability to defects in proportion to the reduced size of the two pieces of which it is composed; being in two parts it may be carried upon two seperate carriages, and by this means be conveyed over bad roads, mountains, swamps, rivers, &c., and so brought up to points altogether inaccessible to other heavy guns.

For marine service this species of ordnance would seem to be particularly useful, forming a long gun of great power for a ship's upper deck, with the advantage of being converted into a large sized mortar in a few minutes ; the two parts being connected or detached with great facility. So that after commencing a cannonading, a shower of shells may be thrown into a town, or fortification, and the cannonading resumed, as circumstances may render expedient.

This invention may probably require some little modification to adapt it for useful service, but in the hands of our skilful artillery-men, I fancy it would be found to be a very formidable weapon, and one which most persons would rather meet in the *Mechanics' Magazine*, than the tented field.

I remain, Sir, yours respectfully,
         WM. BADDELEY.

London, Oct. 5, 1838.

FUEL USED IN THE "GREAT WESTERN" STEAM SHIP—THE PICTON COAL.

Sir,—It is stated in your "Notes and Notices" for the last month, that "the fuel used in the last voyages of the *Great Western* was obtained from the mines of Picton," and "the very fact that the coals of Nova Scotia having been successfully used by the *Great Western* steam-vessel is a matter of much importance." With all deference, I submit, that in a matter of "so much importance," you would have done well to refer to the office before giving publicity to such a statement.

A sense of duty to the public renders it necessary that I should call upon you to put the public in possession of the real facts, which I submit with some reluctance, as it may have happened, that the

Picton coal with which the *Great Western* was supplied, had been exposed, or may not have been a good assortment, or may have been too small, or there may be better qualities. If I were even satisfied of one and all of these surmises, silence might be construed into connivance in misstatements.

The *Great Western* took on board on one homeward passage sixty tons, and upon that little quantity the log contains the following remarks :—

Monday, July 2nd. — " Picton coals are very soft, and will not bear the bare pricking, but burn freely."

Thursday, July 5th. —" Picton coals used forward, the consumpton larger than before, and more difficulty in keeping steam up; a great deal of dirt from them, the clinker soft, and not injurious to the bars. Of the four descriptions of coals used this voyage, the Tredegar is the best for our purposes."

I have only to add, that we have never shipped any second lot of Picton coal; that we have, up to the last voyage out, tried nine different sorts of coal; that we have sent out 500 tons of Scotch (Elgin), and 500 tons of Graigola for homeward voyages; that the last cargo homewards was Liverpool coal, with which, at nearly one-third more consumption than usual, we could not keep steam; and that the entire of the cargo outwards upon, I expect to hear, one of the most trying passages, against heavy and constant gales from the westward, which, as far as we can judge, lasted from the 27th October (her day of sailing) until the 10th of this month, was composed of Tredegar, from Messrs. Homfreys' works, are shipped at Newport, which is the best proof I can give of the value we *at present* set upon it.

Your obedient servant,
CHRISTOPHER CLAXTON.

Bristol, Nov. 15, 1838.

ANCIENT FIRE-ENGINE—HAND-SQUIRTS, ETC.

Sir,—I feel myself called upon to make a few observations on the communication of "J. C. W." (at page 8, of the present volume), wherein mention is made of the ancient fire-engine, which figures in Cyprian Lucar's work. I regret that the sketch, as well as the accompanying description, is somewhat defective; the former is "shorn of its fair proportions," being stripped of all the appurtenances and quaint humour, which, in the original, give life and interest to the scene. "J. C. W." mentions this machine as being "*undoubtedly the first fire-engine ever used.*" Now the fact is, there is no proof given of its being used, or even constructed. Lucar himself goes no farther than to state that it has been "*devised;*" an expression of very common use in his time, for describing proposed schemes, and untried inventions, and there is good grounds for assigning the present contrivance to one of these classes. Not only are we destitute of the slightest proof of its having been used, but from the knowledge we really do possess of what was done in this matter, coupled with the impossibility of making a machine like this to answer its intended purpose, fully justifies us in utterly denying its *existence.*

"J. C. W." says, " it appears (as per pamphlet) to consist of *a hollow cone,* moveable on a wooden frame and open at the vertex, into which might be inserted a long pipe, or even *a hose,* to convey water to any part of a building." This fire-engine was "*devised*" consisted of a cylinder of large dimensions, which formed the working barrel, and terminated in a cone furnished with an adjutage. A piston traversed backward and forward within the cylinder, being acted upon by a screw turned by the winch handle (B. page 7). The cylinder was supported within a wooden frame, mounted on a two-wheeled carriage, and was kept at any required elevation, by means of pins and quadrants at the back. The piston being drawn back, water was ladled into the funnel A.: when filled, the stop cock was turned and the piston screwed up, by which means it was supposed the water would be projected with considerable force upon the fire. I may observe *en passant*, that the leather hose was not known till long after Lucar's time,[*] and even if it had, it could not have been advantageously applied to squirting-engines. There are several cogent reasons, why a machine constructed upon the foregoing plan could not be

---

[*] Flexible leather hose was invented by the brothers Jan Van der Heide, and was first tried by them at Amsterdam, in the year 1672.—Vide Beckmann's History of Inventions.

made "effective in operation;" besides which, I fancy the mechanics of Cyprian Lucar's day, would have found some difficulty in boring out a cylinder of the capacity to hold "an hoggeshead of water, or if you will, a greater quantity thereof,"—to say nothing of fitting the piston.

We are informed by Beckmann, that squirting-engines of considerable power, mounted on wheels and worked by levers, were used in some parts of Germany, so early as 1518.

Similar engines were introduced at London a short time previous to 1633, in which year three of them were employed at a conflagration which took place on London-bridge. These appear to have been of small power and very slightly constructed; however, they were at that time thought very highly of, being described as "such excellent things, that nothing that was ever devised could do so much good; yet none of these did prosper, for they were all broken."

The principal fire-extinguishing apparatus of that period, in London, consisted of leather buckets and brass hand-squirts; which continued to be looked upon as valuable auxiliaries for this purpose, for some time subsequent to the great fire of 1666. I enclose a sketch of one of the *hand-squirts* of that period, on a scale of an inch and a half to a foot;

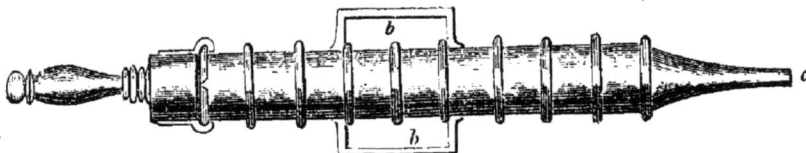

they were made of brass, and when filled, held about two quarts of water; the aperture of the nosel was half an inch in diameter. In using them, a man on each side of the squirt took hold of the handles *b. b.* with one hand, and the nosel with the other, while a third person worked the piston or plunger by the handle *c.* drawing it out while the nosel was immersed in a vessel of water, which filled the cylinder; the bearers then raised the nosel, when the other party pushed in the plunger, the skill of the bearers being employed in directing as much of the water as possible upon the fire. Five of these squirts are still preserved in a rack in the vestry-room of St. Dionis Back-church, in Fenchurch-street, from one of which the accompanying drawing was made.

By an act of common council, passed immediately after the great fire of London, "for preventing and suppressing of fires for the future," each city parish was ordered to be provided with *two* of these brazen hand-squirts; each of the aldermen was to provide his house with *one* ditto; each of the twelve companies was to be furnished with *two*, and the inferior companies were to provide and maintain such a number among them, as the lord mayor and court of aldermen

should direct. So that the number of these hand-squirts originally kept in the city must have been very considerable, but the number at present remaining is exceedingly small.

At the beginning of the seventeenth century it occurred to somebody, that great convenience would arise from fixing one of these squirts in a portable cistern and working the plunger by means of a lever; this idea was carried out, and the valves necessary to ensure its action being supplied, produced what was at that time considered a great mechanical achievement. Most of your readers will, no doubt, recognise in this machine the common tin garden engines, of which a few are still to be met with at some of the agricultural instrument repositories; though these are now almost entirely superseded by the more convenient form of engines, yielding a continuous stream of water; by means of that useful appendage the air-vessel. Syringes are still very extensively employed for horticultural purposes, but they have been made the subjects of very considerable improvements within the last few years.

The squirting-engines projected the water by fits and starts, a cessation of the stream taking place between every stroke of the piston, in consequence of which a great deal of water fell short of the mark and was wasted, and much difficulty was experienced in directing the stream to any required spot. To be at all useful, it was also necessary to place these engines very close to the fire, which exposed the persons working them to imminent danger from the falling of the burning buildings.

Clare, in his " motion of fluids," gives a description of a squirting-engine of the larger class, worked by an arrangement of double levers, which form of engine had just been perfected, when, towards the close of the seventeenth century, Newsham introduced his improved patent fire-engines. These consisted of two working barrels, with an air-vessel for equalizing the stream; which afforded the opportunity of employing leather hose, so as to apply the jet of water with great effect, at a very considerable distance away from the engine. It is a curious but positive fact, that so perfect was Newsham's fire-engine, that at the expiration of upwards of a century, we still find it employed nearly as he left it. Various alterations of convenience, and improvements in the details, of this engine, have been made in the course of that time, but the general character and mode of construction adopted by Richard Newsham have never been surpassed.

I remain, Sir,
Yours truly,
WM. BADDELEY.
London, Nov. 1, 1838.

———

THEORY OF THE SPINNING TOP.
(From the *Liverpool Mercury*.)

Sir,—It is probably not out of your recollection, that, a few months ago, I communicated to you some experiments which I had tried upon the spinning top, in reference to the theory of its motions, and promised, if my communication should be acceptable, to resume the subject, and to state what I conceived to be the true cause of the top's standing. I have now to request your permission to redeem my promise,—a promise which I almost regret, fearing that the subject may be considered too mathematical, and *consequently* too dry for the columns of a newspaper, and that I may not be able to explain my meaning clearly without the aid of a diagram. But, with your permission, Sir, I proceed; and those of your readers who consider the subject uninviting, may easily, in your ample space, find something more congenial to their taste.

In my last letter* I attempted to show the futility of the causes usually assigned for the standing of a top,—first, centrifugal force; and, secondly, Dr. Arnott's supposition, that it is the width of the point. I then stated what are *actually* the motions of the top, and how these may be made to illustrate the movements of the planets,—the rotation, the conical revolution of the axis causing the precession of the equinoxes; the mutation of the axis, and the relative motions of a primary and satellite. I also showed the somewhat remarkable effect produced on the second of these motions by altering the relative positions of the centres of motion and of gravity. I come now to the remaining part of my subject, viz., the cause why a top in motion does not fall. The explanation which you pointed out in the *Mechanics' Magazine*,† I consider not at all satisfactory, since the direction of the diagonal line, upon which the whole rests, is not made to depend either upon the force of gravity or upon the velocity of the top, and, consequently, the top would stand, however slow its motion. The true cause I conceive to be this :—Suppose the top already inclined a little to one side, it will then have a *tendency* to turn over towards that side which is the lowest, in doing which the lowest point of the circumference would of course fall, while the highest point would necessarily rise. But the lowest point, in *beginning* to fall, at the same time is carried forward by the rotary motion, conveying that tendency to fall with it, so that the actual fall takes place at a point in advance of the lowest, which point in its turn becomes the lowest: at the same time the highest point, *beginning* to rise, carries that rise forward by its rotary motion to a point in advance of the highest. Now, before proceeding further, let any one take a top in his hand, stop its motion, hold it inclined to one side, and then with his finger depress a point a little in advance of the lowest, and at the same time raise the side which is immediately in advance of the highest, he will find that this movement tilts the top over (if I may so express it) aside from its former inclination, making the axis now lean towards the side immediately in advance of the formerly lowest point, and, if continued, produces a conical revolution of the axis, and an accompanying slow revolution of the lowest and highest points, corresponding

———

* See Mech. Mag. vol. xxviii, p. 153.
† Ibid. vol. xxviii, p. 279.

exactly, as I said before, to what is termed is astronomy the precession of the equinoxes, except that it is in the *same* direction with the rotation. *Into this conical movement, then,* I contend *the tendency to fall is converted;* but still my demonstration is as yet incomplete, for although I may have shown that the point, which was the lowest at first, will no longer be the lowest, unless I can also show that the *new* lowest point will not be lower than the *old* one, the top *will fall* in spite of this secondary preserving motion. How, then, can it be proved that it will not be so? It can *not* be proved generally; for to do so would be to prove too much; it would be the same error into which the Leamington correspondent (S. W. S.) of the *Mechanics' Magazine* has been led; for a top sometimes *does* fall; but the same theory, a little extended, will show *under what circumstances* it will fall and *under what* it will not. Suppose, for the sake of simplicity, the whole body of the top to be concentrated into one horizontal section, and the whole weight of this section again into one circumference, precisely in the same manner as mathematicians suppose a centre of gravity, a centre of percussion, &c. Then suppose this circle, as before, leaning with its axis to one side; let a parabola be described, touching the lowest point of the circumference with its vertex, and having its ordinates equal to the spaces over which any point in the circumference moves, in its rotation, in given times, and its abscissæ equal to the distances through which the same point would fall, if at rest, in the same times. Also, let distances be set off upon the circumference of the circle equal to the ordinates of the parabola: let the circle, with these points, be orthographically projected, as an ellipse, upon the same vertical plane with the parabola, and, of course, touching it; and let ordinates be drawn from the various points marked upon the circumference, to the conjugate diameter. The abscissæ of the parabola will then show the spaces through which the lowest point would *fall* in the given times, if obeying only the influence of gravity, and the abscissæ of the ellipse will show the spaces through which the same point would *rise*, from the effect of the rotary motion alone, in the same times. If, then, the first abscissæ of the parabola be greater than the first abscissæ of the ellipse, the lowest point will descend still lower, and the top will fall; but if less, the lowest point will become higher, and the top will rise towards an upright position. Now, since the form of the ellipse is constant, while that of the parabola widens or contracts as we increase or diminish the velocity, it is evident that such a

velocity may be given to the top that any abscissæ of the parabola shall become less than the corresponding abscissæ of the ellipse, and that, when this is the case, the top shall stand. or, rather, gradually approach a vertical position.

For the sake of simplicity I have taken the ordinates of definite lengths; but to those who have studied mathematics it will be evident that, in order to be strictly correct, the second point in each curve must be taken in *immediate* succession to the first, making the first ordinate and abscissæ infinitely small. In this case their relative magnitudes *may* be calculated by means of the differential calculus. The following will be found to be the simple result of the whole :— *When the radius of curvature of the ellipse at the lowest point is greater than that of the parabola at its vertex, the top will rise; when less, it will fall.*

This theory will be found equally applicable to the case in which the centre of motion is above the centre of gravity, and will show that the conical motion of the axis must then be backward, precisely as experiment exhibits it.

It occurred to me, that *if the theory were correct* it might be extended further, and that the same effect ought to be produced if the top were in the form of a *ring*, with an attractive power towards the centre substituted for the downward attraction of gravitation; and that, if the velocity were sufficient, it would avoid falling in towards the centre, by an evasive movement similar to that by which, in the other case, it avoids a fall; in short, that it would produce an eccentric revolution of the ring exactly corresponding to what modern refined observations prove to take place with the ring of the planet Saturn. I suggested the probability of this in my former letter, though I then had some doubts of the possibility of proving it experimentally. In that, however, after some difficulty, I have at last succeeded, and have satisfied myself, by ocular demonstration, that the revolution of a ring, with sufficient velocity, round a centre of attraction, gives a *stable* equilibrium. But Laplace has demonstrated mathematically, in his *Mécanique Céleste*, that a *uniform* ring, revolving round a centre of attraction, is in *instable* equilibrium, so that if in the least displaced from its concentric position, the ring would inevitably be brought into collision with the planet or other attracting body; and all other astronomers, I believe, have implicitly received his theory, so that to account for the *actual* stability of Saturn's ring, they have resorted to what I can scarcely avoid regarding as the clumsy expedient of supposing it to be *loaded on one* side, so as to bring it under the law of the usual elliptic planetary movement.—(See *Brewster's Encyclopædia*, vol. 2, p. 646, and Sir J. F. W. Herschel's admirable volume on Astronomy, in *Lardner's Cyclopædia*, art. 444.) Do not be afraid, Sir, that I am going to venture on the arduous task of encountering Laplace's figures, although I have some suspicion of the source of his error (as I suppose it to be). I leave that to more experienced mathematicians ; but an experiment is an experiment, and any one may be excused for believing what he sees with his own eyes.

I have unavoidably trespassed far upon your space, Sir; but, if I have done nothing else, I think I have established this much, at least, that the spinning top may be made something more than a child's plaything.

I am, Sir, respectfully,

J. ELLIOT.

35, Brownlow-street.

REPORT OF THE COMMITTEE OF THE FRANK-LIN INSTITUTE ON THE EXPLOSIONS OF STEAM BOILERS, OF EXPERIMENTS MADE AT THE REQUEST OF THE TREASURY DE-PARTMENT OF THE UNITED STATES.—PART II. CONTAINING THE REPORT OF THE SUB-COMMITTEE TO WHOM WAS REFERRED THE EXAMINATION OF THE STRENGTH OF THE MATERIALS EMPLOYED IN THE CONSTRUCTION OF STEAM BOILERS.

[From the *Franklin Journal*.]

(Continued from page 18.)

Apparatus for High Temperatures.—The general arrangement of the parts of apparatus expressly designed for experiments, at the highest temperatures, is presented in plate I., (See *Mechanics' Magazine*, No. 792, front page,) where F is the portable furnace for charcoal, suspended by an iron ring which is fastened on near its upper edge, and attached by means of pins on the opposite sides, to the two ends of a semi-circular fork on one arm of the lever *x*. The weight of the furnace and its contents is counterpoised by the weight *c*. About the centre of the lever is a slit, 6 or 7 inches long, through which passes the end of a screw suspending-rod. The nut belonging to this screw, is furnished, on its upper surface, with an elevated ridge, serving the purpose of a knife-edge, which applies at pleasure to any one of several transverse notches on the under side of the lever along the slit. This serves, for the time, to fix the lever and to prevent its sliding endwise unless when lifted from its bearing.

As this nut revolves freely on the screw, a horizontal motion about the screw as an axis is readily given to the lever, while, to raise or lower the furnace in order to regulate the temperature, the knife-edge on the nut affords ample facility. The top of the furnace rises between the two guide-rods *s*, *s*, and between the two iron blocks *b'* and *b''* within which the bar to be tried, is confined. When not in immediate use, the furnace is swung round beneath the beam of the frame M, and placed under the cap H, the pipe of which is supported on the end of the crane R, adapted for its reception, and passing into a chimney at *g*. The thermometer *t*, suspended from an arm projecting beyond one of the uprights which support the pulley *p*, is lowered, when in use, into a bath of hot liquid through which the specimen under trial passes, as described below. The details of the arrangement are seen in plate III., where a plan is given of so much of the machine as may be necessary to comprehend the manner of fixing the bars and of applying the heat. In plate IV., a vertical section through the length of the bar, is exhibited ; the references in both these plates, being, so far as they apply to common objects, the same as those in plate I. Thus in plates III. and IV., *i'*, *i'* and *i''*, *i''*. are the straps of iron connected with the blocks *b'* and *b''*, which by means of their projecting arms repose on the two guide-rods *s*, *s*. F is the furnace, *t* the thermometer seen immersed in the bath of hot liquid B, and *x* is the lever supporting the furnace. The bath is composed of an elliptical copper or sheet iron cup, 4½ inches long, 3½ wide, and 4 inches high, with two lips or channels, in the direction of its shorter diameter, each one inch deep, and the same in breadth, to admit the passage of the bar through them, and to contain the packings *w*, *w*, adapted to retain the hot liquid, and cause it to cover the bar *a*. These channels extend each about 1 inch beyond the sides of the cup, affording room for the straps *y*, *y*, passing beneath them and rising on each side, near the tops of which are placed two cross-heads, *c*, *c*, and through these pass the tightening screws *n*, *n*, employed in pressing down the packings *w*, *w*. The manner in which the bars are held by the blocks *b'*, *b''*, is seen at *w*, *w'*, where the dove-tailed form of the holes into which the wedges pass and the arrangement of teeth on the steel face of each wedge, are particularly indicated. In adjusting the bars in their place for these experiments, it became necessary to form perfectly secure joints at *w*, *w'*, where they pass through the channels before mentioned. In most of the experiments below 600 degrees this was affected by means of loosely spun cotton, wrapped about the bar, for an inch at each point where the screws were to be applied. For temperatures above 600° a packing formed of fibres of iron scraped from wire in the manufacture of weavers' reeds was adopted. This being formed into mats, and rolled in

a powder of oxide of tin,—constituted an impervious barrier to the melted metal, particularly after being duly settled and condensed into place, and then firmly compressed by the screws.

Below 600°, the fluid commonly employ-

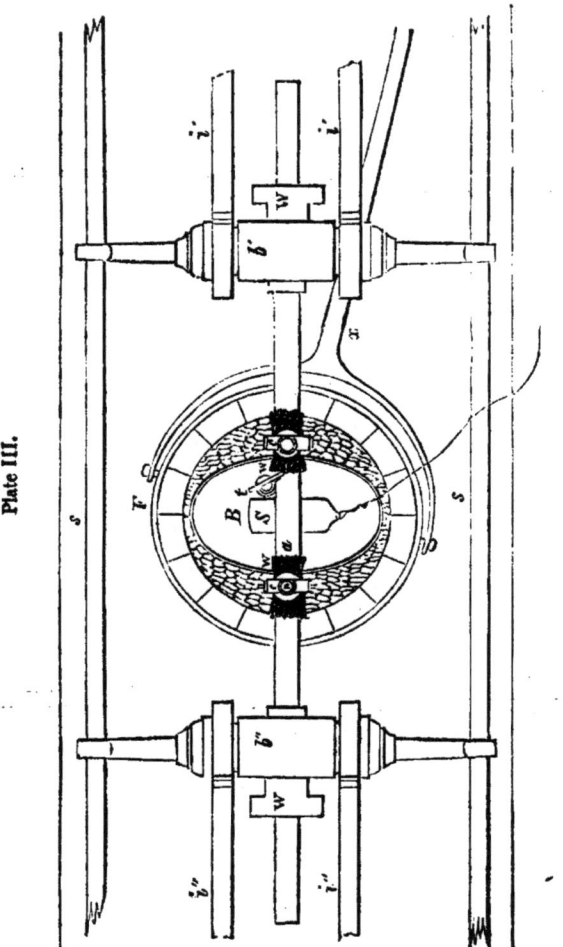

Plate III.

ed was olive oil, and for higher temperatures a mixture of tin and lead. In some cases, however, between the melting point of *tin* and the highest temperatures applied, the latter metal only was used.

The source of heat for moderate temperatures was either a single or a double-wick spirit lamp, of Dr. Michell's form. When the latter proved inadequate to supply the desired temperature, the furnace of charcoal

F was substituted, and, by means of notches *o, o*, at its upper rim, it was raised so high as completely to embrace the bath B as represented in plate IV.

Standard of High Temperatures.—The standard of temperatures below the boiling point of mercury was the mercurical thermometer. Above that point the instrument adopted by the committee, was the steam pyrometer described in the *American Journal of Science*, vol. xxii. page 96, by Professor W. R. Johnson. At S, plates III. and IV., is the standard piece of wrought iron laid horizontally beneath the bar *a*, and kept in its place by the buoyancy of the mixture of tin and lead, the superior density of the latter metal serving to float the iron, and the higher specific heat of the former,

keeping the temperature of the bath more steady than it would have been if lead alone had been employed.

As several improvements have been made in this pyrometer since the date of its first publication, which are conceived to be important in point of accuracy and despatch, it is proper that we should present a view of its structure as used in these experiments, together with a concise statement of the mode in which the latter were conducted.

The instrument is founded on the supposition, that from a mass of water already in ebullition, a weight of vapour may be generated, by immersing a solid, of known weight and capacity for heat, which shall be proportioned to the temperature of such solid above the boiling point of water. In this

Plate IV.

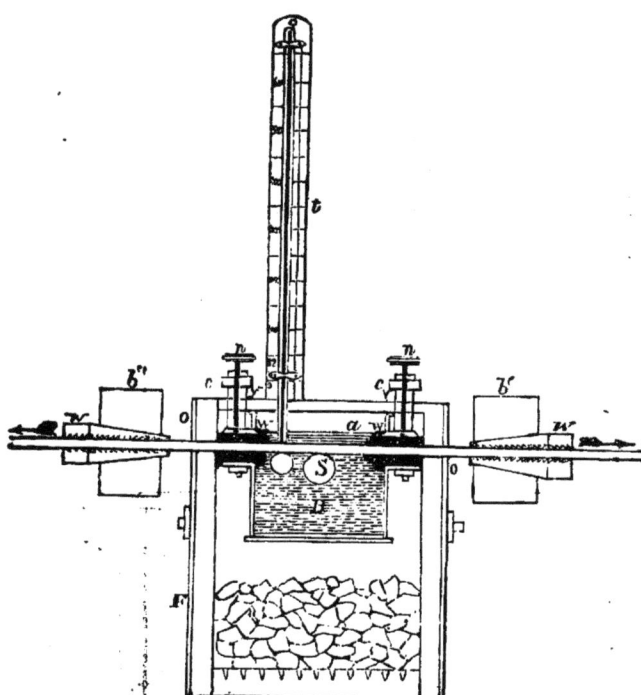

supposition it is, of course, implied that the specific heat of the solid is constant, or that we allow for its variations;—that during the experiment no heat is received by the water from any other source than the hot

solid immersed, and also that vapour ceases to escape, as soon as the solid has been cooled, by vaporization, from the initial temperature, down to that of boiling water. It is further requisite that no heat from the immersed solid be expended in any other way, than the production of vapour by passing from a sensible to a latent state. A convenient apparatus for ascertaining the weight of vapour thus expelled by the hot body, used as a standard piece, is another requisite.

In almost all the usual processes of

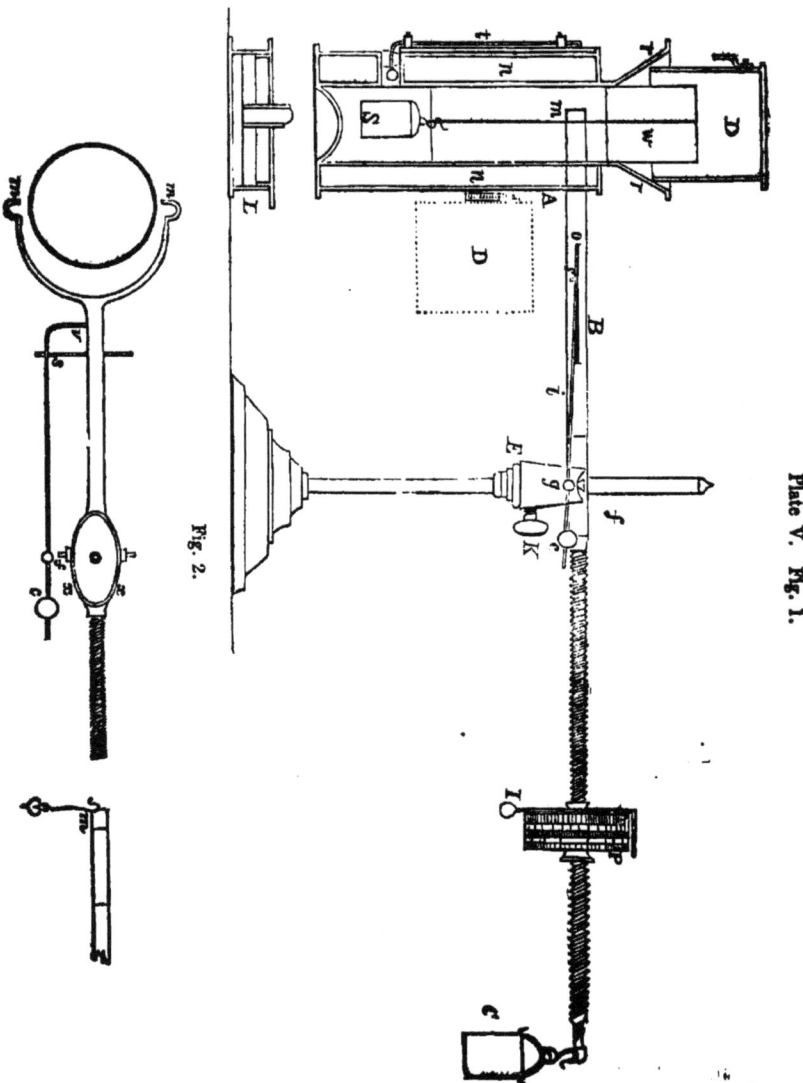

Fig. 2.

Plate V. Fig. 1.

weighing, the time occupied in making an adjustment of the counterpoise, is too considerable to admit the supposition that no loss of vapour from an open vessel of hot water would take place, between the time when the solid attains the boiling point, and that when the equilibrium would be produced. To answer the conditions above indicated, the steam pyrometer is constructed in the form represented in plate V. A, fig. 1, is a cylindrical boiler, 12 inches high, constructed of two concentric cylinders of tinned sheet iron, between which is a stratum n, n, half an inch thick, of dry charcoal-dust, (lamp black,) to serve as a non-conductor, and preserve the water within from loss of heat by radiation during the experiment.

The bottom is formed of a single sheet of the same metal connected with the lower edge of the inner concentric cylinder, and rising in the form of a segment of a sphere to the height of $\frac{3}{4}$ths of an inch, in order to present an enlarged surface to the action of the lamp L, which keeps the water in ebullition until the moment when the solid is immersed. t is a thermometer bent at right angles $\frac{3}{4}$ths of an inch from the bulb, and passing along a tube opening into the boiler. A packing around this part of the stem prevents leakage, and the bulb being wholly immersed in the water, serves at all times to indicate its temperature, and particularly to mark the moment when that of the solid has descended so low as to cease generating steam of *atmospheric tension*.

Mode of ascertaining the weight of vapour expended. The mode of suspending the boiler to the beam of the balance, is seen at m where the dotted prolongation of the beam B forms a forked curve rather greater than a semicircle, each extremity of the arc m, m, (fig. 2.) being turned inward so as to stand at right angles to the direction of the beam ; this brings the two bearing edges which supports short hooks attached to loops on the opposite sides of the boiler, into the same line parallel to the main axis or knife-edge f, at the central part of the beam. These parallels are exactly 12 inches apart. The opposite arm of the balance beam is cylindrical, and cut into a fine threaded screw to within an inch and a half of the fulcrum f, where the beam is divided for a certain space into two portions (x x, fig. 2,) between which passes the upright rod of the supporting stand. The beam used during a considerable part of the experiments contained $150\frac{1}{4}$ threads in 1 foot of the length of the screw, and was $\frac{1}{2}$ an in inch in diameter. At the highest temperatures which the committee had occasion to measure, the number of threads passed over in one experiment, did not exceed $11\frac{1}{4\tau}$, or so much as to measure from 1100 to 1200 degrees of temperature.

By a careful measurement of different numbers of threads, selected at various parts of the screw-arm, (which was 16 inches long) it was ascertained that, though at the two

extremities the threads differed slightly from their mean length, yet at the middle portion, where the revolving counterpoise P is represented in the figure, no appreciable difference could be detected, and as this was the part always occupied by the weight, the instrumental error from this source may be considered altogether unimportant. In fact the extremes of the variation just referred to taken most unfavourably, could not have in the highest temperatures, amounted to more than 7 degrees Fah., which at points so elevated as 13 or 1400 degrees, would not be deemed a very important inaccuracy.

But even this was avoided by occupying that part of the screw where the threads were of equal length ; and by making an adjustment of the balance and weighing any given body placed on the boiler, with the counterpoise in different parts of the range selected for the experiments, it was easy to verify the accuracy of the measurements and determine precisely the error, if any had existed. But this method, when tried, only served to confirm the result of the other.

The revolving counterpoise.—On the screw already described, was placed a revolving counterpoise P, which, together with its index I, placed on a neck projecting beyond the base of the cylinder, weighed 10517 grains ; consequently, as each thread of the screw measured 0.07872 of an inch, the motion of the counterpoise over one thread was equivalent in effect to a weight of 100 grains applied at that end of the beam where the boiler is suspended, or $\frac{1}{100}$ part of a revolution marked a difference of one grain in the weight of water in A.[*]

---

[*] The method used in determining, by calculation, the true adjustment of parts and the graduation of the scale of the steam pyrometer is equally applicable, whatever may be the length of the weighing beam, or of the threads of the screw, and whatever the nature of the material employed for a standard piece, the latent heat of vapour, the kind of liquid from which it is produced, or the scale of thermometer to which we refer the indications, marked on the revolving counterpoise.

Thus, if L be the length of the arm from f to m where the boiler is suspended, and n the number of threads of screw on an equal length of the opposite arm ; then will $\dfrac{L}{n}$ the length of a thread.

Putting P = the weight of the counterpoise, and v = the weight of vapour which must escape in order that an equilibrium, destroyed by its loss, should be restored by making P move one revolution, that is one thread nearer to f, we shall have $\dfrac{L}{n} : L :: v : P$ or $\dfrac{PL}{n} = Lv$; whence $v = \dfrac{P}{n}$. This may be termed the *mechanical relation* of the instrument to the vapour produced.

To determine the *physical action* of the standard piece, if x be supposed = its specific heat ; l = the latent heat of vapour from the liquid at its boiling point ; i = the weight of the standard piece (ex-

The standard piece employed to produce vapour after having been heated in the bath of melted metal surrounding the bar under trial, was formed of wrought iron of the figure seen at S, its greatest length $2\frac{3}{4}$ inches, its diameter one and $\frac{7}{10}$ inch, and its weight 6336 grains Troy. This was sus-suspended by an iron wire $\frac{1}{30}$ of an inch in diameter to the centre of a wire gauze cap w, the lower and smaller end of which entered the mouth of the boiler A, at the base of the funnel r. The upper diameter of this cap is $2\frac{1}{2}$ inches, and its height $2\frac{1}{4}$ inches, giving an area, including the top and sides, and of more than $13\frac{1}{4}$ square inches, or more than three times as much as the section of the boiler at its mouth. The object of employing this cap is to prevent the dashing out of water by ebullition—an effect which is, however, only liable to happen near the close of an experiment when the iron has descended to the temperature of *maximum vaporization*,* and when the boiler contains too much water.

The condenser.—In order to prevent all escape of vapour after ebullition has ceased, a cylindrical cap of tinned iron D, is placed over that of wire gauze, the instant that the

pressed in the same denomination as that of P) ; l— the number of degrees to be placed on the circumference of the revolving weight ; P—the degrees belonging to the same scale as those in which l is expressed ; and v=(as before) the weight of vapour counterpoised by a single revolution of P ; then the efficient cause of vaporization while the standard piece cools through l degrees, will be represented by itx, and the *effect* produced must be expressed by vl ; hence, is derived the equation $lv = itx$, or $v = \dfrac{itx}{l}$. Comparing this value of v with that obtained above, we get $\dfrac{P}{n} = \dfrac{itx}{l}$

Hence, the weight of the standard piece $= i = \dfrac{Pl}{ntx}$, and the five following formulas will give *either one* of the other quantities, when the rest are assigned, viz. $P = \dfrac{nitx}{l}$; $x = \dfrac{Pl}{nit}$; $n = \dfrac{Pl}{itx}$; $t = \dfrac{Pl}{nix}$, and $l = \dfrac{nitx}{P}$. In practice it was found most convenient to make $t = 100$ degrees Fah. But by assuming t equal to the distance *between the freezing and the boiling point of water* under mean atmospheric pressure, a single standard piece would be sufficient to render the instrument universal in its indications. The curved surface of the revolving weight would only require to be divided into as many separate bands as there were different scales to be placed upon it, and graduating each band into as many equal parts as the particular scale comprehends degrees between the two points above mentioned. Thus we should have on the band marked Reaumur, 80 degrees ; Centigrade, 100 degrees ; Delisle, 150 degrees ; and Fahrenheit 180 degrees. It may not be improper to remark that in applying the above formulas, the numerical value of i must also vary with the thermometrical scale. Thus, if for Fah. it be 1037 degrees, it will be for Centigrade, $576\frac{1}{9}$; for Delisle, $863\frac{1}{3}$, and for Reaumur $460\frac{8}{9}$.

* See *Amer. Journal of Science*, vol. 21, p. 304.

boiling point is attained. In general this cap is kept suspended at one side of the boiler A, as exhibited in the outline at D'. The lower rim of the condenser is furnished with an exterior welt or hem of silk, sewed to the tin by means of fine punctures near its edge. This serves effectually to prevent the escape of steam, and, besides allowing the operator to attend deliberately to the adjustment of the counterpoise, will admit, when necessary, the postponement of this process for a considerable length of time. The counterweight c, is to balance the standard piece S, with its suspending wire, and the wire-gauze cap W. As long as the water is kept boiling by the action of the lamp L ; C is removed from the beam and is replaced only after the condenser has been transferred from D' to D. The support E of the beam may be elevated or depressed on its sustaining rod by means of the tightening screw K. Immediately below the fulcrum f, is a small hole drilled horizontally into E, to receive a brass tap carrying a ball $\frac{3}{4}$ of an inch in diameter, through which passes the small index-wire i, so adjusted by means of the screwed counter-weight c, as to be accurately balanced on the tap g, as an axis on which it turns with no other resistance than what is due to the friction produced by its own weight.

Near the extremity v of this wire, it is bent at right angles, and the pointed extremity directed to the side of the beam where, at o, is a straight line $\frac{1}{4}$ of an inch long, serving to guide the eye in reproducing the level after an experiment. A little below this line, is a transverse hole through the beam in which slides the register s (figs. 1 and 2,) —a wire about $\frac{1}{10}$ of an inch in diameter, and $2\frac{1}{4}$ inches long. While the water in A is kept boiling by means of the lamp, the boiler continues to rise, and as the register now projects out beyond the index i, it lifts the latter, keeping the point v always directed a very little above the line o. But when W has been inserted in its place, with S suspended in the water, the additional weight, destroying the equilibrium, depresses the boiler, the base of which rests on the flat surface of the lamp L, the concavity in the bottom serving as an extinguisher to the flame. The index i, is, in the mean time, left at the level attained by the register at the moment before the immersion. While ebullition is proceeding, the operator pushes back the register so as to project but little from the interior side of the beam ; then observing the thermometer t, takes the condenser from the position D', and, at the instant, the ebullition ceases, covers the boiler with it, as at D, letting the standard piece remain in place ; and having attached the counterpoise C, proceeds to bring down the boiler end ef the beam, (which at first rises above the index i,) by causing P to revolve in the direction towards f.

The number of revolutions being counted so many hundreds of degrees, he has only to add to their number 212°, in order to obtain at once the temperature by Fahrenheit's scale.

As both the latent heat of vapour and the specific heat of iron enter into the calculation, in constructing the steam pyrometer, and as on both these points considerable discrepancy prevailed among writers who had treated of these subjects, it was thought important to attempt a direct solution of the question as presented in the particular case of this investigation.

Two methods offered themselves, of verifying the calculations respecting the instrument. The first was, to heat the standard piece to any known temperature above 212°, and in that state plunging it into the boiling water to ascertain whether the amount of water vaporized weighed as many parts measured by hundredths of a revolution of C, as the standard piece had been heated in degrees above the boiling point.

This method being the most direct, was first resorted to by the committee.

As the standard piece S was at first made one or two hundred grains heavier than was supposed to be necessary, trial was made in the way just indicated, and as an excess of vapour was found to have been obtained, the standard piece was proportionally reduced in weight to that which has been already stated.

The other method consisted in determining separately by direct experiment, both the latent heat of vapour, and the specific heat of the standard piece. The researches on these subjects were made in the manner and with apparatus described hereafter.

(To be continued.)

AMERICAN FIRE-PROOF PAINT.

In November, 1837, a patent was granted to Mr. Louis Paimbœuf, of Washington, for a fire-proof paint, to prepare which he gives the following instructions in his specification, published in the *Franklin Journal*. Whether this American paint is in any way similar to 'the " composition for protecting wood from flame,'' lately patented by Mr. Davies, in England, which forms the basis of the Fire Preventive Company and with which various experiments have been made throughout this country, as it appears, has also been the case with Mr. Paimbœuf's paint in Washington, we of course, are unable to say, as the specification of the English patent has not yet been enrolled :—

" My fire-proof paint," says Mr. Paimbœuf, " may be prepared by grinding and incorporating the ingredients used, either with oil, or with water, as may be preferred. That prepared with water or other aqueous fluid, however, has one advantage over that prepared with oil ; namely, it dries very rapidly, and affords the desired protection immediately ; whilst that prepared with oil will not harden until the lapse of several weeks, depending upon the season of the year.

" To prepare my paint, I take the best quick lime, such as when slacked forms an impalpable powder, and slack it by the addition of so much water, only, as is requisite to produce that effect, performing this operation in a trough or vessel, which may be covered over, so as to retain the vapour and heat as perfectly as possible, as upon this procedure I find that much of the effective strength of the composition is dependent.

" When the slacking has become perfect, and the mass has cooled, I, in order to prepare my water paint, add either water or skimmed milk, or a mixture of the two, to the lime, in sufficient quantity to give to it the consistency of cream, or that of ordinary paint. When milk is not used, I add to the water a quantity of rice paste, made by boiling rice in water to a proper consistence, using about eight pounds of rice to every hundred gallons of the prepared paint. To every hundred gallons of this prepared lime mixture. I add twenty pounds of alum, fifteen pounds of potash, and one bushel of common salt. These are the essential ingredients, and the proportions such as I have found to answer well. If the paint is to be white, I find it advantageous to add to these ingredients about six pounds of prepared plaster of Paris, and the same quantity of fine white clay. When the paint is not required to be white, I substitute clean, well-sifted, hard-wood ashes for the potash ; about two bushels being sufficient for the above quantity ; in this case, also, I frequently add three or four gallons of molasses.

" After mingling these ingredients, I first strain them through a fine strainer, and afterwards grind them together in a paint-mill, when the paint is ready for use. When roofs are to be covered, or when crumbling brick walls are to be coated, I mix with my paint a quantity of fine white sand, in the proportion of about one pound to every ten gallons of the paint, as this addition will cause the paint to petrify, preventing leakage in roofs, and binding the crumbling particles of disintegrating brick-work.

"In applying this paint, excepting in very warm weather, it will be advantageous to use it as warm as it can be conveniently kept; and particular care must be taken that it be not allowed to freeze whilst drying, as its binding property would thereby be destroyed, or much impaired. Three coats will be sufficient in all cases. In putting on the first coat, the paint should be more diluted than with the others. It can be managed by any person used to the paint brush.

"When the oil paint is to be prepared, I take forty gallons of good, boiled, linseed oil, and to this I add such quantity of the fine dry-slacked lime as is requisite to bring it to a proper consistence for paint, and to this add two pounds of alum, one pound of pot, or pearl, ashes, and eight pounds of common salt. In this paint, good wood ashes may be substituted for the potash, eight, or ten pounds being used.

"This paint is to be used in the same manner as other paint, taking special care that the first coat is perfectly dry before the second is applied. Under the same circumstances, the addition of a portion of white, fine sand, will produce a like good effect as in the water paint.

"With these paints any of the ordinary pigments may be used, so as to obtain any colour which may be desired.

"I do not claim to be the first who has applied the above-named ingredients to the purpose of rendering wood incombustible, either separately, or, to a certain extent, in combination with each other; but what I do claim as my invention, or discovery, is the combining together of lime, potash, alum and common salt, substantially in the manner herein set forth; whether the same be in the proportions herein designated, or in any other which will produce a like effect, and whether the other ingredients named, or any similar substances, be added to the water, or the oil paint, prepared as above.

LOUIS PAIMBŒUF."

Remarks by the Editor of the Franklin Journal.—When the foregoing composition was first introduced in the city of Washington, it attracted considerable attention, in consequence of some public exhibitions made for the purpose of showing its efficacy, and which were, so far as they went, satisfactory. The small, wooden erections upon which these experiments were performed, had been recently, and very carefully, covered with the composition, and the question was one of much importance, how long will the protective effect continue? Will the paint, or work, stand exposure to the weather, or will it wash, or shell off? To these questions the reply of the experimenter was, that he had roofs covered with it in the West Indies, and that it was good after a lapse of five years. This assurance induced a number of persons to make essays with the material, and it was soon found that it would both shell and wash off at an early day, and that it was, consequently, of no value whatever. Prior to this, the nature of the composition had not been made known, but an application for a patent necessarily called forth the recipe.

Those who are acquainted with the many experiments which have been instituted for the purpose of rendering timber fire-proof, will not meet with anything substantially new in the foregoing plan; all the essential ingredients, as well as many others, analagous in their properties, have been essayed, and the result made public. The propriety of granting a patent was questioned, but it was considered as limiting the patentee to his own particular combination of ingredients, by which the public interest would not be invaded. Twenty compounds equally good, and some better, may easily be prepared, were they worth preparing. Had that in question been faithfully made, by a proper incorporation of the ingredients, and the employment of a sufficient portion of those which were known to be the most beneficial, the eventual failure would at least have been postponed, and would have been less complete.

RYAN'S INSTRUMENT FOR DRYING SILK IN THE LOOM.

The Silver Isis Medal and £5 were presented last year, by the Society of Arts, to Mr. James Ryan, weaver, of Hare-street, Bethnal Green, for his instrument for drying silk in the loom. We extract the following particulars from the *Transactions:*—

Silk thread, like all other vegetable or animal substances composed of fine fibres, is hygrometric; that is, when exposed to a damp air, it absorbs moisture, from which it cannot afterwards be freed except by raising the temperature sufficiently to convert the water into vapour, or to dissolve the water by the action of very dry air. Silk, in contact with damp air on all sides, becomes more moist, and in a shorter time, than similar silk wound into a ball, or on a roller. A silk warp while in the loom, and partly woven, is wound on two rollers, the one, that on which the whole warp is first wound, and from which it is unwound and transferred to the second, or cloth-roll, in proportion as it has passed through the process of weaving, and is become cloth. The space between these two rollers may be divided into two portions, separated by that part across which the shuttle passes; the portion between this part and the cloth-roll being already woven into cloth, and the other being that part of the warp which has been wound off the roll, and is coming up through the harness to the shuttle. This portion is called the porré, and consists of parallel threads exposed both above and below to the air. In damp, cold weather, and during the winter generally, when the weaver leaves off work at night, the air of his workshop becomes colder and damper, till its state nearly approaches to that of the outer air; and, therefore, when the weaver is desirous of beginning his work again in the morning, he finds the porré has become damp; the adjacent threads have, therefore, a tendency to rub hard against one another in making the shed, and the work proceeds heavily and slowly till the fire of his shop has become powerful enough to evaporate the moisture, which often will not be effectually done in less than two hours. Besides the loss of time hence arising, the work done under such circumstances feels loose and spongy, and is very liable to cockle, from certain parts of the threads being more moist than the others during the weaving, especially when two or more kinds of silk are employed in the same warp.

" Another disadvantage is, that the brighter dyes now used, especially for spring wear, are many of them so fugitive, that they are much injured by mere dampness, to which they are particularly liable, as they must of necessity be woven during the winter. These difficulties and imperfections affect both the weaver and his master in so serious a degree, as to render their removal a matter of no small importance.

" Mr. Ryan, after several unsuccessful attempts, at length hit on the simple instrument about to be described, with which he has always obtained the most satisfactory results.

." It is a hollow quadrangular prism of tin-plate, with a round neck at one end, closed by a cork; the length of the prism is about equal to the length of the porré, and on the under side are two handles, to render it more manageable. The prism, being filled with hot water, is applied first to the under, and then to the upper surface of the porré, till, by means of it, all the moisture is evaporated. This is often so considerable, as to cause a visible steam while it is passing off. The instrument is then applied to the

harness; and the whole is thus made dry, and brought into a state proper for working in from ten to twenty minutes, according to its previous state of dampness.

" From the evidence of practical weavers, who have made use of Mr. Ryan's instrument, it appears that the injury caused by dampness, both as regards the work and the more fugitive colours, is not at all exaggerated, and that, by the use of the instrument, they have constantly obtained the advantages attributed to it by the inventor."

NOTES AND NOTICES.

Modern domestic warming Stoves.—When small domestic stoves are used, with very low combustion, as has been recently proposed, upon the score of a misjudged economy, there is great danger of the inmates being suffocated or asphyxied, by the regurgitation of the noxious burned air. The smoke doctors who recommend such a vicious plan, from their ignorance of chemical science, are not aware that the carbonic acid gas, of coke or coal, must be heated 250 degrees F. above the atmospheric air, to acquire the same low specific gravity with it. In other words, unless so rarified by heat, that gaseous poison will descend through the orifice of the ash-pit, and be replaced by the lighter air of the apartment. Drs. Priestly and Dalton have long ago shown the co-existence of these two-fold crossing currents of air, even through the substance of stone-ware tubes. True economy of heat, and salubrity, alike require vivid combustion of the fuel, with a somewhat brisk draught inside of the chimney, and a corresponding abstraction of air from the apartment. Wholesome continuous ventilation, under the ordinary circumstances of dwelling houses, cannot be secured in any other way. Were these mephitic stoves, which have been of late so ridiculously puffed in the public prints, generally introduced, the faculty would need to be immediately quadrupled to supply the demand for medical advice; for headaches, sickness, nervous ailments, and apoplexy, would become the constant inmates of every inhabited mansion. The phenomena of the grotto of Pausilippo might then be daily realised at home, among those who ventured to recline upon sofas in such carbonated apartments; only instead of a puppy being suffocated *pro tempore*, human beings would be sacrificed, to save two-penny worth of fuel *per diem.*—*Dr. Ure's Dic. of Arts, &c.*

Exhibition of Industry at Madrid.—Our own "national exhibition of arts and manufactures" has somehow or other died a natural death—or rather an unnatural one; for an exhibition of that kind, if well conducted would, in this country, be sure to be instructive, and ought to be popular. It is now proposed to establish one in, of all countries in Europe the very last where it could be expected. Señor Annibal Alvarez, the son of a distinguished Spanish sculptor, has published a project for an exhibition of the products of national industry at Madrid. To a foreigner, the national industry of that country appears at present to consist chiefly in mutual murder.

Railway Insurance Company.—The Dresden and Leipzig Railway Company are beginning to count their winnings. At the end of October the receipts amounted to 90,000 dollars, of which half was paid away in interest, and 30,000 of the remainder to the first subscribers. Some increase to the revenue is anticipated from the establishment of an insurance for luggage on the part of the company. A passenger who desires to save himself the trouble of looking after his luggage, will have it in his power to surrender it to the officials on the payment of about three halfpence for every pound weight, and will receive in return a card, on giving up which at the end of the journey the luggage will be re-delivered him. This insurance is to the value of a dollar a pound, by paying double, a double insurance may be effected. It is supposed that few persons will prefer to "take their chance," amidst the inevitable confusion arising from the simultaneous progress of four hundred passengers.

More Rings of Saturn.—Among the other evils of our foggy climate, must for the future be enumerated, one to which much attention has not hitherto been given—its inaptitude for astronomical observations. Signor Decuppis, of the Observatory at Rome, in a recent communication to the Academy of Sciences at Paris, gave an account of the results of some late observations in Saturn, which place this fact beyond a doubt. It was thought for a long time that Saturn possessed only one ring—it was not till Herschel brought into exercise the powers of his gigantic telescope that it was discovered to possess two. In the pure atmosphere of Rome, with the assistance of a powerful telescope, it was discovered on the 17th of June last, that the number amounted to four—and on the 18th of June, Signor Decuppis, to his astonishment, clearly made out five. Having read with how much difficulty Herschel ascertained the existence of the smaller three of the seven satellites of the same planet, the Roman astronomers were amazed to find, that to them they appeared with the greatest distinctness and brilliancy. With such a climate as theirs, it is no wonder that the Italians should be able to claim the discovery of an asteriod by Piazzi of Palermo, but it reflects no honour on them as cultivators of science, that the others should have been introduced to the world of science by Harding and Olbers and that the only planet added to the solar system should have been first detected on the South Parade in Bath.

South Eastern and Dover Railway.—We have paid our periodical visit to the works of this railway, to mark their progress. Our readers are aware that the galleries and shafts of the important tunnel at Shakespeare's Cliff are already complete, and we have now to notice that the two headways are progressing most satisfactorily. The chalk hitherto excavated is of eatraordinary firmness, and fully realises the expectation of the engineer.—that the roof of the tunnel will in no part require any extraneous support beyond the natural chalk. About four weeks hence, should the working continue favourable as hitherto, we may congratulate our readers upon the possibility of walking through the bowels of this noted cliff. The cuttings beyond are proceeding rapidly, and still further, the second tunnel has been commenced under very favourable appearances.—*Dover Chron.*

Proposed Railway from London to the North.—A new line of railway has just been projected, to be called the North Trunk Railway. It is proposed to commence at Islington, passing from thence between Highgate and Hampstead, by Finchley, Barnet, St. Alban's and Bedford, between Higham Ferrars and Wellingborough, and thence to Leicester, there to join the Midland Railway. This route would form a portion of a direct line from London to Edinburgh.

Newcastle and North Shields Railway.—Not-

withstanding the very wet and untoward state of the autumn, the various works on the Newcastle and North Shields Railway are progressing favourably. We understand that the Company intend to apply to Parliament, in the ensuing session, for a renewed extension of their line to Tynemouth, and should they succeed in their praiseworthy undertaking, it cannot fail to be one of the most useful railways in the United Kingdom; it will be the means of opening out to the public, at a cheap rate, one of the best bathing stations on the eastern coast of England. As far as we can learn, the Company mean to carry their line through the town of North Shields by means of a tunnel similar to the one now in use under the town of Liverpool; by this arrangement the present features of the town will be perfectly preserved; and, in order to accommodate the inhabitants, it is proposed by the Company that the locomotive engines shall stop short at the entrance of North Shields, and the trains to be thence forwarded through the tunnel to Tynemouth by horses every half hour, or oftener, if required.

Danube and Maine Canal.—The canal to counect the Danube and the Maine, recently commenced was not a favourite speculation in Germany, It was alleged that the country it was to cross was so sandy, that water would only be attainable at an enormous expense, or not at all, and that the canal would, in consequence, be at a stand still in summer. To the astonishment of every body, the engineers of the canal included, water has turned out uncommonly plentiful in the midst of the sand—that portion of the bed of the canal already completed is partly filled with water 'on the voluntary system;' and it is found that some brooks, whose course it was in contemplation to divert, will not be wanted at all.

Birmingham Railroad.—Return between the 2nd October and 5th November inclusive, to and from Birmingham:—Travelled, 3,553,061 miles. Number of persons, 56,816. Duty paid to Government, £1,850 11s. The foregoing is the substance of the declaration of the Return made before the magistrates, Mr. Twyford, and Dr. Robinson, a county magistrate.

Bottle-Washing Machine Premium.—In answer to numerous inquiries respecting the bottle washing premium, we beg to say, that we are in communication with the firm on whose behalf the prize was offered, and expect to be able to announce the award of the prize in a week or two.

Metropolitan Railway Map.—On the 1st of December will be published vol. xxix of the *Mechanics' Magazine*, price 9s., illustrated with a *Railway Map* of the Metropolis, taking in a radius of 15 miles from the Post-office. Encouraged by the extensive sale which our *Railway Map of England* has commanded, the *Metropolitan Railway Map* has been executed at a very great cost; the utmost exactness has been observed in reducing it from the Ordnance maps, and all the railways projected, up to the day of publication, have been distinctly and accurately marked from actual survey. The limits of the twopenny and threepenny post deliveries are also shown in the Map. The *Metropolitan Railway Map* alone, stitched in a wrapper, price 6d., and on fine paper, coloured, 1s.

The *Railway Map* of England and Wales continues on sale, in a neat wrapper, price 6d.; and on fine paper, coloured, price 1s.

LONDON: Printed and Published for the Proprietor, by W. A. Robertson, at the Mechanics' Magazin, Office, No. 6, Peterborough-court, between 135 and 136, Fleet-street.—Sold by A. & W. Galignani, Rue Vivienne, Paris.

Mechanics' Magazine,

MUSEUM, REGISTER, JOURNAL, AND GAZETTE.

No. 799.]　　　SATURDAY, DECEMBER 1, 1838.　　　[Price 6*d.*

PISTRUCCI'S IMPROVED BANKER FOR SCULPTORS.

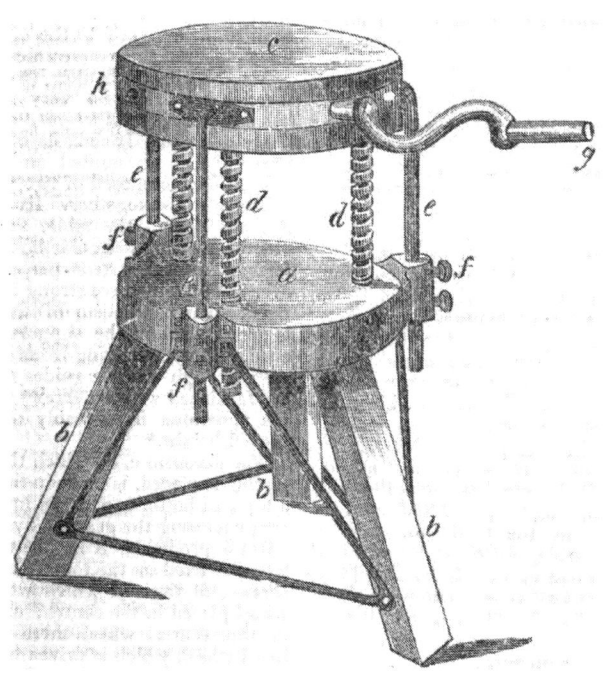

PISTRUCCI'S IMPROVED BANKER FOR SCULPTORS.

Sir,—The splendid talents of Mr. Pistrucci, Chief Medallist of her Majesty's Royal Mint, as "a cameo-gem engraver" (as he has sometimes disparagingly been called), are pretty generally known, and almost universally admitted. His cameos have long been considered scarcely inferior to the best of the antique; and have frequently been mistaken, by some of the most approved judges, for ancient productions. Of his superior excellence as a medallist we have had several recent proofs afforded; but it is not generally known, that to first-rate artistical skill, he adds mechanical abilities of a very high order. Ten years ago he invented, and has since sent to Mexico, a very powerful coining-press of wrought iron, which some of the first manufacturers in Birmingham are now beginning to appreciate and to copy.

Mr. Pistrucci is well known, by every person in the royal mint, to be fully equal, if not superior to, all others, in the judicious selection, and also in the very delicate and difficult art of hardening and tempering steel dies.

His recent and highly important invention of casting *thin iron plates* from models, to be mounted on steel beds, and used in lieu of engraved steel dies, has been already described in your pages (p. 36, No. 767); and his *studio* abounds with simple and ingenious contrivances for facilitating the execution of his multifarious works of art.

During the long protracted (and, by him, undesired) period of his "*official leisure*," Mr. Pistrucci resumed the profession of a sculptor; he fitted up two workshops in the mint, for private works in marble, entirely at his own expense. In one of these he executed his fine colossal bust of the Duke of Wellington, and in the other, that of Prince Pozzo di Borgo; both of which have been greatly admired.

In illustration of the remarks which I have just made respecting Mr. Pistrucci's ingenuity in devising expedients to facilitate or assist his labours, I send you herewith a description of his improved "Banker," which affords a much readier and more convenient method of changing the position of a heavy mass of marble, &c., than any hitherto employed for that purpose. There is both novelty and ingenuity in the present apparatus; and I observe that a patent was obtained by Mr. William Brindley, of Birmingham, in December last, for the application of a precisely similar arrangement of mechanism to the construction of presses of great power. Mr. Pistrucci's banker was made at the well-known manufactory of Messrs. Heaton, Brothers, of Birmingham; and whether Mr. Brindley obtained his idea from that source, I cannot pretend to say, nor is it of any real consequence; for a Mr. Penny exhibited a press, constructed on the same plan, in London, upwards of five years since.

I may hereafter return to the subject of these presses; but, for the present, I shall confine myself to the more legitimate object of this communication.

Mr. Pistrucci's banker (to use commercial phrases) is a very *substantial* one; it carries *great weight*, and is not likely to *break*. It consists of a circular board of oak *a*, mounted on three legs of ash *b b b*, composed of two pieces, and strongly bolted together. Iron ties connect the legs to the table, and to each other. A second circular board or platform, *c*, is supported perpendicularly over the first, by three strong iron screws *d d d*. The parallelism of the machine is preserved, and the support strengthened, by three steadying-irons *e e e*, which pass through suitable guides *f f f*: these are furnished with set screws, for fixing the apparatus immoveably at any approved height.*

The platform *c*, on which the block of marble is placed, is made in three parts: a top and bottom, with an intermediate ring enclosing the gearing, by which the effect is produced. A horizontal toothed wheel is fixed on the top of each of the screws, all three of which work into a wheel placed in the centre; on the axis of this central wheel there is a large bevel wheel, which is driven by a pinion on the winch-handle *g*; so that, by turning the handle, the three screws are made

---

* In this machine the steadying-irons have been placed opposite each of the screws; but, in Mr. Pistrucci's plan, they are placed more judiciously, intermediately, so as to have six equidistant points of support instead of three.

to revolve uniformly; and the platform, with its load, is raised or depressed, according to the direction in which the handle is turned.

By this arrangement, the power required to elevate the platform.c is rendered so trifling, that a child may effect it with ease, even when loaded with two or three tons weight. So great is the mechanical advantage, that when a block of marble, weighing about two tons, was standing on the platform, I could scarcely tell, by the resistance offered in turning the handle, whether I was raising or lowering of it. The space moved through in a given time is, of course, proportionably small; but then the range required is, in these cases, never large.

In addition to the vertical motion thus elegantly and conveniently obtained, a horizontal movement is afforded, on placing a short lever in the hole *h*; when the upper part of the platform revolves on its centre, being supported by a metallic bearing near its outer edge.

The stability and complete efficiency of this apparatus can hardly be conceived except by inspection; and those persons who feel interested in the employment of such a machine, will find it well worth while to pay a visit to the mint—where Mr. Pistrucci's free and open disposition will readily afford them the means of forming their own conclusions on the merits of this invention.

I remain, Sir, yours, respectfully,
WM. BADDELEY.

London, Nov. 7, 1838.

ORGANIC MATTERS THE PROPER AND ONLY SOURCE OF ANIMAL FOOD.

Sir,—I am sorry to observe in your last number (797), that "Junius Redivivus," instead of candidly acknowledging himself in error, makes an ineffectual attempt to vindicate his former statements (page 36, No. 793), setting forth the future probability of *nutritive food* being prepared from inorganic matter. Although fully sensible of my inferiority to "Junius Redivivus" in argumentative powers, I feel that he is so decidedly wrong, and so completely opposed to all recorded facts, in the position he at present endeavours to maintain, that I must beg to offer a few brief observations in reply to the matters broached in his last communication.

"Junius Redivivus" misquotes me at the outset, for no apparent reason but that of introducing school-boys performing the exploit of swallowing *live* frogs —not as a means of subsistence, but out of mere sport—and the bolting of *dead* oysters, by grown up gourmands, as a matter of palatic gratification. He then goes on to assert that "the marble rock is but a congeries of living beings or their exuviæ." It is true that whole mountains and extensive districts in various parts of our globe, appear to be composed almost entirely of animal remains: but such mountains, and *marble rocks*, are two distinct and widely different things. Butchers, with their wives, and sons, and daughters, are next exhibited as fattening on the effluvia of meat: and "Junius Redivivus" says, "yet surely these gases are not living matter." No, but they are most undoubtedly the produce of what was once living *organic* matter; and I shrewdly suspect, by the bye, that if the meat itself was withheld, the fattening influences of the gases would be of a very unsatisfactory character. We are, however, requested to return to the opinion of the philosophers, that is, of one—Humboldt, "who has recorded that certain tribes of Indians, during certain periods of the year *subsist* on a white earth, which they dig up," and "Junius Redivivus" supposes this to substantiate the fact—not that man may derive *nutriment* from, but that he may subsist upon, inorganic matter. In the first place, that this white earth consists *entirely* of inorganic matter is "not proven;" in fact, its peculiar colour strongly induces a belief in the presence of organic (either animal or vegetable) matter, else whence the preference given it over other earths? Secondly, if it was, the mere subsistence for a short time on a substance devoid of nutriment, proves nothing; because in some diseases, mesenteric obstructions for instance, although food is regularly taken into the stomach, and many of the functions of life continue to perform their office, no nutriment is derived by the system, and the food might almost as well be white or any other coloured earth; yet the poor sufferer lingers for a considerable period — perhaps several months—and at length, nature having fairly exhausted all the repositories of self-subsistence,* he sinks a victim to starvation.

Just as well might an Englishman fatten on the *red* earth constantly seen on the banks of his favourite Thames (the colouring matter of which is innumerable animalculæ), and boast of his subsistence on inorganic matter! The mushroom and the olive, brought forward by "Junius Redivivus"—being both organic substances—by no means assist his argument: he seems to depend greatly upon his assertion that "we scarcely know where the stone ends and the vegetable begins, or where the vegetable ends and the animal begins." It has long been a favourite notion with speculative naturalists, that organised beings might be arranged in a continued series, every part of which, like the links of a chain, should be conencted with that which preceded and that which followed it. Linnæus was even impressed with the idea, that nature, in the formation of animals, had never passed abruptly from one kind of structure to another. But the idea of a chain, or continuous gradation of being, was cherished with enthusiastic ardour by Bonnet, who, assuming man as the standard of excellence, attempted to trace a regular series, descending from him to the unorganised materials of the mineral world. Many other writers have adopted this fanciful speculation; but none have carried it to a more extravagant length than Lamarck, who blends it with the wildest and most absurd hypothesis that was ever devised, to account for the diversities of animal structures.

The fact is, the parts, which by their assemblage constitute an organised body, when compared with inorganic matter, exhibit in their chemical, as well as in their mechanical characters, the most well-marked and striking contrast. The solids and fluids of which organic structures are composed, differ very materially in their chemical constitution from the productions of the mineral kingdom. Their elements are combined by a much more complicated arrangement, and united by less powerful affinities; or rather, the balance of affinities, by which they are held together is more easily destroyed, and thus proneness to decomposition is constantly present.

It is an incontrovertible law of nature, that a constant supply of *nutritive* matter is essential for the continuance of life; and when we come to analyze the proximate principles from which *animal nutriment* is derived, we find them reducible to the following: namely, fibrin, albumen, oil, gelatin, and sugar; together with a few others, such as osmazone, which are of minor importance. Perhaps the most exact classification is that of Magendie, who refers all alimentary substances, whether animal or vegetable, to the following heads: namely, farinaceous, mucilaginous, saccharine, acidulous, oily, gaseous, gelatinous, albuminous, and fibrinous. These are adapted to the purposes of nutrition by the functions of assimilation, absorption, circulation, and respiration.

When any of these proximate principles are found to exist in *inorganic matter*, then—and not till then—will man find nutriment therein. "Junius Redivivus" may perhaps go on to ultimate elements; but this will advantage him nothing, as they form organic substances only in definite proportions.

It is too frequently the practice with flighty theorists, though both unphilosophical and unjust, to push their speculations beyond all reasonable limits; and, when challenged, to reply, that "philosophy is not yet a perfect science;" and when met by some insurmountable difficulty, to content themselves by declaring, that "our chemistry is at fault."

I would merely ask, was it not by similar baseless arguments, and such idle speculations of old, that the reign of ignorance was so long perpetuated? and shall we—after the lessons of a Bacon, and the example of a Newton—suffer imagination to run wild, regardless of all recorded experience.

I trust there are none who would assert that our knowledge is perfect, even with reference to the points now under review; but this may be admitted, that whatever further insights the perseverance and ingenuity of man may obtain, into the mysterious operations of nature, we are even now in possession of certain facts, and have acquaintance with several governing laws, that will continue to operate, unchanged, to the end of time. We have positive data to proceed from, however limited our range beyond; and

* The peculiar office of fat as a magazine of nutriment, for the support of the body during the period when food is scarce, is strikingly exemplified in numerous instances, especially in all the hybernating animals.

unprofitable indeed must be our specuations, if, despising these, we start in ignorance—proceed in darkness—and end is chaos.

I am, Sir, yours respectfully,
WM. BADDELEY.
London, Nov. 21, 1838.

SMOKE-CONSUMING STOVE.

Sir,—I inclose you an extract from the *Engineers and Mechanics' Cyclopædia*, by which "Junius Redivivus" will perceive that he has been forestalled in his proposal to make the two side jambs of stoves a repository for coals, for the purpose of effecting the combustion of the smoke. Were the facts stated not beyond all question, still I am sure his natural gallantry would not allow him to dispute the invention with *a lady*.

It is stated in the description, that the hobs or lids "should shut down closely; if air-tight the better." Now, it would be altogether impossible to make them do so; and in that case the gas, smoke, &c., would escape unconsumed into the chimney, and a loss, instead of economy, of fuel, would result. This might be prevented by using a luting; but it would be so frequently required to be renewed, that this is quite out of the question. It appears to me, that by making the side of the stove between the fire and the coals a receptacle for water, in connection with a larger reservoir behind or elsewhere, the coking process would be confined to the strata of coal in contact with the ignited fuel, and thereby prevent the evil consequences before adverted to.

I am, Sir, yours respectfully,
WM. BADDELEY.
London, Nov. 20, 1838.

MRS. SMITH'S STOVE.

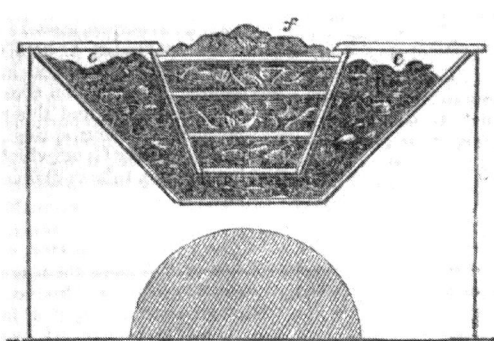

"A plan of a stove designed for burning its own smoke, was communicated by Mrs. Rachel Smith to a periodical journal, which seems susceptible, by its simplicity of construction and soundness of principle, to be made effective for the object intended. The stove is made exteriorly of the usual form, excepting that the fire-part, *f*, is of greater length. or height than is common, and the spaces under the hobs are made into reservoirs to receive the coals, as shown at C C, for supplying the fire. The hobs are upon hinges, and form lids, which shut down very closely; if air-tight the better. The cheeks of the grate are open at the bottom, so that the coals lying upon the inclined planes of the reservoirs descend by their own weight, and occupy the lower part of the grate; as the fuel is consumed, or raised by the poker, a fresh portion of coals enters from either or both of the reservoirs, and fills up the space. In this manner the fuel is constantly supplied, occasioning little or no smoke. The reservoirs should be of sufficient capacity to hold enough coal for the day's consumption."

MODE OF ATTACHING A FRESH CARRIAGE TO A RAILWAY TRAIN WITHOUT STOPPING THE TRAIN.

Sir,—The delay and inconvenience, the wear and tear of machinery, which arise from frequent stoppages, as well as the advantages both to proprietors and travellers, which a solution of the problem above stated would afford, are all so obvious, that they need not be enlarged upon. Two methods of solving this problem present themselves.

1st method: Suppose the last carriage of the train furnished with a cylindrical box, whose axle is parallel to those of a carriage, and containing a spiral spring (like the main spring of a watch), let $l =$ length of the spring; $r =$ radius of the axle on which it coils; $t =$ thickness of the spring; $n =$ number of coils; $\pi =$ 3.14159, then $2\pi \left\{ r + \dfrac{t}{2} + r + \dfrac{3t}{2} \cdots \right.$

$\left. \cdots r + \dfrac{t}{2} + \overline{n-1}.t \right\} = l$ nearly, or

$2\pi . \dfrac{n}{2} .(2r+nt) = l$, or $n^2 + \dfrac{2r}{t} .n = \dfrac{l}{\pi t} \therefore n = -\dfrac{r}{t} + \sqrt{\dfrac{r^2}{t^2} + \dfrac{l}{\pi t}}$. Similarly, if $R =$ radius of cylinder; $T =$ thickness of a *flat* rope coiling upon it and upon *itself*; $L =$ length of rope, that runs out while the spring takes n turns; then $n = -\dfrac{R}{T} + \sqrt{\dfrac{R^2}{T^2} + \dfrac{L}{\pi T}}$.

Example.—Suppose the diameters of the cylinder and axle to be 3 and 1 foot, $l = 93$ feet, $t = \frac{1}{4}$ inch, then in feet $r = \frac{1}{4}$ $t = \frac{1}{48} \therefore \dfrac{r}{t} = 30 \dfrac{l}{\pi} = \dfrac{93}{3.1} = 30$, and

$\dfrac{l}{\pi t} = 1800 \therefore n = -30 + \sqrt{2700} = 22$ nearly; and if $T = 3\,t = \frac{3}{4}$ inch, $L = 3\,l = 279$ feet; and if the train is moving at the rate of 30 miles an hour, or about 40 feet per second, $\dfrac{L}{40} = 7$ seconds nearly; abundance of time for communicating motion to the new carriage (C).

If the last carriage (L) be furnished with several coils of spring, as many must be put in gear as are thought necessary, according to the weight of C.

Let C be placed *near* the rails, and as L passes the rope becomes attached to C, and begins to run out; it remains to get C upon the rails as soon as possible after L has passed. This may be done in various ways.

1st way: Let C be placed on a frame *parellel* to the rails, and be *set on* the rails by a strong spring, which L releases in passing, and by a motion perpendicular to that of the train. By this way, however, at least 1″ of time is lost.

2nd way: Let C be placed on a side rail slightly curved and leading into the railroad, then motion commences the *instant* L passes, and this motion may be aided by a strong spring released by L and *propelling* C; or (which might be still better) by a tube containing compressed air.

Fig. 1.

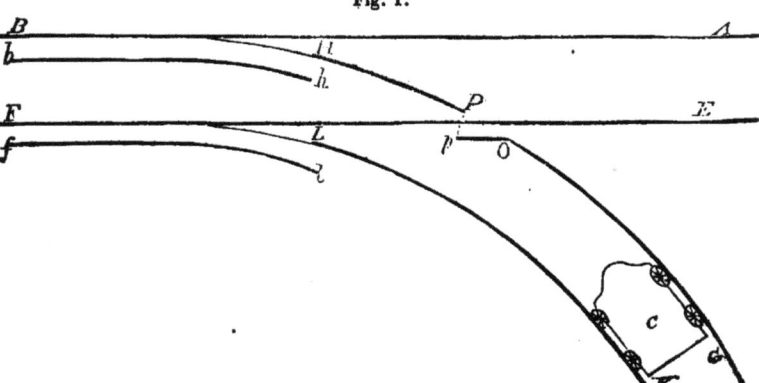

Fig. 1 *explains the construction of the side rail.*—A B, F E, the railroad; G H B F L K, the side rail elevated 6 or 8 inches above A F, and descending into it at B and F. Let the portions B H, F L, be in the positions $b\,h$, $f\,l$, and P O in the position J O moveable round a hinge at O. When L comes to it, it becomes attached to C, and releases the spring which propels C. On passing O, another spring puts p O in the position, P O, completing the rail G P H. When L reaches B F, the portions *b h*, *f l*, assume the positions B H, F L, and C moves into the railroad.

3rd way: Let C descend from above down an inclined rail. Here C may be both attached and detached the instant L passes *under* it; the motion may be aided by a spring as before, as well as by gravity; and if the space through which C descends be regulated according to the speed of the train, a much smaller coil of spring need be carried by L than when the motion of C is wholly produced by the coil.

Fig. 2.

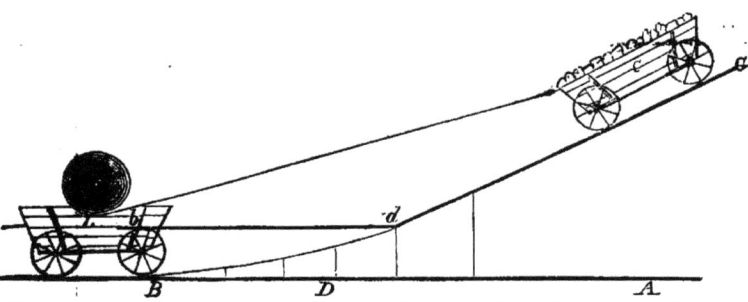

Fig. 2 explains the construction of the inclined rail.—This rail must be supported by a frame extending a little beyond the rails A B F E, so as to allow the trains to pass. A portion, *d b*, of the rail is supported in a horizontal position, and just high enough above A B to allow the wheels of the train to pass under it. When L comes to B the supports are removed, and *b d* falls into the position *d* B, and the next instant C descends down it.

When the new carriage descends from above, the cylinder and coil may perhaps be dispensed with.

Fig. 3.

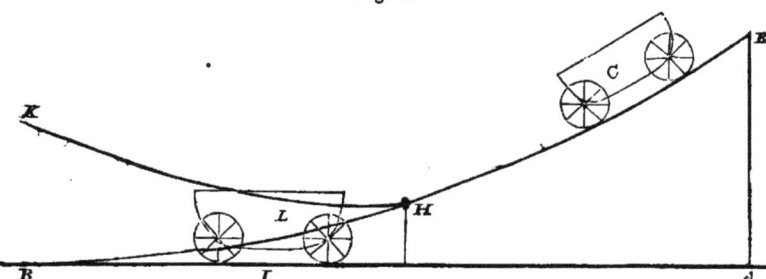

E B, Fig. 2, the rail (slightly curved) down which C descends. A portion, H B, of this rail is moveable about a hinge H, and can be raised into the position H K, so as to allow the wheels of the trains to pass under it. The *instant* L passes B, K H falls, or is *set* in the position H B. The *instant* H B is *so set*, C reaches H, and begins to descend down H B. This will be effected by a trigger, at I, released by L, and having its distance from A properly adjusted. If, then, the height of E, from which C begins its descent be properly assumed (allowing for friction, &c.) C may be made to arrive at B with a velocity greater than that of the train. C will then overtake the train, and by a simple device become attached to it.

This method has the advantage of being very simple, and in deep cuttings the necessary apparatus would be erected at a small expense. On plain ground, and where land can be easily obtained, the tractrix would probably be preferable.

'Several other ways might be mentioned, but perhaps the above are sufficient.

It will be observed that every thing being done by machinery, the times may be calculated to the greatest exactness, so that, although the attachment is effected in a very few seconds, no danger or derangement is likely to ensue.

Fig. 3.

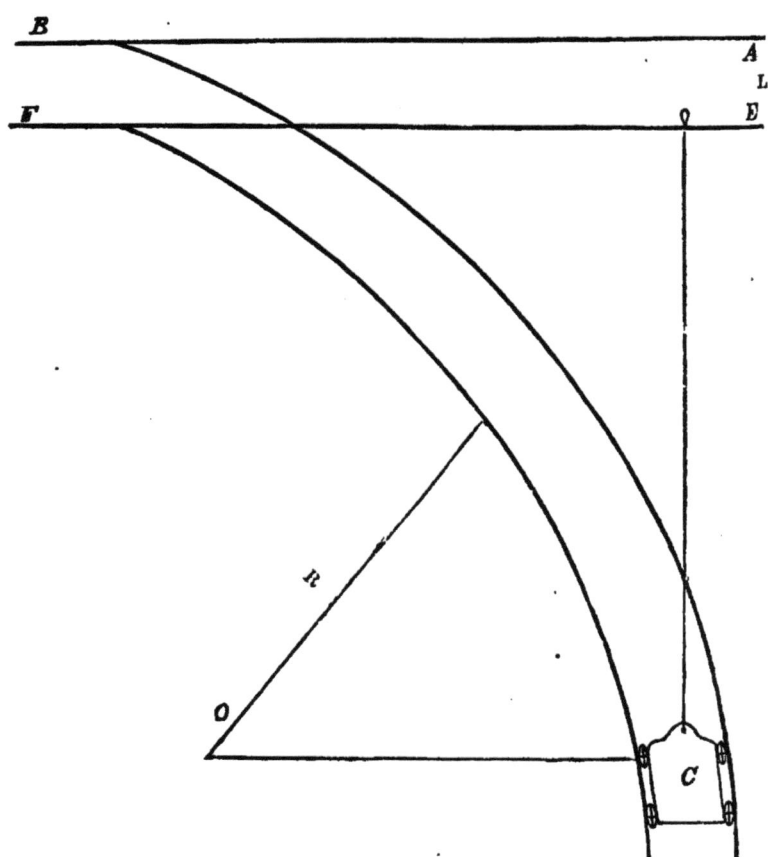

2nd method: Here the cylinder with its coil of spring is dispensed with; C begins to move in a direction *perpendicular* to the motion of the train, and thus it acquires motion without violence.

Let C move on a branch rail forming the quadrant of a circle, whose centre is O, and radius R, and leading into the rail-road, as in fig. 1. E C, a rope = in length to R, and $\perp^2$ E F; as a carriage from the direction L passes, it catches a noose at E, and draws C along the quadrant C F B.

Example.—Suppose 40 feet per second, the rate of the train, then $\frac{40^2}{R}$ = force stretching the rope at the first instant of motion = gravity, suppose = 32 feet, that is, let the rope E C be just able to support the weight of C, then

$$R = \frac{40^2}{32} = 10 \times 5 = 50,$$ or the circle must be at least 50 feet radius. As the rope E C tends to draw L off the rails, this may be prevented by a side rail

parallel to A B, and a small horizontal wheel, or by having another carriage on the other side the rails attached at the same instant with C.

If C were suffered to move freely, it would describe a well known curve called the tractrix. Experience would soon decide whether a slight modification of this curve or the quadrant would answer best in practice. The latter would require the axles of C to converge to O. A spring might be released, which should set them parallel the instant C came upon the straight rails, A B, F E.

G. R.

Cambridge, June 28, 1838.

MR. WHITE'S QUESTION IN SURVEYING.

Sir,—I return "O.N." my best thanks for the simple directions he has given me for solving the trigonometrical question which I proposed in your highly useful Journal (No. 760): I found it quite an easy task to make out an arithmetical solution from "O. N." 's rules. I was somewhat surprised to find that the said question had attracted the notice of a gentleman at the Cape of Good Hope, who has given a geometrical construction, and an analytical solution of it, in the *South African Commercial Advertiser* for June 27th.* I have made out an arithmetical solution of it from his investigations, and I find the required answer exactly agrees with that which I obtained from "O. N." 's method. As the gentleman at the Cape of Good Hope considers the question, and all others of a similar nature, of very high importance in trigonometrical surveying : such being the case, Mr. Editor, perhaps you will republish his solution in your scientific Journal.

"To the Editor of the *South African Advertiser*."
"Sir,—In turning over the leaves of a late number of the *Mechanics' Magazine*, my attention was drawn to the following notice addressed to the editor of that periodical."

[Here Mr. White's letter is quoted, which, of course, we need not repeat.]

"This is one of a numerous and interesting class of questions connected with the subject of trigonometrical surveying, which are all reducible to the well-known

---

* We also have received a copy of this newspaper from a correspondent in Africa.—ED. M. M.

---

problem originally proposed by Townley; by which, from three objects given in position, and the angles which their mutual distances subtend at a station in the same plane, the relative position of the station is determined. They are all of practical importance, both to the military and land surveyor; and in this colony especially, where in the Survey of Grants *every point of importance ought to be determined by triangulation*, they should find a place in the note-book of every surveyor. With the view of drawing the attention of our colonial surveyors to the subject, I am induced to request the favour of a corner in your next paper, for the following solution of the above question.

"Yours, &c.   I.

At H, in the known distance WH, lay off the angles W H Q, WH R respectively equal to the given angles W S C, WCS; and in like manner at W, the angles HWQ, HWR equal to the given angles CS H, SCH. Draw Q R.

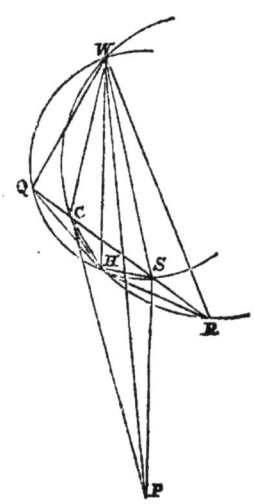

Through the points W, Q, H, describe a circle intersecting Q R in S; and through the points W, R, H, describe another, cutting the same line in C. By this construction, the position of S and C, where the angles were observed, are determined.

Join S H, C H, and on S C describe the triangle S P C, having the angles at

S and C equal to the given angles CSP, SCP. Join PW.

1. In the triangle WHQ there are given the side WH, and the angles to find WQ, QH, which may be computed from the following expressions :—

$$WQ = WH. \frac{\sin. \, WSC}{\sin. \, (WSC + CSH)},$$

$$QH = WH \frac{\sin. \, CSH}{\sin. \, (WSC + CSH)}.$$

2. From the three points W, Q, H, thus determined, and the angles they subtend at S which are given, the distance SW may be found by means of Townley's Problem, of which the following is an analytical solution.* :—

Let s=semiperimeter of triangle WQH
h, w=sides WQ and QH
Q=angle WQH
H, W=angles QHS and QWS
m, n=angles HSQ and WSQ.

1. $\text{Sin.} \, \frac{1}{2} \, Q = \text{sq. rt.} \left(\frac{(s-h) \, (s-w)}{h.w} \right)$

2. $\text{Tan.} \, Z = \frac{w \, \sin. \, n}{h \, \sin. \, m}$

3. $\text{Tan.} \, \frac{1}{2} \, [H-W] = \text{tan.} \, [Z-45°]. \, \text{Tan.} \, \frac{1}{2} \, (Q+m+n)$

4. $W = \frac{1}{2}(H+W) - \frac{1}{2}(H-W),$ where $\frac{1}{2}(H+W) = 180° - \frac{1}{2}(Q+m+n)$

5. $SW = \frac{h \, \sin. \, (n+W)}{\sin. \, n}$

3. In the triangle SWC there are now given SW and the angles; hence SC may be found as in (1); so also may SP in the triangle SCP, of which the side SC, and the angles at S and C, are known.

4. Lastly, in the triangle PSW, from the two sides SP, SW being known, and the included angle given, we have for the distance sought

$$PW = [SW + SP] \cos. \, Z$$

where Z is an auxiliary angle, such that

$$\sin. \, {}^2Z = \frac{4 \, SP. \, SW. \, \cos. \, {}^2 \frac{1}{2} \, WSP}{(SP + SW)^2}.$$

"N.B. Had the geographical position of W been given, and the azimuth of H on its horizon, there would have been sufficient data to deduce the latitude and longitude of P. Quere—what is the solution?"

I am, Sir, your obedient servant,
A. B. WHITE.

Bath, Nov. 1, 1838.

---

* This would be the indeterminate case of Townley's Problem but for the point C, which is determined from the intersection of QR by the other circle.

ON THE CONVERSION OF WATER INTO STEAM, IN THE HIGHER DEGREES OF TEMPERATURE, AND THE BURSTING OF STEAM-BOILERS. BY DR. CHARLES SCHAFHAEUTL.

It is well known that water can, in all degrees of temperature, be converted into steam; nevertheless the vaporization becomes rather difficult in some cases, particularly when the solid conductor which is to communicate the caloric to the water, possesses rather a higher degree of temperature than that under which the liquid is to be converted into steam.

In support of this statement, we need only refer to a long known experiment, which was first made by Leidenfrost, in 1756,—I mean that of letting a drop of water fall into a platina crucible made into a white heat. The water globule spins rapidly round without ebullition, and the evaporation is slow; in proportion as the temperature of the vessel is high. In an experiment made by Klaproth, six drops of water were allowed to fall into a vessel of white hot iron, during the time it was cooling in the air. The first drop required 40 seconds to evaporate; the second, 20; the third, 6; the fourth, 4; the fifth, 2; and the sixth immediately evaporated. Yet, if such a drop, which has remained some seconds in a white hot platina vessel, be turned quickly into the hand, it will be found scarcely warm. The vaporization of the drop of water proceeds quicker in proportion as the temperature of the crucible falls, and in the not yet ascertained degree of temperature above 212, the drop is rapidly dispersed.

In addition to Klaproth, Doebereines, Berzeleus, Muncke, Laurent, and Tomlinson have repeated these experiments, with various alterations, and have given different explanations of it. The reason that the drop of water did not get hotter and evaporate, they ascribed to the circumstance of its being repelled from the hot surface, and for that reason could not remain in contact with the heated body.

Notwithstanding this hypothetical repulsion of the drop of water, by which means it was prevented from getting hot, they found it necessary to account for its rotation, by the circumstance of steam being generated where it touched the hot surface, which forced it away again.

In the following few lines we will endeavour to ascertain the cause of this phenomenon, and bring it in connection with similar ones not yet known or described.

Let us try to heat a fluid at a certain part, i. e. to communicate caloric* to one part of its external surface, and to disperse the caloric, from this point, through the entire mass.

In doing this we must bear in mind two forces, on whose relative action depends the dissemination of the heat, from the point where it is first applied, to the interior. The first of these is the force of cohesion, with which one molecule attracts another in infinitely small distances. The second is the force of gravitation, by which each molecule is attracted by the earth, in the inverted square of the distance.

If a drop of water be allowed to fall on a base which possesses no attraction towards the molecules of which it is composed, that is to say, on a greasy surface, or in the case of a drop of quicksilver on glass, the drop assumes a spherical shape, that is to say, the force of cohesion with which one molecule attracts another is so overpowering, that the force of gravitation, which likewise affects each particle of the molecule, has not the power to spread the drop over the surface, and occasions only a slight depression on the side upon which it falls, all the molecules in the globule being naturally in a perfect equilibrium, occasioned by their own cohesive force.

The circumstances are otherwise with a body of water, which, to preserve its equilibrium, must be enclosed in a vessel. If we imagine this body of water to be composed of single drops, as before described, one drop at the bottom of the vessel has to support the weight of all those in a perpendicular line above it, the force of cohesion in this drop is, therefore, very soon overpowered by the force of gravitation of the drops above it, and would immediately expand, were it not prevented by the vessel it is contained in.

Now the only part of this body of water, which is in perfect equilibrium

without receiving support from a solid body, is the surface. In a drop of water, on the contrary, the whole is in perfect equilibrium without any foreign assistance.

To apply heat to one point of a drop of water, it must naturally be applied to the surface of the globule; to apply heat in a similar manner to the body of water contained in the vessel, we must likewise apply the heat to the surface.

Let us now, for example, put the bulb of a thermometer to the bottom of the vessel, and pour upon the surface of the water contained in it some ether, then set fire to the ether, and as it is consumed continue to supply fresh. Were this continued, for a whole day the water would never boil, and the thermometer scarcely rise, provided the vessel did not get hot, and by this means communicate caloric. In like manner it will be equally impossible to make a drop of water boil in the centre, by applying heat to its external surface. That part of the surface of a drop of water which is touched by an ignited body, is immediately converted into steam; if we now consider the drop of water in the hot platina crucible where it touches the white hot surface, steam is immediately developed with explosion on the point which touches, the elasticity of which immediately elevates the drop, till by the force of gravitation it again touches the crucible on another part, and is forced upwards by another explosion. At the same time the hot atmosphere in the crucible creates a cloud of steam around the drop. One part of this expands and disperses, but the other part of it is still retained by the molecular attraction of the drop, and prevents the further immediate touch of heated air.*

As it was therefore found impossible to conduct heat from the free surface of a liquid to the interior, it being, as it is well known, proved by Count Rumford, that the radiation of heat from one mole-

---

* Melloni, in his interesting experiments on the radiation of heat, has shown that the higher the temperature of a radiating body, the less will the rays of heat be absorbed by transparent bodies.

* By this I am reminded of a similar oft-repeated experiment of a man sitting on a heated oven without receiving any injury, although a fowl was at the same time cooking in it, but the heat was only rendered supportable to the man, so long as he continued to drink pretty freely. The reason of this is obvious. The first attack of heat caused a profuse perspiration, which surrounded the human body with an atmosphere of steam, which being a bad conducting power for caloric, prevented the body from absorbing the heat, which soon decomposed the fowl in which animal life was extinct.

cule to another in liquids was =0; in order to heat a fluid, it is necessary that each single molecule of the fluid should come in absolute contact with the source of heat. It is scarcely necessary to say that this is effected with a quantity of liquid contained in a vessel in which the force of gravitation is unchecked by the force of cohesion; by applying heat to that part which bears the most pressure, which is naturally the bottom.

That part of the water which comes in contact with the heated bottom expands, becomes lighter, and is immediately forced up to the surface by the unheated parts of the water, which are naturally heavier, which process continues until each part of the fluid has come in contact with the heating body. If you apply heat to the drop of water, the outside layers first get warm, and are then of less specific gravity than cold water, and will, therefore naturally always remain on the outside of the water sphere, as was recently shown in an experiment made by Tomlinson, who placed, on some oil heated to 450 or 500 degrees of Fahrenheit, a drop of water coloured with ink, and a drop of ether, whose density in comparison with water =0.7155. Both drops were immediately amalgamated, the water being of the greater density occupied the centre, the ether forming the exterior of the globule.

That the attraction of the vessel towards the water ceases to be perceptible in a degree of temperature in which the water cannot longer exist in its liquid state, is self evident; and we have no reason to have recourse to repulsion, where the evolution of vapour is sufficient to explain all the phenomena of which we have just spoken. If, therefore, it thus appears impossible, under the above-mentioned circumstances, to convert a drop of water, even by an immense quantity of accumulated heat, into steam, it nevertheless is easily accomplished by the following methods :—

We have seen that each time the drop of water came in contact with the hot surface, a certain portion of the water has always been immediately converted into steam; that is, the portion of the surface which came into immediate contact with the vessel. Now, to convert the whole drop at once into steam, it is necessary to divide it into so many small parts that each particle may simultaneously touch the vessel.

In a somewhat similar manner a ball of gunpowder when ignited requires a long time to be consumed, and would not be capable of forcing a bullet out of a gun-barrel; yet if we divide this ball of gunpowder into small grains, of which the common gunpowder is composed, the whole mass when ignited immediately explodes, and propels the bullet with great force to a long distance.

The division of the drop of water above alluded to, is to be effected—first, by mechanical means—viz., by violently propelling it against an incandescent vessel. The drop is thus, by the concussion, scattered into very small particles, which are simultaneously converted into steam. On this principle I constructed a steam-engine, which worked effectually with a cylinder of the diameter of 6-8th parts of an inch to a half-horse power.

The second method is accomplished by chemical means, by balancing the cohesive forces, on which depends the globular form of the drop by the capillary action of a body which has just been ignited. To elucidate this point, let us heat the bottom of a sand-bath to a dark red heat, and cover it about two lines deep with fine well-washed sand. This mass of sand forms an aggregate of small speridical bodies, which, with the interstices between them, act with great capillary force on liquids. As soon as we let a drop of water fall on the heated sand, each grain of the surface of the sand which comes in contact with the water globule, imparts its heat to an equal particle of the surface of the water globule, and thereby each of the grains is perfectly cooled, as the particles of water are converted into steam. This cooled layer of sand imbibes immediately the remaining portion of the water globule, and a further repulsion of it is therefore quite impossible.

The greater height from which the drop of water falls, that is to say, the greater force with which the water is driven into the sand, the more water of the drop is converted into steam, and therefore the greater is the explosion produced.

Let us now fill the sand-bath about two inches deep with the same description of sand, and take an evaporating dish with a semi-circular or spheridical bottom, and press it carefully about an inch and a quarter into the sand, so that the impression in the sand corresponds exactly with the spheridical curve of the bottom of the dish. Then dip the dish the same depth into cold water, and take it out so carefully that as much water as possible remains hanging on the dish, which will collect in the middle by its own gravity into a semi-circular drop with a broad base.

Now, when we carefully put the dish into the corresponding impression in the sand, so that only the extreme part of the drop of water attached to it touches the sand, nothing is heard but a slight hissing noise, which is only momentary. But if we allow the dish by its own weight to sink fully into the impression in the sand, so that the drop of water is flattened between it and the dish, a violent explosion immediately ensues, and the dish is forced several inches out of the sand. After the experiment, we find the bottom of the dish covered with a layer of wet sand, undulating in rays from the centre to the periphery, and which has precisely the same degree of temperature as the water previous to the introduction of the dish into the sand.

This mode of instantly transferring caloric from one small solid sphere to another liquid, forms the introduction to another way in which steam is generated. In all the former experiments the caloric has been transferred from a solid hot body to a liquid. We now proceed to a method of instantaneously generating steam, which, as far as I am aware, has never been described before. I mean the transferring of caloric from fluxed incandescent bodies to the water. We have seen in the experiment with hot sand, that the sound of the explosion has possessed a character of fulness and thickness like that produced by the explosion of a quantity of gunpowder in the open air, which is always a sign that the whole of the water has been converted into steam in a certain measurable space of time. But in the case now under consideration, the incandescent fluxed body imparts its caloric to the water in the same moment that it becomes crystalized, and consequently the instantaneous crystalization of the one body, and the simultaneous volatilization of the other, creates a sharp loud report, similar to that occasioned by the

explosion of fulminated silver or quicksilver.

A simple trial will render this quite clear.

Let us put some drops of water on the smooth surface of a blacksmith's anvil; then heat a bar of iron about 1½ inch broad on one end to a white heat in a forge near the anvil; apply this heated end of the iron with the broad side upon the drops of water, and give the iron a sharp blow with a hammer. A loud explosion, similar to the report of a fowling-piece, will follow.

Let us now repeat the same experiment with this difference, that before we put the white hot end of the iron upon the drops of water, we free it from the slack by a few strokes of a large file, then put the iron on the drops of water, and, as before, strike it with the hammer. No report will then be heard, although we shall find a great portion of the water evaporated. Now, let us repeat the experiment as in the first instance, and surround the drops of water on the anvil with a wall of sheet-iron or paper, except where the iron is to be introduced; after the explosion we shall find a considerable portion of black sandy powder adhering to the anvil and sides of the paper. On a close examination we shall find this powder to consist of minute globules of slack, which, if dissolved in acid, does not evolve even the least trace of gas, and which possesses exactly the same chemical properties as the slack taken from the bar of iron when just out of the forge, and that it consists of part oxide and peroxide, with a little silicious acid.

We find the surface of the iron bar, after the report, quite free from slack on the spot where it came in contact with the water, surrounded by radiations which are caused by the escape of the crystalized slack.

By this we see, that the report is not occasioned by the conversion of the water into steam by means of the white hot iron, which parts with its caloric too slowly to produce that effect; but, on the contrary, by the stroke of the hammer the cohesive force of the fluxed slack on the surface of the iron bar, and of the water on the anvil, is overcome; and that they, by their mutual resistance, are both at the same moment divided into small particles like grains of sand, each of which, in the process of crystalisation, imparts caloric in a greater quantity, and in a much quicker way, to the adjacent particles of water, than the red hot grains of sand would do.

We will proceed to confirm our observations still further.

In iron-works the hearth of a boiling furnace, after boiling pig iron, is covered with fluxed white-hot slack, which is similar to that on the bar of iron alluded to in a former experiment. Upon this white-hot fluid slack we may pour a pint of cold water, without any danger of an explosion. The water will run in globules, as in the white-hot platina crucible, or in large flat circular masses, upon the surface. These *masses* are continually in a waving vibrating motion, occasioned by the generation of steam where they come in contact with the slack. It is very difficult to force one of these water globules or masses into the slack with an instrument, the drop always escaping sideways between the instrument and the liquid slack; but if you succeed in forcing a small quantity of water into the slack, so that both are divided by the pressure and become mixed, a tremendous explosion immediately follows, by which the slack is scattered in all directions, and the furnace not unfrequently destroyed. Yet again, you may safely allow a portion of this slack to run in a basin of cold water. The water will slowly generate steam, and you can perceive the slack for some minutes at the bottom of the water in an incandescent state. In both cases the cohesive force of the drop of water, and that of the drop of slack, prevents their mixing without external violence. But when the cohesive force of a drop of water is partially counterbalanced by the attractive force of another body—for instance, if you put a drop of water on a bar of cold iron, and dip it as quickly as possible into the slack—an explosion immediately takes place. And likewise, although you may safely allow the liquid white-hot slack to run into water, if you let the same hot slack run on wet sand, or even on the damp earth, a violent explosion will be the consequence.

When instead of free water you use water which is chemically fixed to any body, as in the case with water which combines with salts, when they crystalize you will easily cause an explosion.

For example : when you put a crystal of Glauber's salts (sulphate of soda) on the surface of the fluid slack, it melts into a viscous mass, and develops very slowly its water of crystalization. During this process you may easily press the crystal into the slack and cause an explosion. In the same manner as melted slack, act all white-hot fluxed metals, such as copper, zinc, tin, &c. &c.

It is well known, that in the casting of copper, a single drop of water is capable of blowing up a whole casting house; and that the moulds in which copper is to be cast require to be perfectly dry, or a violent explosion will be the consequence of neglect.

Cast iron is an exception. When in a state of fusion it is an extremely slow conductor of caloric. In consequence of its being combined with a large quantity of carbon and silicum, it crystalizes extremely slow; and when brought in contact with water, while in a state of fusion, it decomposes it, combining with one of its constituent parts, and generating carburetted hydrogen very slowly. Therefore iron may be cast in wet moulds without danger.

Concentrated sulphuric acid, which contains some substances in its solution which are not volatile, may be freed from these impurities by the process of dis-

---

* When a pile of iron, in iron forges, is brought out of the reheating furnaces, it is immediately passed in a groove between two revolving iron rollers, in order to weld it and draw it into a smaller shape. Cold water is constantly dropping on the rollers, to keep them cool. The iron, when taken out of the furnace, is naturally covered with liquid slack, as in the case of the iron bar above alluded to, and when the iron is passing between the rollers, drops of water frequently fall upon it and are drawn with it through the groove. When this happens a loud report immediately takes place, and the slack is forced out from between the rollers. But when the iron becomes free from slack, no report will ensue under the same circumstances.

* A phenomenon which frequently occurs, and which has some relation to the foregoing observations, must not be allowed to pass unnoticed. When some salts are dissolved in fluids to saturation, and the crystalization of the fluid takes place instantaneously instead of gradually, a great quantity of caloric is freed; and thus it often happens that a saturated solution of Glauber's salt, which is cooling on a place free from motion, begins, by the slightest shaking, to crystalize immediately, and the vessel becomes at the same moment too hot to hold in your hand.

tillation.; but it is well known that this cannot be done with sulphuric acid, which contains lead in its solution. When this impure acid begins to become concentrated, sulphate of lead begins to chrystalize and fall to the bottom of the retort; and this process creates such a violent disengagement of steam, that the strongest retorts are invariably broken to pieces. This is likewise the case with silicia which has been made soluble in water by melting it with alkali. When this solution is neutralized by acids, and evaporated, the silicia contained in it very often disengages itself in a sort of flocky crystalization, and at the same time steam is so suddenly disengaged that not only a great portion of the liquid is thrown out of the open evaporating dish, but the concussion is so violent that the dish itself is often thrown down from its stand. The same sudden disengagement of steam occurred when the solution was kept boiling in a glass bottle under a pressure of $1\frac{1}{2}$ atmosphere, and never failed to break the glass. Some few seconds before the explosion occurred, the escape of steam from the safety valve became partially impeded, and made a peculiar vibrating noise, combined with a weak hollow sound, which was always followed by the explosion of the bottle. When in the same manner I boiled pure water until the last drop was evaporated, I never could distinguish any change in the sound of the escaping steam, even when the boiling water was agitated. The vibrating motion before alluded to is always a certain forerunner of a rapid crystalization of the bodies held with the water in solution.

Before I proceed to apply the foregoing observations to the explanation of the causes of the bursting of steam boilers, I will first show, that this can in no instance be caused by the decomposition of water.

I welded a gun-barrel 28 inches long, air-tight at the stoutest end, and, by means of a bore, cleared 4 inches of the interior nearest the welded end from oxide, &c., occasioned by the process of welding, to procure a clean metallic surface. At the other end of the barrel a piston was moving air-tight. This piston was perforated in the centre its whole length, and closed at the end with a cock; the piston was likewise so arranged that it might be fastened in any position by means of a screw. At the extremity of the piston rod was attached a dish to put weights upon, to balance the pressure of steam in the interior of the barrel, and keep it in any desired position. I then introduced into this barrel several drops of water; the piston was afterwards brought down near the end of the barrel, and the cock closed. Two inches of the welded end of the gun-barrel was then introduced vertically into a small blast furnace, by an aperture just sufficient to admit it, and heated as quickly as possible. The piston began to rise rapidly, and, by means of the weights placed upon the dish, was kept just 22 inches from the welded end. As soon as the end in the fire became red hot, the piston stopped rising; and when the barrel became white hot, it did not alter the position of the piston. The piston was now screwed tighter in its position, and the apparatus taken out of the furnace, and cooled as quick as possible.

After it was sufficiently cooled, the apparatus was opened under some quicksilver, and the remaining gas was tested. Of three experiments, the results were as follows;—

In the first experiment, the contents were found to consist of one volume azote to eight volumes of hydrogen; in the second, the proportion was as one to six, and in the third as one to five. In neither was there any trace of oxygen to be found. The interior of the barrel was found crusted with oxide, which increased in thickness with each experiment. In the same apparatus, one volume of hydrogen was mixed with two volumes of atmospheric air, and kept, during the process of heating, under a pressure of two atmospheres. As soon as the welded end began to acquire a red heat, the mixture of gases exploded, as might be expected.

Now, to ascertain the influence of steam mixed with this gaseous compoud, and likewise the influence it exercises over the exploding properties of the gases, I introduced into the shorter end of a glass tube, bent in the form of a syphon, the mixture of one volume of hydrogen and two volumes of atmospheric air. The gases were then confined in the tube by means of quicksilver, and brought under a pressure of $1\frac{1}{2}$ atmosphere. The water, whose weight had been nicely calculated for the required volume of gas in the experiment, was allowed to be absorbed by a texture of asbestos, and then, by means of a wire, conducted through the quicksilver into the gases contained in the tube. The whole apparatus was now placed in an iron cylinder, filled with quicksilver, kept in a degree of temperature about 236 degrees of Fahrenheit.

The closed end of the glass tube was then carefully heated by means of a blow-pipe, to a red heat, for the purpose of causing an explosion.

The results of the experiments were as follows :—

One volume of the fulminant gas, mixed with 0.1 volume of water gas, exploded and broke the tube. The explosion was almost as violent with 0.2 of water gas ; with 0.3 the detonation was considerably weaker; with 0.4 the explosion was still weaker; with 0.5 a feeble explosion took place, just as the heated glass tube began to be blown out by the pressure of the gas; and with 0.6 parts, only one explosion took place, though the experiment was tried six times; with 0.7 to 2.0 volumes of water gas, there was no explosion.

Now, from these experiments we may conclude, that if even an exploding mixture were formed in the steam-chamber of a boiler, it must at least be double the quantity of the steam to cause an explosion—which, under any circumstances, is highly improbable.

On considering the circumstances which cause the bursting of steam-boilers, we will pass over those occasioned by the non-acting of the safety valve, or in consequence of its improper size; and proceed at once to those which are not so obvious.

A cylindrical boiler, made of the best charcoal plate, 0.2 of an inch thick, two feet long, and nine inches in diameter, mounted as usual with a safety valve, and connected with a small steam-engine model, and an injection pump, was in this way connected with the flue of a puddling furnace, so that the flame enveloped this boiler in the usual way, and kept constantly working day and night. The boiler, during the first day, worked with 20 atmospheres pressure; and, during the rest of the week, at three atmospheres. After this the injection pump was stopped, and the water kept an inch below the usual level; so that the sides

were allowed to get red hot. When the sides were thus heated, the water was kept in motion, as much as possible, by moving the float, under a pressure of 10 atmospheres. After remaining three hours in this state, the injection pump was again put on, and the boiler worked with three atmospheres pressure to the end of the week.

The first day of the following week, the boiler worked with the same pressure, and with the usual height of water. The safety valve was now gradually loaded to 10 and 11 atmospheres. The weights were added with the utmost care, so as not to impede the action of the valve. When another weight was added to increase the pressure to 12 atmospheres, the boiler burst with a tremendous explosion, and the whole of the upper part was carried almost as high as the large steam-engine chimney, which was sixty-six feet.

On inspecting the edges of the fracture, I found that the part of the boiler which had been red hot, had undergone a remarkable change. A slip of the upper part of the boiler, 1 inch in breadth, was bent to nearly a right angle, by applying a weight of 3 cwt. ; whilst a slip exactly of the same size, cut out of the part which had been red hot, broke by an application of 2 cwt.

The fractured end of this slip was cut off, and polished with a fine file. The greys which thus appeared, held in comparison with the sound slip, which was bent into an angle ; on being placed under a microscope, a dark elliptical form, increasing in number towards the outside, which, to a considerable depth, exhibited a combination with sulphur.

On a careful inspection of various fragments of other steam-boilers which had burst, I found the fractured edges always changed in a similar manner ; and I found that the edges of a piece of iron which was shown to me as a fragment of the boiler which burst on board the Hull steamer, were so brittle, that a slight stroke of the hammer was sufficient to break them off; and, when put into a retort, and diluted in chlorhydric acid, a great quantity of sulphuretted hydrogen was perceptible, the exact quantity of which I was enabled to ascertain by collecting in a solution of nitrate of lead.

By this experiment, it appears to me

clearly demonstrated, that the iron plates of a boiler which, by the pressure of steam contained in them, are always kept in a certain degree of extension by the flame acting upon them, especially by the flame of pit coal, which always contains a quantity of sulphureous acid, are considerably weakened, as the texture of the iron by this means becomes loosened; and the more pyrites the coal contains, the worse it will be, particularly when the boiler, by the sinking of the water level, is often allowed to get into an incandescent state, and afterwards, by the rising of the level, cooled again.

The expansive force of the steam always tends to separate the fibres of the iron; which, in addition, is kept in a state of expansion by the caloric which it imbibes by being allowed to get into an incandescent state. Afterwards, the irresistible contraction of the doubly-expanded metal, caused by its coming in contact with the water, must act upon it in a very powerful and detrimental manner; and this we consider the fundamental cause of most steam-boiler explosions. The slightest irregularity in the generation of steam, which occasions its development by sudden starts, must naturally tend to burst such a weakened boiler, whether the sides be in an incandescent state or cool.

On trying the strength of boiler plates, too little regard is very often paid to their capacity of resisting a sudden shock, which ought always to be borne in mind, and the plates constructed accordingly; for a long series of experiments, too tedious for detail, have proved to me the great difference which exists in trying the absolute strength of such plates by the gradual force of an hydraulic press, and by a sudden and violent

application of the same force. The most fibrous irons are generally the least calculated to resist a sudden check; and the more round and separated, or more stringy the fibres appear, and the more greys that are to be seen on the surface of the polished plate, the less fit is the iron for the construction of a safe boiler. The best plate for this purpose is that which appears to consist of a series of extremely *thin leaves* of a light greyish colour; and the edges of which are not much jagged, but present a somewhat even appearance.

The sudden starts of steam which sometimes occur in a boiler, are generally caused by a saturation of the water with certain earth salts, of which sulphate of lime, silicia, and alumina, are usually the most prominent.

We have shown, in one of the foregoing notes, not only how all porous bodies quicken the development of steam, but likewise that the shock occasioned by steam being generated in such sudden starts, by the crystallization of the bodies which the liquid contained in solution, is alone sufficient to break such a weakened boiler, particularly when the water by such shock is forced against the red hot sides.

The incrustation which accumulates, and is generally chiselled out of the boilers, consists, in most cases, of distinct layers of sediment, often of considerable thickness; which clearly shows that these layers have been quickly formed, and at certain intervals. Water saturated with earth salts, contained in steam-boat boilers with narrow cells is, in the process of boiling, converted into foam; the incandescence of the boiler is, in consequence, hardly to be prevented; and the cocks used for the purpose of ascertaining the level of the water, are not to be depended on.

A sudden alteration in the sound of the steam escaping from the safety-valve is, in my opinion, always a certain sign that a crystallization or sediment is rapidly forming in the boiler; and one of the witnesses (who, if I remember rightly, was a woman), in giving her evidence respecting the late fatal accident on the Hull steamer, particularly mentions, that, previous to the explosion, the escape of the steam from the safety-valve was accompanied by a strange and peculiar noise.

---

* The iron slabs out of which the plates are rolled, ought to be very small and close-grained; so that, when the plate is completed, the granulation of the iron is just beginning to assume a leafy appearance.

If the slabs are too open-grained, before the plates are rolled to their proper thickness, the granulation becomes too loose, and weakens it materially. It is still beleived that all metals become closer-grained and of more specific gravity by rolling: but this is not the case with iron, or any metal which, during the process of rolling, assumes a fibrous appearance; for Broling, who was ascertaining the specific gravity of tilted and rolled iron, was astonished to find, that the more the iron was rolled, the more it lost in specific gravity; which was accounted for by the particles of iron repulsing the water in which it was weighed.

K

We will now conclude these few remarks by repeating, that boilers whose sides are frequently in a state of incandescence, are, even under the most favourable circumstances, never perfectly safe. Yet that the use of distilled water in such boilers, instead of river or sea water, would tend much to lessen the danger, and, in many instances, prevent those distressing and melancholy accidents, too frequently attended with the sacrifice of human life.

ALLEGED DEATH FROM THE USE OF HARPER AND JOYCE'S STOVE.

We were, we believe, the first to point out the danger which would attend the use of the prepared fuel, patented by Mr. Joyce, of Camberwell, and to caution the public against the fatal effects which experiment led us to believe would be the result of its use in confined places. It seemed to us that a knowledge of the first principles of chemistry could not but confirm this conclusion in the minds of any one who chose to take the trouble to examine the subject. It appears to us most extraordinary, that men of high standing in science, are found so reckless of their fame, as to give puffing reports of one-sided experiments upon inventions submitted to them, as in the present instance.

For the free expression of our conviction, we were threatened by Messrs. Harper and Joyce with an action for damages for libel, unless we retracted or explained away our remarks; and more particularly with reference to a note, most applicable on the present occasion. We asked (Vol. 29, p. 208)—" If the death of any party had taken place, in consequence of full reliance on the assurances of the prospectus of the new stove, as to the innocuous nature of the prepared fuel, *would not* the patentees be indictable for manslaughter—*at the least?*" As public journalists, we thought we had done no more than our duty in warning our readers of the threatened danger; and, as we were convinced of the correctness of our opinion we declined either to retract or explain away our published opinions. Messrs. Harper and Joyce have been better advised than to appeal to the judgment of a jury, up to this time; and after the fatal event, a report of which

follows, and which we have abridged from the papers of the day, we hardly think they will be so rash as to venture further in their attempt to stop the mouth of honest criticism. We shall, in our next number, publish Mr. Joyce's specification of his method of preparing the charcoal for fuel.

Report.

On Monday evening, Nov. 19, a lengthened inquiry took place at the Jamaica Coffee-house, Michael's-alley, Cornhill, touching the death of James Trickey, aged 64, who was supposed to have been suffocated by the noxious effluvium of Joyce's newly-invented stove, placed in St. Michael's church. Between 4 and 5 o'clock on Saturday afternoon, the stove was placed in the centre aisle of St. Michael's church, for the purpose of heating it. Mr. Harper, the inventor's partner, suggested that some person should be in the church to watch the stove in its effect, in consequence of which the deceased was appointed for that purpose. Between 6 and 7 o'clock on Sunday morning, the church was found full of a sulphurous smoke, and the deceased was lying on his face in the centre aisle of the church, about a yard and a half from the stove, quite dead. There were two gas lamps burning in the church during Saturday night, but there was no smoke or smell from them.

Messrs. Pugh, South, Ball, and Smith, surgeons, were of opinion that death was produced by apoplexy, and not from suffocation.

Mr. Blinkharne, surgeon, of Gracechurch-street, was decidedly of a contrary opinion. He considered that the deceased's death was the result of his having inhaled some noxious effluvium.

Mr. Harper, of King William-street, London-bridge, stated, that the fuel burnt in the stove was prepared by reducing the common charcoal to a white heat, before which it was sprinkled with a portion of ' chrystallized carbonate of soda.' When it arrived at a white heat it was placed on dampers, and being again sprinkled with the same soda, was put up for use. Witness had been accustomed to have one of the stoves burning in his counting-house, as well as bed-room, with the most pleasing effect; it could be regulated at any degree of heat.

Several gentlemen in the room said that they had used the stove, and a variety of evidence was given in its favour; but, there being a great difference of opinion amongst the jury, the inquiry was adjourned till the following Friday.

At the adjourned inquest, on Nov. 23rd, the coroner said, since they had last met, at the request of Mr. Harper, he had given orders that a stove should be placed in the church with closed doors, from 5 o'clock on the morning of that day, until the time of the jury re-assembling at 6 o'clock in the evening, to test the atmospheric air in the church in which the stove had been burning, at the expiration of the 13 hours. The experiment was, however, spoiled, by the doors of the church being opened, and by Messrs. ooper, Brande, and Golding Bird, entering.

Dr. Golding Bird's evidence was altogether misrepresented in the report of the inquest. The following is the Doctor's statement, as corrected by him, and published in following day's *Times* :—He stated, that when he entered the church at 4 o'clock, p.m., in company with the churchwardens, there was lying near the floor in the middle aisle, a stratum of air sufficiently impregnated with carbonic acid to be, to the best of our belief, exceedingly injurious, and perhaps even fatal, to any one in a recumbent position and exposed to its influence. The carbonic acid evolved during the combustion of the charcoal would, on escaping from the apertures in the stove, divide its caloric with the circumambient air, and then, becoming specifically heavier (in the proportion of 3 to 2), would fall towards the lowest parts of the church. Mr. Cooper, on the contrary, stated, that the carbonic acid, being heated to a considerable extent, would rise towards the ceiling, and that the largest quantity would be found there, until, by gradual intermixture with the surrounding atmosphere, it would become so far diluted as to be no longer injurious. "This gentleman did not take specimens of air near the floor, where I maintain the greatest proportion of carbonic acid would be found. This I did in the presence of the churchwardens, and in the very spot where the unfortunate man's head was found lying, and in this I detected a considerable portion of carbonic acid. The test of lime-water was objected to, on account of the carbonic acid always present in the atmosphere (about 1-1000th part); but this was begging the question—for every chemist must know, that this reagent requires many minutes' exposure to the air in its ordinary state before any opacity appears, whereas the limewater had been scarcely poured into an open vessel placed on the floor, about six feet from the stove, and on a plane below it (on account of a step being situate between this spot and the stove), before it became quite milky. And, whilst stooping to perform this experiment, I unavoidably inhaled this lower stratum of air, and was immediately attacked with severe throbbing of the temples, weight over the eyes, and a disposition to syncope so great that I had considerable difficulty in reaching Mr. Blenkarne's house, where some of the air I had collected was examined. These symptoms are similar to those experienced in crowded and ill-ventilated apartments, especially in rooms in which a chauffeur has been burning."

Mr. J. T. Cooper, Professor of Chemistry, deposed, that he had made several experiments with the stoves, and was convinced from them, that if the stove was properly adjusted to the size of the apartment there was no danger to be apprehended. The quantity of carbonic gas contained in the stove at full heat, never exceeded quite 1 per cent. St. Michael's Church contains about 100,000 cubic feet. The quantity of charcoal which the stove contained at the time of lighting on the Saturday night, was 49 pounds. "If it was possible that such could be consumed in a moment, it would not reach $1\frac{1}{2}$ per cent. 10 per cent. of carbonic gas would be sufficient to throw a person into an insensible state ; 12 per cent. of the gas would, in my opinion, be enough to destroy life. I have tested carbonic gas in the crowded gallery of the theatre, and found it to be 4 per cent. I have also found the same per centage in crowded chapels. At the present time I should say, that there is near 4 per cent. of carbonic gas in this room."

Dr. Brande was next examined ; his testimony was a corroboration of that given by Mr. Cooper.

Mr. Harris, one of the churchwardens of the parish, said, when he went into the church that afternoon, with Messrs. Brande and Cooper, and Dr. Bird, he was seized with a headache and giddiness, and made his way out again as soon as possible.

After some further evidence in favour of the stove, the jury retired about twenty minutes, and then returned a verdict :— 'That the deceased, James Trickey, came to his death by apoplexy, accelerated by inhaling impure air.'"

We may add some further remarks on the subject, by Dr. Golding Bird, which accompanied his communication to the *Times*, correcting the report of his evidence :—

"Considerable stress was laid on the excessive heat evolved by the stove, exciting deleterious vapour from the wooden coffins n the vaults beneath the aisle. This I conidive to be quite unfounded ; for after the consumption of 49lb. (I believe) of charoeal, which became entirely consumed in 12 hours, instead of 18, as has been stated (producing about 1,500 cubic feet of carbonic acid gas), the temperature of the air near the stove was considerably under 60 degrees, and in many parts of the church but little above 50. Afrer a considerable time, by the law of the mutual diffusion of gases, the carbonic acid gas would, indeed, mix with the superincumbent atmosphere, and become innoxious from the dilution. The well-known instance of the Grotto del Cané illustrates the position I have assumed: there the acid gas lies in a stratum reaching up to the knees of a man walking in it, his mouth and nostrils being far above the stratum ; he suffers no inconvenience, whilst a dog or other small animal becomes immediately asphyxiated, his head being immersed in the stratum in a manner analogous to that which I conceive to have been fully possible in the case of the late James Trickey. Another still more remarkable instance of this kind is found in the celebrated Valley of Death in Java—all tending to prove that a considerable time is required before the carbonic acid gas becomes so far diluted as to be respirable ; and if it be objected that, in the latter instances, there are perpetual fountains of the irrespirable gas supplying that lost by intermixing with the air, I would say, that the same remark applies with nearly equal force to the large stove in St. Michael's church, where we have a continued supply of carbonic acid and nitrogen as long as combustion continues.

" In conclusion, allow me to observe, that the remarks now made, as well as the evidence I gave at the inquest, do not apply to the very elegantly constructed stove of Mr. Harper in particular, but to all, under whatever form, in which charcoal is consumed, where there is not a thorough and proper means of ventilation, by means of a sufficiently large chimney, open doors and windows, or otherwise.''

We hardly know what to think of men, standing high in the chemical world, searching for carbonic acid near the ceiling, or in the upper part of a room, instead of at the floor ; it is well known, that this gas is so dense that it may be poured out from one vessel to another. Most of our readers will, doubtless, recollect instances of persons committing suicide by confining themselves in a room in which they ignited a pan of charcoal; and, that they might watch the approach of the deadly fluid, placing lights at various heights from the floor, which were extinguished, one after the other, as the stratum of gas became higher and higher, until it reached the head of the stoical

self-destroyer. To say that one has slept in a room with one of Joyce's stoves in action therein, just amounts to this —that he did not sleep long enough— that the gas from the stove had not had time to rise to the level of the sleeper's head. As well might you aver that a man will not drown in water; and, as a proof of it, place him in a bath, in which the water is not allowed to rise above the man's head.

RECENT AMERICAN PATENTS.
(From the *Franklin Journal* for September.)

MACHINE FOR CUTTING AND DRESSING GRANITE AND OTHER STONE : *Wm. C. Poland, and Earl Blossom, Portland, Maine.*— This machine, we are told, consists essentially of four parts, " 1st. A gang, or set, or sets, of drills, to aid in reducing the face of the stone to an even surface. 2nd. A set, for breaking the pieces left between the holes made by the drills. 3rd. A gang of cutters to do the work usually done by chisels. 4th. A gang of finishers.''

The drills operate much like those used for hand-drilling, in blasting rock, but their shafts are to be lifted by cams, and they are to be gauged so as to enter the stone to a given depth ; provision is made to cause them to turn to a certain distance between each stroke. Rows of holes are thus to be successively made. When this is completed, the set which is to cut away the portion between the holes, is made to operate in a curved direction, something like a cooper's adze. The other instruments are afterwards made to pass over the surface thus prepared, for the purpose of finishing. The claims are to " the combination and arrangement of the drills, set, cutters, and finishers, in the manner described. The form of the set, extending quite across the face of the stone to be dressed, in combination with the other parts of the machine. The method of vibrating the drills as described.''

Since the publication of the account of the successful operation of Hunter's patent stone planing machine, in Scotland, a number of similar machines have been made the subject of patents with us, but hitherto we have not received any authentic history of their successful operation. The reason, we apprehend, is the intrinsic difficulty of causing such machines to operate well upon stone of the texture of good granite. Such stones have been actually dressed by the machines, but we apprehend that the difficulty of keeping the cutters in order has been found to be very formidable, and that it will so continue.

We have, in more than one instance, been

-assured that we should see the machines in .use at the public buildings in Washington, but are still in waiting. We are using, as well as some granite, much miserable sand-stone, which they might cut, although they would occasionally meet with a pebble that might require more humouring than would comport with the steady progress of a ma-chine; our granite, we fear, will still bid them defiance, in which case the walls of our public buildings may continue to be built, as they have heretofore, of *filtering stones.*

IMPROVEMENT IN THE MODE OF CON-STRUCTING DOUBLE CENTERED JOINTS, BUTTS, OR HINGES: *Egbert Hedge, Connecticut.*—These hinges resemble card-table hinges, to which, as well as to other pur-poses, it is intended to apply them. The principle object of the patentee, however, is to use them as rule joints, as in this case the rule when closed has no projecting joint at its end, but all is flush with the end of the rule. As ordinarily made, however, with two joint pins, they would not open and close with truth, and to insure this, they are made with cogs, or teeth, on the con-necting piece and on the joint part of each of the brass straps, which for this purpose are finished off to a curve, of which the joint pin is the centre. The claim is to "the cogs or teeth in double centered joints, to insure the accurate and equal turning upon each centre, in the process of opening and closing the joint, and to keep the parts of the joint in their due positions when at rest either closed or opened."

This construction is ingenious in its ap-plication to rules, but we are apprehensive that, when its greater complexity, and lia-bility to be out of order, are taken into ac-count, it will be found to be more curious than useful.

IMPROVEMENTS IN MANUFACTURING OF COLOURING MATTER, APPLICABLE TO DYEING, STAINING, AND WRITING: *Henry Stephens, of Charlotte-street, Marylebone, Great Britain.*—My improvements (says the patentee) in manufacturing colouring matter, and rendering certain colour or colours more applicable to dyeing, staining, and writing, consists, in the first place, of im-provements in making or manufacturing the ferro prussiates (that is, prussiates of pot-ash and soda); and, secondly, in rendering Prussian blue soluble, and thereby more applicable than heretofore to the purposes of dyeing, staining, colouring, and writing. My first improvement consists in converting certain gaseous products arising from the present mode of making prussiate of potash or soda from animal matter (which are now commonly allowed to escape into the atmos-phere) to the purpose of making prussiate of

potash or soda, so that an increased quantity of prussiate of potash or soda, may be ob-tained from a given quantity of animal matter. For the better explanation of this part of the invention Mr. Stephens refers to drawings, but as the figures are of that apparatus ge-nerally used by chemists and manufacturers, the process may easily be understood with-out them. An iron pot, vessel, or retort, is charged with alkali, and animal or other matter, containing azote, or yielding ammo-nia, which vessel is to be heated to a low red heat. This pot, or vessel, has a moveable cover, which is to be luted on when under operation, but may be removed and placed upon another pot, by disconnecting the joint in a pipe, the joint allowing the head of a pipe to be carried round with a connecting pipe. The first pipe is for conducting the gaseous products arising from the decom-position of the animal matter in the pots into an iron, cylindrical shaped vessel, heated by a furnace below. This vessel is to be charged with alkali, and to be kept at a full-red heat during the operation. There is a pipe leading from the cylindrical vessel to a closed vessel containing a solution of alkali. This vessel is furnished with a jet pipe, or burner, which is merely intended as a gauge cock, to ascertain the state of the gas within; there are furnaces under the pots. The gas generated in the retort passes by the connecting pipe to the cylinder, where, meeting with the alkali in a state of fusion, the effect will be, that the gas becomes com-bined, to a certain degree, with the alkali, and forms prussiate of potash or soda. But there may be portions of the said gas which do not combine or commix with the alkali; these will pass off by a pipe to the closed vessel, and if any of the gas thus passed off should be capable of combining with the al-kaline solution, it may do so in the closed vessel, and that portion which does not combine with the alkaline solution is allowed to pass off by the gas jet-pipe; the state of operation may be ascertained by burning the jet gas from the end of this pipe, for when it ceases to burn freely, the connexion between the pot and cylinder should be disconnected, and the head and pipe be removed round and luted on the pot, which is to be pre-viously charged with animal matter and alkali, the distillation of which will proceed as before described. When the gaseous pro-ducts of several charges have been passed through the cylinder containing the alkali, the cylinder may be opened, and the charge (which will now consist of crude prussiate of potash or soda—or "metal," as it is commonly called in the trade,) be withdrawn into an iron vessel, and when cold be lixi-vated in cold water in the usual manner. The further decomposition of the charge of ani-mal matter in the pot may now be conducted in the ordinary manner of making prussiate in the open vessel by increasing the heat, the contents to be agitated as usual. This process may be repeated alternately in the two pots—the completion of the decompo-sition of the charge of one being effected, while that of the other is subjected to the lower heat, and the operation of distilling off its vapours and passing it to the retort, or vessel.

A similar effect, viz. that of taking up the gaseous products so as to produce an addi-tional quantity of crude prussiate of potash, or "metal," may be obtained in an open, conical chimney, having a false bottom, or grate, or perforated plate, upon which dry potash, or soda, is placed, so that the gas generated in the pot below may pass through the stratum of alkali in the chimney.

A grating or perforated plate is placed under the chimney, or open cone, usually placed on the top of an ordinary pot for making prussiate of potash, in order to con-vey the flame upwards. Upon this grating, or perforated plate a stratum of dry potash, or soda, is laid, and as the gas passes up-wards through this stratum, a portion of it will become combined with the alkali. The chimney, with the stratum of the alkali, may be removed when the flame begins to burn weak, and it may be set aside and applied to future charges, or put into the pot and worked off with the charge in the usual manner of making Prussiate of Potash or Soda.

My second improvement (continues the patentee) viz. the mode, method, or process of treating, or operating upon, Prussian blue, so as to render it more perfectly solu-ble, or more readily disposed to be acted upon by the subsequent process of solution, than when manufactured in the usual way, and in order that the same may be more ap-plicable to the purposes of dyeing, staining, colouring and writing, I effect in the follow-ing manner :—

I take the Prussian blue, whether produced from a combination of prussiate of potash and salts of iron, or the Prussian blue of com-merce as commonly manufactured, and I put this into an earthen vessel, and pour over it a quantity of strongly concentrated acid, sufficient to cover the Prussian blue; muriatic acid, sulphuric acid, or any other acid which has a sufficient action upon iron, will do, but I prefer the muriatic acid. If sulphuric acid is used, it should be diluted a little—that is, a quantity of water equal to about its bulk, at the time when the mass burns white after the Prussian blue is put in. The Prussian blue is to be allowed to

remain in the acid from 24 to 48 hours, or longer. I then dilute this mixture with a large quantity of water, stirring it up at the time, for the purpose of washing from it all the salts of iron. When in this state of dilution, I suffer it to stand until the colour has subsided, when the supernatant liquor is to be drawn off with a syphon, and more water added to it, and I continue the repe-tition of this process until I judge that the acid, with the iron, has been completely washed away, and this is known by testing it with prussiate of potash, which will show if it yields any blue precipitate ; if not, it is sufficiently washed ; I then place it upon a filter, and suffer it to remain until the liquid has all drained away.

The Prussian blue thus prepared is re-duced to a state, as I conceive, containing less iron than the Prussian blue of com-merce, in which state it is more readily acted upon and rendered soluble, than in any other condition. This Prussian blue may then be placed in evaporating dishes, and gently dried. To form the Prussian Blue so operated upon into a solution, I add to it oxalic acid, and mix them carefully together ; after which I add cold water (cold distilled water is best) a little at a time, making it into a dense or dilute solution, according to the colour required. The quantity of oxalic acid may vary according to the quantity of water used. It will be found that the Prus-sian blue that has undergone this process of digestion, as described, requires but a small quantity of oxalic acid to dissolve it. About one part of oxalic acid will dissolve six parts of Prussian blue, (the weight taken before digesting in the acid). This will answer for a concentrated solution, but for a dilute so-lution, more acid will be required. Prussian blue that has not undergone digestion in acid in the way above pointed out, will re-quire a much larger proportion of oxalic acid, from twice to three times its weight, and then it will be greatly liable to precipi-tation after standing ; but when treated in the way described, it is not liable to preci-pitate, but remains a permanent solution.

The chief obstacle to the general employ-ment of the beautiful colour obtained by means of the Ferro Prussiates to the pur-poses of dyeing in the silk, cotton, linen, or woollen manufactures, and also to the pur-poses of staining and writing, has been its hitherto supposed indissoluble nature ; but by means of oxalic acid, (whether obtained by the usual process of mixing or distilling saccharine matter in combination with nitric acid, or from vegetable, or other substances containing oxalic acid, or from combinations of oxalates, whether metallic, earthy, or al-kaline,) I obtain the above perfect solution of the Prussian blue, which is applicable to dyeing, colouring, or staining, in the various manufactures of woollens, silks, linens, cotton, paper, and such other substances as are required to be dyed or stained, and which solution is also available to the purpose of writing, or forming a writing fluid, or ink, to be used with steel, quill, or other pens.

In conclusion, I desire it to be understood, that I do not claim any of the apparatus or machinery described, nor the calcination of animal matter in close vessels, but I do claim the method of obtaining a product of prussiate of potash, or soda, from the gases evolved from the distillation of animal matters, or any other matters that yield azote and carburetted hydrogen—such, for instance, as coal, by means of passing those gases into the mass of alkali in a state of ignition, and into a solution of alkali contained in separate vessels, either closely, or distantly, | connected with the distillatory apparatus.

Secondly—I do not claim the use of acids for the purpose of brightening, or improving the colour of Prussian blue in the ordinary manner. But I do claim the use of strong acids for the purpose of digesting dry Prussian blue of commerce, in order to render it more easily soluble in the oxalic acid than it would be without such a digestion. And I further claim the use of oxalic acid, however obtained, as a solvent for Prussian blue generally, but more especially, as a final process for making a perfect solution of the Prussian blue which has been prepared and digested in the manner above described.

VESSELS TO BE USED AS LIFE PRE-SERVERS. *John Mackintosh, New York.*

I take canvas, or other flexible material, and render it impervious to water by means of a solution of caoutchouc, or in any other of the known ways of effecting this object; and of this flexible material, so saturated, I make my vessel, which is to contain the persons or things intended to be buoyed up, and conveyed upon the water. Such a vessel may be made to assume a variety of forms, dependent upon the purpose for which it is to be used, whether for one or more persons, the transportation of troops and baggage, and for other objects. The manner in which I intend, most commonly, to construct such a boat, or vessel, is the following:—

I take a square piece of canvas, or other material, saturated as above stated. The edges are to be turned over, in the manner of forming a wide hem, so as to leave what, when filled with air, will become a tube, or air-chamber; the turned-over edges being securely cemented down, taking

care that the juncture is air-tight. The material is then doubled over, so as to bring the opposite edges together; and the edges of the doublatures are united by cementing, or otherwise, as are also the edges of the tubes, or air-chambers, so as to cause them to form a continuous air-tight rim, when, if the sides are separated to some distance apart, it will constitute a vessel resembling a boat. A small hose, or tube, furnished with a stop-cock, leads into the air-chamber, which may be inflated in a few moments by applying the mouth to the stop-cock. Instead of a single air-chamber, there may be two or three, one immediately under the other; when, should one be accidentally ruptured, no inconvenience would result therefrom.

It will be evident that a vessel so constructed would float in the water from the buoyancy derived from the air-chamber, and that its lower, or bag part, will also remain at the surface, or nearly so; but if persons, or any weighty articles, are placed upon this part, it will sink so as to displace a portion of water equivalent in weight to itself, if its specific gravity be not greater than that of water, and that in this way, it may be loaded, whilst the tubular part, or air-chamber, will remain at the surface, occupying the situation of the gunwale of a boat.

To form a covering to the persons, or things, contained in the vessel, pieces of air-tight canvas, or other material, may be attached round the air-chamber, and may be folded or drawn over the persons or things contained in the vessel. In some cases, it may be found desirable to leave an air-opening in the covering, which opening may be surmounted by a conical tube, or other device, for admitting air, and keeping out water.

Oars or paddles may be used to give a direction to such vessels; and where, for the conveyance of troops, or for any other purpose, a number of them are to cross any water, a tow-line may be carried by the first, and employed to draw the others over. For the purpose of using oars, there should be thongs at suitable places along the edges, which, when tied together, will form loops through which the oars may pass. Other devices for propelling may be used; as, for example, a triangular or other float-board, having a line attached to it, in the manner of a log line, may be thrown out by a person in the vessel, when, by drawing the line, the vessel will be propelled or drawn towards the float board.

By means of thongs attached to the edges of the air-chamber, the sides of the tubular air-chamber may be made to approach each other in any part desired, and any required form be given to the outline of the vessel, by merely tying these thongs together.

Where it is desired to apply the principle to ships, steam-boats, &c., the bag part need be but little larger than will suffice to contain one or two persons only, and such articles as they may desire to have with them; this may be effected by having the berth matrasses of any ship, steam-boat, &c., cut into two parts lengthwise, and covered with caoutchouc, or with other water-proof flexible material, as aforesaid, to take the place of tubes, with the bag of the aforesaid material, placed between the two parts of the matrass; and in this case, it will be found convenient to attach flexible legging to the bottom of the bag to receive the legs and feet. A person may then carry the whole in his hands, walk about readily, and jump from a vessel or wharf into the water, and when there may use his feet and legs to enable him to swim backwards.

With an apparatus of this kind, a covering may be used which may be drawn round the neck, over the head, or under the arms of the person, as may be desired; and, indeed, this and other parts of the apparatus are susceptible of numerous modifications, which, as they are dependent upon the judgment or the fancy of the person using it, it would be impossible to enumerate.

When the matrass of the vessel is used to form the gunwale of the life boat, such matrass may be made in two thicknesses, which, when used as a matrass, lie upon each other, but when opened out will form the gunwale, the bag part depending from its lower edges. Or the matrass may be cut into two parts, along its middle, so as to consist of two narrow matrasses of half the usual width, which lie side by side when in the berth, but when used as a life boat then open out, the bag as before depending from their edges; in this case, as the two parts of the matrass are not continuous, they are to be connected by water-proof ends, consisting of cloth which may be drawn up in any convenient way. This mode of using the matrass I prefer; the gunwale part being, in this case, of half the width and double the thickness of that first described, which I find to be advantageous.

THE METEORIC SHOWERS OF NOVEMBER.

The night of the 12th of November being the time for the periodical return of an unusual number of meteors, I determined to leave London, on account of the obstruction buildings offer in this crowded city, to locate myself in the delightful village of Richmond, Surrey, for the purpose of observing the annual return of this remarkable phenomenon.

In the afternoon of the 12th the sun sank below the horizon, as to predicate a clear night. When below the horizon, there was a grand display of a rich profusion of red, orange, and rosy-coloured hues.

The wind blew fresh from N. N. E., and the night was, in consequence, very cold. At 10 hours, 13 min. p. m. (clock time), a meteor, without train, fell from the star α Lyra, and took its direction across the Milky Way. At 11 hours another meteor fell from a star north of the Pleiades. At 11 hours 48 minutes a large meteor with train fell from γ Cassiopeia, and crossed the Milky Way at an angle of near 90 degress.

From 12 o'clock (midnight, 12th) till 3 hours 25 minutes on the 13th, nine meteors fell, crossing the galaxy at angles of from 70 to 80 degrees; six were without trains, and three with trains. I now began to despair of witnessing the "grand display" seen in former years; recollecting, however, on some occasions the "shower" did not commence till near 4 o'clock, I continued to direct my attention to the N.N.E., whence the greater number of meteors had fallen.

At 3 hours 35 minutes nothing could exceed the beauty and grandeur of the heavens; from E.N.E. to N. meteors fell like a shower of bomb-shells in a bombardment, and in such rapid succession as to defy every attempt to watch their particular direction and course among the stars, or to ascertain their number. The whole visible heavens were illuminated by the light such a prodigious number of meteors diffused in their descent towards the earth, and a more beautiful and magnificent sight cannot possibly be conceived. At 3 hours 35 minutes the "shower" ceased, and after 4 o'clock all traces of meteors were gone; the stars, however, shown without diminution in number or brightness, and the atmosphere was remarkably clear.

The shower of meteors appeared to take their direction from N.N.E. and N., as if the direction of their trains had been occasioned by the wind, which was blowing fresh from the former point. The total number of meteors could not have fallen short of from 400 to 500, the *maximum* number at 3 hours 45 minutes. We may therefore conclude the "meteoric shower" was equal in interest and splendour to any of a former year.

At Paris it rained till 11 o'clock at night on Monday, while at Richmond it was remarkably fair and clear, and at 5 o'clock on Tuesday morning, the 13th, my thermometer marked 26 degrees of Fahrenheit.—*Correspondent of the Times.*

REPORT MADE TO THE ACADEMY OF SCIENCES, ON VARIOUS FILTERING APPARATUS, AND PARTICULARLY THAT OF HENRY DE FONVIELLE. BY M. ARAGO.

The Academy has charged MM. Gay Lussac, Magendie, Robiquet and myself, with an examination of the filtering apparatus of M. Henri de Fonvielle. The question of filtration is so important and so keenly agitated at the present time, that high authorities, the municipal administration of our chief cities, as well as private individuals, frequently consult the Academy on this subject, so that it has appeared to us to be proper to consider the problem in all its bearings. It is, besides, the best mode of suitably appreciating the new method on which we are appointed to decide.

Mankind use for drink, for cooking, for cleanliness, and for the useful arts, cistern water, well water, spring water, and river water. These four kinds of water have one common origin—rain. Rain is, in general so pure, that foreign matters can be detected in it, only by means of very delicate chemical reagents.

Cisterns, constructed with well chosen materials would therefore be the best means of obtaining excellent water for drinking, if the rain all fell directly into them, and did not bring with it dirt, dust, insects, accumulated in dry weather on the roofs and terraces over which it flows. In certain localities, as at Venice, for example, the inconveniences now alluded to are felt to such an extent, that they found the necessity of causing the rain water, before it reached the great cistern of the Ducal Palace, which was much resorted to by the public, to pass through a thick bed of porous materials, in the interstices of which the foreign matters held in suspension might be in part deposited.

Wells may be compared to cisterns, only they are not supplied by channels of brickwork, stone, or metal. The water of the clouds reaches them, if we may so speak, drop by drop, through the common capillary openings of the soil. It is rare that in this long and difficult trickling in fine streams, the water does not meet with soluble materials, which it dissolves in greater or less quantities. It is not, therefore, strictly speaking, rain water that we draw from our wells : it is generally as clear and limpid ; but it contains, almost always, matters dissolved, whose chemical nature varies with the geological constitution of the country.

The same remarks are applicable to springs. The water which they distribute is also rain water, which, having passed through more or less of the crust of the earth, is returned to the surface by a siphon stream, or in other words, by pressure through streamlets of water from a more elevated situation. The nature and the proportion of foreign matters in spring water, depend also on the extent of the streamlets which feed the source, and the nature of the rocks through which they percolate. When these rocks are of a certain kind, the country will abound in mineral springs. If the vertical descent of the fluid is of a certain extent, the water will issue in a thermal (warm or or hot) spring.

Every river bears to the sea the waters of some principal spring, and those of a certain number of others of minor importance, which unite with the main one in its passage. In chemical composition the water of a river might thus seem to be a medium between those of all the springs of the surrounding country ; but it must be observed, that at the time of freshets or heavy rains, (and if the valley of the river be extensive, these may occur very often,) the fluvial waters do not sink into the earth in large proportions, but flow over its surface in great abundance and with great rapidity ; and that in this superficial flow they can dissolve but a very small portion of foreign matter, compared with that which they will take up when divided into minute streams, and pursuing an underground course, during which the particles are so constantly in contact with soluble materials. To these considerations in favour of the purity of river water, must be added the fact, that carbonate of lime is dissolved by aid of an excess of acid, and that this excess is dissipated during a long exposure of water to the air, in consequence of which the carbonic itself is precipitated.

These remarks, moreover, are to be considered only in a general point of view. It would not be difficult, in fact, without departing from the known laws of geology, to imagine, and even to find arrangements of strata, whose wells and springs would furnish pure water, while the neighbouring rivers, on the contrary, might contain a strong saline impregnation. All that we aim at is, to explain how the reverse of this generally takes place—how the water of the Seine and of the Garonne, for example, are notoriously purer than the waters of most of the springs and wells of the countries through which these rivers respectively flow.

But the advantage of greater purity in river water, chemically considered, is more than lost by their habitual want of limpidity. At each heavy rain, the little torrents are precipitated into the stream, loaded with vegetable soil, clay, gravel, and all sorts of detritus which they tear up from the land; and these heterogeneous ingredients are driven along, until they are gradually deposited in the river's bed.

The proportions of these foreign mixtures held in suspension during freshets, are not the same in different rivers, as might well be expected. In the Seine, this proportion rises sometimes to $\frac{1}{1000}$. He, therefore, who should drink, in the course of his day's work, three quarts of the unfiltered water of the river, at the time of its highest flow, would load his stomach with more than a scruple of sand and mud. What effect must not this, in time, produce upon his health? The question has been much discussed, and it has left physicians and hydraulic engineers very much divided in opinion. For want of exact experiments, both parties agree to leave the question where they found it. We shall certainly not be considered too severe in our judgments, if we add, that one of the declared partisans of these troubled waters, rests his opinion on the alleged observation, that animals, cattle especially, do not begin to drink from the pools which they meet on the way, until they have well stirred up the mud with their feet!

But, every consideration of health aside, it is certainly very disagreeable to drink water charged with dirt. At all times and in all countries, limpidity has been regarded as a necessary quality of the water destined for human beverage; and on this account, long before the invention, or at least the perfection of the means of filtration, the ancients deemed it necessary to dig deep wells, at great expense; or to bring in, by magnificent aqueducts, the water of natural springs, even when their towns were situated on ample rivers.

It is by its rapid motion through, or over, the ground, that water becomes loaded with mud. By repose this is precipitated, and the fluid resumes its natural transparency. Nothing is certainly more simple, than this mode of clarification; it is, unhappily, excessively slow.

From the very interesting experiments and calculations made at Bordeaux, by M. Leupold, we learn that after ten days of absolute repose, the water of the Garonne, taken at the time of a freshet, had not returned to its natural limpidity. At the commencement, it is true, the larger particles subside very fast, but the finer go down with a slowness which would put all patience at a stand.

Simple repose, then, cannot be resorted to as the means of clarifying the water destined for the supply of a large city. Who does not perceive that 8 or 10 separate basins would be necessary, each of sufficient capacity to contain all the water necessary for a day's consumption? Add to this, that in certain places, and at certain seasons, water exposed in a stagnant condition to the open air during 10 consecutive days, would become foul and taste badly, either on account of the putrefaction of innumerable insects which would fall into it from the atmosphere, or in consequence of the vegetation which would begin to take place on its surface.

Repose, however, may be considered as one valuable means of getting clear of the grosser particles which are held in suspension. It is under this point of view, *only*, that basins and reservoirs have been contrived and established in England and France.[*]

Science, or rather chance, brought to light the means of hastening considerably, or rendering almost instantaneous, the precipitation of earthy matters held in suspension by water. This means consists in adding powdered alum to the turbid water. It is averred that at Paris, the gross slime brought down by the Seine, collects into long thick strings, which are very promptly deposited when alum is added. The theory of this operation ought to claim the attention of chemists. It is not at present sufficiently certain, to justify us in affirming, that the same effect would take place in the sediment of every river. Some doubt seems admissible from the fact that the clarification by alum is not always complete; that certain very fine particles escape the action of this salt, remain in suspension, and render the liquid somewhat cloudy (louche), when all the stringy portions have disappeared. If it is true that water, after having been alumed, still requires filtration, we can easily conceive why the employment of alum, as a means of clarification, has not become general. Besides, in the large way, the price of the salt, in addition to other means, might be objectionable. Another more serious objection is, that it affects the chemical purity of river water,—that it introduces a salt which it did not before contain,—that in supposing this salt wholly inactive in certain proportions, consumers might fear that at times these proportions might be very materially exceeded, and that this might easily happen through the negligence, or mistake, of a workman. One of the committee (the reporter) was speaking one day of the aluming of water, to an English engineer, whose

---

[*] * The author appears to forget that reservoirs are indispensable, as the means of insuring a regular supply when the water has to be forced to a greater height, by means of machinery, in order to bring it to the requisite hydrostatic elevation, as in the water works of Philadelphia, Wilmington, &c.

extensive experience had given him much practical acquaintance with the habits and feelings of the public, and who was lamenting the imperfection of the means now in use for purifying water,—"What are you proposing?" said he immediately, "water, like Cesar's wife, ought to be beyond all suspicion."

This, in terms perhaps singular, but true, is a pointed condemnation, of every means of clarification which would introduce into river water any new substance, that it does not originally contain; and therefore the most recent trials of engineers have all been directed to the employment of inert materials, or those which cannot add anything to the water. These materials are gravel of different sizes, sand of different degrees of fineness, and pounded charcoal.

The idea of applying gravel and sand to the clarification of turbid waters was certainly suggested by observing so many natural springs issuing from sandy bottoms with remarkable limpidity; hence the practice is very ancient, and hence do we ascertain it to have been in vogue in the Ducal Palace, at Venice. A bed of fine sand appears to act, in the clearing of water, only as a mass of sinuous capillary tubes, through which the liquid molecules may pass, while the earthy matters in suspension are arrested, in consequence, simply, of their greater dimensions.

From the experiments of Lowitz, Berthollet, Saussure, Figuier, MM. Bussy, Payen, and some other chemists, it is now known by almost everybody, that charcoal has the property of absorbing the matters resulting from the putrefaction of organic bodies. The part which charcoal acts, therefore, in the purification of water, cannot be doubtful.

Theoretically considered, the art of the clarifier appears to be nearly complete; but this is very far from being the case with respect to its economical and successful application, especially when the object is to conduct the operations on a great scale.

Very extensive filtering apparatus have been put into operation by our neighbours on the other side of the water, and especially at Glasgow. The cost of these essays must be counted by millions (of francs). Nevertheless, they have not been successful; but, on the contrary, they have occasioned the ruin of several powerful companies.

Those who are engaged in the amelioration and extension of the useful arts, may certainly find excellent guides in natural phenomena, but on the express condition that they do not allow themselves to be seduced by imperfect similitudes. Such has been, we venture to affirm, the principal

origin of the errors committed in Scotland. Certain springs, it was said, flow uniformly without interruption; they have for ages furnished the same quantity of transparent water; why should not the same result follow from an artificial fountain, under analogous circumstances. But, in the first place, is it certain that these natural springs, of which so much account is made, have experienced no diminution? Where are the wooden conduits by which they have been measured? who has compared their issues, cautiously, year by year, with the quantity of rain which has fallen? Moreover (and herein it is that the Scotch engineers have particularly erred), in an artificial fountain, the filtering strata must always be of limited extent, while the waters of a natural spring are clarified, sometimes, by beds of sand which spread over whole districts, and which act upon a fluid which is but little troubled. In the first case, the capillary tubes of the filter will soon become foul; while in the second, the effect will scarcely be visible.

The result is, that no artificial method of filtration can be successful, unless prompt, economical, and certain means are at hand, of cleaning or renewing the filters. Only one of eight large companies in London, that clarify the water, viz. the Chelsea Company, has attained its object. This has been done by the construction of three large basins communicating with each other: in the first two, the coarser terrene particles are deposited by repose; in the third, the water traverses a thick bed of sand and gravel, whereby it becomes definitively purified. When this basin is empty, the filtrating mass of sand is exposed; at which time, workmen immediately remove by rakes the superficial layers which the sediment has rendered foul, and replace it by fresh sand.

A thought here suggests itself. It is not, certainly, without a good reason, that the able engineer of the company has made his filtering bed six feet thick; the superficial layers, which the workmen remove from time to time, act, without doubt, more effectively than the others; but those below must, nevertheless, have some influence, and must also by degrees become engorged with the matter arrested, daily become less efficient, and in time the whole must require to be changed. The necessity of this, when anticipated, would require the agency of a fourth basin, like the third, and, like it, of the extent of an acre of ground. The total expenditure in these works has amounted to from 300,000 to 400,000 francs; and the manipulations of the filter, which cost annually not less than 25,000 francs, must be continually increasing.

Is it surprising, that in the view of such

heavy expenses encountered by the Chelsea Company for the filtration of 10,000 cubic metres of water per diem, corresponding to about 500 square inches of main pipe, the other English companies should all, in an examination before Parliament, declare that, if compelled to filter the water of the Thames, their rental prices would have to be raised 15 per cent.

The system which Robert Thom, a civil engineer at Greenwich, introduced in 1828, has the advantage over that at Chelsea, of a self-cleaning operation, to which the whole filtering mass is subjected. This mass forms a bed five feet thick. The water is admitted into the basin, filled with sand, either above or below, at pleasure. If the filtration, for example, is by descension, as soon as it is perceived that the filter is obstructed and becomes effete, the water is, for a while, introduced below; and, in its ascensional movement, it drives the sediment from the upper surface into a discharging pipe destined to receive it.

Filtration has not hitherto been attempted in France on a very large scale. In several valuable establishments in Paris, at which it is performed, a large number of small boxes, lined with lead, open at top, are provided, and contain at bottom a bed of charcoal between two layers of sand. These are, in fact, the old filters patented by Smith, Cochet, and Montfort. When the waters of the Seine and Marne arrive at Paris, very highly charged with silt, and undergo depuration in those boxes, it is found necessary to renew the strata, or at least the upper one, every day, and even twice a day.

Each superficial metre of filter gives about 3000 litres (nearly 800 gallons) of clarified water every twenty-four hours: hence it would require 7 square metres, or 7 cubic boxes of one metre in the side, for every inch of fountain pipe; and 7000 such boxes would be requisite for the service of a town, where the consumption would demand 1000 inches.

There is a very simple method of increasing the product of these little boxes : it is, to close them hermetically, and to cause the water to pass through the filtering mass, not by its own weight merely, or by a simple charge, but by strong pressure.

This, Gentlemen, is one of the improvements in the filtration of water, which is proposed, and which has been realized, by the author of the memoir committed to our examination.

The filter of Henry de Fonvielle, at the Hotel-Dieu, though it has not one metre of superficial extent, yields daily, by a pressure of 88 centimetres, (= 34.6 inches of mercu-

rial pressure, = 1⅓ atmospheres.) 50,000 litres (= 13,200 gallons) at least, of clarified water. This amount, deduced from an examination of the various services of the hospital, is a small part of what the apparatus might furnish if the feeding pump were constantly in operation. At certain times we found, in fact, by direct experiment, that the filter would yield as much as 95 litres (= 24 gallons) per minute. This would be nearly 137,000 litres in 24 hours, equal to about 7 inches of pipe. But the quantity first named is 17 times greater than by the methods commonly in use.

Since M. de Fonvielle presented his memoir, and especially since the results at the hospital, several persons, and among others M. Ducommun, have claimed the invention of filtering by increased pressure. In mathematical strictness these claims might perhaps be sustained; for to a greater or less extent, it is unquestionable that in every machine existing, or known only by patent, and particularly those that filter by ascension, there is a pressure, it may be, of some inches; but, regarded in a practical point of view, the question is a very different one. It is whether any one, before the author of the memoir, proposed to effect the filtration of water in vessels *hermetically closed*, allowing nothing to escape from the pressures which the locality or the machine can produce; whether any one, prior to de Fonvielle, had arranged a filtering apparatus in such a manner that *strong* or high pressure would not derange or confuse the different layers; whether, in fact, any one before the experiments at the hospital, had proved that a rapid filtration would give a fluid so limpid as to be perfectly satisfactory? In all these respects the rights of M. de Fonvielle appear to us to be incontestible. From the parliamentary inquiry before alluded to, we learn, that engineers had not been unmindful of the possibility of effecting filtration under moderate pressure,—and that some had adopted this mode in a manner which involved them in hydraulic errors. In France we find every where, and especially at the beautiful mineral water establishment at Gros-Caillou, a fine disposable high pressure, entirely neglected. We see, in fact, M. Ducommun, whose name is so honourably known in this department of the arts, using, at the Hotel-Dieu, three cisterns to clarify 15 hectolitres in 24 hours; while a single one of these cisterns, modified by Fonvielle, yields, in the same time, agreeably to the report of M. Desportes, steward of the hospital, 900 hectolitres of water, perfectly filtered, in lieu of the 15.

But the employment of high pressure is practicable only in combination with another process, of which no one contests the invention with the author of the memoir.

We have seen in time of freshets, a filter of one square meter, requiring to be cleansed once at least in 24 hours, although it would clarify only 3,000 litres of water. It would seem, at the first view, that the filter of M. de Fouvielle, which clears 17 times more, must require cleaning ever hour. Such, however, is not at all the case. No more attention is requisite than in ordinary filters.

The explanation is simple enough when we remark, that under a feeble pressure a filter acts, as it were, only at its surface,— that the mud scarcely penetrates it; while, under great pressure it may, or must, sink deeper. No one will deny that if more turbid water passes in a given time, there must be a proportionate deposition of feculent matter; but if this be found disseminated through a greater depth of sand, the permeability of the filter will not be more changed by it,—the cleaning merely will be more difficult; it is in this respect, above all others, that the new process is worthy of attention.

We have already stated, that at Greenock, the engineer, R. Thom cleans the mass of sand by a rapid counter-current, viz., from bottom to top. This mode may suffice when the filters are choked only at the surface; but the filters of M. de Fonvielle require more powerful means. This the author finds in the action of two counter-currents—in the shock, and sudden shaking and stirring which result from them. In cleaning the hermetically closed filters of the Hotel-Dieu, the workman, whose business it is, opens suddenly, and almost simultaneously, the cocks of the tubes which connect the bottom and top of the apparatus with the elevated reservoirs, or with the body of the feeding pump. The filter is thus tumultuously agitated by two cross currents, by which it is acted upon in a manner not very unlike that thich a garment undergoes in the hands of a washerwoman. These currents have, in every case, the effect of detaching, from the filtrating gravel, the foreign matters which would otherwise remain adhering to it. We have no doubt of the great utility of these conflicting currents; for after having cleaned the filter of the Hotel-Dieu, agreeably to the method of engineer Thom, i. e. by an ascending current, after assuring ourselves that this ascending current came out limpid,—as soon as the two other cocks were opened, the water rushed out from the filter in a very filthy condition.

We may add the passing remark, that the patients who witnessed the operation expressed their great surprise at seeing, after an interval of a few seconds, the same fountain furnish, first a yellow mass as thick as soup, and then water as clear as crystal.

We may add to these numerous details, that the process which you have charged us to give an account of, has received the sanction of time. For more than eight months it has been in operation at the Hotel-Dieu; for more than eight months the same bed of sand, of at least a square metre in surface, has performed its functions without intermission; that there has been no occasion of renewing it; that the Seine, nevertheless, within this period has been extremely foul; and that, at the lowest estimation, 12 millions of litres of water (12,000 cubic metres,) have passed through the apparatus. From these various circumstances we have deemed it unnecessary to make any trials of the further advantages which the author of the memoir expects to derive from a division of the thick filtering body now in use, into three beds, separated from each other; and in confining ourselves exclusively to what we have sufficiently examined, we do not hesitate to say, that in showing the possibility of clarifying large quantities of water with a very small apparatus, M. Henry de Fonvielle has made an important advancement in the arts.*

———

LIST OF ENGLISH PATENTS GRANTED BETWEEN THE 31st OF OCTOBER AND 26th OF NOVEMBER, 1838.

Paul Chappé, of Manchester, spinner, for certain improvements in the means of consuming smoke, and thereby economising fuel and heat in steam-engine and other furnaces and fire-places. October 31; six months.

Luke Hebert, of Staples-Inn, for certain apparatus and process for storing, cleansing, and preserving grain. November 3; six months.

Abraham Bury, Esq., of Manchester, for certain improvements in the mode of printing, colouring or dyeing cotton, and other fabrics, and in the mode of producing certain acid, or acids, applicable to these or other purposes. November 3; six months.

Jacob Telton Slade, of Carburton-street, gent., for certain improvements in pumps for liquids or aeriform fluids. November 3; six months.

Joseph Fraser, of Halifax, railway contractor, for certain improvements in the apparatus or machinery to be employed as centerings or supporters in the construction of bridges and arches, and in tunnels and other mining operations. Nov. 3; six months.

---

* May we not take the liberty of suggesting the probability, that if the ascending current of water at Fair Mount, were made to pass through a tight box, containing the requisite filtering materials, agreeably to the admirable contrivance of De Fonvielle, it might furnish a simple and unexpensive mode of clarifying the water of the river, which, at certain seasons, is so turbid as to be almost past endurance.

Horace Cory, of Harrow-street, Limehouse, bachelor of medicine, for improvements in the manufacture of white lead. Nov. 3; six months.

Charles Callis, Baron Western, of Rivenhall, Essex, for an improvement in drills, for the purpose of drilling corn, grain, seeds, pulse, and manure. November 3; six months.

William Morgan, of New-cross, Surrey, gent., for improvements in the generation of steam. Nov. 3; six months.

Adolphus Henri Erneste Ragon, of Great Portland-street, professor of literature, for improvements in the manufacture of glass, and in the production of other vitrified matters applicable to architectural purposes. Nov. 3; six months.

Edward Cooper, of Piccadilly, for improvements in the manufacture of paper, being a communication from a foreigner residing abroad. Nov. 3; six months.

Charles Flude, of Liverpool, chemist, for improvements in applying heat for generating steam, and for general manufacturing and other useful purposes, where heat is required ; and also for an improved mode of supplying steam-boilers with hot water, the said improvements having for their object the economy of steam. Nov. 3; six months.

Jerome Deville, of Crutched-friars, coach-builder, for improvements in railroads, and in carriages used thereon. Nov. 3; six months.

James Bernington, of Charles-place, Shoreditch, veterinary surgeon of cavalry, for improvements in knapsacks. Nov. 3; six months.

William Henry James, late of Birmingham, but now of Lambeth, for improvements in apparatus for heating, generating, and cooling fluids, and in engines to be actuated by such fluids, parts of which improvements are applicable to the raising and forcing fluids. Nov. 6; six months.

Robert Beart, of Godmanchester, Huntingdon, miller, for improvements in apparatus for filtering liquids. Nov. 6; six months.

Luke Hebert, of Bristol-road, Birmingham, for a new or improved process or processes for embalming the dead, and for preserving corpses for anatomical purposes, being a communication from a foreigner residing abroad. Nov. 6; six months.

Moses Poole, of Lincoln's-Inn, gent. for improvements in apparatus or machinery for obtaining rotatory motion, being a communication from a foreigner residing abroad. Nov. 8; six months.

John Juckes, of Shropshire, gent., for improvements in steam-engine boilers, and in apparatus for feeding furnaces and fire-places, and for the more effectual combustion of the smoke and gases arising therefrom. Nov. 8; six months.

Bryan I' Anson Bromwich, of Clifton-on-Teme, gent., for improvements in machinery to be worked by the application of the expansive force of air or other elastic fluids, to obtain motive power. Nov. 8; six months.

John Small, of Old Jewry, London, merchant, for improvements in filtering liquids, being a communication from a foreigner residing abroad. Nov. 8; six months.

Henry Huntley Mohun, of Regent's-park, M.D., for improvements in the composition and manufacture of fuel, and in furnaces for the consumption of such and other kinds of fuel. Nov. 8; six months.

Thomas Mayos Woodyatt, of Cookly, screw manufacturer, and Samuel Harrison, of Birmingham, for improvements in the manufacture of wood screws. Nov. 8; two months.

John Browne, of Castle-street, Oxford-street, esq., for improvements in paving roads and streets. Nov. 8; six months.

Felix Macartan, of St. Martin's-lane, gent., for improvements in treating the waste matters resulting from the washing of wool and woollen fabrics. Nov. 8; six months.

William Watson, jun., of Leeds, manufacturing chemist, for certain improvements in the manufacture of materials used in the dyeing of blue and other colours. Nov. 8; six months.

John Winrow, of Greenthorpe, Nottingham, mechanic, for certain improved means of, and apparatus for, destroying weeds and insects on land. Nov. 8; six months.

James Drew, of Manchester, for certain improvements in the means of consuming smoke, and economising fuel in steam-engine or other furnaces or fire-places. Nov. 8; six months.

Hugh Ford Bacon, of Fen Drayton, clerk, for an improvement or improvements in the construction of the glass holders and glass chimneys of gas-burners. Nov. 10; six months.

John Holmes, of St. John's terrace, Worcester, engineer, for improvements in forming moulds for castings in metal studs, buttons, nails, tacks, and a variety of other articles. Nov. 13; six months.

George Smith, of the Navy Club-house, Bond-street, captain in the royal navy, for certain improvements in vessels to be propelled by steam or other power, and in the construction and arrangement of the machinery for propelling. Nov. 13; six months.

Anne Bird Byerley, of 147, Strand, widow, and James Collier, of the same place, C. E., for certain improvements in obtaining motive power. Nov. 13; six months.

Sally Thompson, of North-place, Gray's-Inn road, for certain additions to locks or fastenings for doors of buildings, and of cabinets, and for drawers, chests, and other receptacles for the purpose of affording greater security against intrusion by means of keys improperly obtained. Nov. 13; six months.

Edward Samuell, of Liverpool, merchant, for improvements in the manufacture of soda. Nov. 13; six months.

Joseph Eden Macdowall, of 257, High-street, Borough, watchmaker, for an improvement in the manufacture of escapements for chronometers, clocks, and watches. Nov. 15; six months.

Thomas French Berney, of Morton Hall, Norfolk, esq., for certain improvements in cartridges. Nov. 15; six months.

William Thorp and Thomas Meaking, of Manchester, silk manufacturers, for certain improvements in looms for weaving, and also a new description of fabric, to be produced or woven therein. Nov. 20; six months.

William Watson, jun., of Leeds, manufacturing chemist, for certain improvements in the manufacture of liquid ammonia, by which the same may be made applicable to the purposes of dyeing, scouring, and other manufacturing processes. Nov. 20; six months.

Harrison Grey Dyar, of Mortimer-street, Cavendish-square, gent., for improvements in the manufacturing zinc. Nov. 20; six months.

John Wilson, of Liverpool, lecturer on chemistry, for certain improvements in the process of manufacturing alkali from common salt. Nov. 22; six months.

Fanquet Delarne, jun., late of Deville, near Rouen, France, but now of Manchester, calico printer, for certain improvements in the process of printing, or otherwise applying and fixing the colouring matter of madder upon cotton, silk, linen, and other fabrics, without dyeing, and producing by these means permanent colours. Nov. 22; six months.

John George Bodmer, of Manchester, esq., for certain improvements in machinery, tools, and apparatus for cutting, planing, turning, drilling, and rolling metals and other substances. Nov. 22; six months.

Abraham Cohen, of Islington, esq., for certain improvements in the construction of railway carriages, and in the modes of connecting and retarding railway trains. Nov. 26; six months.

LIST OF SCOTCH PATENTS GRANTED BE-
TWEEN THE 22nd OCTOBER AND THE
22nd NOVEMBER, 1838.

Edwin Bottomley, of Aldermanbury, York, clothier, for a certain improvement or improvements applicable to power and hand looms. Sealed 29th of October, 1838; four months to specify.

Laurence Heyworth, of YewTree, near Liverpool, merchant, for a new method of applying steam power directly to the periphery of the movement wheel for purposes of locomotion, both on land and water, and for propelling machinery. October 29.

Thomas Evans, of the Dolwas IronWorks, Glamorgan, agent, for an improved rail for railway purposes, together with the mode of manufacturing and fastening down the same. October 31.

Pierre Armand Leconste de Fontainemoreau, of Charles-street, City-road, Middlesex, for certain improvements in wool combing, being a communication from a foreigner residing abroad. November 2.

James Milne, of Edinburgh, brass-founder, for an improvement or improvements in apparatus employed in transmitting gas for the purpose of light and heat. November 6.

John Hemfrey, of Weymouth Terrace, Shoreditch, Middlesex, engineer and machinist, for certain improvements in the manufacture of hinges or joints, and in the machinery employed therein. Nov. 6.

Charles Flude, of Liverpool, chemist, for improvements in applying heat for smelting, or otherwise working ores, metals, and earths, and for heating steam boilers, and for general manufacturing, or other useful purposes where heat is required; and also for an improved mode of supplying hot water to steam boilers, the said improvements having the economy of fuel for their object. Nov. 6.

Christopher Nickels, of York Road, Lambeth, Surrey, manufacturer, for improvements in machinery for covering fibres applicable in the manufacture of braid and other fibres. November 7.

Thomas Trench Berney, of Morton Hall, Norfolk, Esq., for certain improvements in cartridges. November 8.

Michael Wheelwright Iveson, silk spinner, of Gilmore-place, Edinburgh, for an improved method for preparing and spinning silkwaste, wool, flax, and other fibrous substances, and for discharging the gum from silks, raw and manufactured. Nov. 9.

Moses Poole, of Lincoln's Inn, gent., in consequence of a communication from abroad, for improvements in apparatus or machinery for obtaining rotatory motion. November 14.

Thomas Mellodew, of Wallshaw Cottage, Oldham, Lancashire, mechanic, for certain improvements in looms, for weaving various kinks of cloth. Nov. 14.

Christopher Binks, of Newington, Edinburgh, manufacturing chemist, for certain improvements in the process or processes, for obtaining or manufacturing certain substances, or compounds, applicable in bleaching, and for rendering useful certain products which result therefrom; also improvements in the apparatus employed therein, and in bleaching, and for the application thereto, of a certain agent, not hitherto so employed, which improvements, are also in whole, or in part, applicable to other uses. November 15.

LIST OF IRISH PATENTS GRANTED IN
OCTOBER, 1838.

John Hanson, of Huddersfield, York, for improvements in Machinery, or apparatus for making or manufacturing pipes or tubes from metallic and other substances.

Edward Davey, of Fordtonman, Crediton, merchant, for improvements in saddles and harness.

Robert William Sievier, of Henrietta-street, London, for improvements in looms for weaving, and in the mode, or method, of producing figured goods or fabrics.

Arthur Dumas, of Stamford Hill, for improvements in the manufacture of soap.

George Whitmore, of Austin Friars, London, for a new method of combining, by means of machinery and adhesive composition, all kinds of materials, such as cotton, silk, hemp, tow, furs, wool, hair, &c. into manufactured articles, which may be applied to the purpose for which paper, pasteboard, still board, papie machér parchment, vellum, leather woven fabrics, felt, haircloth, tarpaulins, and the skins of animals, are used.

Edward Shaw, of Fenchurch-street, law stationer, for improvements in the manufacture of paper, or paper boards.

Samuel Hall, of Basford, Nottingham, civil engineer, for improvements in steam engines, heating or evaporating fluids or gasses, and generating steam or vapours.

Peter Fairbairn, of Leeds, machine maker, for improvements in machinery or apparatus for roving, spinning, doubling, and twisting cotton, flax, wool, silk, or other fibrous substances.

David Cheetham, jun., of Chester, cotton spinner, for improvements in the method of condensing smoke, and thereby economising fuel and heat in steam engines, and other furnaces, or fire-places.

Laurence Heyworth, of YewTree, near Liverpool, merchant, for a method of applying steam power directly to the periphery of the movement-wheel for purposes of locomotion, both on land and water, and for propelling machinery.

William Neale Clay, of West Bromwich, Safford, manufacturing chemist, and J. D. Smith, of St. Thomas's Hospital, for improvements in the manufacture of glass.

Job Cutler, of Birmingham, and T. G. Hancock, of the same place, for an improved method of condensing the steam in steam engines, and supplying their boilers with the water thereby formed.

NOTES AND NOTICES.

High Chimney.—The new chimney, recently erected at Mr. Muspratt's chemical works, at Newton, has recently been put into operation. It is stated to be the highest chimney in England, measuring no less than 132 yards 1 foot and 4 inches, from the base to the summit.

Italian Association.—The Italians have at length determined to imitate the Germans, in holding an annual scientific assembly. The first is to be held next year at Florence, in the month of October. The principal promoter of the scheme is Charles Buonaparte, Prince of Musignano, son of Lucian Buonaparte, Prince of Canino, and well known to the world of science as a zoologist of distinguished merit; Charles Buonaparte's principal work is in the English language, a continuation of Wilson's *American Ornithology*, which, begun by a poor Scotch weaver, obliged to fly his country for jacobinism, was finished by an Italian prince, the nephew of the mightiest despot of modern times. The Italian Association is like to do good; Italy being, of all countries, the one most in need of centralization. The hint should have been taken long ago; but the Italians have still an absurd habit of looking down on their neighbours over the mountains, as a sort of barbarians, from whom nothing good can be taken—but their money.

Russia and her Bones.—The Emperor Nicholas has published an Ukasse forbidding the further exportation of bones. The Russians are adopting the new manure; and, having tested its virtues, are commanded to keep, in the words of the proverb, "their own fish guts to their own sea maws." But the trade of grinding bones to fatten turnips will go on in spite of them. Russia is not all the world; and every one knows the power of British gold in drawing hidden stores to light, the bare existence of which was not even suspected. In talking over this subject with Mr. Maxwell, of Gribton, he suggested one patent mode of correcting deficiency, should deficiency arise. Our whalers, in visiting the frozen North, leave behind immense magazines of bones, which may yet form a great article of traffic. Not unfrequently the ships return clear, or only half filled with blubber; then why not complete the stowage with bones, whenever room is left—or deposit them, when the ship is otherwise full, in some place of safety, for future contingencies. The bones of whales and seals may not be quite so good as those of land animals; but the difference is not so great as to mar the traffic, should deficiency of the latter threaten marked enhancement of price. The art of manuring is as yet in its infancy; and practical chemistry has revealments to make in this departments, which will supply more and more what the Emperor Nicholas is pleased to withhold.

Musical Notation. — It is a curious fact, that while the ordinary hand-writings of the natives of various kingdoms of Europe differ materially, the musical notes are written in nearly the same form by all professors of the science throughout the continent, with the exception of some placing the dot before the stem, and others after it. In England the latter mode prevails in regard to crotchets and quavers; but the reverse with minims, generally speaking—*Musical World.* [In the *Archæology of Wales* there are, in the third volume, 175 pages of the musical notation used by the ancient British minstrels in the eleventh century, which are remarkably curious, and consist of letters from the Bardic alphabet, with a variety of characters (now unknown), but not upon lines.]

French Locomotives.—On the 25th October, the first locomotive ever built by French engineers with French iron, was tried on the St. Cloud and Paris Railway. It is, say the accounts, from 40 to 50 horse power; and able to draw 100,000 kilogrammes, or 20 laden waggons. The trial was perfectly satisfactory : it made the journey from Paris to St. Cloud in 16 minutes, and the journey back in 13½, which is at the rate of about 33 miles an hour. Its name is the *Alsacé*; and it is curious to observe, that notwithstanding the boast of its being built by French engineers, it owes its origin to a province, politically French, indeed, but in language, manners, and character, still essentially German. It was built at the manufactory of Messrs. Stehelin and Huber, at Bitschwiller, in the department of the Upper Rhine, in Alsace — a manufactory large enough, it is said, to supply twelve locomotives a year—under the immediate superintendence of Mr. Stehelin, to whose talents it is indebted for a peculiar lightness and elegance of construction. The iron is more indubitably French, being supplied from the works of M. Muel-Doublas, at Abainville. The price of the locomotive is said to be no higher than that of an English one; as, though the cost of the iron is, of course, greater, the difference is made up by the less amount of wages paid to the workmen.

Berlin and Potsdam Railway.—A Berlin correspondent writes to the *Hamburgh Correspondent* newspaper, a flaming account of the sensation occasioned in the Prussian capital, by the recent opening of the railway to Potsdam. At first, he says, the great mass of the public felt unwilling to face the fancied dangers of the new mode of conveyance; but as soon as the rapturous accounts of those who made use of it came to be generally spread, and the non-occurrence of any accident negatived the idea of danger, a rage for railway travelling sprung up, which was daily on the increase. In the course of the first week, 11,400 persons took the ride; and in the second, from 2,200 to 2,300 was the daily average. Double trains were obliged to start every hour. The Berlin correspondent makes no allusion to one cause of popularity for a railroad in Prussia, which, if travellers are to be believed at all, must be no trifling one. The roads there are so deep in sand, that by all description, it is a purgatory worse than Dante's to be dragged along them in a German *eilwagen*, or haste-carriage, so named and so mis-named. In a country of such roads, there must be a peculiar zest in railroad travelling of which we can form no idea.

Railway to Cologne.—At the meeting of the Council of Management of the railway from Cologne to the Belgian frontiers, which took place at Aix la Chapelle on the 9th of November, the directors engaged to open in the spring, one section beginning at Cologne (the greater part of the rails are already laid down), and in the course of the summer a second section, beginning at Aix la Chapelle. The directors again announced, that hitherto the expenses of the land purchased, and the works executed, have been below the estimates; and they congratulate themselves on having, from the very foundation of the society, followed the principle of making the estimates too high, rather than raise illusions, which in other places have been so cruelly dispelled. The execution of the tunnels proceeds rapidly, and is not attended with the difficulties that were expected.

Mr. Imlay, of Philadelphia, has constructed several cars for travelling on railroads, with seats that at night are opened, and made into sleeping berths, as in a ship.

Great North of England Railway.—The workmen have now commenced laying the foundation of the fifth and last bridge of the journey of the Great North of England Railway, near North Allerton, a little to the south of the town, which will cross over the high road leading to Boroughbridge, near to the 220th mile-stone from London.

Metropolitan Railway Map.—Early in December will be published vol. xxix. of the *Mechanics' Magazine*, price 8s. 6d., illustrated with a *Railway Map* of the Metropolis, taking in a radius of 15 miles from the Post-office. Encouraged by the extensive sale which our *Railway Map of England* has commanded, the *Metropolitan Railway Map* has been executed at a very great cost; the utmost exactness has been observed in reducing it from the Ordnance maps, and all the railways projected up to the day of publication have been distinctly and accurately marked from actual survey. The limits of the twopenny and threepenny post deliveries are also shown in the Map. The *Metropolitan Railway Map* alone, stitched in a wrapper, price 6d., and on fine paper, coloured, 1s.

The *Railway Map* of England and Wales continues on sale, in a neat wrapper, price 6d.; and on fine paper, coloured, price 1s.

☞ *British and Foreign Patents taken out with economy and despatch; Specifications, Disclaimers, and Amendments, prepared or revised: Caveats entered; and generally every Branch of Patent Business promptly transacted. A complete list of patents from the earliest period (15 Car. II. 1675,) to the present time may be examined. Fee 2s. 6d.; Clients, gratis.*

LONDON: Printed and Published for the Proprietor, by W. A. Robertson, at the Mechanics' Magazine Office, No. 6, Peterborough-court, between 135 and 136, Fleet-street.—Sold by A. & W. Galignani Rue Vivienne, Paris.

𝕸echanics' 𝕸agazine,

MUSEUM, REGISTER, JOURNAL, AND GAZETTE.

No. 800.]　　　　SATURDAY, DECEMBER 8, 1838.　　　　[Price 3d.

HOLEBROOK'S PATENT PADDLE-WHEEL.

Fig. 1.

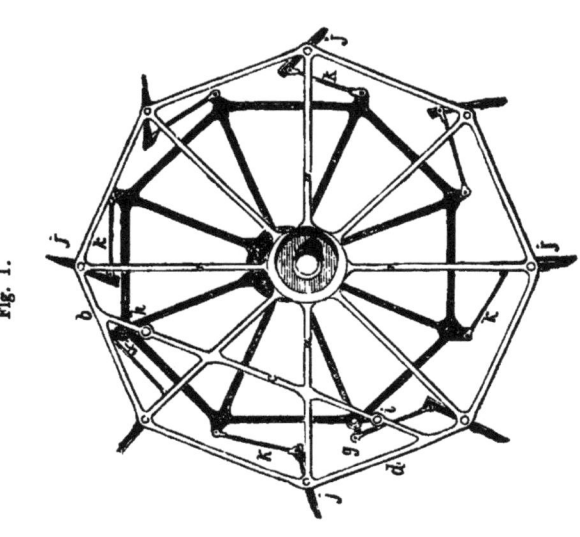

Fig. 2.

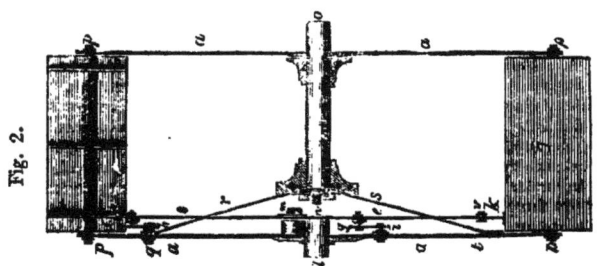

HOLEBROOK'S PATENT PADDLE WHEEL.

Amongst the numerous patents which have lately been obtained for improvements, or rather alterations, in paddle-wheels, and the modes of propelling steam-vessels, the specifications of which appear on the rolls of the Court of Chancery, there is not one which appears to have been more the result of close study and calculation than that patented by Mr. J. P. Holebrook. At first sight, or even upon a cursory examination, it would appear to be very similar to Morgan's, Galloway's, and Seaward's paddles, and, by many, would as such be instantly dismissed from the mind, as a most glaring plagiarism. On examination, however, into the details, and careful tracing of the working of the radial rods, arms, and paddles, through their various positions, it will be seen that there is considerable difference,—and that, in that difference, great superiority lies. That it is a modification of the same principle as that upon which Galloway's or Morgan's wheel is founded, no one can deny; but Mr. Holebrook is as free to come into the field, or rather on to the water, with his arrangement of this principle, as Mr. Morgan, or the many others who before and since Mr. Galloway's patent essayed to improve the common wheel, by applying moveable paddles, guided into their several positions by an eccentric or crank.

In a pamphlet which Mr. Holebrook has circulated upon the subject of the paddle-wheel now under consideration, (and of a shifting paddle-wheel, which we shall take another opportunity of publishing,) he gives us the following introductory summary of the advantages generally considered to be attending upon the use of feathering paddles :—

" It has been suggested, as regards the carrying on of steam navigation by sea, and when the immersion of the propelling apparatus necessarily must be great, that it would be desirable that the paddles should enter and leave the water at angles, other than those obtained by radial positions of the paddles, in order that less power may be consumed by the paddles when entering and leaving the water, and a greater proportional beneficial return be obtained even for the diminished power consumed at such points by the paddles, and in order, further, that a greater consumption of power may take place at parts at which the employment of power is attended with greater beneficial effect. It is demonstrable that such greater effect can be produced by a properly proportioned combination, in which the paddles should enter and leave the water at angles, obtained by positions between radial and vertical ones, agreeably to circumstances, than by paddles fixed radially, as in the common paddle-wheel; but it is essentially necessary that the machinery in use should be of such a construction as to render the employment of it prudent, on the score of strength and lightness, for, otherwise, it will be a matter generally of little doubt that the simpler and stronger machinery will be preferred, even with a certain persuasion of a loss of power, to the more complex and less secure instrument although the employment of it be as certainly attended with increased effect : but, when the parts of such machinery are put together with proper attention to strength and lightness, and the employment of it attended with increased effect, it will scarcely be wise to resist its use on the score of its parts being moveable, and, in consequence, more complicated, because the employment of more complex, and, at the same time, more economical machinery, for the attainment of a desired end, is always a proof of a greater approach to perfection in the means of attaining an object in view. The steam-engine, in its most economical, and in its more simple, form, is a proof of this assertion. Indeed, it may generally be considered that complexity of machinery, while it is attended with increased effect, is no bar to its employment, provided there be no objection to the machinery in point of strength, security, or too great complexity.

" In his improved method of propelling vessels, the inventor has carefully studied to obtain, by the application of a new plan of moving the paddles, such positions of the paddles as are considered desirable, with all due attention to strength and lightness, and within the space usually allowed. He particularly calls attention to the construction of the skeleton of his paddle-wheel, which, though more complex than the framing of a common paddle-wheel, can scarcely be said to be less strong. He also desires an attentive consideration should be given to the construction of his paddles, by which, though they may be guided by parts near their extremities, it will be seen that it is scarcely possible to conceive a case in which they can be twisted and put out of form.

" Of the skeleton of this paddle-wheel, it may further be said that, from its peculiar form, it can, in case of accident, by the addition of common float-boards, be readily converted into a common paddle-wheel;

and, by this, it will be perceived that, in cases of accident to its moveable parts, it may be made equally effective with the common paddle-wheel, while, until such cases of accident occur, the extra benefit of its peculiar positions of paddles may be obtained. That accidents such as those to which allusion has just been made, never have happened with wheels properly constructed with moveable paddles, has been proved by the employment of such wheels, for some years past, in the government navy.

We shall now proceed to describe Mr. Holebrook's wheel, and in so doing, we cannot convey a clearer idea of the plan than by quoting his specification.

" Figure 1 is a side view of a wheel constructed on this plan ; and fig. 2 a sectional view, showing more clearly the essential parts of the wheel. The wheel consists of five principal parts, with their connections and appendages ; namely, an octagonal framework, in two parallel pieces ; the paddles between these frames ; a guide-wheel, to give the desired position to the paddles ; a crank, to which the octagonal framework and guide-wheel are connected in common ; and what I call a star-wheel, which transmits the power directly, from the end of the paddle-wheel shaft, to the outer circle of the paddle-wheel. *a a* are the arms of the outer circle of the paddle-wheel, terminating, on the outer side, in the octagonal framework, and, on the inner, in a nave, which turns loosely on a part of the crank before mentioned. *b c d* is a bar, which passes from one part of the octagonal framework to another, intersecting two of the arms *a a* between the points *b* and *d*. *e e* are the arms of the guide-wheel, which are connected, on the inner side, with the crank before mentioned, and project, on the outer side, a little beyond its octagonal frame.

" The centre of the guide-wheel is placed a little above the centre of the paddle-wheel. The paddle-wheel and the guide-wheel are connected together by means of two metal straps, moving upon four pivots ; two of which are fixed in the guide-wheel, at the points *f* and *g;* and two at the points *h* and *i*, of the bar *b c d*. The object of the bar *b c d* being to obtain the two positions . *h* and *i* for two of the pivots of these straps ; it is so placed, in the paddle-wheel, that if a circle were drawn from the centre of the paddle-wheel, of the same size as a circle drawn from the centre of the guide-wheel, passing through the points *f* and *g* of the guide-wheel, and two radii drawn from the centre of the paddle-wheel, intermediately between the radii of the paddle-wheel, between

which radii the parts *h* and *i* are placed, the points of intersection, of the circle and these two radii, would be *h* and *i*. From the angles of the octagonal framings and the extremities of the arms of the paddle-wheel, pass eight strong rods, *p p p p,* which brace together the inward and outward circles of the paddle-wheel. The paddles *j j,* with the stems fixed to them, are placed upon these rods, and connected with the guide-wheel by means of the rods *k*, which I call guide-rods, one end of which is attached by means of pins, on which they turn, to the projecting ends of the arms *e* of the guide-wheel, and the other end works upon pins placed at the extremities of the stems of the paddles. Each paddle consists of a plate of iron, an iron half-pipe with almost entirely closed ends, and three iron stems. A superficial view of one of these paddles complete, is given separately in fig. 5, and, an end view, in fig. 6. A superficial view of the half-pipe is also given separately in fig. 7, and an end view in fig. 8. The half-pipe, it will be observed, is placed by its edges against the iron plate, and upon this half-pipe and plate are placed the three stems ; the whole being firmly fastened together.

" From the extremity *u* of an arm of the guide-wheel, a guide-rod proceeds to a stem of the uppermost paddle, fig. 2, while, from another extremity *v* of another arm of the guide-wheel, a guide-rod proceeds to the stem of the lowermost paddle ; both stems having holes in them for pins for the connection of their respective guide-rods. Any direction given to either of these guided stems will be mainly transmitted, by the half-pipe, to the other stems and the other parts of the paddle ; and, thus, any twisting of the paddle is scarcely possible of occurrence. It is not necessary to have three stems to a paddle, nor, is it absolutely necessary, to have the paddle constructed of iron, as I have supposed it to be ; because it may be made of wood and iron in various ways ; but the method I have described is one which I think to be, at once, useful and strong. From what I have stated, it will have been seen, that the paddle-wheel and the guide-wheel are attached and made to revolve together, upon the principle of a well known method, namely, by means of straps or cranks, the distance between whose working points is equal to the amount of the concentricity of motion of the paddle-wheel and guide-wheel, and whose same points are in a direction parallel to the direction of the centres of the paddle and guide-wheels. Every paddle, in the course of one revolution of the wheel assumes in turn each of the positions which the different paddles exhibit in fig. 1, or other positions corresponding therewith.

K 2

" In the sectional view, fig. 2, *n o* represents part of the paddle-wheel shaft ; *a a a a* the arms of the paddle-wheel connected together at their extremities by means of the rods *p p p p*, upon which the paddles, *j j*, turn ; the inward arms, *a a*, of the paddle-wheel, terminating inwards in a nave, firmly fixed to the paddle-wheel-shaft ; and the outward arms, *a a*, of the paddle-wheel, terminating inwards, in a nave which revolves loosely upon the lower part of the crank *l m*. Hitherto, the only connections I have shown between the outer and inner circles of the paddle-wheel are the rods *p p p p ;* but, in *q r s t*, fig. 2, and shown, separately, in figs. 3 and 4, is represented the star-wheel before mentioned, which I have introduced for the purpose of further connecting the outer circle with the paddle-wheel shaft.

" I call it a star-wheel because it resembles in form a wheel without a periphery or bounding lines. It is firmly attached, by a nave to the extremity of the paddle-wheel shaft, and, by the extremities of its arms, (which are made of a bent shape for the purpose,) to parts of the arms, *a a*, of the the outer circle of the paddle-wheel. By means of this star-wheel the outer circle of the paddle-wheel is more directly moved, by the paddle-wheel shaft, than by means of the rods *p p p p*, upon which the paddles move. It should properly be stated, in this place, of this star-wheel, that the peculiar bent form given to it, in fig. 2, is not its only form ; because, according to circumstances, it may be bent differently, and may be superseded by a wheel with a rim ; the arms of which wheel need not be bent, but the extremities of which arms may be attached, by means of rods, connecting other formed framings of the paddle-wheel, and this then modified apparatus together ; but, I would here observe, that the form which I have given to this apparatus in fig. 2, is that which I consider generally to be most desirable. It should be here noticed, that this apparatus is not seen in fig. 1 ; because, being placed behind the paddle-wheel, and its arms coinciding in direction with parts of the arms of the paddle-wheel, shown in that figure, it could not be exhibited in such a view of the wheel as given in that figure. In a paddle-wheel, such as is represented in figs. 1 and 2, the measure of the eccentricity of the combination is equal to about one-half of the distance, between the centre of the stem of a paddle and the point at which the stem is guided ; but, it is hardly necessary to observe, that this measure of eccentricity may be increased ; and that, in proportion as it is increased, the paddles, which

are at the bottom of the wheel, will assume positions more nearly vertical ; and *vice versâ*. The converse of this proposition must be equally evident ; namely, that, in proportion as the distance between the two before named points of the stem of a paddle is diminished, the paddles will also assume positions more nearly vertical ones ; and *vice versâ*. From fig. 1, it will be perceived, that the arms of the guide-wheel, upon which the two lowest guide-rods are placed, are equally distant from the lowest arm of the paddle-wheel ; and this will be found to constitute an important novelty in my plan of construction ; for, it is by this, or some like relative position of these two parts, that I am enabled to obtain such positions of the paddles as are shown in the figure. I have used a crank, as a point upon which the guide-wheel may revolve, and, I have placed the guide-wheel within the paddle-wheel, but it is not absolutely necessary, either, that the crank should be used, or, the guide-wheel so placed ; because, a guide-wheel, upon a large eccentric centre might be used, and, the guide-wheel might be placed outside of, and, on either side of, the paddle-wheel, though not, in my judgment, to so much advantage.

The principal points which distinguish Mr. Holebrook's from Morgan's wheel, are very clearly set forth in the following communication which we have received from Mr. Holebrook :—

" The objects sought in the construction of my paddle-wheel, are precisely similar to those obtained in that modification known as Morgan's paddle-wheel ; these objects being a more economical application of power with less swell and less vibration than is attained by the use of the common paddle-wheel. It would be wide of my present purpose to enter upon an illustration and demonstration of the advantages resulting from the attainment of these objects, inasmuch as these have been so elaborately treated upon in the late splendid edition of *" Tredgold on the Steam-Engine,"* &c. : It will, therefore, perhaps be deemed allowable that these advantages should be considered as settled. This being granted, I shall proceed to point out the improvements in my wheel upon Morgan's. The first improvement I shall notice, and, indeed, the most important one, is that in the construction of the framing of the paddle-wheel. In Morgan's wheel it will be recollected, by those who have a knowledge of it, that the power is transmitted, directly to the inner framing of the wheel, from the end of the paddle-wheel shaft; which shaft terminates with the inner framing, leaving the power, which is necessary to turn the outer framing of the wheel, to be transmitted through the ties of the wheel alone. Now, then this mode of transmitting the power, scarcely any method can be be more objectionable; for, it is clear, that the resistance offered to the passage of a paddle, operates as much upon the outer, as upon the inner framing of the paddle-wheel; while the outer framing derives, under this plan, all its strength from the ties which connect it with the inner framing; and, as a consequence, this inner framing must either be made much stronger than the outer framing, or the outer framing must be much less capable of bearing a resistance than the inner one. It is true, that as much ingenuity has been called into exertion, in endeavours to obtain, by a good disposition of the ties of the wheel, as much strength, with as little expenditure of metal, as, probably, the plan of transmitting the power will admit. In my wheel, it will be perceived, that the shaft is continued nearly to the outer framing of the paddle-wheel; and that, then the extremity of the shaft is directly connected with the outer framing by means of an apparatus, which I have denominated a "star-wheel," i. e., "a wheel without a periphery or bounding lines." By this means, I dispense with a complicated system of ties, and obtain what, mechanically considered, is a far stronger form of wheel: in fact, it needs but little examination to perceive that, as far as the framing of the paddle-wheel is concerned, it is scarcely, mathematically considered, less strong than the framing of a common paddle-wheel. That this improved form of framing is an improvement upon the framing of Morgan's wheel, I think very few will deny; and also that, considering that my star-wheel performs the part of the complicated ties of Morgan's wheel, my construction is far less expensive of manufacture.

"The next improvement I will notice, is that resulting from the absence of bent stems. The mode of actuating the paddles in my wheel being essentially different from that of Morgan's wheel, I am enabled to dispense with the bent stems of that modification; I mean those arms projecting at nearly right angles from the surfaces of the paddles, by which the paddles are guided in Morgan's wheel. The absence of these stems in my wheel would necessarily make its cost less than that of Morgan's, while it also obviates a chance of accident from the breaking of these stems. In my wheel, it will be perceived, that the paddles are guided by the extremities of those stems of the paddles which are employed for the purpose of strengthening them.

"The last improvement which it may be worth while, now, to notice, is another dependent upon the difference of my mode of guiding the paddles, as compared with that pursued in Morgan's wheel. In the passage of a paddle through a revolution, it will be seen, that, in Morgan's wheel, in consequence of the guide-rod of a paddle approaching at times the centre of the paddle, it comes within its sphere of action; it, therefore, becomes necessary to cut a slit in the paddle, in order to allow of the proper playing of the guide-rod. Taking this into consideration, I think the absence in my wheel of the necessity for thus slitting every paddle, will be considered as another improvement upon Morgan's combination.

"Having now disposed of the improvements in my wheel upon Morgan's, I propose to make a few observations upon what may seem a similarity, resulting from its being a new application of the same principle of actuating the paddles; I mean, that, because I obtain in my wheel the same positions, or nearly so, that Morgan does in his, it may be inferred, that, while improving the construction of Morgan's wheel, I have been essentially applying a method of guiding the paddles which belongs of right to Mr. Morgan: that I have not thus applied his method it will be my endeavour now to prove. In order to this proof, I shall notice the points of similarity, and points of difference, between mine and Mr. Morgan's wheels.

"The most striking point of resemblance between the two wheels is that of the same positions of the paddles, or nearly so, being attainable in both: in respect of this, I have to say, that I am not aware that Mr. Morgan claims the positions of the paddles, but rather the means by which such positions are obtained; and, as I propose presently to show, that the means employed by him and by me are very different, we may dismiss this point of similarity, and go to the next superficial, and, yet, very apparent resemblance between the wheels—I mean the polygonal form of the wheels. If a charge of imitation were to be preferred against me on this score, it is very clear, that it could be obviated by at once adopting circular framings to my wheel, which would, of course, nearly as well answer my purpose as the polygonal ones; but, it is impossible, that any one could seriously prefer a charge on such an account: and yet, this form of framing, together with the positions of the paddles, I am persuaded, will make many a superficial observer imagine my wheel and Morgan's to be similar. The next point of similarity is the adoption of the crank within the wheel, as a centre for the guide-wheel: upon this point I have only to say, that Mr. Morgan has no claim to this application of the crank, inasmuch as it was published before he employed it. This answer I may also make to the only other point of resemblance, viz., the employment of the rods which brace the extremities of the framings of the wheel together, as centres upon which the paddles may move. I shall now notice the points of difference between the wheels, and, I think, I shall show some very important ones.

" In the first place, upon a comparison, it will be perceived, that the positions of the working points are in as opposite directions in the two combinations as it is possible for them to be: in my wheel, the centre of the guide-wheel is placed *above* the centre of the paddle-wheel, while, in Morgan's, it is *aside*; in my wheel, looking to the lowest paddle, it will be seen, that this paddle is guided at a point *above* its centre, while, in Morgan's, the same paddle is actuated at a point *aside* of its centre; in my wheel, the part of the guide-wheel to which the guide-rod, which moves the lowest paddle, is attached, is situated, as regards the point by which the same paddle is actuated, *aside*, while, in Morgan's, the position of a similar point of the guide-collar is, as regards a similar point of the same paddle, *above*. If the geometrical construction of both wheels be considered, it will be perceived, that *five* working points are essentially necessary, to the proper motion of a paddle in my wheel, while, in Morgan's, *four* working points are sufficient, though, for purposes of convenience, *five* are always employed for every paddle but one, which one has only *four*; this paddle being that one whose bent stem forms the dragging link of the combination. As a consequence of this difference in the necessary number of the working points, the guide-wheel in my combination cannot well be lessened, while, in Morgan's wheel, the guide-collar may be dispensed with altogether, and a pivot substituted in its place, as a point upon which all the guide rods may revolve; indeed, if I were to dispense with my guide-wheel, and resort to such a pivot, my combination would become unstable, and, whatever positions the paddles had, they would be in some degree similar to those obtained in Udny's combination. From my wheel, when thus rendered unstable, Udny's might be obtained, by merely giving an inclination from a vertical direction to the lowest paddle, sufficient to ensure stability of action; and Morgan's could also be obtained by merely turning Udny's a quarter of a circle round, by considering one-half of the paddles to represent the direction of the stems, by entirely removing the other half of the paddles, and by affixing the paddles to the stems at an angle nearly equivalent to that produced by the change in turning the combination partly round. In the mode of connexion of the paddle-wheel with the guide-wheel, or guide-collar, in the two combinations, much difference exists. In Morgan's wheel, one of the bent stems of the paddles serves for a connecting link between the parts, while I am compelled to use two metal straps for the purpose of connecting these parts; and, were I to adopt Morgan's more simple, but not more safe, one link, the paddles in my combination would have very different positions to those which I obtain by my means of connexion. In respect to the mode of arriving at the position of the centre of the guide-wheel or guide-collar, it may be observed, that the well-known method, which applies, when used for this purpose, to Morgan's wheel, is altogether wholly useless, and inapplicable when applied to mine : indeed, this is a natural consequence of the different modes in which the paddles are guided in the two combinations.

"Sufficient has now been stated respecting the properties of Morgan's wheels, and mine, to show, that however similar the results of both combinations may be, the means by which these results are obtained are as dissimilar as it is well possible for them to be. It must not be imagined, however, that all the points of difference have been described, but, only, that it is thought that the remarks which have been made are sufficient to remove any impression of similarity of principle, which a cursory glance at the two wheels might have left upon the mind of any one".

HARPER AND JOYCE'S PATENT STOVE AND PREPARED CHARCOAL.

We have carefully examined the specification of Mr. Joyce's patent for " Improved apparatus for heating churches, warehouses, shops, factories, hothouses, carriages, and other places requiring artificial heat, and improved fuel to be used therewith," which is enrolled at the Rolls Chapel office, and we now present our readers with an abstract of its contents from memory, and which we believe will be found to be substantially correct.

Mr. Joyce commences by stating that by means of his peculiar construction of stove, he is enabled to regulate the combustion of his fuel, so as to produce "any required degree of temperature."! This is an evident plagiarism from a worthy stove-doctor, of some note, whose *ipse dixits* on " warming and ventilation" we some time ago—to use a vulgarism— " hauled over the coals." The doctor could make the temperature of his stoves " as much higher or lower (than 60° or 65°) as he liked", and Messrs. Harper and Joyce can " produce any required degree of temperature." The strictures which we have referred to were, however, more particularly directed against the doctor's book, than his stoves, which are certainly good as far as they go; and he has the merit of bringing them into extensive use by the aid of his name and pen, and of his prior reputation, if he cannot be considered the inventor. The thermometer, or self-regulating fire-places, at all

events, *are not* stoves whose action is deadly, whatever the effects of the constant use of those, or any other hot-air warming machines, may be on the general health of their users.

Mr. Joyce proceeds to describe his improved stove, and it is essentially the same as we have already described, and figured in our 767th number. He describes various modifications as applicable to various situations; amongst others, one in which the stove is surrounded with water, and another in which a chimney is added to it, to conduct the carbonic acid evolved to the outside of the building to be warmed—a very necessary precaution, but one which would so diminish the warming effects of the stove, (which result from the mixture of the products of combustion with the air of the apartment,) as to render it almost useless.

We come now to the most important part of the patent, the prepared fuel. Mr. Joyce states that this fuel consists of charcoal (of which he prefers that made from oak or beech), prepared with alkalis.

To prepare the charcoal, he makes it red hot in an air-tight oven or furnace, and saturates it with a solution of " caustic or carbonated alkalis, alkaline earths, or their salts." The solution is composed of about three pounds of soda, or other alkali, to twelve gallons of water. He states that he generally uses the caustic soda of commerce, but that other alkalis, especially lime, will answer the purpose. The charcoal may be either saturated before or after being made red hot ; where the latter mode is adopted, the alkaline solution is poured on to the charcoal as it is cooling which must, of course, be allowed to become dry before it is used.

MESSRS. SEAWARD'S IMPROVED MARINE ENGINES ON BOARD THE " GORGON."

In a former number (776) we gave a description of these engines, and at that time ventured to predict, that the extensive adoption of a plan in which those cumbrous appendages, sway beams, side rods, cross heads, &c. were done away with, was sure to follow its successful working, and was likely to create

a great change in steam marine affairs.
The result of four months' cruising on
the coast of Spain has fully established
the superiority of the *Gorgon's* engines,
as regards perfection of operation, wear
and tear, and, more important than all,
economy of fuel; these engines of 320
horse power, consuming on an average
15 cwt. of coals per hour, according to
the various reports made to the Admiralty. Their success has been, indeed,
so complete, that the Lords of the Admiralty have ordered Messrs. Seaward and
Co. to fit the splendid frigate *Cyclops*,
of 1300 tons, now building at Pembroke,
with engines upon the same plan. The
Russian, Lubeck, and St. Petersburgh
company's splendid ship now building
in the Thames, of 900 tons, is also to be
fitted by Seaward and Co., with similar engines, of 240 horse power, as
well as a new Man of War for the
East India Company of 240 horse power.

As a remarkable proof of the com-
pactness of this plan of engine, the
engine-rooms in the two latter vessels
are only 45 feet long, with stowage for
120 tons of coals; whereas, with common beam engines of the same power,
60 feet would have been required; making a clear saving of 15 feet in the widest
and most valuable part of the ship.

By the naval papers we find, that
Lord John Hay has left the *Tweed* frigate, and hoisted his broad pennant on
board the *Gorgon*. We believe this is
the first time that a Commodore's pennant
has been hoisted on board of a steam
Man of War.

We perceive by extracts of letters
from Lord John Hay, that the *Gorgon*
has frequently had as many as 2,000
Christino troops on board at one time,
to transport to different parts of the
coast of Spain.

JUCKES'S MODE OF VOTING BY BALLOT.

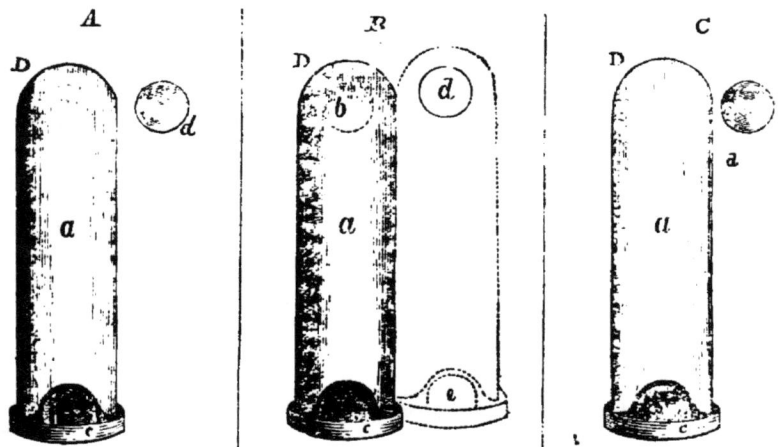

Sir,—Being an ardent admirer of the
principles of voting by ballot, and, looking forward to the time when our representatives will be chosen by such means,
my attention has been directed to devise
a method of giving the vote, and I think
the following plan will be most effec-

tive. If not, I trust some better idea
will be proposed.

I propose, first, that two adjoining
apartments be provided, one for the
candidates and one for the voters, exclusive of the hall or space occupied by the
public. That a lobby, or third apart-

ment, be provided, for putting the requisite questions as to the qualification of each voter, as he presents himself; such questions and answers to be attested. That the voter be then passed to the voting room, where he is not to be seen by, or to see, the candidates. I will suppose there are three candidates, two of whom are to be returned; in such case I would have three balls, coloured respectively, say blue, red, and yellow. Six tubes, or bags, through which such balls will pass, are to be fixed in the partition between the two apartments; three of these tubes to be fixed to convey the balls from the voters' into the candidates' room; and the other three, so as to return the balls to the voters' room, into a receiving box. The mouth of each tube, or bag, to be coloured respectively the same as the balls, and the name of each candidate affixed to his colour; the voter, on entering the room, takes the two coloured balls he chooses, and passes them through the corresponding coloured tubes, or bags; or, if he votes for one only, then he uses one ball; he then passes out by a separate doorway, and another voter comes in, and finds the balls precisely in the same situation as when the first voter came in, they having been returned to the voters' room through the return tubes, by some official, in the interim.

I have made for amusement a model of this simple plan, which is left to be seen at the Old George, George-passage, Snowhill; and having described it to some hundreds, who have been much pleased, I have, in consequence of their approbation, been led to make the plan public.

I am, Sir, your obedient servant,
JOHN JUCKES.

Description of the engravings.—A B C, three compartments, one for each candidate; *a a a,* three tubes, or bags, either on the voters' or candidates' side of the partition. The dotted lines in the middle compartment show the position of one of the tubes on the other side of the partition; *b,* dotted lines, showing the position of the hole on the other side; *c c c,* trays at the bottom of the tubes, to receive the voting ball; *d d d,* three holes, by which to return the voting balls to the tray on the other side of the partition; *e,* dotted lines, showing position of tray on the other side.

IMPROVEMENTS IN BUILDING.

Sir,—It has often struck me, when looking at the public buildings of London, that the stone used is quite unfit for the purpose, owing to its great liability to be affected by the weather. St. Paul's, for instance, in a couple of centuries will not have one of the external sculptured ornaments remaining; even now, the wind-worn angles are quite a deformity; so, also, is Somerset House, Blackfriars and Westminster bridges, and, in fact, all the buildings where the Portland stone is used. It has always appeared to me to be the height of absurdity to bestow great labour on such a worthless material. We have Antidry-rot Companies, why not fall on some plan of preserving stone? An Anti-stonerot Company would be a good speculation for this age of speculations. I am not aware of what stone the new Exchange and new Houses of Parliament are to be built, but I do think it is a subject worthy the attention of the nation, to choose a proper material for these costly and national buildings. Craig Leith stone appears to me to be the best stone I am acquainted with; its colour is excellent, stands weather equal to granite, and can be easily worked. Stones of any size can be had, even up to 100 tons, without a flaw.

I have another subject to mention, equally important. The present method of working stone, I believe, is the same as was employed at the Pyramids of Egypt and Solomon's Temple. Why not employ steam power for all plain hewing? There is a Mr. Hunter, of Arbroath, who has been working stones by steam power for several years. I am certain, if steam is employed, there will be a saving of at least half a million to the nation, in the construction of the Houses of Parliament alone, taking the probable cost at *three times* the architect's estimate, which I have always found to be a good datum.

I am persuaded great economy would be experienced in all richly-sculptured work, for instance, a Corinthian column, if there were cast-iron moulds made, and in short lengths, so that a man might go inside, and give the mould a coating of mastic, and the centre of the pillar might be filled up with brick-work and liquid mortar; a set of pillars for a colonnade might be made at small expense, and would last ten times longer than the stone used at the good West-end houses of the nobility, a wall of which can be completely perforated with the point of an *umbrella;* as witness the entrance-gate to Hyde Park, as a proof.

Your publishing these hasty hints will much oblige

AN OBSERVER.

METHOD OF FINDING THE CURVES OF THE TEETH OF WHEELS, RACKS, THREADS OF ENDLESS SCREWS, ETC. FOR MILLWORK.

Sir,—The teeth of racks are commonly formed by lines square to the pitch line, and those of the pinions which work with them by radial lines within the pitch line, and the teeth of both by curves without the pitch line. Other forms might be given to them, but they would render the direction of the force more oblique, and are therefore objectionable. Curves for the above-shaped teeth may be found in the following manner :—

For the rack, fig. 3, draw two straight

Fig. 3.

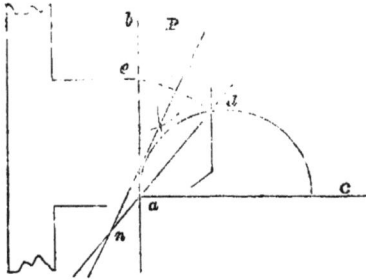

lines, *a b* and *a c*, at right angles to each other, intersecting at the point *a*, and consider *a b* the pitch line of the rack.

Upon the line *a c* draw a semi circle of half the diameter of the pinion, and touching the line *a b*.

Draw a straight line representing the ends of the teeth parallel to *a b*, and cutting the semi-circle at the point *d*.

Upon the line *a b* set off *a e*, equal in length to the arc *a d*.

Draw a straight line through the points *d* and *a*, and raise a perpendicular P to a straight line drawn through the points *d* and *e*, so that P may be midway between those points.

The point *n*, in which P cuts *d a*, is the centre from which to strike the curved part of the tooth, with a radius *n d*, or *n e*.

For the pinion fig. 4, draw two lines

Fig. 4.

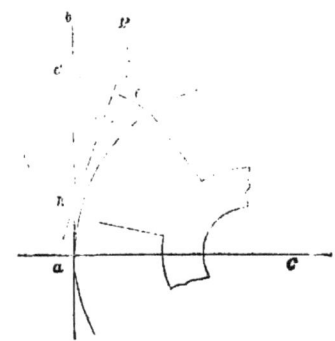

a b and *a c*, at right angles to each other, intersecting at the point *a*.

Upon the line *a c*, draw the pitch line of the pinion touching the line *a b*.

Draw a circle representing the ends of the teeth, cutting *a b* in the point *d*.

Upon the pitch line set off an arc *a e*, equal in the length to the straight line *a d*.

Raise a perpendicular P to a straight line drawn through the points *d* and *e*, so that P is midway between those points.

The point *n*, in which P cuts *a b*, is the centre from which to strike the curved part of the tooth, with a radius *n d* or *n e*.

It is to be observed that, if the arcs are set off correctly, the points *d* and *e* are in the epicycloid in figs. 1 and 2 (see No. 797); in the cycloid in fig. 3; and in the involute in fig. 4 ; and that these are the true curves for the respective cases.

It also appears—1st, That the pitch is not a measure of the radius of curvature of the teeth. 2nd, That the external curve of the teeth of a pinion working with a wheel *twice* its diameter, should be struck from points *n* upon the pitch line. 3rd, That if a pinion works with a wheel *less* than twice its diameter, the points *n* should be within the pitch line (see fig. 1): the rack belongs to this case. 4th, That if a pinion works with a wheel *more* than twice its diame er, the points *n* will be without the pitch line:

the pinion working with a rack belongs to this case.

I am aware, that by taking a line between the pitch line and the end of the tooth, and using it precisely as I have used the line which represents the end of the tooth, a circular curve might be found nearer to the true curve than by the method I have described; but there are objections to this plan,which induced me to reject it.

There are cases in which the points n (see figs. 1, 3 and 4,) will fall between two teeth, where the compass point will not find support; in such cases, the radius with which the teeth are struck may be increased until the compasses get a point of support somewhere upon the perpendicular P; the position of the point from which the tooth is struck being of much consequence, the radius with which it is struck of comparatively very little, so that is not too small.

It cannot be too clearly understood that, *the diameter of wheels of the same pitch should be in exact proportion to their number of teeth,* whatever tables, &c. exist which lead to a different conclusion. Thus, a pinion of four teeth, to work with a rack of one inch pitch, should be 4 inches *circumference;* or between $1\frac{1}{4}$th and $1\frac{9}{37}$ diameters at the pitch line.

If a square threaded screw which is used as an endless screw, were cut by a plane passing through its axis, the section of the thread ought to be of exactly the same form as the teeth of a rack (see fig. 3); and the teeth of the wheel such an endless screw drives, (if the ends of the teeth are parallel to the axis,) should exhibit the same form as those of a pinion (see fig. 4); each suited to the other as directed for the rack and pinion. And the angle each tooth of the wheel makes with the axis, should be derived from the circumference of the *pitch* line of the screw, and its pitch, or rake, and applied to the teeth of the wheel at their *pitch line.* But I had, perhaps, better speak of endless screws, and their wheels, in a separate paper, if you think remarks on the subject worth inserting.

I am, Sir, yours, &c.

C. G. JARVIS.

Nov. 29, 1838.

———

ON THE ACTION OF LIGHT AND AIR, AND THE DISTRIBUTION OF THE ATMOSPHERE OVER THE EARTH'S SURFACE.

Sir,—The object of this paper is to show, that the effects ascribed to the sun's heat are produced by the combined action of light and air; that the distribution of the atmosphere over the earth's surface is effected and regulated by the earth's rotation on its axis; and that the degree of the sun's heat at any given point is in proportion to the intensity of the light, and the density of the atmosphere at that place. And some remarks are also made on the nature of heat and light from compression of air.

To the effect produced by the burning glass, whether it be the burning of a piece of wood, or the melting of a piece of iron, both light and air contribute. It is certain that both these agents are present in the operation, and we do not know that any other agent is present; and in the absence of either of them the burning-glass will not burn. If a light from a lens, which sets substances on fire, be thrown into the same substances in an exhausted receiver, it will produce no perceptible effect. It will not ignite gunpowder. If a burning-light be thrown upon gunpowder in a glass receiver well exhausted, it produces no perceptible effect so long as the vacuum is tolerably good. As the air returns into the receiver (and it cannot long be kept out), the grains of powder begin to smoke and jump about, but the light will not inflame them until the receiver is replenished with air. Hence it is certain that gunpowder cannot be inflamed in a vacuum. A burning light, thrown upon paper immersed in water, produces no perceptible effect, because the water excludes the action of the atmosphere, and not, as is commonly said, because the water absorbs the heat before it can act on the paper. This is evident from the fact, that the light will burn the paper when passed through the water to it. Thus, take a flat transparent glass phial, an inch or two thick, and two or three inches square, fill it with pure water, and place it between the lens and the paper, so that the light must of necessity pass through the bottle of water, and it will then burn the paper. In this case, allowing for a little dispersion which takes place in passing through the water, the light acts with undimi--

nished energy. If the water absorbed the heat, it would absorb it in both cases, and the rays of light in each case would be equally powerless. But it is not so; and the reason is, that in the former case the paper is protected from the action of the air, and in the latter it is not. If a burning light from a lens be thrown upon the hand immersed in water, it causes a pricking sensation; if it be thrown upon sealing-wax so immersed, it makes a crackling noise, very minute globules, resembling air, arise from it, and the parts which the focus has touched are pitted and made rough as if a fluid had burst from them. This is the incipient process of burning, which, when exposed to the action of air, that element completes. These facts show that the effects of the burning glass are produced by the combined action of light and air. Moreover, the summits of the tropical mountains, which, while smitten by the sun's intensest ray are wrapped in everlasting ice, indisputably prove that the sun's ray or light alone will not produce the effect of heat. In all the cases above mentioned, it is the insufficiency of air at the point of action which deprives the sun's ray of its usual power.

Solar light and air, in producing the effects ascribed to heat, appear to act after this manner : the light, proceeding in straight lines and being the more subtle fluid, penetrates substances, and makes way for the admission of air into them; and when the light is sufficiently concentrated, the particles composing bodies are, by the power of these agents, separated, and set at liberty, and the appearances called combustion and melting are produced; an operation, the nature of which is more easily apprehended when we take into consideration the extreme subtility of atmospheric air, and remember that it presses with a force of fourteen pounds to the square inch on objects at the earth's surface. Indeed such are the solvent powers of light and air, when their action is combined and sufficiently concentrated, that few things, if any, can resist them.

That light penetrates solid substances is a fact of which the burning-glass itself furnishes abundant proof. It is evident that all the convergent rays pass through the glass. But this still more plainly appears by throwing the concentrated light of a lens through a transparent phial of water, when the light will be seen as it passes through, and a beautiful sight it is.

The expansion of heated substances is also undoubtedly caused by air which has entered them, and which, when they cool and contract, they give out, or squeeze out; and it is the air so pressed out of cooling substances, that causes the repulsion and hissing noise perceived when water or any other liquid is thrown upon them. If a piece of red hot iron or any other red hot substance is plunged into water, the air which had entered it will be seen issuing from it in innumerable bubbles. Hence, atmospheric air is the most active agent in the melting of metals. Aided by light, and as light or flame, it enters into substances, and separates the particles composing them.

The agitation of boiling water is caused by air which penetrates the substance of the vessel, and not by any conversion of the water into vapour, or gas, or steam, of which the following facts in connexion with those foregoing afford abundant proof.* The boiling of water commences by the formation of small bubbles on the bottom and sides of the vessel exposed to the action of the fire, which, as that action increases, enlarge and rise to the surface, and others in their places are again formed, and rise with continually increasing celerity, until they produce the ebullition called boiling. And if a common drinking glass be put in a vessel of water sufficiently deep to cover it, and turned with the mouth downwards, care being taken to turn it under water, so as not to admit any air in the act of turning, and the vessel of water then be put on the fire to boil, the air will gradually take possession of the glass, and drive the water out. The operation will be plainly seen by watching the glass. If, when the glass is full of air, the vessel be taken off the fire, and allowed to stand, such is the subtileness of the air under the glass, that it will escape through the expanded substances enclosing it, and disappear. But if a portion of the hot water be poured out of the vessel, and the remainder cooled, by adding cold water to it, or if the whole of the water in the vessel when taken off the fire be cooled by adding cold water to it as quickly as may be, without breaking the glass, a portion of the air inside the glass will be retained; and hence it is probable that if by any means the top and upper parts

* See *Mech. Mag.* vol. xxvi, pp. 239, 330, 383,

of steam-boilers could be kept constantly cool, it would not only increase the working power, but by counteracting the expansion of the metal greatly diminish the risk of those dreadful explosions which so frequently happen.

Again, if a tin vessel containing water be partially immersed in a vessel of boiling water, the water in the tin vessel may be raised to boiling heat, but it will not boil, because the water in which the tin vessel is immersed interferes with and interrupts the action of the air. The air penetrates and impregnates the tin vessel and the water it contains, but the surrounding water prevents it passing through in such a current as to produce continued ebullition. It is evident, therefore, that the ebullition of boiling water is caused by air which penetrates the substance of the vessel.

And still further to illustrate this theory, to show the sort of action which is constantly going on about all fires, and how they produce the effects ascribed

to heat, and that those effects are produced by an aerial fluid of which the atmosphere is the source; I may mention an occurrence which lately came under my observation. On the 20th of January last, at noon, the ground being covered with snow, about one inch deep, there being a keen frost, a brilliant sun, which made no impression on the frost, a calm atmosphere, (or if disturbed at all it was by breathings from the north-east) and the thermometer standing at 14 degrees Fahrenheit, a brasier of burning coke was placed upon the snow in the open air, which made upon the ground a radiated figure of this form,—the dark parts being thawed. The brasier was a circular one, and occupied the centre of the figure. In order to ascertain if the radiation resulted from any local cause, the brasier was moved into different places, but it produced in all of them the same figure. The radiation clearly indicates the action of an aeriel fluid forciby emitted in all direc-

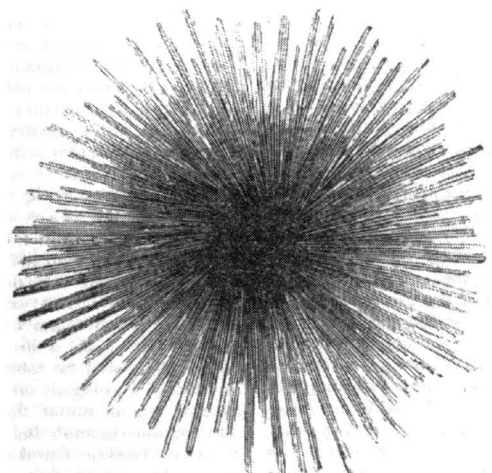

tions from the fire; and it undoubtedly is the action of this fluid which produces the effects ascribed to the heat of the fire. Air, which has performed the office of combustion, and the light and airy particles of matter which it has liberated, are immediately displaced by the pressure of the surrounding atmosphere, and this appears to be the source of the action indicated in the figure.

That the distribution of the atmosphere over the earth's surface is effected and regulated by the earth's rotation, is the necessary consequence of the natural arrangement of matter. The atmosphere being a fluid resting on the face of the earth, and the earth being a sphere turning with great velocity on its axis, the centrifugal force arising from the earth's rotation will inevitably cause an accumulation of atmosphere at the equator, which under the influence of the same force will gradually decline towards the poles.* And a certain consequence of the accumulation will be an increased density at the equator, which likewise will gradually diminish with the accumulation from thence towards the poles. The distribution of the atmosphere results from the order of nature. And that such an atmospherical accumulation and density at the earth's equator do exist, is proved by the fact, that within the tropics, the barometer does not descend more than half as much for every two hundred feet of elevation as it does beyond the tropics. The retardation of the pendulum and the decrease in the intensity of gravity in the equatorial regions, are also attributable to the increased density of the atmosphere in those regions, and not to the earth's centrifugal action or to any dimunition of density. Bodies fall or move with more facility and speed in vacuo than in air, because all resistance to the moving body is removed; through air, than through water, because the air offers less resistance than the water: and it follows as a necessary consequence, that they move with greater facility and speed in rarified than in dense air; and that the resistance to the motion of bodies increases with the density of the atmosphere in which they move. I refer, therefore, to the equatorial retardation of the pendulum and decrease in the intensity of gravity as additional proofs of an accumulation and increased density of atmosphere at the equator. But to my mind the dip of the magnetic needle affords still more striking evidence of such an arrangement of the atmosphere. From the equator to the poles, the dip gradually increases. Suppose then that the density of the atmosphere at every step in the same progress decreases, and we have the reason for the increased dip in the diminished density of the atmosphere, and the diminished resistance consequent thereon. The dip increases as the resistance to it diminishes. And further, the attractive power seated in

the earth which causes the needle to dip, combined with this distribution of the atmosphere will give to the needle a general polarity. It must necessarily point in the direction where it meets the least resistance.

These facts are submitted as evidence of an atmospherical accumulation and density at the equator. And proof of the extreme intensity of the solar light in the equatorial regions is unnecessary, that being a fact well known and universally admitted. We have then such an arrangement of atmosphere and solar light as proportions the degree of solar heat at any given place to the intensity of the light and the density of the atmosphere at that place, and as gives, just such a distribution of natural heat as, (with the exception of the anomalies presented by the Hymalayas) is found to exist. According to this theory it would be greatest and ascend highest within the tropics where the light is most intense and the atmospheric accumulation and density are greatest; and from thence its elevation would decline and its intensity decrease towards the poles. And this is the order we find in nature, there is the greatest heat, attaining the highest elevation, within the tropics; which (with the exception referred to) from thence gradually declines, and decreases in intensity, towards the poles.

The anomalies of the Hymalayas may, perhaps, be adduced as subversive of the theory here advanced; those anomalies, however, appear rather to strengthen than to weaken this theory. They are of that class of exceptions which prove the rule. This range of mountains, situate between 28° and 36° N. lat., extends in one continuous chain from 97° to 67° E. long., a distance, including sinuosities, of about 2,000 miles, presenting, throughout that immense range, a barrier towards the north of the mean height of 17,000 feet. And this is but the average elevation of the ridge, from which rise numerous lofty peaks, some of which attain upwards of 26,000 feet, and few of which rise less than 20,000 feet, above the level of the sea. On some parts of these mountains, the inferior line of perpetual congelation does not descend lower than 17,000 feet above the level of the sea; whereas the inferior line of perpetual congelation at the equator, is only 15,500 or 15,700 feet above

---

* The waters of the ocean under the influence of the earth's centrifugal force are known to be greatly elevated at the equator. That seems to be the power which obeyed the Almighty fiat, "let the waters under the heaven be gathered together." This appears to be the power which first established and still maintains the level of the sea.

that level. But the most remarkable, and to our purpose by far the most important, of these anomalies is, that the heat, and the inferior line of perpetual congelation, appear to attain a much higher elevation on the northern than on the southern side of the Hymalayas; at least this inference fairly may be drawn from the statement, that "the extreme height of cultivation, on the southern slope of these mountains, is 10,000 feet, the height of habitation, 9,500. While on the northern slope, villages are found at 13,000 feet; and cultivation at 13,600 feet." Now, the earth's rotation and centrifugal force, will necessarily cause, and at all times keep up, a strong bearing or pressure of atmosphere from the poles towards the equator, which coming in contact with these mountains, will rush up, or be forced up, by its own momentum, to a considerable elevation against the northern side. And thus it will compress and condense itself, and produce the anomaly of the inferior congelation line attaining a higher elevation on the northern than on the southern side.

With regard to heat from compression of air, it is well ascertained that the temperature or action of atmospheric air is proportioned to its density, increasing where the density increases, and diminishing where it diminishes; and that its density is increased by compression, whether the compression be by mechanical means, or by the superincumbent weight of the atmosphere itself. Hence the reason why the temperature or action of the atmosphere diminishes as we ascend from the earth's surface. Air, by sudden compression, is made to ignite combustible substances. The temperature at the bottom of deep mines is greater than at the top. The deeper the mine, the greater the heat. The heat gradually increases with the depth. This is caused by the superincumbent pressure of the column of air which descends the pit. Proceeding from the poles to the equator, the temperature of the atmosphere increases. Ascending from the earth's surface, it decreases. We know that the density of the atmosphere decreases as we ascend from the earth's surface, and we have seen that it increases as we approach the equator.

By compression air also gives light, or becomes light. The heat and light from collision are caused by the compression of air in the pores of the substances. All solid substances immersed in air must become thoroughly saturated with air. If they were so immersed in water, which is a much less subtile fluid, it would penetrate and saturate them. It would so saturate wood, iron, stone, or almost any other substance. It is evident, then, that the violent collision of substances must cause a sudden compression of air in the parts coming in contact, which is greater or less according to the flexibility of the substances used. Stone and iron in collision, being rigid substances, give out sparks. The ordinary kinds of wood, being softer and more pliable, do not. It is a difficult operation to give one proof of these views which is desirable, namely, the collision of substances in vacuo. It is not an easy matter to maintain a vacuum for such a length of time, as to allow an experiment like this to be properly made. In order to try the experiment satisfactorily, the substances to be struck should lay for some time previously in vacuo, in order to extract the air out of them. Another, and a better proof, is to strike the substances in water, which not only displaces the air from the outward surface, but in some degree follows it into its retreats, the pores of the substances, and drives it out. This will give convincing evidence that neither light nor heat can be obtained by collision, in the absence of air.

Heat from friction is caused by the compression of air in the pores of solid substances; the compression being effected by the action of the substances in air. The axles of machinery would not heat, if the air could be excluded from them Perhaps the nearest approach to a perfect exclusion is to make the axles run in liquids. By friction, in air, substances may be heated red hot; but they can neither be heated red hot, nor heated in any considerable degree, nor, I think, heated at all, in water, because the air is excluded from them.

The light arising from collision and friction, also affords evidence that atmospheric air, under some modifications, gives light, or becomes light; and atmospheric air, being also (as no doubt it is) the fluid which produces dissolution in all cases of combustion, and which, modified in the process of combustion, becomes flame and gives light, we have hence good reason for believing, that all artificial lights are but modifications of atmosperic air.   I am, Sir,

Yours, with much respect,

W.

NOTES AND NOTICES.

Death of Joseph Lancaster.—This celebrated individual, founder of the Lancasterian system of education, died at Williamsburg, near New York, on the 24th instant, in consequence of his being run over the previous day. He was in his 61st year.

Iron Steam Ship for America.—The British and American Steam Navigation Company have this week contracted with Mr. John Laird, of Liverpool, (the builder of the iron steam vessel *Rainbow*, belonging to the General Steam Navigation Company)) for an iron steam ship of 1,200 tons, to be called the *Atalanta*, and intended to run between this country and the United States, in conjunction with the *British Queen* and the *President*. From the experience Mr. Laird has had in this description of naval architecture, and the speed he has already attained in the vessels he has built, those well able to form an opinion on the subject, confidently predict that this vessel will reduce to ten days the average passage between Liverpool and New York.

Sulphur.—M. Maravigno, the Professor of Chemistry in the University of Catania, who possesses a very numerous collection of the crystallized sulphur of Sicily, refers the formation of this substance to the period of secondary rocks. He disputes the assertions of Professor Gemellaro, who pretends that sulphur owes its origin to the decomposition of mollusca, an assertion which has been reproduced in Germany by Professor Leonhard, of Heidelberg. He thinks, that, whilst the secondary formations were being deposited, the currents of acid hydrosulphuric gas, from the interior of the earth, came in contact with the blue marl, held in suspension in water, and that the acid, in decomposing, produced deposits of sulphur, which are still found mingled with the marl. He notices the deplorable system still used in Sicily for extracting sulphur, in which he says that 17 parts are lost out of 18. He then describes the different forms which the crystals present, the first of which has been discovered by him : it is that of a straight rectangular prism, the solid angles of which are truncated, and replaced by triangular facets.—*Athenæum.*

Savory's Clock.—Sir,—As no notice has been taken of a mistake in Nautilus's last paragraph (page 102), allow me to direct his attention to it. He says " It is evident there would be no difficulty in affixing a light index at the *back* of the transparent dial, &c.,"—thus having " *two* indices as in an ordinary clock, one for hours, the other for minutes." Now, " *it is evident*" that he is in error, forgetting, that as the arbor of the hour-hand is to be fixed to the transparent dial, therefore the works must be carried round *it* ; consequently, at the expiration of an hour the framework of the watch will have made 1-12th of a revolution round the arbor of the hour-hand. Now, in the same time, the minute-hand arbor has made *one revolution to the frame of the watch*, which frame having made 1-12th of a revolution to the fixed dial, it follows that the minute-hand will be five minutes in advance of the time. To correct this, it will be necessary that the motion between the two hands should be as 1 to 11, instead of 1 to 12th, which can be effected by making 88 teeth on the great wheel (instead of 96) working into a pinion of eight leaves.

I am, Sir, &c.   ω

St. Pancras, Dec. 5, 1838.

Steam Navigation to India.—The Indian Steam Ship Company have announced that their first vessel, the *India*, will be launched on the 3rd of January next, and be ready to take in stores in the Thames by the middle of April. The vessel is building by Messrs. Scott and Sons, and the engines by Messrs. Scott and Sinclair of Greenock ; she is of 1,200 tons burthen, with accommodation for 80 cabin passengers, and 400 tons of goods. She is provided with a safety apparatus (against boiler bursting, we presume), and built with two strong bulkheads of plate iron across the engine-room, in order to confine accidental fire, and prevent a leak sprung in one division from spreading to another. It is also announced, that another vessel of 1,500 tons burthen is on the stocks, and that a third will be ready within eighteen months, and that three more are about to be commenced. With this number of vessels, it is expected that twelve voyages out and twelve voyages home will be performed in each year, allowing fifty-five days to accomplish the distance from Plymouth to Calcutta, by the Cape of Good Hope.

Steam-vessel Inspectors.—At a late meeting of the Town Council of Edinburgh, a letter was read from Mr. John Cook, W. S., of Moray Place, containing some suggestions for carrying into immediate operation the principle of an Inspector of Steam-boats. The plan proposed by Mr. Cook is this—1st, That the Council should appoint a Committee for this business. 2nd, That the Committee should appoint an Inspector of Steam-boats. 3rd, That the Committee should intimate to all the Proprietors and Captains of Steam-boats sailing from Leith, Newhaven, and Granton, that their Inspector will inspect such vessels ; and that the proprietors of each vessel will have to pay the cost of the inspection. 4th, That the Committee, after receiving the report from their Inspector, should publish a list of the steam-vessels which they approve of and consider safe, as well as a list of them which they do not consider seaworthy and the names of those the owners of which decline to permit them to be inspected, that the public may know what they are about. And, 5th, That the inspector should be renewed as often as the Committee should think fit.—Some of the members of Council expressed doubts whether they had the power to adopt the recommendation of Mr. Cook, and others thought that it might be a very good subject for the Council to petition Parliament upon. In the mean time, however, the letter was ordered to lie on the table.

Metropolitan Railway Map.—On Saturday next, 15th December will be published vol. xxix. of the *Mechanics' Magazine*, price 8s. 6d., illustrated with a *Railway Map* of the Metropolis, taking in a radius of 15 miles from the Post-office. The limits of the two-penny and threepenny post deliveries are also shown in the Map. The *Metropolitan Railway Map* alone, stitched in a wrapper, price 6d., and on fine paper, coloured, 1s. The Supplement to vol. xxix, price 6d.

The *Railway Map* of England and Wales continues on sale, in a neat wrapper, price 6d. ; and on fine paper, coloured, price 1s.

☞ *British and Foreign Patents taken out with economy and despatch ; Specifications, Disclaimers, and Amendments, prepared or revised : Caveats entered ; and generally every Branch of Patent Business promptly transacted. A complete list of patents from the earliest period (15 Car. II. 1675,) to the present time may be examined. Fee 2s. 6d. ; Clients, gratis.*

LONDON : Printed and Published for the Proprietor, by W. A. Robertson, at the Mechanics' Magazine, Office, No. 6, Peterborough-court, between 135 and 136, Fleet-street.—Sold by A. & W. Galignani, Rue Vivienne, Paris.

Mechanics' Magazine,

MUSEUM, REGISTER, JOURNAL, AND GAZETTE.

No. 801.]　　　　SATURDAY, DECEMBER 15, 1838.　　　　[Price 3d.

HEARLE'S PATENT SHIP AND HOUSE PUMP.

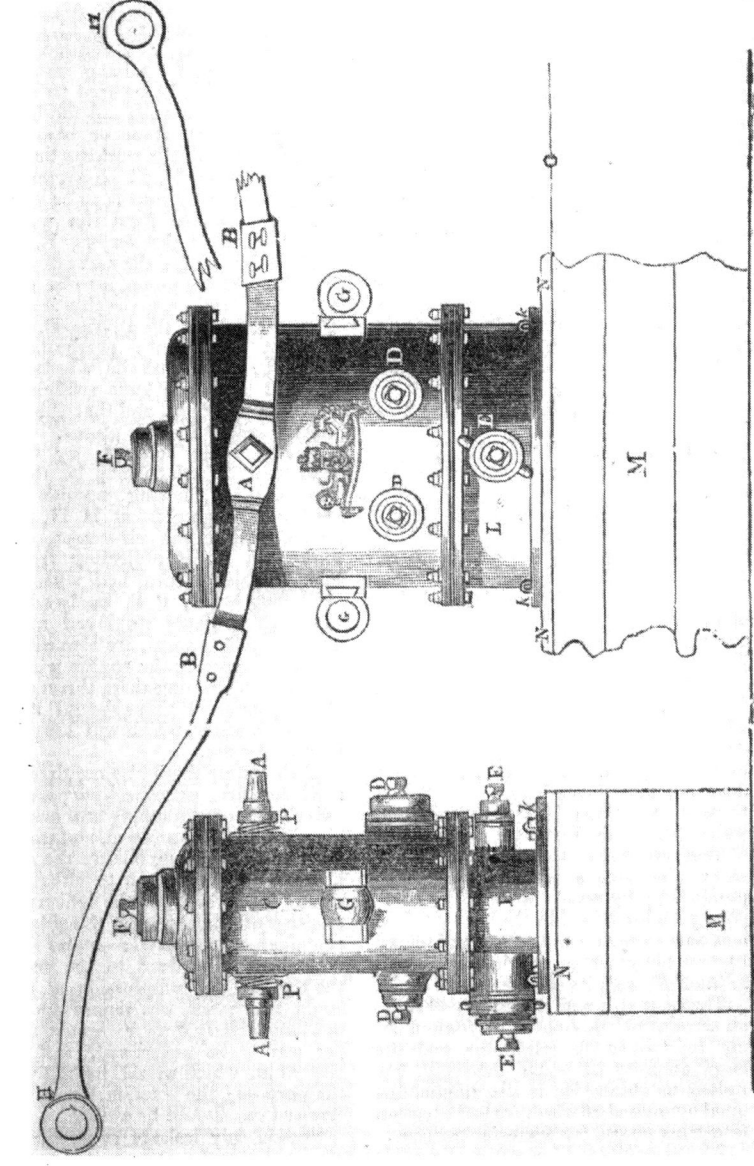

Sir,—Perceiving that one of the correspondents of your highly valuable miscellany (No. 767,) has drawn your attention, and also that of your numerous readers, to "Hearle's patent ship and house pump," I beg leave to forward you a sketch of the same, with a few explanatory notes, that its peculiarities may be the better understood.

I am, Sir, your obedient servant,

JOHN MARE.

Foundry, Plymouth, July 31, 1838.

Description of engravings in the front page.—A, the arbor, working through two stuffing boxes, carrying, on the middle of its length, a lever or beam, to the ends of which are attached the bucket rods in the usual way.

B B, The handles, fitting on the square end of the arbor, at A, meeting at B.

C, the air-vessel, containing the two cylinders.

D D, plugs, removeable at pleasure for attaching in their stead the suction hose.

F, plug, removeable at pleasure to introduce water into the engine, when, if by laying by unused, the bucket leathers, &c., may have become dry.

G G, openings, cast for the reception of the wooden handles, to be used when carrying the engine wherever it may be required.

This engine, it will be readily seen, does not, like the usual fire-engines, where the plunger is used, fill the cylinder, to force back the large body of water to be delivered. Therefore, the water does not pass through those multiplied tortuous windings of a plunger-pump, which materially increases the amount of friction, and consequently of labour also; this being a simple lift and force pump, and, by reason of its being completely enclosed within the air-vessel, it is at once exceedingly portable, and placed beyond the reach of accident or design to interfere with its efficiency.

The ease, also, with which it is worked, on account of the amount of friction being reduced to the minimum, and the large quantity of water delivered, has united to obtain for it the unqualified good opinion of all who have had occasion to use his pump, for whatever purpose.

Its portability is no inconsiderable feature.

An engine on the above construction, having two 6¾-inch cylinders, weighs but four cwt. when complete, and delivers about a hogshead of water per minute.

Directions for using the pump.—The pump, c, when required to be used for domestic purposes, as a common house or lifting pump, for raising water to a great height, is to be secured to its situation by four iron studs and forelocks, k, strongly fixed into stone or wood, in a similar manner to the working board, N, or cover of case, M, which accompanies it. A metal swivel-nut is to be soldered to the supply pipe from the well, and fixed to one of the screws E, in the bed L of the pump; the ascending pipe, if used as a lifting pump, is to be secured in the same manner, to the side screws D, or top screw F, of the air-vessel, as may be most convenient; particular care should be taken that the whole of the caps and screws be hove sufficiently to keep them air-tight, and that the leather washers are all in their places. Motion is given to the axis, A, by the two handles, B B, (which may be detached when needful), having wooden staves passing through them at H H, the two pumps within the air-vessel, C, are worked by levers projecting from the axis, A, which is furnished with glands or stuffing boxes, P P, easily adjusted. The wooden staves employed at H H, in working the pumps, are also made use of for transporting the engine with great readiness, by passing them through G G. The engine at all times should be kept filled with water, which will keep it in a fit state for use.

If the pump should be required to be used for fire, or other purposes, the swivel-nuts of the supply and ascending pipes are to be unscrewed, and the pump fixed on the working board; the suction hoses to be screwed on to either of the screws in the bed, and the delivery hoses to either of the screws in the side of the container or air-vessel;—on no account is the hose to be fixed to the screw on the top, as no compression of the air could take place, but should the engine not immediately draw its water, the upper cap to be unscrewed, and water poured in, which would instantly effect the purpose; the suction hoses, before screwed on, should be wetted.

Should any leakage arise, after long use, at the stuffing boxes, in the side of the container, and through which the axle works, by heaving up the screw a few turns the defect will be remedied. If the valves at any time require inspection, by removing the nuts and screws in the lower part, the bed and container will separate, and the whole of the work immediately be seen; the two metal cylinders may be unscrewed from the bed if required, two notches being cut in the upper part of them for that purpose; but the lower valves may be replaced without removing the cylinders; care should be taken that on screwing together either of the flanges, that the lead which is placed between them be well set in after the screws are hove closely up, in order to make the vessel air-tight. When the engine is used for pumping water from the shore into casks in boats, the leather hoses are to be screwed on to the engine, and the canvass hoses fixed to the leather ones; in deep water, if an oar, or any other buoyant thing, be lashed to the swivel-screws of the hoses, it will prevent them from sinking. If the engine be used for pumping water from any part of the ship, the cap on the top is to be taken off.

Hearle's patent engine pump has met with the most entire approbation of the naval authorities, and it has been adopted in 35 ships of war, with the names of which Mr. Mare, of Plymouth, the patentee's agent, has supplied us, as well as in numerous other vessels, whose names cannot be ascertained. The testimonials of the various officers, who have either been appointed to try the efficiency of the engine, or have had opportunities of witnessing its operation, are extremely satisfactory and decided. Capt. Lockyer, of H. M. S. *Stag*, even goes so far as to say, that "it ought to be supplied to all men-of-war, as it is available for so many purposes."

OBJECTIONS TO MR. PENNY'S CANNON AND MORTAR COMBINED.

Sir,—Your invaluable and talented correspondent, Mr. Baddeley, gives, in your last number, the description of " Mr. Penny's improved cannon," in the course of which he honours my name with favourable mention. In consequence of this quasi appeal to my opinion of the merits of Mr. Penny's suggestion, I feel in courtesy bound to offer a few words on the subject.

The idea of Mr. Penny is ingenious, and his method of connecting and securing the two halves of his cannon, is simple and efficacious. But there is a matter to be considered, which I should suppose that neither Mr. Penny or Mr. Baddeley have had any adequate opportunity of appreciating; that is, the wonderful effects of the explosion of *large* charges of gunpowder. The fitting parts of a musket, or a very small piece of cannon, might possibly be ground so exactly one into the other, as to prevent the entrance of the blast between them, but as we increase the calibre, the charge increases in the ratio of the cube, and so also the effects. There can be no doubt but that, after a few rounds (say with a 24-pounder and 8lbs. of powder), the blast will have, more or less, made its way between the surfaces of the two pieces,—maugre any practicable exactness of fit, or screwing-up whatever. At last, the wear and separation will become so considerable, that, favoured also by the perpendicular abutments offered just before the trunnions (*f*, fig. 1), the anterior half of the cannon will be blown away. I do not take into account the diminution of strength in the cylinder of the piece—and so far aft, too—caused by the interruption of continuity in the substance of its parcites. Fire-arms have long since been made to load at the breech, some of which separate into two pieces at about the same point of Mr. Penny's cannon. That in our days patented and improved by Collins, is good, but a hundred years old. In 1802, a M. Poli, of Paris, produced another, which had a deserved " run." But the best of all that I have seen, is Bevan's, which was patented in 1820 or 1821. This I have used, both with ball and bird shot, and cannot perceive any escape of power from the powder.

For such duplex pieces as Mr. Penny proposes, to be free from the evil I have indicated, they must be constructed in utter opposition to the axiom of the dialectitians who say that "*omne majus contenit minus.*" Now, taking the breech end of Mr. Penny's cannon to be the *major*, it, or its internal cylinder, should be contained within the anterior portion of cylinder, or the *minor*. But then we lose the varied faculty given to this major, upon disjunction, of containing a ball or shell, double the diameter of that which fits the minor cylinder. In fact, were we to reverse the order of the drawing, and suppose the breech and charge of powder and ball, as placed in the half marked D, then, with a little further modification of the connecting insertion, should we arrive at the only plan upon which the cylinders of *bouches à feu* can be formed of two pieces.

There may be some convenience in separating a *bouche à feu* into two pieces, to facilitate transport ; but the efficiency of such would be very little, if at all, diminished, if the anterior half were to be lost by the way. The use of solid shot, in any branch of warfare, either by land or sea, cannot much longer be upheld, even by all the prejudice and apathy of official routine, or by the sly pretended objections put forward by the tacticians, who secretly feel and know a great portion of their " occupation gone," or sadly simplified, when shells and hollow shot* are horizontally projected from ships, batteries, and ramparts, in lieu of the " good old," comparatively inert, solid shot !

For the last twenty years I have preached and written on this important substitution, until I am quite tired of it. At last, I see that some of the British government war-steamers are provided with guns to throw horizontally 68-lbs. hollow shot; they should be 100-lbs. shells.

Mr. Penny, or any one else, will find that for the attack of an enemy's arsenals or towns, large shell and carcass-rockets far preferable to bomb shells. The former lodge in the roofs and upper stories of buildings, and set them on fire ; the latter are apt to bury themselves in the cellars, where they can do but little harm. Moreover, a vessel of the smallest size can throw fifty of the largest rockets for one shell, and that without any *bouche à feu*, or recoil, or strain upon her whatever. We shall one day see rockets far more generally adopted, than the stupid spirit of prejudice has yet allowed them to be. In the field I should like to have 1,000 men, with each six six-pounder rockets (equal in range and power to a six-pounder cannon shot, besides the shell and shrapnell capacity) to meet 6,000 muskets ! The precision of aim is certainly not so great with the

rocket as with the gun ; but, for example, the 6,000 six-pounder rockets *can* be let fly (keeping within three feet of the ground) in vollies of 500 or 1,000 rounds per minute, all concentrated from, and towards, any given point, and merely by laying the rockets on the ground ! If tubes are used, 330 will produce this enormous projection ! Now merely to move a battery of 330 six-pounder guns will require 1,320 horses, and to put these guns in line of battle will occupy a space of almost three miles. Wherever the foot of man can reach, from thence may be discharged annihilating vollies of rockets, each equal in range, but far superior in penetration and destructive effect, to cannon shot of equal weights. The powers of the rocket are as yet unappreciated. It is not a projectile that can be used sparingly. Its very nature requires abundance—and *then* it will sweep all before it. The advantages of large shell-rockets* in naval warfare will be enjoyed by those who first have the sense to use them. Rockets with conical cast-iron heads full of gunpowder, weighing from 100 to 1,000 lbs., will be discharged from the smallest vessel, without recoil or shock, or the use of any *bouche à feu,* save a sheet-iron tube or trough, weighing, at most, 100 lbs. Such projectiles act admirably over the water ; skimming along the surface ; rising a few feet only at every touch, and finally entering the enemy's ship low in the hull, at the same time that large 1,000 or 2,000-pounder rockets are projected in duck and drake fashion. I will take care that, on the first opportunity, vollies of my 6, 12, and 24-pounder prehensile compound rockets shall be sent into the rigging, so that fifty such will be simultaneously burning, with unextinguishable and undetachable fire, throughout the sails, ropes, yards, &c., establishing fifty conflagrations, nourished and maintained by fifty more in another minute of time. Each of these rockets is equal to a shell of equal weight in its penetrating and bursting quality, so that such as strike a mast or yard will do good service ; but the novelty is in the conflagrating power, which no water-engine can extinguish, even could it reach it, and the conflagration of the best wild-fire (" carcass composition") yet invented, endures twenty mi-

---

* Hollow shot is the name given to the smaller natures of shells, fired horizontally from long guns or carronades, 24, 32, 48, or 68-pounders. The 24-pounder, five and a half inches diameter, holds about twelve ounces of powder.

* See *Mech Mag.* vol. xxviii, p. 265.

nutes, projected to the four points of the compass.

The professional officials have the simplicity to tell you, that abundant means of burning ships, &c., are already known to them,—shells, carcasses, red-hot shot, Congreve rockets, &c. &c.; but they affect to forget that any fire excited in the hull of a ship can easily be extinguished if there be any crew on board, and they be not all asleep. Their water-buckets, swabs, and engines, may not extinguish the *composition* of the carcasses, shells, Congreves, &c., but, they will put out the fire communicated to the timber. But let me send fifty per minute of *my prehensile* rockets into the *rigging* of a man-of-war, a surface of 15,000 square feet, which I will do from a distance of 1,500 yards, and we shall see how the gentlemen of the navy will stop the flames, before sails, ropes, yards, and masts are all burnt out. Oh! but we will steep our sails in alum, say they. Good; but then you must not let it rain upon them to wash it out. And what will you do to your ropes, saturated with tar, on each of which a burning "devil" will be irremoveably fixed? We will have our sails of woollen, and our ropes of iron wire, chain, &c. *I* have pointed out a far cheaper remedy than that, upon my offering these same rockets, the screw-popeller (now patented by Captain Ericsson), my spiral percussion shells, shot-proof steam-ship, and a dozen other novelties, to the Lord High Admiral and Admiral Owen, in 1827. But his Royal Highness declared that he "would not, for the world, favour the introduction of such means; why, they would, for a time, put the weak upon a level with the strong! *We* are the strong at sea, and need no such expedients." "But," said I, "your Royal Highness may one day think it expedient to prevent "*the weak*" from being possessed of them." Alas! "a wink is as good as a nod to a blind horse." So I was politely bowed out of the august presence, with my propeller, paddle-wheel, shells, and rockets under my arm!

Many plans for burning or destroying ships have been suggested, by imaginative fire-side inventors. Many unauthenticated, unintelligible stories are told of the "Greek fire," of that of Duprés, under Louis XV. Much was said of Fulton's torpedos, and the catamarans used with such ill success by the English at Boulogne, in 1803. Sub-marine boats were to attach copper globes, full of gunpowder, to the bottom of an enemy's ship. Colonels Paixhans and Ravicchio were to send a powder-boat against the enemy, propelled by large rockets acting on the water; but all these contrivances, however good if they could be applied to the object to be acted upon, can only be compared, in reality, to the well-known plan of catching birds by putting salt upon their tails. But I have extended this letter to an inconvenient length, so will conclude by repeating the remark of one of our present Lords of the Admiralty, who, after examining the construction, and seeing the conflagrating action of one of my prehensile rockets, exclaimed,—"*There is an end to sailing vessels, if attacked by these. No ship of the present rig can stand against any enemy provided with these rockets!*" And yet, no further notice is taken of the matter!

I have the honour to be, Sir,
Your obedient servant,
F. MACERONI.

Dec. 1, 1838.

TRIGONOMETRICAL QUESTION AND SOLUTION—CONJUNCTION OF THE PLANETS.

Sir,—In vol. xv, page 333, I gave a demonstration of a mathematical question, from which I deduced four theorems, which form the basis of analytical trigonometry; by some alterations in the diagram I find that these important theorems may be obtained in a much simpler way.

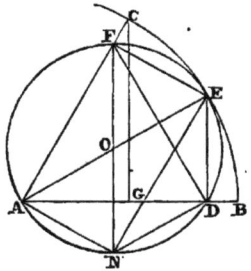

Let B E and E C be any two arches of a circle, whose centre is A. Join A C, A E, A B, and draw E D perpendicular to A B, A C, then the four points A, D, E, F, are in the circumference of a circle of which A E is a diameter; draw the diameter F D, intersecting A E in O; join F D; A N, N D, N E, and draw C G perpendicular to A B. By comparing the triangles A C G, T N B, we find C G = F D, and A G = N D; also A N E F is evidently a rectangle ∴ A F = N E, and A N = F E, calling the arc B E = A, E C = B. ∴ B C = A + B. Let A E, the radius of the circle B E C = unity. Then, E D, E F, and C G, are the sines of the arcs B E, E C, B C; and A D, A F, A G are their cosines; also because, A D E F is a quadrilateral inscribed in a circle A E, F D = E D. A F + T E, A D. But T D = C G and A E = 1; hence sin. (A + B) = sin. A cos. B + cos. A, sin. B. In the same way we obtain from the quadrilateral A N D E, cos. (A + B) = cos. A. cos. B — sin. A sin. B.

Or, by assuming the arc B C = A, C E = B ∴ B E = A — B. Then in the quadrilateral F N O E, by a similar process we obtain sin. (A — B) = sin. A, cos. B, — cos. A sin. B.

Lastly, in the quadrilateral A N D F, cos. (A — B) = sin. A. sin. B + cos. A. cos. B. The writers on anlyetical trigonometry, assume the sine of an arc from 0 to 180° affirmative, but negative from 180° to 360°, also the cosine affirmative from 0 to 90°, negative from 90° to 270°, and positive from 270° to 360°. From this seeming arbitrary notation, if E C is greater than 90°, but B C less than 180°, then sin (A + B) = cos. A, sin. B, — sin. B, cos. A. Now this change in the value of sin. (A + B) is at once attained by making a corresponding change in the diagram; and then it will be found that the quadrilateral first made use of, instead of having T D and E A, for its diagonals, and A D, D E, E T, T A for the sides; the diagonals become A D, T E, and the four sides A F, T D, D E, A E. Then, from geometrical principles, we have, sin. (A + B) = cos. A. sin. B, — sin. A, cos. B. In the same way all the changes that can occur in the values of sin. (A + B), sin. (A — B), cos. (A + B) shows cos. (A — B) may be made out from geometrical principles without any aid from algebra. I may here add, that in the higher analysis, a knowledge of analytical trigonometry is indispensable, and particularly so in physical astronomy.

In speaking of astronomy, I am inclined to think, Kinclaven has made a slip in his last article. He finds that T y = t x, or T : t : : x : y. Now, suppose T is a mean tropical period of the earth, and t that of a planet, x° the precession of the equinoxial points of the earth for the time T, then y° must be understood to be the precession of the equinoxial points of the earth for the time t; by some oversight Kinclaven has supposed that y is the precession of the equinoxial points of the planet whose tropical period is t. Now, on the supposition of uniform circular motion, also that the precession of the equinoxial points of the earth are in proportion to the times, then there can be no doubt, but that $\frac{P\,p}{P-p} = \frac{T\,t}{T-t}$. It may, however, be remarked, that neither of these expressions gives the true synodic for any given revolution. Thus calculating the synodic period of the moon, P and p being the periodical and tropical periods of the earth, we find it to be $29^d . 12^h . 40^m . 3^s$. This period not true for any single synodic period of the moon, it may be two or three hours above or below some true synodic periods of the moon. It is, however, the average period of a great many true synodic periods, the same observations apply to the planets; nor is it true that the precession of the equinoxial points are in proportion to the times; thus, when we say 50″.1 is the precession of the equinoxial points of the earth, this is not true, for any particular tropical period, but the average precession of a great many. With regard to Mr. Utting's period of conjunction of all the planets and satellites, it is all nonsense. Lalande has computed that it requires no less than seventeen millions of millions of years to separate the epochs of contemporaneous conjunctions of the six great planets!

I am, Mr. Editor, yours, &c.
GEORGE SCOTT.

11, York-Buildings, Nov. 15, 1838.

SIR JAMES ANDERSON'S STEAM-BOILER AND CARRIAGE.

Sir, — Being desirous of knowing something with respect to the *portability* and *useful effect* of Sir J. Anderson's forthcoming boiler, I made application to my friend, your correspondent, " Scrutator," who is a user and manufacturer of boilers and steam-engines of forty years' practice, enclosing him the drawing and description of the boiler in your Magazine, No. 775, of June 16th, 1838. The following are his observations, which, if you think worth a place in your valuable publication, are at your service.

I am, Sir, very respectfully, yours,
T. T.

Sir,—I have carefully examined the engraving and description you sent me, of Sir J. Anderson's boiler.

It appears to contain about 750 feet superficial of boiler plate; which ought not to be less than three-sixteenths of an inch thick, even if it is the best charcoal plate.

	tons	ct.	qr.
This, at 7.5 lb. per square foot, will weigh	2	7	0
It is apparent that the iron frames, to which the plates are riveted in 3-inch rows, together with the rivets, angle-irons, bolts, nuts, &c. &c., cannot weigh less than the plates; and comes to	2	7	0
And assuming the steam-chamber, the fire-bars, blowing-apparatus, &c., to weigh only	0	6	0
We have weight of boiler, without water, equal to..	5	0	0
The water required to fill the spaces intended, amounts to 360 feet superficial: which, at 2 inches thick, amounts to 60 cubic feet; and which, at the medium gravity of 62.5 lb. per cubic foot, comes to 3750 lbs., or, in round numbers, to	1	13	0
Now, in addition to this weight of	6	13	0
There is, besides, the weight of the supply tank, and its weight of water; the pumps, eccentrics, rods, steam-engines, fuel, carriage-frame, and wheels: all of which may fairly be rated at	2	10	0
Making altogether a weight of	9	3	0

As to such an enormous weight ever being allowed to work on four wheels, or even six, on common turnpike-roads, if it could do so, appears to be very doubtful. The boiler appears to contain more flue space than necessary, as the farthest part is 60 feet from the fire; therefore, that portion of the water is too far off, to be kept in a proper state of ebullition, except at the expense of the heat abstracted from the water nearer to the fire. The boiler appears to have about 7 feet of fire-grate, with 18 feet of direct impingement; and nearly 400 feet of up-and-down flue and side surface; one-half of which appears to be of very questionable utility, as before observed.

In boilers set in, and nearly surrounded by, a mass of brick-work, which, in some measure, conserves the heat, we usually allow 9 feet of fire and flue surface per horse power.

Now, this boiler being, on the outside, exposed to the condensing effects of the atmosphere, and considering that nearly all the caloric would be expended in the first 30-feet run of the flame and gas from the body of fire, this boiler cannot reasonably be rated at more than twelve horses power, on account of the very great disproportion between the *fire-grate* and *flue surface;* which latter, rising parallel with the flame and gas, instead of being vertical to it (whereby its impingement would be greater), is the reason why no more power could be produced.

All the power from this boiler would, therefore, be required (as a locomotive on common roads) to move itself and carriage; leaving nothing for the propulsion of waggons, or any other vehicles.

I am, Sir, your most obedient,
SCRUTATOR.

Dec. 8, 1838.

ON ROTARY STEAM-ENGINES.

Sir,—The plan for constructing a rotary engine, which you did me the favour of inserting in No. 728, Vol. 27, was one of two I have in my possession; my reason for not sending both was, because the one appeared to me better suited for the purpose; the other plan was of working the slides in a fixed cylinder, having a centre wheel revolving with them, and these slides alternately acted upon by means of the guides.

Some time since I had a model of a rotary engine in the Adelaide Gallery, with the drawing; the slides in this were placed within a centre wheel, similar to one patented by Messrs. Bramah and Richardson; excepting this, that the former had a fixed guide, or tappet, acting upon the projecting arms of the revolving slides, instead of pressing continually upon the inner circumference of the cylinder chamber; this, and the foregoing plan, I abandoned, for the following reasons:—

1. Because of the great friction produced by the circumference of the inner wheel.

2. Because that the slides, not having any support within the chamber of the cylinder, were liable to strain.

3. Because the stress of power rested upon the opening through which the slides passed.

4. Because the slides required to be packed, which increased the friction.

5. Because (in the last of the above plans) the pin, or roller, struck the guide with such abruptness, that it might prove destructive to the machinery when in rapid motion, besides producing an unpleasant concussion and noise.

I perceive, from your monthly Part 137, No. 784, that a patent has been obtained by a Mr. Rowley, of Manchester, for a rotary engine, so similar to the foregoing, that I cannot refrain from noticing it, especially as the guides, or "racers," and the method of using the "pins" or rollers to run within them, are so particularly named. I would beg your correspondent, Mr. Evans, to refer to No. 728, p. 244, and he will there find that these parts are specified as original.

Now, as I have a slight variation in my own plan, not sent you, which, if put into practice, might be considered an improvement, and which, without the use of the guides or "racers" I cannot apply, I would ask—am I to be debarred from using them, because of a subsequent patent?* The plan and arrangement were given to the public; and I consider I have still the privilege of the invention, equal to others.

* The distinctive features of most rotary engines are so faintly marked, that it is hazardous to answer this question without an attentive examination of the two plans. Whether Mr. W. can use the "rollers" and "racers" referred to, altogether depends on the method in which he has applied them to effect his object.—ED. M. M.

The contending opinions, as to the effectiveness or non-effectiveness of the rotary engines, will, I trust, very soon be at an end, and practical results take precedence of theoretical suggestions, placing their merits beyond the influence of conjecture or surmise. The difficulty of rendering the different joints steamtight has been stated to be a principal objection, as it might very well have been 20 years since; but, I would ask, where are all the improvements in mechanical science, the correct adjustment of metallic packing, or the boasted advance and perfection of machine making, if such is to continue the exclusive plea for relinquishing the use of engines, the advantage of which is, on many accounts, admitted?

There was a time, in the infancy of science, when the power of steam was applied without the aid of the reciprocating engine; and no one can now assert, that the reciprocating engines will alone keep the field;—an engine having few parts, and the arrangement of them such as can be easily adjusted to keep it steam-tight, being the only requirements to bring into competition a rotary engine to maintain the rivalship. For, where the saving of room or the increasing of velocity is an object, there is little chance but that the former will give place to the latter; as the power of steam may be applied at so small a cost, and used in so many ways, where it could not be conveniently arranged upon the reciprocity principle.

I have felt more assured, that the opinions that the wear and tear to which the rotary engines are said would be subject, and that the evils arising from this, would prove more than equal to any benefit that could possibly arise under every consideration—are erroneous, from this circumstance being continually brought to my recollection by the asserted experience of those who have engines in use that have metallic packing,—viz., that the more they are worn, the truer and better they work, that this should be a plea, therefore, in favour of one and against another movement, is a paradox which can only have originated in assertions made without consideration, and cannot hold good for the one more than the other.

In the plan of the model in the Gallery, the rotary principle produces the reciprocating, and, in the old system, the reciprocating produces the rotary; now, where the one movement is more required than the other, that engine will of course be the best suited which gives that motion without a needless increase of machinery, friction, and expense. Should any person be desirous of carrying out the principle, with or without the improvement I have before suggested, it will afford me much pleasure in satisfying their inquiries; at the same time I must apologise for so lengthened a paper upon the subject. Trusting I have not trespassed needlessly.

I am, Sir,
Yours respectfully,
JAMES WOODHOUSE.

Pilton, Barnstaple, Dec. 1, 1838.

SETTING OUT CURVES ON RAILWAYS.

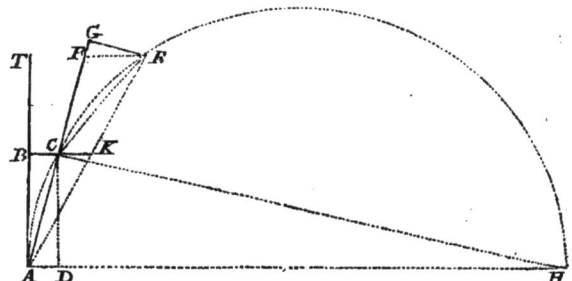

Sir,—Permit me, through the medium of your Journal, to call the attention of engineers on railways to the very erroneous method many of them adopt in setting out the curves, which, if intended to be segments of circular arcs—ought certainly to be so.

Of course it must be generally known, from the great radii of even the least of the curves on railways—it would be impracticable to strike them as in the usual manner from a fixed centre; but from the known relation between the sine and versed sine of an arc, the same object may be simply and expeditiously effected; —for instance, let A T be a tangent to the required curve at the point A. Upon A T measure off any convenient length, A B, and from B draw B C perpendicular to A B = to the versed sine of an arc, whose sine is equal to A B, the point C is the locus of the curve at that point; for it is obvious, that A B = C D the sine of the arc A C; and B C = A D its versed sine; and this operation may be pursued to any convenient distance by continually increasing the sine and versed sine, thereby obtaining so many more points in the curve required. The formula for determining the versed sine A D is of easy application, viz. A D = $\text{Rad} - \sqrt{\text{Rad}^2 - \sin^2}$. And, moreover, since the curvatures of circles vary inversely as their radii, when we have the versed sine of one arc, we may easily obtain the versed sine of arcs of any radii without even the repetition of the above formula. Now, however simple this process may appear, many, in endeavouring to arrive at a higher degree of simplicity, adopt the following inaccurate method, by which the curves are any thing but what they are intended to be, and this I will endeavour to show :—

Let A C E be the proposed circular arc of a given radius, A T its tangent at the point A. Upon A T is taken any part A B, and from B is drawn B C B= $\left(\dfrac{\text{A B}^2}{\text{diam.}} \right)$ perpendicular to A B, then the chord A C is produced to G, making C G = to A B, and from G is drawn the perpendicular G E = 2 B C, and the points A, C, E, are assumed as being in the circumference of the same circle, and other points in the circumference are found by producing the chord C E in the same manner as was before done with the chord A C, and drawing new perpendiculars each = to 2 B C.

Now it may easily be proved, that not only the data by which the value of B C is assumed are incorrect, but that the curve thus determined is very far from being circular, on the contrary, one in which the radius of curvature is continually increasing.

Produce B C to K, and draw F E parallel to B K, join C H, A E and C E.

1st, It is evident that $\frac{A B^2}{\text{diam.} (\overline{A H})}$ is not equal to B C or A D, for (Cor. 8th Euc. 6th).

$$A H : A C :: A C : A D$$
$$\therefore A D \times A H = A C^2$$
$$A D \text{ or } B C = \frac{A C^2}{A H} .$$ And A C

is greater than A B, being the hypothenuse of the right angled $\triangle$ A B C. However, here the error is very trifling, if the sine be small in comparison with the rad.; but strictly speaking, it is incorrect, and in circles of small radii would be serious.

2ndly, Where G E is assumed to be = to 2 B C is decidedly wrong. For (Euc. 32nd 3rd b.) < B A C = < A E C; but in $\triangle$ G C E the < G C E = two interior and opposite angles C A E and C E A (Euc. 32nd, 1st b.) $\therefore \angle$ G C E = $\angle$ B A E. Now the two $\triangle$' B A K, G C E are similar and equal; wherefore, (Euc. 26th, 1st b.) G E = B K. Again, since in $\triangle$ A F E, C K is parallel to F E A F : A C :: F E: C K (Euc. 2nd, b. 6th); but A F is less than 2 A C $\therefore$ F E is less than 2 C K —much more than is G E or its = B K less than 2 C K (since F E is the hyh. of the right angled $\triangle$ G F E). And as B K is divided into two parts B C, C K, if the whole B K be much less than twice one portion C K, it evidently must be much greater than twice the other B C $\therefore$ B K or G E is much more than 2 B C; therefore, admitting the locus of the point C to be on the circumference of a circle of the given radius, the point E is not, if G E is made = to 2 B C—Q. E. D.

Trusting that this communication may not be without interest to some of your readers, I beg to subscribe myself, Sir, yours faithfully,

Δ.

Oct. 5, 1838.

———

REMARKS ON THE LATE ACCIDENT IN ST. MICHAEL'S CHURCH, CORNHILL. — FURTHER EXPERIMENTS WITH HARPER AND JOYCE'S CHARCOAL STOVES.

" This is the patent age of new inventions
For killing bodies, and for saving souls,
All propagated with the best intentions."
BYRON.

Sir,—In our uncertain climate, artificial warmth is an indispensable comfort of life, and various methods are continually resorted to for obtaining it. The extravagant cost of fuel in most parts of our island, renders due economy of heat, and its several sources, a subject of paramount importance. In seeking with too much eagerness after this object, it too often, unfortunately, happens that another circumstance, equally essential to our health and comfort, is altogether lost sight of,—viz., pure and wholesome air.

Our forefathers, of the last two centuries at least, were feelingly alive to the importance of perfect ventilation, both in their streets and houses; but there exists a very strong disposition at the present time to disregard their wise admonitions on this head, confirmed as they are by all subsequent experience, and human life must be sacrificed "to save two-pennyworth of fuel *per diem.*"

The opinion of Dr. Ure upon the modern innovations in "domestic warming stoves," inserted at page 127 of your 798th Number, is deserving of the utmost attention; the pernicious, and slowly fatal effects of "mephitic stoves" (of all kinds) is ably pointed out, and a sufficient caution given against the evil consequences of over-heated and unventilated apartments.

The late lamentable catastrophe in St. Michael's church seems likely to direct a proper degree of attention to the subject of ventilation, and to the avoiding of stoves and fuel, liable to be productive of so much mischief. Under all the circumstances of the case, I apprehend the verdict of the coroner's inquest on the body of James Trickey—("poor fellow! his was an untoward fate,")—must be considered satisfactory and just; though, certainly, not quite in accordance with the weight of evidence brought forward by Messrs. Harper and Co. The jury happened to be sufficiently well-informed to see through the misrepresentations made to them, and mainly supported by what you have very properly designated as "one-sided experiments." There can be little doubt in the mind of any reflecting person, that if poor Trickey had remained in an erect position, either sitting or standing, he would have suffered little or no inconvenience, from the noxious effluvia of the burning charcoal. But there is every reason to believe, in fact, the medical evidence proves as much, that he was seized with an attack of apoplexy—to which, from his full habit of body, he might have been considered exceedingly liable—and fell prostrate upon the pavement of the church, when he instantly became immersed in a stratum of carbonic acid gas, which quickly completed the work of death.

In consideration of the evidence given by Dr. Golding Bird and Mr. Blenkharne, I have been informed that one of the gentlemen who was present at the inquest, has since performed several experiments with Harper and Joyce's patent prepared charcoal in one of their stoves, and he has found, that although *he* was in no way inconvenienced by the effluvia proceeding from it while burning, yet, on placing some small birds on the floor of the apartment they were very soon suffocated. A favourite dog had also very nearly fallen a victim to the experiment, and it was a considerable time after being taken out of the room that he recovered his usual faculties; fully proving the existence of carbonic acid gas, in considerable quantity, upon the floor of the apartment.

It is most disgraceful to the age in which we live, and derogatory to the character of real science, that men should be found, eminent for the supposed extent of their learning, and yet base enough to sacrifice "the truth of science" at the shrine of friendship or of interest. It appears to me, that persons of this class too often imagine that their office, like that of members of the legal profession, is to take up and stoutly maintain whichever side of the question they are hired to defend; to supply *opinions* "to order," with *experiments* "to match," and so to mistify and quibble, or to conceal or misrepresent facts, as to make, if needs be, "the worse appear the better cause."

In criminal matters, the greater the rogue and the more palpable his villainy, so much the greater credit is given to the pleader who shall get him off "scot free." So in the world of science it would seem to be imagined, that the clearer the fact and the more incontrovertible the evidence—so much the greater praise is due to him who shall most effectually bamboozle his hearers and persuade them that black is white in spite of the evidence of their senses. I have understood that a very strong feeling of disgust prevailed with respect to the evidence of some "men of high standing in science" upon the melancholy inquiry before alluded to, and it has been stated, that had the jury been sitting elsewhere, and not composed of friends and neighbours of Mr. Harper, certain learned professors would have been put to open shame; but it was felt, that in dealing with them according to their deserts, Mr. Harper's cause would be seriously injured, and therefore they were permitted to detail experiments and make statements, at which the veriest tyro in chemistry knew to be fallacious and untrue.

The idea of a heavy fluid like carbonic acid gas, which requires to be heated about 250 degrees of Fahrenheit above the atmospheric air to acquire *the same specific gravity*, being found at the top of a capacious building, the air of which, originally at 50°, was only raised 10° in twelve hours by the combustion of 49 pounds of charcoal, is absurd beyond credence.

The tale about the *heat descending* into the vaults beneath the church and extracting deleterious vapours from the coffins, must have been got up by some wag as an excellent satire upon the statements of the learned professors. With quite as much truth, and with more apparent reason, might it have been asserted that the carbonic acid gas had descended into the vaults and suffocated the corpses in their coffins!

I remain, Sir, yours, respectfully,
WM. BADDELEY.

London, Dec. 6, 1838.

MR. PRITCHARD'S "MICROSCOPIC ILLUSTRATIONS."*

The microscope is of about the same antiquity as the telescope; but until within the last twelve years it remained an instrument of comparatively small

* New edition, emended and enlarged. 1838. 248 pp. 8vo., with numerous plates. Whittaker and Co.

utility. The achromatic principle which was at once adapted to the telescope with astonishing success, was productive of no such happy results on its first application to the microscope. Swammerdam, Lewenhoek, De Torre, Euler, and others, applied themselves with great industry to the discovery of some means of remedying the defects of spherical and chromatic aberration to which the simple microscope is subject, but missed the way to the object of which they were in search, by assuming it for granted that the smaller the angle of aperture the better. To such an extent was this notion pushed, that Lewenhoek had lenses with a focal length of not more than 1-700th or 1-800th part of an inch. To the acute, lively, and enthusiastic Dr. Goring it was reserved to discover, in 1827, that the road to perfection lay quite in the opposite extreme—that on the number of rays of light which can be collected by the object lens from every point on the surface of the objects examined, depends the efficiency of the microscope. And thus, to use the words of Dr. Brewster, it has " become quite a new instrument; and promises to be the means of disclosing the structure and laws of matter, and of making as im-

portant discoveries in the infinitely minute world, as the telescope has done in that which is infinitely distant."

Dr. Goring's ablest fellow-labourer has undoubtedly been Mr. Pritchard, whose numerous contributions to microscopic science, both instrumental and literary, we have, more than once, had occasion to notice with commendation in our pages. The volume now before us is an " emended (Query amended?) and enlarged edition" of a work under the same title, which appeared in 1829, and helped greatly to make the value of Dr. Goring's discovery known to, and duly appreciated by, the scientific public; but it is an edition *so much* amended and improved as to be, in fact, almost an entirely new work.

As many persons—not unlearned even —have been heard to express themselves at a loss to understand how it is that an increase in the length of aperture is so efficacious, we shall take this opportunity of extracting the very satisfactory statement of *the sufficient reason* here given by Mr. Pritchard:—

" Every one who has considered the subject at all, will understand that by a series of glasses of different media, aberrations, &c. may be corrected, and almost entirely

Fig. 1.

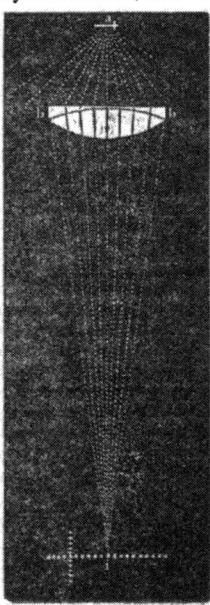

Fig. 2.

dismissed; but it is not so readily perceived how a compound microscope, having a series of glasses, can, with less illumination, give a brighter and more vivid picture, with more of the detail or minute structure of an object, than can be obtained with greater illumination, by a single lens. In this particular, however, consists the main advantage of an achromatic.

"Let me premise that, in order to render any object visible, it is necessary that rays of light should proceed from it, either by reflection from its surface, or by transmission through it, to the eye. Again, if the number of rays be insufficient, the object cannot be seen, notwithstanding we employ a microscope for the purpose. Bearing this in mind, I will endeavour to explain how an increase of angular aperture in an object-glass, independent of any increase of 'its magnifying power, will admit a greater quantity of light from any given point on the surface of an object to pass through the lens, so as to render the structure of the object visible.

"Let A and a represent two objects, in all respects alike, and let us employ two microscopes, of equal magnifying powers, for the purpose of viewing them. Suppose that we are going to look at some spot on the surface of A, or a, which we will imagine to be a delicate tissue. By a well-known law of light, the rays proceed in right lines, in all directions, from this spot, in the manner shown by the dotted lines in both figures. Suppose B B, and b b to be two object-glasses, of equal focal lengths; the former a single lens of the best construction, such as was used in the old compound microscope, and the latter a lens of the newest form, termed an achromatic. Now, these object-glasses will form their respective images at I, and i, and they will be of equal dimensions. But if the number of rays proceeding from A, and falling upon the single lens B B, is not enough, when collected at I, sufficiently to stimulate the eye, any minute pore, striæ, or other marking at A, will not be rendered visible; whilst, from the increase of aperture in the achromatic lens, b b, allowing much more light from a to fall upon it, and to be transmitted through it and collected at i, every marking, &c. at a, will be clearly represented at i; and the eye, being powerfully acted upon by this increase of light, will become highly sensible of it.

"The angles B A B, and b a b, are the angles of aperture of the respective object-glasses; and the quantity of light collected and transmitted by each will be as the squares of B B, and b b, the focal lengths being equal. Hence it is that the power of a microscope, or that faculty it possesses to

render the structure of an object visible, depends upon the angle of aperture of its object-glass, and not upon its magnifying power alone.

"But it may be supposed, perhaps, from this reasoning, that if we throw a greater quantity of light upon an object, more may be collected by the object-glass, we shall be the better able to define its structure; which would probably be the case if the additional light could be thrown only upon those minute parts which we wish to examine, and not upon the whole object. But as we cannot do this—as the increase of illumination cannot be made to increase the *relative proportions* of light which proceed from these minute parts, the intended advantage will not be derived."

Mr. Pritchard mentions a curious "case in point," to show the obstinacy of the ignorance which still prevails, among optical instrument-makers, with respect to the necessity of applying the achromatic principle to the microscope as well as telescope:—

"I have in my possession, at this moment, a triple object-glass, evidently made in imitation of one of Mr. Tulley's achromatics. In constructing it, however, the optician, a person of great respectability, was so unacquainted with what an achromatic is designed to effect, that he actually placed a stop behind the lenses; so that, notwithstanding the focal length does not exceed half an inch, an angle of aperture of only 7 degrees is obtained. This object-glass, of course, is inferior to a common lens: and hence has originated the erroneous notion that the introduction of achromatics has been no improvement to' microscopes."

Every one who has a taste for microscopic investigation, and is desirous of having the best instruments for the purpose, would do well to apply, in future, to Mr. Pritchard himself,[*] whose improved microscopes and appendages are even better than his books upon them. It is here worthy of particular record, that the *Microscope-gnomeometer,* invented by Mr. Pritchard to measure the angle of aperture of microscopic object glasses, has been found of such universal utility that it has been not only employed for reading off the divisions in the trigonometrical survey of Ireland, now in progress, but has been since used by astronomers for similar purposes, in the microscopes attached to transit circles; the advantage of which'

* 162, Fleet-street.

is, that observations can be read off much longer in the twilight of an evening without artificial illumination—thus causing less fatigue to the eye than is experienced by looking alternately "at objects illuminated by different coloured lights."

Much stronger proofs, however, of the beneficial effects produced on science in general by the new stride made in microscopic instrumentation through the discovery of Dr. Goring, and its successful applications by Mr. Pritchard, and other less distinguished followers, remain behind.

The following passages from Mr. Pritchard's "Introductory Remarks" to his present volume are extremely striking. They presented much that was new to us; and will be found, we doubt not, to contain much that is equally new to the majority of our readers:—

" The fact is, that since the modern improvements the microscope has undergone, it is being brought to the assistance, and is at the present time furthering the progress, of almost every branch of natural science. To the Geologist it may be said to be a new instrument. But what has it not even now effected for him ? In his study of organic life and structure, it unfolded to him the precise characters of divers animals and plants which inhabited and clothed our earth in ages which have long passed away. Look at the discoveries of Agassiz on the fossil creatures of the deep ! By a microscopic investigation of such portions of them as have withstood the destructive power of time, namely, their scaly covering, he has been able so to group and class them, that the characters and habits of the genera belonging to each distinct era are clearly demonstrated. A microscopic examination also of the testaceous remains of sundry Entomostraceans found in slate-clay formations, now elevated much above the level of the sea, prove them to have been at some time or other imbedded in the waters. And the Naturalist may even determine by his inspection of the shell, whether the species were the inhabitants of fresh or salt water, and consequently whether the strata themselves were the indurated beds of the sea, or of some river or lake.

" The *most perfect* animal remains which the microscope has disclosed to us, are the various loricated Infusoria of the division Bacillaria.* These minute creatures are

---

* See Mr. Pritchard's Natural History of Animalcules, p. 59.

so inconceivably numerous that they cover many miles of surface with several feet of thickness; as instanced in the polishing-slate and rotten-stone of Bohemia. In Tuscany whole mountains consist almost entirely of the silicified shells of these creatures; thus combining with each other in infinite numbers, to counterbalance, as it were, their individual minuteness, and to teach the unthinking this useful lesson, that Nature, in all her operations, is never employed in vain, and that, what are apparently her most insignificant productions, fall not beneath the notice of the profoundest inquirer after truth.

" To the botanist the aid of the microscope is indispensable. In the investation of our fossil-flora, what does it not exhibit to us ! How beautiful and delicate is the structure of the envelope of some of the fossil-fruits; those, for instance, of our London clay, when viewed under this instrument; and how important is it, that, by its assistance, we can determine with accuracy the natural orders, genera, and sometimes the very species of the trees and plants of former epochs ! How, beyond all question, is now demonstrated the vegetable origin of our coal ! Preserved within a bituminous lump of coal, which has been deposited for thousands of years deep in the bowels of the earth, you may discern not only the wooden fibre, its arrangement, and the disposal and form of the medullary rays, but even the most delicate of the vegetable organs, such as the spiral vessels, and the beautiful termination of those vessels ! These are as distinctly discoverable as in the finest preparations of a recent plant. And what can be more amusing and instructive than the examination of the silicified woods, when formed into sections no thicker than the paper of a bank-note ? Thus rendered pervious to light, the organic structure of the wood becomes plainly distinguishable. And emanating from this, what can be a more interesting subject than the inquiry into the mode in which the silicifying process has been carried on—by which the constituent elements of the inmost and minutest portions are changed—whilst their form and situation and colour remain the same ? In investigating also that extinct genus of plants, the Lepidodendra, a similar idea is raised in the mind, as to what must have been the particular state of the earth with respect to atmosphere and temperature at the period of their growth, and what the changes which have since taken place, in order to bring it to its present condition.

" In our physiological inquiries into the animal and vegetable productions of the present time, the assistance of the microscope is essentially requisite. When Dr. Harvey made his grand discovery of the circulation of the blood, and first lectured upon it, in St. Bartholomew's Hospital, in 1619, he was ridiculed, and lost his practice, through maintaining what was then supposed to be so absurd and wild a theory. The idea was suggested to his mind by reflecting on the valves of the heart and veins, which were evidently so planned as to allow a fluid to pass but one way. All the philosophical reasoning, however, of this celebrated man, could not establish, what appears to us so plain a truth, until it was evidenced in the circulation of cold-blooded animals, by means of the microscope, and thus placed beyond a doubt. Discerning, as we can do, the very forms of the globules of that fluid, as they flow through the capillaries from the arteries to the veins, in obedience to the laws impressed upon them by the Almighty Creator—viewing this most sublime phenomenon, by which life itself is diffused throughout, and sustained in every part of the system—who can resist conviction of the great truth?

" Nor is it a matter of less importance in a scientific point of view, or less interesting, that by the same means we can perceive the fibrous structure of the muscles and nerves, the form and arrangement of the canals by which the internal cavities of the bones are lubricated and nourished, the glandular structure of that beautiful and complex apparatus by which the secretions * are carried on—all, and each of these, requiring but the aid of one of our improved microscopes to render them distinctly visible. Again : how admirably developed, by means of the microscope, are the curious and complex structures of the eyes of insects, the crystalline lenses of those of fish, birds, &c.†, and many of the other parts of the visual organs.‡ The eye—that useful and delightful portion of us which furnishes all the endless variety of objects from which we derive so great enjoyment—resembles, in its peculiar formation and arrangement, an achromatic optical instrument. And if we descend to the lower classes of animals— nay, I would hardly say lower, lest some, perhaps, might imagine that in their small forms they do not evince as much perfection as is discoverable in beings of a higher scale, and have not all the functions which are necessary to life as full in operation as even man himself—if we enter upon an investigation of their minute structures, we can determine absolutely nothing without the microscope ; and our knowledge of the

* Nouvelles Recherches sur la Structure de la Peau, par M. Breschet.
† Philosophical Transactions, 1833.
‡ See Langenbeck on the Eye.

very existence of many highly-organised and active creatures is wholly dependent upon it.

" Vegetable organography, upon which the modern botanist depends so much for his systematic arrangement, and with which the student is so greatly interested and amused, owes almost its very existence to the microscope. This observation will be found to apply in an especial manner both to the cellular and vascular tissues of plants. The membraneous cellules of cellular tissue are sometimes not more than 1-1000th of an inch in diameter ; and those of the ordinary size are about 1-200th or 1-300th. How, then, is it possible that we could become acquainted with their forms and arrangement, but by the aid of the microscope ? And so with respect to vascular tissue : it is absolutely indispensable toward acquiring an accurate knowledge of the structure and forms of these membraneous tubes, and of the spiral or annular fibres which surround them.

" A knowledge of the fructification, if I may so express myself, of that numerous and curious class of plants, the Acrogens, could not be obtained without it ; nor could the existence of many of them, such as the Fungi, Lichens, Algæ, and some of the Musci, be proved. By its powers even the ashes of vegetables may be seen to contain the decisive characteristics of organic structure ; and the long-debated question of the antiquarian, as to whether the ' fine linen of Egypt,' in the times of the Pharaohs, were of linen or cotton fibre, seeing the latter is now indigenous to that country, is for ever set at rest.

" In many of the larger portions of plants, such as the cuticle of their leaves, the stomata, &c., which require but a shallow magnifying power to display them, there is as great a difference manifested when these are viewed under an achromatic microscope, or under the old compound, as is perceptible between the most highly-finished miniature, where the most delicate features and even the down on the skin are correctly depicted, and the mere black and white profile, where we see but the rude contour of the face. Surely, then, as works of art merely, instruments which can effect so much as this are justly entitled to a due share of consideration, even from the most refined and polished minds."

Again :—

" In the study of crystallography, which science may be said to have been for a long time almost at a stand-still, a very extensive field of research appears to be now opening, by the adaptation of polarized light to a microscopic examination of minute crystals—thus eliciting a great variety of curious and beautiful properties, entirely unknown to the world before. This subject being as yet completely in its infancy, it would be unfair to expect an elucidation of it at the present time. To convey, however, some general idea to the reader, of the additional degree of interest which attaches to the phenomena of crystallization by this happy contrivance, it is only necessary to state, that we have now displayed to us minute crystals, with a brilliancy and richness of colouring that is quite inconceivable. We see the smallest difference in their thickness marked by some exquisite change of colour; whilst the beautiful black cross in the circularly arranged crystals of zanthate of ammonia, and the cross with the coloured rings in the compound of phosphoric acid and borate of soda, &c. &c., excite our admiration beyond all bounds. These newly-discovered phenomena, after a patient investigation, may lead to results highly interesting, and of great importance to science."

NOTES AND NOTICES.

American Locomotives.—The Austrian railway from Vienna to Raab in Hungary, the scene of the famous battle in which Eugene Beauharnois defeated the Austrians, is now in progress of execution. The capital is supplied by the well-known Baron Sina, the great Vienna banker, whose name is familiar, it is said, to all the oriental world, and is thought to belong to one possessed of more wealth than all the magicians of the Arabian Nights. The Baron, who enjoys such a peculiar fame in the East, has lately directed his attention to the west, and ordered a number of locomotives for this railway from America. Some of them have just arrived at Vienna, and if accounts from that capital are to be credited, are " of quite a peculiar construction, different from that of the English locomotives, and much more simple and durable." The question of durability had perhaps better be left to be decided by experience, to which, however, none of the excellencies of the new machines appear as yet to have been subjected, as we are told, after much warm praise of their superior advantages, that a trial is intended to be made of them as soon as convenient. American steam carriages, built by Norris, of Philadelphia, have been introduced on the Brunswick and Wolfenbuttel Railway, and appear to have given great satisfaction.

American Steamers.—A line of " grand steamers" is about to be established from Bordeaux to New York, but on what plan is not yet precisely known, several projects being under consideration. The capitalists of Havre are beginning to regret that they did not, as was expected, make a beginning, but they may console themselves with the reflection, that in England, also, the port which was expected to take the lead has been thrown completely into the back ground. In the notice of the project in the French journals, the word " steamer" is made use

of as an expression requiring no explanation. It is worth while to notice what English words have been introduced into the French language of late years. " Steamer,"⁴ " budget," (that of the Chancellor of the Exchequer) and " meeting" (as in " public meeting"), are three that appear to have been completely naturalised. To each of them there " hangs a tale."

New Mines.—The Pacha of Egypt may find some consolation for the loss of his monopoly in general speculation forced on him by the treaty between England and Turkey, in the recent discovery of gold mines in Upper Egypt, the product of which is reputed to have turned out even more rich than was originally supposed. A vein of silver, extremely rich, is stated to have been discovered at Königsberg, the old capital of Prussia. The " Precurseur," a journal published at Antwerp, announces a mine of iron ore has been lately discovered at Capellen, in the province of Antwerp, of a quality superior to that imported from England. It is being worked with activity, and some boat loads have already been sent to Charleroi.

New Parchment.—M. Pelouze states, that if a piece of paper be plunged into nitric acid at 1.5 of density, and left in it a sufficient time for saturation, say two or three minutes, and immediately washed in plenty of water, a species of parchment is produced, which is impervious to damp, and is extremely combustible; and that the same change takes place in cotton and linen stuffs. They owe this property to the Xyloïdine, which M. Pelouze has found in starch, when treated with nitric acid and water.

Proposed Bill for Taxation of Steam-boat Passengers.—The heads of a bill to be brought before Parliament has been settled, by which it is proposed, that " all vessels engaged in the conveyance of passengers by water, when impelled by any other means than sails or manual labour," shall not be employed without a license, apportioned in its price to every 100 tons, nor without paying a duty " for and in respect of every passenger, at and after the rate of ⅛d. for every four miles, or fraction of four miles."—*Mining Journal.*

Jones's Machine for Moulding Bricks.—The earth in its descent, is forced into the moulds by great pressure as they pass under the Pug-Mill, and is delivered therefrom in perfect bricks upon pallet-boards ready to be removed, the whole of which is done by the horse attached in the usual mode to the Pug-Mill, producing from 1000 to 2000 bricks per hour. The earth also being moulded with only one-half the usual quantity of water will take considerably less time to dry, consequently there may be nearly double the quantity of bricks made upon the same space of ground. A machine was at work last week on three successive days, at Messrs. Webb's brick-field, near Ball's Pond Church, Islington, and performed the work admirably.

Metropolitan Railway Map.—Published this day, vol. xxix. of the *Mechanics' Magazine,* price 8s. 6d., illustrated with a *Railway Map* of the Metropolis, taking in a distance of 30 miles from the Post-office. The limits of the two-penny and three-penny post deliveries are also shown in the Map. The *Metropolitan Railway Map* alone, stitched in a wrapper, price 6d., and on fine paper, coloured, 1s. *The Supplement* to vol. xxix, price 6d. *The Railway Map* of England and Wales continues on sale, in a neat wrapper, price 6d.; and on fine paper, coloured, price 1s.

☞ *British and Foreign Patents taken out with economy and despatch; Specifications, Disclaimers, and Amendments, prepared or revised: Caveats entered; and generally every Branch of Patent Business promptly transacted. A complete list of patents from the earliest period (15 Car. II. 1675,) to the present time may be examined. Fee 2s. 6d.; Clients, gratis.*

LONDON: Printed and Published for the Proprietor, by W. A. Robertson, at the Mechanics' Magazine Office, No. 6, Peterborough-court, between 135 and 136, Fleet-street.—Sold by A. & W. Galignani, Rue Vivienne, Paris.

Mechanics' Magazine,

MUSEUM, REGISTER, JOURNAL, AND GAZETTE.

No. 802.]   SATURDAY, DECEMBER 22, 1838.   [Price **3d.**

RIVINGTON'S SELF-ACTING RAILWAY SWITCHES.

Fig. 1.

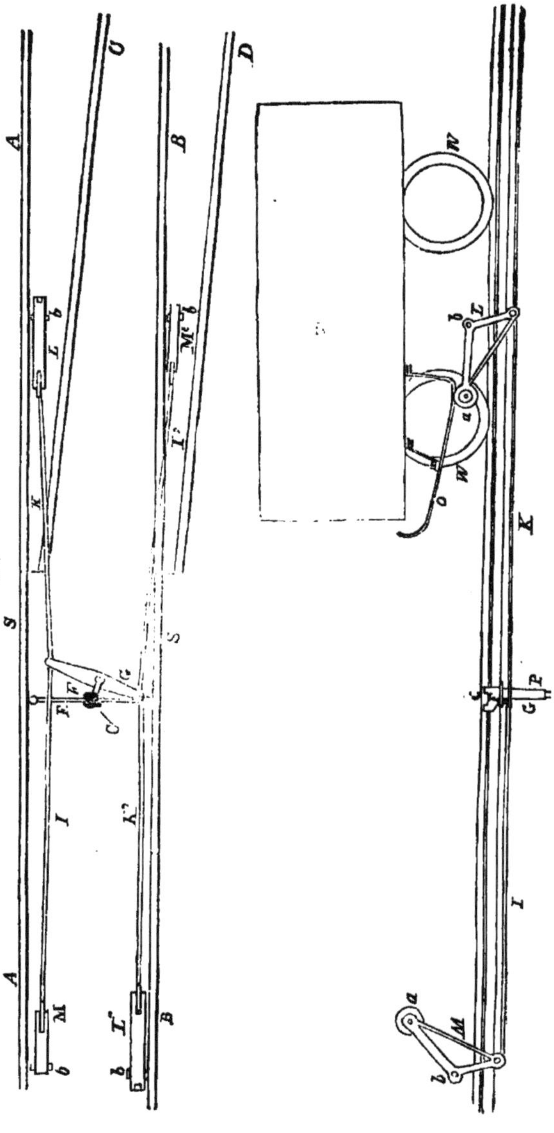

Fig. 2.

SELF-ACTING RAILWAY SWITCHES.

Sir,—A wish to render less frequent one class of accidents to railway trains, has induced me to trouble you with the following plan;—

Fig. 1 is a plan, and fig. 2 a side view. The switches S S are connected by a bar E jointed to them, carrying a frame in its centre, in which the pin* c of the crank F works up and down. This crank is formed on the end of the shaft P, which carries a cross lever G, the ends of which are acted upon by the rods I I' K K' jointed to the lower arm of the bell-cranks L L' M M'. These turn on pivots b, restings in bearing, which, to avoid complexity, are not shown in the engraving. T may be the engine-tender of an advancing train. The same letters of reference are used for both figures.

Now, supposing the figures to represent the portion of railway where the Brighton and Dover lines separate, A B, fig. 1, will be the Dover rails, and C D the Brighton. The tender T, fig. 2 (of a Dover train), has, by the action of the inclined plane O, upon the roller a, depressed the upper arm of the bell-crank L, drawing towards it one end of the lever G by the rod K, and throwing the switches into the position shown in fig. 1. The action of the cranks L and L' will be precisely similar, as also that of M with M'; and it should be observed, that the inclined plane must be fixed on the right side of the Dover trains, and on the left of the Brighton. In delineating the rail in fig. 2, I have supposed continuous wooden sleepers to be adopted.

This plan would, of course, not be applicable to chief stations, where carriages are turned about to all the points of the compass; nor are such deliberate movements attended with danger: it is for trains of passengers travelling at high velocities that such contrivances seem especially necessary, substituting the unerring action of machinery for the carelessness of human beings. Should the above idea be carried into practice, it might be advisable to lengthen the inclined plane considerably, to make its action on the bell-cranks as gradual as possible.

By inserting this in your Magazine, you would oblige, Sir, yours, obediently,
JOHN RIVINGTON.

Sydenham, Nov. 6, 1838.

REMARKS ON VARIOUS PLANS FOR MAKING TIME-PIECES WITHOUT APPARENT MECHANISM.

Sir,—I beg to be permitted to add a few observations to the many which have already appeared in your valuable Magazine, on a subject which has excited, among the curious, some degree of attention, I mean the time-piece of Savory, a representation of which is given on the front page of your 795th number.

That it is no difficult task to construct, on a larger scale, such a time-piece as Savory's, will appear from the three following modes which I now give :—

The first and most coarse method of constructing a time-piece of this kind is as follows: Fix in the centre of a large glass plate, upon which are painted the hour figures, any watch, having first removed the pendant, the glass, and the hour and minute-hands from it; and then affix, upon the hour-hand arbor, a large index, with a centre of sufficient size to cover from sight, when directly viewed, the case of the watch. By this method it must be clear, that the same power which moved the common hour-hand of the watch, would move, under altered circumstances, the larger index of the time-piece, though, of course, not so easily; but it is also clear, that, if a person were to examine the back part of the glass plate, he would at once see the body of the watch, and all illusion would cease.

The next method is a neater mode of attaining the desired end. Take a watch, deprive it of its casing, dial, and hands; then fix the watch in a flat cylindrical casing without a top, leaving only a small part of the hour-hand arbor projecting beyond the hollow of the cylindrical casing; fix this casing in a hole in the centre of the glass plate, with the back of the casing flush with the back face of the glass plate; and, then fix upon the projecting part of the hour-hand arbor, a large index, with a circular centre large enough to cover exactly the cylindrical casing of the watch. Under this method, the means of rotating the hand will not be near so visible as under the first plan.

The third, and last method which I now give, is the most effective, for purposes of illusion, of all. Deprive a watch of its casing, dial, and hands; and, in the centre of the back part of the framing of the watch, affix a small steel pin; then place, on the hour-hand arbor, a large index, with its centre in the form of a cylindrical casing or cap, of such a depth as to contain within its hollow, when thus placed, all the works of the watch; afterwards slip, on the small steel pin, a plate, large enough to cover the open back part of the casing, and lock this plate by some contrivance to the casing: finally, fix the small steel pin, with the watch, casing, and index appended to it, in a small hole, drilled in the centre of the glass plate—this hole being only so large as to allow the pin to fit tightly in it. When these directions have been followed, and the machine is in motion, every part will appear to revolve, with the exception of the pin which affixes the movement to the dial; and this pin may be readily mistaken for a purposely fixed support on which the index may revolve.

In all these methods, it will be seen that the movement of the apparatus is always firmly fixed to the glass plate; and that it is only a large, instead of a small index, which revolves; and, it will also be seen, that a minute-hand might, with very little alteration, be added; but I apprehend, the introduction of a minute-hand would destroy the illusion, which is probably the best feature in such a toy as Savory's time-piece, and, therefore, I should think such a hand would be best omitted.

With respect to the construction of the time-piece originating these observations, I beg to observe, that, as far as an inspection from the street has enabled me to form an opinion, I consider the movement to be partly placed within the centre of the index, and partly within a small flat cylinder fixed in a hole in the centre of the glass plate; I think that the mainspring is attached to the interior surface of the centre of the index, which centre performs the part of what I believe, in some movements, is called the *going-barrel;* that a square end, which projects from the centre of the index, is one extremity of the arbor of this barrel, to which arbor the spring is also fixed, and by turning which, the works are wound up; that the retarding part of the movement is contained within the flat cylinder in the middle of the glass plate; and, finally, that the hand, *i. e.,* the going-barrel, is connected by means of a small tube to the parts of the movement within the cylinder in the glass plate. It may be objected to these suppositions, that it is impossible, within the space taken up by the movement supposed to be placed as I suppose it to be, this space not exceeding the measure of a cylinder of a fifth of an inch in diameter, by a fifth or sixth of an inch in depth, to place the works of a watch. To such an objection, I have to say, that it should be borne in mind that, as this instrument only indicates the hour, all the works of a watch are not necessary; and that instruments were, as I have been informed, made on the continent many years since, which were capable of indicating the hour, of such a size as to be affixed to finger-rings, in the manner of signets; and that such instruments have also been made and placed in breast-pins and broaches. If, therefore, many years ago these instruments could be made of a size fit for the purposes I have mentioned, it is not an unfair conclusion to suppose that, by the application of greater skill and more simple movements, arising in course of time, the constructor of the time-piece under consideration has been enabled to condense, within the limited space I have mentioned, all the works necessary to make an index mark the hours of the day. Possible, as I think the construction which I have hazarded, the workmanship, under such a plan, must be surprisingly beautiful, and, it would appear, that time-piece making could scarcely pass much beyond the limits attained in this instrument.

I beg to conclude these observations, with many apologies for trespassing to such a length upon your valuable time, and to subscribe myself, with much respect, Sir,

Your most obedient servant,
J. P. H.

168, Devonshire-place, Edgeware-road,
Dec. 11, 1838.

ON THE GENERATION OF STEAM.—
DR. SCHAFHAEUTL'S EXPERIMENTS.

Sir,—The article by Dr. Schafhaeu "On the conversion of water into steam &c.," has afforded me pleasure and instruction, and only increases the list (already a long one) of obligations which we owe to the people of North Germany whose productions are marked by that originality of thought and earnestness of pursuit which characterise Dr. Schaf-haeutl's paper.

Dr. S. has referred to my investigation of the phenomena of drops of liquids on highly heated surfaces, as detailed in my *Manual of Natural Philosophy*, page 553; in an incidental notice of calorific repulsion; but he does not seem to be aware of the existence of an article on this subject contained in the 703rd Number of your Journal, wherein I offer an explanation of the phenomena somewhat similar to the one which Dr. S. himself proposes.

Dr. S. also states "that we have no reason to have recourse to calorific repulsion, where the evolution of vapour is sufficient to explain all the phenomena." Now, I think, the following experiment proves that repulsion is an element which cannot be omitted in attempting to generalise these remarkable phenomena :—

If a platinum capsule be drilled with an immense number of holes, as in a coffee or wine strainer, it will not, of course, retain water for a single minute : but, if brought to a white heat by means of well-prepared incandescent charcoal, so that fire only, without flame, be produced, and water be dropped into it, a large globule may be collected, and it will exhibit all the phenomena, just as if a common platinum crucible had been employed.

This capital experiment is due, I believe, to Mr. Perkins.

I am very glad to observe that Dr. S. advocates the only real scientific reason that can be assigned for the bursting of steam-boilers, viz., the elastic force of the steam itself at high temperatures. This, combined with the secondary causes of the weakening of the boilers by corrosion, &c., which Dr. S. so well illustrates, is quite sufficient to account for the frequent disasters that occur; without trying to form explosive mixtures under circumstances which make chemistry appear ridiculous.

I remain, Sir,
Your obedient servant,
C. TOMLINSON.

Salisbury, Dec. 13, 1838.

NEW MODE OF LAYING A RAILWAY.

Sir,—To show your numerous readers that the subject of railways is not exhausted, allow me a place in your Magazine for one of great simplicity, but which I think overcomes many difficulties, to which some of the present railroads are subject.

Lay down cross beams of granite, or any other hard stone, bedded in rubble or gravel, well rammed by means of Col. Maceroni's ingenious portable monkey on wheels; let these beams be in a proportion suitable to their breadths; place them at 8 feet asunder, well levelled. These are my chairs for the rails, which I think no ordinary pressure could sink, as it would be applied equally at each end at the same moment; on the ends of these cross blocks, I repose my rails secured at their union by staples; these may be of Kyanised timber, or oak painted, in beams of eight or nine inches square, according to their length, into which are sunk, in a groove four inches deep, a portion of an iron bar, that is seven inches broad by one inch in thickness, three inches of which rises vertically above the beam, and is either square or round at the top, as may be found most convenient; this blade or bar, is to be well cemented, and well-fitted into the beam; on this blade I run my carriage wheels, the edges of which are to be formed like a pulley, with a groove $2\frac{1}{2}$ inches deep, having a square or round bottom, as shall be suitable to the edge of the rail, and flanching off at its sides in an angle, we will suppose of 60 degrees, whose sloping sides, in case of any lifting (a thing nearly impossible from the form of the blade), on which neither snow nor gravel could rest, would bring it back to its groove; hence no scraper would be required, for the wooden beams should be a little sloped, to throw off water from the place where the blade or rail is inserted.

As to the wheels, I would bring the body of the carriage a little over them for their protection, by which I gain breadth for my carriage, and secure them from accidents on the way side, whose naves should be well supplied with oil, from a circular cavity in their centres, surrounding the axle-tree. Such is my idea of an improved railroad. If the expense of the wooden beams be objected to, on account of their cost and liability to decay, we might then substitute iron rails, whose section would form the letter ⌐L thus placed, but that would add considerably to the expense.

I am, Sir, yours, &c.
GEORGE CUMBERLAND.
Bristol, Dec. 9, 1838.

ON THE ESSENTIAL QUALITIES REQUIRED IN A GOOD GUN.

Sir,—I shall be obliged if you will allow me the opportunity of laying an inquiry before your readers, which may possibly elicit some information that may be of use to others besides myself, who, to a fondness for; mechanical pursuits, occasionally add the use of one of the most improved contrivances of the present day—I mean the gun.

Quære—What is the best method of *proving* the goodness of a gun's shooting ?

In the ordinary method of firing at a mark, it remains a *matter of opinion* whether you have made a good or bad shot, the correctness of that opinion being proportionate to the experience of the person giving it.

On referring to Col. Hawker's work, he says, " A man may be taken in with a horse or a dog, but *never with a gun,* after being simply told how to try it." But, unfortunately, *the instructions* that follow *only* enable you to judge *comparatively* the qualities of several guns ; and the *best* out of half-a-dozen bad guns may be far *inferior* to the worst of half-a-dozen good guns ; and throughout the work I can find no example of what a common double gun *ought to do.*

I want to know some method of ascertaining, to a certainty, the goodness of a gun, supposing I have *only one;* that is to say, having given :—

The weight of the gun ;
Length of barrel and gauge ;
Weight of powder, shot, and No. of shot ; and
Distance ;

what number of shot, or what proportion of the whole charge, should be put into a target of a foot square, or of any other number of square feet?

Surely the answer ought, by this time, to be an established point, or rule, admitted by all gun-makers, sellers and buyers.

Another, and concomitant proof, that of strength, is more difficult to obtain. This is sometimes measured by the number of sheets of brown paper the shot will penetrate ; but this is uncertain, no two persons may use the same kind of paper, and different thicknesses will often be found in the same quire ; some shot of the charge will penetrate much deeper than others, and of many it will be difficult to ascertain whether they are through any particular sheet or not. The penetration into wood is still more uncertain ; if deal, many of the shot may strike the harder part of the grain of the wood, and not penetrate at all, or it may happen that the greater part of the charge may hit the softer part, and be quite buried.

There was a letter some years ago in your Magazine, from the celebrated Mr. Nock, on the comparative advantage of flint and percussion, but he omits several particulars, and speaks of going through 24 sheets of brown paper—this appears more than I can do.

Should this matter interest any of your readers, and they can suggest any accurate proof of the strength and goodness of a gun's shooting, they will confer a benefit upon all purchasers of guns, as well as upon,

Sir, your obedient servant and
Constant reader,
CRITERION.

[Having referred our correspondent's letter to Col. Maceroni, as a gentleman better qualified than any with whom we are acquainted to give the information required, we have been favoured with the following clear rules and instructions upon the choice and essential qualities of a good gun. There are few country gentlemen who will not thank the intelligent Colonel for the valuable information which he has here enabled us to publish. —ED. M. M.]

Sir,—Certes, it is not by merely " troubling you with a few lines" that I could possibly make an adequate reply to the comprehensive queries of your correspondent " Criterion," concerning fowling pieces, their qualities, powers, defects and improvements. All these depend on the existence of certain mechanical and chemical conditions—and on their right understanding and application. Without exact attention to the positions of " Criterion's" questions, but keeping my eye on them, I shall endeavour briefly to em-

brace them in a due and natural conse-
quency. In doing this I must state facts,
which some may think I ought to support by reasons, proofs and authorities;
but such a course would involve me in a
volume, instead of a column or two, in
your excellent miscellany.

1. The interior of a tube for the propelling of heavy substances to a distance,
by means of gunpowder, ought, in principle, to be perfectly cylindrical and
smooth throughout its entire length.
Some *"great"* gunsmiths enlarge the
cylinder slightly towards the breech, in
order that a certain degree of foulness,
deposited by the powder, principally in
that locality, may not oppose the free
descent of the ball or wadding. This is
an error, as I could easily demonstrate,
which has caused many old, thin gun-
barrels to burst; and which makes our
new and heavy ones to recoil more than
need be, although their strength does
save them further evil than a loss of
power from the powder *when they are
clean.*

2. It is evident that the shape of a
mass of gunpowder which can be ignited
in the most simultaneous manner, is the
spherical, and that the best point for ignition would be the centre of that sphere.
The shape next approaching to compact-
ness is the cylinder of equal diameter
and length, so that if we place the touch-
hole of a gun half way between the
breech pin and the top of the charge of
powder, we shall evidently cause the most
instantaneous ignition,—save and except
we could conveniently introduce the fire
through a small tube into the centre of
the charge. This I have tried by way of
experiment, and find it to produce the
effect anticipated. But it is *not* always
expedient to make the charge of powder
ignite so instantaneously as we are supposing. A heavy body requires *some*
fraction of time to receive motion, so
that in pieces of long cylinders, such as
"long guns," fowling-pieces, muskets,
&c , less recoil and equal range is produced by the powder being fired at the
anterior extremity, so that it may be said
to act upon the projectile as it proceeds
l ong the cylinder. " Patent breeches,"
therefore, are no detriment to fowling-
pieces, &c., which are even so much as
orty calibres, or more, in length; but a
mortar, which is of only two or three calibres, should have a conical chamber

with semi-spherical base, and the charge
ignited by a tube reaching to the centre
of the powder. Passing by a crowd of
other axioms and illustration, I must
proceed to—

The most convenient and proper shape
for the projectile to be acted upon by the
powder in the cylinder, which is a sphere.
A like charge will propel a ball of stone
further than one of wood—a ball of iron
further than one of stone, and so on,
according to density, to lead, gold, and
platina. With a cylinder absolutely true,
a ball equally spherical, and also exactly
fitting the cylinder, the range is perfectly
free, barring the effect of wind. But as
these conditions are impossible to achieve
in general practice, the rifle has been invented, which, giving to the ball a rapid
rotation on the axis of its flight, inces-
santly corrects the aberrations, which
would be otherwise produced by its irregularities of shape and unequal friction,
or bounding within a smooth cylinder.
The direction of the flight of all bullets,
cannon shot and shells, becomes less and
less liable to deflection, in proportion as
their diameter increases; because, the
larger the sphere, the less is its surface
and its irregularities of shape in proportion to its mass. For example, I have
irregularities of shape in bullets of one
ounce weight, amounting to the sixteenth
of an inch, and having measured some
well-cast English 24-pounder shot, the
spherical deviation did not amount to
any more. Now, a 24-pound shot weighs
384 oz., so that in the bullet the irregularity is greater as 384 is to one, with
regard to mass, and ten to one, with reference to diameter. Hence the advantages to be gained by rifled *cannon*, supposing such an adaptation practicable,
would not pay the trouble, as the precision with which artillery, with well
fitting, and especially with "*bottomed*"
shot, can be directed, is quite sufficient
for all ordinary purposes. Indeed, cannon so loaded shoots truer than a rifle,
when the w nd blows across the line of
flight. As spheres increase in diameter,
so their mass augments in the ratio of
the circle. The smaller the sphere, the
greater is its surface of resistance to the
air, in proportion to its mass and momentum. A 24lb. shot (were it of lead)
would contain about 512 times the mass
of an ounce musket-ball, whereas its surface. resisted by the air, is only about 40

times that of the latter. The range of projectiles increases in a ratio greater than that of their diameters; for instance, a 32-pounder, which is about 6 diamater, will, with a proportionately less charge of powder, range more than one-third further than an 18-pounder, about 4 inches diameter. I have loaded a small pocket pistol, percussion lock, carrying a ball of exactly 128 to the pound, with twelve grains of the best powder (the strong charge for a percussion duelling pistol), and firing at an octavo volume of above 200 leaves, at six paces distance, the ball, as well as I can remember, penetrated or broke into about thirty leaves. I then loaded a duelling pistol of thirty-two balls to the pound ($\frac{1}{2}$ oz.) with exactly the same charge of twelve grains of powder; when, from the same distance, the ball passed through, or indented, above a hundred leaves of the same book.

The extent and precision of the ranges of projectiles depend on the mass and density of the spheres; the quantity and quality of the gunpowder; the method of its ignition; the loss of its blast by the vent, and especially by its escape around the ball,—a space called " windage." Further, much depends on the true spherosity of the ball, the correctness and smoothness of the cylinder of the piece.

No kind of soft wad or envelope will remedy the defect of a ball being too small. That is, the powder pays no respect to it, but passes by. Thin sheet lead will partially succeed with a rifle.

The escape of powder through the touch-hole of flint lock-pieces is very considerable, though proportionately less in cannon. However, Marshal Augereau, when a sergeant of artillery, had his thumb blown away whilst holding it over the vent of a field-piece, which exploded unexpectly in loading.

Any impediment to the exit of the ball beyond the unavoidable and beneficial friction of a good fit, curtails the power and range. I remember a naval officer showing me a beautiful pair of pocket pistols, of the old screw-barrel conformation. He spoke of their power, &c., when I, who had remarked how much the ball exceeded the bore into which it was so tightly screwed, and the large touch-hole to the flint-lock,—told

him that he could not put a ball through an half-inch deal at twelve paces. After a hearty laugh at my *ignorance*, he fired at the deal bulwark of the ship,—and back came the ball, harmlessly rolling at our feet, with scarce an impression made upon the plank. Half the power of the powder is lost in forcing the ball through the too-small barrel; another quarter, at least, escapes through the touch-hole of these stupidly-imagined pistols, which, I see, are now falling into disuse.

An absurd and vulgar error prevails, even amongst educated people, that the ball from a gun or pistol, commences its career by *rising* above the line coinciding with the cylinder of the piece. One day I had a warm dispute upon this subject, with my comrade, Colonel Wolff, whom I have mentioned in the lately-published portion of my memoirs. His pistols were by the celebrated Kukenrighter, and like all pistols of thirty years ago, were much thinner at the muzzle than at the breech. Thus, as is a cannon, the exterior "line of metal" is not parellel to the interior cylinder, so that by aiming along the former, we give a "real elevation" to the pointing.

" Criterion" will ask, what has all that I have yet written to do with his queries? I reply, that without a due understanding of such fundamental theorems, and their numerous colatarals, no man can know the use of a fire-arm !

But now to come to the projection of charges consisting of many balls, *i. e.*, buck-shot or bird-shot. The proper impingement of these depends upon the same points as decide the range of single balls. The proper proportion of powder to shot, and the application of the greatest possible portion of the blast of the former to the propulsion of the latter. To be brief, the full charge of powder, now so superior to what it was a century ago, is for cannon one-third the weight of the ball; but with regard to muskets, &c., with percussion locks, which admit of no escape through the vents, one-third less powder is required than with a flint-gun.

In projecting buck or bird shot, the most important consideration is the wadding placed over the powder, which should not allow of any of the blast escaping, but send forth the shots as one ball. For sporting purposes I have found no better wad than corks, cut cylindrical, then greased, and divided into pieces about half an inch thick. The cork being elastic *laterally* will go down with a tight fit, whether the gun be clean or foul. On this subject I have already written in your pantechnical pages. I have fired a charge of twenty-one buck-shot, of 120 to the pound, from a musket charged with two drams of fine powder, but with only an ample wad of strong brown paper over the powder. The object was a park paling of oak, at fifty yards: not a shot pierced it; but when I used a wad of cork half an inch thick every shot went clear through. Another good wad is the metallic, introduced, I think, by one Butler. But he has made notches on the edges, for the imaginary need of allowing so much space for the escape of the air from the barrel, which notches afford far too much passage to the blast of the powder, which so gets into the shot, besides being wasted. But by using two such metallic wads over the powder, so that the notches shall not correspond, this inconvenience is obviated.

Another point in the use of bird-shot is the due proportion of it to the gun and powder. Most shooters overload their piece with shot. The rule is, that the shot should be equal to the weight of the ball that would fit the gun. If ever my gun gets so foul (in dry weather), that the wads decend with difficulty, I put half a charge, or less, of powder into it, and then a turnip leaf or wet grass, and working it about with the ramrod, fire it off, so as to clean a great portion of the barrel.

Whilome, a most erroneous fancy prevailed regarding the weight of guns and pistols, which often were boasted of as being so beautifully *light*. A gun or pistol, still more a rifle, cannot be too heavy, until too inconvenient to carry: the heavier the piece the less the recoil, and the steadier the aim. That portion of the charge affected to *moving* the piece backwards is taken away from the range of the projectile forwards. I can load a common fowling-piece so as to send partridge-shot, No. 6, clean through a copper penny-piece, which another person shall cast up in the air over my head. I have done it often.

If the truth of the hasty and disjointed positions I have hastily sketched be admitted, and I am sure they must be upon investigation, it is an indisputable fact, that a clean town musket, bating the flint lock, and its coarseness of parts, must shoot as well as any "Joe Manton," or any other Joe or Jack amongst the multitude of "*artistes*" whom I well know have their pieces, "sure to kill," supplied them by the wholesale makers, instruments, brass bound, case and all, complete, for 20*l.*, and sells them to the west-enders for from 50*l.* to 80*l.*

Before, therefore, I can tell "Criterion" how to prove a gun, I must know many things concerning this said instrument and the way in which he uses it. To make sure of the identity of several aims the gun must be secured in a vice immoveably. The gargon about one maker's gun shooting further, another "closer," another "sharper," another "wider," is all invented by the sellers. A twenty-five shilling gun, with straight barrel and correct smooth cylinder, will shoot as well as a twenty-five pounder, if equally well pointed. But the green-horns are made to believe that Mr. A, or B, or C, can sell them a costly gun that *must kill*, whether they load or aim properly or not!

As to the wretched twaddle of Colonel Hawker upon guns, percussion, and flint, &c., it is far beneath my criticism. What he says about dogs and the proper mode of *finding* game may be all gospel for aught I care. We had a fine specimen of the acumen of the Colonel, and of our military masters, when about two years ago the latter thought fit to consult the former on the substitution of percussion for the flint locks of the troops, &c.! We have seen the vast increase of power derived from the increase in the size of the shot; yet this Colonel Hawker tells us, that the *smaller the shot* the surer we are to kill our game, because, forsooth, "as a game-keeper observed, the smaller shot penetrate farther in *like pins and needles !!*"

The charge of shot used by most sportsmen is about double what it ought to be, whilst the windage allowed by their improper wadding deprives it of a portion of the powder's blast, besides causing it to be unduly dispersed. Less than the weight of the ball that fits the gun, is better than more; and the wad is all important. Over the shot a bit of paper or oakam sufficient to retain it will be enough. Two drams of powder, one ounce of shot, *a cork, or two metallic* wads, will do good service at sixty to eighty yards; so will 100 duck-shot kill well at 100 yards.

Bird-shot, being smooth spheres, upon being pressed together act upon each other just like the particles of water. The hindmost act like wedges between the foremost, so as to cause a tendency to expand and press laterally on the sides of the barrel. This is what causes the "-leading" of a gun-barrel. I have found that instead of rubbing the shot in black-lead, (carburet of iron), to make them smooth and shining, as practised by the sellers, something of the friction on the sides of the barrel is taken away by mixing with the shot a small quantity of impalpably powdered chalk or bone dust. But this is a *finess* in its way, and also has the inconvenience of bringing the dust into your eyes when the wind is against you. I have made cartridges of duck and of buck-shot thirty years ago in this way; and also by pouring liquid plaister in Paris, and also melted tallow amongst them: I also tried very fine sand and brickdust, but loose shot *with such wads as I have named* will answer every purpose of the sportsman.

From all that I have predicated it is, I trust, clear, that I cannot furnish " *Criterion*" with the specific answers he requires without a minute account of the weight of gun, powder, shot, wads he uses, and the distance from which he proves his gun. As he very properly observes, sheets of paper may be thicker or thinner; boards may be harder or softer, in whole or in parts. But I must have tired you out, so I conclude by subscribing myself,

Sir, your obedient servant
F. MACERONI.

PLAN FOR REVERSING A RAILWAY TRAIN WITHOUT TURNPLATES.

Sir,—I herewith forward you a drawing and description of a plan for reversing the trains at the terminus of railways without the aid of machinery or turnplates, which are, I believe, only applicable to single carriages. I am not certain that the plan is novel, but I have never seen it used, and it may be interesting to some of your readers.

I am, Sir, yours respectfully,
R. P. C.

September, 1838.

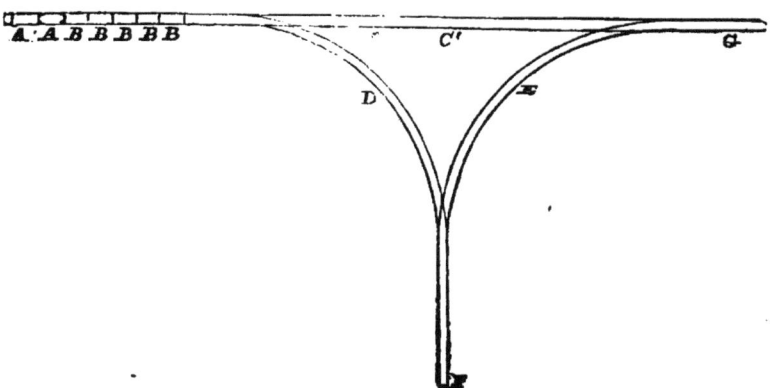

Description of the Plan.—A A, The engine and tender.

B B B B B′, the carriages.
C, the line of railway which the train

has traversed to arrive at the terminus. D E, are curved rails.

The train is supposed to have arrived in the direction from C to A. In order to proceed upon a second trip, it is evident the engine should again precede the carriages, and that the second-class carriages should precede the first class, or, in short, the train should again proceed in the same order that it arrived. This I propose to effect in the following simple manner :—

The train is to enter the curved way D, and proceed until the carriage B arrives at the point F. They are then to enter the other curved way E, and proceed until A (the engine) arrives at G ; the train will then be adjusted for starting.

THE ARKWRIGHTS AND STRUTTS—PRODUCTS OF THE SPINNING JENNY.

We would hardly search for mechanical matter in a work under the title of " *Music and Friends ;*"—in a pleasant book under this title, however, by a Mr. Gardener, we meet with the following particulars, most interesting to all inventive and mechanical aspirants:—

" The private wealth of the present Mr. Arkwright has grown to such an enormous sum, by his unostentatious mode of living, that excepting Prince Esterhazy, he is the richest man in Europe. A few years back I met his daughter, Mrs. Hurt, of Derbyshire, on a Christmas visit at Dr. Holcomb's, and she told me that a few mornings before, the whole of her brothers and sisters, amounting to ten, assembled at breakfast at Willsley Castle, her father's mansion. They found, wrapt up in each napkin, a ten thousand pound bank-note, which he had presented them with as a Christmas-box. Since that time I have been informed that he has repeated the gift, by presenting them with another hundred thousand pounds.''

Of the Strutts, the early associates of Arkwright, we are told—

" John, the son of Mr. George Strutt, who resides at Belper, possesses a refined musical taste, and has rendered his neighbourhood as famous in that science as any district in Germany. * * An idea of the magnitude of their concerns may be gathered from the following circumstances :—About the year 1820, wishing to retire from business, they proposed to any persons who would purchase their works at a valuation, that they would give the parties a bonus of 150,000*l.* To give a higher taste to the work-people at Belper, Mr. John Strutt has formed a musical society, by selecting forty persons, or more, from his mills and workshops, making a band of instrumental performers and a choir of singers. These persons are regularly trained by masters, and taught to play and sing in the best manner. Whatever time is consumed in their studies, is reckoned into their working hours. On the night of a general muster you may see five or six of the furgemen, in their leather aprons, blasting their terrific notes upon ophicleides and trombones. Soon after the commencement of the music-school it was found that the proficients were liable to be enticed away, and to commence as teachers of music. To remedy this, the members of the orchestra are bound to remain at the works for seven years. Mr. Strutt has ingeniously contrived an orchestra, with the desks and boxes containing the instruments, to fold and pack up, so that, with the addition of a pair of wheels, the whole forms a carriage, and with an omnibus for the performers, he occasionally moves the *corps de musique* to Derby or the surrounding villages, where their services are required for charitable occasions. The liberality with which this musical establishment is supported is as extraordinary as its novelty. As an incentive to excellence, when he visits town, he occasionally takes half-a-dozen of his cleverest people with him, who are treated to the opera and the concerts to hear the finest performers of the age.''

ON BRICKS AND BRICKWORK.

(From Hints on Construction, by George Godwin, Jun., F.S.A. and M.I.A. : in various numbers of the *Architectural Magazine*.)

A student, at this time, has no excuse for an ignorance of the leading principles of any of the sciences. Fifty years ago, a man who hoped to attain the extent of knowledge here implied must have toiled earnestly and unceasingly, wading through the theories and contradictions of centuries, with which truth was overlapped ; and gaining thence, if he were sufficiently persevering, some few of its illuminating rays to guide him in his inquiries. It was not to be hoped, then, that individuals, who had specific occupations to employ them, could do more than acquire such information as was actually necessary for the practice of their trade or profession. But now, when men are at the corner of every street, if we may so speak, waiting to deliver verbally, in an hour, the results of the investigations of others during years, to be received by hundreds at the same time, passively, without trouble ; when powerful minds are engaged on all sides applying the hydraulic press to accumulated bales of knowledge (arranging, simplifying, condensing), and spreading the proceeds abroad, in all shapes and in every direction ; to acquire this general notion of all the sciences is not the labour of a life, but the recreation of leisure hours. To possess this knowledge is, therefore, no longer a distinction, but to be without it is a disgrace   *   *   *   *

A brick, we know, consists chiefly of pure clay and flint in the shape of sand, mixed with water ; in other words, of alumina and silica, the latter the basis of glass ; each of them an oxide of a metal (or compound of a metal and oxygen) ; the former consisting, as we learn, of ten parts of aluminum (the metal) and eight of oxygen, the latter of eight of silicium and eight of oxygen. When the two, constituting ordinary brick earth, are mixed together with water, they form a tough, tenacious, and plastic mass ; but if it be heated, if the water by which the admixture was effected be driven off, it loses its plasticity, never to be again acquired, and becomes a solid substance, a silicate of alumina, strong and endurable in proportion as the admixture of the two components is complete, and the burning sufficient. To show the value of properly kneading the earth, and bringing its particles into close connexion, it may be sufficient to say, that, by the bestowal of additional labour in this respect, bricks may be made, capable of resisting twice the amount of pressure which would destroy others prepared less carefully, of similar earth. As regards the effect of perfect burning, and entirely driving off the water, it is but necessary to examine a brick clamp when opened, and note the difference observable between the bricks where the fire has exerted its due influence, and those where it has not done so, in order to comprehend its importance. Where both these points have been attended to, the brick is usually of a bright clear colour ; has a metallic sound when struck ; and, if it be broken, does not crumble to powder, but presents a sharp ragged fracture. Those bricks which have not received the full action of the fire will not fulfil these conditions, and form what are called *place*-bricks : they will not resist the weather for any length of time ; are crushed by a trifling superincumbent weight, and, consequently, should never be used where durability is regarded. The terms place-bricks and stock-bricks are merely

disguises ; they are but other words for *bad* bricks and *better* bricks : and one might reasonably suppose that no person would knowingly use bad materials, to effect a trifling temporary saving, when better might be obtained ; and, therefore, that place-bricks would never be used : unfortunately, however, the reverse is too frequently the case.

If the clay contains too great a proportion of alumina, the bricks contract greatly by burning, and are liable to crack in the operation ; and, if of silica, the bricks will be very brittle. In theory, we may say that bricks which are found to be less than those of ordinary size are, other things being equal, of imperfect composition. For practical purposes, this may, or may not, be worthy of notice ; but there is a real evil attending the use of small bricks, which certainly should be mentioned ; namely, that as, in a given amount of work constructed with them, there will necessarily be a greater number of mortar joints than in the same quantity of walling for which large bricks are employed, it will settle down more, and be, at all events for a time, less stable than in the other case. The heavier a brick is when dry (and this, in a great degree, is regulated by the amount of labour bestowed on the kneading of the materials), the better it is, the more solid, the more impervious to water.

Silica and alumina, when mixed, do not melt on being exposed to the action of heat, unless there are other substances present, such as lime, for example, in which case fusion is easily effected, and a vitrified mass results. The glazing sometimes given to the surface of bricks, for various purposes, depends on the vitrifiability of silica when mixed with certain substances. The ordinary mode adopted is, to throw upon the bricks, when heated, common salt, which we may term a muriate of soda : this is decomposed, and the soda, uniting with the clay, induces fusion of the surface. Glass, we know, is nothing more than silica and soda fused by heat ; therefore, in fact, by this operation the bricks are *glassed* over.

After what we have said, it is hardly necessary to remark that bricks made of sand and clay, containing any portion of lime or other flux, may not safely be used in situations exposed to violent heat, inasmuch as they would readily fuse. Fire-bricks specially so termed consist, for the most part, of pure clay, mixed with a certain quantity of old fire-bricks, or other burnt clay reduced to powder, which fulfils the office of sand, but is less liable to fuse if accidentally brought into contact with ordinary fluxes.

The colour of bricks, although so various

as it is, depends chiefly on the oxide of iron, which all native clay contains; the effect being modified by the substances with which it is combined, or circumstances of which we are ignorant. It is the same with the natural gems, or jewels (many of which consist, too. of the like materials as brick, namely, silica and alumina); for these, although for the most part quite different from each other, owe their colour to the presence of oxide of iron; as, for example, the lazu-lite, which is blue; and the obsidian, which is black; the yellow topaz, and the red garnet.

Concerning the processes of brick-making, although by means beneath the attention of of the architect, we shall say nothing other than to advise the student to inspect them for himself, in some of the numerous brick-fields to be found in the immediate neighbourhood; nor shall we here enter upon the history of bricks, which should probably commence at a time when the first man, Adam, was alive, and would include mention of nearly every known country in the world. England is especially dependent on brick as a building material; and there are numerous excellent examples remaining of brickwork executed many years ago, to show how well it may be performed. We may notice several houses in Lincoln's Inn Fields, and No. 43, in St. Martin's-lane, which display, as indeed do many others even in a greater degree, ornamented pilasters and entablatures formed in the same material with great nicety. At this time, however, in consequence, among other things, of inattention on the part of architects, the system of competition pursued, and the general use of cement as an exterior facing (which naturally induces the men to do their work carelessly, knowing it will be covered, and engenders bad habits), good brickwork is seen but seldom; and, it would now, perhaps, be a matter of difficulty, to find a dozen workmen in London capable at once of imitating some existing specimens. We say at once, because we are perfectly satisfied that there could not be a demand for any amount or sort of skill which England could not supply; and that, if such work were required, and were properly paid for, men would speedily arise equal to the task. We propose to consider some of the various modes of executing common brickwork.

Every treatise on the art of building contains an assurance that brickwork carried up in *English* bond, is stronger than that which is executed according to the *Flemish* mode. Every brick-layer who has had experience will say the same thing, if he be asked; and a careful examination for oneself of the two methods, which every one who would arrive at a sound conclusion is called upon to make, will confirm the fact. Now, in spite of all

this, we still find the Flemish bond adopted in five out of six of the new buildings that are to be observed every day springing up around us with almost dangerous rapidity; and an inquirer would naturally seek for the cause of this singular disagreement between precept and practice. It is manifold and close at hand. Workmen have become accustomed to the latter mode: a good appearance can be produced with less trouble than when English bond is used; and, what is more important than all in the estimation of the speculator, it allows him to use inferior bricks (bricks which, by the mere operation of carting, have become broken into pieces); insomuch as bats may be as advantageously employed for it, so far as appearance goes, as whole bricks.

If custom insist upon retaining the appearance afforded by Flemish bond, a part of the evil may be avoided, it has been suggested, by using English bond withinside, and casing the outside, as it were, with Flemish bond. The weight of the greater quantity of timber in a building is usually upon the inside half of the walls; and the evil, therefore, may probably be lessened by this course, when the walls are thick enough to admit of it; and, by using more whole headers, in the place of bats, the danger of a separation might be prevented. The greatest care would be required, however, so to bond the whole together, that no division could possibly take place.

To enter here on an elaborate description of the two methods might be deemed uncalled for and tedious, and we shall therefore do little more than allude to them.

Flemish bond, as generally performed, implies the arrangement of headers and stretchers *alternately in the same course*, which said headers are, for the most part, merely bats; and thus a wall so constructed often consists of two separate leaves, if we may so speak, very slightly connected together; and the possibility of their separation or dislocation is obvious.

In English bond, on the contrary, each course consists alternately wholly of headers (or bricks laid in the direction of the thickness of the wall), and wholly of stretchers (or bricks laid in the direction of the length of the wall), so it that is bonded together throughout equally, and cannot easily suffer any disruption, unless the bricks themselves be broken by the force exerted. We should recommend the student to obtain a score of bricks, or, better still, some small wooden models of bricks, half bricks, and quarter bricks, or closers, and essay for himself the different combinations which may be produced.

It would seem needless to say that the angles of a building require to be well bonded, and that the architect, or the clerk of the works, if there be one, should have a watchful eye in this respect. It cannot, however, be repeated too often; insomuch as workmen frequently fail to give that additional degree of care which, in order to make sound work, is there required; and unsightly settlements, even if nothing more serious occur, are the certain results. Pieces of thin iron hoop may be advantageously used in some situations, as an additional precaution. We may remark that, if it be intended to cover the exterior of the building with cement, the necessity for care to prevent settlements is not lessened, but increased; insomuch as the slightest disruption produces a crack, which is always strikingly visible, and which *cannot be repaired* without first cutting down a large portion of the cement-work on each side of the fissure, and even then not always effectually.

Many walls, which externally appear to be well bonded together, are in reality defective, through want of proper bond in the horizontal, as well as the perpendicular, joints. In walls not less than two bricks and a half in thickness, especially for foundations, or where they are required to resist great thrust, it will be found a good precaution to introduce occasionally two courses, one over the other, of diagonal or herring-bone bond, which may be done without interfering with the appearance presented by the work externally. In the lower course of a two and a half brick wall, carried up in English bond, for example, there may be on the outside a line of stretchers; then a course of bricks placed diagonally, forming an angle of about 45°, and having the interstices filled up solid; and against them a line of headers, constituting the other face of the wall. In the upper course the operation would be merely reversed: a line of headers would form the outside, and one of stretches the inside, face; and the diagonal bricks would be placed in a contrary direction to the last. Even in a two-brick wall diagonal bond may be introduced; but then it must be in single courses only, between courses of ordinary English bond, as otherwise the face of the wall, not being tied in, would be liable to budge.

We have said, when speaking of diagonal bond, that the interstices should be filled up solid. This should be done in all cases, never allowing the use of bats where whole bricks can be introduced; nor of mortar where *closers* (that is to say, half bats), will serve the purpose. Every crevice in the one course should then be filled with mortar or flushed up, and the whole made perfectly sound and level to receive the next. When brickwork is treated in this manner, provided the weather be not too dry, and the mortar good, (on which head more elsewhere), no further steps are necessary; but as, unfortunately, in the greater number of instances, it is *impossible* to insure the care thus required, the bricklayer's specification should in all cases contain a direction "to grout the work with hot lime and water every eight courses in height," which, *en passant*, we may suggest, does not mean to slobber the face of the wall with the mixture (we know no better word), as some workmen seem to think it does, but to fill up every crevice left by the bricklayer in the interior of it, through carelessness. It likewise serves another useful office, as we think, which may be mentioned. Bricks are often used perfectly dry, and covered with dust, and sometimes even when heated by the sun; in consequence of any of which circumstances the mortar round about each brick is rapidly dried; it is not permitted to be absorbed into the substance of the brick, and sets quickly, without perfectly adhering to it. Indeed, even the imperfect adhesion which does take place is afterwards interfered with by the mere operations of the bricklayer in regard to the next courses; insomuch as mortar, once set, begins to indurate in the form taken, and this form cannot afterwards be changed without destroying the value of the mortar. The best mode of proceeding is, to soak each brick in lime-water before laying it; but, when this is not done, grouting, judiciously performed, may serve as a partial substitute, preventing the too immediate drying of the mortar, and inducing a more general and perfect union of the whole. When it is required that brickwork should dry quickly, grouting is, of course, inexpedient. This would be the case, for example, when building late in the season, and in fear of frost, when we should use the mortar quite hot, and much less fluid than ordinarily, so that the water might be quickly driven off; as otherwise it might become frozen, and, expanding, as water always does in freezing, (being an exception to the general law in nature, that bodies are rendered smaller by the abstraction of heat), cause the mortar to crumble to dust. In such a case as this, we say, grouting should be omitted; but the whole of such a proceeding, although oftentimes expedient, would unquestionably entail the sacrifice of a certain degree of stability.

Pilasters, rusticated quoins, and other projections, whether to be covered with cement or not, should be arranged, as regards their width, so as to bring in whole or half bricks, and will then materially assist to strengthen the walls. The core for all proposed decorations in cement, such as cornices, string courses, and sills (which latter, however, should in all cases be of stone, if practicable), should be *built* with the walls, and not stuck on afterwards, or dubbed out, as it was termed, which is too commonly the case; and for this purpose, when the projection is large, cement should be used instead of mortar, to prevent accidents, notwithstanding that an overlaying course of stone may have been fist bedded on the wall to receive the bricks.

As may be inferred from what we have already written under this head, every thing should be arranged so as to enable brickwork, when executed, to settle down equally. The bricks should all be of the same size, and perfect in form; the mortar joints should be small, or, if necessarily large in part, should be made so throughout the whole extent of the building horizontally. A mixture of old and new bricks in the same wall is, therefore inexpedient; insomuch as the old bricks, being chipped and broken, would require more mortar, to give the work a tolerable apearance, than is necessary for the new, and unequal settlement would, of course, take place. The only way in which old bricks can be intermixed without injurious effects in this respect, is to introduce a continued course or courses of them throughout the building, at certain intervals; the fewer the better. No portion of a brick building should be carried up more than 4 ft. in height above the adjacent parts, which are also to be raised, as it would settle down previously to the erection of the latter; and the consequence would be that, after the remainder was built in conformity to the first erected part, this, in its turn, would afterwards settle down, and, in doing so, shrink away from that which had previously become solid, and so produce a disruption. In reinstating old walls, underpinning foundations, and, indeed, wherever it is necessary that the work should not shrink, but retain its first position, Roman cement should be used instead of mortar, as that sets immediately on being applied, and afterwards admits of no change of form. This fact teaches us not to use Roman cement and mortar indifferently in the same work. We once saw a bricklayer constructing a high wall, whereof the facing bricks were laid in cement and the rest in mortar; and the result was, as might have been anticipated, that the latter shrank away from the facing, producing a general split, and that the wall, becoming perfectly useless, was ultimately taken down and rebuilt. The same evil has been known to occur from the use of facing-bricks, which, being somewhat larger in size than the common bricks forming the wall, were bedded in small layers of fine mortar, while with the latter a large quantity of ordinary mortar was employed.

A little cement may, however, be advantageously used under the ends of bressummers and girders supporting weights, or in other situations where the mortar would be more than ordinarily depressed.

These, we are aware, are every-day matters, and, with many other points even less so, are often omitted by writers on the subject, under the impression that mention of them would be deemed an insult to their readers. We are led to believe, however, that, like other things constantly beneath our own observation, they are frequently overlooked from that very cause, and deem it expedient, therefore, to refer to them, although but cursorily. Our object is to induce inquiry and reflection. Who knows what a single sentence sometimes leads to?

FALLING STARS IN AUGUST AND NOVEMBER.

[From the *Vienna Official Gazette.*]

The phenomenon of an extraordinary abundance of falling stars, about the middle of November, has been again observed this year, and in our part of the world more decidedly than ever. On the 10th of November, when we watched from eight in the evening till one in the morning, we counted about nine such stars in an hour, the sky being rather hazy. On the 11th of November, during five hours after six in the evening, the sky being clear, we counted about 20 in an hour, so that the phenomenon was increasing. On the 12th of Vovember, the sky being quite cloudy, no observation of the kind could be made. On the 13th of November the sky suddenly cleared up half an hour before midnight, and remained perfectly serene till daybreak. During these six hours we noted 1002 falling stars, of which by far the greater part were of the first magnitude, with a long horn of light, and casting much shade, like the moon. The phenomenon decidedly increased from the beginning of the observation till about four in the morning, when it seemed to have reached its culminating point; from that time till daybreak, it decreased, as the following shows:—

In the 1 hour of observation there were 32 falling stars.

—— 2	....	52 ——
—— 3	....	70 ——
—— 4	....	157 ——
—— 5	....	381 ——
—— 6	....	310 ——

Unfortunately the state of the atmosphere on the following night was such, that it did not allow any further observation, so that the duration of this remarkable phenomenon could not be determined. At the beginning of August, there was an unusual number of falling stars, though very far from that in November. On the 7th and 8th of August we counted about six in an hour, on the 9th fifteen, on the 10th sixty, on the 11th and

12th, thirty, and on the 18th, all which day it was again cloudy, only ten in an hour; so that the 10 of August must be considered as the day on which the phenomenon was as its height, since it increased till that day, and sensibly decreased afterwards. We mean to give further particulars respecting the place of this phenomenon in the heavens, and of further peculiarities which result from our observations during the present year, as soon as the necessary comparisons shall be completed.

KARL VON LITTROW.

NOTES AND NOTICES.

Progress of Civilization.—The indefatigable Sultan Mahmoud has a new plan on foot. He is about to establish, in one of the palaces at Constantinople, an "institute for education," on the European plan, in which Europeans only are to be the teachers. A project has been brought forward to establish a theatre in Pera; but its chance of success is doubtful, though it appears the exhibition of a circus of French horse-riders in the Turkish capital has been quite the rage. We hardly know whether it ought to be regarded as evidence of the march of civilization or of uncivilization that, on the 5th of November, a theatre was opened at Warsaw, for the performances of Jews exclusively, in the language known as Judæo-Polish, a barbarous mixture of Hebrew and German, or rather a debased German, written in Rabbinical Hebrew letters, which is generally understood by the Jews of Germany and Poland. This miserable jargon, which produces the same effect on the eye and mind as the ex-Moor dialect, written in the characters of the Greek alphabet, is now it seems to be elevated to the rank of a stage language! The first performance was a drama in five acts, entitled Moses, the author of which, a Vienna Jew of the name of Schertfpierer, sustained the principal character. He is also the manager of this singular theatrical speculation, to which we cannot say that we wish much success.

Leipzig Railway.—The shares in this undertaking were lately lowered a little in value, in consequence of a singular circumstance; Brockhaus, of the great printing and publishing establishment, received an order to print off a number of certificates of shares. Some of them were struck off so badly that they were laid aside to be burned. A man about the office stole a number, forged the signature of the secretary of the company, and set about 200 of them in circulation. When this was first discovered, the alarm of the shareholders was naturally excited, it not being known at the time how many of them had got abroad; the company, however, soon got possession of all, by repaying what money had been advanced on them, and who proposes to commence an action against Brockhaus for the sum lost by his negligence. The evil being found not so great as was anticipated, the shares have nearly regained their former level, being at 94½ per cent., and it is said in the German newspapers that after all, the company will probably sustain less injury by the transaction, than by the retirement of Mr. Worsden, the Englishman, who has hitherto superintended the building of their carriages.

Nuremberg Railway.—The third anniversary of the opening of the railway from Nuremberg to Fürth has just been celebrated. Since the 7th of December, 1835, no less than 1,357,285 passengers have been conveyed on this railway, though the population of both towns between which it runs does not exceed 60,000. This makes seven journeys each year for each inhabitant,—an astonishing result. During these three years no life has been lost on the railway, no serious accident of any kind has occur-

red, and the railway has received no material injury.

Austrian Railroads.—The rails in the Austrian railroads are now made of iron from Styria, which are said to be found more durable than those supplied from England.

Railways in Prussia.—An important law on the subject of railways has just been promulgated by the Prussian government. The most important points are these:—A railway must be approved of by the government, not only in its general plan, but in its details—to prevent undue speculation, any one who takes a share must pay 40 per cent on the nominal capital before he can dispose of it, nor can he dispose of it at all without the consent of the company, who may refuse to take any one else in his place,—at the end of thirty years the government may, if it pleases, take the property into its own hands, on paying to the shareholders a certain amount proportioned to the profits of the preceding five years. After the railroad has been four years in action, the government has the right of naming other persons or companies besides the company of which it is the property, who shall have the right of running carriages upon it, on paying a certain sum fixed by government. The post is to be conveyed gratuitously, and as many persons as the Post-office sends with a free pass.

Sir James Anderson's Steam-carriages.—In a letter from Sir J. Anderson to an *Irish* paper, he gives the following account of the wonderful power possessed by his steam-carriages, which are intended to run on common-roads :—" Subsequent trials of the carriage in question have given the following results:—One hundred weight of coke per hour produces 7500 gallons of steam per minute, driving the engine at a speed of 12 or 15 miles per hour, at a pressure of 50lbs. to the inch, and giving about 4000 gallons of steam per minute, beyond the required consumption. In other works, this immense power is obtained at a cost of one penny per mile." Such *wonderful* results having been obtained by Sir James in his factory—we presume it will not be long now before we have a public demonstration on a common road of these performances. The roads are now in excellent condition—shall we say New Year's Day for a public experiment—or wait till the 1st of April?

Eastern Counties' Railway.—The centring of that beautiful structure, the River Lea Bridge, was struck yesterday, when the depression of the crown of the arch was ascertained, by accurate levelling, not to exceed three-tenths of an inch. Perhaps such a circumstance was never before recorded of a bridge of 70 feet span, constructed of mere brick and mortar.—*Railway Times.*

Presentation of Plate to Dr. Gregory.—Our readers are aware that Dr. Gregory resigned his appointment as Professor of Mathematics, at the Royal Military Academy, Woolwich, in June last. We have just learnt that the company of gentlemen cadets, who are educated in that institution, have done great honour both to themselves and the Doctor, by presenting to him an ornamental piece of plate, value 150 guineas, as a testimonial of their affection and respect. It is an elegant silver vase, with a suitable inscription, and appropriate emblems, including the arms of the Cadet Company as well as the Doctor's arms; the vase, which is 13 inches in diameter, is supported by a cubic pedestal of the most tasteful work. The entire design, which is of a beautiful classical nature, is due, we understand, to one of the senior cadets. The circumstance of this presentation must be the more gratifying to Dr. Gregory on retiring from his long career of official duty; as, during the whole history of the Royal Military Academy, extending over a period of nearly a hundred years, this is the only testimonial of the kind which has ever been presented to a retiring professor.

The "Liverpool" Steam Ship.—After all that has passed in reference to the steam-ship *Liverpool*, you, and all your readers who are interested in the issue of the steam trans-Atlantic enterprise, will be gratified to hear of her arrival in this port this morning, at nine o'clock, having made her passage from Cork in a little less than seventeen days. Of course, all the world knows that the passage *westward* (not eastward) is at all seasons " the *rub*." The average passage of the packets *out* is about thirty-three days, to twenty-two days *in*; and even the fare is proportionally increased. In writing to you from Cork, I gave you what is now confirmed as a correct account of our mishaps. Be it understood, then, that after a comparatively trifling re-alteration in the engines, at Cork, chiefly consisting in a diminution of the draft of the chimnies, as I understand it, the consumption of fuel has been reduced from 40 tons daily, to 26 tons and 17 cwt., so that we had remaining this morning fuel probably for three days or more, although the 150 tons of cargo remained also in the boat. We made nearly 190 miles average progress daily, or between 7¾ and 8 miles the hour. The boat, I must repeat, is no fair specimen of what a candidate for this navigation should be. The company have expended about 75,000*l.* on her, I hear, including 1,100*l.* in fitting up the after-cabin—but she was never intended, and is not fitted for the route, being too long, and wanting beam; strong and staunch enough, but very wet—we had not a dry deck for a single day. Our arrival here is the great event of the times. Much anxiety has been felt for us, as we apprehended. A government brig was even sent out to cruize for us two days since. On the whole, the result of this voyage has done not a little to *strengthen* the Atlantic enterprise in the public mind. It cannot be supposed that an effort to cross the ocean will ever be made under circumstances more conclusive than ours.—New York, Nov. 23, 1838. *Correspondent of the Athenæum.*

The Cottage Fire.—A plan of warming a house, from the back of a kitchen-grate, has been adopted by Sir Charles Menteath. A cast-iron back, an inch thick, is fixed to the grate, and another plate of sheet-iron, placed at a distance of one or two inches from the cast-iron back, shows a species of stove, which serves to warm the under-ground story of a house: and, by means of a circulation of air passing between the two iron plates, a current of warm air, by means of a pipe from the hot chamber between the iron plates, is carried to the next floor above. The air is heated to 190 degrees, by this simple economical method. The wall is hollowed out to the passage or room behind the kitchen-grate. The placing a thin plate of sheet-iron behind the fire of a cottage-grate, adds much to the comfort of the inhabitant. All cottages should consist of two rooms, with a wall in which the grate of the cottager is placed; so that the back of his grate warms the room behind, and dries his clothes.—*Nottingham Mercury.*

Progress of Gas-lighting.—The town of Kirkwall, in Orkney, is now lighted up with gas, thus carrying the comfort and brilliancy of this beautiful invention to the *Ultima Thule* of her Majesty's British dominions.

The Gaudin Light.—The splendid improvement in the Drummond light, which was recently exhibited to the Parisian Academy of Sciences, by M. Gaudin, and was received with much rapture by that illustrious body, is, it seems, claimed by two countrymen of our own. Mr. Keene has addressed to the Academy of Sciences a " reclamation" of the priority of invention, in favour of himself and Mr. Gurney.

Time-pieces without apparent mechanism.—Sir.—I am obliged to your correspondent " Omega," for having, in your minor correspondence of last week, pointed out a circumstance that had certainly escaped my notice. The main object of my communication, in your 797th number, respecting Savory's clock, was to account for the motion of the *hour-hand*; the paragraph in which the possibility of a *minute-hand* also was alluded to, was rather a crude, half-formed hint, than a specific description. But "Omega," in his correction of *my* oversight, has committed one himself, fully as great. The minute-hand being at the back of the dial, must have an apparently reverse or retrograde motion, and would require the figures on the minute-circle to be placed the contrary way to those on the hour-circle; the effect, therefore, of the circumstance pointed out by him, would be, that the minute-hand would *lose*, instead of gain, in respect to the hour-hand; so that the proportion of the motion in the two hands should be as 1 to 13, and not as 1 to 11. I must, however, admit, in candour to those who preceded me in an attempt to guess at the construction of that which, for want of a better name, is called Savory's clock, that I may have been under a mistake with respect to its size; having supposed the latter to be equal to that of the dial on the Polytechnic Institution, no dimensions having been given in the description in the 795th number of the *Mechanics' Magazine.* I have since seen one which was imported from France (where they seem to be common enough) in which the diameter of the dial did not exceed four inches, nor was the hand larger than that in a pocket chronometer. I, of course, admit, that to a dial of that size my explanation is quite inapplicable. I have the honour to be, &c. Dec. 10, 1838. NAUTILUS.

Preservation of Stone, &c.—Sir,—" An observer," who, in your last number, writes on the " improvement of buildings," if he will turn to your volume for 1825, will find several papers of mine on the subjects he now alludes to. Appended to my " Hints to paviors," he will see my suggestions and experiments on the application of coal tar to the formation of paths and roads; on the great hardening effects of oleaginous and of animal gelatinous fluids conjoined with earths and stones, whether applied to soft stone or to cement-covered buildings, or to the formation of artificial stones by mechanical compression. I trust that " An Observer" will also honour me and benefit the public, by supporting me in my exhortations of the same date, to increase the daylight and remove the sombreness of our London streets, by annually white-washing the houses. In 1825, I published these papers in the shape of a pamphlet, and in 1832 Effingham Wilson gave out a second edition, in which is included a paper on the preservation of iron structures, and on oxidation. I have the honour to be, &c.,
F. MACERONI.

Metropolitan Railway Map.—Now published, vol. xxix. of the *Mechanics' Magazine*, price 8s. 6d., illustrated with a *Railway Map* of the Metropolis, taking in a distance of 30 miles from the Post-office. The limits of the two-penny and three-penny post deliveries are also shown in the Map. The *Metropolitan Railway Map* alone, stitched in a wrapper, price 6d., and on fine paper, coloured, 1s. *The Supplement* to vol. xxix, price 6d. *The Railway Map* of England and Wales continues on sale, in a neat wrapper, price 6d.; and on fine paper, coloured, price 1s.

☞ *British and Foreign Patents taken out with economy and despatch; Specifications, Disclaimers, and Amendments, prepared or revised: Caveats entered; and generally every Branch of Patent Business promptly transacted. A complete list of patents from the earliest period (15 Car. II. 1675,) to the present time may be examined. Fee 2s. 6d.; Clients, gratis.*

LONDON: Printed and Published for the Proprietor, by W. A. Robertson, at the Mechanics' Magazine Office, No. 6, Peterborough-court, between 135 and 136, Fleet-street.—Sold by A. & W. Galignani, Rue Vivienne, Paris.

Mechanics' Magazine,

MUSEUM, REGISTER, JOURNAL, AND GAZETTE.

No. 803.]   SATURDAY, DECEMBER 29, 1838.   [Price 3_d._

Printed and Published for the Proprietor, by W. A. Robertson, No. 6, Peterborough-court, Fleet-street.

ARMSTRONG'S IMPROVED WATER-WHEEL.

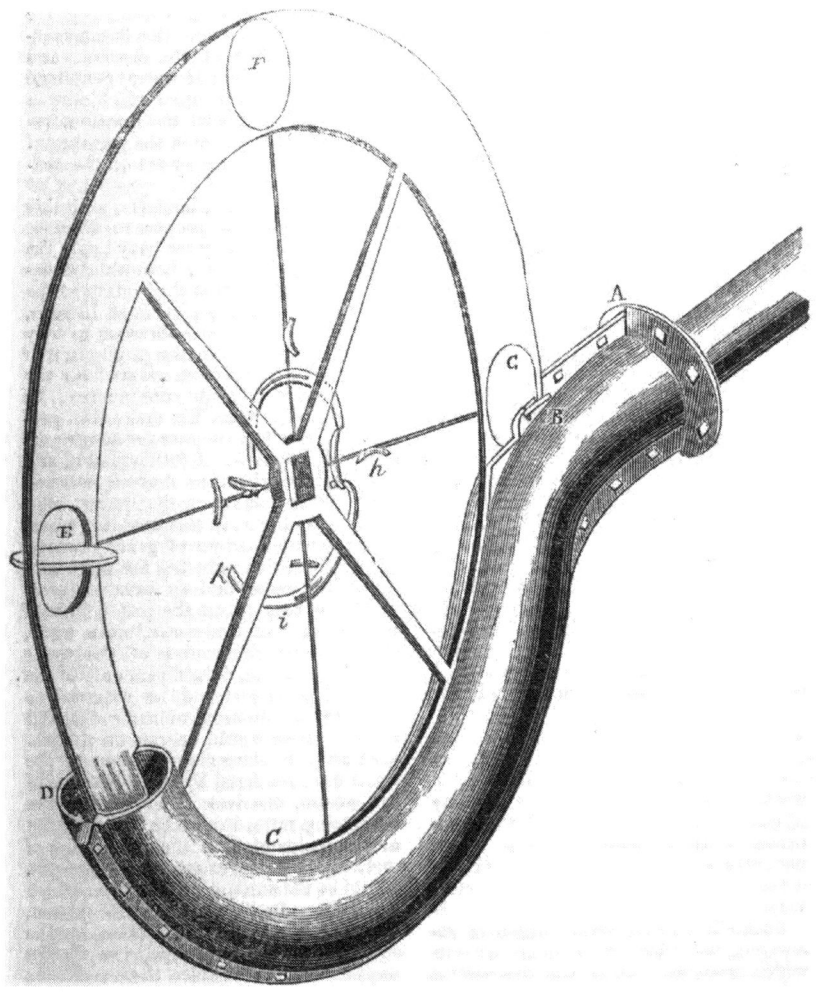

ON HYDRAULIC POWER—NEW HY-
DRAULIC MACHINE.

Sir,—The overshot water-wheel is the only machine in general use, upon which streams of water operate by their gravity, in the course of their descent; and it is only the power exerted by a stream during its descent from a very limited elevation, that can possibly be made available through the medium of this machine; because, in the first place, an overshot water-wheel requires a perpendicular fall, which can seldom be obtained from a very considerable altitude, by artificial means; and, in the second place, because in the few instances in which a very elevated fall is attainable, it is found practically impossible to construct a wheel of a corresponding diameter.

The overshot water-wheel is therefore a very inefficient machine when applied to streams of rapid descent, and the consequence is, not only that such streams are very imperfectly employed, in cases where a water-wheel is used, but wherever the quantity of water is insufficient to turn a wheel with effect, the stream lies wholly neglected, although in a theoretical point of view, its great descent may be much more than equivalent to the deficiency in the quantity of water.

Hilly and undulating districts abound with little rivulets, which flow down the sides of declivities from elevations of from one to three hundred feet, and which, on account of their diminutive size, are at present entirely disregarded. Such streams as these, however, would become most efficient sources of mechanical power, if they were made to operate by their gravity throughout the whole, or the greater part of their descent, instead of the very limited portion of it which a water-wheel is capable of employing. Streams of this description are to be met with on the banks of many of our navigable rivers, and in a multitude of other situations well suited for mills and manufactories, where hydraulic power would prove of the utmost value.

With the view, therefore, both of increasing the efficiency of many streams which are already in actual use, and of bringing into valuable employment an immense number of others, of which at present no use can be made, it is of the utmost importance that some method should be devised, of concentrating the power developed by rapid streams during their descent from very elevated situations, and of applying such power to the purpose of propelling machinery.

The first step towards accomplishing this object, must be to get quit of the necessity of having a perpendicular fall, and the only conceivable way in which this can be done, is by conducting the stream in a pipe from the commencement to the foot of the descent, and bringing the column of water contained in the pipe, to bear upon machinery at the bottom. But still the question remains, *in what manner* is the pressure of the water to be made to act on the machinery?

Before proceeding, however, with this consideration, it is proper to observe, that in whatever way we may apply the pressure of the water, it would be extremely important that the motion of the water through the pipe should be *slow*, otherwise much of the force of gravity would be expended in the production of motion, and the power exerted on the machinery, would, in consequence, be greatly diminished. To make this perfectly intelligible, suppose the lower part of the pipe to be fitted with a piston, and conceive the piston to move forward in the pipe as rapidly as the water could possibly follow it; in this case it is plain, that the whole action of gravity would be expended in producing the mere motion of the water, and no force whatever would be exerted on the piston; but if the velocity of the piston were such, as to *retard* the course of the water through the pipe, then a part only of the force of gravity would be required to generate the motion of the water, and the remainder would operate on the piston; and the slower the motion of the water were rendered by the resistance of the piston, the less in a very rapidly decreasing ratio, would be the quantity of force expended in the production of motion, and the greater, in consequence, would be the amount of the power which would remain to act upon the piston. But the less rapid the motion of the water through the pipe, the larger would the pipe require to be made, in order to carry off the same quantity of water. Hence, therefore, the pipe for the conveyance of the stream would have to be very capacious, as compared with what would be requisite, merely to draw off the whole of the water, if its escape from the lower extremity of the pipe were perfectly free.

It is scarcely necessary to remark, that the diminution of effect, consequent on the expenditure of power in producing the motion of the water, would not be peculiar to a machine propelled by a pressure to be thus obtained by means of a pipe, for it is very well known, that in the case of an overshot wheel, considerable power is lost in generating the motion of the water, during its descent in the buckets of the wheel.

And now, with respect to the manner of applying the pressure of the water to the machinery at the foot of the pipe: the mode which most naturally suggests itself is by means of a cylinder and piston, the same as in the steam-engine, and this method I understand has been recently put in practice at some lead mines in the north of England, but with what success I am not particularly informed. It is easy, however, to perceive that this system of employing the pressure must be subject to very serious objections, for unless the passages for the admission and escape of the water into and out of the cylinder, were many times larger than the corresponding passages in the steam-engine, either the motion of the piston would be excessively slow, or the water would acquire such an accelerated velocity on entering and escaping from the cylinder, as would neutralize a very great part of the pressure; and if, on the other hand, the passages were made sufficiently spacious to obviate these disadvantages, the valves would have to be enlarged to such an extent, as would render the difficulty of working them very considerable. Another objection to this mode of employing the pressure is, that the column of water contained in the pipe would be brought to a state of rest at the termination of every stroke of the piston, and considerable power would be expended in renewing the motion of the water at the commencement of each succeeding stroke.

We ought, therefore, to endeavour to effect the object in view, by some contrivance which would be free from contracted passages, and in which valves might be dispensed with, for without contracted passages, valves could not be employed without inducing great loss of

effect from friction. It is also of importance, that the plan we adopt should admit of the motion of the water through the pipe, being continuous and uniform, and not intermitting or varying.

All these advantages would be possessed by the machine, which is represented in the engraving (see front page), and of which the following is a description.

The pipe for the conveyance of a stream from some considerable elevation, communicates with the upper end of the tube A B C D, which is of somewhat larger diameter than the pipe, and of a shape which will be best understood by reference to the drawing. In the upper side of this tube there is a slit or opening, extending from the point B to the mouth of the tube, by means of which slit, the rim of the wheel, E F G, is admitted into the interior of the tube. The breadth of the rim being somewhat greater than the diameter of the tube, the inner edge of the rim, or that which is nearest the centre of the wheel, remains outside of the slit, while the outer edge of the rim extends to the opposite side of the interior of the tube, and there falls into a groove adapted to receive it. The thickness of the rim is exactly equal to the width of the slit, so that that slit is accurately filled up by the insertion of the rim. Under these circumstances it is evident that the wheel would be capable of revolving with part of its rim constantly immersed in the tube, while no opening would be left through which the water could escape except the mouth of the tube.

The rim of the wheel contains four equidistant circular apertures, three only of which are seen in the drawing, the remaining one being situated in that part of the rim which is concealed in the tube. The diameter of each of these apertures is exactly equal to the interior diameter of the tube, and a circular plate or disk, of the same thickness as the rim, turns upon an axle in each aperture, so as to be capable of assuming a position either at right angles to the rim, as represented at E, or in the same plane with it, as represented at F and G, in which latter position the disks would accurately close up the apertures, and form, as it were, part of the rim.

Now suppose the disk belonging to

the aperture which is concealed in the tube, to be at right angles to the rim, the same as the disk which is represented at E; in this position the disk in question would intercept the passage through the tube, and form a piston, which would sustain upon its upper surface the whole pressure of the superincumbent water, and the consequence would be, that the disk would be pushed forward in the tube, and the wheel would be put in motion. The pressure, however, upon the disk would of course only continue until it reached the mouth of the tube; but in the mean time the disk which occupies the aperture at G, would have entered the tube, and if as soon as it were admitted, it were caused to turn upon its axis, so as likewise to assume a position at right angles to the rim, it also would constitute a piston, upon which the water would begin to operate, the instant it ceased to act upon the other. All then that would be requisite to maintain the constant rotation of the wheel, would be, to cause the disks, or pistons, as I shall now call them, to shut up into the apertures after quitting the tube, and again to open out and assume the requisite position as soon as they are admitted within it.

Many ways may be conceived of changing the positions of the pistons in the order required, but the method which is represented in the drawing, and which I shall now proceed to explain, would probably be as simple and efficacious as any other.

The axles upon which the pistons are fixed are prolonged to the centre of the wheel, and upon each prolonged axle two cross-levers are fixed, one at right angles to the piston attached to the axle, and the other in the same plane with it. The levers and the pistons being thus fixed upon axles common to both, the pistons would necessarily obey any change of position to which the levers might be subject. In order, therefore, to reverse the position of each piston when it has entered the tube, we have only to reverse the direction of the levers with which the piston is connected, and this would be effected by the following contrivance.

h, i, k, is a slide which may be fixed to the frame in which the wheel revolves, or to any other stationary object, and which slide is so adjusted that when the wheel is in motion, the levers which are at right angles to the pistons may strike in rotation with one end against the head of the slide, as they arrive at the point at which the pistons are to be turned. By this arrangement the position of each lever would be changed from a direction perpendicular to the slide, into a direction parallel with the slide, in which latter direction the lever would be retained by the subsequent action of the slide, until the piston in connection with it reached the mouth of the tube. Thus each piston would be turned into the requisite position, exactly at the moment required, and in that position would be maintained so long as it remained in the tube.

A similar slide would have to be fixed on the opposite side of the wheel, in the situation indicated by dotted lines, so as to operate on the other set of levers, and by a similar process, to shut back the pistons into the apertures, preparatory to again entering the tube.

In order to render the pistons capable of turning round in the tube, it would be necessary to make their edges of a spherical and not of a cylindrical form, and the sides of the apertures would require to be similarly shaped to make them correspond with the edges of the pistons. It will be perceived, however, that in consequence of so shaping the edges of the pistons, it would only be a single line round each piston, that could be in contact with the sides of the tube, on which account the pistons could not be made to fit *perfectly* water-tight. If the machine were propelled by the pressure of steam, the leakage which would be thus occasioned would constitute a serious defect, but in a machine propelled by the pressure of water leakage would be, comparatively, of very slight importance, because the quantity of water which would escape through a crevice, under a given pressure, would be extremely insignificant, as compared with the quantity of steam which would escape under similar circumstances.

The mouth of the tube should be turned upwards, sufficiently to render every part of the opening higher than that part of the tube which is situated at C, in order that by the action of the air, the tube might be kept constantly full, below as well as above the piston, which for the time being would be in operation. The advantage of this would be, that on the principle of suction, the water contained in the tube below the piston would operate on the machine with the same effect as if the weight of the water so situated in the tube were added to the pressure on the upper side of the piston.

The width of the tube would of course be regulated by the magnitude of the stream, and the intended velocity of the machine.

With respect to the size of the wheel, it would not require to be very large, probably a diameter of six or eight feet would be as eligible as any other.

The interior of the tube should be lined with brass or copper, and the pistons and rim of the wheel should be cased with a similar material to prevent corrosion.

The friction attending the operation of this machine would be much less than at first sight appears, for it will be observed that the friction of the whole apparatus would be merely that of *juxta-position*, and not of surfaces pressed forcibly against each other.

In the over-shot water-wheel much loss of effect is occasioned in consequence of the impossibility of retaining the water in the buckets, until it arrives at the lowest point in the revolution of the wheel; but from this, or any analogous defect, the machine I have described would be entirely free; and on the whole I think we may fairly conclude, that the relative loss of effect attending the operation of the machine in question, would not exceed that which is experienced in the case of an overshot wheel. It follows, therefore, that by means of the proposed machine we should be enabled to increase the efficiency of a stream in the same proportion that we should increase its available descent.

On an average the fall of an overshot wheel certainly does not exceed 25 feet in height, while by the machine I have suggested, a rivulet might frequently be made to operate from *eight or ten* times that elevation. In such a case, therefore, we should derive from the stream *eight or ten times* the power which under ordinary circumstances would be attainable by means of an overshot wheel, or, what is much the same thing, we should obtain a given effect, from one-eighth or one-tenth of the quantity of water which is at present generally requisite to produce an equal effect.

W. G. ARMSTRONG.

Newcastle-upon-Tyne, Nov. 24, 1838.

STEAM NAVIGATION OF THE UNITED STATES.—CONSTRUCTION OF STEAM VESSELS.

Sir,—Mr. Stevenson's sketch of the *Civil Engineering of North America*, has recently come under my notice, and has much interested me, particularly that part of it which relates to the steam navigation of the United States, having long felt with him, that it is strange we should hitherto " have received so little information regarding it, especially as there is no class of works in that comparatively new and still rising country, which bears stronger marks of long-continued exertion, successfully directed to the perfection of its object, than are presented by many of the steam-boats which now navigate its rivers, bays and lakes."

In 1824, the French Government sent M. Marestier to examine and collect information relative to the steam navigation of the United States. The report which he made on his return was soon afterwards published in Paris. It was noticed at some length in " Papers on Naval Architecture," edited by Messrs. Morgan and Creuze, and in your valuable miscellany, for 1832 or 1833, a short abstract of its contents appeared from the pen of an individual, who at that time contemplated publishing an English translation of it, which he had just completed; circumstances prevented this design from being carried into effect. M. Marestier executed his task with great ability, and his memoir contains a vast amount of practical information, and evinces great tact in presenting the facts to the mind in such points of view as to indicate the general principles upon which the American steam-vessels were constructed. It is to be regretted that the attention of persons conversant with the science and practice of naval architecture in that country has not been more closely directed to the subject of steam navigation, with a view to its general improvement and the elucidation of correct principles of construction.

Our steam-vessels of the present day are undoubtedly superior to those of 1824, but the advance is not so great as might have been reasonably expected in fourteen years. Few of our builders have seriously directed their attention to the subject, and those who have done so have been regarded rather as schemers

almost deserving our pity, than encouraged to persevere in their laudable
efforts to improve, what must at no very
distant period become an important element in our naval greatness. The want
of some medium of communication expressly devoted to this object, may perhaps be one cause why the progress in
improvement has not been more rapid,
but it is to be hoped that Mr. Stevenson's remarks, and the recent republication of "Tredgold on the Steam Engine," with a valuable appendix on
steam navigation, will prove highly beneficial and productive of great improvement in this art. Mr. Stevenson, whilst
he admits the important improvements
which have been made by the American
builders in the construction and action of
their vessels says, (page 119) that "on minutely examining the most approved American steamers, I found it impossible to
trace any *general* principles which seem to
have served as guides for their construction. Every American steam-boat
builder holds opinions of his own, which
are generally founded, not on theoretical
principles, but on deductions drawn from
a close examination of the practical effects of the different arrangements and
proportions adopted in the construction
of different steam-boats, and these opinions never fail to influence in a greater
or less degree the built (build) of his
vessel, and the proportions which her
several parts are made to each other."

Now this course of proceeding is precisely that which is the most likely to
lead to the greatest degree of improvement, and we are inclined to the belief,
from an examination of the drawings of
the different American steam-boats which
have come under our notice, that some
general principles will be found to pervade all fast going steam-vessels, although the builders may not have been
able to communicate the theory of their
construction. The greatest improvements in machinery have been the result of the system of "*trial and error*"
adopted by all practical men, and to this
system it is suspected we shall be indebted for further improvements in the
construction of steam-vessels.

The plans of the two steam-vessels
given in Mr. Stevenson's work are constructed upon the same principles below
the water as to velocity, above the water
they only differ so as to render them

better suitable to their intended purpose.
They may possibly have been furnished
to the author as new designed, but they
are not really so, for the principle upon
which they are constructed, has been
long known and applied by the English
and Scotch steam-boat builders with
success.

The shallowness of the American steam-
boats, their great length in proportion
to their breadth, together with the small
draft of water, are precisely the elements
required for swift propulsion. To these
vessels they have applied machinery of
immense power, moving at a much
greater velocity than is customary in
European marine engines. Thus the
velocity of the piston in the *Rochester* is
equal to 540 feet per minute, whilst in
the English marine engines it rarely
reaches 210 feet per minute. When the
Rochester piston is working at the rate
of 540 feet per minute, and the circumference of the wheel moving at the rate
of 23,13 miles per hour—if the velocity
of centre of effort of the paddles be calculated, and compared with that of the
boat, it will be seen that whether the
boat be going at the rate of 16,55 miles
per hour, and making 27 strokes per
minute, the piston travelling through
540 feet in the same time—or whether
she be going at the rate of 14,97 miles
per hour, making 25 strokes per minute,
and the piston travelling through 500
feet in the same time, that the difference
between velocity of the wheel and the
boat will agree within ,13 of a mile per
hour, which might be occasioned by circumstances not noted by Mr. Stevenson.

At least this practical conclusion may
be arrived at, that the amount to be deducted from the velocity of the wheel is
constant, and does not materially vary
with different velocities of the same
boat in ordinary circumstances.

In the valuable appendix to "Tredgold's Steam Engine,' are some calculations, by which it is attempted to be
shown, that the velocity of the boat is
to that of the wheel, as 2 to 3. This
appears too much as applied to the
Rochester, there the ratio stands thus :—
when going at a velocity of 16,55 miles
per hour, the circumference of the wheel
travels at the rate of 23,13 miles—difference 6,58 miles. When going at a
velocity of 14,97 miles—the circumference of the wheel travels at the rate of

21,42 miles—difference 6,45 miles. If the centre of the paddle be assumed as the centre of effort, which I am aware is not quite in accordance with theory, but, I think is nearly so with fact—the differences are respectively 4,17 miles, and 4,21 miles. And if applied to the *Narragausett*, another steam-vessel, described by Mr. Stevenson, it would give her a velocity of 15,375 miles per hour under the same circumstances that the *Rochester* performed 16,55 miles. Whether this is consistent with the fact, we have no means of ascertaining, Mr. S. not having given her speed, only remarking that she performs her voyages with certainty and speed. The greatest speed of the *Rochester, with* the tide, is described to have been at the rate of 16,55 miles—and her greatest velocity *against* the flood tide, 14,22 miles; which, supposing the velocity of the tide to be constant, would give 15,37 as the velocity of the vessel through the fluid; but Mr. Stevenson states, that for the first 27 miles, her speed was only 12,36 miles, although the current was in her favour. This diminished speed is attributed to the shoalness of the water and the narrowness of the channel, which would certainly occasion a great loss of speed.

The shallowness of her draft of water (4 feet) and the immense power of her engines, as shown by Mr. Stevenson's calculation (772,3 horses), ought to have produced a greater speed if her form had been the best adapted to the purpose. I am inclined to the belief, that if the weight on board our river boats were reduced, so as to bring them to the same draft of water in proportion to their power and arrangements made for driving the wheels with the requisite velocity, that they would equal, if not beat the *Rochester* in speed. The English boats of the same dimensions as to length and breadth, are from one-half to two-thirds deeper in the hold than the American boats—this, as every naval architect is aware, is against them in point of speed, although it renders them better sea-boats. Mr. Stevenson has fallen into a very pardonable error, page 149, in which he speaks of the diverging waves, which *invariably* follow the steamers in this country, and break on the banks of our rivers with considerable violence—not invariably, as I can state from my own daily experience—for it is

here, as in America, that the swiftest boats produce the least disturbance of the fluid, and our own fast boats leave scarce a trace behind; whilst the slow cargo vessels raise such a terrific swell as to affright the luckless cockney from his propriety who happens to be caught by it in a Thames wherry. Apologizing for trespassing so much on your pages,

I am, Sir, yours truly,
GEORGE BAYLEY.
Nov. 26, 1838.

NICKELS AND COLLINS' CEMENT BOOK-BINDING.

The superiority of Mr. Wm. Hancock's patent method of binding books with India rubber, has given rise to various endeavours to effect the same purpose by some other cement. Mr. Nickels, we believe, laid claim to the invention of binding books with India rubber in a specification of a patent for "certain improvements in the manufacture of caoutchouc"—but upon an action to repeal this patent by *scire facias*, upon the ground of the specification claiming more than was set forth in the title, to save the remainder of his patent Mr. Nickels disclaimed, amongst other things, the method of book-binding. He has since, in conjunction with a Mr. Collins, taken out a patent for attaching the leaves of books together with a kind of cement; but it must be evident that the very properties which rendered the natural gum so applicable, are wanting in this artificial substitute, time, however, will be the best test of both inventions. We have seen some very beautiful specimens of the caoutchouc book-binding at Mr. Jennings's, in the Poultry—all Mr. Heath's superb annuals have been bound by Mr. Hancock—and they are most favourable specimens of the invention.

Messrs. Nickels and Collins state in their specification that the flexible or elastic cement, which they propose for the purpose of combining the leaves of books together is cheaper, more expeditious and better than India rubber. They then proceed to state the composition of the cement and the method of using it.

They dissolve a pound weight of isinglass, or of the best glue, in three quarts of hot water, and incorporate about a quarter of an ounce of linseed-oil with a quarter of a pound of dry coarse sugar, and when the sugar has taken up all the oil, it is added by degrees to the dissolved isinglass or glue, stirring it until well mixed; the whole is then to be boiled together, until it is of that consistence that it may be laid on when hot or in a fluid state with a brush. The book is then to be rounded at the back, either in sheets or in single leaves, and put into a press, leaving the back protruding, and a coat of this cement is to be laid on hot, or rather warm, upon the back, and well rubbed in, that the back-edges may be well saturated therewith. A piece of calico, or any other texture or fabric is to have a coat of the same cement, and to be pressed over it, to confine all the leaves together when dry, which in a warm room will be in a short time; the book is then ready to be boarded and finished off in the usual way. The above is the cement which Messrs. Nickels and Collins state they believe to be the best for the purpose, but variations may be made, provided gelatine is a constituent part, either incorporated with albumen or the mucilage of vegetables.

SIR JAMES ANDERSON'S STEAM-BOILER.

Sir,—Having seen in a late number of your Magazine, a letter from me to one of my friends on the boiler of Sir J. Anderson, I take the liberty of sending you another on the same subject direct from myself; and which I am induced to do, owing to the following very extraordinary assertions published in an Irish newspaper:

" In a letter from Sir J. Anderson to an Irish paper, he gives the following account of the wonderful power possessed by his steam-carriages, which are intended to run on common roads. Subsequent trials of the carriage in question have given the following results :—*One hundred weight of coke per hour produces 7,500 gallons of steam per minute*; driving the engine at a speed of twelve to fifteen miles per hour, *at a pressure of* 50lbs. *to the inch*; and giving about 4,000 gallons of steam per minute beyond the required consumption : in other words, this immense power is obtained *at a st of* 1d. *per mile.*"

Now, Mr. Editor, I shall endeavour to show, by the assistance of a little plain arithmetic, how far the above astounding declarations coincide with real practice.

First, he says "7500 gallons of steam per minute is produced, of the elastic force of 50lbs. to the inch." Now, then, I will try to show how much water is required to produce that steam. It is well known that water in volume being 1. steam at the boiling point will be 1711: but steam at the density of 50lb. per inch will be 542; because that the greater the heat applied *under pressure*, the more *water will* be *held* in *suspension* by the *caloric*; as I before observed, in my letter concerning the bursting of the Victoria boilers, in your No. 776; and which your correspondent N. S. construed erroneously to mean an "*assigned vacuum*," in No. 780. Therefore 7500 gals. ÷ 542 = 13·8 gallons of water per minute, which must be boiled away under the pressure assigned of 50lbs. to the inch, to produce 7,500 gallons of steam *of that density and elastic force*.

Again, 13·8 gals. water per minute is = 1,380lbs., and is per hour = 8280lbs., or 3 ton. 13 cwt. 3 qr. 20 lb. Now we are told this is effected with the small expenditure of only "112 lbs. of coke per hour." The usual allowance of coke for boilers seated in brickwork is 4·97, say 5lb. per cubic foot of water at common low pressure; but where the pressure is increased, as in the present case, to 50lb. per inch, 7·2lb. of coke is necessary; and if the combustion is urged by a fan or blower, or by the waste steam, as in locomotives on common roads, then 10lb. of coke will not be to much. I shall, however, endeavour to keep as low as possible, and assign 7·2lb. per foot, we then have as before water per hour 8,280lbs, ÷ 62·5lbs. = 132·4 cubic feet, and which × 7·2 lbs. coke = 953lbs. coke required per hour instead of 112lbs.; which is *just seven-and-a-half times more than Sir J. Anderson says his boiler requires*; and instead of 1d. per mile comes to 8½d. per mile. Now if he can do this, why his boiler is a miracle! but as I am no believer in miracles, and am like Thomas the Apostle a little dubious, I shall not believe without seeing and examination, for the following additional reasons, which I will put in juxta-position, that they may appear more clearly to your readers :—

A double-acting condensing, or low pressure Steam-engine, working expansively, with a boiler of 180 horse-power, requires per hour,	A high-pressure engine boiler to produce steam at 50lb. per inch, and 7,500 gallons per minute, requires per hour,	Sir Jas. Anderson's *Ne plus ultra* boiler is said to produce 7,500 gals. steam per min. of 50lb. pressure, and requires,	A high-pressure boiler of 15 horse-power, (not locomotive, but fixed), requires, per hour, to produce 826,33 gals. steam of 50lb. per in.,
Water, in cubic-feet, 133	Water, in cubic-feet, 132·4	Water, in cubic-feet,	Water, in cubic-feet, 15
Coke, in lbs., 960	Coke, in lbs., 953·0	Coke, in lbs., 112	Coke, in lbs., 112

Mind, I do not say, Mr. Editor, that this Baronet's statement is untrue; but this I do say, that if he can do what he states, many of my compeers in this nebulous island would be very glad to see it as well as—your obliged servant,

SCRUTATOR.

Battersea, Dec. 24, 1838.

COMPETITION BETWEEN THE BOOT AND SHOE TRADE OF BRITAIN AND FRANCE.—THE FRENCH SYSTEM OF BLOCKING.

The preference which is generally given to French boots and shoes over English, is generally set down by would-be patriots, as the result of a morbid prejudice in favour of foreign manufactures, and the superiority of the article is denied by nine out of ten of "the trade." In a pamphlet* which we have received, which is remarkable for the liberal views the writer adopts upon the subject, as well as for the clear, though somewhat redundant style in which these views are expressed, the true and rational cause of this preference is candidly stated; the British manufacture was at one time the best, but the French workmen, in emulating to equal, have excelled ours, and consequently the products of their labour are preferred by the consumer.

"Immediately following the settlement of the last continental war, and for many years after, we neither felt any opposition, nor feared it; and yet the charges made at that period in our trade upon the consumer, were much heavier than those latterly, or at present made. The English boot and shoe was then, generally speaking, the first article of its description in the world, and so there was nothing to apprehend, while the master kept up his prices, and the journeyman his wages. In the progress of time, however, and that, too, of no very extended length, ths healthy condition of things began to alter; over-trading commenced, panics were felt, public credit seemed to reel, and having an abundance of hands ever ready for the market of labour, and nearly as abundant a number of shop-endeavourers, anxious to enrich themselves through the profits, or by the depression of that labour, our competition became more reckless as to character, each doing what best he could for his own interests; while in the mean, the employer and the workman of France saw their advantage, imitating our workmanship, and excelling where they could, till at length by their exertions, their attentions, and their ingenuity, and helped as they were, and still are, in particular by the richer and softer qualities of their leather, and in another way, by the superior fabric of their satin, their merits forced themselves into notice, were favoured, and are now, day-after-day, still in the way of being more highly favoured than ever.

' This new view of the case then, will lead us once more to the still great fact of our having in France in the boot and shoe trade, as other manufactures have in many other instances, an efficient, a persevering, and an intelligent rival; the award therefore, rests with the money spender, be he British, French, or of any other country, as to what article of his consumption he may please to give the preference. His pounds, shillings, and pence, are his own, and his likings are generally seen to accommodate themselves to his benefits. For this reason it is, that the English consumer cares no more for an English *as English*, than for a French boot or shoe; he looks upon the matter in option, as in no ways a test of his patriotism, and makes his selection accordingly, purchasing what best satisfies, or may best serve him, and taking no other heed."

The superiority of the French boots and shoes, Mr. Devlin states, arises as well from the superiority of the leather, as of the workmanship. With regard

* The boot and shoe trade of France, as it affects the interests of the British manufacturer in the same business with instructions towards the French system of blocking. By James Devlin. London: Steill. pp. 47.

to quality of leather we shall have occasion perhaps to say a few words hereafter; suffice it to say at present, that from information with which we have been supplied, we have no doubt but that such an article will shortly be put into the hands of our workmen as will give them an advantage over their Gallic neighbours upon this score, and which will render unnecessary the importation of foreign leather recommended by Mr. Devlin in a subsequent part of his pamphlet, and which we quote; and it will then be our workmen's own fault if they allow their foreign competitors to retain their superiority upon the point of workmanship. We shall do our part in the matter by laying before the English bootmaker Mr. Devlin's observations and instructions upon the French system of blocking. In quoting so largely from Mr. D.'s pamplet, we think we shall be excused by him, the more so, as he declares in his preface that it is written "with a thorough conviction of its utility, and with the desire and aim to assist the well-being of the trade."

"A fact to be closely attended to, is the one, in the boot department, of our very inferior manner of blocking, or turning the front piece of our common Wellington boot; in this we are far behind our neighbours. Take up one of our boot-fronts so prepared, and compare it with a front coming from France, and the difference is as perceptible as lamentable. How stiff, how dead, and how forced is the one; and how easy, moist, and elastic the other. The first, to one unskilled in the operation, might seem to be baked, rather than gently moulded when wet into the position it has received; and then catch it by top and toe, and pull it ever so tenderly back, and, lo! at once its crabbed beauty is gone! and though you may press, shove, or contract it again into something of its original form, still can it never be made to look the same thing as before. Now, do the like to the French front; nay, more, you need not pull it "tenderly," but at your might—apply your strength to the two extremes—force it, as it were, straight; and then, letting it go again, lay it on your board, and by a little application of the hand it will nearly look as well as ever—no puckerings, no looseness, and still possessing the requisite curve. * * * * * *

"After the blocking comes the cutting, and then the closing; but in neither of these have we ought to dread. In our closing, particularly, we stand secure; and, it were a shame did we not—seeing that with us, for

these 30 or 40 years back, we have made of it a separate qualification, taking apprentices to it, and making it our ambition to perfect it in the highest degree.

"The make of our lasts comes next to be considered, and here I must say, that I think Paris has in this matter some degree of superiority over London, that is, generally speaking. The entire bearing of the lasts of the Parisian, from heel to toe, have a straighter cast than those of London; are not deficient in good seats, and being longer in the fore-part, from the fashion of not fitting-up so short as we do, they consequently throw off a boot which has not the snubby or stunted character of the English boot, but stretching forward more taper and slender-like, give the foot altogether a handsomer appearance.

"Lastly, we come to the making;—and here, as above said, if our closing be superior, yet, nevertheless, I know not, as it is followed by us, but it is the cause of one inferiority. The French maker being also his own closer, his opportunities to fit and properly mould his own counters, are better than what the English bootman can possess. In the matter of these he is always careful; he keeps everything—bottom of back, counter, and stiffener—smooth, tight, and hard; working, when he puts his boot on the last, the loose paste out of every corner and crevice, and then afterwards, by the tapping of his hammer, and much close and diligent rubbing, he finishes this part of his labour in a manner unusual in England, where such care in this particular, is seldom or ever employed. But in this we are wrong: a tight, firm, and hard counter, being of much more sterling advantage to a boot, than all our gloss of rand, and prettiness and beauty of bottom.

"The French maker, too, in the entire lasting of his boot, neglects no pains—shows himself always sedulous about the propriety of its position, and the lie and tearing away of his linings; putting these always to rights when displaced, and repairing in the hemming every fracture which may be occasioned by the forcing of his pincers. In the building of his heel, he, likewise, is highly accomplished; in this instance, he has taught ourselves to despise the use of the cobbler's plaster-ball, and in so doing, has torn away the mask from many a spumy, yawning, and ladder-like lift; an imperfection, at one time—and that, too, not very long ago—so disgraceful to our trade. Indeed, in this matter, we may yet do better.

"Now, upon a summing up of these items or indications, what do they lead to? Why, briefly this. *First*:—that our leather is not equal to the French leather; and therefore it behoves the master-shoemaker, at the earliest period—for his interests are fearfully hanging thereon—to enforce, if possible, in this particular, from the only proper quarter, that full respect to his expostulations and wishes, which the importance and exigencies of the case require; and if he succeeds not to the necessary satisfaction, then he is to combine his endeavours with the endeavours of his fellow master-shoemaker, and enter into arrangements for the getting from abroad that superior quality of materials which he has been refused at home.

" Secondly.—And this, as before has been explained, is a very important matter indeed —he must do something to better the character of his blocking. *   *   *   *

" In these countries the shoemaker is his own blocker, while in France, the cumbrure, or blocker, is a person solely in the employ of the currier, 15 or 20 hands being, in some large establishments, wholly engaged from day to day at this work, with a sort of overseer or master blocker, whose duty it is to take care that the procedure be properly done, he being the chief responsible party. Under this system then, as may be supposed, the art in question has had more attention in all ways bestowed upon it, than among ourselves, and, as a consequence, has attained a much higher perfection.

" The blocks themselves in France are of a better make, are considerably thicker at the back portions, or where the tacks are put in, and have not so much of round immediately beside or behind the front or blocking edge, and thus are they easier and better to be wrought on; this roundness impeding as it does, the easy draft of the leather in the operation of the pincers, as also when the leather is being forced back by the pane of the hammer, the blocking iron, or the stick. Again, these blocks are not near so wide, either across the foot-part or up the half-leg, as ours are; the French blocker uniformly putting his tacks in at the under, and not as with us on the top edge; and thus, there being nothing of those rows of tacks in his way, as is the case with the English blocker, he can give a full force to the use of his hammer, blocking iron, or stick, without any danger of breaking down a single tack, and thereby, beside the injury to this particular article, often occasioning unsightly rents in the leather, and loosening, moreover, those parts of the work which had already been accomplished. Against all this the French blocker has easily and judiciously provided, and we must do so likewise. To effect this, however, thoroughly, we must bring in requisition sets of blocks, that is, if we wish to trench our leather for the fronts with economy, the smaller measure

requiring the smaller piece of leather, and that again the smaller block, and the same in the larger, vice versa. This will be evident, and is important; for while we save in one case, we are enabled to give the proper sufficiency for the other, and thereby in the latter instance, are not under the compulsion, in the cutting out of our boots, of leaving those fractures and holes up the side of the front, which is too often observed, and which not seldom are even dangerous, besides their offensiveness to the eye. So far then, these precautions, though very useful, are very simple, being no more than having trenching patterns to the different necessary sizes, and blocks again to suit these trenchings, the smaller blocks being appropriated to the smaller front leg-pieces, and the larger and intermediate blocks, to the larger and intermediate front leg-pieces, thus enabling the work to be done better, from the greater fitness of the instrument to its purpose; no leg-front in being put through the operation of blocking, either leading into difficulty or error from its wasteful abundance, or its confined poverty, two things which should be always avoided.*

The utility of sets of blocks therefore, must be evident, as also our own general neglect in this particular. In many of our petty shops especially, two or three pairs of blocks, are at most the usual compliment in use, while in others, where there are more, these again are so heedlessly assorted—if indeed they be assorted at all—as to be scarcely of greater advantage to the full purposes of blocking; beside, as before said, blocking with us is mostly but no better than a sort of job work, a pair of fronts being put on now and then either by a clicker or a cleaner-up, or by the master himself; or the fronts are bought ready prepared as blocked by some one at the currier's, though these generally are more faultily blocked still, or rather merely shoved in, and then pasted up and thickly gummed over, so that the paste and gum, at the least application of the fingers, breaks and falls off, and thus on closer inspection, discovers the really disreputable nature of the trick altogether, as well in the intrinsic character of the work itself, as in the worthlessness of the mask with which it has been decorated, cheating the eye by an appearance of beauty not at all stable, and by which also, probably, the

* Perhaps I should say, that blocks of the French form are now making, and by the time this pamphlet will be printed, may be had at the Grindery establishments of Mrs. Dennis, George Court, Piccadilly; William Dennis, Little Pulteney Street, Golden Square; and John Dennis, Vere Street, Clare Market.

most inferior sort of leather is sought to be palmed off as being of first-rate quality.

" Blocking, too, is often with us too much the affair of the moment, in all small shops especially. The order is given, the fronts trenched out, then hurried on the blocks, and probably afterwards suddenly dried before the fire—the stamina being thereby in many cases so scorched or burnt out of the leather, as to render the article when finished almost next to valueless.   *   *   *   *

" But to proceed with these instructions ; and now to the more immediate process of blocking itself.

" The front, in the first place, is to be sufficiently wetted in cold water, and never—as is sometimes done—in warm or hot water, a foolish and most dangerous habit. A little soaking, and afterwards a short and brisk rubbing between the hands, with an occasional dip or two, will do for this purpose. On being thus wetted, it is to be doubled and laid on a board, and there drawn out with the pincers as much as possible at the two heel or joint corners, as also at the front-top of the leg, in an upward outward position. After this, it is to be strained, somewhat in the way a jockey-boot tongue is strained, though in a contrary direction, the straining hand in this case being the *left* in place of the *right* ; that is, the downward or foot part is to be endeavoured to be turned, instead of the upward, or the leg part. The straining leather or cord will be fastened or tacked down, therefore, at the *top* of the front ; and as a stronger exertion will be here requisite in pulling with the left hand, it may be proper to recommend as a help to this, that a piece of cross stick or other holding be attached to the end of the strainer, so as to command a greater power while in the act of yielding as much curve to the leather at the instep part as may be possible, though this, at the best, will be no great deal. Such curve, however, as shall be gained this way, will be highly serviceable, and much expedite the after process. When this is done, the fellow-front, or series of fronts, are to be done so by likewise, and then the blocking may commence.

On beginning to block, the curve of the front is to be taken in the right hand, and placed directly over the curve of the block ; then the two corners are to be forced with the fingers as far out as they possibly can be got, and next slightly pincered out and tacked down, at or near the block corners, each corner with two or more tacks, as may be seen in all French blocking.

The instruments or tools to be used are, besides the pincers, only two :—one is a sort of wooden rasp, with rounded back,

made often by the workman himself, the teeth blunt and standing at about a quarter of an inch from each other, the wood of a fine, smooth, and hard description, and the length of the tool altogether being from fourteen to sixteen inches, the rasp portion extending to about ten inches in the middle, and the two untoothed ends, used as the handles. Sometimes also, a circular notched stick, made by the turner, is preferred by certain blockers, though I believe the other kind is thought the best by the best workman, the top portion serving occasionally as as leeking, or what in the trade is technically called a *long-stick*.

The other tool or instrument is the blocking iron or *blocking-knife*, with a heavy deep handle, the whole width of the hand, and cased at each end with a plate of steel, to serve as a hammer in striking in the tacks, it being no matter how the tool, in the hurry of work, may be taken up, each end equally serving the purpose ; the *knife* or under part is of a blunt edge, and in depth about two inches, and near upon four inches wide, with rounded corners.

The working bench of the French blocker ought also to be described : this rises a little above his knees as he sits at work, and upon it he keeps his tacks, pincers, wooden rasp, and blocking knife ; it has a sort of front ledge, between two and three inches in height, nailed or otherwise fastened on, at about the same distance from the near edge of the bench itself against which the workman lays his block for a support, in the various necessary directions, the purchase or hold he thus gets over it, giving him a much greater power in the use of the stick just described, than otherwise he could command.

With these tools then, and being thus placed before this kind of bench, the blocker may proceed. He has already fastened his corners, after which, by the action of his fingers, he may ease and work back the leather on both sides the block as well as he can, using occasionally his knife or his rasp, though yet to no great effect ; two tacks at each side the vamp, at its fullest part, may then be put in, and next, after a little more working of the rasp or knife, one tack at each side, above the turn on the leg ; the top of the front is now to be drawn up to its proper extent, (which practice will better teach, than any description,) and then at both sides tacked, as also, with one tack at the middle or fold, or one tack at each side the fold ; the latter the preferable, as being the firmer method.

The front being thus, in these different places, temporarily secured, namely—at the corners, the foot, and the top,—the wrinklings or foldings of the general pursy leather are next to be conquered. And now it is, that the rasp is brought into efficient operation; being taken between the two hands, and pressed downwards and crossways on each side the block, bruising, confining, and flattening the wrinkles into smaller dimensions, the higher portion of the stick, where are no notches, to be occasionally rubbed over or along the block at its front angle, to keep the leather here level and clear. The blocking-knife is next to be employed, to force these wrinkles to their utmost, or entirely from observation.

Much art and care, however, is required in the working of this tool, as inducing to one of the principal perfections of good blocking. In England, we generally, for this purpose, take the pane or flattened end of our hammers, and force each wrinkle heedlessly in a *direct* line to the position of the working hand—either across, above the turn, from the angle to the sides, or downwards, below the turn, right towards the toe, and to the tacks we put in along the lower part of the blotk;—but in France they have discovered the better system. The French blocker, holding the edge of the block to his person, as it lies on his knees, and the foot to the right, and the top part to the left, pressing inwards to his side, he, in the first place, begins with his knife at the *turn*, and compelling the wrinkles in a sweeping or outward semicircular direction, pincers and tacks out the edges, as he proceeds, for the space of about three inches; then turning the block, he does the same by the other side. He next begins with the wrinkles on the foot or vamp portion, giving likewise the curving direction to the action of his hand, commencing at the turn and making the sweep of about four inches in extent, less or more, according to the length of the foot of the block. In no case does he press the wrinkles right down the foot, and only makes use of the tool in forcing right across the foot when the wrinkles have disappeared, and he is about to pincer the lower portion of the vamp leather to its utmost extent.

There is another operation which he performs in his blocking, that is never practised in England, and which, besides, is of the most positive utility. Before the blocking of the leg and foot parts be entirely finished, he draws his foot tac s, and taking the leather at the extreme end in his pincers, there forces it over the toe at his utmost power: the effect of this being, that when the front comes to be taken off the block, the high state of tension it has been put through, will give it, on being pulled back, that extraordinary degree of tightness at the turn, which is so very remarkable a feature in the superior character of the French to the English blocking; and which also, even through the making and the wearing it uniformly is observed to retain, conferring at once a beauty and a perfection of the most necessary description.

It is needless, I think, to amplify more on this matter, or to tell here of such simple things as all should, at the least, be already well aware of, namely—that in blocking, the tacks should be put in regularly, each tack as opposite to its fellow-tack on the opposing side as may be possible; and that all soft, flabby, or foul leather should be carefully drawn out, even to the top; that in cuts opening out at the edges to the eyes of the blocker in the time of his occupation, he should make it his endeavour to do the best by these he can, helping the good leather as far as he may over to the side of the damage, so that, in the cutting out of the boot, these blemishes may be taken away. It seems to be needless to state these matters, or to caution about them, though one direction I must add, as being of considerable value. This is, that when the boot-front is blocked as before described, and yet in its wet state, a portion of that sort of grease called *dubbing*, should be in all cases, rubbed along the sides of the foot or vamp part—but not such dubbing as may be bought at our curriers' shops, raw and innutritious as it is—but dubbing made from a mixture of the best tallow fat and good oil, an article which, I have been told by a person closely connected with one of the most extensive currier's establishments in Paris, is a chief means to the production, in the dressing, of that superior fulness and softness in the French leather over our own; and if so, or even but partially so, how lamentable it is, that a remedy of such trivial cost should so culpably be overlooked, our thin and rank stuffing oozing out at the blocking and in the wear like mere water, and leaving the meagre starved leather to fret, harden, and perish long before its proper time, not only to the loss of the consumer, but to the constant mortification of the shoemaker himself—the still blamed, though, in this case, the still innocent party.

"These instructions towards blocking having taken up more space than I had thought would be necessary, I mnst now hurry over the remainder of this summing-up, though, as I hope, to no loss; the leather and the blocking being the two most important points which, as resting with ourselves, it devolves upon our own prudence and exertions to effectuate an improvement in.

"As to the *cutting* and the *closing*, these are safe. And now—

"*Thirdly*—comes the character of our

lasts, which, to an intelligent and capable employer, is a matter of no difficulty, scarcely requiring anything more than another change of taste, assisted, as this always must be, by a strict observance of the foot as to its differential points and bearings. Many of our English bootmakers already fit excellently; though, when this is done more generally, it will be both better for themselves, and the great stable character of our trade, as the errors of a majority tend evermore to the injury of the whole.

"*Fourthly*, and in conclusion, we have now only to notice the *making*, and where only a little more care is requisite;—I mean care as to *principles*—such as lasting, and every other real perfection of this most valuable branch of our trade, in contradistinction to that loss of time which is now so irksomely wasted in tracing pretty devices, in pretty colours, along the soles, which the first five minutes' walk will wholly obliterate; and which, then, too often to the cost of the purchaser, are found but a very poor equivalent indeed for the neglect of the more needed perfection.*

LIST OF ENGLISH PATENTS GRANTED BE-
TWEEN THE 26th OF NOVEMBER AND
THE 24th OF DECEMBER, 1838.

John Small, of Old Jewry, merchant, for improvements in the manufacture of thread, or yarn, and paper, by the application of certain fibrous materials not hitherto so employed. Patent sealed Dec. 1; six months to specify.

Peter Taylor, of Birching Bower, Lancaster, rope-maker, and slate merchant, for improvements in machinery for propelling vessels, carriages, and machinery, parts of which improvements are applicable to raising of water. Dec. 1; six months.

Ambrose Bowden Johns, of Plymouth, artist, for improvements in colouring or painting walls, and other surfaces. Dec. 1; six months.

James Hartley, of Bishop Wearmouth, glass manufacturer, for improvements in the manufacture of glass. Dec. 1; six months.

Theodore Cotelle, of the Haymarket, civil engineer, for improvements in extracting the salt from sea or salt-water, and rendering it pure or drinkable, and in purifying other water. Dec. 1; six months.

John Player, the younger, of Longhorneur, Swansea, Glamorgan, for improvements in furnaces and fire-places, for consuming anthracite and other fuel for generating steam, evaporation, smelting and heating iron and other metals. Dec. 1; six months.

William Pontifex, of Shoe-lane, London, copper smith, for improvements in apparatus and materials employed in filtering and clarifying waters and other liquids. Dec. 1; six months.

* It is but right that I should state a great injustice which some masters are in the habit of inflicting on their own native workmen, in looking for or insisting on this valueless loss of time, as occasioned by the figuring and the painting here alluded to, though on those French or other foreign workmen in their employ, the same be not exacted; foolishly, perhaps, thinking as they do, that that boot which the French workman makes, because *he* makes it, must be equal to the really French boot, which is not at all the case, the operative part, as far as he

John M'Curday, of Tonbridge-place, New-road, Esq., for an improved method or methods of generating steam, and applying the same to the evaporation and boiling of fluids, which method, or methods, is, or are applicable to steam-engines, and other purposes, where steam is, or may be applied. Dec. 1; six months.

Stanislaus Darthez, of Austin Friars, London, merchant, for certain improvements in the construction and arrangement of axles, axle-trees, and the naves of wheels for carriages. Dec. 1; six months.

John Shaw, of Glossop, brass-worker, for certain improvements in the arrangement and construction of wind musical instruments. Dec. 1; six months.

Luke Hebert, of Camden Town, C. E., for an improved mode, or modes of fastening trowsers, and other parts of dress or apparel, being a communication from a foreigner. Dec. 1; six months.

Miles Berry, of Chancery-lane, for improvements in the means of, and apparatus for manufacturing gaseous liquids, and for filling bottles and other vessels used for holding the same, and retaining the contents therein, and emptying the same when required; being a communication from a foreigner. Dec. 6; six months.

James Carson, of Liverpool, doctor of medicine, for a new mode of slaughtering animals intended for human food. Dec. 6; six months.

Thomas Robinson Williams, of 61, Cheapside, C. E., for certain improvements in machinery for spinning, twisting, or curling, and weaving horsehair, and other hairs, as well as various fibrous substances. Dec. 6; two months.

Henry Count de Crony, of Picardy, France, now residing at 14, Cambridge-street, Edgeware-road, for certain improvements in filtration; being a communication from a foreigner. Dec. 6; two months.

John Alexander Elsear Degrand, of the Boulevart du Temple, Paris, now residing in Paul's Chain, London, C. E., for improvements in the production of motive power, and in machinery for applying the same to useful purposes. Dec. 6; six months.

Daniel Chandler Hewitt, of Store-street, Bedford-square, professor of music, for certain improvements in musical instruments. Dec. 6; six months.

John Chisholm and Maria Hyppolite Bellemoir, of Pomeroy-street, Old Kent-road, manufacturing chemists, for improvements in treating massicott, litharge, and other compounds of lead, for the purpose of obtaining therefrom silver, and certain other products. Dec. 6; six months.

Godefroy Cavaignac, of Tavistock-row, Covent-garden, gent., for improvements in apparatus for transporting materials for various purposes, from

is concerned, being scarcely of any use towards the perfection of the article. This ignorance, or whatever else it be called, is as condemnable as lamentable, there being no reason why the time of the native workman should be so wasted, while that of the foreign workman is treated with more respect; and, merely, as it would seem, because he is *foreign*. I have actually seen boots made in London by these two classes, and in the one employ, where the difference in the worth of the work, as measured between them in the time it exhausted, was equal to one-third of the wages received by the individual workman. Nor is excellence to be attained in this way;—boots properly being made to *wear*, and not to be made *pictures* of—a shop-window exhibition —and which it were well if the purchaser knew to value as such exhibitions usually deserve.

Men's time should not be thus sacrificed. If they are to become painters in place of shoemakers, let them in their leisure display their abilities on its proper ground—the canvass, and not the hard hide leather which is to be trod in the dirt. Be them, in their available opportunities, painters, poets, or philosophers, any thing, in short, but the mere dupes of a diseased paltry passion, born in whim, and followed up in oppression.

one place to the other, particularly applicable to road-cutting and embankments. Dec. 6; six months

Thomas Sweetapple, of Cotteshall Mill, in Godalming, paper-maker, for an improvement, or improvements in the machinery for making paper. Dec. 6; six months.

Frederick Neville, of Pancras-lane, in the city of London, gent., for an improved method, or process of manufacturing coke, whereby the sal ammoniac bitumen, gases, and other residuous products of coal are at the same time separately collected, and the heat employed in the process is applied to various other useful purposes. Dec. 6; six months.

James Gardner, of Banbury, ironmonger, for improvements in cutting Swedish turnips, mangle wurzel and other roots used for food for sheep, horned cattle, and other animals. Dec. 12; six months.

Thomas Vaux, of Woodford, land surveyor, for improvements in tilling and fertilizing land. Dec. 12; six months.

Crofton William Moat, of Putney, for an improved mode of applying-horse power to carriages on ordinary roads. Dec. 12; six months.

Barclay Farquharson Watson, of Lincoln's-Inn Fields, solicitor, for improvements in crushing or preparing New Zealand flax (phormium tenax). Dec. 12; six months.

Edwin Edward Cassell, of Mill Wall, Poplar, for improvements in lamps. Dec. 12; six months.

Job Cutler, of Lady Pool-lane, Birmingham, gentleman, for improvements in combinations of metals, applicable to the making of tubes or pipes and to other purposes, and in the method of making tubes or pipes therefrom, which improved method is applicable to the making of tubes or pipes from certain other metals and combinations of metals. Dec. 12; six months.

James Lees, of Salem, near Oldham, Lancaster, cotton spinner, for an improvement in the machinery for spinning, twisting, and doubling cotton, silk, wool, hemp, flax, and other fibrous materials. Dec. 17; six months.

John Hawkshaw, of Manchester, C.E., for certain improvements in mechanism or apparatus applicable to railways, and also to carriages to be used thereon. Dec. 17; six months.

Benjamin Goodfellow, of Hyde, Chester, mechanic, for certain improvements in machinery or apparatus for planing or cutting metals. Dec. 17; six months.

John Roberts, of Manchester, machine-maker, for certain improvements in machinery or apparatus for planing or cutting metals. Dec. 17; six months.

John Radcliffe, of Stockport, machine-agent, for the application of an improved covering for the rollers used in the several processes of preparing, drawing, slubbing, roving, spinning, twisting, and doubling of wool, cotton, wool flax, silk, mohair, or any other fibrous material or substance, or so many of such rollers as require, or are deemed to require covering for such several processes, or any of them. Dec. 17; six months.

Joseph Zambeau, of St. Paul's church-yard, chemist, for improvements in rotatory-engines, being a communication from a foreigner. Dec. 19; six months.

Andrew Smith, of Prince's-street, Leicester-square, engineer, for certain improvements in apparatus for heating fluids and generating steam. Dec. 19; six months.

Samuel Parker, of Argyll-place, London, imp maker, for improvements on stoves. Dec. 19; six months.

Carl Augustus Holm, of Mincing-lane, engineer, and John Barrett, of Vauxhall, printer, for certain improvements in printing. Dec. 19; six months.

Daniel Stafford, of 25, St. Martin's-le-grand, London, gentleman, in pursuance of the report of the judicial committee of her Majesty's privy council, for certain improvements on carriages, being an extension for the term of seven years from the 24th ay of December instant, of former letters patent.

LIST OF SCOTCH PATENTS GRANTED BETWEEN THE 22nd OF NOVEMBER AND THE 22nd OF DECEMBER, 1838.

Robert Beart, of Huntingdon, miller, for improvements in apparatus for filtering liquids. Sealed 27th of November, 1838; four months to specify.

Auguste Victor Joseph, Baron D'Asda, of Millman-street, Bedford-row, Middlesex, in consequence of a communication made to him by a certain foreigner residing abroad, for improvements in producing or affording light, which he intends to denominate a solar light. November 29.

John Barnett Humphreys, civil engineer, for improvements in marine and other steam-engines. November 29.

Richard Lamb, of David-row, Southwark, gentleman, for improvements in apparatus for supplying atmospheric air in the production of light and heat. November 29.

James Timmins Chance, of Birmingham, glass manufacturer, for improvements in the manufacture of glass. November 29.

Paul Chappe, of Manchester, spinner and manufacturer, for certain improvements in the means of consuming smoke, and thereby economising fuel and heat in steam-engine or other furnaces or fire-places, which improvements are also applicable in preventing the explosion of boilers. November 30.

Samuel Seaward, of the canal iron works, Poplar, Middlesex, engineer, for certain improvements in marine steam-engines. November 30.

Henry Davies, of Stoke Prior, in the County of Worcester, engineer, for certain improved apparatus or machinery for obtaining mechanical power, also for raising or impelling fluids, and for ascertaining the measure of fluids. December 7.

Joseph Bolton Doe, of Hope-street, White-chapel, Middlesex, iron-founder, for improvements in apparatus used in the manufacture of soap. December 7.

Fanquet Delarue, junior, late of Daville, near Rouen, France, but now residing at the London Coffee-house, London, gentleman, for certain improvements in printing and fixing fast, red, black, and other colours upon cotton, silk, woollen, and other fabrics, without the usual process of dyeing. December 11.

Theodore Cotelle, of the Haymarket, Middlesex, civil engineer, for improvements in extracting the salt from sea or salt water and rendering it pure or drinkable, and in purifying other water. December 14.

Henry Adcock, of Mount Place, Liverpool, civil engineer, for certain improvements in the raising water from mines and other deep places. December 14.

William Thorp and Thomas Meakin, of Manchester, silk manufacturers, for certain improvements in looms for weaving, and also a new description of fabric to be produced or woven therein. December 14.

William Crofts, of Radford, Nottingham, machine maker, for improvements in the manufacture of lace. Dec. 14.

Joseph Green, of Ranelagh-grove, Chelsea, Middlesex, gent., for an improvement in ovens. Dec. 21.

Thomas Nicholas Raper, of Greek-street, Soho, Middlesex, gent., for improvements in rendering fabrics and leather waterproof. Dec. 21.

John Howarth, of Aldermanbury, London, manufacturer, partly in consequence of a communication from a certain foreigner residing abroad, and partly by invention of his own, for certain improvements in machinery for spinning, roving, doubling, and twisting cotton, and other fibrous materials. Dec. 21.

LIST OF IRISH PATENTS GRANTED IN NOVEMBER, 1838.

Joseph Rock Cooper, for improvements in fire-arms. Nov. 6; 1838.

Jean Leandre Clement, for improvements in apparatus for ascertaining and indicating the rate of vessels passing through the water. Nov. 6.

Richard Thompson, for improvements in making a certain spirituous liquor, which he intends to denominate "Thompson's British Wine Brandy." Nov. 8.

John W. Fraser, for improvements in diving, or descending, and working in water, and for raising or floating, sunken and stranded vessels, and other bodies. Nov. 16.

John Henfrey, for improvements in the manufacture of hinges and joints, and in the machinery employed therein. Nov. 17.

Richard Eese, for certain improvements in drying corn, or other grain, seeds. Nov. 21.

Frederick Joseph Burnet, and Hippolyte F. Marquis de Bouffet, for certain improvements in the manufacture of soap, &c. Nov. 29, 1838.

William Neale Clay, for improvements in the manufacture of iron, &c. Nov. 26.

Fauquet Delarue, jun., for certain improvements in printing. and fixing fast red, black, and other colours upon cotton, silk, woollen, and other fabrics without the usual process of dying. Nov. 29.

William Rattray, of Aberdeen, chemist, for certain improvements in the manufacture of gelatine size, and glue. [Omitted in the list for July.]

NOTES AND NOTICES.

Duty on Bricks.—"I wish you would stir up architects to get the duty off bricks: even the double duty taken off would be a boon in favour of the extension of taste. A meeting should be got up in London, to draw up a petition to parliament, which would soon be followed by the rest of the kingdom. In fact, there should be a regular agitation. I have seen the Marquis of Tweedale's 'brick-maker,' and think highly of it."—*W. Thorold. Arch. Mag.*

Steam-boats on Canals.—The Rev. J. W. M'Gauley, Professor of Natural Philosophy to the Board of Education, we understand, has at length succeeded in fabricating a machine for propelling boats on canals without raising a surge, which has been very detrimental to the banks, causing a considerable annual outlay to keep them in repair. The power will be derived from a steam-engine; but instead of the usual paddle-wheels, there will be *a machine immerged in the water underneath the centre of the boat,* the working of which will not cause the least ripple on the surface of the water. There will be a public test of the invention on the Grand Canal about a fortnight hence, with a boat fitted up under the immediate inspection of the Rev. gentleman.—*Dublin Post.*

New Jetty at the London Dock.—A great improvement has been lately made in the London Dock, by the erection of a magnificent jetty, supported on massive piles, extending from the south-west quay, 800 feet across the large basin, affording a quay frontage on both sides for the loading of outward-bound ships of 1,600 feet. The jetty is 62 feet in width, and three lofty sheds, each 208 feet long by 48 feet wide, for the reception of goods and merchandise for exportation, are in the course of erection; one of these store-houses is already completed. There will be a space of 7 feet clear on each side of the warehouses. The erection of the jetty is said to have cost the London Dock Company not less than 69,000l., and it will afford great accommodation to the shipping, and particularly to the Sydney and Hobart Town ships. There are now eight large vessels bound to those places lying alongside the new jetty. They will all carry out a great number of emigrants.

Supplying St. Pancras with Water from Artesian Wells.—We have been much surprised to see by the newspapers that this subject has been seriously thought of, and discussed in meetings at which some persons were present eminent for scientific knowledge. We thought it had been generally known that the sources which supply the London basin, ample as they are, are still limited. As a practical proof of this, it is only necessary to mention that the two great breweries which draw their supplies from wells which penetrate to the chalk, the one on the Middlesex, and the other on the Surrey side of the river, cannot both pump on the same day, and, by agreement, pump on different days. If a part of the Thames water above Richmond, where it is tolerably pure, could, by means of a deep shaft, be made to run into the basin, then, no doubt, the whole of London might be supplied from it, cheaper than is now done by surface-pipes. But, supposing this mode to be adopted, it would only prove sufficient for a century or two; for such would be the quantity of sand and mud carried down by the water of the Thames, that, unless it were filtered before it entered the shaft, it would, in time, solidify the under stratum. Among all the plans that have been devised for supplying London with water, we have no doubt whatever that the present mode by surface-pipes is the best, provided the water be drawn from pure sources. By being brought in in pipes covered by earth, the water is delivered at a lower temperature in summer, and a higher temperature in winter, and free from all those impurities to which an open watercourse is liable: witness, for example, the New River. How to question. Perhaps the real object of the St. Pancras meeting was to hold the Artesian system *in terrorem* over the advocates of the surface system, in order to keep the water companies within bounds.—*Mr. Loudon.—Arch. Mag.*

London and Greenwich Railway.—On Monday the whole line of this railway was opened to the public, and the number of passengers far exceeded the usual average. The distance to and from Greenwich was performed throughout on an average of 18 minutes. Several of the directors and principal shareholders in the company went to Greenwich and back.

Metropolitan Railway Map. — Now published, vol. xxix. of the *Mechanics' Magazine,* price 8s. 6d., illustrated with a *Railway Map* of the Metropolis, taking in a distance of 30 miles from the Post-office. The limits of the two-penny and three-penny post deliveries are also shown in the Map. The *Metropolitan Railway Map* alone, stitched in a wrapper, price 6d., and on fine paper, coloured, 1s. *The Supplement* to vol. xxix, price 6d. *The Railway Map* of England and Wales continues on sale, in a neat wrapper, price 6d.; and on fine paper, coloured, price 1s.

☞ *British and Foreign Patents taken out with economy and despatch; Specifications, Disclaimers, and Amendments, prepared or revised: Caveats entered; and generally every Branch of Patent Business promptly transacted. A complete list of patents from the earliest period (15 Car. II. 1675,) to the present time may be examined. Fee 2s. 6d.; Clients, gratis.*

LONDON: Printed and Published for the Proprietor, by W. A. Robertson, at the Mechanics' Magazine Office, No. 6, Peterborough-court, between 135 and 136, Fleet-street.—Sold by A. & W. Galignani, Rue Vivienne, Paris.

Mechanics' Magazine,

MUSEUM, REGISTER, JOURNAL, AND GAZETTE.

No. 804.] SATURDAY, JANUARY 5, 1839. [Price 3*d.*

Printed and Published for the Proprietor, by W. A. Robertson, No. 6, Peterborough-court, Fleet-street.

DUCHEMIN VICTOR'S ROTARY STEAM-ENGINE.

Fig. 3.

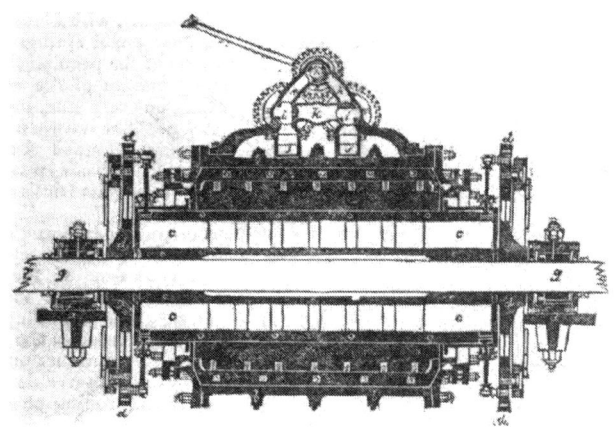

Fig. 4.

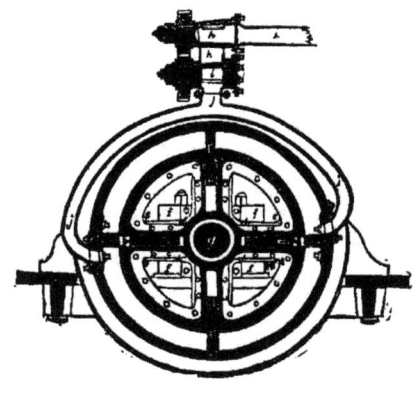

VICTOR'S ROTARY STEAM-ENGINE.

A patent was granted for this engine on the 19th of March, 1838, to Mr. Duchemin Victor, of Boulogne-sur-Mer, and the following description has been supplied to us by a friend of the inventor:—

This rotary engine, to be worked by steam, or other aeriform fluids, has four moveable pieces or pistons, entering, by means of an exterior arrangement of machinery, into an interior and concentric cylinder, upon which the pressure is always equal, since it always presses simultaneously upon equal and opposite surfaces; its great cylinder is inwardly cylindric in all its parts, and is not cut for the passage of any piece, which advantages secure to it all the strength of the metal; it can also have a great diameter, and a great height from one base to the other. This engine can be made use of at any pressure, as well as with condenser and air pump, and its peculiar construction exempts it from those defects which are necessarily attached to engines of the ordinary construction.

Description of the engravings :—

Fig. 1, is an elevation of the engine.

Fig. 2, a side view.

Fig. 3, a section by C D.

Fig. 4, another section by A B, A C.

a, in fig. 1, 2, 3, and 4, is the foundation plate.

b, is the exterior or great cylinder of the engine; it is shut at each extremity (fig. 3) by a ring, which forms, at the same time, the fixed and uneven part of a stuffing-box. The capacity between this cylinder and the interior cylinder (fig. 4) is divided into two equal parts by separations or abutments, which resist the pressure of the steam; these separations or abutments are furnished with blades, whose length is equal to the height of the interior cylinder, and these blades are squarely finished, in order that the part which, by means of small springs, will gradually approach to replace the part worn out, may be always of the same length.

c, is the interior and concentric cylinder; it is fixed upon the axle-tree. The four arms of this cylinder (fig. 4) are prolonged on the outside (fig. 1 and 3), and are grooved in their full length, sufficiently deep (fig. 3) to permit the pieces receiving the impulsion of the steam, and communicating the motion to the axle-tree, to lodge themselves in, when they pass before the separations or abutments (fig. 4). This cylinder is so arranged at each extremity (fig. 4) as to allow of a ring (fig. 2 and 3) being fixed to it; this ring forms the part of the stuffing-box, whose surface is polished, and which moves itself with the interior cylinder. Small plates, fixed at the extremities of the arms (fig. 1, 3, and 4), complete the closing of the grooves. The moveable pieces acting as pistons (fig. 3 and 4) are furnished, as are also the separations or abutments, with blades, constantly pushed by small springs; these blades, by means of the peculiar arrangement of the extremities of the cylinder (fig. 3), prevent, on every side, the leakage of the steam. The outward motion of these blades is limited by small pins (fig. 3); small cavities in the arms (fig. 3 and 4) diminish the friction of the moveable pieces.

d, shows a cross, fixed upon the axle-tree at each extremity of the cylinder; the middle of each arm of the cross, placed precisely opposite the moveable pieces, serves to guide a sliding piece (fig. 2 and 3), which receives the motion on one side by a roller, and communicates it by the other to the moveable pieces, by means of a rod rolling through a small stuffing-box.

e, shows the pieces in which the rollers run. These pieces are fixed upon the foundation plate. The rollers, in the parts placed opposite the separations or abutments, run in parallel guides towards the axle (fig. 2), and approach it sufficiently to allow the moveable pieces to pass before these separations or abutments without touching them.

f, are the boxes, with bearings, in which turns the axle-tree of the engine. Regulating screws acting upon these bearings keep the axle-tree always, and in every way (notwithstanding the wearing out), in its proper position.

g, is the axle-tree of the engine, by which the motion is transmitted.

h, are the pipes and cock, for the admission of the steam into the cylinder.

i, shows two cocks (fig. 1 and 3) permitting, alternately, the entrance of the steam into the cylinder, whilst its going out takes place by the other, according to the direction in which the engine works.

j, are two forked pipes (fig. 2 and 4),

Fig. 1.

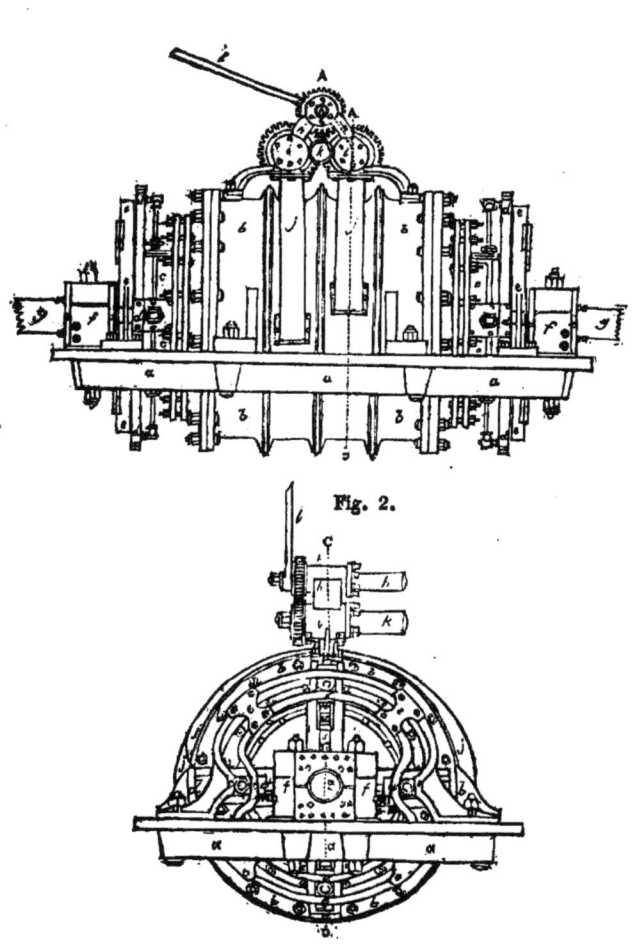

Fig. 2.

serving alternately for the entrance or going out of the steam, which is introduced simultaneously upon equal and diametrically opposite surfaces, and acts constantly upon two of the four moveable pieces. It is to be remarked that at each time the steam goes out, it is only the quantity which was contained in the capacity between two moveable pieces.

k, is the pipe for the exit of the steam.

l, is the handle (fig. 1, 2 and 3) by which the engine can be put in motion, in one direction or in the other, or can be stopped with the greatest facility, since this handle operates at the same time upon the three cocks (fig. 1 and 3), by means of three tooth-wheels, one of which has thirty teeth, and the others forty. In the position where the handle is (fig. 3), the steam is introduced by the pipes on the left, and it goes out by those on the right. To reverse the motion, it is only necessary to place the handle, which is now a sixth part of a circle, on the left, in a similar position on the right. To stop the engine, the handle must be placed in a perpendicular direction.

In conclusion it is to be remaked, that this engine is extremely simple ; all its parts can be easily examined by so arranging the foundation plate as to admit of one of its extremities being taken down, which would allow the exterior cylinder to be taken off.

PHILOSOPHY OF ROAD MAKING.—IRON FRAME AND WOODEN BLOCK ROAD.

Sir,—Amongst the mechanical arts tending to the progress of human civilisation, the first in importance, after that of printing, is the art of road-making. Printing circulates ideas, and roads circulate men and women, the originators of ideas; and they moreover, quicken and increase to an enormous extent the circulation of printed ideas. Viewed thus, it seems strange that, with the single exception of the process of macadamisation, all road-making in England —rail-road-making inclusive—is most unphilosophically performed.

Leaving out of the question the operations of levelling or preparing the ground for road-making, the first principle to bear in mind is, that the whole surface be so bonded together that no part of it can be forced down below the level of the rest, by the heaviest load, concussion included, which may pass over it. Or, if the nature of the material or its mode of preparation preclude this, then every separate portion of which the surface is composed should possess sufficient breadth of bearing to ensure it against sinking by the passing weights. The heaviest load is probably a coal-waggon, weighing nearly two tons, and carrying in it a load of four tons. This will give a ton and a half to each wheel; add the momentum, and the frequent unequal bearings, with the falls from prominences into hollows, the wheels will frequently press with a weight of upwards of four tons each.

Let us now examine how far the various roads in use comply with the essential principles laid down.

A road may be bonded together either chemically or mechanically, or in both modes combined. An exemplification of the chemical bond is the concrete formed with hot lime and broken stone. A very hard and white road may be thus made ; but the disadvantage of this bond is, that when once broken it will not again unite.

The conjoined chemical and mechanical bond is exemplified in the roadways of squared granite blocks, grouted with hot lime. The disadvantages of this is that the mechanical bond is imperfect, the stones being a series of wedges, only touching each other at their upper edges. The pressure of the loads soon breaks the chemical bond, and the stones are forced one below another, when the road becomes rapidly useless.

The mechanical bond may be subdivided into two kinds—that of pressure united with friction of surfaces—and that of adhesion, owing to the plastic nature of the material. The common Macadamised road of small pieces of granite is an example of pressure and friction united. The rough facets of the fragments of stone, being shaken into something like order, fitting against each other, are in that state firmly pressed downwards and laterally, by the passing loads and become a white mass. This road can only be destroyed by positive wear, and it is repaired with perfect facility, a few days of passing loads restoring it to its level condition.

But when worn into inequalities and out of repair, this road gives an undulating motion to wheel carriages of a very unpleasant kind. The imitations of this road done in flint have no bond whatever. The surfaces continually slide past each other : and the material being brittle is crushed into powder.

Another example of pressure and friction is the ordinary granite paving for roadways, consisting of squared blocks, or rather wedge-shaped blocks. These are laid side by side, on a soft bedding of gravel, with their upper edges bearing against each other. In this state they are rammed with a hand rammer. They are supposed to constitute an arch, but in fact they are only a series of teeth, liable to sink whenever they are pressed with a greater weight than that which originally rammed them. When one is forced down below the level of the others, the surrounding ones soon follow, and a receptacle for water is formed, which helps forward the work of rapid destruction

Clay and chalk on roads are examples of mechanical adhesion. In dry weather they pulverise rapidly into dust ; in wet weather they lose their form under pressure. and become mud. Clay especially is unfitted for a road, even as a foundation, and where it occurs, care should be taken to make the covering waterproof, by either the great depth or other impermeability of material.

Of late a new feature in road-making has appeared—the use of bitumen under the name of " asphalte." This promises to be valuable, though sufficient experience is yet wanting for confirmation. It is an example of mechanical adhesion, not acted upon by water, nor liable to have its bond broken by violence, like the chemical bonds of water cements ; at least I imagine so, unless it shall prove to become brittle and friable with frost. The authorities who rule in Marylebone, have very wisely arranged their disputes as to which is the best mode of road-making, by agreeing to lay down various samples side by side in Oxford-street.

" And like that best whose merits most shall be."

In one mode bitumen is worked up with granite fragments into conglomerated blocks, about a foot square and six inches thick. These are laid in lines at right angles with the street, and the joints are run in with hot bitumen mixed with coarse sand. Another mode is by laying these squared blocks diagonally with the line of the street, cemented together.

A third mode consists in preparing blocks of asphalte mixed with coarse sand, some twenty inches in length and about six inches wide and thick.

A fourth mode consists in laying down squared blocks of granite, of the usual paving size, and cementing them together with sand and bitumen in a heated state.

In addition to these, a specimen of Macadamisation is laid down, and a sample of what is called " wood paving."

The latter consists of hexagonal blocks of fir, about twelve inches deep, and grain up and down, set side by side on the earth.

It has yet to be proved whether the lateral bond of bitumen and sand will so firmly unite the blocks together as to prevent separation by the concussion of the passing loads. If they continue to adhere perfectly, owing to the elasticity of the bitumen, then they will form the most perfect road, as the water will not penetrate them, and the under surface will not squeeze up. They will moreover, permit carriages to roll over them with less percussion than any mere stone road. But I incline to think that the samples laid down will scarcely prove sufficiently thick to resist separation. Once separated, the road will be a most wretched one.

The sample of wood paving laid down seems so far a mere waste of material, from the defect common to the whole experiments—an unfirm and insufficient under stratum. Had a level been first made with concrete, from eight to ten inches in depth, a most durable road might have been made on it, either with bitumen or wood. It seems strange that this essential matter should still be so little understood.

For general road-making there can be little doubt that the macadmising process is the best; but for the streets of a town of much traffic, the wear of it is too rapid. Cementing it with bitumen, if not too costly, is a great improvement. But there is one serious objection, the necessity of frequently taking up the street pavement to get access to the sewers, water, and gas pipes.

As there is little chance of our attaining subways for our water and gas pipes,

—though in all new streets they ought to be as scrupulously provided as common sewers—we may assume, that of a necessity, the under-stratum of our streets must always be loose and unfirm. The roadways should, therefore, be constructed so as to suit this condition. Every portion of the road should be capable of sustaining itself on the soft ground without sinking, and we must contrive this in the same mode architects contrive to sustain buildings on unfirm soils, by breadth of bearing.

Let the ground be first levelled and made as firm as usual for paving. Let it then be covered with a series of cast-iron frames about four feet square in surface, and nine inches deep. For the sake of strength, the bottoms should be cross ribbed. The sides of these frames should abut against each other. When first laid down the hollows should be filled up with squares of bitumen, like the present experimental roadway, or better still with the hexagonal wooden blocks, caulked with the bitumen. Over this the carriages would run with little noise and little percussion.

It must be evident, that in such a road there could be no partial sinkings. No weight in common use could sink a paving block 4 feet square, and the wear of wood endways would be trifling. There is one objection only, which strikes me as to the use of wood, the chance of its under surface rotting and producing miasma. But probably the iron rust would correct this.

If required to take up the paving, the removing and relaying of these masses, with a proper tackle would be a very simple affair—and the road-way would be in as good a condition immediately after relaying, as it now is after long consolidation.

The expense of the cast iron is the only consideration worth attending to. But if the durability and saving of expense be equivalent to it, then the first cost is not a consideration. I should be glad to hear through the medium of your publication, that an experimental piece of wood in this plan were added to the list of the others ordered by the Marylebone authorities.

Colonel Maceroni must now agree that I have made the *amende honorable* for joking about his wooden pavements. He has often argued on the necessity of efficient ramming for the streets. He will, perhaps, give your readers some hints for improving the plan I propose of extended surface bearing.

Mr. Baddeley is wrong in supposing that I wish to uphold an argument in the face of facts. I want only to invite to the pursuit of new genius and new gleanings in physical science, and even were it proved futile to attempt to use inorganic matter for food,—which I hold to be not yet proven—even then I cannot agree with him that it would be time lost, for our chemistry would give numerous new facts. I do not argue with those who laugh at the hunters after the philosopher's stone and the perpetual motion. They had at least an incentive to pursuits by which they explored mines of knowledge for the world, and which they would never have discovered, but for their aims at which they considered higher things. Were the pursuit after inorganic food to lead to the knowledge of better modes of preparing organic food, it would still be a great gain. Be it a phantasy or not, I cannot reconcile it to my nature to believe that we are to be for ever the slaughterers of the lower animals, to sustain our own lives in common with the most savage tribes in the remotest periods of the worlds history,

I remain, Sir, yours obediently,
JUNIUS REDIVIVUS.
December 25, 1838.

COLES'S FRICTION-WHEEL CARRIAGES.

Sir,—Passing Charing-cross a few days since, my attention was caught by a single-sheet pamphlet, exhibited in the shop-window of a truss-maker, headed "Coles on Railroads." Having procured a copy "for the small charge of twopence," I discovered that this method had been resorted to by Mr. Coles, to endeavour to ward off the consequences of some remarks, which appeared in No. 776 of your Magazine, on Mr. Coles's "patent friction-wheel carriages," and which Mr. Coles is pleased to describe as, "calculated to annihilate his invention."

In a quotation from *"The Times"* appended to his pamphlet, Mr. Coles is described as "a gentleman of some reputation as a practical mechanician, though not a professional engineer;"

whatever may be his talents in this line, certes they will be assisted mighty little by his literary labours, if I may judge from the specimen before me, in which equal disregard is had, both to facts and rules of grammar.

Mr. Coles commences his lucubrations as follows :—". As railway or locomotive travelling is found to be the most easy and delightful, affording advantages alike to the sportsman, the merchant, and the man of business, it is a subject which need only be made known to the world to become universal; where there exists a possibility of *erecting them* (*i. e.* the sportsman, the merchant and the man of business) the means can always be attained !" Following this, is an account of Mr. Coles's delight at the idea of a railroad in the vicinity of this great metropolis ; of his early visit to the London and Greenwhich railway on its opening ; and on his rapturous travelling backward and forward until he fairly rode— not his hobby's tail off—but all the grease of the wheels, which began to " assail his ears with a most disagreeable noise." This being altogether unbearable a remedy was sought for, "and in about six months," says Mr. Coles, " I had my patent specified for large anti-friction wheels working over the axles of the ground wheels to relieve them from friction." If Mr. Coles is to be believed, his patent only applies to models upon a scale of one inch to the foot, for referring to the drawings at the top of his pamphlet, he says they are " from *models* of an engine or four-wheel carriage and two two-wheel carriages, made upon a scale of one inch to the foot, for which patents have been obtained for the United Kingdom, France, and America."

Having explained the construction of his models at some length, Mr. Coles describes the circumstances attending their exhibition in different towns, &c. These models were exhibited for three days in Liverpool before many engineers and influential men of the place ; " but nothing occurred there worthy of remark !" At Birmingham, Mr. Coles was more fortunate, his models having been "examined minutely in all their parts by—Dr. Church! whose opinion was perfectly in accordance with that which had previously been given by one of the directors of the Greenwich railroad, George Walter, Esq., viz.: that the first

railroad company that adopted Coles's patent—would run away from all the rest; and the others must follow !"

Mr. Coles's models next underwent a three days' scrutiny in Manchester, and here it seems something did occur "worthy of remark," inasmuch as they were seen by Mr. Roberts, certainly one of the first practical mechanics of the day, and his opinion was most decidedly unfavourable to Mr. Coles's plan. Mr. Roberts having made several experiments on a *large scale*, with anti-friction wheels as applied to railway carriages, which proved unsuccessful, he was not to be gulled by the specious performances of " miniature models." Mr. Roberts, " could not be persuaded" that Coles's anti-friction wheels " would have any other effect than to retard the progress of the carriage; and he did not think the railroad companies would *gain* any thing by adopting my patent."

Mr. Coles states,

" In Dec.1837, I find in the *Mech. Mag*. that Mr. Roberts was so anxious to get rid of every obstacle which impeded the progress of his engines and train, that he was almost ready to quarrel with Dr. Lardner, who made no allowance for the obstruction which the wind caused to locomotive travelling, whereas he found that a peg-top, by being simply lacquered, would not run so fast as when it was not lacquered, in illustration of which *they* (*i. e.* Mr. Roberts and Dr Lardner) spun out a very lengthy article, and actually gave a drawing of *his* peg top. Verily there are men in the world who will strain at a gnat and swallow a camel ! !"

The straining and swallowing of Mr. Coles is exceedingly amusing, but unfortunately for his veracity, neither Mr. Roberts nor Dr. Lardner ever wrote a line—or as Mr. Coles elegantly expresses it—" spun out a lengthy article" on the subject. A very interesting paper on this topic appeared in your 747th Number, furnished by your esteemed correspondent, Mr. Richard Evans of Swansea, to which, I suppose, Mr. Coles blunderingly refers.

But of all Mr. Coles's statements, perhaps the following is the most outrageously absurd : he says " when I submitted my invention to Mr. Robertson, the secretary of the Eastern Counties Railway, be promised to procure for me an introduction to Mr. Braithwaite, their engineer, which was promising probably more than he was able to perform ; *therefore, to get rid of my importunity,* an attack was made on my invention in the *Mechanics' Magazine*, No. 776 "!

As the writer of the letter here referred to, and most unjustly characterised as an *attack*, I will just make a few remarks in reply, to Mr. Coles's most illiberal representation. Not happening to be in communication either with Mr. Robertson or Mr. Braithwaite, I know nothing at all of what had passed between them and Mr. Coles; and the article which has so grievously offended the latter gentleman was my free and unbiased opinion of Mr. Coles's plan, given without favour or affection to any man, and partly provoked by the appearance in a previous number of the *Mechanics' Magazine* of what Mr. Coles himself now states to have been, an *" exaggerated statement."* *

It is due to myself, and to your readers, that I should be permitted most unequivocally to deny the truth of the imputation cast upon me by Mr. Coles. *I have never been employed* either to attack or to uphold the plans of any person, in your pages, but have always followed the dictates of my own judgment in the spirit of free inquiry to which your journal is ever open. I have carefully perused my paper in your 776th number, and would not wish, at this time, to change a single word; since that number was published I have continually sought for, and obtained much, information on the subject of anti-friction wheels, but all tending to prove the correctness of my views on the subject. I am still open to conviction, and really feel so much interest in the question, that had I the means I would gladly furnish Mr. Coles with the opportunity of testing his plan on a proper scale, and for ever set at rest all doubts upon the subject.

I know not what Mr. Braithwaite's opinion of Mr. Coles's plan may be, and although upon *some subjects* we may differ as widely as the antipodes, yet I shrewdly suspect our conclusions on Mr. Coles's plan have some slight resemblance. At any rate, Mr. Braithwaite

* This statement was supplied to us by, and printed from the manuscript of, Mr. Coles himself—the only alteration we made being the necessary grammatical and orthographical corrections, and the omission of *his* opinions upon the immeasurable importance of his patent. Had he supplied us with a correction of the error complained of in his pamphlet, we should have immediately published it.—ED. M. M.

is a gentleman to whom Mr. Coles would have found no difficulty in gaining access—*sans* introduction; but it is highly absurd to suppose that the secretary of any railway company should be unable to effect an introduction to *their* engineer!

Mr. Coles puts a suppositious case, and imagines a *possibility* of his anti-friction wheel-carriages being tried, when he says " probably a host of such men as the Robertsons, Roberts, and Baddeleys, would raise their voices and ply their quills to prevent it, crying like Demetrius, " our craft is in danger." I must confess I feel myself highly honoured by this undeserved association, and beg to return Mr. Coles my best thanks, at the same time I have to inform him, that unfortunately *my craft* is in no way either directly or indirectly connected with railways, or carriages; try back Mr. Coles! Mr. Coles seems to comprehend all possible applications of anti-friction wheels within the limits of his patent-right; he says " the principle is equally well adapted for trucks, for warehouses, or for water-carts when *manual labour* (unfeeling man) would be found quite adequate to work them." He also asserts their fitness " for *rollers*, for roads, gardens, or agricultural purposes;" but I must freely confess myself so " very untutored in mechanical science," as not to comprehend their utility in the latter case where no weight is to be carried, nor any very high velocity to be obtained!

Not a single *practical* result has yet been adduced by Mr. Coles in support of the boasted superiority of his plan; neither has he obtained a favourable opinion of his plan from any one person of acknowledged experience in railway engineering—if I except, perhaps, Colonel Landmann.

Mr. Coles is, doubtless, very angry at his ill success, in not being able to persuade any body to use, what he has so carefully secured to himself by *six patents*. The contents of his pamphlet are

" A fine sample on the whole
Of rhetoric, which the learn'd call *rigmarole*."

Before he " goes to press" again I would suggest, in a spirit of kindness, that Mr. Coles should learn the use of *inverted commas* to mark any quotations he may have occasion to make, so that his readers may not be at any loss to know what was said by Mr. B., and what by Mr. C. A few lessons on the proper arrangement of the *personal pronouns* would also be of great service. There are several other matters to which I had intended to call attention, but I find I have already " spun out" such " a lengthy article" that I must forbear.

I therefore beg to subscribe myself, yours respectfully, but independently,

WM. BADDELEY.

London Dec. 20, 1838.

P. S.—I deem it necessary to enclose the copy of Mr. Coles's treatise, that you may be assured I am not joking with you.

TRIGONOMETRICAL QUESTIONS PROPOSED.

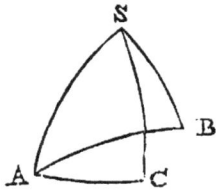

Sir,—I return you my best thanks for republishing the solution of my question in your highly valuable Journal, as given by I., in the *South African Commercial Advertiser*, of June 27. Perhaps you can find a place for the two following questions, which have severely puzzled me, and no doubt some of your able contributors will give the best possible solutions for. The first is from *West's Course of Mathematics*, the second is, I believe, a new one.

I am, Sir,
Your obedient servant,
A. B. WHITE.

Bath, Dec. 18, 1838.

Question 1st, The elevation of a tower at three several stations (no two of which are in the same straight line with the foot of the tower) are 50° 45′, 58° 15′, and 46° 45′; also the distances of the first and second station is 24, the distance of the second and third 38, and the distance of the first and third 50 yards?

Question 2nd, In the two right-angled spherical triangles A S P, A S C, right angled at P, and so are given S C and S P, also the spherical angle P A C, to determine A C and A P?

SETTING OUT CURVES OF RAILWAYS.

Sir,—In the *Mechanics' Magazine* of the 15th December, there is an article signed △, pointing out an erroneous method in use for setting out circular railway curves, and substituting a true one. Your correspondent is right in condemning the erroneous method which I believe is made use of by many surveyors through ignorance, but the one he substitutes, though true in principle, is impracticable, except on level ground. If he refers to a paper of mine in *Weale's Scientific Advertiser* for July last, he will find that he has been forestalled in his condemnation of the erroneous methods in use, and a very easy, practical and true one laid down, and demonstrated, together with a rule for projecting curves on unlevel or hilly ground. The curves projected on erroneous principles, are unshapely, even when the surface is level, but they become palpable to the most superficial observer, on hilly ground, which is the case on the London and Southampton Railway, the one alluded to in my article in the *Scientific Advertiser*. In order to detect the errors in the London and Southampton Railway the following easy method was adopted. I took a point on the outside rails, and measured a straight line, touching the convex side of the inner rails, and terminating at the outside rails. This line is double the size of the arc comprehended between its extremes and the distance of the rails (which is given) is the versed sine. The line thus measured varied at different parts of the same curve, and in every case gave a radius widely different from that to which is professed the curves belong. In truth the curves alluded to are neither circular, eliptical, parabolic, or belonging to any regular order of curves whatever. It is to be hoped, for the credit of engineers and surveyors, that the false methods will be abandoned, and others substituted more in accord with truth and science.

I remain, Sir, yours,
FOSTER CHARLTON.

Dec. 24, 1838.

HERAPATH AND COX'S IMPROVED
MODE OF TANNING APPLIED TO
DRESSING LEATHER.

Sir,—I know you feel interested in
every improvement in manufactures,
particularly when great national advantages are likely to arise from them, and
as your excellent Magazine has already
contained a discussion upon the merits
of Messrs. Herapath and Cox's patent
process for tanning butts or sole leather,
you will not perhaps consider me troublesome in sending you the results of
my first experiments upon dressing leather, under license with their apparatus.

I first took 50 Danish hides, weighing
in the salt 33 lbs each, and such as by
the old process took me more than three
months, in the same tanning solutions;
they were taken off the expressing rollers
in 24 *days*.

I send you a sample as curried; it is
very good; and if the hides had remained
on a few days longer, they would have
been nearly equal to calf, as you will observe they are not sufficiently tanned.

The next trial was upon 50 Buenos Ayres
horse-hides, weighing 24 lbs. each in the
salt, which also would have required between three and four months; they were
taken off in 28 days: and I think you
will agree with me that the specimen
which I send (curried) is of very excellent quality, and will bear the scrutiny of
any one who may see it.

I am now at work upon hides of a
different description, and shall nearly
complete those (*the third set*) *upon the
same apparatus, in about three months*, a
thing unparalleled.

From this experiment I have no doubt
of the patent reducing the period of tanning dressing leather to one-third of
that usually occupied, while the quality
of the leather will be very superior.

I assure you my experiments are with
cold liquors, and bark only, of the Hamp-
shire growth; and I would observe, the
leather does carry a most astonishing
quantity of oil, or rather what is termed
"stuff" (a mixture of oil and tallow), and
you perceive it does retain a good colour,
particularly the cordovan, which is tanned
through.

Part of the hides from which the samples are taken I have sent to Messrs.
Tuckett and Rake, leather factors, Bristol. The patentees are almost strangers
to me, yet it is but justice to them to say
that it is the most important improvement in tanning since the days of Seguin,
and is just that which was so long sought
for in vain by Sir H. Davy and others,
being simple and exceedingly economical.
I have no doubt of its ultimate success.

I am, Sir,
Yours, very respectfully,
E. WILKINS,
Tanner and Currier,
Southampton and Romsey.

67, High-street, Southampton, Dec. 26, 1838.

[The foregoing communication is part
of the information which we last week
stated that we were in possession of, upon
the subject of improved dressing leather,
when reviewing Mr. Devlin's pamphlet
on the boot-making of Britain and France.
Whether we are borne out in what we
stated by the opinions of persons in the
trade we shall be glad to know; and to
afford those who are interested an opportunity of judging, we beg to state that
the samples referred to by Mr. Wilkins
may be examined at our publishing office.—ED. M. M.]

DESGRAND'S MODE OF MAKING PAPER
FROM WOOD.

In a recent number we announced
amongst the Notes and Notices, that the
Messieurs Montgolfiers, the French paper makers, had succeeded in substituting
wood for rags in the manufacture of
paper. A specification of a patent for
this object, taken out by a Mr. Desgrand,
of Size-lane, has lately been enrolled,
which, as it states the invention to be a
"communication from a foreigner residing abroad," is doubtless the Montgolfiers process. Pure wood, of the fir or
poplar kind, alone is employed in the
process, all external bark or epidermis
being carefully removed and excluded.
The process prescribed is as follows :—

The trees must be entirely deprived
of the bark, and cut into logs four
or five feet long, which are split into
lengths a few inches square; these lengths
of wood are carefully assorted in their
various shades or colours, whereby you
may make at less expence white paper
out of the whitest pieces, and coloured or
darkened paper out of the dark or shaded pieces, without any addition of colouring matter; and by the addition of
colouring matter, all sorts of shades requisite may be produced with the white
and with the coloured pieces of wood.

The lengths are chopped into small chips of about two to four inches long, and one to three inches thick, taking care to chop them so as to make as many openings or splinters in the chip as possible, the more open they are, the more freely they admit the liquor, after spoken of, which divides and separates the different fibres forming the texture of the wood. All knots of the wood, all decayed parts, and all parts of the fibres which are not straight, must, at this stage of the process, be picked out and rejected.

A quantity of chips of one shade, about half a ton, are deposited in a pit, water-tight, in which there is an outlet for water, and there is poured over them, until they are covered, lime water, or water saturated with as much lime as it will take up; the chips remain in this for a greater or less time, according to the temperature; in the south of France, from three to six weeks were necessary to complete the action of the lime-water, or dissolve the glutinous parts of the wood which keep the different fibres of the wood united together. The bath will have produced its full effect when all the chips have sunk to the bottom of the pit. When the chips are sufficiently dissolved and saturated, the lime water is let out and replaced by a sufficient quantity of the pure water, so as to wash away all the lime water, or as much as is possible. In that state the fibres of the chips may be easily separated by submitting them to the action of stampers or fulling mallets to pound them, which opens, divides, and flattens the fibres of the chips, and facilitates the further process they have to undergo in the usual paper engine or cylinder to reduce them in pulp. When the pulp is sufficiently reduced by the paper-engine it is in a fit state to make paper or paste-board, either by itself or mixed with the usual material generally employed for manufacturing paper or paste-board, either by a paper machine or by frames.

In using the paper-engine, as now in operation for pulping rags, or other usual materials, care must be taken to employ the same sort of tackle, but use it very blunt or dull. If for white paper, the fibres, after they have been pounded, and flattened, and divided, are submitted to the action of the several chemical products which have the property of bleaching vegetable produce.

———

APPLICATION FOR EXTENSION OF PATENT.

Privy Council, Wednesday, December 12.

The petition of James Russell, Esq., of Handsworth, came on for hearing before the Privy Council this day. The Lords present were Lord Brougham, Mr. Justice Parke, Mr. Justice Bosanquet, Sir Stephen Lushington, and the Hon. Mr. Erskine. The petitioner prayed for the renewal of a patent " for certain improvements in manufacturing tubes for gas and other purposes," assigned to him by Cornelius Whitehouse, a workman employed by him for the purpose of carrying into effect the manufacture of tubes by machinery. Mr. Cresswell (in the absence of Sir William Follett) detailed the history of gas-tubing from the period of the application of old gun-barrels for that purpose. The present invention had arisen in consequence of the petitioner having been prevented by the combination of his workmen from meeting the demands of the gas companies for tubes, the supply afforded by manipulation being limited, and too dependent upon the caprice of his workmen to allow him to enter into contracts. The great demand for the patent tubing, and its great superiority over the hammered tube, had led to numerous infringements, which, coupled with the enormous extent of litigation consequent on such piracies, had deprived the petitioner of the fair and adequate remuneration he ought to have obtained for so valuable an invention, without which gas-lighting could not have been carried on to its present extent, and which had led to several inventions of great utility. Angier March Perkins, Esq., civil engineer, proved that he was the inventor of an improved method of heating buildings, which had been adopted in the British Museum, Milbank Penitentiary, and in many churches, houses, and other buildings, both public and private, to a great extent, and that his patent was entirely dependent upon the patent tubing of Mr. Russell, without which he could not have carried out his invention. Francis Bramah, Esq., civil engineer, proved that he had inspected Mr. Russell's works, and was delighted with the beauty of the invention. That he had for some years used the patent tubing, and had submitted it to a pressure of a ton upon the square inch. Mr. Bramah also spoke to the great reduction in price effected by the patent, and its utility for hollow axles, spindles for machinery, and a variety of mechanical purposes, independent of its value for transmitting heat, gas, or fluids. The Lords of the Council having intimated their opinion that the value of the patent was in some degree proved by the numerous decisions of the courts of law in its favour, the accounts were put in and verified. Mr. Fletcher, of Dudley, as the solicitor to the petitioner, produced the original patents and other documents, and proved that he had complied with the regulations promulgated by the Privy Council. The Attorney-General then addressed their lordships on behalf of the Crown, and stated that he was fully acquainted with Mr. Russels patent, having been employed in opposition to it in different courts of law. He could, however, fully attest the value and utility of the invention; and if their lordships should be of opinion that sufficient remuneration had not been afforded, he should, under the peculiar circumstances of the case, rejoice if their lordships would grant an extension, in order that adequate reward might be given to an invention of great public benefit. Lord Brougham.—Their lordships having taken the whole of this matter into account, retain the opinion which they have had impressed upon their minds from the very beginning, that this is an invention of extraordinary merit, doing the greatest honour to the inventor, conferring great benefit on the community; founding in this eminent merit not merely the application of a known principle, embodying it in new machinery, and applying it to practical purposes, but involving the discovery of a new, curious, and most important principle, and at the same time applying that principle to a most important purpose. Their lordships have, on the same side of the question, taken into account (which it is material to mention) Mr. Russel's merit in patronising the ingenious and deserving author of this invention, in expending money till he was enabled to complete this invention, and in liberally supplying the funds which were requisite for the purpose of carrying the invention into execution. On the other hand, their lordships have taken into mature consideration (which they always do in such cases), the profit made by the patentee, Mr. Russell, standing in the place of the inventor. They find that it is not a case, as in claims of other inventions of great ingenuity, and and certainly of great public benefit, of actual loss in some, and of very scanty if any profit at all realised in others—but that a considerable profit has been realised, and, upon the whole, no loss. It is to be observed, that the profit is not perhaps very much greater, if at all greater than the ordinary profits on stock to that amount employed without the privileges and extra profits of a monopoly. It is proper to consider that one great item of deduction from those profits involves great pain and anxiety and suffering to the party, namely, the litigation to which he has been subjected, and which is generally found to be in proportion to the merit and usefulness of a patent, namely, he temptation to infringe it, and to set at

nought the right of the patentee, both in the Court of Chancery, when he applies for protection by injunction, and afterwards in a court of law, when he comes to claim compensation for damages; the temptation being, as I have stated, in proportion to the benefit and the demand for the invention. That is an item which has to a considerable degree attracted the attention of their lordships in this profit and loss account which has been laid before them in the course of these transactions. Taking the whole of the matter into consideration, the merits of the patentee, the merits of Mr. Russell, and the loss that has been sustained in the litigation —and setting against those, on the other hand, the profits which have been made. their lordships were of opinion the term ought to be extended, and upon due execution being given to the undertaking, which has been just given by Mr. Fl-tcher on behalf of the inventor, that the term ought to be extended for the period of six years.

OXFORD-ST. EXPERIMENTAL PAVEMENT.

The importance of ascertaining the best species of pavement for the carriage-roads of the metropolis is some excuse for the confusion, accompanied by the smoke and offensive odours from the cauldrons, which have prevailed at the east end of Oxford-street for the last two months. The inhabitants have, however, been great sufferers thereby; but we may now congratulate them that, at last, all the ground is assigned and set out for the different varieties, while many of them are completed, and the rest are in progress. Commencing at Charles-street are the *asphaltum blocks of Robinson*, one half laid straight, the other diagonally. This is followed by *granite pavement* nine inches deep, jointed with Claridge's asphaltum; then is to succeed a granite pavement of stones only 4½ inches deep, also to be joined with the same substance, Mr. Claridge being of opinion, and desirous of proving, that his cement is sufficiently strong to bind even these shallow stones into one solid mass. To this succeeds the *Bastenne Company's* portion; the blocks in this part are in the form of bricks, but somewhat larger; they have been laid both ways, straight and diagonally. Next follows the *granite pavement*, laid by the parish, which is undoubtedly one of the finest specimens of work of its kind to be found in London. It consists of three parts:—1. Stones laid in the ordinary way, on a well-formed bed of concrete. 2. Similar stones laid diagonally on a bed of the same material; the joints of both these portions are filled with a grouting of lime, sand, &c. 3. Stones laid in the usual manner, but on the earth without any official bed, and the joints are filled in with fine gravel. The whole of this work has a good curved surface, and the regular thickness of the stones has evidently been carefully attended to. The next experiment, going towards Tottenham-court-road, is what is called the *Scotch asphaltum granite* (said to be a patented article). This composition has the appearance of stone, and the blocks are about six inches thick, nine inches broad, and eighteen inches long on one face, while the other is only thirteen inches long. In laying them (which is done with Parker's cement), every alternate block is reversed, so that every second block lies solid on its base, or longest face, while the others fit in between them as keystones, and when joined each may be said to support the other. The next division is the *wood pavement*, composed of blocks of fine timber Kyanized, they are of a hexagonal form, seven inches diameter, and fifteen inches deep, part is laid on a bed of one and a half inch planks. Then follows the *Val de Travers Asphalte*, which will occupy the remaining portion of the street devoted to the experimental pavement. This last article consists of blocks, about ten inches square and five inches thick, formed of a bitumen thickly studded by broken pieces of granite; so that, when laid, it may be looked upon as a sort of macadamised road, where, in lieu of earth, for filling the interstices between the broken granite, and making the whole of a solid mass, a strong binding composition has been employed. *C. E. and Arch. Journal.*

EXPERIMENTS ON THE POWER OF MEN. BY JOSHUA FIELD, V. P. INST. C. E., F.R.S.
(From the *Trans. Inst. C. E.*)

In this paper are recorded the results of some experiments made to ascertain the working power of men with winches, as applied to cranes. The experiments were undertaken with a view of ascertaining the effect men can produce working at machines or cranes for short periods, as compared with the effect which they produce working continuously.

The apparatus, a crane of rough construction in ordinary use, and not prepared in any manner for the experiments, consisted of two wheels, of 92 and 41 cogs, and two pinions, of 11 and 10 cogs; the diameter of the barrel, measuring to the centre of the chain, was 11¾ inches, and the diameter of the handle 36 inches. The ratio of the weight to the power on this combination is 105 to 1.

The weight was raised in all cases through 16½ feet, and so proportioned in the different experiments, as to give a resistance against the hands of the men of 10, 15, 20, 25, 30, and 35 lbs. *plus* the friction of the apparatus. * * * *

In order to compare these experiments with each other, these results must be reduced to a common standard of comparison, and it is very convenient to express the results of such experiments by the pounds raised one foot high in one minute, this being the method of estimating horses' power. The number is in each case obtained in the following manner. I will take the first experiment.

Here 1,050 lbs. was raised 16½ feet high in 90 seconds; this is equivalent to $(1,050 + 16\cdot5) = 17,325$ lbs. raised one foot high in 90 seconds, which is equivalent to $(17,325 \div 1\cdot5 = 11,550$ lbs. raised one foot high in one minute. In this case then the man's power = 11,550.

The same calculations being pursued in the other cases, give the numbers constituting the last column of the following table:—

No. of Experiment.	Statical resistance at handle.	Weight raised.	Time in seconds.	Time in minutes.	REMARKS.	Man's power.
I.	10	1050	90	1·5	Easily, by a stout Englishman	11550
II.	15	1575	135	2·25	Tolerably easily, by the same man........	11505
III.	20	2100	120	2	Not easily, by a sturdy Irishman	17325
IV.	25	2625	150	2·5	With difficulty, by a stout Englishman....	17329
V.	30	3150	150	2·5	With difficulty by a London man	20790
VI.	35	3675	132	2·2	{ With the utmost difficulty, by a tall } { Irishman..................... }	27562
VII.	..	..	150	2·5	{ With the utmost difficulty, by a London } { man, same as Experiment V....... }	24255
VIII.	..	..	170	2·83	With extreme labour, by a tall Irishman..	21427
IX.	..	..	180	3	{ With very great exertion, by a sturdy } { Irishman, same as Experiment III. .. }	20212
X.	..	..	243	4·05	With the utmost exertions, by a Welshman	15134
XI.	..	..	35	....	Given up at this time, by an Irishman	....

We may consider experiment IV. as giving a near approximation to the maximum power of a man for two minutes and a half; for in all the succeeding experiments the man was so exhausted as to be unable to let down the weight. The greatest effect produced was that in experiment VI. This, when the friction of the machine is taken into the account, is fully equal to a horse's power, or 33,000 lbs. raised one foot high in one minute. Thus, it appears, that a very powerful man, exerting himself to the utmost for two minutes, comes up to the constant power of a horse, that is, the power which a horse can exert for eight hours per day.

MANUFACTURE OF GUN FLINTS.

An important application of siliceous substance is in the formation of gun flints, for which purpose it must be cut in a peculiar manner. The following characters distinguish good flint nodules from such as are less fit for being manufactured. The best are somewhat convex, approaching to globular; those which are very irregular, knobbed, branched and tuberose, are generally full of imperfections. Good nodules seldom weigh more than 20 pounds; when less than two they are not worth the working. They should have a greasy lustre, and be particularly smooth and fine grained. The colour may vary from honey-yellow to blackish-brown, but it should be uniform throughout the lump, and the translucency should be so great as to render letters legible through a slice about one-fiftieth of an inch thick, laid down upon the paper. The fracture should be perfectly smooth, uniform, and slightly conchoidal; the last property being essential to the cutting out of perfect gun-flints.

Four tools are employed by the gun-flint makers.

First, a hammer or mace of iron with a square head, from one to two pounds weight, with a handle 7 or eight inches long. This tool is not made of steel, because so hard a metal would render the strokes too harsh, or dry as the workmen say, and would shatter the nodules irregularly, instead of cutting them with a clean conchoidal fracture.

Second, a hammer with two points, made of good steel well hardened, and weighing from 10 to 16 ounces, with a handle 7 inches long passing through it in such a way that the points of the hammer are nearer the hand of the workman than the centre of gravity of the mass.

Third, the disc hammer or roller, a small solid wheel, or flat segment of a cylinder, parallel to its base, only two inches and a third in diameter, and not more than 12 ounces in weight. It is formed of steel not hardened, and is fixed upon a handle six inches long, which passes through a square hole in its centre.

Fourth, a chisel tapering and bevelled at both extremities, 7 or 8 inches long, and 2 inches broad, made of steel not hardened; this is set on a block of wood, which serves also for a bench to the workmen. To these 4 tools a file must be added, for the purpose of restoring the edge of the chisel from time to time.

After selecting a good mass of flint, the workman executes the following four operations on it.

1. *He breaks the block.*—Being seated upon the ground, he places the nodule of flint on his left thigh, and applies slight strokes with the square hammer to divide it into smaller pieces of about a pound and a half each, with broad surfaces and almost even fractures. The blows should be moderate, lest the lump crack and split in the wrong direction.

2. *He cleaves or chips the flint.*—The principal point is to split the flint well or to chip off scales of the length, thickness, and shape adapted for the subsequent formation of gun flints. Here the greatest dexterity and steadiness of manipulation are necessary; but the fracture of the flint is not restricted to any particular direction, for it may be chipped in all parts with equal facility.

The workman holds the lump of flint in his left hand, and strikes with the pointed hammer upon the edges of the great planes produced by the first breaking, whereby the white coating of the flint is removed in small scales, and the interior body of the flint is laid bare; after which he continues to detach similar scaly portions from the clean mass.

These scaly portions are nearly an inch and a half broad, two inches and a half long, and about one-sixth of an inch thick in the middle. They are slightly convex below, and consequently leave in the part of the lump from which they were separated a space slightly concave, longitudinally bordered by two somewhat projecting straight lines or ridges. The ridges produced by the separation of the first scales must naturally constitute nearly the middle of the subsequent pieces; and such scales alone as have their ridges thus placed in the middle are fit to be made into gun-flints. In this manner the workman continues to split or chip the mass of flint in various directions, until the defects usually found in the interior render it impossible to make the requisite fractures, or until the piece is too much reduced to sustain the smart blows by which the flint is divided.

3. *He fashions the gun-flints.*—Five different parts may be distinguished in a gun-flint. 1. The sloping facet or bevel part, which is impelled against the hammer of the lock. Its thickness should be from two to three-twelfths of an inch; for if it were thicker it would be too liable to break ; and if more obtuse, the scintillations would be less vivid. 2. The sides, or lateral edges, which are always somewhat irregular. 3. The back or thick part opposite the tapering edge. 4. The under surface, which is smooth and rather concave. And 5. The upper face, which has a small square plane between the tapering edge and the back, for entering into the upper claw of the cock.

In order to fashion the flint, those scales are selected which have at least one of the above mentioned longitudinal ridges ; the workman fixes on one of the two tapering borders to form the striking edge, after which the two sides of the stone that are to form the lateral edges, as well as the part that is to form the back, are successively placed on the edge of the chisel in such a manner that the convex surface of the flint, which rests on the forefinger of the left hand, is turned towards that tool. Then with the disc hammer he applies some slight strokes to the flint just opposite the edge of the chisel underneath, and thereby breaks it exactly along the edge of the chisel.

4. The finishing operation is the *trimming*, or the process of giving the flint a smooth and equal edge ; this is done by turning up the stone and placing the edge of its tapering end upon the chisel, in which position it is completed by five or six slight strokes of the disc hammer. The whole operation of making a gun-flint, which I have used so many words to describe, is performed in less than one minute. A good workman is able to manufacture 1000 good chips or scales in a day (if the flint-balls be of good quality), or 500 gun-flints. Hence, in the space of three days, he can easily cleave and finish 1000 gun-flints without any assistance.

A great quantity of refuse matter is left, for scarcely more than half the scales are good, and nearly half the mass in the best flints is incapable of being chipped out ; so that it seldom happens that the largest nodules furnish more than 50 gun-flints.—*Dr. Ure's Dictionary of Arts and Manufactures.*

PLAN FOR DETECTING THE DISPLACEMENT OF RAILS, AND OF PREVENTING ACCIDENTS ON RAILWAYS.

Captain Smith, R.N., suggests the following plans for giving additional safety to railway travelling :—For the purpose of detecting the displacement of a rail on any line of road, that the policemen stationed for its protection should be desired to pace from one station to the other, once or twice during their watch, drawing after them a *staff along the edge of the rail ;* the staff being fitted with a hook on the end, so formed as to *fit the edge of the rail.* It appears to me, that by this simple and inexpensive plan the slightest derangement would be detected with much greater certainty than if left to the vision of the men, though aided by a lantern, especially during heavy rain or snow.

The application of this detector would also serve to discover any thing that might be thrown on a rail by design, or driven on it by a powerful gust of wind, &c.

The application of the plan would ensure the patrolling of the line of road from one end to the other, the men being desired to pass each other a few yards before they returned to their station-box, which they should do by the other rail, drawing the staff or detector after them, as on the first. The men might be required to exchange their staff (which should be numbered), as a proof they had communicated with each other on their beat.

I ventured to suggest, some time since, that a gong might be used to advantage, if attached to the *last* carriage in each train, to be struck for the purpose of warning a train that might be expected to overtake one that was detained on the road by accident or otherwise[*]

I am aware that powerful lights are used for these purposes ; but as fogs are occasionally exceedingly dense, and people's eyes are not always open, I feel assured that the sound of a powerful gong would be found much more useful in peculiar cases—and it might also be used as a means of calling the attention of the engineer, should the guard at the opposite end of a train require to do so, in the event of a separation in the train or other casualty.

It is also suggested for consideration, that in order to enable the trains to run without loss of time when the rails are slippery, that the engine should be fitted with a box or compartment to contain sand, from which two pipes should lead before the first wheels, and immediately over the rails, to be arranged so that the engineer could with ease cause sand to be sprinkled lightly on the rails, whenever he found them too slippery to proceed as fast as he wished. The sand to escape on the principle by which some seeds are sown. The first or the last car-

---

[*] An accident occasioning the loss of *lives*, is just announced in the papers as having occurred in America in consequence of a dense fog.

riage might brush the sand off (if thought requisite), as it followed the engine, to prevent the rails being unnecessarily worn by the increased friction. [The use of sand, &c., has been already suggested by Colonel Maceroni.]

NOTES AND NOTICES.

Papier Maché Ornaments, for the "Actæon," Liverpool and Glasgow Steam-ship.—We have been favoured, within the last few days, with an inspection, at the manufactory of Messrs. Jennens and Bettridge, of a set of panels, in *papier maché*, intended for the decoration of the "Actæon," Liverpool and Glasgow steamer; which, as works of art, have not, we believe, been surpassed by anything of the kind ever produced at this celebrated establishment. The panels are 28 in number, four of which are very large, and consist of historical subjects, some original, and others copies from the works of celebrated masters. The first represents the triumphal entry of Alexander into Babylon; the second exhibits a view of a Grecian sea-port, and the arrival of a victorious fleet: the third describes the Olympic games, the combats of gladiators, &c.; the fourth gives a representation of the Hippodrome, the Temple of Victory, and chariot races. Each of these subjects is depicted by the artist with the vividness and freshness of life. The various groups of Grecian, Egyptian, and Persian figures, the richness and brilliancy of the costumes, the colossal statues, temples, and columns, in their architectural grandeur and beauty, furnish a vivid representation of the barbaric pomp and magnificence of by-gone ages. The smaller panels are divided into the classes, devoted to the illustration of particular subjects. The first series represents full-length figures, emblematic of Victory, Commerce, and the Arts and Sciences, surrounded with beautiful ornamental work, drawn in imitation of *alto-relievo;* the whole surmounted with the arms of Liverpool and Glasgow. The second embraces mythological subjects, representing the triumph of Neptune, Juno, and the Graces, Actæon, &c.; the whole adorned with an emblematic framework. The third comprises mosaic heads, and emblems ornamented with arabesque foliage, birds, flowers, and fountains. Viewed separately, each of these paintings is an exquisite specimen of the advanced state of this department of our manufactures and the fine arts; and, as a whole, they form unquestionably one of the most unique and splendid collections of the kind ever produced. The panels will not, we believe, be removed for a few days from the show-rooms of the manufactory, where artists and other visitors may have an opportunity of inspecting them.—*Birmingham Herald.*

Railway Missionaries.—The Bishop of Bath and Wells has appointed the Rev. F. Campbell, M.A., as a missionary among the navigators employed on the Bristol and Exeter Railway. The Railway Directors, and the Church Pastoral Aid Society, have liberally contributed, and it is hoped that the same well-judged efforts will be made elsewhere to reclaim them from their present heathen state.—*Bath paper.*

Sleep on Railroads.—The following notice of an invention of accommodation "to sleep," as if at home, while travelling on railroads, appears in a recent number of the *Baltimore American:*—" The introduction of the newly-invented sleeping cars on our railroads makes that kind of travelling almost perfect—all that is wanting now is a dining car. The

sleeping cars will soon be placed on the railroad between this and Philadelphia, so that traveller s leaving here in the seven o'clock train may go *to* sleep in this city, and not be disturbed till they reach Philadelphia. These cars are fifty feet in length, and the seats, which are sideways, can, by a simple movement, be converted into *berths;* in each car forty-eight passengers can be accommodated with berths."

Heart of Oak.—One of the piles used in the foundation of the old bridge at Lancaster was taken out a short time since, and found to be " as sound as an acorn," although it must have been under water at least 900 years.

Preservation of Sculpture.—Experiments are in progress at the École des Beaux Arts at Paris, with some oily substance to be used for the preservation of marbles and works of art, which suffer so much in the wear and tear of a great metropolis.—*Galignani's Messenger.* [This is one of our neighbours reinventions of old English inventions; some years ago Henning, the Sculptor, used a coating of wax on the triumphal arch in the Park, and hitherto complete with success. Colonel Maceroni has also repeatedly proposed similar plans of preserving stone.]

Zincography.—The *Baltimore American* states, that a method has been invented of drawing on zinc, said to be very superior in effect to lithography. The mode of preparing the metal so as to fit it for the purpose is said to be a secret unknown but to one person in the country. In the process of stamping, a delicate pink tint is conveyed to the paper, by which the engraving is made to assume the appearance of drawing on chalk.

Galvanized Iron, is the somewhat fantastic name newly given in France to iron tinned by a peculiar patent process, whereby it resists the rusting influence of damp air, and even moisture, much longer than ordinary tin plate. The following is the prescribed process: Clean the surface of the iron perfectly by the joint action of dilute acid and friction, plunge it into a bath of melted zinc, and stir it about till it be such as is used for making tin plate. The tin forms an exterior coat of alloy. When the metal thus prepared is exposed to humidity, the zinc is said to oxidise slowly by a galvanic action, and to protect the iron from rusting within it, whereby the outer tinned surface remains for a long period perfectly white, in circumstances under which iron tinned in the usual way would have been superficially browned and corroded with rust.—*Ure's Dictionary of Arts, &c.*

Chanter's Smoke Consuming Furnaces.—Some time ago (No. 771) we described a locomotive engine furnace, invented and patented by Messrs. Chanter and Co., the principal feature of which was the placing the furnace bars at a considerable angle, so that the fuel being supplied at the top of the incline, would gradually descend as that beneath it was consumed. A deflecting plate, at a corresponding angle with that of the furnace bars, caused the smoke from the newly added fuel to pass over the fuel in a state of incandescence, and be thus consumed. This plan has, we are informed, answered well in stationary engines, but the great draught of the locomotive furnace forced the fuel too quickly down the incline at whatever angle it was placed, and the plan was, in consequence, for the time abandoned. Messrs. Chanter and Co. have, however, since made such improvements as completely (according to experiments which have been detailed to us) to obviate this difficulty, and to effect an immense saving of fuel. We are promised particulars of the invention as soon as the patent right is secure.

LONDON: Printed and Published for the Proprietor, by W. A. Robertson, at the Mechanics' Magazine Office, No. 6, Peterborough-court, between 135 and 136, Fleet-street.—Sold by A. & W. Galignani, Rue Vivienne, Paris.

𝔐𝔢𝔠𝔥𝔞𝔫𝔦𝔠𝔰' 𝔐𝔞𝔤𝔞𝔷𝔦𝔫𝔢,

MUSEUM, REGISTER, JOURNAL, AND GAZETTE.

No. 805.]   SATURDAY, JANUARY 12, 1839.   [Price 3*d.*]

Printed and Published for the Proprietor, by W. A. Robertson, No. 6, Peterborough-court, Fleet-street.

NATIONAL SWIMMING SOCIETY'S MACHINE FOR TEACHING SWIMMING.

NATIONAL SWIMMING SOCIETY'S MACHINE FOR THE TEACHING OF SWIMMING.

Sir,—I beg to transmit for insertion in your widely circulated journal, a drawing and description, of a machine belonging to the National Swimming Society, invented for the purpose of facilitating the teaching of swimming. Before describing the machine in question, it may be well to observe, that this Institution has been founded for the purpose of promoting the noble art of swimming throughout the British empire, with a primary view to the preservation of human life, and a secondary object of affording to all classes an opportunity of enjoying a delightful, healthy, and invigorating exercise.

As the utility of the knowledge of swimming cannot be questioned, more especially when we reflect on the many distressing circumstances which daily occur from the loss of life by drowning, which a knowledge of this art might have prevented, the Committee have spared no trouble or expense in perfecting their plans, one of which forms the subject of the present paper.

It was found by the swimming-masters connected with the Society, that in teaching youth, or others, to swim, they must at once discard the old system of bladders, preservers, &c., as being both dangerous and impracticable; and it was also found that the practice of standing half in and half out of the water while giving instructions, was very injurious to the health of the teachers: hence, the present machine was invented with a view to combine safety to the pupils, with facility in teaching, together with comfort and convenience to the masters. So far as the attainment of these objects are concerned the Committee have had their utmost wishes realised, as the machine has answered the purposes for which it was constructed to the fullest extent; and has been daily in use on the Surrey canal during the last summer.

The machine (see front page), consists of two pontoons, formed at the ends in the manner of a barge, each 18 feet long, 2 feet wide, and 2 feet deep. These are placed parallel, at the distance of 12 feet apart, and support a stage the lower part of which is 2 feet above the upper sur-

face of the pontoons, and is supported by the iron stancheons, braces, &c., shewn by the drawing. The stage is 14 ft. 4 in. long by 4 ft. wide, framed together with braces 8 inches deep, and surrounded by a strong rail. The end rails are permanently fixed into iron sockets; but the side rails are hinged so as to be let down when the machine is in operation, (as exhibited in the drawing), thus making the whole width 9 ft. The rails when let down are supported at the ends by the transverse braces of the frame-work, which are projected beyond the middle part of the stage for that purpose. In the side of each rail thus let down are fixed three strong iron rods about 14 inches long, radiating to the centre of the stage, and having brass pulleys at their ends. From these the swimmers are suspended by strong cords attached to a swivel-ring, which is fixed to a "harness" placed on the upper part of the body,—the cords passing through the pulleys to others at the bottom of the mast, thence to pulleys at about three feet above the floor of the stage. The ends of the cords are fixed to clamps fixed round the mast. It will be seen that the swimming master can adjust the cords to the greatest nicety, and give the pupils more or less of "rope," as circumstances may require, or even raise them out of the water if necessary. The points of suspension are 4 feet apart, which leaves sufficient room for the limbs to strike out. The heads of the two rows of swimmers are placed in the same direction; so that the pupils while learning to use their limbs are also progressing forwards, drawing the machine along with them; and as the points of suspension across the stage are 11 feet 4 inches apart, the feet of the first row are sufficiently in advance of the heads of the second. The floor of the stage is boarded longitudinally, leaving an opening of one inch between each for the escape of water. At the ends of the stage are boxes, shewn in the drawing, which serve both as seats in dressing, and places to hold the clothes or other property of the swimmers.

The lower part of the mast, which is hexagonal, is firmly fixed to a transverse beam in the stage, and the upper part is fixed into the lower part, in the manner of a pivot and socket screw, and can be unshipped at pleasure.

A temporary awning has been placed round the machine, for the purpose of privacy in undressing and dressing, which, however, has nothing to do with the invention.

The drawing of the machine, with the swimming master and his assistant on the stage, together with the representation of the pupils in exercise, will, I trust, render further description unnecessary.

It is the intention of the Society to establish stations throughout the kingdom, at each of which a machine will be placed; and as the invention is perfect and complete in itself, the machine can be removed in its respective locality to the most approved stream or piece of water, without the disadvantage of openly exercising, a stationary bath in still water, or, the expense of paying for a covering in addition to the expense of teaching.

The machine in full operation, with the flag of the society unfurled, mast high, presents rather a novel appearance, in addition to its utility, and has afforded much amusement as well as gratification to numbers who have seen it on the Surrey canal.

The machine is the invention of Mr. Strachan, the founder of the Society, and was built under his direction by Mr. Thomas Ower, of Pimlico.

I remain, Sir, yours, obediently,

J. Robertson,

Hon. Sec. to the N.S.S.

Society's House, Green Man Tavern, Berwick-street, Oxford-street, Dec. 29, 1838.

It is with great pleasure that we publish the preceding communication from our old and esteemed correspondent. The institution in which he now appears to take so much interest is one of great national and social importance, and if judiciously managed cannot fail of bringing about most beneficial results. Bathing, in a sanatary point of view, can never be properly and pleasurably enjoyed without that confidence and self-possession which is only felt by the practised swimmer. We think the plan of the National Swimming Society good, far superior, indeed, to that of "L'Ecoles de Natation," which glitter in gaudy paint, like immense fair-booths, on the Seine at Paris. The portable and inexpensive nature of the English apparatus will most certainly tend in a very great degree to its general adoption in all localities where there is a convenient river or sheet of water. One of the machines, upon a smaller scale than above described, would be, we should think, a beneficial addition to the public swimming baths which are established in the neighbourhood of London.

In addition to the information contained in our correspondent's letter, we add the following extracts from the introductory notice to the rules and regulations, written by Mr. J. Strachan, stated to be the founder of the Society, who we are happy to see, has the full merit of his honourable endeavours awarded to him by his associates.

"From my earliest years I have entertained an opinion that an Institution for teaching the Art of Swimming would be desirable, as the acquirement of that art is, in many cases, so essential to the preservation of human life. * * * There are few individuals who, at some period of their lives, are not exposed to marine dangers, and it is truly gratifying to know that those who are proficient in the art, when thus unhappily situated, exhibit a courage and tranquillity which confidence inspires; whilst those who are without this attainment have their courage shaken, and are instantly overwhelmed.

"As swimming is acknowledged by all who have participated in its enjoyments to be an individual benefit, and a national accomplishment; so, likewise, it is allowed, that the want of a knowledge of this valuable acquisition has been the chief cause why so many of our worthy fellow-countrymen, the brave defenders of our nation, have while fighting its victorious battles met with a watery grave. Therefore, as an inducement to all classes to learn this most useful exercise, it is the intention of this Society, in the extension of its usefulness, not only to teach as many as will come under its banners, but likewise to encourage by example and rewards, the most expert swimmers. By teaching this art to youth, we not only increase our individual safety, but likewise our public power; inasmuch, as all will be taught not only to save themselves, but also to rescue others in the time of danger— which must tend in the highest degree to strengthen both our Navy and Army."

We cannot better dismiss this subject for the present, nor more forcibly recommend it to the attention of our readers, than by quoting the eloquent lines of Thomson; albeit, the season is not the most tempting in which to commence a pupilage in the art.

"This is the purest exercise of health,
The kind refresher of the summer heats ;
Nor, when cold Winter keens the brightening flood,
Would I weak shivering linger on the brink.
Thus life redoubles, and is oft preserv'd,
By the bold swimmer, in the swift illapse
Of accident disastrous. Hence the limbs
Knit into force ; and the same Roman arm,
That rose victorious o'er the conquer'd earth,
First learn'd, while tender, to subdue the wave.
Even, from the body's purity, the mind
Receives a secret sympathetic aid."

THE RESISTANCE OF THE ATMOSPHERE TO GREAT RATES OF SPEED ON RAILWAYS.

Sir,—I take the liberty of handing you the following paragraph which I have extracted very recently from an evening paper. Excepting, certainly, the somewhat shy and sarcastic hint conveyed in the parenthesis, it has every appearance of being derived from the first authority, and may therefore be considered as authentic information.

"Dr. Lardner has discovered by experiments recently made on the Liverpool and Manchester Railways, that the atmosphere is an opponent more formidable than has ever been suspected. At 32 miles an hour the resistance it offers is nearly 80 per cent. of all that the steam power has to encounter, and it increases in a proportion so enormously greater than the speed, that there is not the slightest possibility of any such velocity of transit being gained, as some (and among them none more ardently than Dr. himself) have anticipated. It is certain, that even 40 miles an hour, cannot be maintained, except at a cost which amounts practically to a prohibition."

Dr. Lardner is peculiarly unfortunate in his opinions on practical matters, whether in regard to what can, or what cannot be performed, but in this he is not singular among mathematicians, the tendency of whose minds and studies, as contrasted with men of practical views, is not to invest, but rather to divest things of all those physical circumstances and adjuncts, which give to them character and reality. Hence the surprise which Dr. Lardner, and others who are mathematicians rather than philosophers, manifest at the enormous resistance which the atmosphere offers to great locomotive velocity, is not participated in by practical men, who have long been aware of this circumstance, and have long since in their anticipa-

tions, approximated to that estimate of it which is now announced as being the result of experiments. This time Dr. Lardner has had the satisfaction of being himself instrumental to the correction of his former opinions, the experiments by which it has been effected, being those, I presume, which the British Association directed to be undertaken, and for which they granted a sum of money under the designation of ascertaining Railway Constants.* But even the men of abstract science should not thus have been found so short sighted, for no very long time has elapsed, since they had a warning and a lesson upon the subject of atmospheric resistance, when the mathematical theory of projectiles sunk in ruins before it, and when the experiments of Dr. Hutton, taught them, that in order to their investigations being possessed of an iota of value in a practical point of view, the dignity of science must bend, even before such common place practices as observation and experience—that it must submit to have the certainty and exactness of its results, alloyed and lowered by the modifying influence of the data so obtained, and its severe and beautiful simplicity disfigured, by the correcting formulæ of a mere empyrical process.

The mathematicians have been exceedingly busy with their various and opposing theories of locomotive action, as among other places the pages of your Magazine testify, and now, to the great amusement of practical men, they are all at once shaken back in the progress of their fine spun speculations, by discovering, that the resistance of the air is so rude and so powerfully a disturbing force, that, abstracting its influence, their conclusions are altogether useless, and have ended only in vanity and vexation of spirit. Who would have thought it, say they, that it should prove to be so important an element in the calculation ? What cruel havoc has it made among many an imposing display of our art, such as strikes the uninitiated dumb— creates so powerful an illusion in our favour,—imparts such a prestige even to the silliest of our lucubrations, and enables us to *look down* your mere prac-

---

* Our correspondent is mistaken. The experiments which have led Dr. Lardner to his *new* theories are those made under the direction of Mr. Wood upon the Great Western Railway, to try the efficacy of Mr. Brunel's system, some account of which we shall give in our next.—ED. M. M.

tical and mechanical man, and to *symbolize* him out of countenance.

But the practical men can say, that they thought of it, and told them of it too, in those communions which they have the high privilege to hold with them, in the humble section G of the British Association for the advancement of science. There they forewarned them of the mighty resistance which it would be proved that the air opposes at high velocities ; offering at the same time illustrations of it, and urging that experiments should be made to decide the point. There also, these opinions of practical sagacity received the support of such among the men of science present, who are philosophers as well as mathematicians. What the result of the experiments directed by the Association to be made, has been, the paragraph above announces, and fully does it justify the far seeing anticipations of practical skill and discernment. And yet methinks, from the manner in which it is worded, and in which I doubt not many more paragraphs on this subject will from first to last be indited, the public will be led to imagine, that the men of science, emphatically so called, have again stepped forth to be the guides and instructors of that lamentably ignorant and unmathematical class of men, the civil engineers of Great Britain*— that they have once more generously unfolded to them novel and important views, concerning even their practical pursuits ; and have with noble condescension, undertaken the task which high science only could inspire, and for which closet speculations alone could qualify, of correcting the crude ideas, and of moderating the extravagancy of expectation, which must needs spring up in minds, untutored in the colleges of the Universities of London and of Durham.

The subject of atmospheric resistance at high locomotive velocities, is no stranger to the pages of your Magazine. So long since as the year 1833, in the course of the controversy occasioned by

that most ludicrously absurd scheme of an undulating railway, brought forward by Mr. Badnall, and so quixotically championed by all your mathematical correspondents, * I adverted to the enormous resistance of the air in the following terms : — "Mr. Badnall, in common with many more, appears to regard very lightly the resistance of the air. I am disposed to think that in our rage for speed, and provided corresponding improvements take place in the formation of the road line, the stability and solidity of the rails, the diminution of percussion, the bestowing of needful elasticity, and in the more minute details of the wheels, bearings, &c., we shall not rest content, until, on a level, the air will oppose *nineteen parts in twenty* of the total resistance." Since that time those improvements in railway mechanism at which I hinted, have in part taken place, and now we are informed, that " at 32 miles an hour, the resistance which is offered by the air, is nearly 80 per cent. of all that the steam power has to encounter," which is coming pretty close to what I anticipated, and which, at the proposed velocity for the mail of 40 miles an hour, would fully realise my expectations. The atmosphere, therefore, is *not* "an opponent more formidable *than has ever been suspected.*" In bringing this topic to bear as an argument on the subject of the controversy, I observed, that " Mr. Badnall gives us alternate fits of high and low velocities, and makes the resistance of the air greater than necessary, or than would be encountered with the average of those velocities; for he should remember, that this kind of resistance increases with the *square* of the velocity," and now we are told that " it increases in a proportion *enormously* greater than the speed." (*Query*—In a greater or less proportion than the square ?) I also inquired in the same communication (No. 524) " whether Mr. Badnall had ever calculated how many hundred horse power an engine must be in order to produce a velocity of 100 miles in an hour, in opposition solely to

* Some observations were lately made in the presence of a distinguished mathematician and pupil of "l'Ecole Polytechnique," on the deficiency of mathematical knowledge among our civil engineers, when the Frenchman, with a significant shrug of the shoulders, candidly remarked that there was not much to be regretted on that head, for that mathematical attainments, except of course such as are of routine acquirement in a professional education, are but poor qualifications for an engineer.

* Mr. Badnall always claimed Dr. Lardner as being his supporter in this notable affair, but in justice to the Doctor it ought to be known, that in section G. of the Dublin meeting of the British Association, he emphatically asserted, either that he had not countenanced, or that he did not [*then*] countenance the undulating railway principle; but which was the expression he used I do not distinctly recollect.

this resistance," and in a subsequent paper I stated that such an engine must be of 312 horse power.

I would not have brought forward my own statements to show that there is no novelty in the results now elicited by Dr. Lardner's experiments, but that I was at the time most unmercifully assailed on account of them by your mathematical correspondents. They were all up in arms on the occasion. "Kinclaven," the redoubtable mathematical Mentor and oracle of your pages, was particularly forward and zealous; and even the celebrated lady "M. S.," was roused to give her opinion on the subject. It will scarcely be credited by your readers, unless they refer back to your 20th vol., that "Kinclaven" had the hardihood to assert, "that for any safe velocity, the resistance of the air is an element that may altogether be rejected," and "that instead of more than 300 horses, it would not require the one-twelfth part of the force of a single donkey to overcome the resistance, allowing two donkeys to equal the strength of one horse. Here then, as the event has proved, is a most instructive example of the little value of even the highest mathematical attainments, unless under the guidance and correction of sound common sense, such as in general preeminently distinguishes our civil engineers,* and which practice and experience, if not the only means of acquiring, are at least the most efficient means of improving. It ought to teach our mathematicians to be a little more modest, in regard to the opinions which they are too apt most dictatorially to propound to practical men, even upon those practical matters about which the one party is necessarily the most and the other the least conversant; it ought to teach them to abstain from, or at least to moderate, that officious interference in other men's professional affairs, to which, incited by an over-weening conceit of their own superior wisdom, they have of late years been so remarkably prone; and it ought, in connection with other cases of a similar kind, to put the public and the legislation on their guard, against confiding, I had almost said in

any degree, in statements conveying expectations, either of failure or success, which have only chamber practice and mathematical skill, as a guarantee for their probability and trustworthiness.

It is probable, that among a certain class of thinkers, opinions will now take as strong a turn against the practicability of greatly increasing the velocity of locomotives beyond the present rate, as heretofore they were, in the same unreasoning manner, extravagantly in favour of it.* It will become, I have no doubt, a favourite topic, and have quite a run; for these persons betray their want of discrimination and judgment, equally by the denunciation of what is feasible, as by outrageous speculations as to what is practicable. Such, however, is the ingenuity and perseverance of our engineers, that it is not at all improbable, by improvements which may yet be made in the interior construction of the locomotive, so as to diminish the velocity of the moving parts, and by an attention to the exterior form, both of the engine, the tender, and one accompanying carriage, so as to diminish the effective area of resistance, they will, after all that can be said respecting the opposition of the air, be able at least to double the present speed, if any case, such as the transit of the mail, should authorise such a profuse expenditure of power.

Between two and three years after this subject was discussed in the pages of your Magazine, Mr. Herapath introduced it in his *Railway Magazine*, and took the same view of it as I had done; only that, instead of using Dr. Young's formula for calculating the resistance of the air, or availing himself of Dr. Hutton's experiments on Smeaton's table, he derived it in his usual *à priori* manner, from certain theoretical considerations, peculiar to his own whimsical hypothesis of the constitution and phenomenon of matter. He thus brought out the resistance, rather less per square foot than I had done; but in other respects, such as the area of resistance in a train being taken at thirty square feet, and the law of the resistance being in the ratio of the *square* of the velocity, he

---

* My own profession is, I imagine, sufficiently well known, to shield me from the imputation of including myself in the above eulogium upon that distinguished body.

* The writer in the Quarterly Review just out, anticipates nothing less than a velocity of 100 miles an hour.

took the same data as I had adopted. He, however, treated the subject, as though the views which he gave of it were entirely new, and had originated with himself. The observations in which he indulged, expressive of his contempt for engineers, were wholly uncalled for. He may be assured, that the subject of atmospheric resistance had not escaped their attention, though mathematicians, and men of science, so called, had failed to notice it. It was very late in the day before he himself had recognised it, and then only because the circumstance was forced upon his attention by his mathematical formulæ giving results so widely differing from practice and experiment.

Our civil engineers can better employ their time than in talking, or writing, or lecturing. Posterity will appreciate their minds in their deeds, and pass into oblivion the scientific twaddle of their would-be instructors. At the next meeting of the British Association, this atmospheric resistance will afford a capital subject for high science to seize and descant on, for the especial instruction, but real amusement, of the civil engineers who may be present. There will be such big looks of wig-wisdom on the occasion,—such an air of oracular decision,—such a condescending attention to begin at the A, B, C, of the subject, so as to bring it perfectly down to the comprehension of that mere practical order of intellects which is to be addressed,—such a cool attempt to attribute novelty and originality to the views and opinions then propounded,—and such a subdued but self-complacent tone of triumph, on this account, over the men of clods of stones and of iron, although they were the first to court the investigation, and to anticipate the results; that, altogether, there will be exhibited as pretty a piece of solemn scientific foolery as a man would wish to behold on a fine summer's day—and " may I be there to see." But I hope that the threatening symptoms of a secession of the more eminent of our practical men from this section, will not be further exasperated; but they will subside under the healing influence of those more numerous as well as more distinguished members of the Association; who being themselves not merely the expounders, but the real working parties, and the pioneers of the body scientific, are better able to appreciate, and more willing to allow, the great talents and the energy displayed, by those who labour in the departments of practical knowledge and enterprise. I hope also, Mr. Editor, you will be able to give us a better report of the proceedings in Section C, than you will find in the *Athenæum;* for this important section has never had justice done to it in that publication.

Some of your readers may be curious to know, and certainly it will not be uninstructive to show, how so expert a mathematician as " Kinclaven" could continue to blunder so egregiously on this question. The fact is, that, like many other mathematicians whose minds are cramped in mathematical ways, he attends more to the signs than to the things signified. Thus the entirety and the mutual play of the physical relations are lost sight of in those which are purely mathematical. Kinclaven had taken the formulæ, which I had adopted for estimating the resistance of the atmosphere, and for determining the power of an engine necessary to cope with it at a velocity of 100 miles in an hour; and having symbolised them, and so manouvered the symbols as to bring forth results, in two distinct cases of velocities, in the ratio of the *cubes* of the velocities, he characteristically jumped to the conclusion, that this ratio represented in my calculation the law of the resistance of the air. In such an absurdity as this would undoubtedly have been, he assumed the whole affair to be involved—and thus Q. E. D. This was an amusing instance of that sharp, thimble-rig practice in which an algebraist is sometimes outwitted by his own cleverness. In not attending to what, in point of physics, was really embodied in his algebraical expressions, "Kinclaven" overlooked the fact, that, by adopting the engineer's estimate of the power of a horse, namely, as being equivalent to 33,000 lbs. raised one foot *in a minute*, the consideration of the *expenditure* of power, *in a given time*, became thereby involved in the question, and therefore, of necessity, became included in the ultimate ratio of the results. Now, any tyro among the civil engineers—who, "Kinclaven" says, ought always to be mathematicians—would be ashamed not to know, that when the resistance is, as in the case of air, in the ratio of the square of the velocity, the supply of power must necessarily be *in the ratio of the cube.* Thus the results which were to have frightened me out of my propriety, were just what they ought to have been. But "Kinclaven" was dreaming all the while of the degree of force merely equivalent to a certain resistance. He threw out of the question its own velocity, which, of course, must be taken into the account in determining the capabilities of an engine. When his eyes were *pulled open* to the proper consideration of the subject, did he candidly acknowledge his error? Oh! no—a mathematician always stands by his order; he never lowers its dignity by striking to a plain, practical, unmathematical sort of man: he, therefore, covered his retreat in smoke, raised by a page full of the most illiberal banter, to which the fitting reply it received was silence.

I am, Sir, yours, &c.
　　　　　　Benj. Chiverton.

ROE'S PATENT WATER-CLOSETS.

Sir,—As some remarks have been made in your valuable work on the merits of my water-closet basin, suffer me to add a few words.

To your correspondent "Z." who writes in No. 796, page 83, I beg to express my thanks; and to "Z. Z." also, in 797, p. 104, I have acknowledgments to offer, for his admissions of the superiority of my plan, though these appear so reluctantly given as almost to awaken the suspicion that he is not altogether an uninterested party.

A gentleman connected with the Society of Arts called a few days since at No. 69, Strand, to examine my closet, and if other scientific gentlemen would do the same, they would be as satisfied as he was pleased to express himself to be—"that my basins, fixed either upon Braham's valve, or Hardcastle pan closets (which are the common closets in use), would make them more simple, and cheaper in fixing; and that they are altogether a great improvement upon every article of the sort he had met with."

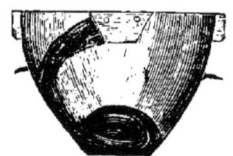

The annexed engraving is a section of the basin, shewing the water passing the fans, and the recess or rim, overflowing as it does when the handle is up. This recess, when the handle is down, answers the purpose of the *lead box*, &c. in the cistern,—viz., supplying the bottom of the basin, or making what is technically termed the "water joint."

Your correspondent "Z." states that "any number of closets with my basin may be supplied from one cistern, situate in any part of the building." This may certainly be done if the basin be supplied with a cock, so made that it will neither work stiff nor leak in any reasonable time (and if it did leak it could do no damage, as the water would run into the basin or the trap; and even if it should leak, after a few years use I would make it equal to new for half-a-crown). I have lately added an oil cup to the stop cock, from which it supplies itself with oil.

The size of the water-way is varied according to the pressure.

Those that object, as "Z. Z." does, to the stop cock and charged pipe, may not be aware that a closet with my basin can be supplied with a valve on top of pipe, without a lead box, &c., in cistern. It can also be supplied by means of a small pump from a cistern beneath.

A charged pipe, covered with a nonconductor of cold or heat—such as hemp or wool—will receive no injury from frost.

"Z. Z." is mistaken in the intimation he gives that the cock at my closet is suddenly closed; no danger of bursting occurs on this account.

　　　　　　Yours, &c.
　　　　　　Freeman Roe.
Windmill-place, Camberwell.

FRENCH PORTABLE FIRE-LADDERS.

Sir,—At page 5 of your 791st number, Mr. Felix M. Simeon has given a description of the portable folding-ladders carried with the fire-engines by the Sapeur Pompiers of Paris; by means of which, they are said, to ascend from window to window of a burning house, until the highest story is attained. I fancy it will be apparent that this must always be, as designated by Mr. Parry, at page 28, a very "complex" mode of proceeding, and one that is fraught with difficulty and danger. In the event of flames issuing from the lower windows, this plan of ascending cannot be resorted to. It will very likely have occurred to many of your readers as somewhat singular, that in a corps so decidedly military as the Sapeur Pompiers of Paris, the scaling ladders of the army should have been overlooked! This, however, is not the fact; in addition to the folding ladders described by Mr. Simeon, others are pretty generally employed in Paris, known as the "Italien Ladder," very closely resembling those which for the last few years have been so successfully employed in various parts of England.

Fig. 2.   Fig. 1.

d e, armed with triangular pieces of iron.

Fig. 3.

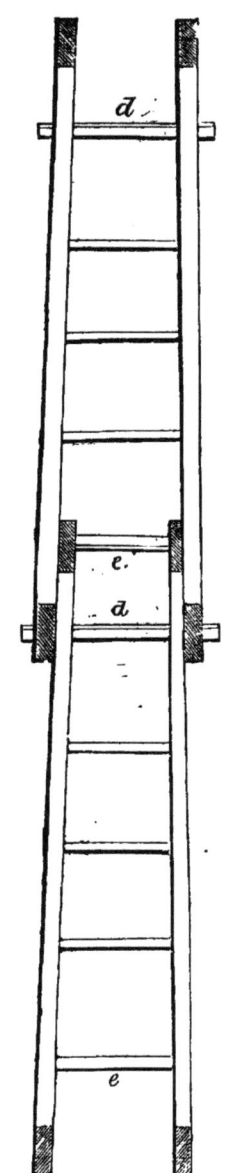

The accompanying sketch of the "Italien Ladder," will show its construction, and exhibit the points of resemblance as well as of difference, to those already so fully described in your pages. Fig. 1 is a front view, and fig. 2, a side view of one length of the "Italien Ladder;" fig. 3 shows two of these lengths as united. Each ladder consists of two side-pieces of ash A. B., tipped at either end with iron, hollowed out both at top and bottom as shown at C *c*, fig. 2. Each ladder is six feet long, and is furnished with five steps; the middle three being cylindrical, and the upper and lower ones, to fit the cavity C. *c*. The upper step of each length, *d*, passes through the side-pieces and projects beyond them far enough to enter the hollow of the next ladder and take a proper bearing. These ladders are intended to fit one into another, *ad infinitum*, until any required elevation is attained; they are six inches shorter than our ladders, and somewhat heavier. It will be seen that the principal difference between the two, is in the mode of making the joint. In the French ladders the steps merely fit into grooved cavities in the ladder ends, while in the English ladders addi-

tional security is obtained by the use of iron straps which make a firmer and a better joint. The Parisians might with great advantage adopt the pair of small wheels on the top ladder, and also the escape-belt for lowering females or infirm persons, &c., which are here now considered an indispensable appendage to this elegant and convenient form of ladder.

I remain, Sir, yours respectfully,
WM. BADDELEY.

London, Dec. 7, 1838.

ON THE SUPPLY OF GAS—METHOD OF READING OFF GAS-METERS.

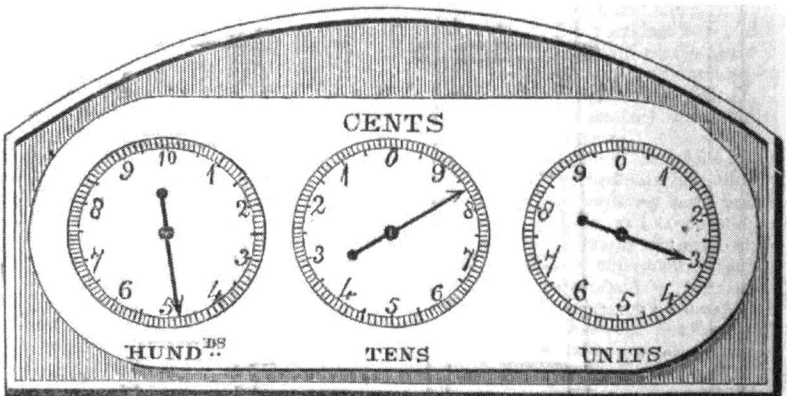

Sir,—The present augmented and daily increasing consumption of gas, from its extended employment in culinary operations, for manufacturing processes, and as a convenient source of household warmth, renders every circumstance connected with its diffusion, of more than ordinary interest.

In a former communication, I expressed a confident opinion that it would be equally advantageous, both to the manufacturers and the consumers of gas, if it was in all cases supplied *by measure only*. Its general employment for the above purposes, as well as a source of light, will eventually render it absolutely necessary to make the supply unintermitting. Thus, for instance, in Red Lion-street, Clerkenwell, where a number of artisans, jewellers, &c., required gas for soldering and other purposes, upon application being made, a continuous supply has been afforded

them, and the same thing has already been accomplished in some other districts. The supply of gas by meter, although at this time very considerable, is not yet become *universal;* this may arise from various causes. In some cases it may originate in misconceived notions of economy on the part of the consumer, who forgets that the charges *per light* are so regulated, as to protect the Gas-company against the wilful waste of the extravagant, and the unfair consumption of the fraudulent. These rates, therefore, are in most cases much beyond what the fair trader would be called upon to pay, who chose to economise his gas-light, in the same way that he husbands his oil or his candles. The non-employment of meters may sometimes be traced to unacquaintance with, and consequently want of confidence in the apparatus, by means of which the passage of the gas is demonstrated.

Besides, another circumstance tends considerably to diminish the satisfaction, which the use of the meter should give to the consumer, viz. : inability on his part to read off the quantities registered by the machine; he is therefore very apt to fancy himself too much at the mercy of the inspecting clerk, whose duty it is to take an account from time to time, of the state of the meter.

The beautiful and simple mechanism of the gas-meter has already been explained in one of your former volumes, and a description of it may be met with in almost every treatise on gas-lighting; I shall not, therefore, enlarge upon this point, but proceed briefly to explain, what has not hitherto been made generally intelligible, viz.: the method of reading off the quantities of gas registered by the index of the machine.

At the upper part of the meter, in front, a small tin door opens, and exhibits, within a glass, an enamelled plate with three dial-faces and pointers; above the middle dial stands the word *cents*, indicating that hundreds of cubic feet are registered. The dial to the reader's right-hand is marked *units*, (or hundreds of feet); the middle dial *tens* (or thousands of feet); and the left hand dial *hundreds* (or tens of thousands of feet.) Each division from figure to figure of the *units* dial, is one hundred feet: these divisions are subdivided into ten parts, each of which is equal to ten feet of gas consumed. One complete revolution of the pointer on this dial indicates the passage of one thousand feet of gas.

The middle dial, marked *tens*, is also divided into ten parts denoted by the numerals, each of these spaces being equal to one thousand feet of gas; these are again subdivided into tens or hundreds of feet.

The last dial, marked *hundreds*, is divided like the two former, the numeral divisions being ten thousands of feet, and the sub-divisions one hundred feet each.

The axle of the first, or units pointer, is connected by wheel-work with the drum revolving in water within the meter, so as to move from figure to figure while one hundred feet of gas are passing through; upon this axle is placed a pinion of six leaves. The axle of the middle pointer carries a wheel of sixty teeth, which, working into the pinion on the first, makes one revolution by the time that has completed ten. A similar wheel and pinion communicate motion to the last pointer, so that the quantity of gas registered by each dial is tenfold that of the preceding. The consequence of the foregoing arrangement is, that the central pointer travels in a direction opposite to the motion of the others, and the numbers therefore read the reverse way. From what has been stated it will be evident, that when the last dial has made one complete revolution, the middle pointer will have made ten, and the unit hand one hundred revolutions.

If the last dial was made sufficiently large to admit the subdivisions distinctly, all the purposes required would be answered, but it has been found more eligible in practice to employ three dials, and to dispense with the *subdivision*, marking the ten points only on each.

In reading the meter, first note down the units, which in the accompanying example we find at stated 3. = 300 feet. The central pointer denotes 8 = 8,000 feet; while the last hand marks 4 = 40,000, —shewing that 48,300 feet of gas have passed through the meter. It is always necessary to take the figure last passed, the reason for which will appear on inspection.

In this case the subdivisions on the *tens* dial, in fact, show the quantity = 8,300, and those on the *hundreds* dial 48,000 ; but as I said before, the subdivisions are not employed, from the greater distinctness and diminished liability to error resulting from the use of three dials.

As the motion of the pointers is continually progressive, it is always necessary to deduct from the present quantity shewn by the meter, the quantity noted at the previous settlement; the difference between them, being the additional quantity consumed, since that examination was made. Thus, for instance, supposing the quantity denoted by the meter and charged for at Michaelmas last, to have been 36,000 feet; that quantity being deducted from the present shewing at Christmas of 48,300, leaves 12,300 feet as the quantity now to be paid for.

Trusting that this explanation will be found sufficiently intelligible,

I remain, Sir, yours, respectfully,

WM. BADDELEY.

London, Dec. 21, 1838.

ON THE REQUISITES OF A GOOD GUN.

Sir,—Living in the country, I receive your Magazine in monthly parts. In the December part (received yesterday), there is an inquiry, by " Criterion," relative to guns, &c., and a reply by your very ingenious correspondent, Colonel Maceroni. During my life, I have (as all my friends know) indulged a sort of mono-maniacal propensity towards guns and pistols, and have expended many a hundred weight of shot and balls in experimenting therewith. With this prelude, I bear testimony to the correctness of Colonel Maceroni's statements " *en masse ;*" but, in detail, there are one or two points which take me by surprise—such as admitting the maximum weight of shot to be the exact weight of a leaden ball that would fit the bore of the gun—and the possibility of driving leaden shot through a copper pennypiece.

These are matters which it would ill become me to deny, or even to doubt, after the assertion of Col. Maceroni; but they do astonish me. .

Amongst very numerous experiments, I have tried long barrels with small bores, and short barrels with large bores, and every thing intermediate, and with every size of shot; and I do not hesitate to affirm, let the description of fowling-piece be what it may (provided the barrel have a " patent breech," and the bore be cylindrical), that if the length of the barrel be not less than 42 times the diameter of its bore, nor greater than 47 times the diameter of its bore, the barrel will, if properly charged, " shoot well." I prefer the length to be 45 times the diameter of the bore; not because it happens to be about the medium between the proposed extremes, but because I have one barrel of these relative proportions, which " shoots" better than any gun I have ever possessed or tried, although its miniature dimensions, viz. 25⅝ inches barrel, and $\frac{9}{16}$ inch bore, would rather indicate a child's toy than a serviceable implement. I do not recommend so small a barrel for ordinary purposes, because a few discharges produce a very considerable alteration in its calibre and efficiency, which would be still more apparent were the barrel longer.

Eleven-sixteenths, or two-thirds, of an inch bore, is good for any purpose;

and either of these diameters, multiplied by 45, will give a very satisfactory length of cylinder. A barrel of these proportions, smoothly bored, and of good iron, with a " patent breech," *n'importe* who may have been the maker, or for how small a sum it may have been manufactured, should throw from six to ten shots into half a sheet of common-sized letter-paper, from a distance of sixty yards; and this opinion is humbly submitted to " Criterion," by

Yours, very truly,
CHAS. THORNTON COATHUPE.

Wraxall, Jan. 4, 1839.

———

ON THE REQUISITES OF A GOOD GUN—
BARREL BORING.

SIR,—I have been much pleased to read in your interesting Magazine of the 22d December, a paper by a gentleman subscribing himself " Criterion," enquiring into " the essential qualities of a good gun," and highly gratified to find a reply by Col. Maceroni who has expressed himself so scientifically on the subject, and who has stated facts, the greater number of which I can bear testimony to.

To " Criterion," I would say that a good gun ought to place with regularity about eight shot (say No. 5), in a space of four square inches, at a distance of 40 yards, one particular good quality will be the regularity with which the shot are placed; a gun that will not do this, will be of little use at a distance of 60 or 80 yards-

The hardest and closest shooting gun I have heard of, would pierce 55 sheets of brown paper, and place about 180 shot in the sheet at 40 yards, I can give reference to this same gun if required; but, as " Criterion " has said, this is more than " I can do " with my gun, I am obliged to be contented with something less.

I am glad to find that Col. Maceroni lays it down as a principle, that, the interior of a barrel ought to be a cylinder, it has always been a subject which I have contested more warmly with gunmakers than any other, and I think if gunmakers were to make on such a principle, we should have a complete reformation in our guns, a cylindrical interior must be the easiest, the closest, and the strongest shooting gun, and one which must hold the shooting longer than any other form, as well as being less liable to lead.* we might then have sure data for ascertaining the charge and the length of barrel that the gague should require.

The gun I have just referred to, was, as near a cylinder as it was possible to obtain.

I should be glad to learn of any gentleman, through the medium your of valuable periodical, if the method of boring guns with the ordinary boring bench is to be considered a good one, or one that is, or is not capable of being greatly improved. I am inclined to think (though I wish to observe I give my opinion with much diffidence), that the method now practised is one which is not capable of producing a *true cylinder*, unless by chance, or at any rate, if it is possible, it must be with great difficulty and judgment on the part of the operator, and perhaps this is the reason why many gunmakers object to cylinders.

Draw boring, or lead boring, as now practised by hand, I think equally uncertain.

The method of boring cylinders for steam engines must certainly be a correct one, but I am not aware if the same method might apply to gun barrels.— but think the barrel and the machinery ought to be kept firm, instead of the bit and barrel wabbling about as it now does in the ordinary boring bench.

I am, Sir,

Your most obedient servant,

T. WHISTLER.

COLONEL MACERONI'S PLAN OF ROAD MAKING.—OXFORD-STREET EXPERIMENTS. — PRESERVING STONE AND STATUARY.

Sir,—So sick, sick at heart, am I of "preaching to the desert," and of "throwing pearls to swine," during five-and-twenty years, that although I am called upon, or quoted by name, on three different subjects in your last num-

ber, (804), I almost deem it useless and superfluous to trouble either you or myself with a single line in rejoinder.*

I have read with that astonishment and disappointment which fills us when we see our services and counsel acknowledged, appreciated, confessed, but slighted and discarded, the letter of Junius on the "Philosophy of road making," in which he throws overboard all that he wrote in the "*Monthly Repository*" for Feb. 1834, on the subject of street paving, confirmatory of my views, and what is more, of your own, Mr. Editor, on that socially important subject!

You must remember, Sir, that in 1825 I printed a pamphlet called *Hints to Paviours*, which was copied into your Magazine, and shortly republished, enhanced by a preface, or "*Introductory Review of other paving plans*," written by yourself. In 1832 Mr. Effingham Wilson published another edition, to which was added an essay "On the increase of daylight in London," another, "On the Oxydation and preservation of stone and plaister covered buildings, and on the use of *coal tar*, and of gelatinous or oleaginous fluids in forming paths and roads, and hardening earths and soft stones." All these papers appeared in the *Mechanics' Magazine* in 1825 or 1826. I have distributed gratis upwards of one thousand copies of the pamphlet, but *The Monthly Repository. The Times*, and *The Morning Herald*, were the only papers that took any notice of it. All three, however, carefully and at length analysed the plans, compared them with all the others proposed and tried, and gave to mine the most unqualified preference and approbation. * * *

But to street paving, "Junius Redivivus remarks on the Oxford-street sam-

---

* I am inclined to think that the leading of a barrel much depends on the state of the iron, as some barrels will lead much more than others under the same circumstances. Query—Is it from the iron being too soft.

* Having received Col. Maceroni's communication late in the week, and thinking that it would be preferable to publish it at once, whilst the public interest is excited upon the subject of paving, than to defer it to a succeeding number, we have been compelled, for want of space, to omit a considerable part in which he very justly complains of the apathy and neglect with which his proposals of improvements on many important subjects,—paving and stone-preserving among the rest,—have been treated by the Board of Works and other parties. We may add, that the favourable opinion which we many years ago expressed upon the Colonel's plan of road making, remains unchanged, and we are convinced that ere long its efficiency will be tried and proved, and we hope that, as is too often the case, the inventor will not then be altogether unconsulted or unrewarded. Ed. M. M.

ples, and recommends a more expensive plan, if possible, than any of them.

There is no difficulty in forming a smooth immoveable pavement that will bear any weight, by the mere appliance of *good* mortar and rectangular stones, without any recourse to the ridiculously expensive humbug exhibited in Oxford-street. But, as I have shown in my "Hints to Paviours," although this plan is existing and answering every purpose in Milan, Sienna, Florence, Rome, and Naples, it will not do for London, where it would be daily disturbed for water and gas-pipe laying, or for cleansing or repairing the rotten brick sewers. To illustrate all this, it would be necessary for me to add my entire pamphlet to this long letter. Such of your readers as take any sort of interest in the subject, but who may not like to pay a shilling for the work to Mr. Effingham Wilson, may refer to your volumes for 1825 and 1826. Every one of the plans exhibited in Oxford-street is utterly inapplicable to *London* streets. They are all as difficult to lift and *properly* to replace as are the *horizontal walls* of Puzzolana mortar and Besalt, which cover the streets of Rome and Naples, and which never require to be disturbed. One merit these "inventions" surely do possess, and that is, vast expense!

"Junius Redivivus" recommends a pavement of wooden blocks, placed in cast iron boxes, four feet square, so as to ensure a large "surface bearing." Now, "in for a penny (such a penny!) in for a pound," say I. Iron boxes, covering several hundred square miles, will rust and wear rapidly; let us have them of gold, or of silver, at any rate; what a capital *job* for the silversmiths of Ephesus, I mean of Bishopsgate, and Newgate, and Tyburn turnpike (*tour du baton.)*

The writer in the "*Civil Engineers' and Architects' Journal*" likens that absurd and most costly specimen of the genius of Mr. Macnamara, (who, by the bye), has been "taken up" by a company), of broad pyramidal stones placed side by side, to a flat arch and a congregation of "keystones!"

The scientific gentleman does not tell us what *mechanical* bearing supports every alternate stone which has its broadest side, or base, downward on the soft unequally resisting earth! Pray, good Sir, just go and look at a specimen of this mutually supporting "key-stone" pavement, by the Foundling Hospital, where each of the stones that speedily sunk make a hole sufficient to break an axle or a bone.

"Junius Redivivus" also advises the wooden pavement to "be placed on a level of concrete, eight or ten inches deep," while he objects, very properly, to other plans of blocks joined by bitumen, &c., into immoveable surfaces! He totally forgets his panegyrics upon my most uniquely simple and cheap method of merely hardening the substratum and then beating down the stones, old or new, by a powerful locomotive "monkey." My plan requires no sublayer of gravel, much less another over that of broken granite, and still less any lime "grouting." The addition of a little old brick and mortar "dry rubbish" over the soil, and then, the old stones, which are better than the new larger ones, well rammed down by my machine of five hundred weight, will constitute a pavement that can never budge; but *should* any protuberance appear, one more ramming *on such protuberances* will settle the street for the period of the stones endurance of mere wear from friction on their surface, which, mayhap, is longer than it is worth while to calculate. Here are no new stones; no new material; no difficulty in taking up any part of the pavement, and then, by a mere *local mechanical* appliance, restoring any such portion to any degree of density, according to the number of strokes of the rammer.

After giving the most unqualified approbation and preference to my cheap and "philosophical" system of paving, and praising you, Mr. Editor, for showing its preference to all others, "Junius Redivivus" not only returns to the "concrete" horizontal immovable wall, but advocates the condemned iron-box absurdity, besides giving great credit to "the authorities who rule Marylebone, who have very *wisely* arranged their disputes as to which is the best mode of road-making, by agreeing to lay down various samples side by side in Oxford-street." Why the Marylebone sages have not given a sample of my simple street-paving along with the jobbing, expensive, absurd vagaries he has long ago, conjointly with the *Times, Herald,* &c. declared to be quite inapplicable to the streets of London, "Junius" does not say, perhaps, because he does not know. He is not at all offended, however, at this studied scorn for his opinion! But the fact is, when I sent them a dozen copies of my pamphlet, I asked for no contract,—no reward;—I offered "the authorities who rule Marylebone" to superintend the application of my plan, for common *labourer's wages.* In fine, these great "authorities" have excluded from their notice and trial, the only system which costs not a shilling more than the older one, save the first expense of the locomotive wooden rammer.

The Strand, Holborn, Cheapside, Fleet-street, &c. have been repaved every four months with the absurd layers of gravel, then of broken granite, then of new large stones, then of mortar grouting; in fact, one or other of these great urban arteries is *always* stopped, always paving, always level as the river in a gale of wind! All the vast traffic of carts, waggons, stages, omnibuses is at some time diverted from some portion of the regular track! No sooner is the grouted pavement laid, than scores of men are seen, in numerous places, labouring at taking up parts so sunk as to attract the notice even of the "authorities." Then anon a great board proclaims that by authority of the Lord Mayor, or some other official, no carriages are to pass that way. Thanks to such "authorities," how many a quiet, narrow, dingy lane, has latterly been favoured with the rattling current of three thousand vehicles a day, where nothing but a cart with sugar casks or cotton bales had passed before this era of complex paving *jobs!*

In your last number, Sir, my name is mentioned in connection with the preservation of stone and plaister by the application of oleaginous matter. On this subject I hope to be allowed to say a word.

Being at Rome, in 1811, I was acquainted with the Count Forbin Janson, Intendant-general of the museums of the Emperor of France. At my suggestion he applied oil to all the statues and bas-reliefs exposed to the open air. Subsequently, the fresco painting of the cupola of the Pantheon at Paris, being in a damp and mouldering condition, *my* plan of preservation was applied by the celebrated chemist d'Arcet. A charcoal fire was placed in a grate attached to a long handle, like those used to pitch the sides of a ship. The part, well dried, was then "paid" with a mixture of linseed oil and wax. Then, when cold, polished with a brush and flannel.

In 1810 I had caused a similar preservative operation to be performed upon many of the frescos of Pompeii, which were constantly being rendered less and less vivid and discernable, by the guides throwing water over them for visitants to see them to the best advantage. A coat of good varnish, at my suggestion, rendered the freshness permanent and preserved the paintings. I also caused the beautiful group of the Toro Farnese, and other antique statues, exposed in the Villa Reale, at Naples, to the sea spray, charged with muriatic acid, to be covered with linseed oil, in which bees'-wax was dissolved. Cardinal Forbin Janson, brother to the count above named, will well remember the fact. You state that the builder or artist of the Hyde-park triumphal arch, protected his marble with a similar application. This was years after the publication even of my paper on that subject in your Magazine.

Apologising for having occupied so much of your valuable space, but *not* upon unimportant subjects,

I have the honour to be, Sir,

Your obedient servant,

F. MACERONI.

No. 3, St. James's-square.

KILN-DRYING WHEATS FOR GRINDING.

We have been lately struck with a passage in the last *Quarterly Journal of Agriculture,* recommending the process of kiln-drying wheats, as applied in Scotland and the North of England to oats. "It is evident, observes the writer, "that all grain intended for household consumption can be readily preserved: but that all meal is less easily kept from the attacks of insects, and is with greater difficulty recovered from the deterioration. - We incline to think that the vast superiority of Scotch oatmeal over that which the unfortunate Southerns obtain from English millers, and the great length of time during which the former will remain sweet and good compared with the latter, which will keep useable only a fortnight, is attributable to the method of kiln-drying that the Scotch millers adopt, which not only destroys incipient (or perhaps actual) insect vitality, but imparts that richness of flavour which is wholly deficient in English oatmeal. In fact, the mealmen of the South confess they know not how to give the same appearance, taste, scent, or texture, to their oatmeal, that is so apparent and so delicious in that of Scotland.

"We are firmly of opinion that if wheat could be subjected to a similar process to that which is adopted for oats, the results would be the same, namely, that it would keep a much longer time; it would be secured from insects; its flavour would be highly improved in richness and delicacy; it would be decidedly more wholesome, and would require much less time in cooking. Our reasons for speaking thus positively arise from actual demonstration. It is well known that flour from wheat is rejected altogether by the desptic, 'because it lies heavy on the stomach,' 'because it turns sour,' 'because it ferments,' 'because'— twenty other reasons; therefore it is not a fit aliment for young children, where, from ill health or fortuitous circumstances, a mother is unable to nurse her offspring, and recourse is to be had to extraneous nourishment. Every succedaneum with which we are acquainted has been tried for the food of infants, and not any is so entirely free from objection as this very contemned *wheaten-flour*, when it has undergone the process of either baking or boiling for the space of a couple of hours. All crudity is then subdued —it becomes a compact hard mass, which is to be rolled or grated when wanted for use— requires merely heating, not even boiling, to make it palatable, and is in flavour really delicious; it is also light, nourishing, free from all those objections made by the dyspeptic, and moreover will keep free from insects, shall we say *for ever?* if we do, our readers will see that we mean an indefinite period."

NOTES AND NOTICES.

Royal Exchange Improvements.—The plan suggested for building the new Royal Exchange, it is understood, will be adopted with little, if any alteration. All the buildings immediately contiguous to the late Royal Exchange and West of that building are to be thrown down, to make room for a grand front to the new building facing the West, in the best style of architecture. The side-front, opposite the Bank, will be much larger than the old Exchange, and extend nearly as far as Freeman's-place. Finch-lane, according to the plan, will be flagged, and no longer a passage for horses or carriages, and there will be no building at either side of the new edifice, to obstruct its view. It will be completely insulated, and be a considerable improvement to that part of the city when finished.

Steam-boat Incident.—The following incident is related in the *Buffalo Commercial Advertiser*, as having occurred on Lake Erie, during the tremendous gale which swept over it in the month of November last. In that fearful night, the steam-boat Constitution, Capt. Appleby, was out amidst the terrors of the gale. By the glimpses caught at intervals, when the fitful storm for a moment broke away, the anxious and watchful commander was made aware of the critical situation of his boat, which was rapidly drifting in—under the hurricane power of the gale, which blew almost directly across the lake—toward a dangerous reef, from which escape would have been impossible. He went directly to the engineer, and ordered on "more steam." The reply of the engineer was that there was already as much on as the boilers would safely bear. Again did the Captain seek the deck, to see if his labouring boat was making head-way, and again returned to the engine-room. He explained to the engineer their hazardous situation, and told him that all hope was lost, if no more head-way could be gained, but left the engineer to use his discretion in the crisis. A moment of reflection, and the decision was made. He coolly directed the heads of two barrels of oil to be broken in, and the furnaces were rapidly fed with wool dipped in the highly inflammable liquid, while two men with ladles dashed the oil into the flames. The intense heat which these combustibles created, generated steam with the rapidity of lightning, and soon the resistless vapour forced up the safety valve, and issued forth with tremendous violence, its sharp hissing heard above the wild uproar of the waters and the storm. With a desperate and determined courage, which equalled the most daring heroism that the page of history has ever recorded, the engineer *sat down upon the lever of the safety valve*, to confine and raise the steam to the necessary power required to propel the boat against the drifting waves! In this awful situation he calmly remained, until the prodigious efforts of the engine had forced the Constitution sufficiently off shore to be beyond the threatened danger. This intrepid act was not a rash and vain-glorious attempt to gain the applause of a multitude by a foolhardy exposure of life, in some racing excurion—it was not the deed of a drunken and reckless man, wickedly heedless of the safety of those whose lives were perilled—but it was the self-possessed and determined courage of one whose firmness is worthy of all admiration.

The Thames Tunnel.—In 1838 a distance of eighty feet of this work was completed,—and it has now advanced to within 90 feet of the low water mark on the Middlesex side of the river. We understand that the works continue to progress slowly, but steadily.

Nelson's opinion of Steam Navigation.—In 1800, Mr. Bell purchased a small French vessel, fitted her up with a steam-engine of four horse power, and sailed with her from the Clyde by the Land's End to London, at the rate of seven or eight miles an hour; with the view of submitting his plans to Government. That boat was inspected by the late Lord Melville, and a number of naval gentlemen, all of whom, except Lord Melville and Lord Nelson, considered the invention of no value to the country. The remark made on that occasion by Nelson is worthy of notice. He said, "Gentlemen, if you do not take advantage of this invention, you may rely on it, other nations will."—*Edinburgh Journal.*

LONDON: Printed and Published for the Proprietor, by W. A. Robertson, at the Mechanics' Magazine Office, No. 6, Peterborough-court, between 185 and 136, Fleet-street.—Sold by A. & W. Galignani, Rue Vivienne, Paris.

Mechanics' Magazine,

MUSEUM, REGISTER, JOURNAL, AND GAZETTE.

No. 806.]   **SATURDAY, JANUARY 19, 1839.**   [Price 3*d*.

Printed and Published for the Proprietor, by W. A. Robertson, No. 8, Peterborough-court, Fleet-street.

CHANTER AND CO.'S PATENT STOVE.

Fig. 1.

THE STOVE EXCITEMENT SEASON OF 1837-8—CHANTER AND CO.'S PATENT STOVE.

The winter of 1837 having been one of unusual severity, the inventive faculties of men were exerted in a proportionate degree to devise means of supplying by artificial means that congenial warmth in their houses, places of business and of congregation, of which nature for a time had deprived them. The countless plans which were conceived in the mind, drawn out upon paper, experimented upon in private, or put even into operation in a limited circle, we leave out of the question, but the number which came to the maturity of publication in scientific journals, or of tempting their inventors to the risk of the expense of a patent of monopoly, form a striking indication that thousands of minds were probably directed towards the attainment of the desired object—economical and healthy artificial warmth. In the columns of the *Mechanics' Magazine* alone many plans have been published, as well as numerous hints and remarks upon their efficacy or inefficacy;—and these were hardly a tithe of the communications we received and were compelled to keep back, either from want of space in our pages, or deficiency of merit or originality in the plans. On the records of the inrollment offices within the same time, thirteen specifications have been placed—how many of them will repay their patentees we venture not to say.

The plan which received the greatest share of public attention was that of Messrs. Harper and Joyce. The causes of this popularity were, in the first place, the singularly clever means taken to excite the public attention, at the Jerusalem Coffee House, and in newspaper paragraphs,—and, secondly, in the specious advantages attributed to it of extreme portability and cheapness : a portability which was to render it available under every circumstances of travel, in the cabin, in the carriage, on the railway, in a lady's muff, and it was almost hinted, in a gentleman's waistcoat pocket ! No chimney was to be required—London was to be for ever freed from smoke and the effects of that sombre vapour. The stone of its palaces and its public buildings once cleansed, was to remain white as Parian marble,—and its atmosphere was to rival in clearness that of the cities of the sunny south. In point of cheapness, a hundred people were to be warmed for a song, and a fifty-horse steam boiler fed for the price of a ballad. All this was to be effected by a wonderful chemical discovery ! The monopolizing coal owners were in despair ; the weapon which was to overthrow their combination appeared in the form of an insignificant urn. To buy up and crush this enemy was now their aim, and it was currently reported, that a hundred thousand pounds in sterling money had been subscribed and offered to the fortunate inventor for his secret, or if he would keep it his own, and let it die with him—but was refused ! What delirious visions of greatness must have disturbed the slumbers of the poor gardener of Camberwell —what a prospect of transition from obscurity to splendour, from poverty to riches—unequalled almost in the fanciful events of the Thousand and One. Sadly, however, was poor Joyce disappointed— and most strikingly in this instance was the old adage verified;—a sudden " slip" indeed was there " 'twixt the cup and the lip." We are credibly informed, that up to this day, Joyce has only received sixteen pounds for his discovery, and involved himself, to boot, in the meshes of a chancery suit ! By what arts this deprivation of the inventor of the fruits of his invention has been effected, (whatever its merits may be,) we are not sufficiently informed to venture to state—but this we know, that whilst his name is blazoned forth as a party in an apparently extensively conducted business, the puffings, and advertisings, and manufacturings, of which must be carried on at no trifling expense—he is totally unconnected with, and ignorant of its proceedings—and is a poorer and less happy man than he was before he made his wonderful discovery, or dreamt his splendid dream.

To return to our subject. From the moment we became acquainted with the material to be used as a fuel in Harper and Joyce's patent stove—we felt it to be our duty to set our face against its introduction. So convinced were we of the danger certainly attendant on the combustion of charcoal under any modification, that we should have felt ourselves to be conniving at the wholesale administration of poison, had we been silent on the occasion of an attempt to make its use prevalent—more especially under the guise of ostensible salubrity. A twelve months' experience has had its effect—the excitement has subsided, and the invention is dying away, and will shortly be spoken of as one of the nine days' wonders of an age rife in pseudo-miraculous events.

Fig. 2.

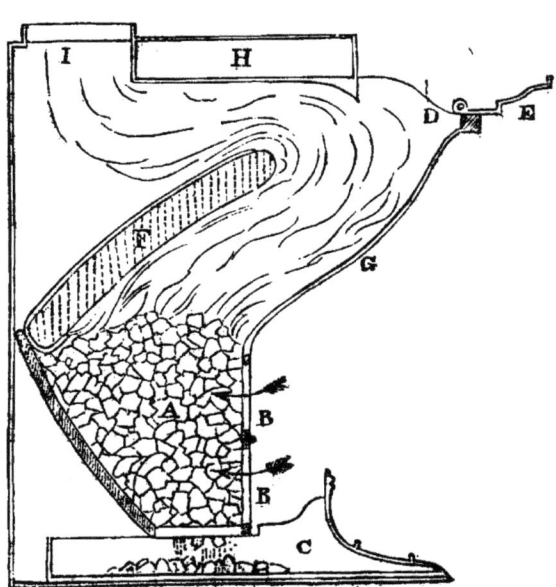

With more solid pretensions to utility as a warming apparatus, Dr. Arnott's stove has kept the field, and is now in very general use. We are still of the same opinion that we formerly expressed upon the culinary capabilities, and the extravagant no-trouble-giving qualities attributed to it by its introducer to public favour. The competition in the manufacture has, also, had a baneful influence upon its character. The invention being public property—no manufacturing patentee's reputation being at stake,—every blacksmith tried to make it cheaper than his neighbour, and its correct action being dependant on the due proportion of its parts, upon its being air tight, and upon the size of its flues; and these essentials being neglected through ignorance or carelessness—many have got into use, which, besides the fault of the smell of the iron, common to all hot air stoves, have that of the sulphurous vapour of the fuel, aptitude to smoke, the air entering very often at the wrong end, and a liability to explosion. In the greater number of those in use, the very distinguishing feature of Dr. Arnott's stove, the addition from which he gave it its name, the mercury *Thermometer* regulator, is wanting, and the admission of the air is left to the care of the hand of man. Is not this as we prophesied? What reason now remains to call these hand-regulated stoves by the name of Dr. Arnott? For the causes we have just stated, it appears to us that Dr. Arnott's stoves will not continue long in demand; but the change once made from the old open fire-place, will lead the way to the trial of various plans by many people before they will consent to return to Rumford's and registers. Whether this will be the result of the insurrection amongst fire-places, we do not take upon ourselves to say;—as movers forward we should hope otherwise.

Amongst the plans which we think well worthy of a trial, is the invention we shall now describe, and which is figured in the first and third pages of this week's number. Messrs. Chanter and Co. are well known in the mechanical and steam-engine world as having patented many plans for the complete combustion of the fuel supplied to fire-places and furnaces, whether in a solid vapoury, or gaseous form. This heating apparatus, is one in which the cheerful appearance of an open fire is combined with all the real or supposed advantages of a hot air stove,

Figure 1, on our front page, is a perspective view of the stove, and fig. 2, on p. 259, is a transverse section; such parts of the section as are seen in the perspective view, are distinguished by the same letters of reference.

A, the visible incandescent fuel, seen through B B, the front grating or fire-bars, and through which also, air enters to the fire; C, the ash-pan, which is moveable; D, the aperture to supply fuel; E, the lid to close the aperture, which turns on a hinge, and which must fit as air-tight as possible; F, a back deflector, the prime feature of the invention, formed of fire-brick; G, another, and front deflector, to throw warmth downwards on the feet; H, the hot-air chamber communicating with spaces at the sides and back; I, the flue, communicating either with an ascending or descending chimney tube.

From this description it will be seen that there is an open fire, fed from the top and front of the stove, and as the coals descend on the back of the inclined deflector, they become coked and and prepared to offer a continual cheerful appearance through the front grating. The back deflector partially throws back the smoke and gases upon the fire to be consumed, the greater portion of which would otherwise uselessly escape. The front deflector throws the heat from the fire down upon the feet of a person near it—a most essential advantage both in point of health and comfort. The stove is made in two cast iron bodies, an ornamental casing, (some are of very elegant design, as may be seen from the specimen on our front page) and a cockle, of which fig. 2 is a section. A considerable space is allowed between the two bodies, in which a large current of fresh air is continually heated and supplied to the apartment which it is destined to warm. Experience, the best test of merit, will prove whether Messrs. Chanter's invention is more worthy of patronage than those of their predecessors or contemporaries. We have done our duty in making it public.

ON THE MANUFACTURE OF PAPER FROM WOOD.

Sir,—I have read with some attention the notice in your 804th number, (page 234), descriptive of Mr. Desgrand's patent mode of making paper from wood, which is supposed to be identical with the process followed by the Messieurs Montgolfier, who are now said to have " succeeded in *substituting wood for rags* in the manufacture of paper." There are two points to be noticed, the *material* employed, and the *process* adopted, and in neither of these can I perceive the slightest claim to originalty, nor the shadow of a pretext for the exclusive monopoly of a patent-right. With respect to the *material :*—it has been long known, that excellent paper can be made from almost every vegetable substance capable of yielding easily an abundance of strong fibre. The bark of the willow, beech, hawthorn, maple, elm, plane tree and some others; the stalks of the thistle, nettle, hop-bines, and straw, have all at different times been converted into paper. Some of your readers may remember the white-and-brown papers made for the late Mr. William Cobbett, from the refuse of his Indian-corn. Wood, in the state of shavings, has been converted in a like manner. The art of making paper, from vegetable matters reduced to a fibrous pulp, was known in China long before it was practised in Europe; and the Chinese have carried it to a high degree of perfection. The different sorts of paper vary in China according to the materials of which they are composed, and the various modes of manufacturing these materials. Every province has its peculiar paper. That of Sechesen is made of linen rags, as in Europe; that of Fo-kien, of young bamboos; that of the northern provinces, of the interior bark of the mulberry; that of the province of Kiang-nan, of the skin which is found in webs of the silkworm; and in the province of Houquang, the tree chu, or ko-chu, furnishes the materials of which paper is made. A few years since a patent was taken out in America for the manufacture of paper from wood shavings, but this had been done long antecedently in England. I have by me a piece of paper made from this material; it is semi-transparent and rather brittle.

From all this, it appears to me that the only circumstance connected with Mr. Desgrand's mode of proceeding that has the slightest pretension to novelty, as regards the material employed, is the use of *log-timber* chopped up for the purpose; and, in this respect, whatever may be the claim to merit on the score of novelty, the plan is exceedingly objectionable as regards the expense. On the contrary, wood in the state of shavings is comparatively valueless, and in so fine a state of division as to be most easily acted upon by the decomposing agents employed for effectually separating the fibres.

With respect to the *process* adopted by Mr. Desgrand, it differs nothing from that which has been followed in China for some centuries, where the method of fabricating paper from the bark, &c., of different trees, and from the bamboo, is as follows:—The whole substance of the bamboo is employed in this operation: the shoots of one or two years standing, which are nearly as thick as a man's leg, being preferred.

They strip the leaves from the stems, cut them into pieces of four or five feet in length, make them into parcels, and put them into water to macerate. As soon as they are softened, which usually happens in about five days, they are washed in pure water, put into a dry ditch, and covered for some days with lime watered for the purpose of slacking. They are then washed carefully a second time, and every one of the pieces is cut into filaments, which are bleached by exposure to the sun's rays. After this they are boiled in large kettles, and then reduced to pulp in mortars of wood: This pulp being mixed with a glutinous substance extracted from a plant called in China *Koteng*, is made into sheets of paper in the usual manner. The paper made from the bamboo is sufficiently white, soft, and closely united, without the least inequality on the surface to interrupt the motion of the pencil. But every kind of paper made from the bamboo, or from the bark of trees, is much more liable to crack than that made from rags. The Japanese employ the young shoot of the *morus papifera sativa*, or true paper tree, from which they obtain the detached fibres by means of long saturation in water, and by subsequent boiling in a ley made from wood ashes. The method of treating wood shavings, as patented in America, is to boil them in water containing from twelve to 18 parts, by weight, of common alkali, which in a very short time effects the decomposition of the wood, and reduces it to a mass of fibres, ready for the subsequent processes of bleaching, grinding, &c., which being finished, the pulp is moulded into sheets of paper in the usual manner. A Mr. Lambert took out a patent a short time since for manufacturing paper from straw. Having cut away all the joints and knots, the pieces of straw are boiled with quicklime in water, for separating the fibres, and extracting the mucilaginous and colouring matters. Or caustic potash, soda, or ammonia, may be employed instead of quicklime for this purpose. The material is then to be washed in clear water, and afterwards subjected to the action of an hydro-sulphuret, composed of one pound of quicklime, and a quarter of a pound of sulphur to every gallon of water, for the more effectual removing of the mucilage and silicious matters. After several successive washings, bleaching, &c., the stuff is put into the beating engine and worked up in the usual manner.

Perseverance and ingenuity will ultimately triumph over the most obdurate elements, and it is quite certain that all the substances I have named, with many others of a similar character, have been converted into paper. But, as a matter of commercial speculation, I am quite convinced that no paper *composed entirely of any ligneous materials*, will ever prove a successful or profitable manufacture. The vast improvements that have been made within the last few years, both in the chemical and mechanical departments of our paper-mills, have enabled our manufacturers to convert the very coarsest linen, and even cotton rags and refuse, into a good and useful paper: by this means rendering the supply of rags, upon the whole, fully adequate to the demand, immense as that is.

The gigantic strides of education in all directions has already created, and will continue to create, an increased demand for books and paper; what the ultimate effect of this may be, is difficult to determine; but, should the supply of rags hereafter prove unequal to the requirements of our manufacturers, it may be desirable to find some substance that being easily and cheaply obtained may be employed *in conjunction with rags*, so as to eke out their quantity without deteriorating their quality.

Some paper-makers now mix a certain quantity of straw with the rags for making *hand* papers, but it is cut too short, and is not previously submitted to the very essential process of fibrous separation. The use of plaister, chalk, &c., in white, and of clays, &c., in brown papers is a species of fraud that is detrimental to the article produced; a fictitious weight and apparent strength is given, while the positive strength is in reality diminished.

I do believe, that a judicious combination of one or more of the foregoing vegetable substances with rags might even now be made with advantage, and prove the means of adding very considerably to the useful character of several kinds of paper.

I remain, Sir,
Yours, respectfully,
WM. BADDELEY.

London, Jan. 10th, 1839.

WATER-PROOFING CLOTH.—MRS. DUKE AND HER FOLLOWERS.

Sir,—About twenty-five or thirty years since, a lady of the name of Duke took out a patent for rendering cloth waterproof, and carried on her business for some time at Holloway. The cloth that had undergone her process was not at all altered in appearance, and when water was poured on it instead of being absorbed immediately it stood in large drops, or ran about like quicksilver; or if you even formed a dish of the cloth and poured water into it none came through, so that Mrs. Duke's discovery became the wonder of the day, and was soon patronised by the army and the public. In the mean time the period for enrolling the specification of her patent arrived, and when every body was curious to know how this wonder was produced the cunning lady disappointed them; and did not deign to specify, but instead of it put them again on the *qui vive* by dashing at another patent; and when the specification again became due the naughty lady again disappointed them; and, if I am not greatly mistaken, carried on the farce a third time, and never specified at all. By this time, however, the army and the public found out that the whole affair was a farce; instead of being water-proof the cloth had only been immersed in some dyer's wash containing animal matter and alum, which having little affinity for water slightly repelled the pearly drops that ran about upon it, and even afforded her customers the gratification of exhibiting in their windows little dish-like pieces of cloth containing water. But, alas! a little wear, and the " peltings of the pitiless storm," betrayed the joke, and all the drizzling dripping drops drained through, to the discomfort and dismay of many a wet-skin weary traveller. About two years since a patent was taken out for a similar process, which of course came to nothing. But now, in the year 1839, the stale joke is in course of being tried again, but not in the tame and spiritless attempt of an individual. Half the tailors in town and country are now in the secret, and soon from the Orkneys to the Lizard Point will poor Sawney and John Bull have their gullibility dosed and tested with this old new invention. Poor Madam Duke! you little thought how many would follow in your ancient footsteps. Some say they follow you so exactly that they are the only possessors of the great secret you would not even entrust to one of the national records of the Court of Chancery; others assert, that you employed a solution of bees-wax; some, a preparation of acetate of lead: but one of your patentee followers says he finds nothing to effect the purpose so well as a solution of soap in water for the first dip, and a solution of alum in water for the second, and that by these dippings dried and repeated the cloth is rendered water-proof. I have no doubt that you noticed your prepared cloth was *not* water-proof, but I don't know that you figured away with this as the most meritorious part of your discovery. Here your disciples have got the start of you immeasurably; the great desideratum in the modern portion of the discovery is that the cloth rendered water-proof by this process is not water-proof. Chafe and rub the water on the cloth a little, say they, and it will come through, which proves, that if it will admit water from without, it will admit the escape of perspiration from within, while cloth really water-proof will not. Oh, Madam Duke, you see what dupes your followers would make of us!

Yours, &c.,
DRYSKIN.

___

SIR JAMES ANDERSON'S BOILER—CORRECTION OF SUPPOSED INACCURACY OF STATEMENT OF ITS POWERS.

Sir,—In a recent communication respecting the quantity of water evaporated to produce 7,500 gallons of 50 lbs. steam, the benefit of the doubt whether 50 lbs. steam pressure, or 50 lbs. load per square inch on the safety-valve was meant has been given to the statement. I own I am not disposed to be so liberal.

$50^{lbs} + 14\frac{3}{4} = 64\frac{3}{4}^{lbs} = 4\frac{1}{4}$ atmosphere steam, the corresponding volume of water is about 450. Consequently 7,500 $\div 450 = 16\frac{2}{3}$ gallons of water $= 1666^{lbs} = 2\frac{7}{10}$ cubic feet of water per minute per 1.866^{lbs} of coke, and nearly 160 cubic feet of water per hour for 112^{lbs}; ten times more than a very full allowance. Is it not probable that an 0 was added by the editor of the Irish newspaper in error?

My corrected reading becomes 750 gallons $= 120$ cubic feet of 50 lbs. steam —rejecting the equal forces on both sides. When it ceases to be a boiler question, we have

$$\frac{120 \times 144 \times 50 \times .6}{33.000} = \text{about 13 horse}$$

power. The omission of the reference to the atmosphere of steam, balancing atmospheric pressure of the air, has

sometimes produced curious results, such as plausible proof of the superiority of high pressure engines, while their actual inferiority to the contrasted low pressure amounted to 25 per cent, while it is equally essential in the comparison of gross power exerted in high pressure engines using steam at a different pressures, since the atmosphere wasted bears a different ratio in such cases to the gross power actually expended.

N. S.

___

LIEUTENANT WATSON'S WIND TELEGRAPH.

The great and melancholy loss of life and property occasioned by the storm on Sunday the 7th instant, was principally caused by the *sudden change* of the wind from one quarter to another. A correspondent has supplied us with the following account from the *Liverpool Albion* of 9th Nov. last, of a weather telegraph to convey intelligence of the state of the wind at a distant point, to vessels about to leave port. It occurs to us that the use of such a plan would perhaps have been the means of preventing in some degree the distressing calamities of the last week. Our republication of the article cannot be otherwise than useful—and we leave it to the consideration of our nautical readers.

"Lieut. Watson has favoured us with a sketch of an apparatus he is about to erect on Bidston-hill, which appears to us of such great importance that we have provided a wood cut, in order that our readers may understand it. The apparatus may be seen from any part of the piers along the whole line of docks from the river. The object is, that masters of vessels and pilots going to sea, may themselves at once ascertain *the point and strength of the wind at Holyhead; the relative rise or fall of the barometer there, compared with the previous day; and also the state of the wind of the Northwest Lightship, or at Bidston.* The advantages of this simple arrangement will be immense. How often has it happened, that vessels with moderately fine weather here, have been induced to go to sea, and, by meeting adverse winds outside, have been either lost, or compelled to put back with damage, when, if the commanders had been aware of the state of the weather outside, they would never have left dock! We instance one case only out of the many that might be brought forward:

On Friday, 23rd Dec., 1836, high water at 11 45 a.m.

"Wind at Holyhead, E.N.E, *blowing fresh;* barometer *falling,* and a *very heavy sea* outside.

"Wind at Liverpool, *North. light breeze.* Amongst the vessels that sailed that tide

The Heyes, B. ship, was totally lost.
.. Sandbach, do., got aground.⎫
.. Caroline, barque,　do.　⎬ More or less
.. Promise, brig,　　　do.　⎬ damaged.
.. Thos. Tucker, do.,　do.　⎭
A barque, name unknown.
A brig, unknown, near Jackson's Buoy, a complete wreck.
Quentin Leitch, B. ship.⎫ In danger near the
Hope,　　　B. brig. ⎬ Banks, with signals of
Canton,　　B. ship. ⎭ distress, &c.
The Race, schooner, and a great many others put back, damaged.

" Now, had it been generally known, that the wind was different at Holyhead, and blowing *much stronger,* with *a falling barometer,* the probability is, these vessels would *not have put* to sea. In future, no pilot need not go round the Rock without knowing precisely the state of the weather outside ; and we have no hesitation in saying, that the plan will be the means of saving both lives and property, and prove such a benefit as calls for some public demonstration to the projector of it. Certainly, amongst all the inventions of the present day, on " *the go a-head principle,*" it never entered into our imagination, that a wooden barometer would give the indications of the weather at a distance of upwards of seventy miles ; nevertheless, this is to be done in less than a week's time.

" We proceed to describe our wood cut.

" The circle at the bottom is about twelve feet in diameter, and is provided with letters marking the points of the compass. It has two hands like a clock, the longer one will show the point of the wind at Holyhead, the shorter one at the Northwest Lightship or Bidston. At the top of the mast is an iron rod, on which a ball works. When the ball is at the top of the rod, it indicates a light breeze ; when in the middle, moderate ; and when seen at the bottom, blowing very fresh. The space between this rod and the circle is for the barometer ; the pointer on the *right* hand side, or *north,* shows the position of the barometer, at Holyhead, *on that day ;* and the one on the *left,* or *south,* the position the *preceding day.* The whole is very simple, and easily understood.

" We congratulate the public at large upon this important addition to the acknowledged advantages derived from the Telegraph, and have no doubt that our merchants and shipowners will duly appreciate Lieut. Watson's unremitting exertions to promote and generally to extend the utility of the establishment he has so long and so ably superintended.

" We are glad to find, that there is every prospect of the new arrangement with the Dock Committee answering his purpose personally. Indeed, with his moderate

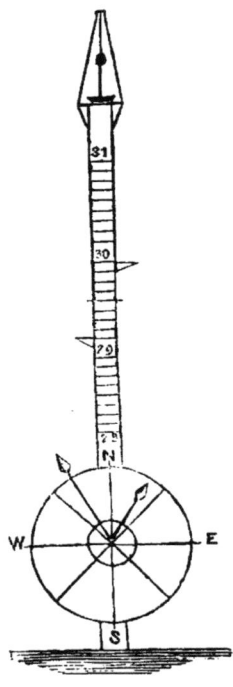

views, it can scarcely be otherwise, when we consider that the poles at Bidston formerly cost, and their inefficiency compared to the present telegraph. If we understand rightly, Lieutenant Watson fixes his subscription at the extremely low rate of 10*s.* per vessel per annum ; but, from firms requiring other information, besides that of their own vessels, he expects an additional sum ; some houses, it appears, having six or seven vessels, have subscribed their ten guineas, for which they are to be furnished with information of all vessels coming from particular ports in which they are interested. We are imformed, that this is *one* only of several equally important improvements contemplated by Lieutenant Watson. We are sure the Liverpool public will not let his merits and exertions continue unrewarded. To no person is the port of Liverpool more indebted than to him for his perseverance, through all manner of difficulties, in accomplishing the great benefits already derived from the telegraph. We think that the underwriters, of Liverpool, London, and elsewhere, ought to substantially acknowledge this important improvement. If the telegraph does occasionally save insurances, even such a loss to them will be far more than counterbalanced by saving many a " good vessel from shipwreck and disaster."

GREAT WESTERN RAILWAY INQUIRY, —DR. LARDNER'S EXPERIMENTS ON THE RESISTANCE OF THE ATMOSPHERE TO RAILWAY TRAINS.

At the general meeting of the Great Western Railway shareholders in September last a large portion of the proprietary expressed great dissatisfaction with the extraordinary departure from all the received rules of construction, which had been made by their engineer Mr. Brunel, (son of the engineer of Tunnel celebrity), in the formation of this most important line. No small part of the dissatisfaction evinced doubtless arose from the implied censure or stigma which this departure threw upon the directors and engineers of other railways, who had adhered to the tried and proved plan in the works which had been placed under their superintendence. Good and rational grounds for setting on foot an inquiry into the matter, however, existed in the startling extent of the deviations made by Mr. Brunel; in the want of sufficient reasons given at the outset for these deviations; and in the practical inexperience of their projector. These same reasons for inquiry, however, existed in equal or greater force in the first instance; and had the matter been then and previous to adoption, investigated, the shareholders would have been more likely to have arrived at the truth, and the engineer would have received greater honour and fame by a favourable result of such an examination, than after the works have been allowed to proceed to such a length that many have thought it would be sounder economy to continue on even a bad system, than to decide on any material change so late in the day. In the heat of the, (we had almost said fierce) controversy which has been carried on by the partizans of the old and new

systems, many men, otherwise eminently qualified to have advised the shareholders, have by ardent approbation or condemnation of the new system previous to the institution of the inquiry, staked their reputation upon the success of one plan or the other. The opposition was principally on the part of the Lancashire shareholders, a class of men better qualified than any other to form a judgment, from their familiarity with railway works; and at their instance the question of the utility of Mr. Brunel's plans was referred to the separate investigation of Mr. Hawkshaw and Mr. Wood, and the meeting of the proprietary was adjourned from September to the present month, to give time for the necessary inquiries and experiments before the Directors' Report should be received, and their line of proceeding thus approved of and confirmed.

Although it was not strictly and formally agreed that the report of these referees should determine the decision of the meeting, yet such was certainly the received understanding; and the repudiation of the reports of these gentlemen, and the disfavour with which they were treated by the Company because they happened to be unfavourable to their preconceived opinions, appears to unbiased minds a rather uncourteous proceeding. The fitness or unfitness of these examiners for the task imposed upon them should have been taken into consideration before trusting them with it, and any objection of such a nature after an unfavourable report is certainly most ungracious, and of suspicious partiality. The reports of Mr. Wood and Mr. Hawkshaw are of too great length, and already too extensively published, to render it expedient for us even to give an outline of their contents, suffice it to say that they are both condemnatory, for various and different reasons, of Mr. Brunel's plans, but notwithstanding this, the shareholders determined by a majority of 7,792 to 6,145 to adhere to the new system. The ostensible ground for this decision appeared to be that Mr. Brunel had with one stroke of a hammer upset all the experiments upon which Mr. Wood's report was founded, by breaking off a bit of the blast pipe of the *North Star* engine with which the trials as to speed were made, and thus improving the draft and enabling it to perform constantly what Mr. Wood had declared was impossible to be done consistently with economy.

Most of the experiments referred to were conducted by Dr. Lardner at the request of Mr. Wood, and a very interesting account of them has been published in the January number of the *Monthly Chronicle*, a well conducted Magazine under his avowed editorship. Upon the point of atmospheric resistance, as our esteemed correspondent Mr. Cheverton plainly proved in our last number, the Dr. has shown himself to have played the part of the tyro who in performing a routine of experiments, falls upon results which he thinks to be new to the world, because they are so to himself, and publishes them as discoveries of his own, which they certainly may be, though any merit attendant on their being brought to light has long before been absorbed by the first discoverers.

Passing over in Dr. Lardner's article on "The Great Western Railway Inquiry," his general sketch of the nature of railway constructions and operations, and the reasons *pro* and *con.*, the old and new plans, as to gauge, sleepers, rails, and other points, we come to his account of "a more extensive and varied course of experiments than was ever before accomplished on railways, or probably than ever could have been accomplished except for the peculiar combination of circumstances which in this case produced it." The greater part of this course of experiments, we must for the present forbear noticing, relating to the deflection of the rails and sleepers, the motion of the carriages, and the descriptions of the ingenious instruments which were invented and constructed to test these points, but which, as most interesting in a mechanical point of view, we shall take another opportunity of adverting to more particularly in our pages : we then come to that part of the subject which Dr. Lardner characterizes as "transcending all others in importance," that is, " *the actual amount of resistance opposed to the moving power at present on railways.*" This resistance is from two causes,—friction, and the atmosphere. That from friction, is dismissed very briefly and doubtless to the Doctor's mind, satisfactorily ; but on the atmosphere he, bestows supreme attention, seeking the "bubble reputation" by proclaiming new discoveries in the laws of its resistance. Like other "castles in the air" the flimsy foundation has been blown away by a few prior discoverers, and the Doctor is left discovery-less.

Notwithstanding the freedom of our speech upon the pretensions of Dr. Lardner to the *discovery* of the supposed laws of resistance of the atmosphere, we by no means think lightly of his experiments upon the subject, which are highly philosophical, and apparently strictly to the purpose ; those upon the Whiston and Madeley inclined planes are very striking, and unless some such untoward circumstance as was asserted by Mr. Brunel to have influenced the experiments with the *North Star* was here in action, (which is not likely), appear to be so far conclusive.

The resistances of friction and the atmosphere are so intimately connected, that instead of confining our extracts to the latter branch of the inquiry, we shall now take the liberty of quoting the whole of the account of the experiments upon resistance given in the *Monthly Chronicle ;* as also Dr. Lardner's introductory and passing remarks, and conclusions. The experimenter's mind was doubtless well prepared for the investigation of the matter by the consideration that he had, previously given to the question, as being the appointee and depository of the grant of money made by the British Association at their last meeting, to ascertain and determine railway constants. The British Association and the Great Western Railway Directors have jointly conferred a benefit on railway science by engaging so ingenious an experimentalist to investigate a subject of such growing national importance, securing to the public materials towards a more complete and satisfactory settlement of the question, than either body singly could have done,—the one for want of time, the other of means

But we keep our readers too long from Dr. Lardner on the resistance of friction and the atmosphere.

" Some years since a French gentleman, M. de Pambour, made a course of experiments on the Liverpool and Manchester

Railway with a view to determine the amount of this resistance. The results he obtained, however, were not satisfactory, nor were his methods of inquiry such as would have afforded correct conclusions.*

" It is not necessary here to notice his calculations more fully, as we shall presently show that the investigation now before us presents the question of resistance altogether in a new light.

" The resistance offered to the tractive power by a carriage proceeding with a uniform motion on a straight and level railway is produced, partly by the friction of the axles of the wheels in their bearings, partly by the rolling of the tires on the rails, and partly by the inertia of the *air* which the carriage displaces in its progress.

" By a degree of accuracy of mechanical construction, which is within the present limits of engineering skill, and by a good system of lubrication, the friction of the axle in its bearing may be reduced to an exceedingly small amount.

" The amount of resistance which attends a rolling motion is small, under the most unfavourable circumstances, as is manifested by the facility with which enormous weights are moved even on the rough surface of the earth, when coarse rollers of wood are placed under them. How insignificant, therefore, that part of the resistance must be which proceeds from the rolling of the tire of a wheel accurately finished in the lathe, on the surface of a not less accurately rolled iron bar laid as truly even and level as art can effect, may be easily conceived.

" The resistance proceeding from these causes has been generally considered to be the same at all velocities; and if such be the case, it would follow that the expenditure of the moving power, in transporting a load over a given distance, would, so far as this source of resistance is concerned, be the same whether it were carried at five miles or fifty miles an hour. Some slight differences, however, on this point have existed between the results of the experiments of those philosophers who have inquired respecting it. Coulomb conceived that his experiments showed a slight decrease of resistance with the increase of speed, while Morin and others maintain that it is quite independent of the velocity. All these series

of experiments were, however, made at velocities so much less than those at which railway carriages move, that any laws of friction established by them should be applied with considerable caution in railway investigations. Some of the experiments made in the course of the inquiry now before us raise a doubt on this point, and suggest a probability that the resistance from friction *decreases* as the speed is increased.

To these sources of resistance, and to these only, have those who have devoted their attention to the practical working of railways, hitherto directed their inquiries. To reduce these to the lowest possible amount by the excellent construction of the carriages and engines, and the exquisite perfection of the road on which they move, has been the object to which the engineering profession has addressed all its powers, and with what signal success it is needless here to say. Such carriages and such roads could never have entered into the contemplation, even of the most sanguine speculator on the progress of art.

" The remaining source of resistance—the *air*—has been overlooked, or, if it received a thought, it was regarded as bearing so small a proportion to the other causes of resistance, that without producing any error of importance, it might be confounded with them; that its effect might be calculated on the same principles; and that the estimate of resistance thereby obtained, would be sufficiently near the truth for all practical purposes. That estimate was, as we have said, at the usual speed of railway trains, from eight to ten pounds per ton of the gross load.

" We shall presently see how far this assumption, and the estimates based upon it, are countenanced by the immediate results of experiments.

" It appears from the report before us, that the method decided on for investigating the resistance upon the Great Western Railway, was the common method of observing the rate at which a train in motion is gradually retarded. If it be admitted (as it has been always assumed to be), that friction is the only, or the principal retarding influence, it must then be admitted also, that the velocity which a carriage will lose when not impelled by any force will be equal in equal times. On this principle, proper formulæ were constructed by Dr. Lardner, in which due allowance was made for the effect of the momentum of the wheels of the carriages in rotation; and in order to obtain as great a number as possible of distinct experiments, from which a mean value of the friction might be deduced, he divided the interval between the moment at which the carriage

---

* In the mathematical formulæ which follow from M. de Pambour's reasoning, and which he uses in his calculations, he has wholly omitted the effect of the momentum of the wheels of the carriages in accelerated and retarded motion, so that his formulæ, in fact, represent the motion of a sledge, and not that of a wheeled carriage. The effect of such an error is far from inconsiderable, where the weight of the wheels and axles is so great as in railway carriages.

was dismissed with a known speed, until it came to rest, into a succession of short intervals, for each of which the velocity was observed. By such means the velocity lost in each of these successive intervals was ascertained, and such velocity formed a datum from which the amount of friction or resistance might be calculated.

"Upon applying these formulæ to a number of the experiments, a result was obtained, which was so unexpected, that in first instance it was deemed to be an error of calculation. It was found, in fact, that the computed amount of resistance for the first interval in each experiment after the train was dismissed was enormously greater than any estimate which had ever been made of that resistance. Thus it was found, that when the train was started with a speed of about thirty miles an hour, the computed value of the friction was about twenty pounds a ton, instead of not exceeding, according to the common estimate, eight or ten pounds ! The idea that this proceeded from any error of calculation or of observation was soon dispelled by finding that a like result followed from every experiment, and every calculation, without exception. It was also observed that the computed value of the resistance was greatly increased where the velocity of the train was considerable at starting. It was further observed that the computed values for the successive intervals until the train was reduced to rest were gradually less, the computed value for the first interval being generally two or three times greater than for the last.

"No doubt now remained that the resistance which was developed in these computations was a real resistance of much larger amount than any which has been hitherto contemplated, and that it has a direct dependance on the velocity, which it is known friction has not.

"The atmosphere of course presented itself at once as the cause of this resistance. It has been established by the experiments of various philosophers, that this resistence within the limits of their experiments increases as the square of the velocity; but their experiments did not extend to railway speed, and therefore could not be assumed with certainty as a datum. It was thought necessary, therefore, to reduce the question to immediate experiment on the railways themselves ; and although such experiments as those just adverted to, computed by the formulæ which were used, gave results which could not be far from the truth, it was considered, that where an effect was indicated by the calculations so very different from what practical men have hitherto supposed to exist, such a result

should, if possible, be deduced more immediately from experiment, and be made more independent of calculation founded on mere mathematical reasoning. For this purpose Dr. Lardner proposed, as an *experimentum crucis*, to dismiss a train of coaches at a high speed down a steep inclined plane, and to observe with precision the extent to which it would be accelerated in its descent by the gravity of the plane. If it were true that the resistance indicated by the above calculations were really that of the atmosphere, and that that resistance increased as the square of the speed, it was expected that in the descent a speed would be obtained which might produce a resistance equal to the gravity of the plane, and that when that happened, no further acceleration would take place, but that the train would move uniformly to the foot of the plane. It was farther proposed to select a second plane less steep than the first, and to make upon it a like experiment ; the gravity upon the latter being less than upon the former in proportion to its inclination, a less speed would produce a resistance in equilibrium with it, so that each plane would have a limit to its accelerating power, depending jointly on the resistance of the air, and on the weight of the train.

"It was likewise proposed to vary the weight of the train upon the same plane, in which case the limiting velocity would be varied in a corresponding manner.

"These experiments were accordingly tried with complete success, the results verifying all that was anticipated from them. The two planes selected for the purpose were the Whiston Inclined Plane on the Liverpool and Manchester Railway, and the Madeley Plane on the Grand Junction Railway, the former descending at the rate of one in ninety-six, and the latter at the rate of one in a hundred and seventy-seven.

"A train of four coaches, loaded with a weight equal to forty-two passengers, was impelled from the top of the Whiston Plane at the rate of about thirty miles an hour. Its velocity was observed to increase for a few hundred yards, when it obtained a speed of thirty-two and a quarter miles an hour, with which it descended uniformly to the foot of the plane.

"The same carriages deprived of their load were started in like manner down the plane, when they were found to attain a velocity of thirty-one miles an hour, which received no augmentation during the descent.

"In like manner on the Madeley Plane a similar train was started, and it gradually attained a speed of twenty-one miles an hour, which it retained until it completed its descent. Each of these experiments was repeatedly tried, always giving nearly the same result.

" Here, then, are facts which, being independent of all theory or calculation, cannot be either evaded or disputed. A load of eighteen tons has a gravitating power down one in ninety-six, amounting to four hundred and twenty-one pounds ; that gravitating power was, it appears, balanced by *some* resistance when descending at thirty-two and a quarter miles an hour. This re-sistance, amounting to four hundred and twenty-one pounds, was of course composed of friction and the atmosphere. If the fric-tion were taken at the common estimate of nine pounds, the friction of this coach train would be one hundred and sixty-two pounds, and it would then follow that the atmos-pheric resistance at thirty-two and a quarter miles an hour was two hundred and sixty pounds !

" But even this would appear too low an estimate of this hitherto neglected opponent to railway speed, for, by comparing the uni-form speed obtained in the descent of the Whiston Plane with that obtained in de-scending the Madeley Plane, assuming that the atmospheric resistance is in proportion to the square of the velocity, Dr. Lardner found that the value of the friction could be obtained, and the value which he obtained for it was by this process a small fraction more than five pounds a ton. If this value be correct, that portion of the whole resist-ance due to friction would be about ninety-three pounds, leaving three hundred and twenty-eight pounds to the account of the atmosphere !

" This very low value of the friction was deduced by a process in which nothing was assumed, except that the resistance of the air is as the square of the speed, and that the friction of the two trains used in the two experiments was the same. The two trains were certainly not composed of the same identical coaches, but they were composed of coaches similar in construction, equal in weight, and equally loaded, and were sup-ported on a similar number of wheels of like magnitude ; and, in short, no reason existed for supposing that the friction could be materially different.

" By varying the load on the Whiston Plane it was also ascertained that the re-sistance of the air did not vary sensibly from the proportion of the square of the speed. If the squares of 31 and of 32¼ be taken, they will be found to be very nearly in the proportion of 15·6 and 18, which was that of the loads used.

" Much on this interesting subject still remains for investigation, and many more experiments will be necessary, before the mean amount of the atmospheric resistance to railway trains can be considered as as-certained with the requisite degree of pre-cision. Meanwhile it is indisputable that this resistance at the common speed of pas-senger trains is of very formidable amount. That part of the resistance which arises from friction has probably been reduced as low as it is likely to be. At all events, what-ever importance may have heretofore at-tached to its further diminution, it can now have very little weight in the economy of railway transport. Even supposing the whole friction annihilated, we should not be relieved from much more than twenty per cent. of the present expenditure of power in passenger traffic. But since it is as impos-sible that this annihilation of friction can take place as that the perpetual motion should be discovered, it may be safely as-sumed that we cannot practically reckon on any increased economy of power worth se-rious attention, by any further improve-ments directed towards the diminution of friction. To what, then, it may be asked, are we to look to for that diminution of re-sistance which appears indispensible for ob-taining the increased speed after which rail-way engineers aspire ? It is an ascertained fact, that every augmentation of speed will produce an augmentation of resistance, not proportional to the increase of speed, but in the vastly greater proportion of the increase of the square of the speed. Thus if the railway train tried upon the Whiston Plane, were required to be moved at sixty miles an hour instead of thirty, the resistance which it would suffer from the atmosphere, instead of amounting, as it did, to about three hun-dred and twenty-eight pounds, would amount to one thousand three hundred and twelve pounds, to which ninety-three pounds being added for friction, would give a total resist-ance of one thousand four hundred and five pounds ! Thus the power of the engine to accomplish this double speed would require to be increased in the proportion of four hundred and twenty-one to one thousand four hundred and five ! If, then, the present engines are cumbrous and unwieldy, and overload, and injure the railway, what is not to be feared from engines capable of pro-ducing a power of energy so enormously greater, and producing that power with double the speed ! We are sure that no sober practical man will differ from us when we pronounce that in the present state of art the accomplishment of such an object is impracticable.

We are happy to observe from this concluding observation, that Dr. Lardner has learnt wisdom by experience ; when the time shall come that a speed of forty miles an hour on a railway will not be "*impracticable*" in an economical point of view, the qualifying words "*in the present state of art,*" will be found to have been an admirable provision for the future

---

DR. CARSON'S PATENT METHOD OF SLAUGHTERING ANIMALS FOR HUMAN FOOD.

In our last month's list of patents there appears mention of one granted to Dr. Carson, of Liverpool, for "an improved method of slaughtering animals for human food." The objects proposed to be effected by this invention are twofold, first, a speedy and therefore merciful method of depriving the animal of life, and second, improvement in the meat, by rendering it more nutritious and of better flavour. The Doctor has in the press a pamphlet which will contain a full explanation of his manner of slaughtering, and the physiological principles upon which the plan is based ; and from which work, as soon as it appears, we shall take the necessary extracts to explain the matter to our readers. In the course of the last week several butchers in Liverpool slaughtered some oxen and sheep by the patent method, and a party of twelve gentlemen supped off a shoulder of one of the sheep, which was pronounced by all who partook of it to be superlatively fine. The butchers of Liverpool are, we understand, adopting the plan very promptly.

In the mean time, as the subject is one of considerable interest, we extract a few passages from a letter which Dr. C. has addressed to the Editor of the *Liverpool Mercury*.

"The object of the mode of slaughtering recommended is, to remove the impediments which nature has set up against the elasticity of the lungs, on the existence of which impediments, life, in a great measure, depends, and, of course, to allow the lungs to resiliate into their natural dimensions, or in other words, to collapse, while the animal is still alive,

"The method of removing these impediments in the circumstances stated, to break fully and at once what may be considered the mainspring of life in all animals with elastic lungs, in the most humane manner,

has been the object of long and anxious consideration and of much varied experiment. The result has been effectual and conclusive.

"The effect of this method of slaughtering is to retain the lymphatic and lacteal fluids, and, indeed, all the finer juices of the body, at the same points and in the same proportion in which they existed while the animal was still alive, instead of being accumulated in the large vessels and discharged out of the body in a mass, and becoming a nuisance.

"The result is an increase of the edible part of a carcase to the amount of at least one-tenth beyond that which it would supply by any mode of slaughtering hitherto in use. The meat thus obtained is more juicy, tender, and far better flavoured. It sets sooner, and, of course, is earlier fit for use. It keeps much longer sweet and untainted. This remarkable and important property is evidently derived from none of the vessels being empty so as to admit the external air, and from an oozing of juice or lymph from the full vessels, when any portion is cut, upon the raw surface of that portion, and, by its tenacity, forming a sealing cement. This property is of great importance to butchers, who lose a great deal of meat in certain states of the weather ; to the public in general ; but particularly to mariners and to the inhabitants of warm climates.

"The time to which this meat will keep in different states of the weather has not been ascertained, for it has always been used before any marks of approaching putrescency have been discovered ; but, in the course of last autumn, this meat continued sweet and untainted for many days after other meat, killed at the same time and placed in the same circumstances, had given such proofs of spoiling as to render it necessary to cook it.

The meat, the produce of slaughtering in this mode, is more economical. In the first place, it never shrinks, but on the contrary, enlarges in cooking ; the fat, being supplied with the juices which enrich the muscular or red portion, is much more savory and is more acceptable to delicate palates, and is, therefore, less wasted. It requires less time to cook. The centre of a large joint is done nearly as soon as the surface. These properties are derived from the meat being less spongy and a better conductor of heat than meat obtained by any of the modes of slaughtering at present in use. It preserves well and requires a much less quantity of salt. Other valuable properties will present themselves to those who use this meat, and the method in question is also attended with advantageous results affecting hides, skins, &c., upon which it is not necessary now to dilate.

"I may, I trust, be permitted to state, that the addition of one-tenth, at least to the edible portion of animals slaughtered for human food, with a high degree of improvement in the quality of the whole, with the property of keeping much longer, and with a diminution of the expense of cooking, as well as less waste in the substance cooked, will not be deemed one of the least important boons which science has, at any period, been found to yield to the exigencies of human life."

ENGINEERING DUTIES OF THE BIRMINGHAM RAILWAY OFFICERS.

[From Lecount's History of the Birmingham Railway.]

The labours of the engineers, it is almost needless to state, commenced long before the ground was broken. In fact, many of them were employed in getting assents to our Bill, from the land-owners who have shown themselves so wise in their generation. Then came the various surveys and levellings required for fixing the line ; then the designing and drawing of bridges and other works in detail, in order that approximate estimates of costs might be laid before Parliment. When the period arrived for executing the works, it was necessary to calculate the time which those of the greatest magnitude would be likely to occupy, so that they might be let to the contractors in such an order, that the whole might be simultaneously completed, as far as possible, with reference to the successive openings of portions of the whole line, which was desirable, not only as a measure of pecuniary interest, but to get the road in good repair, and to drill every one into his particular duty. The order of letting the contracts having been decided, assistant and sub-assistant engineers were appointed, as required, upon the general principle of dividing the whole line into four districts, and each district into three lengths, so as to place about ten miles under the immediate superintendence of one sub-assistant engineer ; thus each assistant engineer had three sub-assistants, being all subordinate to one engineer-in-chief.

When any particular portion of the works was to be prepared for letting, the sub-assistant engineer, under the direction of his superior, had to revise all the Parliamentary surveys and levels with the utmost care, and draw to a large scale very accurate plans and sections of the land, in order that the quantity of excavations and embankments might be obtained as nearly as possible. It was also necessary to make detailed plans and working drawings, elevations, and sections of every bridge and culvert which carried a road or stream across the railway, or which carried the railway over a road or stream. These, being roughly sketched by the engineer on the spot, were sent to the chief office, to be fairly drawn out with full details, and upon a uniform system laid down by the principal engineer ; the object being to put them in such a shape that parties wishing to tender for any of the contracts might clearly understand the nature of the works, and make accurate estimates from the drawings without difficulty. The limits of each contract were defined with reference to the most convenient execution of the works, regard being had to the disposition of the earth works, so that each contractor might make his embankments with the materials yielded by his excavations, as far as it was practicable ; care being taken that the aggregate amount of the contract should not exceed the means of the generality of persons in the habit of tendering for such works.

A contract of 100,000l. was thought a very responsible undertaking ; and the experience of the London and Birmingham Railway has shown that those amounting to or exceeding that sum, have called for extraordinary exertions. Of these there have been seven upon the whole line ; four were very soon relinquished by the parties originally contracting for them, and the remaining three executed with great difficulty.

The drawings being completed, and the limits of the contracts fixed, detailed specifications were drawn up, under the engineer-in-chief's superintendence ; the whole was then submitted to the inspection of parties willing to tender for the works, who, on an appointed day, delivered in their respective estimates ; and the lowest tender was generally, but not invariably, accepted,—regard being always had to the character and means of the parties. The whole of these extensive and important works were let at prices which were under the estimate of the engineer-in-chief.

The original contract drawings were signed by the engineer-in-chief and the contractor, and preserved as documents. Three copies of each, however, had to be made out—one for the use of the committee, one for the engineer-in-chief, and one for the assistant engineer.

When it is borne in mind that the engineering works of the whole railway, in accordance with the above system, were divided into thirty separate divisions, each requiring its own set of drawings, estimates, and specifications, and that all these works, with two unimportant exceptions, were let to various contractors, between May, 1834, and October, 1835, it will be perceived that an extensive and efficient drawing establishment must have been kept at work. Speaking in round numbers, we must say, that for eighteen months, not less than thirty drawings per week, each requiring two days' work from one pair of hands, were turned out from the engineer-in-chief's office.

NOTES AND NOTICES.

A Transparent Watch.—A watch has been presented to the Academy of Science at Paris, constructed of very peculiar materials, the parts being principally formed of rock crystal. It was made by M. Rebellier, and is small in size. The internal works are visible; the two teethed wheels which carry the hands are rock crystal, the other wheels of metal, to prevent accidents from the breaking of the springs. All the screws are fixed in crystal, and all the axles turn on rubies. The escapement is of sapphire, the balance-wheel of rock crystal, and its springs of gold. The regularity of this watch as a time-keeper is attributed by the maker to the feeble expansion of the rock crystal in the balance-wheel, &c. The execution of the whole shows to what a state of perfection the art of cutting precious stones has been carried in modern times

Mr. Wivell's Fire Escape.—At a meeting of the Society of Arts on Thursday, the 10th instant, Mr. Wivell's fire-escape was brought before the Committee of Mechanics. He has somewhat simplified it since it was described and figured in No. 723 of our Magazine, by the substitution of a single for a double ladder, in other respects it is the same. The Society thought that it was the best that had been brought before them, and awarded Mr. Wivell a silver medal, on condition of his leaving a model in their Museum. An exhibition of its operation took place at the Society's house in the Adelphi on the following Tuesday. It certainly proved itself to be as complete an apparatus as could be desired, in all cases where it can be applied,—that is, to windows within a foot or two above or below the exact heights of the ladders. No experiment upon the principal point of objection urged by our correspondent, Mr. Baddeley, was made, that of indelicacy in regard to females. The Society's house-keeper could not be prevailed upon to make the descent in the canvass trough.

Long-going Time-keepers.—The *Revue du Havre* states, that a journeyman watchmaker of that town has invented a new movement, by which he can make a lady's watch go for a year after being once wound up, a gentleman's watch for three years, an ordinary clock for twenty years, and a church or other public clock for 280 years! The cost of one of these watches is only 50 francs.

Paris Artesian Well.—The boring in the Artesian well at the Abattoir de Grenelle has now attained the depth of 1400 feet, but no water has yet been found.

Railway Labourers and Navigators.—There is always one feature which strikingly distinguishes the construction of railways from that of canals, and this is the employment of the surrounding agricultural population. When the reader is informed, that for nearly three years from fifteen to twenty thousand men were engaged on this work, taken almost invariably from the adjacent towns and villages, and that, in actual labour, nearly four millions have been expended in earth-work, brick-work, brick-maikng, &c., among the local population, he would have some idea how this would influence pauperism and the poor-rates; whereas, in the making of canals, it is the general custom to employ gangs of hands who travel from one work to another, and do nothing else. These banditti, known in some parts of England by the name of " Navies," or " Navigators," and in others by that of " Bankers," are generally the terror of the surrounding country;

they are as completely a class by themselves as the gipsies. Possessed of all the daring recklessness of the smuggler, without any of his redeeming qualities, their ferocious behaviour can only be equalled by the brutality of their language. It may be truly said, their hand is against every man, and before they have been long located, every man's hand is against them; and woe befal any woman, with the slightest share of modesty, whose ears they can assail.—*Lecount's History of the Birmingham Railway.*

Magnetic Observations.—A deputation from the Royal Society had an interview with Viscount Melbourne on Saturday, the 5th instant, in Downing-street, to communicate some resolutions of the council, recommending the equipment of a scientific expedition to the southern regions, with a view to magnetic observations and the establishment of fixed magnetic observatories in Canada, St. Helena, Van Dieman's Land, Ceylon, and at the Cape of Good Hope. The deputation consisted of Mr. J. W. Lubbock, Vice-President and Treasurer; P. M. Roget, M.D., and Mr. S. H. Christie, Secretaries; Sir John F. W. Herschel, Chairman; and Major Sabine and Mr. Charles Wheatstone, Secretaries of the Physical and Meteorological Committees of the Royal Society.

Vegetable Weather Prophet.—The attention of scientific men is just now directed to a curious discovery of Professor Stiefel—well known throughout Germany for his Natural Science—the result of which has been the attainment of a more accurate knowledge of those changes to which the atmosphere is subjected, than was possible by the old method. The instruments hitherto in use have been the thermometer and the barometer, but an unerring standard has been considered a desideratum; that is said to have been at last supplied in the shape of Geranium fruit, the awns of which are *in* and evolved by the dryness or humidity of the atmosphere, in obedience to laws so regular and unvarying, that being fixed upon a dial-plate properly graduated, the change from one part of the room to another may be noted with the greatest accuracy. A paper on the subject was to be read at the meeting of German naturalists, held this year at Freyburg. Professor Stiefel is the greatest weather-doctor in southern Germany, and has for many years tabulated all changes in the atmosphere, according to a plan suggested by Goëthe; but he does not venture to predict for more than twenty-four hours at a time, and laughs at our weather prophets. By observation, he says, one may get the *rule*, but not the *exceptions.*—*Athenæum.*

Economy of Fuel in the Smelting of Iron.—It is stated by M. Teploff, mining engineer in Russia, that in the Ural Mountains, where many mines of iron are worked, they obtain 14 lbs. of iron by a consumption of the same quantity of fuel, pro rated both the quantity and rapidity of the air which enters into combustion is properly regulated; but only from four to six pounds when the action of the blowers is badly managed. In an experiment made by order of the Government, it was found that 100 cubic feet of air, under a pressure of two inches of mercury, have produced the same effect as 200 cubic feet of air under a pressure of one inch, but with this difference, that in the latter case the consumption of fuel was double that required for the former. M. Teploff further states that a furnace had produced 22,000 lbs. of iron in twenty-four hours, and which had only consumed 16,000 lbs. of fuel for this operation, whilst before the proper regulation of the blast, double the quantity was required to produce an equal portion of iron. According to the same engineer, the results obtained by this method are superior as regards economy to those produced by means of the hot blast.—*Recueil de la Société Polytechnique, June,* 1838.

LONDON: Printed and Published for the Proprietor, by W. A. Robertson, at the Mechanics' Magazine Office, No. 6, Peterborough-court, Fleet-street.—Sold by A. & W. Galignani, Rue Vivienne, Paris.

𝔐𝔢𝔠𝔥𝔞𝔫𝔦𝔠𝔰' 𝔐𝔞𝔤𝔞𝔷𝔦𝔫𝔢,

MUSEUM, REGISTER, JOURNAL, AND GAZETTE.

No. 807.] SATURDAY, JANUARY 26, 1839. [Price 3*d*.

Printed and Published for the Proprietor, by W. A. Robertson, No. 6, Peterborough-court, Fleet-street.

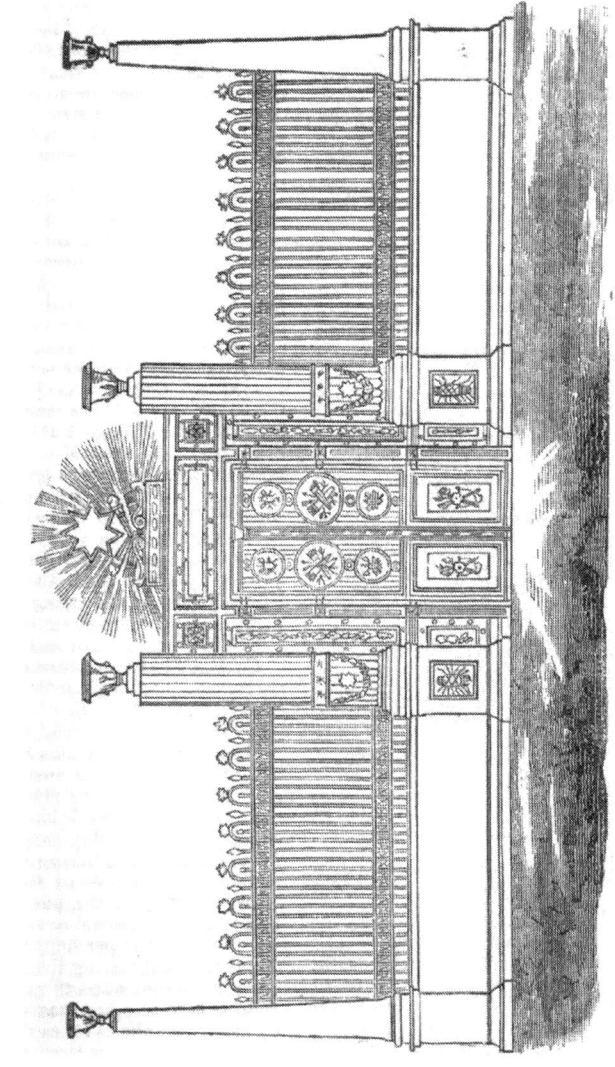

THE GRAND SULTAN'S PALACE GATES, MADE BY MR. DEAN, OF BOLTON.

THE GRAND SULTAN OF TURKEY'S PALACE GATES — MADE BY MR. DEAN OF BOLTON.

Sir,—I beg to send you herewith Mr. Physic's lithographed drawing of the beautiful gates, two pairs of which, with palisading, are now about being finished by Mr. Dean, engineer of Bolton (not as stated in the newspapers, for the Pasha of Egypt but) for the Grand Sultan of Turkey; at whose palace, in Constantinople, upon the banks of the Bosphorus they are to be erected.

The height of the gates, including the center ornaments over them, will be 35 feet, these, as well as all the other ornaments will be richly gilt (a taste you, say) more suitable to the Orientals than to us.

The height of the gates themselves will be 22 feet, and they will have a gateway of 12 feet in width; but including the hanging pilasters, which are formed to represent beautifully pendant vine leaves, the width will be 18 feet 3 inches.

The gates are formed of splendid devices and of admirable castings, and will be supported on each side by marble columns, 3 feet 9 inches each in diameter, surmounted by elegant vases.

The palisades add much to the general effect, they are 23 feet in length on each side, and are fixed in marble basements 7 feet 6 inches each in height, making in the whole 22 feet.

The weight of the two pair of gates will be forty tons.

The cost of them when fixed up, the manufacturer informs me, will be upwards of 20,000l. The patterns 900l. Packing cases 150l.

I am, your's, faithfully,
RICHARD EVANS.

7, Portland-street, Manchester,
Nov. 21, 1838.

ON STREET PAVING—DESCRIPTION OF A NEW METHOD.

Sir,—The subject of road-making interests the public in the present day so much, particularly during the time allowed for the trial of those portions of experimental pavement now laid down in Oxford-street, that I hope you will allow me to say a few words on the subject.

I understand that the time allowed for the trial of the different plans in Oxford-street is until the end of June next; now I contend that this is unquestionably not a sufficient length of time for a fair trial of these experiments, because there can be no doubt that most of them will be found infinitely superior to the old mode of paving the street, and six months is not a sufficient length of time to judge of the merit of any particular system—surely a twelvemonth is the least time that should be conceded, more particularly as these bituminous compositions should undergo the fiery ordeal of the summer's heat as well as the extreme of the winter cold.

I myself have always contended that a street ought to be paved upon a plan purely mechanical and simple, and upon which I shall have a little to say before I conclude this letter. I hold that boiling cauldrons with the subsequent daubing and pasting of the bituminous matter under, over, or between the stones is perfectly useless and unnecessary. When I make use of these terms I acknowledge the superiority of this style of pavement over the old, although I at the same time condemn it as useless. It has always been to me a matter of great astonishment how any men whose business it was to have studied this subject could have persisted for so many years in paving streets upon the old plan the whole system of which, from the removal of the old stones to the laying down of the new ones and completing the road, was one of the most absurd and fallacious.

Colonel Maceroni's plan of having the stones all of one size is no doubt perfectly correct, and without this being the case you can never have a road that will last for any length of time, because a heavy weight passing over a small stone will force it farther into the foundation than the next which is much larger; this fact, coupled with the additional fact of the foundations always having been too soft, has been the cause of the immediate destruction of all the paved streets of London. The Colonel next recommends a very heavy rammer for hardening the foundation and driving the stones down; now this, unless done effectually, will be of no use, or in other words it will be of no use unless the effect of the blow from the rammer be more than equivalent to the pressure from any weight that will pass over the stone; if it should not be so, the weight will force down the stone lower than the blow from the rammer, and hence an uneven surface.

Junius Redivivus recommends his square iron boxes, but these from the expense are totally out of the question; as also for many other reasons which might be mentioned: what, for instance, can be the use of the sides of the iron frame work, supposing even the square frame at the bottom to be of any service in supporting the blocks above? Surely there can be no necessity for the iron sides where the stone or wooden blocks are to be placed close together. Junius says he considers that the greatest weights which pass over the paved streets of London are the coal waggons laden. Now within two days of my reading this statement of his, I saw three timber waggons laden, pass along Pall Mall and Cockspur-street, each of which I will venture to assert weighed four or five times more than any loaded coal waggon he ever saw; they each contained trees of an enormous substance and length, and were with difficulty dragged along by a great many horses each.

With respect to the application of the bituminous cement which so completely fills up the interstices between the stones, let me ask whether considerable inconvenience may not be found in case of the bursting of a water pipe underneath? The water will certainly not be able to make its way up through this solid surface, but I apprehend will flow for a considerable distance until it finds some more porous outlet through which it can force itself; in which case there might be a considerable difficulty in finding the fracture from whence the stream originally flowed.

One of the plans adopted in Oxford-street, of having the stones 18 inches long on one side and 13 on the other, is really most laughable and beneath criticism; one stone having nothing to support it but the foundation, offers a resisting surface of 18 inches to the weight upon it; the adjoining one having the support of the stone next to it on each side as well as the foundation, offers a resisting surface equivalent to 49 inches in length. What will be the state of this pavement in the course of a little time,

when heavy weights have passed over these stones so irregularly supported?

I paid a little attention to the different specimens of pavement in their progress of formation, and passing accidentally one day, I observed that the old solid macadamized road was by the rigorous application of blows from the pickaxes of about a dozen stout Irishmen, broken up and removed for the purpose of making a foundation for the blocks to be placed above; and this new foundation was to consist of an artificial composition or concrete; now even supposing this concrete to form a solid bed, could you possibly have had one more solid or more durable than a macadamized substance of several years standing? It appeared to be as hard and as solid as a rock, and was evidently broken up with very considerable difficulty. Why go to the expense of removing a good ready-made foundation for one which could not at any rate be better, and in all probability not any thing like so good?

The great error in paving has always been an insecure foundation in the first instance, and then placing stones upon that insecure foundation of an unequal size. It strikes me, that if the following plan be tried it will be found to answer, and that most effectually. It is very simple and can be described in few words.

For the foundation let broken granite be laid down as if for a macadamized road, and let the public pass over this foundation until it is formed, into a perfectly smooth and solid mass: we know very well that a road of this description when once made by the public, is capable of resisting any pressure that may pass over it, no waggon-wheel, whatever may be the weight resting on it, makes, or can make the slightest indentation upon it by merely passing over it: very well, if that be the case, I will now ask whether it be possible to have a better foundation than such a mass to support the road itself? I save Colonel Maceroni the expense, the trouble and uncertainty of his monkey or rammer, because I merely lay the stones down and get the foundation completed free of expense by the public traffic. What can be the use of concretes, bitumens, asphalts, and other artificial compositions at an enormous

R 2

expense, when you can have that which is equal to a rock for a foundation, and that by the cheapest and most simple of means. After this macadamized foundation has been reduced by the public use to a solid mass, it is then fit for the support of the granite-stones, which I propose to place upon it. We know very well, that supposing the stones to be all of one size, it is desirable to have them as large as possible, because the larger the stone the greater the resistance, and therefore, the more durable the road. The stones I intend to employ are not to be of an unusual size, but to be so shaped, and so placed, that they shall have the same effect as if they were of large size. I have not yet quite determined upon the exact dimensions, but for the present, I shall say let them be 2 feet long 1 foot broad, and about 8 inches in depth, but made in the shape of a wedge, so that looking at the end of the stone it shows something like a triangle ; let rows of these stones, being all of the same size, be placed on this macadamized foundation close together with their flat sides downwards, and, of course, their sharp edges uppermost; then let other rows be laid between them, fitting in as wedges, the stones being reversed, and with their flat sides uppermost, but let the centre of each of the upper stones be placed exactly over the line of conjunction of the four under stones, so that in fact any one stone having a pressure upon it, will have the support of four others under it; or in other words, a stone 2 feet in length and 1 in breadth will, from the position and the principle upon which it is placed, offer a resisting surface to any weight, equivalent to a stone *four feet in length and three in width*, the stones being all of the same size ; this will be the case in every part of the road ; you cannot place a wheel on any part of it, or on any one of the stones, where you shall not find that the pressure of it extends, or is diffused over a space 4 feet long by 3 in width. If my foundation alone is capable of resisting any weight passing over it when exposed, what will it be when protected by an upper stratum such as I now describe? If you were to place St. Paul's Cathedral upon wheels and pass it over such a road it would make no impression upon it, the only in-

jury that could possibly arise to it, if used by the public, would be, from the friction of the wheels coming in contact with the surface of the stones, and wearing it away by an equal extent in all parts, which sort of wear and tear we know is very trifling even in a great many years.

There are many more points connected with the subject which I ought to touch upon had I time or you more space, but sufficient has been said, I think, to make the general principle intelligible. Independent of its great durability, it will always preserve an undisturbed and level surface ; it will also always be in a comparatively clean state, for no mud can arise from the under surface; it is also with very great facility taken up and replaced, when such an operation shall be required for the purpose of repairing it, making any new sewers, or laying down pipes of any description ; the difficulty of draught over it will also be very much lessened in consequence of its level surface being always preserved without any alteration.

SELYM.

———

ON MECHANICAL DENTITION, AND THE EDUCATION OF DENTISTS — BY JOHN GRAY, ESQ., DENTIST.

Sir,—As it is evident that only a duly qualified surgeon is competent to act as a surgeon dentist, and that an experienced practical mechanic, only, can succeed as a mechanical dentist, so it is equally dishonest for the mere surgeon to assume the mechanical department, as for the mechanic to play the quack in the surgical department ; indeed there is a much greater distinction between them than is commonly supposed.

The mechanical must always precede the surgical education ; for, it has been observed, that he who is not an expert mechanic at the age of twenty, will never afterwards be able to acquire the mechanical dexterity that is necessary for the fabrication of artificial teeth ; whereas, in the acquirement of surgical knowledge, so much more serious thought and riper judgment are requisite, that the student reaps, comparatively, but little benefit from his studies before that age. Hence a mechanic may become a surgeon, but he who is first a surgeon can never afterwards become a mechanic.

It would appear, however, that if a surgical education be not acquired in early life it cannot be afterwards, else many quacks whose impositions have proved successful in a pecuniary point of view would gladly obtain the respectable designation of surgeon. This view of their case is corroborated by the anxiety which they often evince to give their sons a regular surgical or medical education; thus showing how keenly, though secretly, they feel their own degradation, even in the midst of successful imposture.

It is only when the accomplished mechanic has superadded the qualification of surgeon, that he may legitimately assume the whole range of the profession of a dentist with credit to himself and advantage to his patients.

Although a mechanical dentist ought to be a first rate workman yet there is not that in the work itself, which would ever produce a workman.

As a physician must acquire a thorough knowledge of anatomy before he can practice medicine, so must a mechanical dentist acquire his mechanical abilities before he turns dentist, or he will ever remain a " botch." One reason of this is, that the workers in comparatively soft materials do not make their own tools, the fabrication of which is the most essential part of the education of a mechanic. The workers in steel and other hard metals make the greater part of the tools they use, by which means they become critically acquainted with the temper and other properties of those that best suit their peculiar mode of working; and the shades of difference in this respect are so nice, that few workmen can use the tools of others without injuring them—for instance, no mechanic can properly use a *drill* that he has not made for himself.

An employment like clockmaking practically exercises all the laws and principles of mechanics, and thus clockmaking is to the watchmaker and mechanical dentist what anatomy is to the surgeon, Without it the watchmaker can never rightly understand his business, and, consequently, can never become a sound workman. Watchmaking, which is clockmaking in miniature, requires such anxious care and exquisite execution of minute parts that the making of artificial teeth would appear comparatively coarse work to a talented clock and watch maker; all the practical knowledge and mechanical dexterity of a clever general workman as a clock and watchmaker, is requisite for, and can be applied to, the making of artificial teeth; and thus a watchmaker may be a dentist, though a dentist cannot be a watchmaker. It must not be understood, however, that every watchmaker is an Earnshaw, or fit to become a dentist. None but a genius and an enthusiast will ever shine in either character.

As health or indisposition, comfort or pain may be the result of their performances, it is evident that dental artists should be the first mechanics in the kingdom. The importance of their labours demands this, and the liberal price paid for their services should command it. To the ingenious the work will be delightful from its constant variety, which calls into action all the inventive and mechanical faculties and affords them scope. And although the making of artificial teeth has been, hitherto, (with very few exceptions) carried on as a trade by rapacious pretenders, yet it is not a *trade*, but an *art* of a high order, and as such it cannot be performed by proxy. Artificial teeth cannot be "got up" even by clever artists without seeing the person to whose mouth the work is designed to conform, and the functions of which it is intended to assist; as well might an artist be set to paint a portrait without being permitted to see the original. Like other artists the mechanical dentist must execute all, or very nearly all, the work with his own hands or he never can succeed as a dentist, nor ought so to do.

The increasing demand for artificial teeth opens a rich field for the enterprise and encouragement of ingenious mechanics, particularly clock and watchmakers ; and young men brought up in the country as general workmen, will always possess a decided advantage over those bred in large towns, where they generally learn but one branch of watch work, which is very seldom preceded by clockmaking, and consequently they are very little better than *automata* ; parents and young men themselves should endeavour to remedy this fundamental defect in the "division of labour education," which renders them liable to be thrown out of employment by every alteration in their business, and unable to turn their hands to anything else.

I shall conclude this subject with a sketch of what I should consider to be a proper education for a mechanical dentist.

A boy at the age of twelve years with such a developement of faculties as clearly indicates a "mechanical genius," should be placed in a clockmaker's shop till the age of seventeen, if the work carried on in the shop be of a general or mixed nature, which is commonly the case, and if many men be employed, so much the better, provided the principal part of the work done be clockmaking. From the age of seventeen to nineteen or twenty, he should be employed at watch work, either repairing or finishing, in order to "fine down his hand," so that he may never afterwards experience any difficulty with work, on account of its minuteness.

Having now, it is presumed, acquired mechanical knowledge and manual dexterity, he may commence the making of artificial teeth under the best instructor he can procure; and if the education of the dental preceptor has not been equal to that of the pupil, the latter will soon surpass his master. To give entire ease and comfort to the wearer, the mechanical artist must be capable of carving a piece of ivory to fit the gum so perfectly airtight that it shall adhere and remain securely in its place for the purposes of mastication, &c., by the mere pressure of the atmosphere, which renders the wearer almost unconscious that his teeth are artificial.

The young artist is here cautioned against falling into the quackery of constructing artificial teeth on gold or silver plates, which are only manufactured by such persons as have recourse to the hammer because they cannot use the graver. Such botch-work is worse than useless because highly injurious to health and destructive to the other teeth.

If the surgical is intended to be added to the mechanical education, his anatomical and other surgical studies may be commenced simultaneously with those of the dentist without prejudice to either, for they will assist rather than retard each other. At this period the beauties of the interesting field that opens upon the ar-

dent mind of youth may be experienced but cannot be described. At the age of twenty-five his surgical knowledge may be so complete as to procure his admission as a member of the College; and, when a few years of experience in the actual practice of his profession have given him the ease and confidence attendant on ability he will be able to look back with satisfaction on the progressive steps by which he gained his knowledge and present eminence.

A young man of genius and sufficient enthusiasm for the task, may, by his own exertions, acquire the above education, including the surgical part, without any assistance from parents or friends. I mention this for the encouragement of merit, knowing it to have been achieved with comparative ease. Similar spirits may aspire to the same honour. Self acquired advantages are at once the most honourable and valuable to the possessor.

J. GRAY.

26, Old Burlington-street, November, 1838.

STEAM CARRIAGES ON COMMON ROADS.—MR. WALTER HANCOCK'S ESTIMATE OF EXPENDITURE AND RECEIPT OF A LONDON AND BIRMINGHAM ESTABLISHMENT.

Sir,—The subject of Steam Locomotion on Common Roads having been lately discussed in the various mechanical periodicals, and as as no satisfactory replies have been afforded, will you permit me space to give my opinions upon the arrangement and economy of this mode of transit, resulting from actual practice, with which your readers and the public generally can acquaint themselves, by perusing a work I have lately published.*

I will select the London and Birmingham road, having recently been called upon to furnish an estimate for that line by some gentleman interested in it; and as great care was taken in drawing this up, I annex a copy.

* Narrative of Twelve Years' Experiments, 1824-1836, with Steam-carriages on Common Roads. Published by John Weale, High Holborn, and J. Mann, Cornhill.

Although a change of drags midway would be sufficient, I would recommend two stations for that purpose, which would divide the runs into about 36 miles each, which would provide more effectually against delay in case of necessity.

The stations for fuel and water, should never be more than 10 miles apart, as it is not advisable to carry an unnecessary weight upon the drag.

Foul bars do not form any obstacle to my progress, for I have a means of readily changing a floor of bars without lowering the fire, and which I patented in 1833.

Having run the road from London to Birmingham and back, with one of my carriages, in August, 1835, I am enabled to give the above opinions with some degree of confidence as regards the road in question; as well as from the experience of others on which I have run long distances; and from actual working, particularly for six consecutive months on the Paddington road. These data afford me the means of stating truly, the great returns which running steam-carriages for hire, on common roads, would afford to the capitalist. I would bind myself to build drags, which should perform all that I have stated in the annexed estimate.

Railways, of course, afford the speediest means of transit, but a few hours saved, is not such an object with the majority of travellers, as to induce them to forsake the pleasures of the common road. Steam-carriages form the medium between railways and horse-coaches, and had the coach-masters generally been a little more liberal and unbiassed in their views, the public at large would have had demonstration of their effect long since. I have had no body of capitalists to thank for either cash or candid investigation, for want of which support, the public are at this moment quite unaware of the perfection to which I have brought my carriages and the work which they have performed on the public roads:—whilst I have been openly desiring inquiry, the railway engineers have been receiving almost unbounded and blind support; how far a profitable return for which will be made remains to be proved.

Some months since, I offered Lord Litchfield to convey the mails at 20 miles an hour, and if my experience had not borne me out, I should certainly never have made such an offer.

It is true that many mere speculators have disgusted and prejudiced the public against common road steam-carriage conveyance, by sending forth exaggerated accounts of performances, which have been carried out only upon the drawing-board or writing-desk; by these the subject has been sadly overrated; no spare drags have been by them deemed necessary; but I am quite willing to admit, that if three drags up and three down should be required on the London and Birmingham road daily, that I would recommend the whole stock to consist of 12 drags; the railways for a like distance would have 18 locomotives; this provision, in either case, is necessary to ensure sufficient time to clean, inspect or repair.

To have only the exact number required on the road at one time, would be tantamount to driving a set of horses all through. It is only an increase of outlay in the first instance, for certainly a carriage will last longer in proportion, under moderate, than under excessive work.

I think that the velocity being less, and from the peculiar construction of my carriages, that the repairs would not amount to more than one-fourth of those for railway-carriages, running equal distances.

Of course, I am in no small degree mortified, after having spent fourteen years of the prime of my life to perfect this invention, to meet with less of public encouragement, than of clamour and prejudice; when I well know that society would be benefitted, by the general adoption of steam locomotion on common roads; but I have yet hope, that at no far distant time, it will be a favourite mode of transit; and I beg to observe, that this hope is not founded on opposition to railways, but merely to afford the public a wholesome competition.

Certainly, the railway engineers can have no cause for complaint against the constructors of common road locomotives, the former having copied rather copiously from the latter: I would instance their general arrangement of machinery, which is from Mr. Gurney; the extended surface of boiler in equal space, from myself, as well as the adoption of double valves to their pumps; again increasing the draft, by passing the waste steam up the funnel, from Mr. Gurney, with several others; and it may be fairly asserted, that until they copy more from common road locomotives, in many other material points, they will not remove the evils at present so justly complained of.

The subjoined is the estimate referred to at the commencement of this letter:—

Estimate of Steam Power for One Day's Work on the Common Road from London to Birmingham, and back, a distance run of 216 miles; the Passengers' train at 12 to 14 miles an hour, the Goods' train at 10 miles an hour.

Passengers.	£.	s.	d.	Passengers.	£.	s.	d.
Dr.				*Cr.*			
To coke, oil, engineers, guards, stokers, and tolls, 1 drag with 3 coaches to travel 108 miles each way................	26	0	0	By 80 passengers each way, or 160 passengers and luggage at 10s.	80	0	0
Repairs, wear and tear, renewal of engines rent of stations, pay of attendants, and contingencies	22	0	0	1½ ton of goods each way, or 3 tons at ½d. per lb........	14	0	0
	48	0	0		94	0	0
Goods.				*Goods.*			
Coke, oil, engineers, guards, stokers, and tolls, 4 drags and 4 vans, to travel 108 miles each way	52	0	0	32 tons of goods, carried 108 miles, or each way at ½d. per lb.	149	6	8
Repairs, wear and tear, renewal of engines, rent of stations, pay of attendants, and contingencies.................	44	0	0		243	6	8
Daily interest on capital £2,500 at 5 per cent.............	3	8	6				
	147	8	6	Deduct 20 per cent. on 94l. and 10s. per cent. on 149l. for light loads..................	33	14	0
Daily profit.........	62	4	2				
	£209	12	8		£209	12	8

Capital.

4 drags for passengers .... at	£1,600	....	£6,400
9 common carriages .... "	130	....	1,170
8 drags for goods .... "	1,600	....	12,800
7 vans for ditto .... "	80	....	560
Fitting up stations			2,000
Surplus capital for contingencies........................			2,070
			£25,000

Daily profit £62 4s. 2d. on daily expenditure of £147 8s. 6d., is about 42¼ per cent.

I am, Sir, your obedient servant,

WALTER HANCOCK.

Stratford, Essex, January 21, 1839.

P.S. The above estimate does not quite accord with the one in my narrative; the difference is caused by this being calculated for a more extensive concern, whereby the expenses are proportionably diminished, and the profits thereby increased.

W. H.

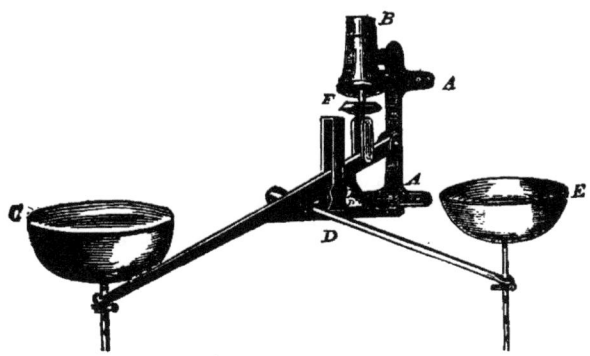

The Society of Arts lately awarded to a Mr. Crockford, a silver medal for his invention of improved ball valves for water tanks, cisterns, and other vessels, of which the following is a description supplied us by the inventor :—

The use of the ball cock as a means of stopping the further entry of the water into the cistern or other vessel to be filled, when it has reached a certain height, is attended with many advantages ; at the same time, however, many disadvantages are attendant on it, which considerably lessen its utility, viz. its uncertainty of action, liability to get out of order, and the length of time which is taken in filling the cistern or other vessel : this latter arising from the circumstance that the moment the water begins to float the ball, the supply of water through the service pipe is to a certain extent cut off by the partial closing of the cock, consequent on the rise of the ball, and the supply, by the continued rising of the ball, is continually decreasing in quantity, till at length,

when the vessel is nearly filled the water merely enters drop by drop. To remedy this, many contrivances have been proposed to allow the cock to remain open till the proper moment arrives for closing it, and then to close it entirely at once ; but of these contrivances, ingenious as many of them are, none that we have hitherto seen are so simple and so well calculated to perform the duty required as the one represented by the above cut, the method of action of which will be easily understood by reference to the following description of the various parts :

The frame A is screwed to the side of the cistern by four screws, the soldered joint to the supply pipe at B, the large ball C is fixed on end of long lever, and is kept down by the trigger D, until the water rising to the small ball E on the short lever releases the trigger, and by the power of the large ball the valve F is immediately closed.

This valve has comparatively little friction, is fixed with little trouble, and possesses certainty of action.

EXPERIMENTS WITH CAPTAIN ERICSSON'S CANAL BOAT PROPELLER.

We are glad to find that the success which attended the first experiments with Ericsson's propeller, (noticed in our 721st, 751st, and 781st numbers) has induced some American canal proprietors to build a steam tow-boat, fitted with it, for the purpose of putting it to a complete practical test. The boat is called the *R. F. Stockton,* and has lately arrived in the Thames from Liverpool, where she was built last summer

by Messrs. Laird, under the superintendence of Mr. Ogden, the American consul for that port. Several experiments have been made, the results of which appear very satisfactory, both in relation to the application of the propeller to inland and to ocean navigation ; and these experiments derive additional weight from the fact of their having been performed and approved of in Liverpool, the grand emporium of shipping and of commerce.

Respecting the speed which it has been asserted may be attained by the new propeller, we have to notice a trial made below Blackwall on the 12th inst., in the presence of about thirty gentlemen, many of whom were scientific and practical men. The result was, that a distance of nine miles (over the land) was passed in 35 minutes, with the tide : thus proving the speed through the water to be between 11 and 12 miles per hour. The propeller was only 6 feet 4 inches in diameter ; the dimensions of the boat are given in the account of the next experiment.

An experiment proving the great power power of this propeller, with an account of which we have been supplied, was made on the 16th instant, between Southwark and Waterloo bridges, the result of which was as follows : — Four coal barges, with upright sides and square ends, as represented in the engraving,

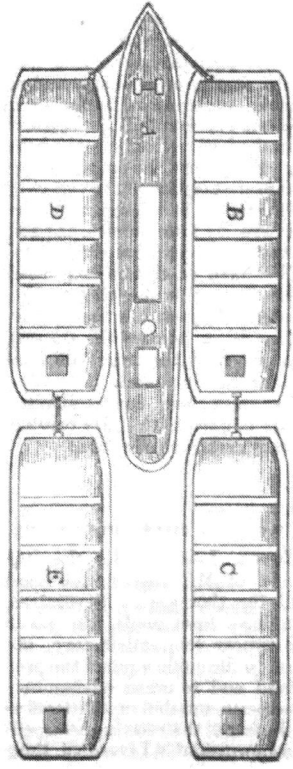

B *Nep*, 15 ft. beam, drawing 4 ft. 6 in. water.

C *Joseph*, 15 ft. 7 in. beam, drawing 4 ft. 6 in. water.

D *Mary*, 15 ft. 2 in. beam, drawing 4 ft. 6 in. water.

E *Ugie*, 13 ft. 4 in. beam, drawing 4 ft. water.

were made fast to the steamer A, which is 70 feet long, 10 feet beam, and draws 6 feet 9 inches water. Steam being set on, full speed was attained in about one minute, and the whole distance between the bridges, precisely one mile, was performed in 11 minutes, the time chosen for the experiment being high water. The number of strokes made by the engines was 49 per minute—the cylinders 16 inches in diameter with 18 in. stroke. The difference in the speed of the propellers being as 9 to 10, and the outside one, revolving at the greater speed, and being attached to the crankshaft directly, it follows that the inside propeller made only 44.1 revolutions per minute. Now, although the circumference of the propeller is nearly 20 feet, the spiral planes are placed at such an angle that, *were resistance of the water perfect*, the boat could only proceed 14,04 feet for each revolution ; hence the distance passed over in one minute could only be $44,1 \times 14,04 = 619$ feet per minute with such *perfect* resistance. The distance actually passed being $\frac{5280}{11} = 480$ feet, it follows that a distance of 139 feet was lost out of 619, which amounts to only $22\frac{1}{2}$ feet less of speed. Considering the *square form* of the barges towed, and that they presented together 59 feet 1 inch beam, with an average draught of 4 feet 4 inches, besides the sectional area of the steamer which is 43 square feet, and, considering that the propeller is only 6 feet 4 inches diameter, occupying less than 2 ft. 6 in. in length behind the stern of the boat ; the result we have now recorded may, in a mechanical point of view, be considered of great importance.

The *R. F. Stockton* has been constructed as a tug-boat for the Ohio and Raritan canal in the United States, whereto she will shortly proceed.

DESCRIPTION OF DR. CARSON'S PATENT METHOD OF SLAUGHTERING ANIMALS.

Since our last publication we have received Dr. Carson's pamphlet upon this important subject. We could not consistently with the character of the *Mechanics' Magazine* enter into a detail of the physiological principles upon which the new method of depriving animals of life is founded. Those who are more deeply interested in the subject than the generality of our readers we may refer to a paper by Dr. C. in the *Philosophical Transactions* for 1820, and to his work published in 1833, entitled *Inquiry into the Causes of Respiration, &c.* The following brief statement must suffice on this head :—

" The physiological principle upon which this discovery is founded, and the existence of which was first ascertained by myself, is, that a power, of great and extensive use to animal life, is derived from the stretch in which the lungs are held both in the living and recently dead animal, in opposition to their elasticity. In consequence of this power existing after death, while the antagonist and controling power derived from irritability was destroyed with life, the blood, lymphatic and lacteal fluids are all drained from the arteries and small veins of the larger circulation, and from the lymphatics and lacteals of the parts traversed by that circulation, into the viscera and blood-vessels within the chest, and into the roots of the larger veins approaching to it."

The following are descriptive particulars of Dr. Carson's patented methods of carrying his invention into operation :—

" It is necessary for the full accomplishment of the object aimed at by this method of slaughtering, that a pressure at least equal to that of the atmosphere, shall be made to rest upon the external surface of the lungs and of the other viscera within the chest, from which surfaces, a part of that pressure is removed during life, and for a considerable term after the death of the animal, by the elasticity of the lungs. This object is accomplished by a free, sufficiently abundant, and uninterrupted admission of air into both the cavities formed by the membrane lining the chest and that forming the external covering of the lungs.

" A pressure, at least equal to that of the atmosphere, being thus brought to rest upon the external surfaces of the organs contained by the chest, the lungs, in obedience to their structure, collapse. When the animal in this situation attempts to inspire, and expands the chest, the air required to occupy the increased dimensions, instead of entering through the windpipe, in which direction it has to encounter the disengaged elasticity of the lungs, is admitted through other passages, provided in the manner about to be described. The transmission of air through the windpipe, in other words, breathing, is rendered impracticable ; and as none of the viscera of the chest will be relieved from any portion of the atmospherical pressure, all suction or draining of the fluids towards the chest is prevented, however extensive that expansion may be.

" The method for admitting air into the cavities of the chest is the following :—

" A tube, sharp at one end, and perforated near the point by several holes, is to communicate at the other end with an air-tight bag, filled with air, and of a size fitted to different animals ; a button is to be placed on this tube, with a slight concavity on the side facing the point, and at such a distance from that point as to admit all the holes in it, to be within the cavity of the chest, into which it is to be introduced. A tube thus armed is to be introduced, by a simple perforation, made by its sharp point, into the cavity of the chest, between the fifth and sixth rib, on each side of the chest, as far as the button will allow it. The concave side of the button is to be filled with some greasy substance, and pressed firmly against the side of the animal, to prevent any air escaping along the outside of the tube. The air is then to be pressed gently out of the bags through the tubes, into the chest ; the lungs instantly collapse ; and in a minute or two the animal expires.

" Not a drop of blood is lost in this operation ; and if the same vessels be divided, which in the case of an ox being killed by knocking on the head or pithing, would discharge, in the course of a minute or two, from forty to seventy pounds weight of blood, they would not, in this case, yield more than a spoonful or two.

" A pressure is required to be given to the air introduced, additional to that of the atmosphere, in order that when the chest is expanded, in the violent efforts which the animal may make to breathe, the air within the chest may still, after being thus rarefied, sustain a pressure equal to that of the atmosphere. This will be easily accomplished, by keeping a constant pressure upon the air-bags. The additional heat acquired by the air, after it has been a few seconds within the chest, will aid this effect. By this means any chance of a portion of the atmospherical pressure being removed from the pectoral viscera, while it rests wholly upon the rest of the body, will be prevented, and, of course, all draining of fluids from the body into the viscera of the chest guarded against.

" For the successful application of this apparatus it is necessary that the animal should be secured against the power of making any extensive movement, or against any accidental change of position.

" Lambs, calves, sheep, or pigs may be sufficiently secured by being placed upon an oblong, shallow cradle, to each corner of which is fixed a perpendicular post ; these posts are fitted with buckles, at different heights, by which a limb of the animal is fastened to each post ; by these means the animal, without being subjected to much pain, is secured against the power of making any movement or change of position. I need not say that these cradles must be of different dimensions, suiting the bulk of the animals to be placed on them.

" For oxen a peculiar machinery will be required. A stall or booth, into which he is to be led, is formed in the following manner :—

" Two broad bars of iron (for it ought to be made of the strongest materials) in length between two and three feet, are to be placed at a distance from each other, equal to the length of the animal. These bars are to be connected at each end, on the same side, by iron rods. At each end of these bars, an upright rod of iron is to be erected, nearly of a length equal to the height of the animal as he stands. These upright rods are to be connected on each side, and also those in front by other rods, at different heights. Chains are to be attached to the top of the upright rods or topmost side rods, on one side, that after the animal is led into it, may pass over his shoulders and buttocks, and be hooked upon the corresponding rods on the other side. The feet of the animal are to rest upon the broad bars, the fore feet on one and the hind feet on the other ; each limb is to be bound or buckled at different heights to the upright rod nearest it. After the animal has been thus secured, the entrance of the booth is to be closed by a chain. Strong leather bands are to be passed under him at the breast and flanks, and secured on each side, either to the upright rods or those which connect them ; these bands are not so much for security, as for preventing the falling of the animal during the short and simple process by which it is to be deprived of life. The head of the animal is to be fixed by a rope.

" This machine, though tedious in the minute description may evidently be made of speedy application to the purposes in view.

" In the position in which the animal stands secured in the stall, a convenient and easy access is afforded for the application of the apparatus by which the animal is to be deprived of life.

" Another method, where such an apparatus as that described cannot be procured, is to throw the animal upon its back, bind its fore feet to its hind feet and fix them in that position by tackling, so as that it may not be able to swerve upon either side.

" Another method still is, to suspend it, either by the fore or hind feet, and then securing its loose feet, separated, to bolts in the floor.

" The method which gives the least preparatory pain is to be preferred ; and on this account, as well as the greater security it affords, placing the ox in a stall similar to that described above, appears to be the best."

RECENT EXPERIMENTS AT PARIS ON RAISING AND HEATING WATER BY SAVERY'S ATMOSPHERIC ENGINE. BY M.M. COLLADEN AND CHAMPIONNIERE.

The note which I present to the Academy of Sciences (says M. Colladen) is a summary of the experiments I made with M. Championniere, civil engineer, with the steam engine on *Savery's construction*.

In these very simple machines, the steam raises the water by its immediate action. The steam is introduced into a vessel, then condensed, and produces a respiration or flowing in of water.

A second admission of the steam drives the water up into the reservoir.

These machines were the first steam movers employed in large works. They were afterwards abandoned for the machines of Newcomen and Watt.

Several manufacturers, especially Manoury D'Hectol, nevertheless have employed them.

As our experiments may serve to fix the value of these engines, and the conditions under which the employment of them may be preferable, we think it may be useful to publish them.

We possess very few estimates of the power of the Savery Machine :—Bradley, Smeaton, Manoury and Girard, have published some memoirs on its effects ; but we find in no publication on the subject, the measure of increase of heat in the water elevated, nor of any other element needful to the theory of these motive powers.

But a very small number of the Savery machines are in existence. We know of but five in operation, there are in the department of the Seine, the fourth in Loire

Inferieure, and a fifth at Lyon. We believe there are none ramaining in England.

We have experimented with the three of the department of the Seine. The oldest is at the *abattoir de Grenelle*, and was constructed by Manoury. The two others are in the Vigier baths ; they were made by Gingembre.

The following numbers were obtained from these three machines in three series of experiments :—

Experiment of the 26th of March, 1833, *on the bath machine of Pont Marie.*

Temperature of the water of the Seine, $6\frac{1}{4}°$
Mean tension of the steam, 3atm.
Water raised per hour, 12.213m.
Height of elevation, 6.6m.
Temperature of the water raised, $10\frac{1}{4}°$
Dry wood burned during one hour, 30.4k.
Duration of a period, 26.0″

Experiment of the 10th of July, 1833, *with the same machine.*

Temperature of the water of the Seine,$19\frac{1}{4}°$
Mean tension of the vapour, 3atm.
Water raised per hour, 12.724m.
Height of elevation, 6.10m.
Temperature of the water raised, $23\frac{1}{2}°$
Dry wood burned in an hour, 46kil.
Duration of a period, 26″

Experiment with the machine of Manoury D'Hectol.

Temperature of the water of the well, $12\frac{1}{2}°$
Mean tension of the steam, 00
Water raised per hour, 15.400m.
Height of elevation, 14m.
Temperature of water raised, $16\frac{1}{2}°$
Charcoal burned in an hour, 13kil.
Duration of a period, 90″

Agreeably to the first and second tables, the machine of Pont Marie gives 2.595 dynam, to a kilogramme of wood.

This is about eight times less than the effective force of a small piston machine of the same force which would work pumps. But the water raised would have to be afterwards heated, so that we must take into account the increase of temperature, which was four degrees Cent. ($=7\frac{1}{4}$ F.) in the first series, in the month of March, and three quarters in the second in July. Thus, in the first case, each kilogramme of wood sent up to the reservoir, by the action of the machine 1702, portions of heat (caloric) and the second 1255. With a more complicated machine than that of Savery, an additional heating apparatus would have been necessary, and this addition would have required the same expense.

Thus, whenever water is to be both raised and heated (and this often occurs in manufactories) the almost forgotten machine of Savery is the most advantageous motive power. It is the least costly at first, the least subject to accidents, and to wear and tear, and the most easily managed.

We will add a few words on the comparative effect of the three machines. In all of them the accession of heat was about four degrees, although the Manoury machine differs essentially from the two others.

The last machine performs more than double the work of those of Gingembre, at the same cost. Agreeably to the public report of M. Girard, in the 21st Vol. of *Annales de Physique et de Chimie*, the Manoury machine gave 20,202 dynam for each kilogramme of charcoal. This result surpasses that obtained by us, whence the increase of temperature of the water must at that time have been at a maximum of 2.8° instead of 4°. This measure is wanting in the memoir referred to.

From the foregoing experiments it results :

1. That the Savery Machine is a very valuable motive power which may be advantageously employed in many of the arts.

2. That the use of it ought to be limited to those cases in which water is to be heated as well as elevated.

3. That the machine of Manoury is the best model for imitation.—*Ann. des Pontes et Chaussees, trans. in Frank. Journal.*

ON FOUNDATIONS UPON SAND, AND ON COATINGS OF MINERAL TAR : BY M. OLIVIER, ENGINEER.

1. *Foundations on sand.*—At the school *des Pontes et Chaussées*, in 1830, it had been pointed out, that foundations on sand might be laid, wherever the earth was compressible, and in no danger of being carried away by floods. The canal of Saint Martin was given as an example. I have several times applied the system thus indicated and always with success. The following are examples : M. Dupuis, one of the conductors in my district, an architect of the town of Pont-Auderner, was employed to erect a building for the mayoralty. Its situation required that the edifice should be founded on the natural soil. This was well, for there is, in the valley of the Rille, a little below the soil, a bed of solid stones, mixed with sand, of about $31\frac{1}{4}$ inches thick. M. Dupuis, feared that the ground under the bed of gravel was not good, and he had it sounded. It proved to be compressible, and when the gravel was removed, it became impossible to lay a good foundation on the earth which it covered. The architect deemed it needful, in consequence, to resort to piles, and these, it was ascertained, must be very long

to reach solid ground. I went to see the work as they were beginning to drive the piles: it was a very expensive undertaking, which I proposed they should avoid, by substituting a bed of water sand, well watered with cream of lime. M. Dupuis, being responsible for the work, could not decide upon taking this advice, and continued the piles sufficiently for the whole front wall; but he adopted for the other walls the plan I had recommended. These were all, of course, united, though resting on different foundations, but they have all remained firm without any movement, or at least it has been uniform.

This furnishes a new proof of the safety of foundations on sand; 1st, since all the erections in the valley of the Rille, founded on the bed of gravel before mentioned, stand very well, though the ground underneath is compressible; 2nd, since the walls placed on sand, resting on soft ground, have not sunk more than those built on piles, driven with the greatest care to a solid foundation.

Another fact. M. Fauquet Lemaitre, is a proprietor at Bolbec of several cotton factories. One of them being burnt down, he extended the other, which made it necessary to connect the new with the old walls : these walls, situated at the foot of a hill, were partly on a mass of chalk and partly on a bottom of green sand, in spaces where no chalk existed. This sand was moistened by infiltrations of water, which could not however wash it away. When a weight was placed on this sand and left at rest, the mass remained firm ; but if a little motion were given it, it became pasty and almost liquid. The builder thought he must have recourse to piles, and several foundations were prepared for their being driven, when M. Fauquet spoke to me about his buildings, and of the position in which he found himself. At this time, the experiment before cited had been made, and I advised him to lay his foundations on sand. I requested him to converse with M. Frissard, chief engineer at the Port of Havre, and he did so. The latter coincided with me, and added that all the masonry of the steam-engine of 60-horse power, was founded on sand, and nothing had moved it. It was not so with the structures on piles ; a side wall, connected with the foundation of the engine, placed on piles driven as deep as possible, had moved so much that the connecting stones were broken, so that they had to saw them off from the engine walls, the level of which had not changed. This accident, it was believed, occurred from the water contained in the sand, having collected more abundantly around the piles ; and the friction of the latter against the

ground, being thus diminished, they sunk until the masonry rested on the sand.

As other walls erected on sand or on rocks, have not moved, this experiment proves that foundations on sand are as safe as those on rocks, while we cannot rely upon the stability of an edifice constructed on piles and driven into sand ; the friction which they encounter induces the belief that they have gone as far as possible, or necessary, and when any cause diminishing this resistance from friction occurs, an accident follows which proves the contrary.

The first experiment was made under my own eyes ; the second I did not witness, but have every reason to believe that a true account was given me.

2. *Employment of mineral tar in structures of masonry.*—It has for a long time appeared to me that mineral tar, which does so well upon wood and iron, might also be used for covering stone and brick work, as a defence against moisture. Four experiments were made which confirmed this apprehension. But it will be well to premise that as mineral tar is obtained by distilling vegetable materials, it would be more suitable to call it pyroligneous tar.

Without touching upon all the cases in which pyroligneous tar may be employed, which we believe to be very numerous, we shall simply cite a few in which we have tried it.

The light house of Quillebuef had become much degraded by North East storms. The rains were very copious, and the water passing into the brick tower, caused the bottom of the staircase to rot. We repaired the masonry, and in the month of May, 1833, painted the tower with pyroligneous tar, which so far has perfectly answered our expectations ; excepting that a few of the pilots pretend that the light-house, being now black, is not seen so well as when it was white,

M. de Cachelu painted with the tar an earthen wall, exposed to the rains so much as to become very wet inside the building. When I saw these walls, the tar had served as a complete defence against the dampness.

Walls much exposed to storms of rain, are commonly defended by a coating of slate or cement, but the above experiments show that these two modes of defence may be advantageously replaced by a coating of pyroligneous tar.

The joints of the wall being well filled up and smooth, the tar is spread over it and it penetrates the wall. When dry, a second coat is applied and immediately powdered over with sand. This, when solidified, is covered with lime white-wash, as thick as can be put on with a brush. This acting on the carbonic acid of the atmosphere, forms a crust of limestone which exists for a long time, and once in two or three years the wall may be re-white-washed.

We have employed this treatment on bridges very successfully.

In courts, and yards, and terraces, the tar coating is now employed with great advantage. When worn or broken it is easily repaired. — *Translated from Annales des Ponts et Chaussées.*

LIST OF ENGLISH PATENTS GRANTED BE-TWEEN THE 24th OF DECEMBER, 1838, AND THE 24th OF JANUARY, 1839.

Samuel Clegg, of Sidmouth-street, Gray's Inn-road, engineer, for a new improvement in valves and the combination of them with machinery. Jan. 3; six months to specify.

Henry Robert Abraham, of Keppel-street, Russell-square, architect, for improvements in apparatus applicable to steam-boilers. Jan. 3; six months.

Thomas Nicholas Raper, of Greek-street, Soho, gentleman, for improvements in rendering fabrics and leather waterproof. Jan. 3; six months.

Abel Morrall, of Studley, needle maker, for certain improvements in the making or manufacturing of needles, and in the machinery or apparatus employed therein. Jan. 3; six months.

Louis Mathurin Busson du Maurier, of Lombard-street, gentleman, for improvements in the construction of springs for carriages. Jan. 3; six months.

Miles Berry, of Chancery-lane, for certain improvements in rotatory engines to be worked by steam or other fluids. Jan. 4; six months.

William Hickling Burnett, of Wharton-street, Bagnigge Wells-road, gentleman, for new and improved machinery for sawing, planing, grooving and otherwise preparing or working wood for certain purposes. Jan. 8; six months.

Joseph Clisild Daniell, of Limpley Stoke, Wilts, for an improved method of weaving woollen cloths and cloths made of wool together with other materials. Jan. 9; six months.

Moses Poole, of Lincoln's Inn, gentleman, for certain improvements in clogs. Jan. 11; six months.

John Howarth, of Aldermanbury, manufacturer, for certain improvements in machinery for spinning, roving, doubling, and twisting cotton, and other fibrous materials. Jan. 11; six months.

John Ashton, of Manchester, silk manufacturer, for an improvement or improvements in manufacturing plush of silk or other fibrous materials. Jan. 11; six months.

John Swain Worth, of Manchester, merchant, for an improved machine for preparing and cleaning wool for manufacturing purposes. Jan. 11; six months.

William Newton, of Chancery-lane, for certain improvements in machines for drilling land or sowing grain and seeds of different descriptions. Jan. 11; six months.

Francis Brewin, of the Old Kent-road, tanner, for certain improvements in using materials employed in tanning, and preparing the same for other useful purposes. Jan. 11; six months.

Robert Logan, of Trafalgar-square, for a new cloth or cloths constructed from cocoa nut fibre, and for certain improvements in preparing such fibrous materials for the same and other purposes. Jan. 11; six months.

William Ponsford, of Wangye-house, Essex, gentleman, for an improvement in the manufacture of hats, and an improved description of felt suitable for hats and various other useful purposes, and improvements in preparing the material or materials chiefly used in the manufacture of such felt. Jan. 12; six months.

Edwin Marten, of the village of Brasted, Kent, plumber, for an improved method of laying covering composed of lead or other metal on the roof of houses or other buildings, with drains, whereby the part of the water falling on such roof which would otherwise penetrate, is carried off, and rolls and seams are rendered unnecessary. Jan. 12; six months.

Joseph Burch, of Bankside, Blackfriars, calico printer, for certain improvements in printing cotton, woollen, paper and other fabrics and materials. Jan. 15; six months.

William Witham, of Huddersfield, machinest, for improvements in engines to be worked by steam-water or other fluids. Jan. 15; six months.

Hugh Ford Bacon, of Fen Drayton, Cambridge, for an improvement or improvements in apparatus for regulating the flow or supply of gas through pipes to gas burners, with a view to uniformity of supply. Jan. 17; six months.

William Holme Heginbotham, of Stockport, gentleman, for certain improvements in machinery or apparatus for propelling boats or other vessels, to be employed either for marine or inland navigation, and to be worked by steam or other power. Jan. 17; six months.

William Newton, of Chancery-lane, civil engineer, for certain improvements in engines, to be worked by air or other gases. Jan. 17; six months.

Oglethorpe Wakelin Barratt, of Birmingham, metal gilder, for certain improvements in the process of decomposing muriate of soda, for the manufacture of mineral alkali and other valuable products. Jan. 19; six months.

Joseph Garnett, of Haslingden, dyer, for certain improvements in machinery or apparatus for carding cotton, flax, wool, or any other fibrous substances. Jan. 19; six months.

Richard Dugdale, of Paris, now residing at Manchester-street, Middlesex, engineer, for a method or methods of increasing the security and strength of beams, axles, rods and other articles made of iron and steel. Jan. 19; six months.

Caleb Bedells, of Leicester, manufacturer, for an improvement in gloves, stockings and other articles of hosiery. Jan. 21; two months.

John Coope Haddan, of Baring-place, Waterloo Road, Surrey, civil engineer, for improvements in machinery or apparatus for propelling vessels and boats by steam or other power. Jan. 22; six months.

George Stevens, of Stowmarket, brewer, for certain improvements in stoves. Jan. 22; six months.

Thomas Dowling, of Chapel-place, Oxford-street, gentleman, for improvements in preparing metals for the prevention of oxidation. Jan. 24; six months.

John Harrocks Ainsworth, of Halliwell, Lancaster, bleacher, for certain improvements in machinery or apparatus for stretching, drying and finishing woven fabrics. Jan. 24; six months.

IRISH PATENTS GRANTED IN DECEMBER, 1838.

John Wilson, for improvements in the process of manufacturing alkali from common salt. Dec. 6; 1838.

Charles Wye Williams, for improvements in the means of preparing the vegetable materials of peat, moss, or bog, so as to render it applicable to several useful purposes, particularly for fuel. Dec. 5.

Peter Fairbairn, for improvements in looms for weaving ribbons, tapes, and other fabrics. Dec. 17.

Dr. Mohan, for improvements in the composition and manufacture of fuel, and in furnaces for the consumption of such and other kinds of fuel. Dec. 17.

John Milling, for improvements in locomotive steam-engines to be used upon railways and other roads, part or parts of which improvements are applicable to stationary steam-engines or to machinery in general. Dec. 21.

John M'Dowall, for improvements in the machinery for sawing or cutting timber, and in the mode of applying power to the same. Dec. 22.

Daniel Stafford, for improvements in carriages. Dec. 26.

NOTES AND NOTICES.

Artificial Suns.—In our 797th number, p. 112, we noticed the proposition of M. Gaudin to give artificial light to the city of Paris from an elevated Drummond light. Our attention has been called by our valued correspondent Mr. W. H. Weekes, of Sandwich, to a similar proposition of his, published in our 15th vol. p. 255, June, 1831, more than seven years ago. Mr. Weekes also states, that "during the greater part of the last year he has been endeavouring to make arrangements for the purpose of carrying an experiment of the kind into effect from the top of one of the church towers in Sandwich, in lieu of a more convenient and elevated situation."

National Swimming Society.—At a meeting of this Society on the 15th inst. a silver medal bearing the emblems of the society, was presented to Mr. John Fennymore, as champion in swimming for the year 1838; and another silver medal was awarded to Mr. Mason, of Furnival's Inn, for the best essay on the subject of swimming.

Pistrucci's Mode of Medal Striking. — The following "protest" against Signor Pistrucci's claim to the new mode of medal striking has been published by M. Caque, Engraver in the Gallery of the Kings of France, at the Paris Mint, in a French Journal. " M. Pistrucci, who holds a station in the Mint at London, is announced as the inventor of a process for striking a matrix with a punch which has never been touched by the graver, and which gives, nevertheless, a medal identically the same as the original model in wax. In this operation, the beauty and perfection of the design are, at a single blow, transferred to the metal, whether of gold, silver, or copper. The process is this: The model being given in wax, earth, wood, or any other convenient material, take a mould of it in plaster, when the mould is dry or oiled : to harden it, take an impression of it in the moulding sand for cast iron, as fine as possible, in order that the points may be sharp, and that the materials may become as hard as tempered steel. The back is to be dressed to a plain surface. This piece, solidly fixed in a piece of steel, becomes the matrix, on which may be struck either the medal itself or a punch, if it be desired to multiply steel matrices of the medal. M. Pistrucci has tried his process on medals three inches in diameter, and with perfect success. The importance of such a discovery is very obvious. Not only medals, but many pieces of jewellery which require to be chased, may be treated in the same manner." Permit me to say, Mr. Editor, that this important discovery, as you may be easily convinced, is two years and a half old. At that epoch I made known at the Royal Mint in Paris, all that M. Pistrucci has just done in London. A plan of the apparatus, and the details of the operation were deposited by me in the Royal Mint, to be placed at the disposal of my fellow labourers in France. They are in the cabinet of the Director of the Mint. Since that time the medal engravers have employed my process, and more than thirty medals of the reign of Napoleon, have been thus reproduced. You may easily, Mr. Editor, obtain a proof of my statement on application at the Mint."

Varnish for common Candles intended as a substitute for Wax Candles.—Take equal parts of the balm of benzoin and resin mastic ; put each of them in a separate vessel of glass or lead, add spirits of wine, and heat them gently till the resinous parts are dissolved. Let each of the solutions remain a while at rest, and then unite them in one vessel. Prior to using this composition, it is adviseable that the fluid be heated to 25 deg. or 30 deg. cent. (= 80 or 90 Fah.) Dip the candle in it from 5 to 10 seconds, then dry it carefully, which will take about 10 minutes. The proportion of the ingredients may vary, but in proportion as the benzoin is diminished and the mastic increased, the candles become more liable to soften by handling. If the benzoin be increased, the candle dries too soon and loses its polish and colour. The quantity of alcohol will vary according to the thickness of the coat to be given to the candle.

To dye Wool and Goat's Hair a delicate Blue.—M. Buisson, apothecary, has communicated to the Royal Society of Agriculture at Lyons, a new and very simple process. The colour is strong enough to resist water, the sun, and even soap ; while that obtained by dye-woods fades much more easily, and is very inferior in brightness in an artificial light. To obtain this colour as pure, fresh, and deep as possible, the water of the bath while cold, must be first saturated with crystallized verdigrease (acetate of copper), then slightly acidulated with acetic or pyroligneous acid ; dip the stuffs in the usual way, then wash and dry them.

Electric Power of the Gymnotus.—At the first evening meeting of the Royal Institution for the session, held on Friday evening, Dr. Farraday delivered a lecture on the Electric Powers of the Gymnotus and the Silirus. He stated that an interesting specimen of the former was now in this country, which has enabled him to make some experiments upon its electric powers. The first part of his lecture was devoted to an explanation of the phenomena of electrical action, and he then explained the apparatus by which similar effects were produced by the fish to those of an electrical machine when in action. They consisted in copper collars, lined with india-rubber, placed near each extremity of the fish, with which conductors were connected, and amongst other phenomena the formation of a helix was very obvious. He also announced that another fish was daily expected in this country, which could afford further opportunity for experiments. One surprising part in the economy both of the electric ray and eel was, that the organs of life constituted but a small part of the animal, the rest being mere electrical appendages, and which it had been proved might not only be removed without injuring them, but that the fish even attained greater vivacity and lived longer. Now this afforded a room for believing, that further experiments would throw greater light upon the nature of the nervous influence, and he suggested as an experiment the removal of these appendages and the trying the effect of passing electrical currents into them. The lecture was illustrated by some of the most splendid apparatus in the possession of the Institution, and the demonstrations were given in the lecturer's most happy manner.

Errata in Col. Maceroni's article On the Requisites of a good Gun. No. 802, p. 198, col. 2, line 7 from bottom, for "circle" read "cube."—p. 200, col. 2, line 4, for "town" read "Tower."

LONDON: Printed and Published for the Proprietor, by W. A. Robertson, at the Mechanics' Magazine Office, No. 6, Peterborough-court, Fleet-street.—Sold by A. & W. Galignani, Rue Vivienne, Paris.

Mechanics' Magazine,

MUSEUM, REGISTER, JOURNAL, AND GAZETTE.

No. 808.]　　　　SATURDAY, FEBURARY 2, 1839.　　　　[Price 3d.

Printed and Published for the Proprietor, by W. A. Robertson, No. 6, Peterborough-court, Fleet-street.

WOODHOUSE'S ROTARY STEAM-ENGINE.

Fig. 1.

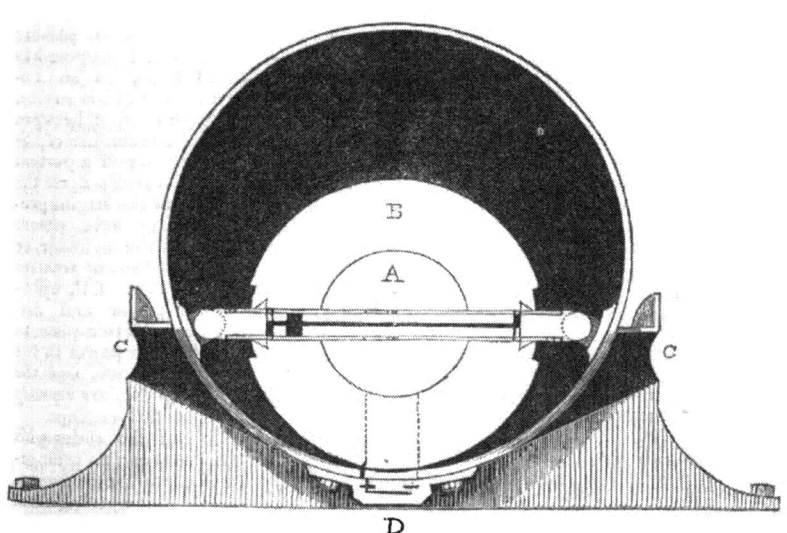

Fig. 2.

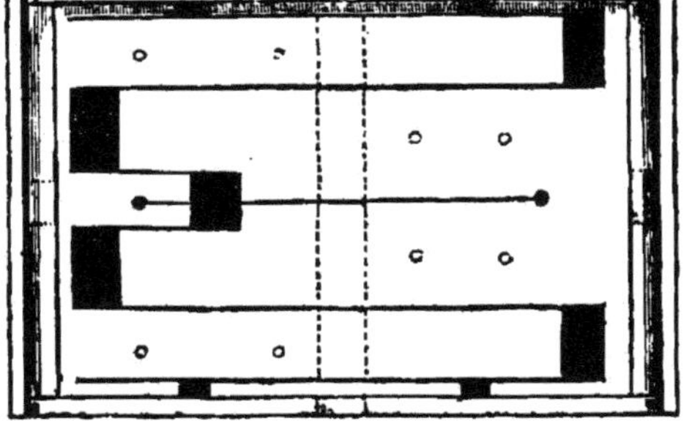

WOODHOUSE'S ROTARY STEAM-ENGINE.

Sir,—The accompanying are sketches of a rotary engine, the arrangement of the slides of which I made more than two years ago, and which is intended as an improvement upon Trotter's, patented many years since, (described in vol. 9, *Repertory of Arts*, 1805,) by avoiding the difficulties which that invention presented in the working parts upon its revival by Earl Dundonald.

The central axis contains two slides passing through it, and alternately closing the steam chamber, to accomplish which, the power is admitted by means of small moveable plugs, or spring valves, which traverse small openings, communicating each way with a small chamber or chambers, between the slides. This produces a requisite pressure of each slide against the inner circumference of the cylinder or case, and is done without the least waste of the power applied. The slides afford a mutual support, and never separate, but alternately work within each other, as shown, fig. 2, being one arrangement of the plan. The *whole* of the foregoing arrangements I *believe* will be found *original,* and are calculated to produce good effects without *needless* friction; for so long as the power is exerted, whatever it may be, the slides will be influenced by its operation, and in proportion to its extent.

The leverage power will more than compensate for its friction, which though more in proportion than the plan of my former engine, published, July, 1837, yet its parts being fewer and more compact, are so far preferable. The power though variable is never reduced to less than one-third. The ingress for the motive power, it will be seen by fig. 1, can be applied at either side of the machine.

The slides are rendered air-tight by means of friction-plates at their sides and ends, and also within the axis, by wedge-shaped plates, as shown; friction-plates are also introduced at the sides, where the inner, (B) exceeds the diameter of the outer portion of the axis, (A). The situation of the axis shows the increase made to the leverage power, and the advantages likely to be gained in reduced bulk.

I have also combined this plan with that of my former engine, acted upon by guides or *racers ;* that is, by making the slides to work *in* the axis, by which means I avoid the use of the cogged wheels and pinion, and make the machine still more compact; but there will be a small addition to the friction.

Fig. 1. is a section of the rotary engine, showing the plan of working the slides through the axis, and of making them to support each other during the whole movement; also one method of introducing the power into the slide-chamber, which, if plugs are used, the openings pass the friction-plates ; but if spring-valves are used this is immaterial. The foregoing arrangement of passing the power into the slides, I proposed to Earl Dundonald in 1835, as an improvement to the working of his engine, by allowing a communication between the inner and outer circumference of the eccentric wheel, and taking off a portion of the friction which its stress upon the vacuum or outlet-side of the engine produced. A, the working axis, passes through the cheeks of the cylinder or case, and may be adapted to any rotative or reciprocatory movement. CC, openings to connect the power and discharge-pipes. D, lower friction-plate, to prevent the passing of the power to the discharging-pipe. The above, and the one acted upon with guides, are equally suitable as water or forcing-engines.

Fig. 2. is a section of the slides with the side-plates off, showing an arrangement of the method of working them together ; the small tubular opening from the plug-opening passes down the centre, midway in the thickness and solid part of each slide, and are here shown.

The arrangement of the packing-plates is such, that the pressure upon them will effectually hinder the escape of the power, as the power itself presses them against the case, in addition to the springs which I should use.

As the subject of the rotary engines has at last obtained notice, it would probably not be unacceptable to your readers, if the arrangement of working Earl Dundonald's engine in the manner I proposed, be sent you, which, although certain difficulties attend that invention, might show one among the many novelties on that head.

I am, Sir,

Your's, respectfully,

JAS. WOODHOUSE.

Pilton, Barnstable, Dec. 5th, 1838.

————

IMPROVED MODE OF CONNECTING MACHINE BANDS.

Fig. 1.   Fig. 2.   Fig. 3.

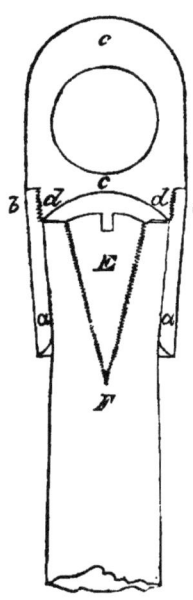 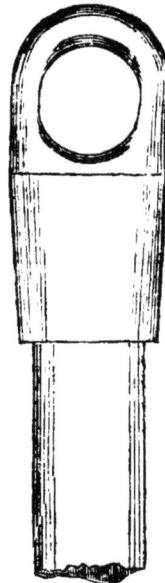 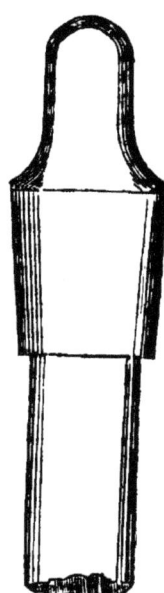

Sir,—I believe all practical mechanics will allow that the common mode of connecting catgut or other bands used for driving machinery, by screwing on to the band a socket having a hook or eye formed in it, is objectionable, inasmuch as the screwed socket is frequently flying off, thereby causing considerable annoyance and loss of time, and to some bands (such as the patent elastic band made by the London Caoutchouc Company), it is entirely inapplicable; for the screw in the socket cuts the fibres so that it actually destroys the parts it is meant to hold by.

The annexed sketches are intended to show a scheme of mine to obviate those difficulties.

Fig. 1, a section and figs. 2 and 3 the external appearance. *a a*, in the section, a conical socket screwed internally as far down as shown at *b b*, *cc* the eye having a short cylindrical screw, as shown at *dd*, made to fit the screwed part of the socket *a a*, E is a conical brass screw having a fine thread cut on it, and F is the end of the band or catgut.

Mode of Fastening.—Force the end of the band through the socket *a a* about an inch or two; screw into the end of the band the brass screw E, then draw it back into the position shown in the sketch, and screw the eye *c c*, into its place.

I have put those fastenings to some very severe tests, among others I hung four half-hundred weights to the end of a piece of patent elastic band (which by the bye is admirably adapted to driving machinery) five-eighths of an inch diameter, and let them fall through a space of five feet, three times, when the band broke some two or three feet from the eye, without in any way injuring it or disturbing its hold of the band.

As I imagine this will be useful to some of your readers you will oblige me by inserting it in your valuable publication.

I am, Sir, respectfully,
Your obedient servant,
EDW. HUMPHRYS.

December 18, 1838.

REMARKS ON MR. FIELD'S EXPERI-
MENTS ON THE POWER OF MEN—
AND ON THE CAPABILITIES OF
SHALDER'S FOUNTAIN PUMP.

Sir,—Of all the very desperate attacks
yet made upon the existence of my
father's patent invention for raising or
conveying water or other fluids—not one
of them is worth notice compared with
the attempt of Joshua Field, Esq., F.R.S.
and V.P., Inst. of C. E.; (*See Mech.
Mag.* p. 237), and if I cannot parry this
attack, (the force of which, or the intention of making, I am convinced this
gentleman is altogether unconscious of,)
I would at once retire from the everlasting scheming and study required in
adapting the invention to its almost
infinite variety of applications.

After the experience of two appren-
ticeships—a time surely sufficient for to
enable any flat to find out his folly—
still flattering myself that the Fountain Pumps are perfect, both in principle and practice, or at least so near
perfection as to leave little chance of improvement—to be told that a common
crane, with its many incumbrances, will
beat it in raising a weight (fluid or solid,
makes no difference) beyond all chance
of competition, is most disheartening.
To prove, however, that we are not so
far wrong, by the favour of Messrs.
Boardman and Harmer, Wharfingers,
Norwich, I have made some experiments recorded beneath.

The following is a table of the effects
produced by a good common crane
expressed in pounds' weight raised
one foot high, by six different men
for minute, and half minute spells;
the resistance at the crank-handle, with
the addition of friction being about 50 lbs.
for the majority of the trials.

Experiments.	Men's weight in lbs.	Height in Inches.	Years of Age.	Weight raised in lbs.	To an elevation, of in Inches.	Time in Seconds.	Weight raised 1 foot high in ½ minute.	Weight raised 1 foot high in one minute.
1	175	70	41	1760	60	30	8,800	
2	138	65	31	"	61	"	8,946	
3	175	70	41	2,016	96	50	gave	up
4	138	65	31	"	72	60		12,096
5	157	67	28	"	99	"		16,632
6	164	66	52	"	87	"		14,616
7	157	67	41	"	56	37	gave	up
8	177	67	32	" .	99	60		16,632
9	175	70	41	"	111	"		18,648
10	138	65	31	1,792	96	"		14,336

In the foregoing experiments, the
height, weight and age of the men are
given from their own statements, which
I had no reason to question, as they
appeared quite correct enough for such
a rough trial.

1st and 2nd Experiments were in
hoisting a butt of currants neat weight,
marked as 15 cwt. 2 qrs. 24 lbs.; for this
short spell and light weight, the difference between the most powerful man
and myself was scarcely discernible.

3rd Experiment—a scale and chains,
weighing ½ cwt., were taken, and 35 half
cwts. A 13-inch crank was used in this
and the former experiments, which was
changed for a 15-inch crank in all the
others.

4th Experiment; here the resistance
which I fixed upon with the men's concurrence, as most suitable for their
strength, proved, as I expected, too heavy
for me for a minute's spell; greater speed
and less power being more beneficial, as
shown in the last experiments.

5th, 6th, 7th and 8th Experiments, by
powerful compact-built wharf and watermen, who all entered into the spirit of
the trials and did their best.

9th Experiment; by the most powerful
men, same as trial the 1st and 3rd Experiments.

10th Experiments, with less resistance;
still I believe I should have done more
work; and for all the men, except the
strongest, perhaps rather less resistance

would have shown a trifle heavier result.

The nearest easy-made practical guess I could give of the friction of the crane was made by hanging a sufficient weight upon the crank-handle, just to bring it down from a little above to a trifle below the horizontal line; and then trying what proportion of this weight would be drawn up again through the same space by the hanging load: the medium between these weights, taking the average at several different elevations of the load, appeared to show about 10 per cent. friction for the movements of all the working parts—they being well oiled before making the trials. It may be as well to observe, the chains and hook to which the chain and weights were attached, on the average, appeared to about balance its own weight on each side the guide pulley.

Comparison between the power expended by myself in working the crane, and the effect I have produced and can produce, again, with a well made and well proportioned Fountain Pump.

Crane.   lbs.
Half Minute Trial.—Actual weight
  raised 1 foot high 8,946
Add the friction of the working
  parts...................... 894
                           ———
Total expenditure of power 9,840

Fountain Pump trial, with heavy lever, suitable for two men to work at for a continuance, for filling road watering-carts.

                             lbs.
Half Minute Trial.—19 full strokes
of 6¼ lbs. weight of water, each
to a clear perpendicular lift of
71 feet from the surface of the
water in the well, is equal to
raising 1 foot high 8,768
Add for the friction of the lever
  bolts, 3 per cent............ 233
                           ———
Effect produced 9,001

                             lbs.
Crane.—Minute Trial. My most
effective spell raised 1 foot high 14,336
Add the friction, 10 per cent... 1,434
                           ———
Total expenditure of power 15,760

                             lbs.
Fountain Pump.—Minute Trial.
34 strokes of 6¼ lbs. each to

71 feet, is equal to raising 1 foot
  high 15,691
Add 3 per cent. gudgeon friction...................... 470
                           ———
Effect produced.............. 16,161

For a minute's spell, I am of opinion that a crank motion, with the addition of a fly-wheel, besides overcoming in its simplest applications to pump-work, at least 5 per cent. extra friction, will enable a man to produce full 15 per cent. more work than a lever, provided both are adjusted for their best effect. I should not therefore despair of making a Patent Fountain Pump, which suits such a 12½ stone man as performed the 1st, 3rd and 9th Experiments, able to raise in one minute's time, either

224 Imperial gallons of fresh water to 10 feet high
112   "   "   "   " 20   "
22½   "   "   "   " 100 "
7½   "   "   "   " 300 "

or in proportion to any elevation or number of such operations. But as the length of the spell increases, the scale gradually turns in favour of the lever power, owing to the smaller size and less distance travelled by the mechanical bearings, and in its superior maintaining position for manual labour; and, judging from many Experiments made with Fountain Pumps, the lever fulcrums of which being fixed about 6 feet high, and the resistance properly proportionate to the men's strength, I have found that sturdy labouring men, of middling height, from 25 to 40 years of age, weighing 150 to 170 lbs. each, can, by making the most of their strength, average the following results, expressed in pounds' weight, raised 1 foot high; each minute for short and long spells:—

                             lbs.
For 1 minute's spell 15,000 to 16,000
" 3   "   " 11,000 to 12,000
" 5   "   " 9,000 to 10,000
" 10   "   " 7,000 to 8,000
" 15   "   " 6,000 to 7,000

Any person, therefore, wanting to know the quantity of water he can have supplied by a man's power, has only to divide any of these sums by the height to which he wants the water raised.

Let any man try whether he can run treble the distance in one minute, or double in 5 minutes of what he can average, through a very long spell, or up a flight of steps with convenient hand-rails, or a mining-ladder, which may be a less disputable criterion :—if he does not find that task tough enough, I am much deceived ; therefore for any man of common dimensions, English or Foreign, to fully double this effect, ac-cording to Mr. Field, appeared to me to be clearly impossible—3,675 lbs. weight raised 16½ feet high in 2⅓ minutes by the Irishman, and in 2½ minutes by the Englishman; the latter, too apparently, very soon after a bye-play exertion that would cause millions of strong men to "kick the bucket" before they could accomplish the same. From the length of the crank given, the diameter of the chain barrel and combination of cogs, I can find no error in the statement of the crane giving 105-fold purchase; conse-quently, the men's hands passed through 105-times 6½ feet, which, for the Irish-man, gives near 84 revolutions per minute with an 18-inch crank, and may possibly have been done with a one-hand load of half the resistance stated. I shall, therefore, conclude by giving a guess, that instead of the weight being raised 16½ feet, it must have been that 16½ feet of chain was wound upon the barrel, the end of this chain passing under a move-able pulley to which the load was at-tached, and up to the hook at the ex-tremity of the crane's jib, which addi-tional pulley would give double the pur-chase, and show only half the effects for all the trials.

I am, Sir,
Your obliged,
W. SHALDER, jun.

Bank-place, Norwich, Jan. 23rd, 1839.

ON AN EXPERIMENT OF PERKINS, TO PROVE THE CALORIFIC REPULSION OF A WATER GLOBULE FROM AN IGNITED METALLIC SURFACE—BY DR. SCHAFHAEUTL.

Sir,—Mr. C. Tomlinson has had the kindness to direct my attention to one of his ingenious researches contained in No. 703 of your valuable Magazine, which I much regret not being earlier acquainted with. He likewise alludes to an experiment made by Mr. Perkins, of which I was perfectly ignorant, and which he seems to consider a convincing proof of the doctrine of calorific repulsion of certain heterogeneous bodies.

The following is the experiment to which Mr. T. refers :—

" If a platinum capsule be drilled with an immense number of holes, as in a coffee or wine strainer, it will not, of course, retain water for a single minute ; but if brought to a white heat by means of well-prepared in-candescent charcoal, so that fire only, with-out flame, be produced, and water be dropped into it, a large globule may be collected, and it will exhibit all the phenomena, just as if a common platinum crucible had been em-ployed."

Now if I am allowed to give an opinion on the subject, Mr. Perkins's experiment tends to prove nothing more or less than the well known fact, that a drop of water evaporates very slowly in an ignited vessel.

We have a certain phenomenon before us by which, through the medium of our senses, we are convinced of the fact, that a water globule in an ignited metallic vessel is extremely slow in evaporating. Now in searching for the cause of this slow evaporation, two different explana-tions have been given of it. One party ascribes the slowly increasing tempera-ture of the drop of water to the repulsion developed by the ignited metal, I, on the other hand, consider it to arrive from the difficulty we experience in heating any liquid by applying caloric to its surface, and to the development of steam by touching the ignited metal.

In both cases the ignited surface is the essential medium through which the effect is produced, viz., the slow evapo-ration of the drop of water, and where this ignited surface is wanting, no effect whatever can be produced.

As an ignited metallic surface is there-fore absolutely necessary, a hole in this surface cannot possess the essential means, and can have no influence what-ever on the drop of water, and if the drop is of smaller diameter than the hole, and is moving in the axis of the hole, it will certainly fall through it. This will be the case whether the vessel be hot or cold, provided the current of air does not throw the drop out of the axis of the hole. If the drop of water be of larger diameter than the hole, then it must of course touch the ignited surface, and the development of steam and the spinning motion of the drop will be the same as if there were no hole. It is further even difficult to press a drop of water through the hole of a plate which is not ignited, if it be of smaller diameter than the drop.

The well-known method of making a water microscope, is by drilling a hole in a blackened plate about one-tenth of an inch in diameter, and then letting a drop of water fall into it, the drop remains very firmly suspended in the hole, and it requires rather a violent concusssion of the plate before it will fall through.

That a fluid will run through a coffee or wine strainer is owing to the gravitation of the upper portion of the liquid pressing on the lower, and thus overpowering the cohesive force of the molicules, but when this cohesive force is predominant, as is always the case in a small drop of water, it is with great difficulty that the drop will be made to pass through the innumerable holes of a clean dry wine strainer.

But Mr. Perkins adduces a far more wonderful and extraordinary experiment, in which the repulsive force of a red-hot cylinder was so great, that steam of many hundred atmospheres was not able to escape through a hole bored in it of full one-eighth of an inch in diameter. But Berzelius himself, who is an advocate of the doctrine of the calorific repulsion between heterogeneous bodies, thinks this experiment requires further confirmation. With this opinion I cordially concur, particularly as all attractive and repulsive forces of the molicules act only in infinite small distances, which amount in fact to an actual touch. But if Perkins' assertion should really be true, we can ascribe it to nothing else than the difficulty of forcing liquids through channels whose sides are not moistened by them; but here again the size of the hole seems to present an insurmountable difficulty.

An experiment which it is more difficult to explain. is alleged by Berzelius in his *Manual of Chemistry*, where he states:—

" If, for example, you melt on the end of a platinum wire a little borax or diphosphate of soda, and afterwards make this end of the wire as hot as possible, the drop will be found to recede from the point containing the greatest heat, and this will be the case not only when you hold the wire horizontally, but also when it is so inclined that the drop is forced to ascend."

Berzelius ascribes this ascension to the repulsive force of the platinum wire, but it is easily seen that no real repulsive force exists in this experiment, for no fixed degree of heat is necessary in performing it. It does as well if the wire is only red hot, as in the highest possible degree of white heat. The only circumstance to be observed is, that the two extremes of that portion of the wire which comes in contact with the drop must have a different degree of heat; and this alone produces the effect alluded to. If you heat the drop and the wire by means of a blow-pipe to the highest degree, taking care that both ends of the wire have an equal degree of heat, the drop will remain stationary. On the contrary, a red heat if unequally applied will make the drop move.

Now if the platinum wire really possessed a repulsive force, it would be the more powerful in proportion as the heat was increased, and if this repulsive force had the power to move a drop of water *contrary* to the direction of gravity, it would be still stronger *in* the direction of gravity, and at all events cause the drop to separate from the platinum wire. But it is just the contrary, for a white hot wire holds the drop of borax so firmly, that the most violent vibration is not able to separate it entirely from the wire This experiment is more satisfactorily explained by the cohesive forces of the molicules of a drop.

The cohesive force of a drop of water is by far greater than appears generally to be believed. A single experiment will confirm this assertion. Let us draw out the end of a glass funnel with rather a wide mouth into a capillary tube, so that water drops from it very slowly. If the level of the water is kept always equal, the drops will be all of the same size, provided the vessel is kept motionless. Let one of these drops fall on a sensible balance and ascertain its weight. Afterwards put the funnel in the sliding ring of a stand, and receive the drop on a board, covered with a thin layer of the pollen of the *lycopodium clavutum*.

The drops on this board, assume immediately a globular shape, similar to one in a white hot crucible, and may be moved with the same facility, though it is perfectly clear that the *lycopodium* does not exercise any repulsive force in this case. I was often able to make one of these drops resist the flame carefully applied by means of a blow-pipe as long as it would resist the heat of a white hot platinum crucible.

Now raise the funnel till the drops of water begin to be divided or scattered in pieces by their own velocity in falling. Then let us measure the distance between the end of the capillary tube and the board upon which the drops fell; and this space with the weight of the drop will give you the final velocity of the drop from which you may easily calculate its momentum, or the power which is necessary to overcome the cohesive force of a drop of water.

In an experiment made with common water from the New River Company, a drop of water weighing exactly 0.651 of a grain, in the course of 20 trials, fell on an average through a space of 2,4 inches before it was scattered into pieces; and this gives, taking the falling of a body in the first second through a space of 16 feet = 3,062 feet final velocity, which shows that a force of 1,993 grains is required to overcome the cohesive force of a single drop of water, and that the force of gravitation to that of cohesion, is as 1 : 3,062.

Now a globule of borax is attracted even by the hot platinum wire, for if it repelled it, the drop would disengage itself by the slightest shake and fall to the ground. But the drop even surrounds the wire which is passing as a kind of axis through the upper part of it, and the form which the drop assumes on the platinum wire depends on three forces:

1st, On the attractive force of the platinum wire.

2nd, On the cohesive force of the molicules of the drop itself.

3rd, On the force of gravitation.

I endeavoured to measure the attractive force of the platinum wire by fixing one end of it, and measuring the angle of deflexion which would be required to separate the liquid drop from it when in a vibrating motion: but as always some portion of the drop remained on the platinum wire, I concluded, that as the attractive force of the wire overcame the cohesive force of the drop, the former must at least be equal to the latter.

All these three forces, when in perfect equilibrium, cause the drop to assume a lengthened form, whose longitudinal section seems to be bounded by a parabolical and catenary curve, which both sections are divided by the wire as a line of abscisses. Now suppose we destroy the equilibrium of the three forces by heating one end of the wire, and thus partially do away with the attractive force of this portion of it, the cohesive force of that portion of the drop thus liberated will begin immediately to act and contract itself at this end by receding from the heated portion of the wire. At the same time, the opposite end of the drop thus freed from the counteracting forces of the receding half of the globule, is moved onwards by the attractive force of the other less heated end of the wire, elongating itself at the same time, until it resumes its original shape, and with it the state of equilibrium on a higher portion of the wire, in a somewhat similar way as water rises in a capillary tube, and as it is thus perfectly clear that the acting forces of cohesion and attraction are always greater than that of the gravitation of the drop, it is natural that the drop will even rise upwards on the inclined wire contrary to the action of gravitation.

The object in this, as in all similar cases, is merely to give a sufficient explanation of a certain fact, which, in our case consists in the drop of water but little heated * in an ignited metallic vessel, and the more we are able to found the explanation of certain phenomena on already known laws, the greater simplicity and truth will be introduced into science. And that extraordinary genius, Young, with his doctrine of interference, by which he has referred the theory of light and sound to the doctrine of waves, has rendered a service to science, which will, perhaps, be only duly appreciated by posterity.

I have, therefore, endeavoured to explain the before-mentioned experiment, by the long known fact, that it is almost impossible to heat liquid bodies by applying caloric to their surface, and this may

---

* I found the actual heat of the drop of water out of a well-polished platina-vessel, to be between 198 and 199 Fahrenheit. Bandrimont (Ann. de ch. et de Phys. LXI, 319,) found the temperature lower than this, and Laurent (Ann. de ch. et Phys. LXII, 327,) found it near the boiling point. From my experiments, I found that the polish of the platina-vessel had great influence on the temperature of the drop.

be ascertained by burning sulphuric ether on a glass of water, which glass of water is nothing more than a section in the radius of a large water globule, the action of the remaining parts being substituted by a wall of a solid body. On the other hand, the advocates of the calorific repulsion in certain heterogeneous bodies,are compelled, in order to explain the before-mentioned phenomenon, to create a new particular problematic force, viz., the calorific repulsion of certain heterogeneous bodies in all their different states of aggregation, with which the so-called repulsion-force of the caloric in a *homogeneous* mass, by which repulsive force, the expansion of bodies by heat is generally explained, stands in no scientific relation.

I remain, Sir,

You most obedient servant,

CHAS. SCHAFHAEUTL.

Cornhill, 22nd Jan. 1839.

DR. LARDNER'S EXPERIMENTS ON THE RESISTANCE OF THE ATMOSPHERE TO RAILWAY TRAINS.

Sir,—After a careful perusal of Dr. Lardner's article on the Great Western Railway Enquiry in the *Monthly Chronicle,* and your criticism on it, I come to the conclusion that the Doctor in estimating the aerial resistance which astounded him, has overlooked the resistance of undue friction, made manifest even by his own report, which curiously contradicts itself in many essential particulars. I shall quote his words, and remark upon them seriatim.

" The tire of the wheels has a conical form, which gives their combination the effect of a wedge tending to force the two rails of the same line asunder, or, in other words, to widen the gauge. It was, therefore, to be expected, that the rails would bend outwards while the wheels were passing over them. It is understood that this was found generally to take place when the instruments were applied to the rails of the Great Western Railway, but on all other railways, the rails exhibited as frequent a yielding *inwards,* and in some instances no outward yielding whatever was indicated."

We have here precisely the effect which a sound practical critic would have foretold. While the rail is maintained in a truly vertical position, as is the case with

the continuous bearing on the Great Western Railway, (i. e. the base is sufficiently broad and solid to prevent the rail being forced down or twisted by the weight pressing on the inner edge,—which is the *direct* tendency of the gravity acting on the coned wheels while at rest as well as in motion,—whereas the widening of the gauge between the rails is an effect of the lateral thrusts from irregular motion. While the bearing is solid,) the lateral thrusts only take effect, tending to widen the gauge, but with the bridge rails or iron bars in piers or chains, the unfirm structure subsides or vibrates immediately under the pressure of the coned wheels, and will continue so to do until the faces of the rails shall have accommodated themselves to the horizontal angle of the cones.

Dr. Lardner goes on to say,—

" The principal irregularity of motion to which railway carriages are liable, is a lateral swinging to the right and left between the rails. * * *

" The application of this apparatus shewed in a very conclusive and satisfactory manner, that the ease and smoothness of a carriage depends upon other circumstances than the goodness of the road or carriage. Thus, a carriage placed in the middle of a train will have less motion than it has when placed at the end of it. Also, the carriages of a train coupled by Mr. Booth's couplings, which convert the train into a column in some degree solid, will show less motion than if coupled simply by chains. A carriage, also, which much overhangs its wheels, shows more motion than one whose wheels are further apart; the end body of a carriage than the middle body ; a carriage heavily loaded less motion than when more lightly loaded. The speed of the motion has also a material influence on its irregularities, the lateral swinging between the rails being greatly increased by the increase of speed ; but *cæteris paribus,* these irregularities are always proportionally small as the road is well constructed. * * * The resistance offered to the tractive power by a carriage, proceeding with a uniform motion on a straight and level railway, is produced partly by the friction of the axles of the wheels in their bearings, partly by the rolling of the tires on the rails, and partly by the inertion of the *air* which the carriage displaces in its progress. By a degree of accuracy of mechanical construction which is within the present limits of engineering skill, and by a good system of lubrication the friction of the axle in its bearing may be reduced to an exceedingly small amount.

" The amount of resistance which attends a rolling motion is small under the most unfavourable circumstances, as is manifested by the facility with which enormous weights are moved, even on the rough surface of the earth, when coarse rollers of wood are placed under them. How insignificant, therefore, that part of the resistance must be which proceeds from the rolling of the tire of a wheel accurately finished in the lathe, on the surface of a not less accurately rolled iron bar, laid as truly even and level as art can effect, may be easily conceived."

The Doctor has committed an error very common to mathematicians, i. e· reasoning from imperfect data. Whence comes the "lateral swinging to the right and left between the rails?" Does it proceed from the "accurately turned wheels," or the "not less accurately rolled iron bar;" or does it proceed from the vehicles thus eulogized in a subsequent sentence?

" To reduce these (sources of resistance) to the lowest possible amount by the excellent construction of the carriages and engines, and the exquisite perfection of the road on which they move, has been the object to which the engineering profession has addressed all its powers, and with what signal success, it is needless here to say. Such carriages and such roads could never have entered into the contemplation even of the most sanguine speculator on the progress of art."

If the Doctor will experiment upon the "wooden rollers" he alludes to, he will find that the "enormous weight" will have a tendency to advance other than in a right line, owing to irregularities of the road. But the friction is not thereby materially increased, as there is no *lateral* resistance. This process, if continued, would ultimately enable the roller to escape from beneath the load, but this tendency is salutarily corrected by the attendant labourers with a blow of a crowbar, which again restores the tendency to the right line. But if the Doctor were to couple the two rollers together by a frame-work to coerce them to an equal motion, he would find the friction become so great that they would be nearly useless for his purpose. On the railway this is done—the two rollers of iron are shackled and forced to run in a groove. If all the work be so good as

the Doctor describes, there ought to be no "lateral swinging." But there is lateral swinging, and in lieu of the labourers with the crowbar, the flanges of the wheels are usually made to rectify the errors, and that not very gently. As an improvement on this, " Mr. Booth's couplings are made to convert the train into a column in some degree solid." The blows on the flanges are lessened, but the restrained freedom of motion must be at an expense of surplus friction, and consequently at an increased expenditure of motive power. We shall be told " friction is nothing on railways," yet Dr. Lardner says, it is twenty pounds per ton instead of eight or nine, as it has been usually estimated, but then chary of the carriages and roads, he adds, " the atmosphere" is to blame. Another sentence seems very strange.

" Some of the experiments made in the course of the inquiry now before us, suggest a probability that the resistance from friction *decreases* as the speed is increased."

Is it not at least as probable that the increasing momentum more than counterbalances it?

Dr. Lardner may be very right in regarding the rolling friction of the wheel tires as insignificant in amount; but there is one important circumstance to be added to the three retarding cause she names, i. e. *the lateral rubbing friction of the wheel tires* during the process described of " lateral swinging to right and left between the rails." If he asserts this also to be insignificant, will he inform us why it is that the average wear of the tires of the tram carriages on the London and Birmingham Railway is only three months, though steadied by Mr. Booth's couplings, the mass of iron exposed to the wear being four inches in width and an inch and a quarter in thickness? Or will the Doctor give us a formula of the time and distance necessary to wear it out by simple rolling friction on a flat iron surface? Will he inform us what the process is which pinches out the tires laterally in small patches as a cook does paste on the edges of a pie, or which sometimes shreds away the solid flange as though it were a cheese rind?

The truth is, the rapid wear of the tires is conclusive evidence of a faulty principle of construction. If the carriages and roads were so perfect as Dr. Lardner describes them there would be no "lateral swinging." An inaccurately constructed carriage or an imperfect road, yields a motion like the *bias* in "bowls," and this cannot be compensated for by so rude a process as that of coning the wheels. On common roads other and better means have been resorted to, or otherwise the work could not have been done. In machinery the fly-wheel is resorted to to equalize the power, and some effectual compensation process must be found for the irregularities of railway motion before we can materially lessen the friction and expenditure.

Such a sentence as the following ought not to have been penned by an experimentalist like Dr. Lardner.

"The two trains were certainly not composed of the same identical coaches, but they were composed of coaches similar in construction, equal in weight and equally loaded, and were supported on a similar number of wheels of like magnitude ; and, in short, no reason existed for supposing that the friction could be materially different."

If a Thames waterman were asked to choose one of two boats, "similar in construction," he would prefer a trial by rowing ; and if an R N. were to choose from two frigates, "similar in construction," he would prefer trying them to the sailing test. And had Mr. Brunel put the "North Star" in perfect order before Mr. Wood made his unerring calculations, that gentleman's observations would have retained some value. The stage-coachmen know, that two coaches rarely run equally light, and it is to be found, that a varying construction extends to railway-coaches as well as railway-engines. At present, Dr. Lardner's experiments are far from conclusive as to what retardation belongs to the atmosphere, and what to friction. One very obvious experiment might have occurred to the Doctor. As the resistance of the atmosphere must of course be in proportion to the superficial front of the train, he might have increased and diminished this frontage, or, removing the bodies from the carriages, he might have replaced them with an equivalent weight of iron bars. It is not stated whether the experiments were performed in a still atmosphere, or with the wind for or

against. These circumstances would make a sensible difference. When a really efficient train shall be tried, I suspect the result will be found as startling as the effect produced by Mr. Brunel on the North Star engine, to the discomfiture of Mr. Wood.

With regard to the increased width of the gauge on the Great Western Railway, there can be no doubt that it much diminishes the risk of overturning, by increasing the base without increasing the height; at the same time the power of the wind on the train is diminished. And supposing that at any future time it be considered desirable to increase materially the height of the wheels, it will be found quite practicable to place the bodies between the wheels instead of above them. The defect of the wide gauge is, that the amount of friction of the present trains—which afford no compensation for irregularity of motion—must be greater than that of the narrow gauge. Of the advantage of the continuous bearing, there can scarcely be a doubt in any rational mind. That the timber work was of too small a scantling to bear the weights rolling over it, is no imputation of the soundness of principle.

High praise is due to Mr. Brunel for the moral courage he has shown in venturing upon an untrodden path at the dictates of reason unswayed by prescription. But for him, the art of railway-making would have remained the same as Mr. Stephenson left it at Manchester; for all that has been done since is merely imitation, varying as little as canals and canal-boats. It was to be expected that the race of imitators would decry without mercy or fairness, as has been the case with every new plan since the world began ; but the change is achieved, and the experiment will be fairly tried in practice. If successful, a new race of imitators will arise, who, in turn, will abuse every other innovation after the same fashion. It is an easy thing to imitate, and the public would be unwise to deal harshly with its few original minds, even if they chance to err in their calculations. Take away the originators from amongst us, and farewell our pre-eminence as a wealthy nation.

COLONUS.

Jan. 27, 1839.

DR. LARDNER'S OPINIONS ON THE RESISTANCE OF THE ATMOSPHERE TO RAILWAY TRAINS—GREAT WESTERN RAILWAY INQUIRY.

Sir,—Your observations on the Great Western Railway Inquiry, and on Dr. Lardner's experiments, in your last number, have tempted me to make the following remarks :—We Bristolians are under some obligations to the learned Doctor for having drawn forth our energies to prove him completely mistaken. He certainly laboured all in his power at the meeting of the British Association in this city, to demonstrate that the *Great Western* would never cross the Atlantic direct to New York, and he now stakes his reputation to prove that no steam locomotion can *profitably* travel on our railway (or any other) at the rate of forty miles an hour. On the first point, at the Liverpool meeting, the Doctor confessed himself in error, and, on the second, at the next in Birmingham, I have little doubt, Mr. Editor, that you will have to report a similar change of opinion; since the one he holds at present, on the resistance of the atmosphere, was, *by himself*, in my presence and that of a hundred others, in the mechanical section at Liverpool, treated as unworthy of consideration. This new opinion is said to be founded on experiment. I therefore beg leave to state some counter-experiments to disprove the assertion that " the extreme mean maximum rate of speed accomplished by the *North Star* engine on the Great Western (the fastest on any railway) was 41.15 miles an hour, when at full speed, and with a load of only 15 tons," this, be it remembered, being the supposed limit of velocity occasioned *by the resistance of the atmosphere*. Marshal Soult and suite were conveyed from Manchester to Liverpool in about 35 minutes. In Canada it is not uncommon, when the St. Laurence is frozen over, for a car propelled by sails to go at the rate of nearly 100 miles per hour, not of course before the wind, for that would outstrip the propelling power and amount to an absurdity—but with the wind, " on the beam," as the sailors say. Many birds will fly at the same rate. Where then is this powerful resistance, or how is it avoided ? In the first place, are the laws of fluids in motion so accurately known as to be the subject of exact calculation ? I doubt it, because I have never yet seen a good explanation of the cause of that apparent paradox—the attempt to blow off a disc of considerable diameter in comparison with the pipe (or quill) on which it is placed, for, if on this disc a little spirits of wine are inflamed, it will be seen that a current is produced tending to fix it the more firmly on the aperture the more violently you blow through the tube; secondly, " how is it to be avoided ?" Clearly in the same manner as it is reduced in all things intended for rapid motion—by shape. The American steamboats, and the *Great Western* also, will pass through still water at the rate of 15 miles per hour; in the rarer fluid of the atmosphere what may not then be expected by attention to *shape* in railway carriages and engines ? By the bye, Mr. Editor, there is a simple mode of setting the question at rest. The late gale of wind at Liverpool and Manchester, on the Monday morning, ought, according to Dr. Lardner, to have stopped all the trains proceeding from the latter place; what effect had it on them ? If you can obtain an authentic reply to this query you will afford valuable information to your readers.

Yours, respectfully,

H. A. M.

Bristol, Jan. 22, 1839.

GREAT WESTERN RAILWAY INQUIRY —MR. WOOD'S EXPERIMENTS.

In the remarks which we made upon the Great Western Railway Inquiry in our 806th number, we stated that the ostensible ground upon which the Company discarded Mr. Wood's experiments, was that " Mr. Brunel had with one stroke of a hammer upset all the experiments upon which Mr. Wood's report was founded." Mr. Wood's prompt and satisfactory reply to this assertion, which we subjoin, shows how easily a public meeting may be led away to the giving of hasty decisions by the feelings of the moment—how " turned by the trifle"—the *trifling* we should say, of Mr. Babbage's eloquence—how " tickled by a straw"—the experiments "of straw," the magic touch of a Brunel's hammer !

The following is the letter which Mr. Wood has addressed to the Great Western Railway directors :—

" GENTLEMEN.—When I undertook the task of reporting upon the Great Western Railway, knowing the nature of the inquiry, I had prepared myself to expect that my opinions would be severely scrutinized by your engineer, and by the public ; and I determined to submit to such criticisms with due resignation, come from whatever quarter they might. With your body, however, it is different ; it is in my opinion imperative upon me to see, which I am sure will meet your concurrence, that the conclusions drawn by you from my repor should be strictly correct, and those only which are warranted by the opinions therein expressed, or by the experiments therein detailed.

" A copy of your report reached me yesterday : generally the conclusions drawn from my report, are therein fairly, clearly, and impartially stated ; but there is one part in which you have been led to an erroneous conclusion, by contrasting with each other, *entirely dissimilar experiments,* and which it appears to me, to be due to the shareholders, to the public, and to my own character, that I should set right. You detail an experiment, personally witnessed by yourselves, with the ' North Star Engine,' *after being improved by Mr. Brunel ;* where that engine took from Maidenhead and back, a train of carriages loaded to 43 tons, an average speed of 38 miles an hour, the consumption of coke being ·95 lbs. per ton, per mile. This you contrast with a consumption of 2·75 lbs., as shown by some of my experiments.

" The consumption of 2·75 lbs. was however, derived from an experiment where the load was only *sixteen tons,* which it is quite preposterous to contrast with a load of *forty-three tons.* Fortunately we have in my report the data afforded for making a correct comparison. In Table, page 17, you will find a set of experiments, (the third,) by the ' North Star Engine' detailed, showing a load of 41·61 tons, conveyed at the average rate of 38·8 miles an hour, the consumption of coke being 1·09 lbs. per ton, per mile. Mr. Brunel states the improved performance of this engine at 40 tons, taken 40 miles an hour, with a consumption of ·9 lbs. per ton, per mile. Comparing these experiments with each other, we have the following results of the performance of that engine, before and after being improved :—

	Load in tons.	Speed in Miles pr. hour.	Coke pr. ton. pr. mile.
Director's Experiments..	43	38	·95 lb.
Mr. Brunel's, do.	40	40	·90
Mean	41·5	39	·925
Mr. Wood's Experiments	41·65	38·8	1·09

" After the above remarkable similarity of effect, it is unnecessary for me to say a single word, as to that part of your Report, wherein you state ' that if I had witnessed the recent performance of the engines, I would unquestionably have changed my opinions.' It would, indeed, have been a grievous waste of money, time, and labour, if these experiments could have been ' blown to the winds' by the ' knocking off the end of a pipe.' The preceding remarkable coincidence of performance, elicited by different experimentalists, and at a different period, is the strongest possible proof of their value and importance.

" I beg you will also allow me this opportunity of more clearly bringing before you and the public an exposition of my opinions as regards the atmospheric resistance, which your Chairman stated to be ' too startling to be true, or at any rate, to be implicitly received without further investigation.' The experiments on atmospheric resistance were brought forward merely to show, that at high rates of speed the atmosphere had considerable influence on the resistance of railway trains ; and I expected that I had in my Report, (which it was my undoubted intention to do), clearly explained *that no standard of the amount of resistance could be drawn from the experiments adduced.* That my opinions on these experiments may not be misunderstood, I beg you will allow me to quote that portion of the Appendix which relates to this subject. After giving the formula alluded to in p. 48 of my Report, I state—

" ' In these formulæ the resistance of the atmosphere to railway trains is assumed as being proportional to the square of the velocity ; this has not yet, however, been ascertained by experiment with sufficient accuracy, to be adopted as the standard of resistance. The figure, outline, and frontage, opposed by railway trains, consisting of a different number and description of carriages, are so various, and the circumstances affecting the resistances so complicated, that until experiments more varied than the preceding are made, no fixed standard of resistance can be safely assumed. For these reasons, therefore, I have not at this time entered upon calculations founded upon these formulæ, to determine the values of f and d, or the relative amount of the friction properly so called, and that part of the resistance which arises from the effect of the atmosphere.

" ' It is my intention to pursue the subject further, with railway trains composed of a different number of carriages, and of all the varieties which are used in practice ; and it does appear to me, that until experiments are made on all the varieties of trains, no practically useful conclusions of the precise amount of atmospheric resistance can be drawn. These experiments are extremely valuable as part of a series, to accomplish this object; their great utility in the present inquiry is unquestionable, as corroborating to a certain extent the results determined by the experiments with the engines ;—viz.. that at high rates of speed the atmospheric resistance to railway trains is much greater than has been generally supposed.' "

" NICHOLAS WOOD."

Killingworth, Jan. 14, 1839.

M. SOLIEL'S NEW EXPERIMENTS ON LIGHT.

M. Arago lately read a note addressed to the Academy by M. Soliel, jun., the object of which was to lay his claim to undoubted priority in all that relates to the construction of chromatic apparatus, intended to exhibit on a large scale the experiments of polarized rays of light through crystalline laminæ. For many years past, those philosophers who have paid particular attention to the optics of polarization, and among whom we will mention Messrs. Barbinet, Delezenne, Pelcet, Pouillet, and Nœremberg, who have given their advice to this able artist, who has himself introduced many useful modifications in the arrangement of these experiments, which are so brilliant, and yet so little known. We ourselves can testify to the magnificent effects produced by these apparatus.

Last week, notwithstanding the inconvenience occasioned by the sun shining very dimly, Professor Pouillet exhibited these beautiful phenomena to the immense audience that is attracted to his course of natural philosophy. By the aid of this apparatus, which is both simple and reasonable in price, fifteen hundred persons were enabled to see and admire the system of admirable fringes produced by the interference of rays, an experiment attributed to Fresnel, and which is but a very happy modification of the coloured rings of that master of all natural philosophers, Sir Isaac Newton.

The instruments of M. Soliel are equally adapted to the demonstration of the laws of double refraction, and of their identity with luminous polarization. By interposing crystalline laminæ in the path of rays, polarized by tourmaline, it exhibits on a large scale the brilliant complementary colours discovered in this case by M. Arago. This same crowd of spectators enjoyed the sight of the surprising results shown by crystals with one or two axes, which throw on a white screen, systems either circular, hyperbolic, or lemnisbates, curves tinted with a thousand co-

lours, and of a brilliancy which no other process can even approach. Thus, all these magnificent images, obtained in France by Fresnel, Biot, and Arago, and in England by Brewster, Herschel, and Airy, are no longer cabinet phenomena, but may be exhibited to the most numerous assemblages.

We will add, that M. Soliel, jun., is not only the first who has made at Paris these beautiful apparatus, but he has discovered a very curious effect produced by the dilatation of carbonate of lead on its optical axes and their superb curves. But these are things that must be seen to be appreciated, and all the descriptions in the world cannot give the slightest idea of the dazzling phenomenon of polarized optics.

OXFORD-STREET EXPERIMENTAL PAVEMENT.

At a meeting of the Marylebone vestry, held on Saturday last, the subject of the experimental pavements laid down in Oxford-street was the subject of discussion. It was proposed, that as some of the plans tried had already most completely failed, they should be removed. " The portion more particularly alluded to was that of the Parisian bitumen, laid down at the commencement of the experiment at Charles-street. The asphaltes altogether appeared to be a failure, but the portion next Charles-street had attained a most deplorable state of dilapidation ; in fact, it had worked into holes at various parts to such an extent, that omnibusses and other vehicles passing over it, from the excessive jerking caused, ran the risk of snapping their springs, and endangering the lives of the public." It also appeared, that some of the competitors had been allowed to repair and renovate their pavement.

It was further stated, " that the only experiment, except the granite paving, that would be successful was the wooden blocks. It was proposed, therefore, to remove the bitumen, and instead of filling up the space with Guernsey chippings, to allow the wooden blocks to be laid down, so as to give them a trial on a more extensive scale. This was, however, objected to as a most unfair course of proceeding, and it was ultimately agreed that the Surveyor should write to the various projectors, forbidding any of the experiments to be touched again, and also that the Committee should take proper steps to remedy the evil complained of.—*Abridged from the Observer.*

MANUFACTURE OF BLACK INK.

Nut-galls, sulphate of iron, and gum, are the only substances truly useful in the preparation of ordinary ink; the other things often added merely modify the shade, and considerably diminish the cost to the manufacturer upon the great scale. Many of these inks contain little gallic acid or tannin, and are therefore of inferior quality. To make 12 gallons of ink we may take,—

12 pounds of nutgalls,
5 pounds of green sulphate of iron,
5 pounds of gum senegal,
12 gallons of water.

The bruised nutgalls are to be put into a cylindrical copper, of a depth equal to its diameter, and boiled during three hours, with three fourths of the above quantity of water, taking care to add fresh water to replace what is lost by evaporation. The decoction is to be emptied into a tub, allowed to settle, and the clear liquor being drawn off, the lees are to be drained. Some recommend the addition of a little bullock's blood or white of egg, to remove a part of the tannin. But this abstraction tends to lessen the product, and will seldom be practised by the manufacturer intent upon a large return for his capital. The gum is to be dissolved in a small quantity of hot water, and the mucilage, thus formed, being filtered, is added to the clear decoction. The sulphate of iron must likewise be separately dissolved, and well mixed with the above. The colour darkens by degrees, in consequence of the peroxidizement of the iron, on exposing the ink to the action of the air. But ink affords a more durable writing when used n the pale state, because its particles are then finer, and penetrate the paper more intimately. When ink consists chiefly of tannate of proxide of iron, however black, it is merely superficial, and is easily erased or effaced. Therefore whenever the liquid made by the above prescription has acquired a moderately deep tint, it should be drawn off clear into bottles, and well corked up. Some ink-makers allow it to mould a a little in the casks before bottling, and suppose that it will thereby be not so liable to become mouldy in the bottles. A few bruised cloves, or other aromatic perfume, added to ink, is said to prevent the formation of mouldiness, which is produced by the ova of infusoria animalcules. I prefer digesting the galls to boiling them.

The operation may be abridged, by peroxidizing the copperas beforehand, by moderate calcination in an open vessel; but, for the reasons above assigned, ink made with such a sulphate of iron, however agreeable to the ignorant, when made to shine with gum and sugar, under the name of japan ink, is neither the most durable nor the most pleasant to write with.

From the comparatively high price of gallnuts, sumach, logwood, and even oak bark, are too frequently substituted, to a considerable degree. in the manufacture of ink.

The ink made by the prescription given above, is much more rich and powerful than many of the inks commonly sold. To bring it to their standard, a half more water may safely be added, or even 20 gallons of tolerable ink may be made from that weight of materials, as I have ascertained.—*Dr. Ure's Dictionary of Arts and Manufactures.*

---

manufacture of thread, or yarn, and paper, by the application of certain fibrous materials not hitherto so employed. Jan. 21.

John Thomas Betts, of Smithfield Bars, London, rectifier, in consequence of a communication from a certain foreigner residing abroad, for improvements in the process of preparing spirituous liquors in the making of brandy. Jan. 21.

Benjamin Ledger Shaw, of Honley, near Huddersfield, York, clothier, for improvements in preparing wool for and in the manufacture of woollen cloth, parts of which improvements are applicable to the weaving of other fabrics. Jan. 21.

John Chanter, of Earl-street, Blackfriars, Middlesex, Esq., and Peter Borrie, of Dundee, engineer, for improvements applicable to steam boilers. January 21.

NOTES AND NOTICES.

Sir James Anderson's Steam Carriage.—We understand from a letter which appeared in the *Morning Herald* last week, from Mr. Shaw, the Secretary, in answer to the complaint of a disappointed Shareholder, in the "Steam Carriage and Waggon Company," that Sir James intends to be in London with his Steam Carriage, at the time of the meeting of Parliament this month.

Joyce's and Arnott's Stoves.—On Sunday last at the parish church of Downham, Norfolk, about the middle of the service, a child fell down, (as it was supposed) in a fit, and was carried out; soon after, several children who were sitting at the chancel felt themselves compelled to leave the church; and toward the latter part of the service a lady fainted, and was immediately followed by several others, who were compelled to be carried home. In all about twenty persons felt the serious effects of some unwholesome gas, generated by the stoves placed in the church for the purpose of warming it. There are two of Harper and Joyce's stoves, and one of Dr. Arnott's, employed.—*Patriot.*

New Method of Cleaning Glass.—Reduce to very fine powder a piece of indigo, moisten a rag, apply it to the powder, and smear the glass with it. Wipe it well with a dry cloth. Very fine sifted ashes applied in the same manner, by a rag dipped in brandy or spirits of wine, will answer well; but spanish white ought to be rejected, as it is apt to take the polish off the glass.—*Journal de bon. Uc. et Prat.*

Improved Crayons for Drawing on Glass.—Take equal quantities of asphaltum and yellow wax, and melt them together. Add lampblack, sufficient to give the mixture the requisite colour, and stir it well, and pour it in moulds for Crayons. The glass should be well wiped with leather, and in drawing, care must be taken not to soil the glass with the fingers. It is sometimes difficult to trim the Crayons with a common knife, for if too sharp, it cuts in too much, and if too dull, it cannot make a fine point; but if the edge be bevelled, like scissors, and very sharp, the point may easily be rendered very fine.—*Rec. de la Soc. Poly.*

Wivell's Fire Escape.—On Wednesday, the 23rd ult., at one o'clock, Mr. Wivell explained the application of his machine at the Guildhall, Bath, (the use of which was granted by the Mayor,) to a considerable number of individuals who were present. The various operations were highly applauded, and so much pleased was Lord James O'Brien, that not content with bestowing barren praise upon the inventor, he generously ordered one of the largest sort, with all its apparatus, complete, at the cost of Fifty Pounds, for the use of his fellow-citizens in the event of a fire—an act, for which, his Lordship is entitled to the public thanks.

Kollman's Patent Railway.—A model of this invention has for some time been exhibited at No. 6, Carlisle-street, Soho. The objects proposed to be effected, are safety, facility in ascending inclines, and turning curves. Mr. Kollman attends in Carlisle-street, three days a-week, for the purpose of exhibiting his models and engine, the latter of which has been beautifully manufactured at an expense of £300, and is on the scale of one and a half inch to a foot. It is worked by steam, and performs various and speedy evolutions upon a model-railway, formed in the shape of the figure 8, "which demonstrates *practically*, (says the *Mining Journal, sed quere,*) its capability of moving round a circle of fourteen feet radius, besides its wonderful power of ascending a hill of one mile in 14 acclivity. This revolution is effected by an additional fore-wheel on either side, of smaller diameter, and concentric with the large driving-wheel, the tire of which is *roughened* to give it the necessary hold on the surface of the rail, which is elevated at the commencement of the acclivity, so as for the smaller wheel to act upon it; being also just sufficiently elevated to raise the larger wheel from the line of rail on which it previously acted. To prevent the possibility of the engine being diverted from its course, there is a centre-rail with two horizontal wheels in front of the engine, which acts as a pole to a carriage, and makes its direction completely subservient to the middle or centre rail, the outward wheels running upon a plain and unconfined surface of iron.

Wind and Steam-Navigation.—The French government steamer, *Veloce*, according to the French papers, has been fitted out on a new principle for working the vessel, with either sails or steam, and is now on her voyage from Rochefort to Mexico, for the purpose of testing this important invention. When fallen in with, of late, by a Spanish ship, north lat. 40., long. W. of Paris, 14., the captain reported that her rate of sailing under top-sails, studding-sails and royals, had been for two days and a half upwards of eleven knots an hour. The papers do not give any particulars of the plan adopted.

Steam Carriage Tram-Roads.—It is said to be in agitation to revive the plan for establishing tram-roads by the side of of turnpike-roads to compete with railroads.

Steam-boat night steering signals.—At the Society of Arts last week a silver medal was awarded to Mr. Jennings, for his invention of night signals for steamers, to intimate to the steersman in what direction to vary the helm. The plan consisted of three differently coloured lamps, fixed on a horizontal rod attached to a swivel which the man on the paddle-box varies in position relative to the steersman's eye, and thus conveying to him the requisite orders.

French Waterworks.—The Municipal Council of Versailles has voted a sum of 400,000 francs (16,000*l.*), and applied to government for a grant of 1,500,000 francs (60,000*l.*), for the purpose of enabling the town to be supplied with an additional quantity of water, by taking advantage of the works recently finished at Bezons, a little above Marly.—*Galignani's Messenger.*

LONDON: Printed and Published for the Proprietor, by W. A. Robertson, at the Mechanics' Magazine Office, No. 6, Peterborough-court, Fleet-street.—Sold by A. & W. Galignani, Rue Vivienne, Paris.

Mechanics' Magazine,
MUSEUM, REGISTER, JOURNAL, AND GAZETTE.

No. 809.]   SATURDAY, FEBRUARY 9, 1839.   [Price 6d.

Printed and Published for the Proprietor, by W. A. Robertson, No. 6, Peterborough-court, Fleet-street.

CHANDOS-STREET FIRE ENGINE STATION.

LONDON FIRES IN 1838.

" The pile decreas'd that lately seemed so high,
And sheets of smoke roll'd upward to the sky :
As humid vapours from a marshy bog,
Rise by degrees, condensing into fog,
That intercept the sun's enliv'ning ray,
And with a cloud infect the cheerful day.
The sooty ashes, wafted by the air,
Whirl round and thicken in a body there."

Dryden's Ovid.

Sir,—The year revolves, and again brings round the period at which I have undertaken to record one class of the calamities of which London's fair city, and its environs, have been the site. The inauspicious opening of the year eighteen hundred and thirty-eight will long be remembered for the appalling number, and the serious extent of its *conflagrations*, both at home and abroad. The year had scarcely set in, which it did with a severe frost, than fires began to blaze around us in every direction, with a frequency that has never been surpassed within the memory of man. Among the other casualties of this period, the Royal Exchange, " one of the greatest glories and ornaments of London," fell a victim to the flames ; at the same time, information reached us of the misfortunes of our continental neighbours. " Three notable fires" says Mr. Leigh Hunt, " have lately taken place in different parts of Europe, as if on purpose to contradict the freezingness of the weather, and scorch old Winter's beard, —one at the Imperial Palace in St. Peterburgh, another at the Italian Opera House in Paris, and the third at the Royal Exchange in London. It has been well observed by a French paper, that the buildings were characteristic of the different nations,—the Palace of Russia, despotism ; the Theatre, of French love of the fine arts ; the Exchange, of English commerce."[*]

The result of this year's record is to show, that the metropolitan alarms of fire have amounted to the astounding number of seven hundred and fifty-five, as set forth in the following table : to which I have added the average of the five years' experience of the London Fire Establishment, so that the difference may be at once perceived :

MONTHS.	Number of Fires.	Number of Fatal Fires.	Number of Lives Lost.	Alarms from Chimneys on Fire.	False Alarms.	Average of last Five Years.
January	63	1	1	11	10	41
February	44	1	5	8	9	38
March	49	0	0	15	8	44
April	37	1	4	8	6	35
May	45	0	0	9	4	45
June	47	1	2	10	7	40
July	34	0	0	5	6	44
August	50	0	0	5	4	43
September......	50	2	2	7	4	39
October	54	1	1	12	10	38
November	50	2	3	6	6	42
December	45	2	3	11	6	47
Total......	568	11	21	107	80	496

The number of fires wherein the premises were *totally destroyed* is................ 33
Very seriously damaged.. 152
Slightly damaged 383

568

* *Monthly Repository*, No. 386, for February, 1838.

Brought forward	568
Alarms which proved to be occasioned by chimneys on fire	107
False alarms, originating in error or design	80
Making the total number of calls	755

The number of instances in which Insurances had been effected on the *building*
and *contents*, was 161
On the *building* only 59
On the *contents* only 128
Neither Insured 220

568

The foregoing list is exclusive of a very large number of fires in chimnies —many of them exceedingly troublesome, and some highly dangerous—which have been attended by the men and engines of the London Fire Establishment. Those mentioned in the table (107) were supposed to be serious fires, and caused a general turn-out of the force. The number of false alarms last year, was somewhat below the usual average; they were of the usual varied character—the greater number originating in error, and some few in design. The kindling of bonfires, and burning of refuse matters in the open air, the escape of large quantities of smoke or steam, unusual appearances of light from gasworks, kilns, &c., are continually giving rise to groundless alarms of fire.

At a quarter after eleven o'clock at night on the 13th of September, a brilliant appearance of the *Aurora Borealis* caused a turn-out of the firemen.

The annual return of this singular phenomenon on the 13th of November, produced similar results. Upon this occasion, one of the most splendid specimens of the "Falling Stars," and other meteoric phenomena, that have ever been witnessed by the inhabitants of London, was observed at an early hour, and attracted considerable curiosity. For some time the appearance resembled that of an alarming conflagration : several Fire-Brigade engines were turned out, and for upwards of two hours were traversing the metropolis in quest of the supposed fire. The engine from the King-street station proceeded as far as Hampstead, before the error was detected; while others from the stations in Baker-street, Wells-street, Crown-street, Holborn, Jeffery-square and Wellclose-square, went to Kilburn, Ealing, Staint John's-wood, Holloway, and other places.

The men belonging to the Society for the Protection of Life from Fire, with their cumbersome "fire-escapes," were running in various directions, as were also parties of the Metropolitan Police.— During the progress of this meteoric phenomenon the atmosphere was remarkably clear, and the stars shone with unusual brightness; the air was rather frosty, with a sharp breeze from the east. It began a quarter before two, and the first object that attracted the attention of observers was, several stars of an ordinary size, shooting from their original spots, and falling apparently to the earth, when it seemed as if they exploded, for immediately afterwards the horizon was brilliantly illuminated by a vivid light; this, within ten minutes, disappeared, but another light of a most splendid description rose from the same quarter, and gradually expanded over the whole hemisphere. At intervals immense masses of crimson vapour appeared, intermingled with branches of silvery corruscations, which at times formed a rich and variegated canopy, covering the entire expanse from the east to the western hemisphere, presenting a most gorgeous spectacle. I was called up on this occasion soon after four o'clock, at which time the appearance, as seen from a narrow street, looking northward, so closely resembled a tremendous conflagration, that any one might well have been mistaken. On reaching an open spot, however, from whence a distinct view of the horizon could be obtained, the character of the phenomenon was at once apparent. It then consisted of two distinct rays of brilliant crimson light, rising divergently from the north. Soon after four, the phenomenon became more faint, but the bright columns of light radiating from it, retained their splendour till hlf-past four , when it entirely disap-

T 2

peared, vanishing at last with considerable rapidity.*

The time when this phenomenon has been observed during the last six or seven years, has been in the interval extending from the 12th to the 16th of November. The same phenomena has been observed, at precisely the same time, in Germany, Russia, Australia, and America, and has created similar consternation and alarm, notwithstanding the periodical return of these singular appearances are now so well understood.

It happens on this, as on most former occasions, that the instances in which the premises have been completely burned down, are by no means the most serious fires. With some few exceptions, which will be noticed in detail, the "*total losses*" have consisted principally of buildings of a very unimportant character; in many cases small old buildings, and for the most part composed of timber. Upon *twenty-one* occasions the buildings were very small, and so completely on fire before it was discovered, that they were wholly enveloped in flames before any assistance could possibly reach the spot. In the *three* following cases, the buildings were filled with such highly inflammable materials, as to defy all attempts to extinguish the fire. The first of these broke out about half-past three o'clock on Friday morning, February the 9th, in an extensive range of timber buildings belonging to Mr. Edgington, in the Old Kent-road, occupied as a tarpauling manufactory. There being at the time a very high wind blowing from the south west, the flames speedily communicated to several adjoining buildings, one of which was the Coach-manufactory of Mr. Ward; another a house in which Mr. Lyon carried on the business of paper-staining; these together with the house of Mr. Richardson, and the stabling of Mr. Wellan, were, with nearly the whole of their contents, totally consumed in less than an hour. It was altogether impossible to control the raging of the flames, in consequence of the immense quantity of pitch, tar, and rosin, deposited in the premises, and which on becoming ignited, gave to the conflagration a fearful

* For a highly interesting description of these appearances, as seen at Richmond, vide page 152 of the present volume.

ascendancy. These buildings were all in immediate connexion, and entirely composed of timber; upwards of an hour and a-quarter elapsed, before a drop of water could be obtained to supply any one of the numerous engines that were waiting the arrival of this necessary element. This conflagration was visible at a great distance in every direction, and presented one of the most appalling sights that has occurred for a great length of time: far exceeding even those of the Royal Exchange and Fenning's Wharf.

The next fire was that of Sir Charles Price's Turpentine Distillery, Mill Wall, Poplar, which was burned down on Monday afternoon, May 14th, from the boiling over of a copper of the inflammable material there manufactured. This was the most extensive establishment of the kind in the metropolis, and occupied seven or eight acres of ground. On the arrival of the firemen and engines, they found the Distillery (a building upwards of seventy feet long by fifty wide) enveloped in one vivid sheet of flame. It contained six boilers, all of which were full; also four large stills, which were filled with liquor ready to be worked off; on the stage of the Distillery were two hundred and fifty barrels of turpentine, ready to be removed to the storehouses. Notwithstanding the presence of ten engines, besides the two powerful floating engines of the London Fire Establishment, and a plentiful supply of water being afforded by the proximity of the City Canal, all efforts to preserve any portion of this building was unavailing. With the stock in the yard, however, the firemen were more successful.

At a quarter before nine o'clock in the evening of Wednesday October 10, a fire broke out in the wadding manufactory of Mr. Yorke, situated in Norfolk-place, Shoreditch; from the nature of its contents, it instantly ignited from top to bottom, and the whole was rapidly destroyed. This building was in the midst of surrounding premises, which at first appeared doomed to a similar fate, but the prompt arrival and energetic efforts of the firemen, effected their preservation. *Seven* of the total losses were at such remote distances from the metropolis, as to preclude the possibility of the London firemen reaching the scene of destruction time enough to render much useful aid towards preserving the building in which the fire had originated; they have, however, often succeeded in saving adjacent premises, which, but for their exertions, would have been included in the general devastation. These fires were as follows:—February 22nd, eight o'clock P.M. at the Snuff Mills of Messrs. Taddy, Attfield and Company, Morden, Surrey, nine miles from the nearest London engine station. Three engines started off, with four horses each, and arrived at the fire time enough to arrest its progress in the dwelling-house, but the manufactory was wholly consumed. The fire originated in the drying stove. On the following morning, at five o'clock, a fire broke out in the house of Mr. Davies, tinman, Camberwell Green, which, with the adjoining one of Mr. Bedwell, tobacconist (both small buildings) were consumed before any assistance could be rendered, the distance being three miles. April 3rd, at two o'clock in the morning, a fire broke out in the loft of Mr. Bailey, sailmaker, Heath-street, Barking, Essex, eight miles from town. The sail-loft was destroyed, and the fire communicated to three other premises, but by the intrepid exertions of the Royal Exchange Assurance Company's resident Engineer, and the arrival of a party of the London Brigade, with two engines from town, the fire was happily arrested in its course. July 25, about a quarter before eight at night, a tremendous fire broke out in the extensive premises of Messrs Gordon, Brothers, and Co., shipbuilders, &c. Grove-street, Deptford, adjoining her Majesty's Victualling and Dock-yards. This establishment was one of the most extensive in Deptford, comprising warehouses, timber-house, workshops, and various covered buildings, occupying upwards of an acre of ground. When the fire was first discoverd, it was raging in a counting-house belonging to the Iron Steam-boat Company, from whence the flames rapidly communicated to the buildings on either side. With the exception of the warehouses, the rest of the buildings were constructed principally of wood, the breakings up of old ships. Within half an hour after its commencement, the conflagration had extended to the warehouses where the mouldings for vessels were manufactured, and although there were strong

party-walls between them, they one by one took fire. The devouring element, notwithstanding the pulling down and cutting through roofs to prevent communication, proceded with frightful rapidity, so that the whole of the premises fell a sacrifice to its ravages. Engines from Greenwich, Deptford, the Victualling Office, and those of the Fire Brigade from town (distant three miles) were brought up with most praiseworthy speed, but unfortunately the absence of water rendered them altogether useless: when there no longer remained a possibility of saving any portion of the premises, a supply of water was obtained. The heat of the burning mass was so intense as to defy a near approach, and several houses on the opposite side of Grove-street, occupied by tradesmen and working people, were set on fire and seriously injured; but, by dint of great exertions on the part of the firemen, they were saved from total destruction.

On the 3rd of September, Croydon was the scene of a conflagration, which occasioned considerable destruction of property. This fire broke out in the workshops of Mr. Ingram, carpenter, &c., but how it originated is not known. The flames extended right and left with extraordinary rapidity, communicating to the Printing-offices of Mr. Langford, the warehouse of Mr. Corker, currier, the workshop of Mr. Restell, watchmaker, and some other buildings. The dwelling-house of Mr. Langford was pulled down to cut off the communication of the fire in that direction, but his printing-office was wholly consumed. The military from Croydon Barracks with their engine, were soon on the spot, but a great scarcity of water paralysed their efforts. This fire was discovered soon after eight o'clock in the evening, and before eleven a party of the London Fire Brigade, with an engine, reached the scene of danger, and rendered most effectual aid in finally supressing the flames.

On Monday evening, October 22, soon after six o'clock, the retired village of Harrow-on-the-Hill was thrown into the greatest consternation by the bursting forth of flames from the residence of the Rev. C. Wordsworth, Head Master of Harrow School. The parish engine, and the private engine of the Rev. H. Drury, were immediately put in requisition, but with so little effect that expresses were sent off for further assistance, and in less than two hours a couple of engines, with a strong force of the London firemen were on the spot, but the extreme difficulty of obtaining water, again materially circumscribed their usefulness. The extraordinary exertions of Mr. Greenhill of Roxheth, and his labourers, with that gentleman's private engine, contributed most eminently to check the speed of the fire; the preservation of Mr. Bowen's premises is ascribed solely to the well-directed efforts of this gallant band. The distance of the West-end Engine-stations from the London Fire establishments is eleven miles.

November 25th. Sunday morning, about half past four o'clock, Ray House, Woodford-bridge, the residence of T. Lewis, Esq. was destroyed by fire. It was at the time under repair, and there is every reason to believe that the accident is attributable to carelessness on the part of some of the workmen employed on the premises. Although the distance is eight miles from the nearest station, the Brigade were on the spot time enough to effect the preservation of the offices adjoining; the dwelling itself was too far gone to afford any chance of preserving it.

The only remaining fires characterised as "total losses," are the memorable conflagration of the Royal Exchange on the 10th of January, and that in Paper Buildings, Temple, on the 6th of March, both of them too important in their character and consequences to be briefly passed by. "We doubt," says a writer in the *Monthly Repository* for February, "if the Royal Exchange is not the very last public building which any one would have expected to be destroyed by fire." It has, however, again fallen before the mighty foe. Soon after ten o'clock on the never-to-be-forgotten night of Wednesday, January the 10th, a fire was discovered to be raging in Lloyd's Coffee-room) at the north-east corner of the edifice. On the first discovery of the fire, considerable difficulty was experienced in obtaining access to the building; the flames appeared to have got to a very great head before the discovery was made, and it was very soon predicted by those persons acquainted with the

building, that the whole of it would be destroyed, and this foreboding was too correctly realised. From the intense frost that prevailed, some difficulty arose in obtaining water, but afterwards an abundant supply was afforded. I was on the spot from eleven o'clock till nearly six the following morning, and having assisted to raise the ladders, was one of the first to enter the south gallery from the balcony in the quadrangle, where a most gallant effort was made to cut off the fire in that direction; the branches of three engines being got up into the passage, and their powerful jets directed point blank upon the advancing flames, but unfortunately with no apparent effect. A similar attempt was at the same time made in the north gallery; the whole body of fire being then confined to the west side of the building; but the building was so intimately connected together with one continuous roof, and with so many narrow intricate passages, that the firemen were completely surrounded and enveloped with fire before they were conscious of their danger, and they did not yield until completely beaten from their respective positions by the flames advancing above their heads, beneath their feet, and on either side of them. Strenuous efforts were also making to endeavour to save the shops which surrounded the edifice on the north, west, and south sides, but without success; the exertions of the firemen on behalf of Mr. Betts's Musical Instrument warehouse, under the north piazza, which fell one of the earliest victims to the flames, gave rise to the following *jeu d'esprit* which appeared in *Bentley's Miscellany* for February :—

"The men of Braidwood's Fire Brigade,
    In water to their middles,
With skill and great precision play'd
    On Arthur Betts's Fiddles,"

It was a touching circumstance, that at twelve o'clock, when the flames had reached the north-west angle of the building, and were rapidly advancing towards the tower, the chimes struck up the air of

"There's nae luck about the house
    While our guid man's awa.*"

---

* The "guid man,"—that is to say, the keeper of the key, for want of which entrance was delayed, and the fire gained a considerable head, was gone to Greenwich.

By this time (twelve o'clock) the whole of the long range of offices belonging to the Royal Exchange Assurance Company, Lloyd's establishment, the coffee room, the captains' room, and the offices of the underwriters, presented one body of flame, which shot up to a great height, illuminating the Bank, St. Bartholemew's, St. Michael's, and St. Mary Woolnoth's churches. At one o'clock the north and west sides of the Exchange were consumed, and the fire was rapidly approaching the clock-tower. The efforts of the firemen appeared hitherto not to have had the least effect; the flames continued to extend rapidly over the building, although upwards of twenty engines were in full work.

As it was pretty evident that if the fire passed into Sweeting's-alley, it would be difficult to say how far its ravages might not extend, this point now became one of vital importance; accordingly, by Mr. Braidwood's direction, eight powerful engines were concentrated upon the eastern side of the building, and by keeping up a continuous torrent of water, happily succeeded in saving the shops in Sweeting's-alley, and effectually stayed the progress of the flames in that direction. At five o'clock the fire was still raging, but all apprehension of danger to the surrounding buildings was at an end.

During the fire, the bridges over the Thames were crowded with people, and the flames were distinctly visible at Windsor Castle, a distance of 24 miles from Cornhill. An Essex farmer states, that the fire was seen at Thoydon Mount, near Epping, 18 miles from London. On the high lands of Surrey, within 20 miles of the metropolis, the progress of the conflagration was observed by the astonished country people, who watched the destruction of the tower, when the flames ascended that part of the edifice. The calamity is supposed to have arisen in the stoppage and over-heating of one of the flues connected with the kitchen of Lloyd's coffee-house. This department was separated from the rest of the building by a thick, but old boarding, which, indeed, formed the only partition between the several offices on the first floor. To this want of a party-wall may be traced the speedy ignition of the premises on all sides, a circumstance

which, with a north-east wind, would have been almost impossible had such a barrier existed. It seems exceedingly strange, that although nobody was allowed to have fire or candle in the building after a certain hour, and a surveyor went round every evening to see that this regulation was not violated, no person was allowed to remain within the building all night to guard it against intruders.

About five o'clock, Tuesday morning, March 6, one of the most destructive fires which has ever occurred in the Temple, broke out in the chambers of Mr. Maule, Queen's Counsel, and M.P. for Carlow. As soon as the alarm of fire was given, the watchmen got out the two engines belonging to the Inner Temple, but they were found to be in a very indifferent state; after a considerable lapse of time messengers were dispatched to procure more efficient aid, and the engines from the various stations of the London Fire Establishment arrived in quick succession. On reaching the spot the firemen found the fire had gained too much ground to be stopped in the first building—in fact, the second was already on fire: in addition to the disadvantages of a *late call*, they had to contend with a great deficiency of water, occasioned by the circumstance of some of the adjacent mains being at the time under repair. Unfortunately it was low water, so that a supply could not be drawn from the Thames. Before seven o'clock, however, a tolerable supply of water was obtained, and sixteen powerful engines discharged their streams on the burning mass. The new and powerful floating fire-engine having been brought up from its station at Southwark-bridge, it was manned by ninety auxilliaries, and upwards of 800 feet of leather hose united, by means of which a copious jet of water was directed on the various points subjected to the action of the fire, with the best effect; the flames being completely beaten in the third building to which they had extended. The chambers of 36 parties were destroyed, and the loss of property, both insured and uninsured was very considerable; but by far, the heaviest loss was the destruction of books, papers, deeds, documents and legal instruments of the greatest importances, of very many of which no duplicate copy existed.

T 4

This fire completes the list of total losses, and such a list as has hardly ever before occurred.

At each of the following fires very serious damage was sustained, and although the firemen were eventually triumphant, it was only by dint of most extraordinary efforts that the flames were vanquished :—January 25, 5 P.M. Mr. Bowe, furrier, 3, Little Carter-lane, Doctor's Commons, upper part of building and back workshops destroyed; ten other buildings more or less damaged. February 6, 5½ A.M., Mr. R. Flack, licensed victualler (sign of the Brown Bear), Lemon-street, Goodman's Fields; building and contents nearly destroyed, and two adjoining buildings damaged. March 8, 4½ A.M., Mr. J. Murfitt, military ornament maker, 26, Porter-street, Newport Market; second floor, attics and roof burned off and three other buildings damaged. Sept. 15, 1¾ A.M., Mr. Thompson, boat-builder, Rother-hithe; part of a ware-room, &c., burned down. Sept. 16, 1 A.M., Messrs. Gale and Son, Love-lane, Shadwell; paint warehouse and oil cellars destroyed; rope ground and four other buildings seriously damaged. Sept. 27, 8¾ P.M., Mr. Edwards, carpenter and builder, Bennet-street, Stamford-street; back workshops and front dwelling-house nearly destroyed; three other premises damaged. Sept. 28, ¾ A.M., Mr. G. Collard, tinman, 33 Wych-street, Strand; house burned all through, and great part of the roof off; two other buildings damaged. October 9, 9 P.M., Mr. White, stable-keeper, White Hart Mews, London-wall; hay-loft and roof burned; three other buildings slightly damaged. October 29, 11¾ P.M., Mr. Llewellyn, cheesemonger, &c., 24, Liquorpond-street; nearly destroyed, and four other buildings damaged. December 1, 7¾ A.M., Messrs. Foster and Smith, rice and oil mills, Shad-Thames, Dock-head; rice mill and about three-fourths of an oil mill and warehouse adjoining, destroyed. Dec. 20, 8¾ P.M., Messrs. Bowman and Co., sugar refiners, Duncan-street, Whitechapel; about two-thirds of the Russia-house destroyed; machinery saved; five other buildings seriously damaged. Dec. 28, 11 P.M., Mrs. Waters, coffee-house keeper, High-street, Shoreditch; this house nearly destroyed, and three adjoining ones damaged. At all these, and several other fires, the firemen have had to encounter much difficulty, from a late call, deficiency of water, or the inflammable nature of the building or contents; and the termination of each fire is highly creditable to their unwearied skill and perseverance. During the past year, nearly 800 buildings have been damaged by fire, in or near the metropolis.

The following list exhibits the occupation of the premises wherein the foregoing fires have occurred, care having been taken to discriminate between fires originating in that portion of the building occupied in trade or manufacture, and those that have happened in, and damaged dwelling houses only.

Apothecaries	1
Bagnios	3
Bakers	9
Booksellers, Stationers, &c.	9
Beer-shops, retail	7
Boat and Barge-builders	1
Brewers	3
Brokers, &c.	6
Builders	3
Cabinet-makers	9
Carpenters, &c.	21
Cement Works	2
Charcoal and Coke, dealers in	4
Cheesemongers	5
Chemists and Druggists	4
Coffee-roasters	1
Coffee-shops, &c.	3
Colour-makers	2
Confectioners	2
Carried forward	**95**

Brought forward	95
Cornchandlers	1
Coopers	3
Drapers, &c.	15
Distillers	1
Drysalters	1
Dyers	1
Eating-house Keepers	6
Feather Merchants	2
Fire-preventive Company (!)	1
Fire-work makers	1
Flax-dressers	1
Founders	6
Furriers and Skin-dyers	2
Gas-works	4
Granaries	2
Grocers	5
Hat-makers	2
Horse-hair Merchants	1
Carried forward	**150**

Brought forward 150	Brought forward 476
Hotels, Taverns, &c. 7	Ships 5
Japanners 2	Ships, steam...................... 4
Lodging-houses 37	Ship-builders 4
Laundries 1	Ship-chandlers.................... 1
Lucifer Match-makers 8	Soot Merchants 1
Manganese Manufacturer........... 1	Sugar-refiners 4
Marine Stores, dealers in 5	Snuff-mills 1
Mills (Steam)..................... 1	Tailors 8
Oil and Colourmen (not manufacturers) 7	Tallow-chandlers, &c.............. 2
Private dwellings.................. 179	Tanners 1
Painter and Glaziers 1	Tarpaulin-makers 2
Painted-baize manufactory.......... 1	Theatres 1
Pawnbrokers 1	Tinmen, Braziers, and Smiths 5
Pipe-makers...................... 3	Toy Warehouses 2
Printing-ink Manufactory 1	Turpentine Distillery 1
Printers, copper-plate............. 2	Unoccupied buildings 1
Printers and Engravers 6	Under repair 9
Pork-butchers 2	Upholsterers 3
Public Buildings 3	Victuallers, licensed 23
Rag Merchants 2	Wadding Manufactories 2
Rope-makers 1	Warehousemen, Manchester 1
Sale Shops and Offices 38	Weavers, silk 1
Sail-makers 1	Weavers, carpet 1
Saltpetre refiners 1	Wine and Spirit Merchant 8
Saw-mills........................ 2	Wood Merchant 1
Stables 13	
	Total 568
Carried forward 476	

The number of fires on each day of the week during the past year was as follows :—

Monday.	Tuesday.	Wednesday.	Thursday.	Friday.	Saturday.	Sunday.
81	87	91	81	84	67	77

Their distribution throughout the 24 hours has taken place in the following ratio :—

	First Hour.	Second Hour.	Third Hour.	Fourth Hour.	Fifth Hour.	Sixth Hour.	Seventh Hour.	Eighth Hour.	Ninth Hour.	Tenth Hour.	Eleventh Hour.	Twelfth Hour.
A.M.	20	37	17	24	17	7	16	12	13	13	12	12
P.M.	15	21	13	19	22	27	39	47	46	42	43	34

The result of careful and diligent inquiries into the causes of each fire—which continues to be greatly facilitated, by prompt attendance of the firemen the moment an alarm is given—furnishes the following epitome of the origin of the greater part of them. A perusal of this list will show that the usual amount of carelessness in the use of fire and candles is kept up; that the same want of caution as heretofore, is still manifest.

Accidents of various kinds ascertained to be for the most part unavoidable . 34	Brought forward 138
Apparel ignited on the person 5	Ditto - ditto to window curtains.. 33
Candles, various accidents with 38	Carelessness, palpable instances of .. 17
Ditto, setting fire to bed curtains 61	Children playing with fire 5
	Fire, sparks from.................. 12
Carried forward 138	Carried forward 205

Brought forward 205			Brought forward 427	
Fires kindled on hearths and other im-			Lucifer Match-making	9
proper places	15		Ovens, overheated, defective, &c.	11
Fires, portable charcoal	2		Reading in bed....................	1
Fire-heat applied to various purposes			Shavings, loose ignited	17
of trade and manufactures	39		Spontaneous ignition of Lime	4
Fire-works	3		Ditto ditto Rags........	3
Flues, stopped up, defective and ignited	58		Ditto ditto Tan	2
Fumigation, incautious	1		Stoves and stove-pipes, overheated, &c.	26
Furnaces, overheated, &c.	15		Stoves, drying	5
Gas, sundry accidents from escape of..	34		Suspicious	8
Gas, accidents in lighting of	8		Tobacco-smoking..................	4
Gunpowder	1		Wilful	6
Intoxication	4			
Lamps, sparks from	9			523
Linen, &c., airing before fire	32		Undiscovered	45
Lightning.......................	1			
			Total.................	568
Carried forward 427				

The writer in the *Monthly Repository* before quoted (Mr. Leigh Hunt),alluding to the relative causes of fire, observes that " a theatre is a place full of night work, lamps, and combustibles; every siege and fire-work in a melodrama seems to threaten it. A church has a vestry-room,and perhaps a fat sexton who sleeps after dinner. Public offices have work by candle-light, and housekeepers who reside in them. In coffee-houses a perpetual household warmth is going on; strangers or visitors are constantly going in and out, and supping, and sleeping; and the greatest wonder, after all, in a metropolis, is, that there are not a dozen or even a hundred conflagrations every night. Think of all the careless people that go to bed every night with candles in their hands,—of the children, the sick, the sleepy, the superannuated, the foolish, (*the fraudulent*), the revengeful, (*and*) the drunken. Think of the food for fire, which exists all about them, —of the narrow escape which curtains and clothes *must* be having, (so to speak), every instant; and then be any bounds, if you can, to the wonder at London's being so very unburnt a city. Doubtless, the partial accidents which occur, and which nobody hears of, must be numerous,* yet still it is a reasonable ground of amazement, that fires are not constantly occurring on the largest and most tragical scale.

The number of fatal fires in the present

report is less than the last, but the number of lives lost is rather greater ; some of last year's fires were attended with most distressing casualties, which I will briefly describe in the order of their occurrene.

On the 8th of February, about half-past one o'clock in the morning, a fire broke out in the house of Mrs. Parke, dealer in rags and marine stores, situated between Charlotte-street and George-street, Gravel-lane, Southwark. The fire was discovered by police constable Wm. Pool, 33 M, who observed smoke issuing from the shop; having forced an entrance, he succeeded in getting out Mrs. Parke and four of her children, who were fast asleep in a room at the back of the shop. He then proceeded up stairs and met several persons coming down ; he asked them if there were any more persons remaining up stairs, and being assured there was not,he concluded they were all saved. On returning into the street he found somebody had pulled down all the shutters,which had increased the draught so much that the fire was rapidly extending all over the house. It unfortunately happened that the policeman had been misinformed; a poor widow and her three children, and a woman named Mary Ryan, were still within the fated building.

On the coroner's inquest, James Robinson, a glass-blower, stated that he occupied the front room on the first-floor; he was awoke by his wife, who said the house was on fire. Having jumped out of bed and opened his window he saw flames coming through the roof of the

---

* These accidents, are, in fact, much more numerous than those, which come to the firemen's knowledge, and are reported.—W. B.

shop which projected some distance from the house into the street. He immediately snatched up one of his children and ran down stairs, followed by his wife with the other child. As he passed through the passage the wainscoat partition was on fire; notwithstanding this he returned to the house for the humane purpose of assisting the widow (Mrs. Sweeney) and her family, who occupied the back room on the second-floor, and any others that remained. When he reached his own room, the flames were coming through the flooring, and he found himself immediately so surrounded by fire, as to leave him no alternative but to throw himself out of the window into the yard, a height of about 15 feet. In falling he came in contact with a water-butt which seriously injured him. When he got upon his feet he heard Mrs. Sweeney calling for help from a window above, and told her to throw her children out of window and jump after them. He had scarcely uttered these words when Mary Ryan fell upon him from an upper window and knocked him down. As soon as he recovered himself, he held up a pillow, calling out to Mrs. Sweeney to throw out her children; she then threw one out, it fell upon the pillow but Robinson was so week he could not support it, and it rolled on the stones. He held up the pillow to receive another, and saw Mrs. Sweeney leave the window, and heard her calling to her children, but she returned without them, when the smoke and heat overpowered her, and she fell over the window ledge with her arms stretched out in a supplicating attitude. There the poor creature remained until the flames reached her, uttering the most heart-rending cries for assistance for herself and children. The little ones were running about the room screaming dreadfully; their piteous cries however were soon stifled by the ravages of the fire, and they shared the fate of their parent.

In the mean time, intelligence of the fire having been forwarded to the several engine-stations, and to myself, by the policemen, I instantly repaired to the spot with a set of Merryweather's portable fire-escape ladders. Although but 10 minutes had elapsed from the discovery of the fire, on reaching the spot I found it had extended so rapidly

that the whole building was one mass of flame.*

My arrival was speedily followed by the engines from Southwark-bridge-road, Waterloo-road, Farringdon-street, and other stations. They were for some time useless, upwards of half-an-hour having elapsed before any water could be obtained; at length a supply having been procured from the main of the Southwark water-works in Union-street, several engines took up their station there, and by means of an extraordinary length of leather-hose joined together, were thus enabled to reach the fire, and check the spread of the flames, which were communicating to the buildings on either side. By great exertion these premises were saved from destruction, but the marine-store shop and dwelling-house were entirely consumed. As soon as the ruins were sufficiently cooled to be entered, search was made for the bodies, when the remains of the mother and two children were discovered, burnt in a shocking manner. The other child, and the woman who fell from the window, received such serious injuries that they expired soon after in the hospital, making a melancholy total of five victims to this disastrous conflagration. When it is known that *twenty-six persons* were asleep in the house at the outbreak of the fire, it is next to a miracle that so many escaped.

The next fatal fire broke out about a quarter past five o'clock in the morning of Sunday, April 22, in the house of Mr. Cockerell, fire-work maker, Paradise-row, Lower-road, Islington. The inmates consisted of Mr. and Mrs. Cockerell, aged persons, their four sons and a daughter, who all slept in the house. The mother and daughter slept in a room on the first-floor, and the father with three of his sons in separate beds in another apartment on the same story, divided from the females' bedroom by a room strongly secured, in which gunpowder and other combustibles were deposited. Mr. George Cockerell, slept in a small apartment on the ground-floor; he stated that he was awoke by a

---

* Mr. Woods, the coroner, was pleased to say at the inquest, "that the public thanks were due to Mr. Baddeley for his prompt attendance, and he earnestly hoped that Mr. Merryweather's excellent fire-escape would be brought into general use."

loud explosion, and he instantly jumped out of bed, and on looking out of the window saw a cloud of smoke issuing from the house ; there was a ladder standing in the garden, which he seized and raised to the window of his mother's apartment, in doing of which he broke a pane of glass which roused his mother and sister who instantly pushed up the sash.   His sister descended the ladder first, she slipped and fell, but was not much hurt; the mother was next got down in safety.   Mr. George Cockerell then attempted to raise the ladder to his father's apartment but was so much exhausted he was unable to do so.   William Harris, police constable, No. 112 N, coming up at the time placed the ladder to the window and went up, Mr. G. Cockerell following him; the window was opened by Mr. Cockerell, sen., and his son and the policeman dragged him out of the room on to the ladder, but unfortunately he fell with great violence to the ground.   While the above parties were attending him, Thomas Easom, a neighbour, ascended the ladder and on looking into the window he saw two persons lying on a bed under the window ; they appeared to have been suffocated and he believed they were then dead ; he caught hold of the hand of one man (John Glenn) and attempted to drag him out but was soon compelled to let go, in consequence of the great heat and smoke that issued from the windows, indeed he was so overcome that he fell to the ground.   All attempts to save the young men proved fruitless, and the flames raged furiously.   The engines were promptly in attendance, but a great deficiency of water prevailed for some time. The knowledge of the dangerous substances in the house prevented a very near approach to the premises which were blown to pieces, by the frequent explosions ; blue, crimson, and other flames ever and anon illuminating the atmosphere.   As soon as the fire was subdued it was found that Elijah Cockerell, aged 21; Henry Cockerell, aged 17; and John Glenn, aged 38 (a step-son) had perished in the flames.   Mr. Cockerell, sen., expired on the following Thursday, in St. Bartholomew's hospital from the severe injuries he had sustained while effecting his escape from his burning habitation.   The coroner and jury complimented Mr. George Cockerell, his

neighbour Thomas Easom, and the policeman Harris, for the intrepidity they displayed in their unsuccessful endeavours on behalf of the three young men.

On the morning of Friday, June 8th, between two and three o'clock, a fire broke out at No. 7, Pancras-place, Pancras-road.   The policeman on duty observed flames in the shop, and immediately gave an alarm, when a most distressing scene took place.   The house, which was an exceedingly small one, was occupied by several poor families, amounting in all to twenty-two individuals, all of whom were in a profound sleep when the fire was discovered.   By the exertions of the police and neighbours, all were rescued except the two daughters of Mr. Pelbean, who occupied the first floor front room.   The eldest was a fine girl between fifteen and sixteen years of age; the youngest was six years old.   The elder girl had nearly escaped, but she returned to the room to save her sister, and had got as far as the landing, when the stairs gave way, and both were precipitated into the burning mass below, from whence the bodies were soon dug out, but so disfigured as scarcely to retain a vestige of the human form.   The house, with all its contents, was entirely consumed, and the timely arrival of the Farringdon-street and Holborn engines, aided by a plentiful supply of water, was the sole means of preserving the adjoining buildings.   These engines were first and second on the spot, and extinguished the fire ere the *St. Pancras engine* made its appearance.

September 14th, Friday evening, about eight o'clock, a serious explosion of gunpowder took place at the shop of Mr. Dallow, oil and colourman, corner of Myddleton-street, St. John-street-road.   It appeared that some gunpowder at the bottom of a small cask had become damp and caked together, and was placed under a counter as useless. The errand boy being left alone in the shop, got under the counter with a candle, and accidentally dropped a spark into the cask, the contents of which exploded with such violence as to force out the shop windows, while the contents of the shop were much damaged by the heat.   The poor boy was picked up in a mutilated state and conveyed to St. Bartholomew's Hospital, where he soon after expired from the severe burns he had received.

On Wednesday morning, September 19th, about a quarter before four o'clock, the vicinity of Tottenham-court-road was the scene of a dreadful calamity, by the breaking out of an alarming fire in the premises of Mr. Holt, No. 19, Pancras street, one door from Tottenham-court-road; the house was tenanted by nineteen individuals. The fire originated in the first floor back room, occupied by a young married woman, named Thornton, whose husband, a currier, was at the time at work in the country. The inmates of the building were all rescued by the courageous conduct of the police, with the exception of the young woman just alluded to, who, there is no doubt, was dead before the fire was discovered. As soon as the fire had been extinguished, the first object that attracted the attention of the firemen on entering the ruins, was a blackened substance lying upon the rafters of the first floor back room, which proved to be the remains of Mrs. Thornton; the body was burnt completely to a cinder; the arms, legs, and portions of the head being burned off. From the spot where her remains were found, it appeared as if she had risen from her bed and was endeavouring to gain the window, when she sunk overpowered by the smoke and heat.

About half-past seven o'clock, Monday evening, November 12th, a fire occurred at the house of Mr. Cole, 10, White Conduit-street, Pentonville. A spark is supposed to have fallen from a candle after a child had been put to bed, and the room was left without its being noticed. On an alarm of fire being given in the street, the room was entered, and the bed and furniture found to be enveloped in flames. The poor child was so severely burned as to die soon afterwards, but the fire was confined to the bedroom, and extinguished by the neighbours and police.

Two days afterwards (November 14th), soon after eight o'clock in the evening, Mrs. Hodges, 1, Durham-street, Strand, ran out of her house enveloped in flames, and screaming in a most heart-rending manner. Several persons instantly flew to her assistance, and having with some difficulty extinguished the flames, conveyed her to the Charing-cross Hospital. During her screams she had several times mentioned a child, and on some persons entering the house, which was filled with smoke, and reaching the first floor, they found a back room in flames, by the light of which a female child, two years' old, was observed near the fire place; a gentleman present rushed into the room and succeeded in rescuing the child from her dreadful situation, and she was conveyed to the same hospital, where these two unfortunate sufferers subsequently died of their burns. The fire was occasioned by a spark from the fire-place having ignited the bed which was closely contiguous to it. By the exertions of the firemen, the flames were eventually confined to the room in which they commenced, the whole of the contents being destroyed.

Another fatal fire occurred on the 19th of December, about half-past nine o'clock in the evening, at the house of Mr. Baxter, No. 4, Brunswick-place, East, City-road. It appeared on the inquest, that the parents had gone to spend the evening at the house of a friend, leaving the children in care of a servant girl, who put them to bed about eight o'clock, and some time afterwards went up stairs to see that all was right. Upon her second visit to the room it is probable she dropped a spark from her candle, for shortly afterwards there was a cry of fire raised in the street, and several persons having been admitted to the house, they proceeded to the room where the two children were lying, and rescued them, but not until they were both so dreadfully burned, as to expire in the course of the night. The room, a double-bedded chamber, was burned out, but the flames were prevented from extending further.

The two remaining fatal fires originated in the wearing apparel of adult females taking fire upon their person, and causing death. Although these two cases only have been attended, and therefore reported, by the firemen, such accidents have been most appallingly frequent during the past year; as many as six per week have several times been reported in the public press. Nearly one hundred and thirty persons—adult females and children—have, to my knowledge, been burned to death during the period under review.

The supply of water upon the whole has been remarkably good, although upon some occasions, as already stated, a most lamentable deficiency has arisen. It is but fair to state, however, that this deficiency has mostly arisen from the circumstance of early intimation of the fires not having been given to the several turncocks, and not from inability on the part of the water companies to furnish a prompt and sufficient supply. The necessity of speedy application to the firemen appears to be perfectly well understood, but the equal necessity for as quick an application to that important functionary, the *turncock*, is not generally perceived.

There is one circumstance calculated to give rise to serious injury, if persisted in. I allude to the frequent habit of cutting down the customary rewards to the turncock and firemen for prompt attendance. In one parish, and a large and populous one too, upon several recent occasions, the turncock having turned out at dead of night, and furnished water to fires which have been happily "nipped in the bud," has been refused any reward, and *called a fool for turning out !* Under such treatment what is to be expected ?

The past year has been marked by the successful introduction of a *fire-proof cement*, patented by Mr. Joseph Davies, which, if generally adopted, seems calculated to reduce very materially the extent of damage from fire. In your 778th number I furnished a brief account of a very successful and highly satisfactory experiment made with this cement, in Dorset-street, Clapham-road, on the 6th of June last. Since this, a further trial of the merits of this composition has been made at Manchester, where its efficacy was even more severely tested. Upon this occasion, a building of two floors in Trafford-street, was prepared, by covering the rafters, joints, lintlets, &c., with the patent cement, in coats varying from one to three eighths of an inch in thickness, as well as the ladder staircases connecting the floors. The cement in the ceilings was laid upon common lath-work, and about one-fourth of the ceiling in each room was finished off with fine white plastering upon the cement. An immense crowd had assembled long before the hour appointed for "lighting up." The apartments having been filled with a mass of old furniture, wood, shavings, and other combustibles, the premises were carefully examined by the high-constables of Manchester, the conductors of the fire-engine departments, the police of that place, as also of Liverpool, Stockport, Bolton, &c., and many of the leading manufacturers, all of whom appeared extremely anxious to ascertain the result. Soon after three o'clock, the materials on the ground-floor were set on fire, and the flames spread rapidly round the room, enveloping the step-ladders, &c., and consuming the window-frames—they being unprotected by the cement. The fire in the ground-floor having expended its fury in about half-an-hour, without having communicated to the apartments above, the incendiary ascended a ladder and ignited their contents. On after examination by the local authorities, it was found, as in former experiments, that while the common plastering was partially destroyed, and all the unprotected wood-work utterly consumed, access to flame had been most effectually prevented, and all the rafters, joints, &c., covered with the cement, were entirely preserved. In a certificate signed by the principal witnesses of this experiment, it is stated that "the quantity of fuel used was excessive, and the flames exceedingly fierce and of long duration." It seems to be an almost universal opinion, that this cement is one of the most effectual preventives to the spread of fire ever offered to the public. The patent has been purchased by a Fire-Preventive Company, who have established a manufactory in Upper-ground-street, Blackfriars-road, from whence builders and others will shortly be enabled to obtain any quantity of this highly useful and eligible material

On the 12th of June I had the gratification of receiving a silver medal from the Society of Arts, for my invention of *a portable dam or cistern*, now in constant use at fires in London, and in other parts of England. Although, intrinsically considered, this may be thought a poor equivalent for sixteen years' unwearied perseverance, and an outlay of upwards of one hundred pounds—yet, being the only acknowledgement I have ever received, I consider myself somewhat fortunate in obtaining this mark of the Society's opinion of my invention.

When the London Fire-engine Establishment first came into operation, they for the most part adopted the then existing engine-stations, so far as they coincided with their plans, and among these was the engine-house of the Westminster Insurance Company, in Bedfordbury. Being a very confined, and otherwise inconvenient situation, the committee sought for some more suitable spot, and at length obtained a piece of ground near the end of Bedfordbury, in Chandos-street, upon which they erected the building, a sketch of which accompanies this communication, from the design of E. H. Browne, Esq., architect. The elevation is designed in the Roman style of architecture, and is divided into two compartments; the lower one consists of a rusticated archway, comprising the ground and mezzonine floors. The upper compartment is composed of two Roman Doric pilasters, with a bold projecting cornice, the metopes forming the windows of the attic story. The first, second, and attic stories are comprised in this division of the elevation. The extremely irregular form of the ground, and its very contracted space, were great obstacles to the formation of a perfect plan, and materially injured the effect the elevation would otherwise possess. Surrounded by these impediments, the architect has done much better than could have been expected, and the building is altogether highly creditable to his taste and skill. In describing the new engine-station in Farringdon-street,* (also by E. H. Browne, Esq.) I noticed the general employment of slate for stairs, &c.; upon the present occasion the architect has again judiciously adopted this material for the whole of the sinks, cisterns &c., which is found to answer extremely well; such of the stairs as could be inserted into the walls, are slate throughout.

Three engines, with a suitable complement of men, are attached to the Chandos-street Station, and when its central position in the populous parish of St. Martin's—its proximity to the Theatres and to the Government Offices, are considered, it will be seen to be a station of very great utility.

In consequence of the unlimited amount of manual power which can always be obtained for money, at every fire in the metropolis, the committee of

*Vide volume 24, page 364.

managers have had several of the Establishment engines increased in size—the six-inch working barrels being exchanged for seven-inch; by which means a larger stream, and much greater quantity of water, is delivered in a given time. This alteration has been found most essentially useful at several large fires, and it is, I believe, the intention to adopt seven-inch barrels generally in future.

In conclusion I have only to observe, that the exertions of the firemen continue to be crowned with such uniform success, as greatly to astonish those persons who are unacquainted wit the admirable system upon which their operations are conducted. In the promptness of their attendance, and in the efficiency of their services, they are altogether unrivalled. During the severe weather of January and February, 1838, the firemen cheerfully encountered all the harrassing duties of that extraordinary period. Their energies seemed to increase with the exigency of the time : fire after fire—watching after watching—required their constant attendance, and many of the men for days never laid down, except in their accoutrements. The uniform exemplary conduct of the Brigade upon this trying occasion, drew forth the warmest encomiums of the public press. On the 12th of February, the Committee of Managers of the London Fire Establishment passed the following resolution, to show the high sense they entertained of the exertions of their servants :

"The Committee have observed with the greatest satisfaction, that during the arduous duty occasioned by the late numerous and extensive fires, the foremen, engineers, and firemen have all, severally, not only done their duty, but with an alacrity and zeal that entitles them to the approbation of the Committee for the same."

Under the protection of the present highly efficient fire-establishment, the comparative security of lives and property is so much increased, as to be a subject of general remark. In a humorous ballad published a few weeks since, entitled "*the lament of the penny-a-liners*," the change effected in this state of things is thus alluded to :—

"Those beggars of the 'Fire-Brigade'
  Involve us in deep cares ;
Confine the flames to one small house,
  Instead of streets and squares !"

Where *all* have acquitted themselves so deservingly, it would be invidious to particularize, but there is one subject upon which I must add a few words, and in some measure deviate from this generally wholesome rule. During the year, it is shewn that fatal fires have occurred and several lives have been lost, but, upon no occasion has a single person perished *after the arrival of the firemen ;* and when it is known how rapid their attendance has ever been, it goes to clear them from the slightest imputation of blame. Nor is this all, upon several occasions the firemen have been the happy instruments in the hand of Providence, of rescuing persons who, but for their timely aid, would have swelled the fatal list of those who perished. I particularly mention these occurrences, because the firemen have over and over again been charged with exclusive attention to *property*, and with neglecting the preservation of *life*.

On Sunday evening, February 11th, a poor woman named Martin, residing in Well's-mews, Oxford-street, having rushed into her burning apartment in search of her child, which had been previously extricated, the door closed upon her, and she sunk, overpowered by the heat and smoke, and would inevitably have perished but for the intrepid conduct of Dowding of the Fire-brigade, who burst open the door and extricated her; her clothes had taken fire, and she was much scorched; the contents of the room were destroyed, but the conflagration was prevented from extending farther.

On the 12th of January, a fire broke out in the ale and wine cellars of Messrs. Collyer and Wilson, in Leman-street, Goodman's-fields. On the firemen from the Wellclose-square station reaching the spot, they found the cellar on fire and filled with dense smoke, which prevented entrance. On being told that a man was in the cellar, however, Thomas Overton and Thomas Loader instantly entered, and after twice groping round on their hands and knees, they at length found the object of their search, apparently dead. Having got him out, he was conveyed to the nearest surgeon ; after two hours judicious treatment, signs of life appeared, and he was subsequently removed to his own residence

in Lambeth-street, and eventually recovered.

Soon after five o'clock in the morning of Tuesday, February 6th, a fire broke out at the Brown Bear public-house, in Leman-street, when T. Loader again signalised himself. Mr. Flack, the landlord, writes as follows: — " Independently of the great activity displayed generally by Mr. Loader during the progress of the fire, too much praise cannot be given him for the promptitude which he exhibited in procuring the ladders whereby the inmates of my house were enabled to escape, when, but for his assistance, they would most probably have all perished; for in consequence of the rapidity with which the flames spread, all retreat was cut off from the lower part of the house, when Loader having brought the ladders, placed them against the wall of a cooperage situate at the rear of my house, and proceeded at the imminent risk of his own life, in rescuing *nine persons*, and was thus the means of saving the lives of so many of his fellow-creatures who, but for his timely aid, would have perished."

On the 30th of November, Thomas Loader was again instrumental in saving life at a fire which broke out in the basement story of a house in Shelton-court, Chandos-street. On the firemen's arrival, the smoke and heat was so great as to render entrance by the staircase impossible. Loader climbed up to the first floor window, and was about to force an entrance, when he heard faint cries from the room above. Four lengths of ladder having been raised, Loader ascended to the second floor, but found it impossible to enter for the smoke; on leaning over the window-sill and groping about, he got hold of a boy, nine years of age, named Wootton, whom he handed to Mr. Mallett, who was with him on the ladder. In a few seconds more the child would have been suffocated.

Loader and Overton were presented with medals at the meeting of the Society for Preventing Loss of Life by Fire*, in April last, for their praiseworthy conduct.

* The proceedings of this Society have been designated by some portions of the periodical press, as the most abominable fraud ever practised upon the British public. By the Society's last report, it appears that of £730 collected last year, there has

Good soldiers and good officers are alike necessary to the formation of an effective army. The numerous good qualities which shine so conspicuously in Mr. Braidwood, the Superintendant, seem to pervade his little corps—little as compared with the magnitude of their duties. Foremen, engineers, and firemen, zealously join heart and hand in the fearless execution of all those hazardous offices which impending dangers render needful. Their unanimity and friendly co-operation is one of the principal secrets of their uniform success. It is but an act of justice to notice here, the valuable assistance that is continually afforded by the firemen of the West of England Insurance Company, under their foreman, Mr. Conorton. Upon three hundred and eighty occasions no engines or firemen have been present but those of the Insurance Companies.

Several of the firemen have encountered severe falls, with many hard knocks, but no fatal injury has been sustained. Courage without conduct is always reprehensible, but with the firemen discretion is not only the better part of, but also the very essence of valour. They are intrepid without being too venturesome, and brave without being fool-hardy; the consequence is, that amid the war of elements and the crash of matter, they escape comparatively unscathed.

" To Braidwood, the King of the plug and the hose,
    Let's fill up again and again ;
To the Fire-Brigade, the fire-king's foes,
    Let us also our cup of thanks drain—
May they all, whilst the tide of old Thames fairly flows,
    Be ever found right in the main."

I remain, Sir.
        Yours respectfully,
                W. BADDELEY.

Wellington-street, Blackfriars-road,
    January 30th, 1839.

been expended upon the legitimate objects of the Society, such as the purchase and maintenance of fire-escapes, the pay of conductors, medals, and rewards, &c. £235. While, to keep up the delusion in the public mind, and remunerate sufficiently those who live upon this speculation, has taken no less a sum than £400 ! So that the public are willingly cajoled into paying, for a thinly-scattered and very inefficient species of protection, a sum of money adequate to provide a nightly efficient safeguard against loss of life by fire throughout the whole of this metropolis. Three years have elapsed since the formation of this Society, but up to this time not one life has yet been saved by the Society's instrumentality.

ON PHANTASMAGORIA, OR ARTIFICIAL AND NATURAL MAGIC.
BY HENRY D——.

Black spirits and white,
    Red spirits and grey ;
Mingle, mingle, mingle,
    You that mingle may.
                MACBETH, Act IV.

Phantasmagoria is a name derived from two Greek works, φαντασμα, phantasma, a phantom ; and γωσιαω, göriaö, tolaugh at. This term is principally applied to those superior exhibitions with a powerful magic lantern, which realise to the fancy its most fruitful and fantastic visions of the world of spirits.

The introduction of scenic shadows to mimic life, and to illustrate the poetry of olden times, may be traced to very remote period of history. For example, the Japanese of our day, in their dramatic representations, though they sometimes employ living actors, yet in what they call their wáyang, they substitute shadows; the origin of which must bear a very ancient date, for they are considered to be representations of history and fable referrable to an early period, and are traceable on very old monuments. Sir Thomas Stamford Raffles, in his History of Java, 1817, says ;—" The different characters in the history are in these wáyangs represented by figures, about eighteen inches or two feet high, stamped or cut out of pieces of thick leather, generally of buffalo's hide, which are painted and gilt with great care and at considerable expense, so as to form some supposed resemblance of the character to the individual intended to be personified. The whole figure is, however, strangely distorted and grotesque ; the nose, in particular, being unnaturally prominent." The character of these figures will be better understood by the following representation of one of them. After noticing their probable great antiquity, he proceeds to describe the mode of using them. "These figures are fastened upon a horn-spike a, and have a piece of thin horn hanging from each hand, b b; by means of which the arms, which are jointed at the elbow and shoulder, can be moved at the discretion of the manager. A white cloth or curtain is then drawn right over an oblong frame of ten or twelve feet long and five feet high, and, being placed in front of the specta tors, is rendered transparent by means of a hanging lamp behind it. The se-

veral figures are made, in turn, to appear and act their parts. Previous to the commencement of this performance, the *Dálang*, who is seated behind the curtain, arranges the different characters on each side of the curtain, by sticking them into a long plantain stem which is laid along the bottom. The *gámelan* then commences; and as the several characters present themselves, extracts of the history are repeated, and the dialogue is carried on, generally at the discretion and by the invention of the *Dálang*. Without this personage nothing can be done; for he not only puts the puppets in motion, but repeats their parts; interspersing them with detached verses from the romance illustrative of the story, and descriptive of the qualities of the different heroes." In another branch of these performances, being

poems founded on ancient fabulous history—" in the course of the entertainment, all the varieties of ancient weapons named in these poems are represented behind the transparent curtain." The exhibitions afford the common people exquisite delight.

We may, many of us, recollect having seen somewhat similar representations at home. About Christmas, when every town is crowded with exhibitions, these migratory shows occasionally form a part of the general stock of amusement. The figures throw a black shadow on a screen of stretched linen, and appear to dance or walk, and perform other feats at the will of the exhibitor; who generally accompanies these ingenious tricks with some comic song.

But we have hitherto been recounting the merits of only opaque and dark figures. We shall now proceed to show how readily ignorance and fear may be induced to conjure up the grossest absurdities respecting the most natural, pleasing, and scientific effects resulting from optical skill.

That division of optical science called *Dioptrics*, is the doctrine of refracted vision: it explains the effects of light refracted by passing through different media, whether air, water, glass, or similar transparent substances. On the subject of Light, Sir David Brewster says:—" When light falls upon any body whatever, part of it is reflected or driven back, and part of it enters the body, and is either lost within it or transmitted through it. When the body is bright and well polished like *silver*, a great part of the light is reflected, and the remainder lost within the silver, which can transmit light only when hammered out into the thinnest film. When the body is transparent, like *glass* or *water*, almost all the light is transmitted, and only a small part of it reflected.

" The light which is driven back from bodies is reflected according to particular laws, the consideration of which forms that branch of optics called *Catoptrics*; and the light which is transmitted through transparent bodies is transmitted according to particular laws, the consideration of which constitutes the subject of *Dioptrics*."

Among other optical instruments which a knowledge of dioptrics will enable us, most satisfactorily, to explain, ranks Phantasmagoric Apparatus; of which the chief is the Magic Lantern, an invention of the celebrated Athanasius Kircher *, about the middle of the seventeenth century. Its name is a sufficient commentary on the opinion which the vulgar would form on its apparent ability to raise and put in action a variety of spectral appearances. These instruments are made of various powers, and to exhibit in different ways. The commonest will throw a large circular field of light on a white wall : and within this enlightened space the coloured objects are represented. In a better kind of lantern, the figures only are illuminated, all the rest appearing dark : thus heightening the magical effect, but still the lantern and its light are not hid. The most improved method is, for the operator to be on one side of a semi-transparent screen of varnished muslin, and the spectators on the opposite side; which can only be effected by partitioning off a portion of the room, or fixing the screen within the doorway. The only difference between the magic lantern and the solar microscope is, that in the latter, as its name implies, the sun affords the light, while in the magic lantern an argand lamp is employed. As the oxyhydrogen light can be applied to the microscope, so it can also be applied to illuminate the transparencies of a magic lantern; and this has been done with considerable advantage.

When proper care has been taken to fit up a chamber for the exhibition of the magic lantern, the instrument will be entirely out of view, and no light will escape from the one apartment into the other. The screen will probably be from five to seven feet square, composed of fine muslin or taffetas, carefully varnished, and entirely surrounded with a curtain, wooden partition, or wall. In one room the spectators await the exhibition; in the other, the operator has his lantern on wheels, placed on a long table or raised platform. The lantern is large, and contains an argand burner placed before a concave mirror on one side, and before a wide projecting horizontal tube on the other. At top it has a bent tubular chimney. Within the horizontal tube are placed two illuminat-

ing lenses which receive the direct light of the lamp as well as the reflected light of the mirror just mentioned, the light thus condensed is thrown immediately on the painted glass slides, which now enter the tube at a distance a little before the anterior focus of one or two magnifying lenses. · It can now be set in motion on its carriage and platform, when, according as it approaches to, or recedes from the screen the figures will be proportionably magnified or diminished. By a simple mechanical contrivance, the outermost or magnifying lens is made to adjust itself according to the required distance from the screen to render the objects well defined, and by another contrivance a lesser screen is let down over the lens to lessen the intensity of the light as the body is rendered diminutive, the effect of the lens being to render it brighter, which, if not thus guarded against, would destroy the optical illusion of great distance, for the representations so displayed, actually appear to approach or to fly off to a considerable distance as the case may be.

In an exhibition of this remarkably ingenious optical instrument, M. Robertson, of Paris, contrived to introduce the living figure into the picture by strongly illuminating the person. But it is said, that the appearance was more natural when only the shadow of the person or other living object was thrown on the screen.

A knowledge of these various, and curious methods of managing optical deceptions at once exposes the influence which similar exhibitions would naturally have on rude, unenlightened and superstitious people. The machinery of ancient necromancy must have been very imperfect, but on the other hand, the reign of ignorance was spread sufficiently wide to give efficacy to an operation shrouded with all the circumstances of mystery, which the magician's dress, his personal appearance, his incense, his wand, his magic circle, and his incantations could concentrate into the absurd ceremony of exorcising spirits. Much curious deception may be made to result from the use of combined plane mirrors. Among the ancients these mirrors were usually formed of flat burnished silver, or other metal. A combination of two plain mirrors, forms what is called the Magician's Mirror from the singular

* Born in 1601.

effect produced when two persons look into it, each seeing the other's and not his own features. However, the most powerful instrument is the concave mirror, the surface of which should be perfectly elliptical; with this the image of the spectator may be made to appear in the air. It may also be used to display the image of some other object that can be illuminated by throwing upon it the rays of a strong light. In this way a visionary skull is often shown floating in the air. Another common and favourite exhibition is the *magic dagger*, which may be made to appear to strike at the spectator's breast, and therefore, seldom fails to excite the utmost astonishment.

Sir David Brewster observes respecting this caster kind of optical exhibition :

" When the instruments of illusion are themselves concealed,—when all extraneous lights but those which illuminate the real objects are excluded,—when the mirrors are large and well polished and truly formed,—the effect of the representation on ignorant minds is altogether overpowering, while even those who know the deception, and perfectly understand its principles, are not a little surprised at its effects. The inferiority in the effects of a common concave mirror to that of a well arranged exhibition, is greater even than that of a perspective picture hanging in an apartment, to the same picture exhibited under all the imposing accompaniments of a dioramic representation."

The same writer notices a phantasmagoric exhibition by M. Philipstal in 1802, in which thunder and lightning with figures of ghosts, skeletons, and known individuals were introduced. By means of combined sliders the eyes and mouths of some of the figures were set in motion. Many terrific figures advanced upon the spectators, enlarging as they approached, and then appeared to sink into the ground.

In an exhibition with the magic lantern performed by a Frenchman some time back in London, several very wonderful feats were produced. Among others, by means of gasses, or what he called dissolving lights, large figures and entire scenes underwent complete and perfect transformations ; thus, a winter piece, in which a redbreast was introduced and appeared to fly about the snow which seemed to envelop every object, gradually disappeared, the cottage showed its thatch, and the trees and earth their verdure. He produced the most perfect imitation of a spirit *walking* into the room, and strange as it may sound, he actually made the appearance of a skeleton with a dart in its uplifted arm stand before each spectator in a threatening attitude. At the time of this exhibition much discussion in the public journals was excited by the strangeness and perfection of the delusion.

The magic lantern, with every improvement that has been yet made upon it is still subject to some impediments. The small transparent pictures, however well painted, are but miniatures, which, as they become magnified, their defects are equally enlarged, and what was scarcely a speck on the glass, becomes an unsightly blot in the coloured shadow. A useful application of the magic lantern for scientific purposes when illustrative drawings are wanted, has lately been introduced, the only disadvantage, of which is the repeated darkening of the theatre or lecture room.

To obviate in some measure the defective magical effects of the lantern in producing optical phantasms, it has been proposed to introduce apparatus for *cata-dioptrical phantasmagoria ;* a combination of reflexion and refraction, using a large concave mirror and a powerful lens, with a screen all properly adjusted.

The illusions we have been considering are all artificially produced, it may be as well, therefore, before concluding, to allude to that curious natural phenomena the *Spectre of the Brocken.* The Brocken is the loftiest of the Hartz Mountains, in the kingdom of Hanover. The spectral appearances usually take place at sunrise. M. Haue, in 1797, visited this lofty mountain thirty times before his curiosity was gratified. Standing on the summit of the mountain, with the setting sun behind, and looking towards a distant eminence he saw a gigantic human figure. He found all his actions were imitated, as when he bent forwards or raised his arms, and calling another person to him, two colossal spectral figures were observed in the distance. But the most singular part was, that a *third* figure afterwards appeared, and all gave imitations to the actions made by the two astonished spectators.

Among other aërial phenomena we

have the *Fata Morgana*, or the *Castles of the Fairy Morgana*—the *Enchanted Coast* as seen by Scoresby in his voyage to Greenland—aërial figures of cows and horses have been seen in the air; as also of vessels, sometimes, therefore, called *Spectre Ships.* It is sufficient for our present purpose thus briefly to notice these natural phenomena, the account of which might easily have been extended. They could not be entirely passed over in treating on *phantasms,* which subject has now been considered first in the simplest form of dark unmeaning shadows, then by optical illusion made to take an apparent corporeal in spiritual body; and to conclude, we have shown that these spectral illusions are also prodigies of the material world, and that while they astonish, they should rather call forth our admiration, than excite mere vacant amazement and groundless alarm.

January, 1839.

ON THE EDUCATION OF MECHANICAL DENTISTS.

Sir,—I am an old subscriber and admirer of the *Mechanics' Magazine,* and I feel real regret when I see its pages desecrated and its usefulness marred by an attempt to convert it into a circulating medium of advertisement, under the specious colour of conveying useful information; yet never was such an attempt more glaringly apparent, or more successful, than in the paper by a Mr. Gray, dentist, in your 807th number. This puff is introduced under the flimsy covering of pretended information as to the best means of acquiring the anatomical and mechanical proficiency necessary for a dentist, such portion of the letter being pure and unadulterated *fudge;* but qualifications as needful as any; namely, the education and manners of a gentleman, are never once hinted at. How must the mass of respectable dentists in England, many of whom by education and *status* in society are fully entitled to rank as gentlemen, shrink abashed at the consciousness of not having had the honour of being reared as journeymen clockmakers!

But even in a mechanical point of view, what analogy, in the name of honest Sam Slick, is there between the work of a clock-maker and that of a dentist—unless it be, that one has the teeth of wheels to file; the other, of human beings—one, to construct pallets of escapement; the other palates of mastication? There is scarcely a single tool common to both, at least not without being at the same time common to half a dozen arts beside; and the very one pointed out by Mr. Gray as the *ne plus ultra* of perfection, the sculptor, (miscalled by him "the graver") peculiarly appertains to the dentist, perhaps to the jeweller, but certainly *not* to the clock-maker.

As to anatomical certificates it is notoriously a vulgar and common absurdity with dentists who have no other means of respectability, to sigh for the distinction of a diploma, that they may be thereby placed upon an equality with some transformed surgeons, who, unable to succeed in their own profession, have recourse to the *easily attempted* one of dentist (where want of cleverness is less quickly detected by the public),and who hold out their "M.R.C.S." as a universal certificate of excellence. I once knew one of these worthies, who, by way of bearing down all opposition, actually had the arms and motto of the College of Surgeons emblazoned on his powder boxes and scraping tools. Now the truth is, that so slight, in comparative importance, is the knowledge of the merely *local* anatomy necessary for a dentist (which any one might acquire in a day) compared with the practical judgment, which long experience and close observation concentrated upon a single subject, can alone confer, that medical men of real eminence invariably refer to the opinion of respectable dentists in all matters connected with the teeth.

But the gravamen of Mr. Gray's offence is his concluding puff, intermixed, as it is, with a gratuitous impertinence to the most eminent men in his profession. By what right does Mr. Gray attempt to stigmatise the construction of gold palates (the universal practice of first class dentists) as "botch work?" Is it because he, from inability to succeed in that most exquisite and beautiful branch of the art—or from the greater cheapness of the material preferred by him—or from a wish to excite attention —or from whatever other motive, may choose to confine himself to one mode of construction out of the many adopted by other dentists as expediency may demand, that he should, like the monkey in the fable, endeavour to cry down that which he himself is unable to exhibit? Unfortunately for really respectable dentists, there is no art of which the uninitiated are more incompetent judges than theirs, and this it is that renders it such a prolific field for puffing, otherwise I should dilate a little upon your bone practitioners; I should unfold the little expedients *to assist the atmosphere,* so easily detected and smiled at by practical men—the wedging between the teeth—the plug of soft wood—the small and almost invisible wire—and a dozen other artifices daily, and unblushingly resorted to by your *soi-disant* capillary attraction, atmospherical pressure, and no-ligature men; to say nothing of the absolute absurdity and humbug, in a philosophical sense, of attributing to "atmospherical pressure" an adherence, which, (when successful) is the result of a good fit, promoted by the clamminess of the saliva, and assisted by the muscles of the cheeks and by the constantly recurring support of the tongue and under jaw, a result well known to dentists, and taken advantage of by them for these thirty years past.

I have the honour to be, Sir,
Your very obedient servant,
ARTIFEX.

January 29, 1839.

ON THE SPEED OF CANAL BOATS.

Sir,—In my letter published in No. 555 of the *Mechanics' Magazine,* I tried to explain why, when a quick passage boat on the Paisley canal is sailing up to the speed which keeps it upon the top of the wave, there is then much less strain on the towing line than there is when the boat is sailing at a rate so slow as just to allow the swell to keep ahead of it; but since writing that letter another method of explaining part of the principle has presented itself to me. In No. 555 of your Magazine it is shown, that the whole length of the wave is shorter than the boat; this being the case it will be evident that when the boat is upon the top of the wave, which has a convex form, the draught of water near to the centre of its length, will be slightly increased, and in this way buoy up the

centre, so as to cause the draught of water fore and aft to be much less than it is when the boat is sailing at a slower rate with a swell ahead; for at any place near to the centre of the boat there is a much greater buoyancy in a given length than there is at the stern and bow. When a quick passage boat on a canal is upon the top of the wave, it is supported by the water, so as to pass through it, and be resisted less, than when sailing with a wave at its bow, or even when sailing on wide water at a speed equal to its best on a canal, and with the water nearly on a level all round its sides—although the water may buoy up or support the boat to the same extent in all these cases; for when sailing on the top of the wave the resistance caused by the action of the bow and stern upon the water, as also the friction of the water on these parts, will be greatly diminished, they being then a very small distance in the water, while the friction caused by the nearly parallel part of the boat passing through the water, will be but slightly increased; as a small increase of the draught of water at the middle part of the boat, where the bulk is great in a given length, will much diminish the draught fore and aft, where the boat has a small capacity in the same length.

A quick passage boat on a canal will be drawn at its best speed as easily as another boat of the same burden, but of a longer and consequently sharper shape, will be drawn at the same speed upon wide water; indeed the passenger boats upon the Paisley canal act much in the same way as longer and sharper shaped boats do upon wide water, for in each of these cases the resistance caused by the inclined parts acting on the water is very small, as the canal boat is mostly buoyed up or supported by the water at a part which runs nearly parallel to the direction in which it is sailing, and no part of the very long boat for wide water is much inclined to its course. From this it would appear that the best form for a quick passage canal boat will be different from that of a boat for sailing fast upon wide water.

The bow and stern of a canal boat are not so deep into the water when sailing at its best speed as when sailing at rather a quick rate but behind the swell, and on this account the swell is not so great in the former as in the latter case.

When a boat is buoyed up, or supported by the wave in the manner already described, it is a very different affair from skipping or rising to the surface of the water; for when sailing upon the wave the vessel is buoyed up at its most bulky part, which is nearly parallel, so as partly to lift its ends out of the water; but skipping upon the surface is caused by the action of the water upon the bow of the boat.

When a canal boat sails upon the top of the wave, less power must keep it in motion than is required when the boat is moving at a quick rate, but behind the swell; for in the first case there is in a manner only one wave formed during a long run, and the minute portion of its force continually destroyed by friction upon the banks of the canal or other resistance, is supplied by the action of the boat upon the water, whereas by moving slower than the swell a great many waves are in consequence formed in the same time; as the particles which form the wave have a motion in the same direction as that of the boat, this also will give some little advantage when the boat sails upon the wave.

A quick passage boat of the ordinary construction, if sailing upon wide water, cannot get advantage of the wave in the same manner that it does when sailing at its best speed upon a canal, for in a canal the wave is confined about the boat, and in wide water the waves are sent away towards the banks.

I am, Sir,
Yours truly,
JAMES WHITELAW.

Glasgow, Jan. 26, 1839.

GENERATION OF GASES IN THE
CASTING OF IRON.

Sir,—Having read in your Magazine, No. 799, an article upon the conversion of water into steam at high temperatures, by Dr. Schafhaeutl, I beg through the medium of your valuable Magazine to correct an inaccuracy of which the doctor has been guilty, relative to the casting of iron in damp moulds, and the gas evolved from them. The doctor states that cast iron may be run into wet moulds without danger, whereas, I know from experience, that such is by no means the case; for many serious ac-

cidents have occurred from this cause. In one instance which came under my knowledge, the upper box of a large casting was blown completely up to the roof of the foundry, and some of the beams set on fire by the melted iron which was spirted up by the formation of steam beneath it. The moulds are always dried by charcoal fires suspended over them, retaining merely sufficient dampness to bind together the sand. He also states that carburetted hydrogen is generated very slowly from the contact of the iron with the water; he is mistaken in supposing that it is carburetted hydrogen which is evolved. The iron combines with the oxygen of the water to form oxide of iron, and a large quantity of hydrogen is liberated, which is permitted to escape through a quantity of vents made in the moulds for that purpose, where it is enflamed, burning with a blueish flame, and emitting a smell of hydrogen wherever it escapes combustion. The small quantity of carburetted hydrogen which is evolved, may be regarded merely in the light of an impurity contained in the hydrogen; as, for every hundred parts of common cast iron which enter into the state of an oxide, there are but 7.066 of carbon set free, which, with perhaps a little carburetted hydrogen obtained from the coal dust with which the foundrymen colour their moulds, forms a very small quantity indeed compared with the gas derived from the water, which, so far from being generated slowly, escapes with very great violence, on which account the moulds are pierced in a great many places to permit of its free exit. With many apologies for having trespassed so long on your paper, I subscribe myself,

AN ENGINEER.

METHOD OF OBTAINING FAC-SIMILES OF
OBJECTS BY THE ACTION OF LIGHT.

Considerable interest has been excited amongst the lovers of the fine arts, by the announcement that a method had been invented by a M. Daguerre of Paris, for fixing the beautiful representations of objects obtained in the *Camera Obscura* upon a surface prepared in a certain manner, so as to form a perfect picture. By some, the discovery of this use of sunshine, was thought to be all moonshine, but we are happy to say that it appears likely to be turned to some practical purpose. We subjoin an extract from the *Constitutionnel*, giving an account of the reception the announcement of the discovery has met with at the *Academie*, and by the Parisian public.

" At the last sitting of the Academy of Sciences, M. Arago announced one of the most important discoveries in the fine arts that has distinguished the present century, the author of which has already acquired universal reputation by his miraculous diorama—M. Daguerre. It is well known that certain chemical substances of chlorate of silver, have the property of changing their colour by the mere contact of light ; and it is by a combination of this nature, that M. Daguerre has succeeded in fixing upon paper prepared with it the rays that are directed on the table of the Camera Obscura and rendering the optical tableau permanent. The exact representation of whatever objects this instrument is directed to is, as everybody is aware, thrown down with vivid colours upon the white prepared to receive them, and the rays of light that are thus reflected have the power of acting in the way above alluded to on chlorate of silver, or certain preparations of it. In this manner an exact representation of light and shade of whatever object may be wished to be viewed is obtained with the precise accuracy of nature herself, and it is stated to have all the softness of a fine aqua-tint engraving. M. Daguerre had made this discovery some years ago, but he had not then succeeded in making the alteration of colour permanent on the chemical substance. This main desideratum he has now accomplished, and in this manner has been able, among other instances, to make a permanent chemical representation of the Louvre, taken from the Pont des Arts. M. Arago, in commenting upon this most extraordinary discovery, observed, that a patent would be by no means able to preserve the rights of the discoverer sufficiently to reward him for his efforts ; and he therefore urged the propriety of an application being made to the legislature for a grant of public money as a recompense. M. Biot, on the same occasion, compared M. Daguerre's discovery to the retina of the eye, the objects being represented on one and the other surface with almost equal accuracy.

" What is the secret of the invention ? What is the substance endowed with such astonishing sensibility to the rays of light, that it not only penetrates itself with them, but preserves their impression ; performs at once the function of the eye and of the optic nerve—the material instrument of sensation, and sensation itself ? In good sooth we know nothing about it. Figure to yourself, says,

a Parisian contemporary, a mirror which after having received your image, gives you back you portrait, indelible as a picture, and a much more exact resemblance. Such is the miracle invented by M. Daguerre. His pictures do not reproduce colour, but only outline, the lights and shadows of the model. They are not paintings, they are drawings ; but drawings pushed to a degree of perfection that art never can reach.

" One has heard of writing by steam, but " drawing by sunshine" (or moonshine) is a novelty for which the world is indebted to M. Daguerre, of Paris, the diorama painter. M. Arago and M. Biot, who have made reports to the Academy of Sciences on the effect of M. Daguerre's discovery, have given up all attempts to define its causes. The complaisance of the inventor has permitted us to see these *chefs-d'œuvre*, where nature has delineated herself. At every picture placed before our eyes we were in admiration. What perfection of outline—what effects of *chiaro oscuro*—what delicacy—what finish ! But how can we be assured that this is not the work of a clever draughtsman ? As a sufficient answer, M. Daguerre puts a magnifying glass in our hand. We then see the minutest folds of the drapery, the lines of a landscape, invisible to the naked eye. In the mass of buildings accessories of all kinds, imperceptible accidents, of which the view of Paris from the Pont des Arts is composed, we distinguish the smallest details, we count the stones of the pavement, we see the moisture produced by rain, we read the sign of a shop. Every thread of the luminous tissue has passed from the object to the surface retaining it. The impression of the image takes place with greater or less rapidity, according to the intensity of the light ; it is produced quicker at noon than in the morning or evening, in a summer than in a winter. M. Daguerre has hitherto made his experiments only in Paris ; and in the most favourable circumstances they have always been too slow to obtain complete results, except on still or inanimate nature. Motion escapes him, or leaves only vague and uncertain traces. It may be presumed that the sun of Africa would give him instantaneous images of natural objects in full life and action."

The invention has since been the subject of discussion at the Royal Institution in London ; and we have been told that in some of the specimens exhibited, so perfect was the resemblance of the picture produced to the original, that the very threads and fibres of a person's garments were plainly shown.

The invention, if it produce results at all approaching to what has been stated, is a most important one ; and we are happy to be able to claim it as of English origin. The French must content themselves with the merit of applying it to practical purposes—no trifling honour in these utilitarian days.

To substantiate our claim we append a letter which we have received from Sir Anthony Carlisle, stating that he performed experiments to the same effect *forty years ago;* also a communication from our correspondent Mr. Oxley, claiming the credit of the idea for himself and others ; but the date of their experiments is far subsequent to those of Sir Anthony.

ON THE PRODUCTION OF REPRESENTATIONS OF OBJECTS BY THE ACTION OF LIGHT.

Sir,—At the evening meeting holden at the Royal Institution on Friday last, several specimens of shaded impressions were exhibited, produced by the new French Camera. The outlines, as well as the interior forms of the objects, were faintly pictured, and hence the application of this method of impressing accurate designs may become disregarded after public curiosity subsides.

Having, about forty years ago, made several experiments with my lamented friend, Mr. Thomas Wedgewood, to obtain and fix the shadows of objects by exposing the figures painted on glass, to fall upon a flat surface of shamoy leather wetted with nitrate of silver, and fixed in a case made for a stuffed bird, we obtained a temporary image or copy of the figure on the surface of the leather, which, however, was soon obscured by the effects of light. It would be serviceable to men of research if failing experiments were more often published, because the repetition of them would be thus prevented. The new method of depicting by a camera, promises to be valuable for obtaining exact representations of *fixed* and *still* objects, although at present they seem only to possess the correct *elements* for a finished drawing.

Few artists of competent skill addict themselves to drawing natural objects, although the value of such designs wholly depends on exactness. For anatomical purposes designs should be faithfully correct, and the new instrument and new method are well suited to those purposes.

Among the many splendid plates devoted to illustrate anatomy, none are so truly executed as those of Cheselden, which were taken by a Camera Obscura.

Your obliged reader,
ANTHONY CARLISLE.

Langham-place, Jan. 30, 1839.

REMARKS ON M. DAGUERRE'S CHEMICAL AND OPTICAL INVENTION FOR PRODUCING DRAWINGS FROM OBJECTS BY THE ACTION OF THE RAYS OF LIGHT.

Sir,—In No. 775 of your valuable Magazine for 16th June, 1838, I showed in the most ample manner how very difficult it is to determine the merits of the originality and priority of inventions and discoveries, and I now again respectfully beg leave to call your attention to this subject. I have just bought a copy of a paper containing an account of M. Daguerre's invention or discovery of a means of producing drawings by the action of light. I certainly congratulate that gentleman on being, as I believe, the first to bring such a project into full operation ; but I cannot compliment him on being, by many years, the first or earliest projector. This honour I believe is due to England and to Englishmen, and I claim it for myself and two other gentlemen, viz. Mr. John Turneau, a very excellent artist and eminent portrait painter, the inventor and patentee of the Liverpool lamps, so much celebrated before the introduction of gas lights ; the other, Egerton Smith, Esq., Editor and Proprietor of the *Liverpool Mercury*. I scarcely need say that both these gentlemen are highly esteemed by all that have the honour of their acquaintance, and universally respected, not only in Liverpool, but generally throughout Lancashire, for their integrity, talents, and urbanity. I was for upwards of eight years, during the whole of my residence in Liverpool, on the strictest terms of friendship and intimacy with them, and we have often met for the purpose of discussing various interesting inventions and discoveries. Mr. Turneau in 1823 or 1824, lent me Senefelder's large quarto work on the Discovery and Practice of Lithography, with which I was much amused and greatly interested ; in fact, I was wonderfully well pleased with Senefelder's discovery, which forcibly reminded me of a project I had conceived a few years before ; for having then a moderate knowledge of chemistry, I was well acquainted with the fact, that certain chemical preparations, especially those of silver, were subject to great changes of colour from the effect of the rays of light. I communicated my ideas on this subject to the gentlemen I have named, proposing that a camera obscura or similar optical means should be made to throw the images of the objects to be delineated upon paper saturated with the chemical preparation. Mr. Smith and Mr. Turneau both told me that they had long before I mentioned it, entertained ideas of the possibility of accomplishing such a thing ; and indeed, I have not the least reason to doubt them, for each of them may be, and is considered as being, a universal genius, familiar with every useful and curious invention. They observed to me, that as many chemical substances were acted upon by light, it might be attended with considerable trouble to find out which was the best to be used for this purpose. I myself proposed using a solution of nitrate of silver, or lunar caustic in water, and to prepare the paper by moistening it with this liquid, for I had many years before used this as a marking ink for linen, and had observed that if this liquid was exposed to the light it became black and less useful for marking with. I considered that the paper might be dried in the dark and kept close shut up until wanted, and that when it had been acted upon, and had received the desired impressions of objects by means of the camera obscura, it should be withdrawn therefrom in the dark, or as little as possible exposed to the light until it had been immersed in an alkaline solution, or a solution of muriate of soda, or saturated with spirit varnish, to fix the tints and prevent further change from the action of the light. As I had no camera obscura, I did not proceed with my intended experiments, and have ever since, from various circumstances, been prevented from following up my project, although Mr. Turneau and myself discussed the matter with so much earnestness as to argue which room in his house would be the best to place the camera, and to try the experiments in. Mr. Turneau then resided in Lord Street, Liverpool, and if he should see this communication it cannot fail to remind him of the whole affair. Thus you see, Mr. Editor, there are three Englishmen to a certainty, and perhaps more of our countrymen who projected making use of the rays of light to delineate drawings or likenesses of objects long before M. Daguerre.*

I am, Sir, yours, &c.
THOS. OXLEY, Teacher.

3, Elizabeth-place, Westminster-road,
February 2, 1839.

IMPROVED SAFETY-VALVE FOR
STATIONARY STEAM ENGINES.

Sir,—It is a remarkable fact, that whilst nearly every part of the steam-engine has been the subject of improvement—whilst we have had improved pistons, air-pumps, buckets, and valves, improved modes of generating steam, and improved modes of condensation—still that most important,

---

* We reserve a further communication of Mr. Oxley's for another occasion.—ED. M. M.

that most essential part of this mighty machine, namely, the safety-valve, remains in statu quo—still remains, in more senses than one, fixed upon its seat. In its most perfect form of construction, it is very little superior to that proposed by its inventor, Papin, whilst in many cases it is decidedly worse. It is very common in this part of the country to fix a safety-valve only two inches diameter upon a ten-horse boiler, and three inches upon a twenty-horse, and so on. Now, taking it for granted that these valves are free to act, (which in many cases they are not, on account of the stuffing-boxes with which some of them are provided), there is still an insufficiency of room for the escape of steam; from the circumstance that a pressure greatly superior to that with which the safety-valve is loaded is required to raise it any considerable distance from its seat. I never recollect to have seen a safety-valve raised one-eighth of an inch, although the steam may have been blowing off at a pressure considerably above that indicated. If we take one-eighth of an inch, for example, and multiply it by the circumferences of the valves whose respective diameters in inches are 2, 3 and 4, we shall have the several quantities ·7854, 1·178 and 1·568, being the respective areas, in inches, of the openings by which the steam is allowed to escape from a 10, 20 or 30-horse boiler, as the case may be. Pretty safety-valves, truly! But perhaps the worst form of safety-valve, and one which is coming into very general use, is that which is loaded by a spring in lieu of a weight; by this contrivance an additional force is operating to prevent the escape of the steam, at the very moment that the contrary should take place; or, in Hibernian phrase, "the more it wants to escape, the more it can't."

A safety-valve, to be worth the name, should allow the free escape of all the steam which can possibly be generated in the boiler to which it is attached, at the very moment it has acquired a certain degree of elasticity; and it was under this impression that the following plan suggested itself to me, which, though perhaps inapplicable to steam vessels and locomotives, might certainly be easily applied to all fixed boilers, whether high or low-pressure. Suppose A, B, C, D, (in the annexed figure) to represent a common low-pressure boiler, the water-level in which is indicated by the line E F; G H being the pipe by which it is supplied with water; of course, there will be an hydrostatic column in this pipe, the height of which will depend upon the elastic force of the confined steam; if this force be equal to 5lbs. upon the square inch, above the pressure of the atmosphere, then the column H I would be about 11 feet 6 inches in height, If we now suppose a small cistern, J, attached by means of the bended pipe, K, to the feed-pipe, G H, it is clear that when the steam had forced the water to the level, I, that it would stand the same

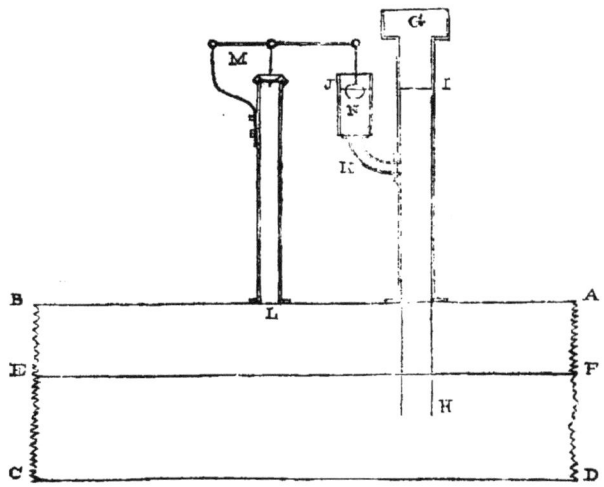

height in the cistern, K; it is equally clear, that if the safety-valve be mounted upon the pipe, L M, and loaded by means of a lever the weight attached to which was suspended in the cistern J, that the moment the pressure of the steam had raised the water into that cistern, a great part of the weight N would be supported by the water so raised, and consequently taken off the safety-valve; further, if this weight be cast hollow, so as to float in water, of course it would be taken off altogether—from the circumstance that the water would rise in the pipe G H, nearly 28 inches for every pound of pressure. It is evident, that a boiler fitted upon this plan would work at any assignable pressure, and yet not the smallest portion of steam escape from the safety-valve; whilst, if that pressure be exceeded only so much as the one-sixth or one-seventh of a pound upon the inch, the safety valve would be as effectually opened as if it were taken out altogether. If there should be an objection to the steam blowing off at (suppose) every time it ex-

ceeds the assigned pressure but a small degree, it is evident that it might be so loaded as to lift, before the water was forced as high as the weight N; whilst, if the valve should happen to stick upon its seat, (a very common case,) or there should be more steam generated than can escape under those circumstances, the increased pressure of steam would raise the water so as to surround the weight N, and consequently release the valve. Of course it is no more necessary that the safety-valve should be fixed so much above the boiler, than that the column, H I, should be water; a well-arranged system of levers in one case, and the substitution of mercury in the other, would answer all the ends required.

The admission of this communication into your valuable columns, (should you deem it worthy of the honour,) might, perhaps, afford a useful hint to *many*, and would certainly be a source of gratification to *one* of your constant readers.

J. R.

Todmorden, Dec. 17th, 1838.

THE MECHANICS' MAGAZINE METROPOLITAN RAILWAY MAP.

Sir,—The Metropolitan Railway Map, which forms the frontispiece to your xxixth volume, is undoubtedly a specimen of lithography, which reflects the highest credit on your artist, and goes far towards proving that some may prove a very formidable rival to copper and steel in the map de-

partment as well as others. As far as execution is concerned, there is no ground for complaint; but may I be permitted to hint that there is some deficiency in that still greater essential, correctness ? Trifling errors cannot well be avoided in a sheet comprising such a number of objects and their names : the greatest care probably would not suffice to expurgate such small and yet puzzling mistakes as the putting "Athyns" for "Albyns," "Borndon" for "Horndon," or "Penae-common" for "Penge-common;" but the faults I allude to are of a graver character, as occurring in a *Railway* map. In this it is of course pre-eminently necessary that some idea should be given of the face of the country, to which end the range, and, if possible, the elevation of the hills should be marked, and particular attention paid to the courses of the streams crossing or accompanying the lines of rail. This is done in some parts of the map in question, but not in others; so as to detract very considerably from the utility of the whole. The great range, or double range of hills in Kent and Surrey *is* shewn, so that the reason of the proposed deviation of the Dover Railway is made plain to every spectator. But on the other hand, Hertfordshire is delineated as having a perfectly flat surface, so that the windings of the Birmingham line appear quite causeless and inexplicable, while the perplexity in this quarter is still further increased by the omission of the rivers Gade and Bulbourne, whose valleys in reality determine the course both of the railway and the Grand Junction Canal. The Gade, as far as can be made out in the map, appears to rise in Cashiobury Park, instead of to the northward of Hemel Hempstead; and as to the Bulbourne, that has disappeared altogether. While mentioning streams, I may remark that one of them is made to flow no-where, another into the Paddington Canal, and a third into the Great Western Railway. This must have resulted from sheer carelessness, doubly mortifying in connexion with a work, which in other respects bears so many marks of patient pains-taking.

Is it quite correct to give other names to the various lines than those they are authorised to bear by Parliament? The practice, at any rate, may lead to confusion, to say the least of it. Besides, if the "Northern and Eastern Railway" is to be styled the "Cambridge," and the South Eastern Railway," the "Dover," why should not the "Eastern Counties," by a parity of reasoning, be called the "Norwich?" Let there be uniformity, at any rate; this is, however, a virtue but too little attended to throughout; for instance, the different "stations" are duly marked on some of the lines, while they are totally omitted upon others, and this without reference to their being in actual operation or not. As to the "Blackwall" and the "London Grand Junction" railways, they are left out altogether, although perhaps quite as likely to be com-

pleted as some that are inserted,—the "Thames Junction," to wit,—and what an omission for a "Metropolitan" map!

I take it for granted that the various lines are delineated correctly in the main, but there is certainly a very material error as to the terminus of one of them. The "Eastern Counties" is represented as entering London immediately north of the Mile-end-road, proceeding close to the north side of Whitechapel, and terminating near the end of Fenchurch and Leadenhall-streets. In reality, this line is intended to proceed much further to the north, through the *terra incognita* of Bethnal-green, to a terminus on the eastern side of Shoreditch. These are points which the artist might easily have ascertained, and which are by no means unimportant to notice. Your artist's line, for example, would be immeasurably more expensive than the real one, from the havoc it would make with valuable property, while he lands *his* railway travellers in a characteristically different part of the town from that which is to be the real site of the terminus. Again, he makes the "Cambridge" railway join the Eastern Counties' only at the depôt common to both. Now, it is understood that the former line will join the latter some miles from town (it is said, I think, at Angel-lane, Stratford) which, however, is on a different bank of the Lea from the works in progress at Tottenham). If the lithographer had shown all the hills, as he ought to have done, it would be plainly perceived that, according to *his* line, the Northern and Eastern (Cambridge) Company would not avoid the high ground about Clapton, which would require a tunnel or deep cuttings at a vast expense, the saving of which seems to have formed their principal inducement for abandoning their original plan, and agreeing to a junction with the Eastern Counties. Surely all information as to these matters might easily have been obtained in the proper quarters, and those errors avoided which now certainly form a blemish in a very beautiful work of art. It may not, even now, be too late to remove that blemish in some degree, by the insertion of these few corrections in the pages of the Magazine which has introduced the work in question to the world.

I remain, Sir, yours, respectfully,

January 30th, 1839.   H.

[Although this communication is rather severe upon ourselves (or rather our artist), and, as we shall show, unjustly so, we readily insert it from a respect for the spirit of usefulness in which it is written. To some of the minor errors noticed, we, on the part of our artist, must plead guilty; it is not a little, however, to his credit, that so sharp-sighted a critic as "H." has, with all his microscopic acumen, been able to discover so few amongst the thousands of names, and representations of roads, and rivers, and towns, and buildings, which cover to almost a needle's point, our superfic'al square foot of paper. From the "faults of a graver character," found by "H." we must defend ourselves. With regard to hills, Hertford is by no means a flat country, but its undulations are, for the most part, of so gradual a nature as to render their representation on a map, even of ten times the scale, almost impossible ; the abrupt ranges of hills of Kent and Surrey, are easily shewn. With regard, in the next place, to the junction of the Northern and Eastern with the Eastern Counties Railway, our correspondent will find plainly marked out on the map *in dotted lines*, the intended deviation of the Northern and Eastern railway at Tottenham, to the Eastern Counties' at Angel-lane, Stratford. These railways are shewn on the map as authorised by Parliament to be constructed (excepting the slight error as to terminus, noticed by "H.") ; and until an act is obtained, authorising the proposed deviation, we do not think we are authorised in treating it as a *de facto* railway. The very fact of "H's" being able to detect so trifling an error, as he mentions with regard to the terminus of the Eastern Counties Railway, upon a map of London occupying little more than a *superficial inch*, is a proof of the general correctness of the plan. The real errors and omissions pointed out shall be remedied in the next edition.—ED. M.M.

INSTITUTION OF CIVIL ENGINEERS.

[Abridgment of the Report for 1838]

The following are the officers elected for the ensuing year : — *President :* James Walker, F.R.S.L. and E.—*Vice Presidents:* W. Cubitt, F.R.S. ; Bryan Donkin, F.R S. ; Joshua Field, F.R.S. ; Henry R. Palmer, F.R.S.—*Other Members of the Council :* F. Bramah ; I. K. Brunel, F.R.S. ; J. Howell; J. Locke, F.R.S. ; G. Lowe, F.R.S. ; J. Macneill, F.R.S. ; M. A. Provis ; Major Robe, R.E. ; James Simpson ; R. Stephenson.—*Treasurer:* W. A. Hankey,—*Auditors :* W. Freeman ; Charles Manby. — *Secretary :* Thomas Webster, M.A. The general meeting of the institution was held on the 15th instant, when the annual report was read. The following are extracts from this document. After alluding to the progress and success of the institution, the council observe :—At the close of the preceding session, the council issued a list of subjects, to adequate communications on

which they would award premiums. The following communications were received : " An elaborate and beautiful set of drawings of the shield at the Thames Tunnel," from Mr. Brunel, and two sets of drawings of Huddart's rope machinery ; the one from Mr. Birch, the other from Mr. Dempsey. The council, feeling this communication and the invention of the shield were entitled to a high mark of approbation, determined on presenting Mr. Brunel with a silver medal, accompanied by a suitable record of the high sense entertained of the benefits conferred by him on the practice of the civil engineer. Feeling, also, that the beauty of the drawings justly merited some mark of approbation, they determined on presenting the draughtsman, Mr. Pinchback, with a bronze medal in testimony thereof. The communications by Mr. Dempsey and Mr. Birch, on Huddart's rope machinery, likewise called for some special mark of approbation on the part of the council. The liberality of Mr. Cotton, in throwing open to the institution the works of the late Captain Huddart, is fresh in the recollection of most present ; with that same liberality, he at once acceded to the wish of the council to allow any person to attend and make drawings of this celebrated rope machinery for the institution. Two young men availed themselves of this liberality, and, with great perseverance, measured and took drawings of this elaborate machinery. The council felt, that to have attempted to distinguish betwixt the merits of these two communications would have been both difficult and invidious ; they have, therefore, awarded a Telford medal, in silver, accompanied by books to the value of five guineas, both to Mr. Birch and Mr. Dempsey. On the other subjects, issued at the same time, the council have not yet received communications of adequate merit ; but they have the pleasure of announcing several to have been promised. These subjects have been again announced, with others, as prize subjects for the present session. But though the council received no communication in which the subject of steam was treated with the generality and comprehensiveness which they desired, they received the following on parts of this great subject, to each of which they awarded a silver medal : — "On the Effective Pressure of Steam in the Cornish Condensing Engine," by Thomas Wicksteed ; "On the Expansive Action of Steam in the Cylinder of some of the Cornish Engines," by W. J. Henwood ; and on the "Evaporation of Water in the Boilers of Steam Engines," by Josiah Parkes.—The council also awarded a silver medal to the communications of Lieut. Denison, "On the Strength of American Timber," and of Mr. Bramah "On the Strength of Cast Iron." The series of experiments by Lieut. Denison was undertaken by that officer when stationed abroad, with the view of establishing some proportion betwixt the strengths of different kinds of American timber, and affording a means of comparing their strength with that of European. The communication by Mr. Bramah is also a valuable addition to our knowledge. These experiments, undertaken with the view of verifying the principles assumed in the work of Tredgold on cast iron, surpass every other series in number, and in the care taken to insure accuracy, since two similar specimens of each beam were subjected to trial. The principles, with the view of establishing which these experiments were undertaken, are, that within the elastic limit the forces of compression and extension are equal; and that, consequently, a triangular beam, provided it be not loaded beyond that limit, will have the same amount of deflection, whether the base or apex be uppermost; and a flanged beam the same deflection, whether the flange be at the top or the bottom. This communication is accompanied by some observations by Mr. A. H. Renton, pointing out the agreement which subsists between the experiments and results of the formulæ of Tredgold. The Council have peculiar pleasure in pointing out the two preceding communications as of a kind on which they conceive the Telford Premiums may be worthily bestowed.—A silver medal has also been awarded to Mr. Green, for his communication "On the Canal Lifts on the Grand Western Canal;"

to Mr. Harrison for his communication "On the Drops on the Stanhope and Tyne Railway;" and to Josiah Richards for his elaborate Drawing of the Rhymney Ironworks. The council have also awarded a silver medal to Francis Wishaw for his "History of Westminster Bridge." The institution received, during last session, from Mr. Rendel, an elaborate and beautiful set of drawings, accompanied by a suitable description of the Torpoint Floating Bridge. The council felt that in awarding a silver medal to Mr. Rendel, accompanied by a suitable record of the sense entertained of the benefit conferred by him on the inland communication of the country, this, the highest acknowledgement in their power to make, is most amply merited. A bronze medal has been awarded to Mr. Ballard for his ice-boat, and a description of his method of breaking ice by forcing it upwards. This simple method is applicable at about one-third the labour of the ordinary ice-boat. A bronze medal has also been awarded to Thomas M. Smith for his drawing and account of Edwards's, or the Pont-y-tu Prydd Bridge, in South Wales. Mr. Smith being in the neighbourhood of this bridge, availed himself of the opportunity to make accurate drawings of its curious and interesting structure. The council have also awarded five guineas to Mr. Guy for his method of making perfect spheres. This great desideratum in the mechanical art has been in a great measure supplied by the ingenuity of this individual, and a simple method furnished of readily producing very accurate spheres of metal or other hard substances.

ON THE CHEMICAL EXAMINATION OF GUNPOWDER.

I have treated five different samples: 1, The government powder made at Waltham Abbey; 2, Glass gunpowder made by John Hall, Dartford; 3, The treble strong gunpowder of Charles Lawrence and Son; 4, The Dartford gunpowder of Pigou and Wilks; 5, Superfine treble strong sporting gunpowder of Curtis and Harvey. The first is coarse-grained, the others are all of considerable fineness. The specific gravity of each was taken in oil of turpentine: that of the first and last three was exactly the same, being one 1·80; that of the second was 1·793, all being reduced to water as unity.

The above density for specimen first may be calculated thus:

75 parts of nitre, specific gravity = 2·000
15 parts of charcoal, specific gr. = 1·154
10 parts of sulphur, specific gr. = 2·000

The volume of these constituents is 55·5, (the volume of their weight of water being

100;) by which if their weight 100 be divided, the quotient is 1·80.

The specific gravity of the first and second of the above powders, including the interstices of their grains, after being well shaken down in a phial, is 1·02. This is a curious result, as the size of the grains is extremely different. That of Pigou and Wilks similarly tried is only 0·99; that of the Battle powder is 1·03; and that of Curtis and Harvey is nearly 1·05. Gunpowders thus appear to have nearly the same weight as water, under an equal bulk; so that an imperial gallon will hold from 10 pounds to 10 pounds and a half as above shown.

The quantities of water which 100 grains of each part with on a steam bath, and absorb when placed for 24 hours under a moistened receiver standing in water, are as follows:—

100 grains of Waltham Abbey, lose 1·1 by steam heat, gain 0·8 over water.

of Hall	0·5	2·2
Lawrence	1·0	1·1
Pigou and Wilks	0·6	2·2
Curtis and Harvey	0·9	1·7

Thus we perceive that the large-grained government powder resists the hygrometric influence better than the others ; among which, however, Lawrence's ranks nearly as high. These two are therefore relatively the best keeping gunpowders of the series.

The process most commonly practised in the analysis of gunpowder seems to be tolerably exact. The nitre is first separated by hot distilled water, evaporated and weighed. A minute loss of salt may be counted on, from its known volatility with boiling water. I have evaporated always on a steam bath. It is probable that a small portion of the lighter and looser constituent of gunpowder, the carbon, flies off in the operations of corning and dusting. Hence, analysis may show a small deficit of charcoal below the synthetic proportions originally mixed. The residuum of charcoal and sulphur left on the double filter-paper, being well dried by the heat of ordinary steam, was estimated, as usual, by the difference of weight of the inner and outer papers. The residuum was cleared off into a platina capsule with a tooth brush, and digested in a dilute solution of potash at a boiling temperature. Three parts of potash are fully sufficient to dissolve out one of sulphur. When the above solution is thrown on a filter, and washed first with a very dilute solution of potash boiling hot, then with boiling water, and afterwards dried, the carbon will remain ; the weight of which deducted from that of the mixed powder, will show the amount of sulphur.

I have tried many other modes of estimating the sulphur in gunpowder more directly, but with little satisfaction in the results. When a platina capsule, containing gunpowder spread on its bottom, is floated in oil heated to 400° Fahrenheit, a brisk exhalation of sulphur fumes rises, but, at the end of several hours, the loss does not amount to more than one-half of the sulphur present.

The mixed residuum of charcoal and sulphur digested in hot oil of turpentine, gives up the sulphur readily ; but to separate again the last portions of the oil from the charcoal or sulphur, requires the aid of alcohol.

When gunpowder is digested with chlorate of potash and dilute muriatic acid, at a moderate heat in a retort, the sulphur is acidified ; but this process is disagreeable and slow, and consumes much chlorate. The resulting sulphuric acid being tested by nitrate of baryta, indicates of course the quantity of sulphur in the gunpowder. A curious fact occurred to me in this experiment. After the sulphur and charcoal of the gunpowder had been quite acidified, I poured some solution of the baryta salt into the mixture, but no cloud of sulphate ensued. On evaporating to dryness, however, and redissolving, the nitrate of baryta became effective, and enabled me to estimate the sulphuric acid generated ; which was of course 10 for every 4 of the sulphur.

The acidification of the sulphur by nitric or nitro-muriatic acid is likewise a slow and unpleasant operation.

By digesting gunpowder with potash water, so as to convert its sulphur into a sulphuret, mixing this with nitre in great excess, drying and igniting, I had hoped to convert the sulphur readily into sulphuric acid ; but on treating the fused mass with dilute nitric acid, more or less sulphurous acid was exhaled. This occurred even though chlorate of potash had been mixed with the nitre to aid the oxygenation.

The following are the results of my analyses, conducted by the first described method :

100 grains afford, of	Nitre.	Charcoal.	Sulphur.	Water.
Waltham Abbey	74·5	14·4	10·0	1·1
Hall, Dartford	76·2	14·0	9·0	0·5 loss 0·3
Pigou and Wilks	77·4	13·5	8·5	0·6
Curtis and Harvey	76·7	12·5	9·0	1·1 loss 0·7
Battle Gunpowder	77·0	13·5	8·0	0·8 loss 0·7

It is probable, for reasons already assigned, that the proportions mixed by the manufacturers may differ slightly from the above.

The English sporting gunpowders have long been an object of desire and emulation in France. Their great superiority for fowling pieces over the product of the French national manufactories, is indisputable. Unwilling to ascribe this superiority to any genuine cause, M. Vergnaud, captain of French artillery, in a little work on fulminating powders, lately published, asserts positively, that the English manufacturers of ‘ poudre de chasse’ are guilty of the ‘ charlatanisme’ of mixing fulminating mercury with it. To determine what truth was in this allegation, with regard at least to the above five celebrated gunpowders, I made the following experiments:

One grain of fulminating mercury, in crystalline particles, was mixed in water with 200 grains of the Waltham Abbey gunpowder, and the mixture was digested over a lamp with a very little muriatic acid. The filtered liquid gave manifest indications of the corrosive sublimate, into which fulminating mercury is instantly convertible by muriatic acid; for copper was quicksilvered by it: potash caused a white cloud in it that became yellow, and sulphuretted hydrogen gas separated a dirty yellow white precipitate of bisulphuret of mercury. When the Waltham Abbey powder was treated alone with dilute muriatic acid, no effect whatever was produced upon the filtered liquid by the sulphuretted hydrogen gas.

200 grains of each of the above sporting gunpowders were treated precisely in the same way, but no trace of mercury was obtained by the severest tests. Since by this process there is no doubt but one 10,000th part of fulminating mercury could be detected, we may conclude that Captain Vergnand's charge is groundless. The superiority of our sporting gunpowders is due to the same cause as the superiority of our cotton fabrics—the care of our manufacturers in selecting the best materials, and their skill in combining them.—*Dr. Ure's Dictionary of Arts and Manufactures.*

NOTES AND NOTICES.

Kynaxing.—It is to be hoped that we shall have no more tampering with dry-rot docters and their nostrums, for the preservation of her Majesty's ships. The steeping of large logs of timber in solutions of any kind, is *perfectly useless;* the solution penetrates only skin-deep, whereas, the real dry-rot commences at the centre, where the fibres being the oldest, first gave way, as is the case in standing tree. The only plausible and promising preservative of timber is the gas of the Kerasote, procured from the distillation of coal or vegetable tar, which, when driven off in the shape of gas, will penetrate every part of the largest logs, and render the wood almost as hard as iron; so hard, indeed, as not easily to be worked. It is understood that in Belgium they are using it as blocks for the railroads. The worm (*teredo navalis*) as proved at Sheerness, will not touch it, while pieces of the same wood, *steeped in corrosive sublimate, sulphureous acid and other active solutions,* were bored through and through. Let our ships be built of good sound English oak, as they formerly were, well seasoned, under cover, and left on the stocks as long as they conveniently can be allowed, and we shall hear no more of dry-rot or wet either. The *Royal William* carried a flag, as guard ship, when nearly a century old; the *Sovereign of the Seas* was burnt when half a century old; and Anson's *Centurion,* always in service, was broken up when nearly forty years old; yet none of these—nor many hundred old ships—were steeped in any quack nostrums.—*Sir John Barrow's Life of Lord Anson.*

Preservation of Flour.—A very strong compression of flour, in rectangular moulds, is said, by M. Robineau, to preserve it both from damp and from insects. The bran must not be separated from it before it is pressed. A cake of flour, thus prepared, was placed by him in a very damp cellar, from which it was taken, at the end of six weeks, without any alteration. Another was put into some flour infected with insects, and after remaining there for eight days, it had acquired the unpleasant smell of the spoiled flour, which it retained for a long time, but the insects had not attacked it.—*Athenæum.*

Improved Saddle.—A saddler of the name of Collinson, an ingenious young man, resident in Burneston, near Bedale, has invented a saddle with an air seat. which, from its comfortable pliancy and elasticity, prevents much of the weariness and all the other usual unpleasant consequences, occasioned by long sitting on horseback. The invention will, in a little time, be fully made known to the public.

New Thrashing Machine.—On Tuesday last, a new hand thrashing machine was exhibited at the market, the first merit of which is its cheapness (its cost is only a few pounds), and its second, that it will thrash about four quarters of wheat daily; two men are employed, in turning it, and two others in feeding it and clearing the straw away. The machine, instead of diminishing manual labour, increases it, by giving employ to those labourers who cannot use the flail, and who in wet days could not be profitably occupied. It is very portable, and occupies but a small space. John Ellman, Esq., a few weeks back, introduced the first machine of this description into the county, and Mr. Cheal, the machinist, has made the one alluded to upon the same principle. As we are led to believe it to be of great utility to the farmer, there is little doubt that it will be extensively used.—*Sussex Express.*

Ericsson's Propeller.—On Tuesday last, Jan. 29, the R. F. Stockton, (described in *Mechanics' Magazine,* No. 707) towed the American packet ship, *Toronto,* from Blackwall to the lower point of Woolwich, a distance of three miles and a quarter, in 40 minutes, against the flood tide, then running from two to two and a half miles, thus towing her through the water at the rate of upwards of six miles an hour. The *Toronto* is 650 tons burden, she measures 32 feet beam, and drew at the time of the trial 16 feet 9 inches water; thus presenting a sectional area of more than 460 square feet. Now the fact of this body having been moved at a rate of upwards of six miles an hour, by a propeller, or piece of mechanism, measuring only 6 feet 4 inches in diameter, and occupying less than three feet in length, is one which scientifically considered, is interesting in the extreme, and in a practical or commercial point of view is of immense importance. We understand a company is about being formed to apply the propeller to a ship of 1,000 tons burden, to be employed in transatlantic navigation; and as her sailing qualities will not at all interfere with her steaming power, it is confidently anticipated that increased safety will be insured and her passage greatly accelerated, at a saving of at least one-half the fuel.—*Times.*

LONDON: Printed and Published for the Proprietor, by W. A. Robertson, at the Mechanics' Magazine Office, No. 6, Peterborough-court, Fleet-street.—Sold by A. & W. Galignani, Rue Vivienne, Paris.

Mechanics' Magazine,

MUSEUM, REGISTER, JOURNAL, AND GAZETTE.

No. 810.]　　　　SATURDAY, FEBRUARY 16, 1839.　　　　[Price 3*d*.

Printed and Published for the Proprietor, by W. A. Robertson, No. 6, Peterborough-court, Fleet-street.

MOTT'S AMERICAN PATENT PARLOUR STOVE.

Fig. 1.　　　　　　　　　　　　　　Fig. 2.

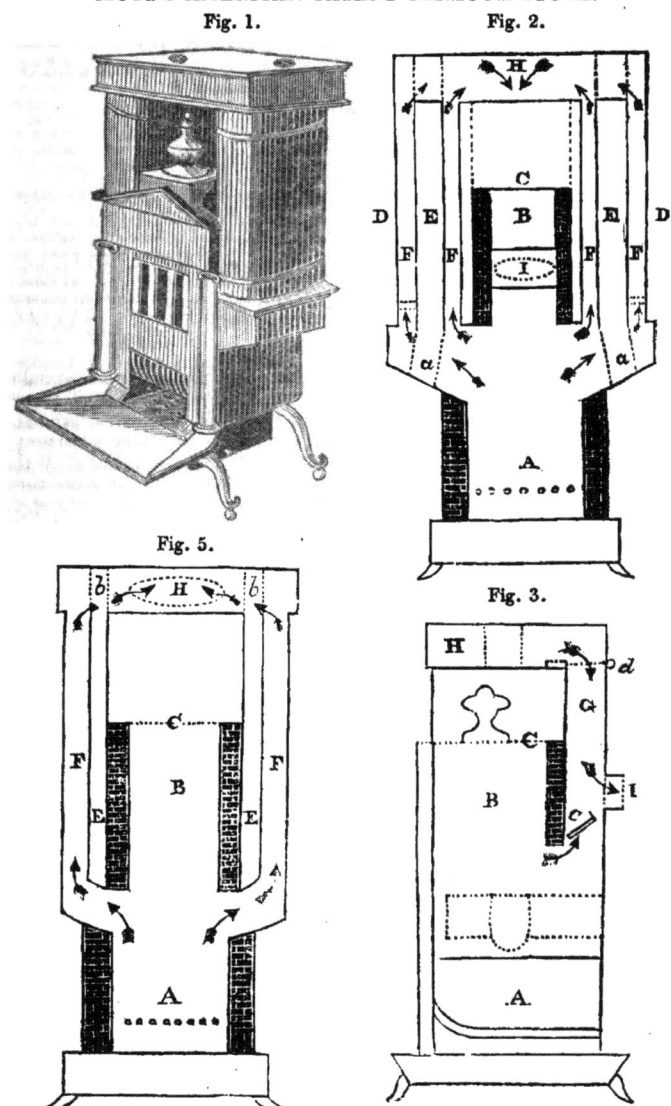

Fig. 5.

Fig. 3.

MOTT'S AMERICAN PATENT PARLOUR STOVE.

Mr. J. S. Mott of New York, lately obtained a patent in America for an improved parlour stove, which he states in a description thereof, published in the *Franklin Journal* for Nov. last, to be "furnished with air heaters, so constructed as to economize fuel, and to supply air of genial warmth, which is not deteriorated in its passage through the heating flues." The following is the description of this par lour stove, as set forth in Mr. Mott's specification :—

"Figure 1, in the accompanying engravings, is a perspective view of the stove ; fig. 2, is a vertical section through the middle thereof, parallel with its front; and, fig. 3, is a vertical section from front to back through the middle.

" In each of the figures, like parts are designated by the same letters of reference.

" A, is the grate or fire-chamber, which is surmounted by B, a reservoir for coal, which has a close fitting cover at C, allowing a considerable quantity of fuel to be supplied at once, and to burn out gradually ; D D are the combined flues and air-heaters, consisting of one cylinder, or rather oval tube, or chamber, within another, the innermost being the air-chamber and the space between it and the outermost, the smoke-flue ; E E, fig. 2, &c., are sections of the inner, or air-flues, and F F, the spaces surrounding them, and forming the smoke-flues. Air is admitted into the interior or air-flue at its lower end, through tubes or apertures leading into it, in any convenient way, as at the part represented by the dotted lines *a a*; these tubes or appertures may be extended, if preferred, so as to admit the air from without the room, but this will seldom be found necessary. The air which is heated in passing through these tubes, escapes into the room through openings at *b b*, in the top of the stove.

" The direction of the draught or passage for smoke and heated air from the fires, is represented by arrows. There are, as in many other stoves, two directions for the escape of the smoke, &c. ; one directly to the escape-pipe, and the other by a more circuitous route, which is to be used after the fuel has become perfectly ignited. There is a flat flue G, extending up from the fire-place to the chamber H, at the top of the stove ; in this flat flue there are two valves or dampers *c* and *d*, by which the draught is governed. L, is the escape-pipe for smoke, and into which it passes directly when *c* is opened and *d* closed ; but when these valves are reversed, the draught is carried through the flues surrounding the air tubes, and down the flue G, to the

Fig. 4.

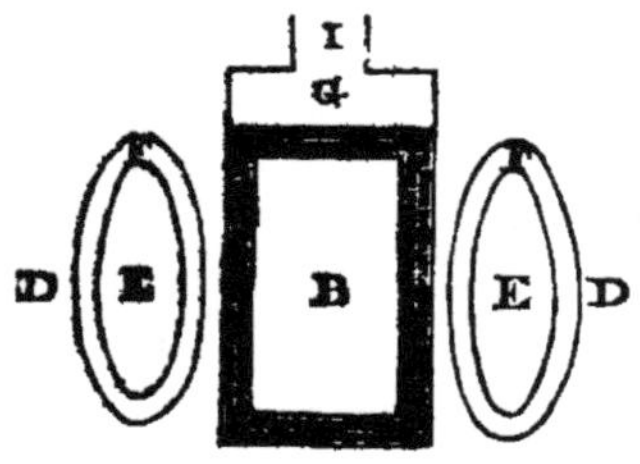

escape-pipe. Fig. 4 is a horizontal section of the stove at the level of the escape-pipe, the parts of which figure are designated by their proper letters of reference. By this arrangement of the flues and of the air tubes within them, the exterior or shell of the stove is directly heated by the heated air, and a free radiation takes place into the room. Fig. 5, shows a modification of this stove, in which the smoke-flues do not entirely surround the air-flues, the latter being heated on the side directly towards the reservoir B; but still, through the intermedium of the brick lining. In other respects, the construction of this stove is identical with that before described."

Mr. Mott claims as new " the air chambers, or tubes surrounded in whole or in part by the smoke-flues in the manner, and located as herein described, in combination with the side openings or apertures that connect the smoke-flue F F, with the fire-chamber ; whether said openings be under or above the surface of the fuel ; whether combined with a stove, such as that herein represented, or with one of any other construction to which such side flue, so arranged and combined, can be advantageously appended." No claim is made to air-tubes surrounded by the smoke-flues, when placed immediately above the fire-chamber.

GENERATION OF STEAM AND GASES IN CASTING IRON, &c.

Sir,—I learn from your valuable journal No. 809. that I have committed an inaccuracy in my remarks on the conversion of water into steam at high degrees of temperature, contained in your number 799; but it seems to me it is more in my manner of expressing myself, than in my allegation of facts, and I hope the kindness of your readers will readily excuse in me, as a foreigner, errors of such a nature.

I stated in the alleged remarks, that moulds in which copper is to be cast should be perfectly dry; and in continuation, said, the iron might be cast in *wet* moulds, because the water which the moulds contain is not so quickly converted into steam, but becomes slowly decomposed.

I perceive the word "*wet*" has here been the cause of the misunderstanding. I used the word *wet* as an antithesis to *perfectly dry*, and to imply, that the moulds in which copper is to be cast should not contain any water which might be evaporated by the usual mode of drying in copper foundries; and that the moulds for cast iron might contain water, because it became decomposed, and not instantaneously converted into steam.

Now, my respected opponent himself, admits that his so called dried moulds, contain water sufficient to bind together the sand, saying "the iron combines with the oxygen of the water," and, again, "a large quantity of hydrogen proves distinctly that water is decomposed." I can in no way ascertain how I am to describe a mould containing so much water as to render necessary a great number of holes to provide for the escape of a large quantity of hydrogen, and also to combine the particles of sand: I perceive I have made a mistake in using the word *wet*, but even the expression *damp* moulds would seem from the remarks of your correspondent, too strong; I shall therefore content myself by mentioning the fact, that many of our iron founders on the continent in rough casting of small things, such, for example, as laundress's box irons, &c., mould in sand sufficiently *wet* to admit of its adhereing together, and the moulds are filled with the melted iron without being previously dried.

I have myself witnessed several explosions in casting iron, but in all such cases, the moulds had been previously carefully dried, and the cause of the explosion appeared to the founders themselves unusual and inexplicable. If I remember rightly, I found, in the treatise on iron and steel, contained in two volumes of *Lardner's Cabinet Cyclopædia*, mention of a similar unexpected and unexplained explosion on a very large scale.

In breaking various moulds which had been carefully dried, I discovered in several of them near the inner surface several cavities, like blisters, which were filled with water, and I am therefore inclined to ascribe such explosions more to the collection of a mass of water in this manner, than to water which is retained by the capillarity of the sand.

In reference to the chemical composition of the gas* developed, I will observe, that I only once analyzed the gas escaping from such a mould, and burned it in Volta's eudiometer with oxygen gas in the well known way; 100 volumes of this gas left 37.26 (after the necessary corrections) volumes of dried carbonic acid gas, a result which coincides with no known atomical combination, but which, nevertheless, shows that carburetted hydrogen is mixed with hydrogen gas in a greater quantity, than can be considered as a mere impurity.

I scarcely need mention, that if we consider 100 parts of cast iron to contain 5 per cent. of carbon, and assume the whole of such carbon entering into combination with the freed hydrogen during the conversion of the metallic iron into the *oxidium ferroso ferricum*, 3.02 pure hydrogen and 6.63 carburetted hydrogen will be developed.

In my own manuscript, instead of "and generating carburetted hydrogen very slowly," the passage stands, "and generating a mixture of hydrogen and

---

* The process of decomposition of water by cast iron, as in this case, is by no means so simple as it appears at the first view, particularly as the real chemical composition of the different sorts of cast iron steel and malleable iron is still enveloped in considerable darkness, notwithstanding the many efforts made by Bergmann, Mouchette, Vauqelin, even up to Karsten, Berthier, &c., as I intend shortly to show, in an elaborate treatise on the combination of iron with carbon and different other metallic bodies, and in which treatise I intend giving a minute comparative chemical analysis of the different sorts of cast and wrought iron of England, France, Germany and Sweden.

slight carburetted hydrogen very slowly:" I cannot undertake to say whether this is a fault of the copyist or the compositor. I here take the opportunity of correcting some other errors in print which have occurred in my remarks, to prevent perhaps other misunderstandings: page 141, line 19 from below, instead of "part oxide," read "protoxide"; and page 144, column 2, line 7 from below, instead of "diluted in chlorhydric acid," read "dissolved in diluted chlorhydric acid"; again, five lines lower, instead of "by collecting in a solution of nitrate of lead," "by collecting it in a solution," &c.

Finally, referring to the remark of your correspondent, "which, so far from being generated slowly, escapes with very great violence." The words slowly, or violently, can only be used in a relative sense; the idea which led me to use the expression "hydrogen generating slowly," was, in comparison with the momentary explosion occasioned by casting or pouring copper into moulds containing the slightest degree of damp or moisture; the reason that no explosion takes place in pouring iron into moulds containing moisture is, that the water is not at once converted into steam, but becomes decomposed; now the quiet development of hydrogen gas during the decomposition of water (continuing at least several seconds) in comparison to the momentary and destructive bursting up of water into steam, I called slowly, and I think no one will blame me for having so done.

Pardon me for having thus far trespassed on your valuable journal, and permit me to subscribe myself,

Your obedient servant,
C. SCHAFHAEUTL.
Cornhill, Feb. 9, 1839.

SWIMMING MACHINES.

Sir,—The "Swimming Machine" described in your 805th number as the "invention" of a Mr. Strachan, cannot lay any valid claim to consideration as a novelty. An apparatus for teaching swimming on the same suspensory principle was used by Captain Clias, the predecessor of Voelker in the introduction of Gymnastics to this country, and is described and figured at full length in his well-known work upon the subject,

published many years ago. It is matter of notoriety that the same system has been long in use in the Government Swimming-schools of Berlin, under the direction of General Pfuel, whose method is described in several English works, among others the *British Cyclopædia*.

The variations introduced by Mr. Strachan are few in number, but, singularly enough, they are all very much for the worse. The Prussian master instructs one pupil only at a time, from a fixed stage placed sideways to the learner. Mr. Strachan, on the contrary, proposes to instruct six at once, though he does not say how the ropes are to be managed, *by one man*, so as to avoid drowning some of the luckless beginners —and certain it is that if the back rank, wisely omitted in the cut, were to be let down as much too low as the front rank are drawn up too high, the master would find, when he had time enough to attend to that half of his charge, that they would never require to undergo the suspending process again. By Mr. Strachan's plan, also, as illustrated by the engraving, the front pupils will always have the master behind them, so that he cannot possibly show them any of the motions they are to go through; while their unfortunate companions behind, (if the rope be long enough to allow any play at all), will, when they make any progress forward, be quite under the stage, where they can neither see nor be seen. The idea of making the stage a floating one is peculiarly absurd, since any movement of the persons upon it will subject the parties below to a mouthful of water at the least, which is generally quite sufficient to disconcert a beginner. "The pupils," we are informed, "while learning to use their limbs are also progressing forwards, *drawing the machine along with them*." Considering the cumbrousness of the machine, this will be no small task, especially as most tyros in the natatory art find it exceedingly difficult to make any way ahead, with nothing but their own bodies to propel. By M. Clias's plan, the scholar, as soon as he had mastered the rudiments, was left to himself for a time, suspended loosely by a cord fixed to a ring going over a horizontal pole of somewhat smaller circumference, so that he was absolutely free to move, while he was sure to be supported as soon as he began to sink; the superiority of this to Mr. Strachan's clumsy contrivance is too obvious to need dwelling upon.

The "portable and inexpensive nature" of Mr. S.'s machine is pointed out as one of its recommendations. A slight consideration of the matter, however, will suffice to convince any one who may think it worth while to take the trouble, that such an apparatus as that referred to as already in use, must be both costly and cumbrous. The carriage of such a machine, *on land*, as proposed, would in a few trips amount to as much as the original cost, although that must have been no trifle.

After all, it may very reasonably be doubted whether any method of learning to swim yet discovered can vie for safety and celerity with the simple plan of beginning in a place known not to be out of depth, and so dispensing with "apparatus" altogether. It is hard to say why a learner, of all people in the world, should venture where, but for artificial support, he would infallibly be drowned. A depth of less than his own height is amply sufficient for the swimmer's purposes.

And I remain, Sir,
Very respectfully yours,
AQUARIUS.

Jan. 23, 1839.

SOLUTION OF MR. WHITE'S QUESTION IN SPHERICS.

Sir,—I send you a solution of Mr. White's ingenious question in spherics, proposed in No. 804 of your Magazine.

I am, Sir, yours, &c.
IVER M'IVER.

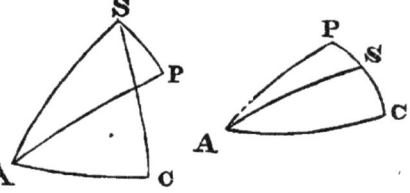

In the right angled spherical triangle A P S; sin. A S: sin. S P : : 1 : sin. S A P $\therefore$ sin A S $= \dfrac{\text{sin. S P}}{\text{sin. S A P}}$. Similarly sin. A S $= \dfrac{\text{sin. S C}}{\text{sin. S A C}}$: assume the spherical angle S A C $= x + a$, a being $\frac{1}{2}$ the angle P A C $\therefore$ S A P $= x - a$, hence, $\dfrac{\text{sin. S P}}{\text{sin. } (x-a)} = \dfrac{\text{sin. S C}}{\text{sin. } (x+a)}$ or

$\dfrac{\text{sin. S C}}{\text{sin. S P}} = \dfrac{\text{sin. } (x+a)}{\text{sin. } (x-a)} = \dfrac{\text{tan. } x + \text{tan. } a}{\text{tan. } x - \text{tan. } a}$;

hence, sin. S C . tan. x — sin. S C tan. a = sin. S P . tan. x + sin. S P . tan. a ;

whence, tan. x = tan. $a \left(\dfrac{\text{sin. SC} + \text{sin. S P}}{\text{sin. SC} - \text{sin. S P}} \right)$

$= \text{tan. } a \left(\dfrac{\text{tan. } \frac{1}{2} \text{ (S C + S P)}}{\text{tan. } \frac{1}{2} \text{ (S C — S P)}} \right) = \text{tan. } a$

tan. $\frac{1}{2}$ (S C + S P). Cot. $\frac{1}{2}$ (S C—S P);

whence, the angles S A C and S A P became known and A P, A C found by the common rules for spherical trigonometry. Among the many applications of the above question, I shall select one.

Suppose the declination of a celestial object is 25° .. 17′ .. 54" N. Latitude 2° .. 26′ .. 15" N. To find its right ascension and longitude of the object, the obliquity of the Ecliptic being 23° .. 27′ .. 44".

$\frac{1}{2}$ (S C + S P) = 13° .. 52′ .. 4½" tan. 9·392487
$\frac{1}{2}$ (S C — S P) = 11° — 25′ .. 49½" cot. 0·694244
$\dfrac{a}{2} = \dfrac{23° .. 27′ .. 44"}{2} = 11° .. 43′ .. 52"$ tan. 9·317345

$x = 14° .. 13′ .. 40½"$ tan. 9·404076

Hence, S A P = 2° .. 29 .. 48½ and S A C 25° .. 26′ .. 32½".

In the right angled spherical triangles A S P and A S C, we have sin. A P = co. tan. S A P . tan. P S and sin. A C =

co. tan. S A C . tan S C : which on calculation will be found to be 77° .. 14′ .. 30½" and 76° .. 18′ .. 28" respectively.

When the point S falls within the angle P A C, by a slight modification we

find, tan. s = tan. a . tan. $\frac{1}{2}$ (S C — S P).
Cot. $\frac{1}{2}$ (S C + S P), or if S P is greater
than S C tan. s = tan. a tan. $\frac{1}{2}$ (S P—S C).
Cot. $\frac{1}{2}$ (S P + S C). Finally, when S C
is greater or less than S P, the angle
S A C will be greater or less than S A P.

It may be added, that the supplements of the above found arches—viz.
(180°—77'°.. 14'.. 30$\frac{1}{2}$″),(180°—76°.. 18'
.. 28″)=102°.. 45'.. 29$\frac{1}{2}$″ and 103°.. 41'
.. 32″, will also satisfy the conditions of
the question.

I. M.

TRANSACTIONS OF THE INSTITUTION
OF CIVIL ENGINEERS.

Although it may be conceded that the
second volume of the Transactions of the
Institute of Civil Engineers* (which has
recently been published), falls short in
some measure of the interest attached to
the first of the series, it must at the same
time be allowed that the papers it contains are of very considerable importance,
and, generally speaking, of no mean degree of excellence. Nearly the whole of
them, as might be expected, are of a decidedly practical complexion : science is
taken up in them at the very point where
it is usually abandoned in the Transactions of older and perhaps prouder associations—at the point, that is to say,
where it ceases to be merely " contemplative," and begins to be also useful.
We need not stop to enquire which of
the two systems is to be preferred ; a sufficient apology for the publication of the
labours of the Institute of Civil Engineers
may be found in the fact of the almost
total neglect of practical science in the
older repositories, and in the consideration that, while mere "contemplative"
science would be sadly at a loss without
the assistance of its practical brother, the
latter *might* manage to make some progress, even in the absence of the former.
There is room enough, however, in the
wide world (of literature) for both; only
it is to be hoped that "practical" will
for the future occupy as much space as
of right belongs to him; or, in other
words, that ere long we shall have a
goodly number of such works of utility,
as this publication of the Society of Civil

Engineers, for instance, to set against
the now overwhelming mass of works of
speculation of the same class, such as the
"*Philosophical Transactions*," and a
host of others. The time seems fast coming when such a consummation may be
looked for, not in vain.

The antagonism of theory and practice
is well illustrated in some of the essays
in the volume before us. Almost the
only paper which is not drawn directly
from a practical source is one by Mr. G.
H. Palmer, on "Steam as a Moving
Power, especially with reference to the
Economy of Atmospheric and High-
Pressure Engines." In this the writer
labours hard to prove that the amount
of duty attributed to the celebrated
pumping engines of the Cornish mines
must have been highly exaggerated ; in
fact, that the amount reported could not
be attained without reversing the *laws
of nature* herself! According to this
gentleman, the maximum weight which
one bushel of coals can raise one foot
high is 44,467,500 pounds, and that, too,
putting all consideration of friction out
of the question. No other member of
the Society seems to have taken the
trouble to attempt the refutal of Mr.
Palmer's objections in a similar form ;
but a little further on in the volume we
come to a paper, of a thoroughly practical character, on the "Effective Power
of High-pressure Engines in the Cornish
mines," by Mr. Wickstead, which entirely
demolishes Mr. Palmer's "contemplations," by setting in array a few simple
facts, which militate against his sweeping
conclusion. Aware of the scepticism of
many of his brethren, Mr. W., on a recent visit to Cornwall, took the opportunity of again testing most closely the
power of the engines, and, as it happens,
he appears to have taken the very precautions against fraud and deception
which Mr. Palmer points out in his
paper,—and what was the result ?—why,
Mr. Wickstead found that, despising the
"limits fixed by the unchanging laws of
nature," the *illegal* engine had raised
more than double the *utmost-possible*
quantity of 44,000,000 lbs. ;—in fact,
upwards of 100 millions !* Nay, more,

---

* Transactions of the Institute of Civil Engineers,
vol. 2. London, 1838, John Weale, 4 to., pp. 246,
(with 33 plates.)

* "The fire under the boiler was worked down
as low as could be without stopping the engine.
The pressure of steam was 40 lbs. per square inch
in the boiler; I took the counter and the time, and
then started the engine. At the end of 2$\frac{1}{2}$ hours the

Mr. W. goes on to show that this "limit" has long been habitually passed; and to produce a pretty strong proof, if any were wanting, that there can be no mistake about the matter, we extract the passage, commencing with a table, which shows in a striking light the astonishing improvements effected in the working of the steam-engine in Cornwall, within the last threescore years and ten:—

Date.	Lbs. raised 1 foot high, with the consumption of one bushel, or 94 lbs. of coals.	Lbs. of coal per horse-power per hour.
1769	5,590,000	33.33
1772	9,450,000	19.70
1786 to 1800	20,000,000	9.30
1813	28,000,000	6.64
1814	34,000,000	5.47
1815	50,000,000	3.72
1825	54,000,000	3.44
1827	62,000,000	3.
1828	80,000,000	2.32
1834	90,000,000	2.06
1836	97,000,000	1.91
1835 (Trial of Fowey Consols engine)	125,000,000	1.48

"Mr. John Taylor, an authority that cannot be disputed, stated, in a lecture delivered by him to the members of the Society of Arts, that in 1829 he procured authentic accounts from the Consolidated Mines, of coals purchased and used in 1799, and also in 1828; from Wheal Alfred Mines, of the coals purchased and used in 1816 and in 1825; from Wheal Towan Mines, of the coals purchased in 1814 and 1826; from Dolcoath Mines, of the coals purchased and used in 1807 and 1817; and the result of his calculations, when comparing the depth of the mines at the different periods, the water raised, and the coals consumed, showed a saving upon the books of the mines, proportionate to the improvements stated to have been made during these periods in the working of the engines."—p. 67.

Now, how are these apparent contradictions between Nature and Art to be reconciled? Are we to suppose that the laws of Nature are suspended for the especial behoof of the miners of the west; or are we to suppose, on the strength of Mr. Palmer's dictum, that his theory is correct, and that the Cornish engineers, Mr. Wickstead, Mr. Taylor, the account books of the mines, and every man of science who has *actually tested* the performances of the engines, are in a conspiracy together? The answer is plain; Mr. Palmer must give up his theory, and yield to the force of circumstances, resolving for the future, it is to be hoped, to rely more upon direct experiment, and less upon mere speculative inferences, which, after all, must be deduced from despised experiment itself. Thus Mr. Palmer's "law of Nature" rests entirely on the credit due to certain anonymous trials of the virtue contained in "seven pounds of the best Newcastle coals;" some years ago, as will be seen from the table, it might then have tallied well enough with the actual performances of the Cornish engines, but now that the latter are reported to have so far transcended the assumed boundary of possibility, it would assuredly have done no injury to Mr. Palmer's reputation for sagacity to have instituted a new series of experiments on the power of coal, previous to troubling his fellow engineers with a long attempt to uphold a theoretical point, in the very teeth of facts. He reminds us much of poor Dr. Lardner, and a multitude of the same *genus*, who rear so enormous a pile of dogmatism, on an insufficient foundation, of data, that, when assailed by plain matter of fact, the huge structure "topples down headlong," and the luckless builder, reputation and all, disappears from view among the rubbish of his own creation. We trust Mr. Palmer is reserved for a better fate than this.

Practice is triumphant throughout the remainder of the volume. Mr. Francis Bramah contributes a paper on the "strength of cast iron," which is well balanced by another, of "experiments on American timber," by Lieutenant Denison. Mr. Harrison communicates a detailed description of the "drops" for loading coal on the Wear, which are of a construction peculiar to the locality; while Mr. John Reynolds relieves the volume from the reproach it might otherwise incur, of containing nothing whatever on the grand topic of the day in the en-

fire was lowering, and it was necessary to have more fuel, the 94 lbs. of coal having been consumed, the engine was then stopped, and the counter again taken. It had made 672 strokes, or very nearly 5 strokes per minute. The weight of water raised was (285.6 lbs. × 672 strokes) 191,823,2 lbs.; the height to which it was raised was (42 fath. 2 ft. 6 in. + 37 fath. 5 ft. 6 in. + 8 fath. 5 ft. 6 in. —) 535 ft. 6 in., the weight multiplied by the height in feet is equal to 102,721,323 lbs. of water lifted one foot high with 94 lbs. of coals."

gineering world, by contributing an article on " railways of continuous bearing." Captain Smith, of the East Indies, publishes a method of mounting a " reciprocating lighthouse," much more simple and economical than that in use, though not universally applicable : Mr. Hays makes known a " machine for cleansing small rivers," likely to prove of great utility, and which has been for many years in use on the Little Storer River, in Kent, on the banks of which its inventor, an honest miller, was a resident; and Mr. Green furnishes the particulars of a boat-lift, or a substitute for a lock, as used on the Grand Western Canal. These, with two or three others we are about to notice, comprise the papers of most interest and importance in the present volume.

The erection of bridges has always been a grand branch of engineering. The first volume of the Institute's Transactions was extremely full in this department, nor is the second one at all deficient. It presents us, in the body of the work, with the details connected with the bridge and embankment—a bold and imposing structure—executed at Youghal, in Ireland, by the late lamented Alexander Nimmo, and with similar particulars respecting a wooden bridge over the Calder, in Yorkshire, by Mr. William Bull. But the most important papers of this class are those which commence and close the volume, the former relating to the bridge thrown over the Severn at Tewkesbury by the late president of the Institute—*Thomas Telford*,—and the latter comprising a highly-interesting account of (what is styled) the " Floating Bridge" over the Hamoaze at Plymouth, contributed by the ingenious inventor and designer of the plan, Mr. J. M. Rendel. It may be objected that the term " bridge" is applied inappropriately in this instance, and certainly very few persons would form beforehand any idea of the actual *thing* from the term employed. But we will let Mr. R. speak for himself, and quote his own description :—

" The bridge is a large flat-bottomed vessel, of a breadth or width nearly equal to its length, divided in the direction of its length into three divisions, the middle being appropriated to the machinery which impels it, and each of the side divisions to carriages and traffic of all kinds. These side divisions

or decks are raised from two feet to two feet six inches above the line of floatation, and by means of strong and commodious drawbridges or platforms, hung at each end of each deck, carriages drive on and off the deck from the landing place, embarking and disembarking thereby. without difficulty, or occasion for the least disturbance of horses or passengers, who remain in their places during the time of crossing the river. To make the passage certain and safe in any weather, and by night as well as by day, the bridge is guided by two chains, which, passing through it over cast-iron wheels, are laid across the river and fastened to the opposite shores, consequently forming as it were a road, along which the bridge is made to travel forward and back from shore to shore as required. Two small steam-engines are employed as the moving power, by turning a shaft, on each end of which is a large cast-iron wheel whereon the guide-chains rest. The peripheries of these wheels are cast with sockets fitted to the links of the chain, so that when the wheels are stationary, the bridge is, as it were, moored by the chains, but, when put in motion by the steam-engines, it is moved in the reverse direction of, and with the same velocity as, the wheels. The landing-places on each shore are simple inclined planes from low water mark to two feet above high-water mark, formed to a slope or inclination of 1 in 12 or 1 in 14, and as the bridge approaches, the drawbridge is lowered on the plane ; the draught of water of the bridge, and the projection of the drawbridge being such, that carriages, &c., are embarked and disembarked dry, or considerably above the water-mark, whilst the bridge is all afloat, and out of danger of grounding or drifting, being held fast by the chains.

' To prevent the chains being so tight as to interrupt the free navigation of the estuary, or to endanger their breaking, instead of being fastened or moored to the shores, their ends have heavy weights attached to them in shafts sunk at the end of each landing place. Of course these weights rise and fall as the strain upon the chains becomes more or less, and prevents the tension ever exceeding the balance-weights, which are considerably below the weights to which the chains have been proved."—p. 215.

From this it will be seen that the " bridge" is of a locomotive character, and, in fact, bears much more resemblance to a ferry-boat than to the stationary article heretofore called by the name of " bridge," to which the absence of locomotion has till now been considered essential. Smeaton maintained the possibility of erecting " a bridge from Dover to Calais ;" but no one ever dreamed that the feat had been already accomplished by the establishment of a line of packet-boats! Mr. Rendel will remind us, indeed, that *they* have no chain to guide them, but this will not serve to bear him out : there are thousands of places whose rivers are crossed by boats moved by means of a rope stretched across, or, —worse still for Mr. R— by a *chain :* yet no one ever dreams of calling one of these ferry-boats a " bridge,"—and, if he did, he would only be laughed at for his pains.—The introduction of chains, (which, when not in use, lie across the bottom of the river) is, it will be perceived, the grand feature of Mr. R.'s plans, which appears to have been a highly successful one, and to merit adoption in the many situations on estuaries near our coast, where such a mode of communication is desirable—whether the boat be called a " bridge," or no:—

" By employing chains as a guide from shore to shore, the passage by the bridge is rendered safe by night as well as by day, and in rough weather as well as fine ; whilst, by the employment of them as a medium through which motion is conveyed to the bridge, a command is obtained over the motion, which enables the man at the engine to start, stop, and move forward and back, with a facility and rapidity that could not be obtained by any other means. These advantages are found of the utmost value in the approach to, and departure from, the landing-places, the chains acting better than any warps, and superseding all necessity for men to attend that operation, as well as for a crew, such as steersman, look-out-man, &c. &c. There being only two persons necessary for the working of this kind of bridge, viz., the man at the engine, and the man at the drawbridges, and to direct the engine-man when to stop, and start. I have before stated, that the speed at which the bridge is worked across the river, is, on an average, 320 feet per minute. This might be considerably increased, if necessary, though I do not think it capable of being made equal to the speed of ordinary steam-boats worked by paddles, still it must not be lost sight of, that in this, as in every other case of travelling, the proper measure for speed is the time taken to perform the journey, or as applied to the instance of crossing a river, the time which is occupied in the passage, from the embarking at one shore to the disembarking at the other. Now by the employment of chains the course is *direct*, the speed

uniform, and may be maintained with safety to the shore approached ; and the delay of backing, warping, &c. &c., which, at some ferries, takes as much time as crossing the river, is entirely got rid of.

" In illustration of this, I need only mention, that the time occupied in crossing by this bridge is seven minutes at low water, and eight minutes at high water (which is 320 feet per minute, the width at high water being 2550 feet) the time being uniform, whether in fair or rough weather, night or day.

" As a criterion by which to judge of the capabilities of the bridge for accommodation, I would state, that I have seen on it at one time three carriages, each with four horses, one carriage with a pair of horses, seven saddle horses, and sixty foot-passengers, and still there was nothing like crowding or discomfort. Though the exposure of the site is such that the sea frequently breaks over the funnel of the engines, I have never yet known the passengers of the Devonport and Falmouth mail, or of the other coaches which regularly cross it twice a day, leave their seats, even from the top of the coach. Of course, in such cases, the roadway on the lee side is chiefly used."—p. 222.

With this extract, which will enable our readers to form their own judgment of the merits of Mr. Rendel's invention, we take our leave of the present volume of the Civil Engineers' Transactions ; only remarking, in conclusion, that the book is excellently got-up in every particular, and that the numerous plates, especially, are executed in a style worthy of the reputation which our engravers, in what may be called the " engineering branch" of the art, have attained, as beyond all doubt the *first engravers in the world.*

PHOTOGENIC DRAWING.

Some Account of the Art of Photogenic Drawing, or the Process by which Natural Objects may be made to delineate themselves without the aid of the Artist's Pencil. By Henry FoxTalbot, Esq. F.R.S.

[From the *Athenæum.*]

1. In the spring of 1834, I began to put in practice a method which I had devised some time previously, for employing to purposes of utility the very curious property which has been long known to chemists to be possessed by the nitrate of silver ; namely, its discolouration when exposed to the violet rays of light. This property appeared to me to be perhaps capable of useful application in the following manner :—

I proposed to spread on a sheet of paper a sufficient quantity of nitrate of silver, and then to set the paper in the sunshine, having first placed before it some object casting a well-defined shadow. The light, acting on the rest of the paper, would naturally blacken it, while the parts in shadow would retain their whiteness. Thus I expected that a kind of image or picture would be produced, resembling to a certain degree the object from which it was derived. I expected, however, also, that it would be necessary to preserve such images in a portfolio, and to view them only by candlelight, because, if by daylight, the same natural process which formed the images would destroy them, by blackening the rest of the paper.

Such was my leading idea before it was enlarged and corrected by experience. It was not until some time after, and when I was in possession of several novel and curious results, that I thought of inquiring whether this process had been ever proposed or attempted before. I found that in fact it had; but apparently not followed up to any extent, or with much perseverance. The few notices that I have been able to meet with are vague and unsatisfactory; merely stating that such a method exists of obtaining the outline of an object, but going into no details respecting the best and most advantageous manner of proceeding.

The only definite account of the matter which I have been able to meet with, is contained in the first volume of the Journal of the Royal Institution, page 170, from which it appears that the idea was originally started by Mr. Wedgwood, and a numerous series of experiments made both by him and Sir Humphry Davy, which, however, ended in failure. I will take the liberty of quoting a few passages from this memoir.

"The copy of a painting, immediately after being taken, must be kept in an obscure place. It may, indeed, be examined in the shade, but in this case the exposure should be only for a few minutes. No attempts that have been made to prevent the uncoloured parts from being acted upon by light, have as yet been successful. They have been covered with a thin coating of fine varnish; but this has not destroyed their susceptibility of becoming coloured. When the solar rays are passed through a print, and thrown upon prepared paper, the unshaded parts are slowly copied; but the lights transmitted by the shaded parts are seldom so definite as to form a distinct resemblance of them by producing different intensities of colour.

"The images formed by means of a camera obscura have been found too faint to produce, in any moderate time, an effect upon the nitrate of silver. To copy these images was the first object of Mr. Wedgwood, but all his numerous experiments proved unsuccessful."

These are the observations of Sir Humphry Davy. I have been informed by a scientific friend that this unfavourable result of Mr. Wedgwood's and Sir Humphry Davy's experiments, was the chief cause which discouraged him from following up with perseverance the idea which he had also entertained of fixing the beautiful images of the camera obscura. And no doubt, when so distinguished an experimenter as Sir Humphry Davy announced "that all experiments had proved unsuccessful," such a statement was calculated materially to discourage further inquiry. The circumstance also, announced by Davy, that the paper on which these images were depicted was liable to become entirely dark, and that nothing hitherto tried would prevent it, would perhaps have induced me to consider the attempt as hopeless, if I had not (fortunately) before I read it), already discovered a method of overcoming this difficulty, and of fixing the image in such a manner that it is no more liable to injury or destruction.

In the course of my experiments directed to that end, I have been astonished at the variety of effects which I have found produced by a very limited number of different processes when combined in various ways; and also at the length of time which sometimes elapses before the full effect of these manifests itself with certainty. For I have found that images formed in this manner, which have appeared in good preservation at the end of twelve months from their formation, have nevertheless somewhat altered during the second year. This circumstance, added to the fact that the first attempts which I made became indistinct in process of time (the paper growing wholly dark), induced me to watch the progress of the change during some considerable time, as I thought that perhaps all these images would ultimately be found to fade away. I found, however, to my satisfaction, that this was not the case; and having now kept a number of these drawings during nearly five years without their suffering any deterioration, I think myself authorised to draw conclusions from my experiments with more certainty.

2. Effect and Appearance of these Images. —The images obtained in this manner are themselves white, but the ground upon which they display themselves is variously and pleasingly coloured.

Such is the variety of which the process is capable, that by merely varying the pre-portions and some trifling details of manipulation, any of the following colours are readily attainable :—

<blockquote>
Sky-blue,
Yellow,
Rose-colour,
Brown, of various shades,
Black.
</blockquote>

Green alone is absent from the list, with the exception of a dark shade of it, approaching to black. The blue-coloured variety has a very pleasing effect, somewhat like that produced by the Wedgwood-ware, which has white figures on a blue ground. This variety also retains its colours perfectly if preserved in a portfolio, and not being subject to any spontaneous change, requires no preserving process. These different shades of colour are of course so many different chemical compounds, which chemists have not hitherto distinctly noticed.

3. *First Applications of this Process.*— The first kind of objects which I attempted to copy by this process were flowers and leaves, either fresh or selected from my herbarium. These it renders with the utmost truth and fidelity, exhibiting even the venation of the leaves, the minute hairs that clothe the plant, &c. &c.

It is so natural to associate the idea of *labour* with great complexity and elaborate detail of execution, that one is more struck at seeing the thousand florets of an *Agrostis* depicted with all its capillary branchlets (and so accurately that none of all this multitude shall want its little bivalve calyx, requiring to be examined through a lens), than one is by the picture of the large and simple leaf of an oak or a chesnut. But in truth the difficulty is in both cases the same. The one of these takes no more time to execute than the other ; for the object which would take the most skilful artist days or weeks of labour to trace or to copy, is effected by the boundless powers of natural chemistry in the space of a few seconds.

To give an idea of the degree of accuracy with which some objects can be imitated by this process, I need only mention one instance. Upon one occasion, having made an image of a piece of lace of an elaborate pattern, I showed it to some persons at the distance of a few feet, with the inquiry, whether it was a good representation ? when the reply was, "that they were not so easily to be deceived, for that it was evidently no picture, but the piece of lace itself."

At the very commencement of my experiments upon this subject, when I saw how beautiful were the images which were thus produced by the action of light, I regretted the more that they were destined to have such a brief existence, and I resolved to attempt to point out, if possible, some method of preventing this, or retarding it as much as possible. The following considerations led me to conceive the possibility of discovering a preservative process.

The nitrate of silver, which has become black by the action of light, is no longer the same chemical substance that it was before. Consequently, if a picture produced by solar light is subjected afterwards to any chemical process, the white and dark parts of it will be differently acted upon ; and there is no evidence that after this action has taken place, these white and dark parts will any longer be subject to a spontaneous change ; or, if they are so, still it does not follow that that change will *now* tend to assimilate them to each other. In case of their remaining *dissimilar* the picture will remain visible, and therefore our object will be accomplished.

If it should be asserted that exposure to sunshine would *necessarily* reduce the whole to one uniform tint, and destroy the picture, the *onus probandi* evidently lies on those who make the assertion. If we designate by the letter A the exposure to the solar light, and by B some intermediate chemical process, my argument was this : Since it cannot be shown, *à priori*, that the final result of the series of processes A B A will be the same with that denoted by B A, it will be therefore worth while to put the matter to the test of experiment, viz. by varying the process B until the right one be discovered, or until so many trials have been made as to preclude all reasonable hope of its existence.

My first trials were unsuccessful, as indeed I expected ; but after some time I discovered a method which answers perfectly, and shortly afterwards another. On one of these more especially I have made numerous experiments ; the other I have comparatively little used, because it appears to require more nicety in the management. It is, however, equal, if not superior, to the first, in brilliancy of effect.

This chemical change, which I call the *preserving process*, is far more effectual than could have been anticipated. The paper, which had previously been so sensitive to light, becomes completely insensible to it, insomuch that I am able to show the Society specimens which have been exposed for an hour to the full summer sun, and from which exposure the image has suffered nothing, but retains its perfect whiteness.

4. *On the Art of fixing a Shadow.*—The phenomenon which I have now briefly mentioned appears to me to partake of the character of the *marvellous*, almost as much as any fact which physical investigation has yet brought to our knowledge. The most transitory of things, a shadow, the prover-

bial emblem of all that is fleeting and momentary, may be fettered by the spells of our "*natural magic*," and may be fixed for ever in the position which it seemed only destined for a single instant to occupy.

This remarkable phenomenon, of whatever value it may turn out in its application to the arts, will at least be accepted as a new proof of the value of the inductive methods of modern science, which by noticing the occurrence of unusual circumstances (which accident perhaps first manifests in some small degree), and by following them up with experiments, and varying the conditions of these until the true law of nature which they express is apprehended, conducts us at length to consequences altogether unexpected, remote from usual experience, and contrary to almost universal belief. Such is the fact, that we may receive on paper the fleeting shadow, arrest it there, and in the space of a single minute fix it there so firmly as to be no more capable of change, even if thrown back into the sunbeam from which it derived its origin.

5. Before going further, I may however add, that it is not always necessary to use a preserving process. This I did not discover until after I had acquired considerable practice in this art, having supposed at first that all these pictures would ultimately become indistinct if not preserved in some way from the change. But experience has shown to me that there are at least two or three different ways in which the process may be conducted, so that the images shall possess a character of durability, provided they are kept from the action of direct sunshine. These ways have presented themselves to notice rather accidentally than otherwise; in some instances without any particular memoranda having been made at the time, so that I am not yet prepared to state accurately on what particular thing this sort of semi-durability depends, or what course is best to be followed in order to obtain it. But as I have found that certain of the images which have been subjected to no preserving process remain quite white and perfect after the lapse of a year or two, and indeed show no symptom whatever of changing, while others differently prepared (and left unpreserved) have grown quite dark in one-tenth of that time, I think this singularity requires to be pointed out. Whether it will be of much value I do not know; perhaps it will be thought better to incur at first the small additional trouble of employing the preserving process, especially as the drawings thus prepared will stand the sunshine; while the unpreserved ones, however well they last in a portfolio or in common daylight, should not be risked in a very strong light, as they would be liable to change thereby even years after their original formation. This very quality, however, admits of useful application. For this semi-durable paper, which retains its whiteness for years in the shade, and yet suffers a change whenever exposed to the solar light, is evidently well suited to the use of a naturalist travelling in a distant country, who may wish to keep some memorial of the plants he finds, without having the trouble of drying them and carrying them about with him. He would only have to take a sheet of this paper, throw the image upon it, and replace it in his portfolio. The defect of this particular paper is, that in general the *ground* is not even; but this is of no consequence where utility alone, and not beauty of effect is consulted.

6. *Portraits.*—Another purpose for which I think my method will be found very convenient, is the making of outline portraits, or *silhouettes*. These are now often traced by the hand from shadows projected by a candle. But the hand is liable to err from the true outline, and a very small deviation causes a notable diminution in the resemblance. I believe this manual process cannot be compared with the truth and fidelity with which the portrait is given by means of solar light.

7. *Paintings on Glass.*—The shadow-pictures which are formed by exposing paintings on glass to solar light are very pleasing. The glass itself, around the painting, should be blackened; such, for instance, as are often employed for the magic lantern. The paintings on the glass should have no bright yellows or reds, for these stop the violet rays of light, which are the only effective ones. The pictures thus formed resemble the productions of the artist's pencil more, perhaps, than any of the others. Persons to whom I have shown them have generally mistaken them for such, at the same time observing, that the *style* was new to them, and must be one rather difficult to acquire. It is in these pictures only that, as yet, I have observed indications of *colour*. I have not had time to pursue this branch of the inquiry further. It would be a great thing if by any means we could accomplish the delineation of objects in their natural colours. I am not very sanguine respecting the possibility of this; yet, as I have just now remarked, it appears possible to obtain at least *some indication* of variety of tint.

8. *Application to the Microscope.*—I now come to a branch of the subject which appears to me very important and likely to prove extensively useful, the application of my method of delineating objects to the solar microscope.

The objects which the microscope unfolds to our view, curious and wonderful as they are, are often singularly complicated. The eye, indeed, may comprehend the whole which is presented to it in the field of view; but the powers of the pencil fail to express these minutiæ of nature in their innumerable details. What artist could have skill or patience enough to copy them? or granting that he could do so, must it not be at the expense of much most valuable time, which might be more usefully employed?

Contemplating the beautiful picture which the solar miscroscope produces, the thought struck me whether it might not be possible to cause that image to impress itself upon the paper, and thus to let Nature substitute her own inimitable pencil for the imperfect, tedious, and almost hopeless attempt of copying a subject so intricate?

My first attempt had no success. Although I chose a bright day, and formed a good image of my object upon prepared paper, on returning at the expiration of an hour I found that no effect had taken place. I was therefore half inclined to abandon this experiment, when it occurred to me that there was no reason to suppose that the common muriate of silver was the most sensitive substance that exists to the action of the chemical rays; and though such should eventually prove to be the fact, at any rate it was not to be assumed without proof. I therefore began a course of experiments in order to ascertain the influence of various modes of preparation, and I found these to be signally different in their results. I considered this matter chiefly in a practical point of view; for as to the theory, I confess that I cannot as yet understand the reason why the paper prepared in one way should be so much more sensitive than in another.

The result of these experiments was the discovery of a mode of preparation greatly superior in sensibility to what I had originally employed: and by means of this, all those effects which I had before only anticipated as theoretically possible were found to be capable of realization.

When a sheet of this, which I shall call "*Sensitive Paper*," is placed in a dark chamber, and the magnified image of some object thrown on it by the solar microscope, after the lapse of perhaps a quarter of an hour, the picture is found to be completed. I have not as yet used high magnifying powers, on account of the consequent enfeeblement of the light. Of course with a more sensitive paper, greater magnifying power will become desirable.

On examining one of these pictures, which I made about three years and a half ago, I find, by actual measurement of the picture and the object, that the latter is magnified seventeen times in linear diameter, and in surface consequently 289 times. I have others which I believe are considerably more magnified; but I have lost the corresponding objects, so that I cannot here state the exact numbers.

Not only does this process save our time and trouble, but there are many objects, especially microscopic crystallizations, which alter so greatly in the course of three or four days (and it could hardly take any artist less to delineate them in all their details), that they could never be drawn in the usual way.

I will now describe the *degree of sensitiveness* which this paper possesses, premising that I am far from supposing that I have reached the limit of which this quality is capable. On the contrary, considering the few experiments which I have made, (few, that is, in comparison with the number which it would be easy to imagine and propose,) I think it most likely that other methods may be found, by which substances may be prepared, perhaps as much transcending in sensitiveness the one which I have employed, as that does the ordinary state of the nitrate of silver. But to confine myself to what I have actually accomplished in the preparation of a very sensitive paper.

When a sheet of paper is brought towards a window, not one through which the sun shines, but looking in the opposite direction, it immediately begins to discolour. For this reason, if the paper is prepared by daylight, it must by no means be left uncovered, but as soon as finished be shut up in a drawer or cupboard and there left to dry, or else dried at night by the warmth of a fire. Before using this paper for the delineation of any object, I generally approach it for a little time towards the light, thus intentionally giving it a slight shade of colour, for the purpose of seeing that the *ground* is *even*. If it appears so when thus tried to a small extent, it will generally be found to prove so in the final result. But if there are some places or spots in it which do not acquire the same tint as the rest, such a sheet of paper should be rejected; for there is a risk that, when employed, instead of presenting a *ground* uniformly dark, which is essential to the beauty of the drawing, it will have large white spots, places altogether insensible to the effect of light. This singular circumstance I shall revert to elsewhere: it is sufficient to mention it here.

The paper then, which is thus readily sensitive to the light of a common window, is of course much more so to the direct sunshine. Indeed, such is the velocity of the effect then produced, that the picture may be said to be ended almost as soon as it is begun.

To give some more definite idea of the rapidity of the process, I will state, that after various trials the nearest evaluation which I could make of the time necessary for obtaining the picture of an object, so as to have pretty distinct outlines, when I employed the full sunshine, was *half a second*.

9. *Architecture, Landscape, and external Nature.*—But perhaps the most curious application of this art is the one I am now about to relate. At least it is that which has appeared the most surprising to those who have examined. my collection of pictures formed by solar light.

Every one is acquainted with the beautiful effects which are produced by a *camera obscura*, and has admired the vivid picture of external nature which it displays. It had often occurred to me, that if it were possible to retain upon the paper the lovely scene which thus illuminates it for a moment, or if we could but fix the outline of it, the lights and shadows divested of all *colour*, such a result could not fail to be most interesting. And however much I might be disposed at first to treat this notion as a scientific dream, yet when I had succeeded in fixing the images of the solar microscope by means of a peculiarly sensitive paper, there appeared no longer any doubt that an analogous process would succeed in copying the objects of external nature, although indeed they are much less illuminated.

Not having with me in the country a *camera obscura* of any considerable size, I constructed one out of a large box, the image being thrown upon one end of it by a good object glass fixed in the opposite end. This apparatus being armed with a sensitive paper, was taken out in a summer afternoon and placed about 100 yards from a building favourably illuminated by the sun. An hour or two afterwards I opened the box, and I found depicted upon the paper a very distinct representation of the building, with the exception of those parts of it which lay in the shade. A little experience in this branch of the art showed me that with smaller *cameræ obscuræ* the effect would be produced in a smaller time. Accordingly I had several small boxes made, in which I fixed lenses of shorter focus, and with these I obtained very perfect but extremely small pictures; such as without great stretch of imagination might be supposed to be the work of some Lilliputian artist. They require indeed examination with a lens to discover all their minutiæ.

In the summer of 1835 I made in this way a great number of representations of my house in the country, which is well suited to the purpose, from its ancient and remarkable architecture. And this building I believe to be the first that was ever yet known *to have drawn its own picture.*

The method of proceeding was this : having first adjusted the paper to the proper focus in each of these little *cameræ*, I then took a number of them with me out of doors and placed them in different situations around the building. After the lapse of half an hour I gathered them all up, and brought them within doors to open them. When opened, there was found in each a miniature picture of the objects before which it had been placed.

To the traveller in distant lands who is ignorant, as too many unfortunately are, of the art of drawing, this little invention may prove of real service; and even to the artist himself, however skilful he may be. For although this natural process does not produce an effect much resembling the productions of his pencil, and therefore cannot be considered as capable of replacing them, yet it is to be recollected that he may often be so situated as to be able to devote only a single hour to the delineation of some very interesting locality. Now, since nothing prevents him from simultaneously disposing, in different positions, any number of these little *cameræ*, it is evident that their collective results when examined afterwards, may furnish him with a large body of interesting memorials, and with numerous details which he had not had time either to note down or to delineate.

10. *Delineations of Sculpture.*—Another use which I propose to make of my invention is for the copying of statues and bas-reliefs. I place these in strong sunshine, and put before them at a proper distance, and in the requisite position, a small *camera obscura* containing the prepared paper. In this way I have obtained images of various statues, &c. I have not pursued this branch of the subject to any extent; but I expect interesting results from it, and that it may be usefully employed under many circumstances.

11. *Copying of Engravings.*—The invention may be employed with great facility for obtaining copies of drawings or engravings, or fac similes of MSS. For this purpose the engraving is pressed upon the prepared paper, with its engraved side in contact with the latter. The pressure must be as uniform as possible, that the contact may be perfect; for the least interval sensibly injures the result, by producing a kind of cloudiness in lieu of the sharp strokes of the original.

When placed in the sun, the solar light gradually traverses the paper, except in those places where it is prevented from doing so by the opaque lines of the engraving. It therefore of course makes an exact image or print of the design. This is one of the experiments which Davy and Wedgwood state that they tried, but failed, from want of sufficient sensibility in their paper.

The length of time requisite for effecting the copy depends on the thickness of the paper on which the engraving has been printed. At first I thought that it would not be possible to succeed with thick papers; but I found on trial that the success of the method was by no means so limited. It is enough for the purpose, if the paper allow any of the solar light to pass. When the paper is thick, I allow half an hour for the formation of a good copy. In this way I have copied very minute, complicated, and delicate engravings, crowded with figures of small size, which were rendered with great distinctness.

The effect of the copy, though of course unlike the original (substituting as it does lights for shadows, and *vice versâ*), yet is often very pleasing, and would, I think, suggest to artists useful ideas respecting light and shade.

It may be supposed that the engraving would be soiled or injured by being thus pressed against the prepared paper. There is not much danger of this, provided both are perfectly dry. It may be well to mention, however, that in case any stain should be perceived on the engraving, it may be readily removed by a chemical application which does no injury whatever to the paper.

In copying engravings, &c., by this method, the lights and shadows are reversed, consequently the effect is wholly altered. But if the picture so obtained is first *preserved* so as to bear sunshine, it may be afterwards itself employed as an object to be copied; and by means of this second process the lights and shadows are brought back to their original disposition. In this way we have indeed to contend with the imperfections arising from two processes instead of one; but I believe this will be found merely a difficulty of manipulation. I propose to employ this for the purpose more particularly of multiplying at small expense, copies of such rare or unique engravings as it would not be worth while to re-engrave, from the limited demand for them.

I will now add a few remarks concerning the very singular circumstance, which I have before briefly mentioned, viz., that the paper sometimes, although intended to be prepared of the most sensitive quality, turns out on trial to be wholly insensible to light, and incapable of change. The most singular part of this is the very small difference in the mode of preparation which causes so wide a discrepancy in the result. For instance, a sheet of paper is all prepared at the same time, and with the intention of giving it as much uniformity as possible; and yet, when exposed to sunshine, this paper will exhibit large white spots of very definite outline, where the preparing process has failed; the rest of the paper, where it has succeeded, turning black as rapidly as possible. Sometimes the spots are of a pale tint of cærulean blue, and are surrounded by exceedingly definite outlines of perfect whiteness, contrasting very much with the blackness of the part immediately succeeding. With regard to the theory of this, I am only prepared to state as my opinion at present, that it is a case of what is called "unstable equilibrium." The process followed is such as to produce one of two definite chemical compounds; and when we happen to come near the limit which separates the two cases, it depends upon exceedingly small and often imperceptible circumstances, which of the two compounds shall be formed. That they are both definite compounds, is of course at present merely my conjecture: that they are signally different, is evident from their dissimilar properties.

I have thus endeavoured to give a brief outline of some of the peculiarities attending this new process, which I offer to the lovers of science and nature. That it is susceptible of great improvements I have no manner of doubt; but even in its present state I believe it will be found capable of many useful and important applications besides those of which I have here given a short account.

MR. MACKINNON'S NEW PATENT LAW BILL.

[From the *Commercial Gazette*.]

Mr. Mackinnon (the member for Lymington) has already given notice of his usual sessional motion on the subject of the present defective law of patents. It will be recollected that he brought forward a similar measure last session, which was frustrated, as many other useful measures were, by the pressure and mismanagement of public business, the balance of parties, and the obstructive proceedings of the Upper House. Amendments have been already made in the old law of patents; but, although they were well-intentioned, they are admitted by everybody conversant with the subject to be utterly deficient in point of remedy. There is no subject at any time of more vital importance to the commercial interests of this commercial country than the "Law of

Patents." But it is more especially so at the present time—when those commercial interests are so widely involved and so seriously impaired. In the protection desirable to be extended by an equitable and improved patent law, we include all the useful results of scientific discovery; all the improvements of mechanical invention; all the amelioration produced in our manufactures by the encouragement of the Art of Design; and, finally, the protection of *Pattern Right* in our great calico-printing, cotton, and silk manufactures. It will be evident that in the last departments of national industry alone—the protection of the *Pattern Right*—(otherwise the copyright of the new design or pattern, upon which the sale of the goods chiefly depends) is a subject of paramount interest to the whole manufacturing class. The importance of Mr. M.'s motion may be inferred from these preliminary remarks. The incompetency of the present Law of Patents to protect invention was one of the prominent grievances established by the evidence before the "Select Committee of *Arts and Manufactures*." Mr. Martin, the celebrated painter, and several architects and sculptors, produced curious and startling examples of the total incompetency of the present "Patent Law" to protect the property of individuals engaged in the improvements of art, or in the discoveries of science. Again; some of the first manufacturers of the country, on whose combined prosperity its commercial prosperity chiefly depends—such as calico-printers, cotton and silk manufacturers, ironmasters, paper-stainers, brass-founders, and Staffordshire potters—all demonstrated that the present Law of Patents was so deficient in adequate protection as to cut up profits and frustrate improvement in the arts of design, by rendering the purchase of patterns, a profitless and useless speculation. We entertain hopes that Mr. Mackinnon's proposal may remove the chief grounds of grievance, the justice of which has been so undeniably established. The great problem to be solved in the reform of the Patent Laws is to reconcile the individual rights of private property with the general exigencies of free trade. Finally, equity demands that those great benefactors of the human species—those "kings of mind," whose victories are bloodless and immortal; those conquerors of natural difficulty on behalf of the whole race who apply the discovery of science to social comfort—who invent useful machinery

—who circulate beneficial ideas, or give impulse and pre-eminence to staple branches of national industry, should be protected in the exertion of their talents, and in the fair enjoyment of the remuneration which that exertion and those talents win and deserve. Reward to such men cannot be too high—too jealously guarded; nor too permanently secured.

NOTES AND NOTICES.

Wide Velvet Weaving.—At a late meeting of the Society of Arts and Manufactures, the most judicious rewards of the season were bestowed upon two Spitalfields velvet weavers, named Hanshard and Cole, for their invention of a mode of weaving wide velvets. It appeared that rather more than a year ago a velvet shawl two yards square was imported from France, and Hanshard hearing of this, devised a means of performing the same work, and offered to his employers to undertake it at his own risk. He succeeded, and delivered a seven-quarter square, for which he was paid four pounds. Cole knowing also that a wide velvet was in demand, and having seen Hanshard's velvet, also devised a means of effecting the object, which turned out to be the same as Hanshard's, and he undertook, and performed the work for two pounds five shillings. Hanshard endeavoured to keep his means secret, but was undoubtedly the first inventor; he was rewarded with a prize of five pounds; Cole, although a subsequent, appeared also to be an original inventor, and made the process known to the trade immidiately: he was awarded a premium of three pounds. The difficulty to be overcome in weaving a wide velvet was this: the width of the fabric being greater than the stretch of a man's arm, he could not pass the wire containing the silk across it, the wire being so thin and flexible. To obviate this, Hanshard put the wire in a small brass tube, pointed at the end, which held the wire stiff, so that it could be passed across. In working, however, the end of the tube was liable to catch in the fabric and break the thread, and this difficulty was overcome by putting a pointed cap upon the end of the tube after the wire had been inserted. Cole, as stated, followed Hanshard in the invention of the tube, but the cap was solely Hanshard's.

New Silk Mill.—An English manufacturer, residing in Turin, is said to have invented a new silk-mill, the mechanism of which is so extremely simple, that it may be worked by children only 10 years old, and yet produces three times the quantity of twist made by the old mills in the same period, and of a much superior quality.

Another Substitute for Steam.—A correspondent informs us that an ingenious mechanic (Mr. William Dupe, of this city), has discovered a substitute for steam for propelling wheel carriages and ships. The invention, which is a very simple one, acts by condensed air. It is calculated that it will not cost more than one-third of the expense to work it, and will take up a much smaller space than a steam-engine. A model of this invention has been shown to several scientific gentlemen of the University and city, who have expressed their decided approbation of it.—*Oxford Chronicle.*

LONDON: Printed and Published for the Proprietor, by W. A. Robertson, at the Mechanics' Magazine Office, No. 6, Peterborough-court, Fleet-street.—Sold by A. & W. Galignani, Rue Vivienne, Paris.

𝔐echanics' 𝔐agazine,

MUSEUM, REGISTER, JOURNAL, AND GAZETTE.

No. 811.]   **SATURDAY, FEBRUARY 23, 1839.**   [Price 6*d*.

Printed and Published for the Proprietor, by W. A. Robertson, No. 6, Peterborough-court, Fleet-street.

MACRAE'S PATENT HYDRAULIC GAS-HOLDER COUNTER-BALANCE.

MACRAE'S PATENT HYDRAULIC COUNTERBALANCE FOR GAS-HOLDERS.

SIR,—May I request the favour of a place in your scientific Gazette for the following description of an improved hydraulic counterbalance.

The necessity for a contrivance of this nature has long been felt by manufacturers of gas; and the simplicity and economy of the plan invented by Mr. Macrae are self evident. Gas proprietors are, perhaps, not generally acquainted with the important fact, that a very great portion of the gas generated, is exposed to continual waste, in consequence of the present defective system of charging their gasometers.

Mr. Macrae particularly alludes to this subject in his specification, and points out very clearly the waste which is daily taking place, from the practice of conveying gas into gas-holders of such ponderous weight as those that are now in general use in the metropolis.

It is set forth in the specification that he has

"proved from an extensive experience in the manufacture, that in large establishments several thousands of feet of gas are totally lost every twenty-four hours, from the employment of gas-holders that are not sufficiently counterbalanced; and his evidence has been amply corroborated by the testimony of a number of the most experienced practical men at present engaged in the trade."

The principle of the invention he assimilates to the action of the domestic bellows:

"This instrument is elevated by artificial power, so as to admit a body of air, and this air is expelled by a reverse application of the same power; my object," he goes on to state, "is to approach as near as it is practicable, the above principle, by regulating the ascending and descending motions of a gas-holder, by means of a hydraulic counterbalance."

The specification explains, that the *heavy* gas-holders used, operate most destructively upon the generating and purifying machinery; and states, that although a heavy gas-holder is indispensable as a means for propelling an abundant supply into the street mains, yet, it is attended by an enormous sacrifice of property—in waste of gas, damage to machinery, loss of time in charging the gas-holders, and an unnecessary consumption of fuel in the furnaces.

The inventor, at considerable length, enters into a practical detail of the *different and opposite effects* produced upon the machinery of the retort and purifying houses, by a heavy and a light gas-holder. The following is an extract from the specification in reference to this point:

"Immediately the process of charging the *heavy* gas-holder commences, a considerable quantity of gas will be observed forcing its way through the luteing of the retort lids, similar escapes are also observed from the damaged parts of the hydraulic and ascension mains; and at the same time, the liquid in the hydraulic main rises in the (out of action) dip-pipes, to an elevation of from twenty to thirty inches; and while those destructive effects are being produced in the retort house, equally injurious results are taking place in the purifying department—for the lime water in the purifiers is depressed from its proper level, and in consequence a great portion of the gas passes from the purifiers without being sufficiently exposed to the action of the lime.

"The heavy gas-holder also, during its rising, prevents the gas from escaping with the required rapidity, from the generating vessels, and being therefore confined in, and exposed to, the intense heat of the retorts, the gas must be deprived of a great portion of one of its most valuable illuminating ingredients, viz.; the carbonaceous base.

"Frequent and dangerous reactions are also produced by the same cause: the destruction to machinery—waste of gas—and danger to the persons of the stokers, resulting from those reactions, are familiar to the intelligent workmen employed in several of our metropolitan establishments."

The frequency of this occurrence during the last two winters, and more particularly within the last three months, in several of the large works, ought, surely, to induce the proprietors to adopt a remedy so simple in its application, and so inexpensive, as the hydraulic counterbalance. Having now alluded to the destructive effects of *back pressure,* caused by gasometers that are not sufficiently counterpoised; I must remark, that there is a total absence of those evils during the process of filling a light gas-holder; the stream of gas, in this case, will flow in a uniform and uninterrupted current, from the retorts to the gasometer. Now, the hydraulic counterbalance will effect this desirable object, by facilitating the ascent of the gasometer, and when it has attained its highest point, by simply discharging the liquid from the tank, the whole and entire weight of the gas-holder will then be allowed to press upon the outlet main, and thereby cause an abundant supply to consumers. I beg to subjoin one additional extract from the specification. It is stated, that

" an immense quantity of gas is consumed during the time the stokers are engaged in drawing the charges ; for it is well known that upon such occasions the whole of the gas contained in the retorts and ascension-pipes is destroyed. Now, in establishments where there are from 200 to 300 retorts drawn four times in every twenty-four hours, the loss to the proprietors must be enormous in the course of twelve months. To prevent, in a great measure, this waste of gas," Mr. Macrae suggests, " that previous to the charge being drawn, the gas-holder should be rendered as light as it will admit of, by adding for that purpose the *required weight* of liquid to the hydraulic tank. If this operation be skilfully performed, a large portion of the gas (which would otherwise be consumed) will be drawn from the retorts and ascension-pipes, and conveyed into the light gas-holder."

Description of the apparatus.—The invention merely consists of a square tank, formed of wood, copper, or iron ; its dimensions must be in proportion to the weight of the gasometer it is intended for.

a, The hydraulic tank ; *d*, the well ; *b*, discharging valve ; *c*, a feed-pipe, which may be supplied from any convenient source.

" Previous to the gas being conveyed to the gas-holder, the tank (which is then at its highest point of suspension) is charged with the required weight of liquid—water, tar, ammoniacal liquor, or whatever is most suitable." The tank being thus charged, the gas-holder is then, by the weight of the hydraulic balance, enabled to rise freely, and without causing back pressure, as it does by the system at present in use. When the gasometer has received its complement of gas, the tank, as shown in the sketch (front page), is then at its lowest elevation. Previous to the gas being sent into the street mains, the liquor in the tank is drawn off through the valve at *b ;* when this is accomplished, it is obvious that nearly the *whole weight* of the gasometer will then be allowed to communicate any required degree of pressure ; and if it is necessary at any time

of night to check or subdue the pressure, or even to increase the supply of gas to consumers, either purpose may be accomplished by simply adding to, or diminishing by degrees, the quantity of liquid in the tank, by any number of pounds weight of liquid.

The specification contains a variety of drawings, exhibiting different methods of attaching and working the hydraulic tank, according to the nature of the ground upon which a gas-holder stands. The drawing prefixed to this description represents the apparatus as working in a well. This method is suggested by Mr. Macrae in consequence of the greater facility it affords for charging and emptying the counterbalance : also by this arrangement the tank will not be affected in windy weather.

I am, Sir,

Yours, most respectfully,

S. Q , ENGINEER.

16, Moscow Road, Bayswater, London.

LAMBERT AND SON'S HEMISPHERICAL JOINTS FOR GAS, STEAM, &c.

Sir,—In the infancy of an art, however simple, the attainment of perfection all at once is hardly to be expected ; improvement is the result of observation and experiment—often very laborious, and frequently extending over a long period of time. Ultimate perfection, or at least a reasonable approximation to it, is the produce of a series of steps or gradations, each following the other in due and stately order.

Perhaps the art of *gas-lighting* affords one of the most remarkable instances that could be adduced, of the progressive nature of improvement in maturing a new manufacture. If we contrast the form of oven—the description of retorts—the mode of separation—and, the purifying apparatus of the earliest gas-makers— with the most approved arrangements of the present day, a striking and important difference will be observed. In the article on "Gas-light" in the seventh edition of the *Encyclopædia Britannica* now publishing, it is observed, that, "so rapid has been the progress of gas illumination, that, in the course of a few years after it was first introduced, it was adopted by all the principal towns in the kingdom, for lighting streets, as well as shops and public edifices. In private

Fig. 4.

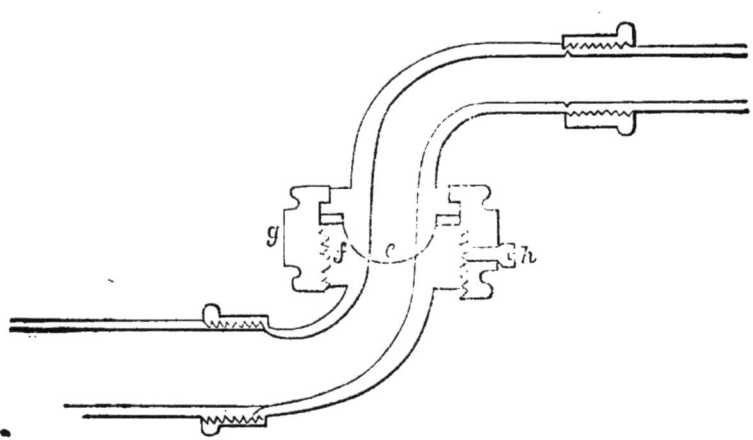

houses it found its way more slowly, partly from an apprehension, not entirely groundless, of the danger attending the use of it; and partly, from the annoyance which was experienced in many cases, through the careless and imperfect manner in which the service-pipes were at first fitted up." It is very certain, whatever might have been the errors and defects of the original *gas-makers*, they were altogether out-heroded by the continual blunderings of the early *gas-fitters*. Their first mistake was in the employment of unsuitable materials for much of their work; and secondly, in so greatly underrating the friction of the gas in its passage, as to employ tubes much too small for transmitting it with sufficient facility to afford a uniform and adequate supply. To these may be added, the want of attention to the necessary accuracy in the several joints and fittings. These inconveniences, however, have been in a great measure, if not wholly removed, by a more intimate acquaintance with the character—and increased experience in the management of gas. The copper tubing of small diameter, has now given way to pipes of iron for service-mains, and to flexible tin tubing for the radial branches; considerable alterations and improvements have also taken place in the burners, valves, cocks, and furniture generally, so as to afford increased economy and safety–frequently combined with great elegance in appear-

ance. One of the latest—if not the very last—improvement in gas furniture, is that which I have now the pleasure of introducing to your readers; viz., the improved hemispherical joints for gas-brackets, pendants, &c. &c., manufactured by Messrs. Lambert and Son, New Cut, Lambeth.

Swing brackets and pendants, as formerly constructed, consisted of a key or plug working in a socket, with a narrow groove around it which left a very contracted and tortuous passage for the gas, seriously impeding its progress and exceedingly liable to stoppage or derangement.

In Messrs. Lambert's brackets, &c., a free and uninterrupted passage for the gas is afforded throughout; the progress being through easy curves; all sharp or right-angled turnings are entirely avoided.

In the accompanying drawings, which are on a scale of three inches to a foot, Fig. 1 represents one of Messrs. Lambert and Son's double-jointed folding brackets. Fig. 2 is a section through the same when fully extended. Fig. 3 is a pendant light; the joints are situated at *a* and *b*; *c* is the stop-cock; *d* the burner. Fig. 4 is a section (full size) of the hemispherical joint; it is composed of a convex hemisphere *e*, fitting into a corresponding concavity on the end of the lower branch pipe *f*, and kept closely in contact by the screw-nut *g*. When

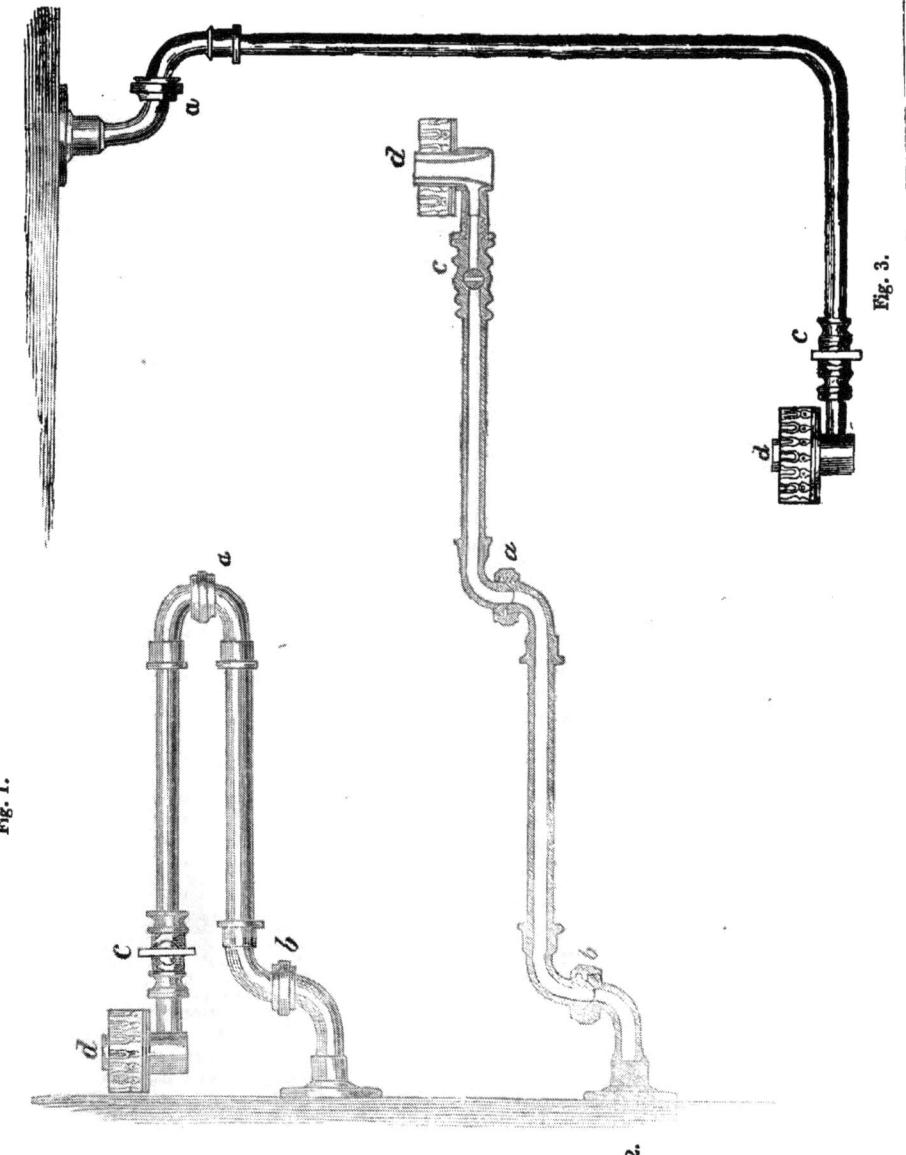

Fig. 1.

Fig. 2.

Fig. 3.

properly adjusted, the nut is fixed immoveably by a small set-screw, shown at *h*. By reference to the sections, figs. 2 and 4, it will at once be seen how freely the gas is transmitted along the whole length of this improved furniture.

In a communication on the subject of revolving slide-rests, inserted at page 267 of your last volume, I alluded to the manufacture of these joints; and I have now only to state, that they are produced by means of an apparatus of this kind, which, for its simplicity of construction—the extent of its powers—and the accuracy of its performance—is, I believe, without a rival in the mechanical world. Although this piece of apparatus, in its first construction, has been somewhat costly, yet by its powers, these joints, *i. e.*, the internal and external hemispheres, are produced with a facility and accuracy that ensures perfection, and at a cost so moderate as to warrant the expectation of their universal adoption in all good work. Some gentlemen connected with one of the principal Metropolitan gas-works, have pronounced these jointed brackets, as manufactured by Messrs. Lambert's, to be the most beautiful things of the kind they ever saw, and have declared these joints to be the greatest improvement hitherto made in gas furniture.

These joints are equally well adapted for connecting tubes employed in the transmission of steam, &c., a flexible joint being thus obtained that will continue steam-tight under almost any possible pressure. I may here observe that the hemispherical joints of Messrs. Lambert and Son, have sometimes been confounded with the old *ball and socket joints*, but this is a great mistake, the difference between them being very marked. In the well-known ball and socket joint, a *limited* range of motion was obtained in *every direction;* in Messrs. Lambert's joints on the contrary, an *unlimited* range in *one direction* only is obtained. Two of these joints being placed at right angles to each other, form an universal joint which is extremely useful for several pneumatic and hydraulic purposes.

By means of their very beautiful rotary slide-rest, Messrs. Lambert and Son can produce metallic spheres for valves with a facility and truth never before attained to, and in some recent trials with the spherical form of valve in one of their double-action pumps,* considerable advantage appeared to attend their employment.

I remain, Sir,

Yours, respectfully,

WM. BADDELEY.

London, Feb. 12, 1839.

— ◦◦ —

CHRONOMETER PREMIUMS—APPARENT ABUSE OF OFFICIAL INFLUENCE.

Sir,—The necessity of strict impartiality and disinterestedness in official persons who may be intrusted in any degree, with the duties of judge, or arbiter of any description, is so generally allowed, that I shall not take up a line of your valuable pages in insisting upon it. The public exposure of all departures from rectitude by such parties is the duty of every one; it may neutralize mischief already perpetrated; it is sometimes sufficient to bring the erring parties back to the right path; and it always awakens a vigilant circumspection on the part of the public with regard to their future conduct.

In the present case, I shall confine myself to the request that you will permit me to take advantage of the circulation of your Magazine among the class more particularly interested in the affair which I am about to mention, and to transfer to your pages a paragraph from the *Times* newspaper of Tuesday the 29th ult., headed with the characteristic "*(Advertisement)*" with which that paper honestly marks this species of paid and pseudo-editorial insertion; I, also, propose to place in juxta-position with this paragraph, for the sake of more effective contrast, some *verbatim* extracts from the "*Appendix to Captain Sir John Ross's (C.B., K.S.A., K.C.S., &c.) Narrative of his Second Expedition to the Arctic Regions.*"

(From the *Times*.)

" (Advertisement).—The importance of accuracy and skill in the manufacture of chronometers is too obvious to require comment, and we have much pleasure in insert-

---

* A description of this pump will be found in your 27th volume, page 82.

ing the following honourable testimonial in favour of the talents of Mr. Caster, of 61, Cornhill, and 207, Tooley-street, from Professor Airy of the Royal Observatory Greenwich."

Royal Observatory, Greenwich, Oct. 22.

"I certify, that since the commencement of the annual trials of chronometers at the Royal Observatory in competition for re-wards offered by the government for the best chronometers, Mr. H. Caster, chronome-ter-maker of Tooley-street, has obtained the greater number of rewards than any other chronometer-maker, and that he ob-tained the two last rewards given by the government.

" I certify, also, that since I have had charge of the Royal Observatory, several chronometers constructed by Mr. Caster, either belonging to the Royal Navy, or on trial for purchase by the government have been rated at the Royal Observatory, and that they have generally been extremely good.

" G. B. AIRY."

Mr. Caster in other advertisements put forth previous to the date of the one above, has constantly stated the number of his successful chronometers, during the twelve years public trials to be four !

———

(From *Ross's Appendix*.)

" Our chronometer, No. 1410, in 1828, for which we then received the premium of Three Hundred Pounds."

" In 1830, one of our chronometers was entitled to the second prize.

" In the trial of 1831, the chronometers made by us obtained *the whole of the three prizes.*'

" In 1832 and 1833 chronometers made by us were entitled to prizes ;—in this trial (1833) sixty chronometers were sent by various makers, and at the termination, the numbers were reduced to ten, *four* of which were made by us."

" Of *eight* chronometers entitled to the prizes during the last three annual public trials at the Royal Obser atory, *five* were constructed by us.

" PARKINSON AND FRODSHAM.'

Simple addition will show that the number of chronometers made by Messrs. Parkinson and Frodsham, which were suc-cessful during the twelve years public trials amounted to *fifteen !*'

It is evidently impossible that *both* of the above statements can be true; and it is not probable that the detailed enumera-tion of respectable persons like Messrs.

Parkinson and Frodsham* of their own chronometers is so *egregiously false,* as to permit that amount of correction which would be necessary to make Mr. G. B. Airy's *true;* the more so, as the state-ment of Messrs. Parkinson and Frod-sham was made advisedly for publication by a celebrated naval officer, of great in-telligence and sagacity, intimately ac-quainted with the subject in question and not likely to suffer a gross deception to be introduced to the public under his patronage.

The painful conviction which must press upon every unprejudiced mind, in this case, is, that the Astronomer Royal has made (apparently at least) a more deliberate, and certainly, a far more dan-gerous, departure from truth, than in a *faux pas* of the same character which he committed in a letter he addressed to Sir James South, and which was pub-lished in the *Times* during a recent con-troversy.

It may be possible that Mr. G. B. Airy has been deceived by misrepresentations, and that the Caster-certificate has been obtained from his *good-nature* while his *discretion* was idle or asleep. Mr. G. B. Airy certainly could not give the first item of this document *from personal knowledge of the facts it certifies,* for he did not succeed to the post of Astronomer Royal until long after the public trials alluded to had ceased. The late Mr. Pond alone could have done this, with any propriety, as he it was who pre-sided during the whole period of the trials; if Mr. G. B. Airy had referred to documents which must still be in his office, he would probably have shaped this part of the certificate very differently if he had thought it consistent with a man of honour to have given it at all.

The *public, undisputed, and pointed contradiction* given to this part by Messrs. Parkinson and Frodsham will assuredly taint the character of the whole certificate, to a degree which must prevent the second part being received with that ready and implicit confidence which ought to follow the official attesta-tions of an Astronomer Royal of England, when committed to writing by himself and relating to facts in practical science

———

* Lately made an F.R.S. for his discovery of " a fundamental principle in all pendulums."

which are within his professional cogni-
zance.

PHILO-VERITAS.

ON STEAM LOCOMOTION ON COMMON
ROADS. BY G. S.

ARTICLE I.

It has been observed that every age
has its peculiar characteristics, which
distinguish it from those that have preceded it. That every fresh era brings
with it a fresh tide of popular result,
and which, from the avidity with which
it has been followed, has been not inaptly
denominated the *rage* of the day. We
have seen the rages for war and volun-
teering, the rages for canals and manufacturing machinery—for foreign loans
and bubble companies—for gas companies—and a great many others, in
which fortunes have been lost and won
with amazing rapidity; and when the
present becomes the past, this will, no
doubt, be designated as the age of locomotion. The rapid strides which have
been made in steam navigation, by which
we are rendered independent of wind and
tide—and the rapidity of transit which
has been effected by means of the railroads, justly entitle it to this characteristic. The newest rage, or that of the
day, is the one last mentioned, which
conquers distance overland, in the same
manner as the previous one does over
water. There is but one thing wanting
to finish the work, to make it perfect and
complete, and crown the whole, and that
is locomotion on the public highways,
at a cheap and expeditious rate, by means
of steam power.

There has been no want of attention,
or niggardliness of application to this
branch of the subject;—eminent, industrious, and ingenious men have applied
the powers of their minds and bodies to
the subject, but hitherto without success,
—discoveries have been announced—patents taken out—companies formed—and
days of starting fixed; but which have
all exploded in smoke and vapours*. We
have seen exhibited for the amusement
of the public, steam waggons, coaches,
and cabs; which have been stared at for
the day, and then consigned to the tomb
of all the Capulets. All this proves,
not that the thing is impracticable, but
that the right way of doing it has not
yet been discovered; not that Nature
has set her face against it, but that we
are not yet sufficiently versed in her laws
to take proper advantage of the vast
powers she has placed at our command
for the purpose. Some progress has,
however, been made even in discovering
that such or such a way is not the right
one; it serves to sharpen the intellects,
and furnishes experience to direct us to
new channels of research, which, though
occasionally unsuccessful, may, eventually, end in the discovery of the object
sought for: like negative quantities in
algebra, which, though nothing, or less
than nothing in themselves, assist us in
the eventual discovery of the positive
quantity sought. The magnitude, and
vast importance of the subject cannot be
denied. The power that would open up
to us an easy, cheap, and expeditious
communication with all the immense
ramification of road, by which the face
of this country is so thickly interlined
and interwoven—bringing locomotive
steam to our very doors, and into every
village to which a road exists, is a consummation devoutly to be wished. Neither can it be denied that the powers of
the steam-engine are competent to all
that is required of it in this direction,
its powers have never yet been known to
fail in any object to which they have been
discreetly applied, and we have no right
to say the case would be otherwise if so
applied here. What then is to hinder
us in the pursuit of an object so very desirable? We have had successive defeats
tis true, but the varieties of application
of which the power at our command is
susceptible, and these varieties of application so very far from being exhausted,
give us every encouragement to go on,
bearing in mind the directions of the poet
who tells us—

" Tho' baffled oft, as oft your task pursue,
  For perseverance ever gains its view;"

and recollecting the object we are in
search of, is nothing less than the perfection of locomotion—the very *acmè* of
travelling. It is no objection to our researches to say that we are already superseded by the railways—that the object
sought has been already obtained through
their means. It has only been *partially*

---

* We must be allowed to dissent *in toto* from
this assertion of our correspondent, so far as it
has reference to Mr. Hancock, whose carriages, if
properly established on a line of road, and fairly
backed by capital and enterprise, would, we have
little doubt, be successful.—ED. M. M.

obtained, and that part a very small one. What we seek is the *liberty* of locomotion—a license for highways and byeways, and for any ways we like, to which, at present, a horse and pair of wheels are eligible. A railroad does not possess this liberty; tied down to its course as completely as its rails are to the sleepers, a carriage on it cannot budge an inch, either to the right or the left without upsetting, or stopping the whole concern.

Besides this, the disadvantages to which railroads are liable, when put in competition with a successful application of steam on the common roads, are numerous and palpable. Amongst the principal disadvantages are the following:—a railroad cannot be efficiently used for short distances, or places in contiguity to each other; it is only effective when communicating with two places at a considerable distance from each other, say twenty miles and upwards, or any extended line, such as from London to Birmingham, Liverpool, &c.

Another difficulty is, that after the trains have once attained their impetus, and got into full swing, they cannot be conveniently stopped, so as to let out a passenger here, and take up another there, who are consequently compelled to go from one station to another, several miles apart, without the possibility of communicating with any portion of the intervening country, and are frequently obliged to pass and leave behind them, or stop short of by many miles, the spot they would otherwise wish to be set down at.

Another difficulty is, railroads cannot be made to accommodate themselves to the route of the principal towns in their line, or choose their course with impunity, from the necessity they are under of due regard being paid to the levels of the country through which they pass; for this reason, they seldom communicate direct with a country town, but only approximate to it as nearly as the levels of the surrounding country will allow. Thus with the exception of the two termini, a passenger, after leaving the train, has to finish his journey by other means than the railway, which has only served to give him a lift on the road, instead of taking him direct to the place which is the object of his journey

Another difficulty is, that the enor-

mous expense attending their construction, renders it necessary to tax their passengers with a charge for conveyance equal to that of the ordinary coaches, consequently, the only saving is in time, not in money; and unless a passenger wants to go to a place at, or in the vicinity of the station when the train stops, (for be it remembered, they do not always stop at every station), it is a chance but what he may have to pay as much more for travelling across the country, as the journey would have cost him direct from London by the stage.

The only real advantage the railroads possess is, as has been stated, a saving of time in long distances; in this respect it is useless to compete with them, further then in attempting to approximate them as near as may be, since no adaptation of machinery on turnpike roads can, by any possibility be made to keep up with them in point of celerity, where the places are a long run distant from each other. But though we cannot expect to equal their speed,—though we have no hopes of attaining that rapid transit which they are capable of giving, we have in view other advantages, equally pleasant and agreeable to make up for it, and to which they are strangers—a liberty to go where we like, which they have not—a liberty of stopping when, where, and as often as we like, which they are deprived of. We have the face of nature, the beautiful prospects of hill and dale,—the delightful views of the country open, before, and around us, while they are winding their way, one half of their time, through doleful cuttings, and dismal tunnels, with nothing but the half-mile posts flitting past them, to break the monotony of the scene, and make their journey bearable.

As we have no hopes of competing with them in celerity, so neither have we any prospects of equalling them in burthen; we cannot expect a common road engine to run away with a train of 20 or 30 coaches, or waggons behind it. But we have a right to calculate on being able to convey from 30 to 60 passengers, at a rate of from 12 to 16 miles an hour, or a waggon heavily laden, at a rate of four to seven miles; and if we do this, we shall be more than a match for them, as we shall be able to do it at a much less charge, in consequence of our having to pay nothing for the making of the roads, except the turnpike dues; while they, before starting, are burthened with a debt of millions, the interest of which must, for a long time, swallow up all their earnings, besides their being placed in the unpleasant predicament of being undersold in the transit market, by a rival, who can afford to do so, and at a a profit too, in consequence of not being subject to similar circumstances.

The importance of the subject being undoubted, it behoves us to inquire into the causes of failure in the numerous attempts which have been made to accomplish it; and by comparing what has been done, with what has not been done, to find out, if possible, what ought to have been done.   This will be treated of in a second article.   G. S.

OBSERVATIONS UPON DR. LARDNER'S EXPERIMENTS TO ASCERTAIN THE FORCE OF ATMOSPHERIC RESISTANCE TO RAILWAY CARRIAGES.

Sir,—Your clever, but very intemperate correspondent, Mr. Cheverton, having in a letter published in your 805th Number, endeavoured to establish the truth of some of the opinions formerly contended for by him, has brought forward for this purpose some recent experiments performed on the Great Western Railway by Dr. Lardner, and seems to think that these experiments are quite conclusive and satisfactory on the point; you will therefore permit me in a few words to point out the inaccuracy of the principle on which these experiments were conducted, and the equally erroneous and futile deductions drawn from them; particularly as it is, I understand, intended to perform a more extended series of experiments of the same description for the purpose of solving the important problem of atmospheric resistance. The great error committed by Dr. Lardner in the manner these experiments were performed, consisted in this, that he calculates the tractive power of gravity of a body moving down an incline plane with great velocity, in the same manner as he would calculate the force necessary to prevent the same train, when at rest, from moving down that plane; by proportioning this force to the weight of the train, as the length of the plane is to the perpendicular height: the distinction in these two cases are very obvious,

and although it escaped the notice of Dr. Lardner, I am astonished how it could have escaped that of your very shrewd correspondent Mr. Cheverton. Mr. Wood himself, in his report to the Directors, does not appear to be satisfied with these experiments on the inclined planes, but he does not hint at where the gross error was committed.

If a body be projected with great velocity in a horizontal direction from the top a of the inclined plane a b, and

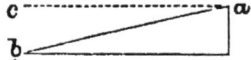

proceed to c with a velocity equal to $96 \times 16 = 1536$ feet in a second, it is well known the force of gravity will make it proceed in the direction of the diagonal b a, and it will not require a solid plane to move on; it could not receive any assistance therefore from the tractive power of gravity on the plane; under these circumstances, and when the body is projected with a great velocity along the plane, this tractive power of gravity will vary under every variation of velocity, as the center of gravity falls thereby with different velocity.   Dr. Lardner might as well attempt to weigh the gravity of a body in a pair of scales, the beam of which would be in constant oscillation, by the ordinary means, as to calculate the force of gravity in the manner he has done. A simple experiment will convince him of this fact: let him take a small incline plane made to move freely on four wheels, let a line be extended from the lower end b, and pass over a pully with a weight suspended to it; now let a be another weight, being in proportion to the former as the length of the incline is to its height, and let this be kept in its position at the top of the plane by a line secured to a hook, and drawn parallel to the incline; under these circumstances every thing will remain stationary, but the moment you cut this latter line the heavy body will descend, and the inclined plane will at the same time move forward. I agree with Dr. Lardner, that when a train of carriages is moving down an incline of 1 in 96 with a uniform velocity of 31 miles an hour, that the tractive power of gravity is the measure of the force of atmospheric resistance and friction; but I do not agree with him, that this power, where a rapidly moving body is concerned, is to be calculated in the same manner which he has done; if the tractive power on a horizontal plane for 15 tons, moving at 31 miles the hour, amounted to the enormous amount of 364 pounds, what I beg leave to ask him, must have been the power of these engines when starting, and the intensity of the engine not reduced by great speed, when the resistance of the atmosphere was trifling; it would more resemble the power of gunpowder and be exerted more in the same way, than the manner we generally find our steam-engines to act.

I am, Sir, &c.

MENTOR.

Dublin, Feb., 1832.

PALMER'S PNEUMATIC FILTER.

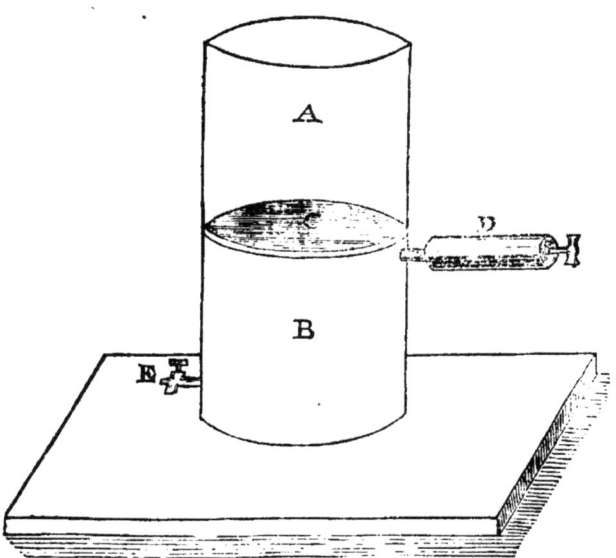

Mr. Palmer, the chemical and philosophical instrument maker, of Newgate-street, has lately made a very useful application of pneumatic pressure, in the filtering of liquids. The following is a description of the apparatus: A B is a tin vessel of any required dimensions, divided by a rim at about the middle. C is a moveable division, formed of a plate of zinc, perforated with numerous holes. This plate is for the purpose of supporting the superincumbent pressure, and above it is laid a filtering fabric of paper, calico, flannel, leather, felt, or other material, according to the nature of the fluid to be filtered. D is an air-pump, attached to the upper part of the lower division. There is a rather heavy loose brass ring to place above the edges of the filtering cloth or fabric to keep the rim close to the side of the filter. E is a cock, to draw off the filtered liquid. The mode of using this filter is as follows:

Lay smoothly on the perforated support, in the centre of the filter, for ordinary purposes a piece of calico cut round to the size of the outside of the loose brass ring; then on the calico lay a piece of filtering paper similarly cut, upon which drop gently the brass ring: pour in the liquid until the vessel is nearly full; then exhaust the air by means of the air-pump, D, when the liquid will quickly filter through, leaving the filter nearly dry. Unscrew the air-pump, and draw off the liquid at the cock, E, and should it not be quite bright with one operation, pass it through again in the same manner without removing the filter.

———

GEOMETRICAL SOLUTION OF MR. WHITE'S TRIGONOMETRICAL QUESTION: BY MR. J. NELSON.

SIR,—I beg leave to send you a solution of Mr. White's first trigonometrical question, proposed in the *Mechanics' Magazine*, No. 804.

I am, Sir, yours, &c.
                    J. NELSON.
Paddington-street, Feb. 15th, 1839.

Construct the triangle A B C, the three sides of which, namely, A B, B C, C A, are 24, 38, 50. Suppose P to be the bottom of the tower; join A P, B P, C P; then, evidently, these lines must be to one another as the cotangents of the angles of elevation. Hence the following construction:

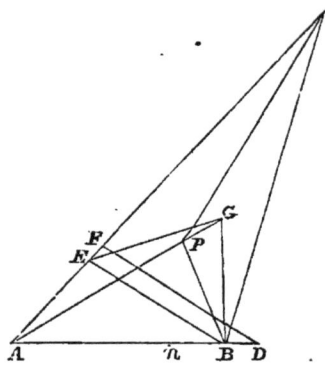

Make A B, A D, A N, in proportion to the cotangents of the angles of elevation taken from the points A B C. Draw B E, making the angle A B E = A C B, and draw D F parallel to B E, and from the centres B and E, with the radii A N, A F, describe arcs intersecting in G. Join A G, E G, B G, and make the angle A B P equal to A G B. Join A P, P C. Then will P be the position of the bottom of the tower.

Demonstration.—The triangles A B P, A G B being equiangular, A P : B P :: A B : B G=(A N) also A B : A P :: A G A B, and A B : A C :: A E : A B ∴ A P A G = A C, A E, both being equal to A B², ∴ the triangles A C P, and A E G, are similar, ∴ A P, P C, :: A E : E G = (A F) :: A B : A D; that is, A P, B P, C P, are respectively as A B, A N, and A D; that is, as the cotangents of the angles of elevation.

Steps of the calculation.—The cotangent of 50° 45′ being assumed = 24, the cotangents of 58° 15′ and 46° 45′, will be 18.1775, and 27.6328 = A N and A D respectively; also A C : B C : A B : B E, or 50 : 38 : 24 : 18·24 = B E; and A C : A B : A D : A F; or 50 : 24 : 27·632 8 : 13·2638 = A F = E G.

1st.—The three sides of the triangle A B C being given, the angle A C B = A B E, will be 27° 35′ 10″.

2nd. The three sides of the triangle E B G being given, the angle E B G will be 42° 43′ 5″; consequently 42° 43 5″ + 27° 35′ 10″= 70° 18′ 15″= to A B G = A P B.

3rd. In the triangle A B G are given two sides and the contained angle, from which we find the angle A B P = 65° 56′ 22″, and the angle G A B = 43° 45′ 22″.

4th. In the triangle A B P are given all the angles, and A B to find A P.

As sin. 70·18·15· ∠ A P B   Logarithm .. 9.973818
To sin. 65·56·22 ∠ A B P 9.960525
So A B 24 .... 1.380211

To A P.. 1.366918
            Tan. 50° 45′ 0.087760

Height of the tower 28.4821 yards...................... 1.454678

The above calculation may be verified by finding the height of the tower from B P,

As sin. 70 . 18 . 15 ∠ A P B...................................... 9.973818
To sin. 43 . 45 . 22 ∠ P A B...................................... 9.839847
So A B 24 .... 1.380251

To B P.. 1.246240
Tan. 56° 15' 208437

Height of the tower 28.4891........................... 1.454677

J. N.

ORAM'S PATENT FUELS.

It is well known that in working coal pits, and in the loading, unloading, and other operations to which coal is subjected before it comes into the market, a great quantity of dust or powder is produced, which, although possessing all the combustible qualities of the larger pieces, cannot be used advantageously in consequence of the minuteness of the particles. Many attempts have been made to bring this article into value by mixing it with various substances, and preparing it in different ways. A correspondent has favoured us with the following particulars of a patent lately obtained by a Mr. Oram for "Improvements in the Manufacture of Fuel," which consists in compounding small coal or dust with other matters, so as the fuel produced is of as great and sometimes of greater value than the large pieces of the same coals. We shall not take upon ourselves to uphold this statement of the patentee or of our orrespondent, nor even the originality of the compound; but enable the reader to form his own judgment by publishing the account supplied to us of the combinations by which Mr. Oram produces his improved fuel.

The materials employed in the making Mr. Oram's fuel are, first, small or dust of bituminous coal; secondly, mud, alluvial deposits, marl, clay, or any other earth containing vegetable matter; thirdly, water; and there are several other substances which may, under certain circumstances, be employed with the above three, but are not absolutely necessary to make a good fuel, such as mineral tar, coal tar, gas tar, mineral pitch, vegetable pitch, rosin, asphaltum, or any other bituminous matter, chalk or lime, sawdust, anthracite or stone, coal, coke or coke-dust, and breeze.

First description of Fuel : — Take thirty pounds of vegetable tar, coal tar, gas tar, mineral pitch, vegetable pitch, rosin, asphaltum, or any other bituminous matter, (the vegetable tar, coal tar, and gas tar, will readily mix with the other ingredients used; but if either mineral pitch, vegetable pitch, rosin, asphaltum, or any other bituminous matter be employed, it should first be dissolved in boiling water, and whilst hot, mixed with the other materials). One hundred and eighty pounds of dry mud (the best for the purpose is that taken from rivers), clay, marl, or any other earth containing vegetable matter, and fifty gallons of water, and mix them together; then add by degrees thirty pounds of powdered lime (stone lime is the best) or chalk, passed through a fine sieve, and one ton of small or dust of bituminous coal. The whole should then be well stirred up with rakes or other suitable instruments, until the several materials are thoroughly combined, or they may be mixed together by machinery, it being necessary to obtain a perfect blending of the materials in order to their adhering together and burning equally. The materials so combined are to be put into moulds of any shape, either square, oblong, or angular) the dimensions of which may be of any size found most convenient, and then pressed either in a screw, lever, or other press.

The lumps or blocks thus produced are to be placed to dry, leaving spaces between the lumps for the circulation of the air; and it will facilitate the drying to place them in a room or shed, the atmosphere of which can be heated, though in warm dry weather this will not be necessary.

Second description of Fuel — suitable for furnaces having a powerful draught. Take ten hundred weight of small or dust of bituminous coal, ten hundred weight of small oven-made coke or coke dust (which proportions will admit of variation), thirty pounds of tar or any other of the bituminous matters before specified, two hundred pounds of dry mud, clay, marl, or other earth containing vegetable matter, fifty gallons of water, and thirty pounds of lime or chalk, and mix, mould, and press them in precisely the same manner as described for manufacturing the first mentioned fuel.

Third description of Fuel:—Take fifteen hundred weight of small or dust of bituminous coal, five hundred weight of breeze (which proportions will also admit of variation), thirty pounds of tar or any other of the bituminous matters before specified, two hundred pounds of dry clay, marl, mud, or other earth containing vegetable matter, fifty gallons of water, and thirty pounds of lime or chalk, mixed, moulded, and pressed in like manner.

Fourth description of Fuel:— Take thirteen hundred weight of anthracite or stone coal, seven hundred weight of small or dust of bituminous coal (which proportions will admit of considerable variation), forty gallons of water, forty pounds of tar or other bitumen, as before, thirty pounds of lime or chalk, and one hundred and eighty pounds of dry clay, mud, marl, or other earth containing vegetable matter, mixed, moulded, and pressed in like manner.

Fifth description of Fuel:—Take fifteen hundred weight of small or dust of bituminous coal, five hundred weight of sawdust (which proportions will admit of considerable variation), forty pounds of tar or other bitumen, as before, two hundred pounds of dry clay, mud, marl, or other earth containing vegetable matter, seventy gallons of water (the quantity of water must be varied in proportion as the quantity of sawdust is used), thirty pounds of lime or chalk, mixed, moulded, and pressed in like manner.

Sixth description of Fuel:—Take five hundred weight of peat turf, peat earth, peat moss, or bog earth, five hundred weight of sawdust, ten hundred weight of small or dust of bituminous coal, thirty pounds of lime or chalk, thirty pounds of tar or other bitumen, as before, two hundred pounds of dry clay,

mud, marl, or other earth containing vegetable matter, and seventy gallons of water, mixed, moulded, and pressed in like manner.

In manufacturing each of the above species of fuel, the lime and bitumen may be omitted; but the use of them not only increases the adhesion of the other materials, but the lime has the effect of neutralizing the sulphureous acid gas contained in the coal, and the bitumen adds to the ready combustion of the fuel. Vegetable tar is preferable to bitumen, and mud (especially river mud, and more particularly such as is taken from the river Thames) to any other earth; stone lime is also preferable to chalk or any other description of lime; and the sawdust from the pine to sawdust of any other description of timber.

GALVANIZATION OF METALS; OR PATENT PROCESSES FOR THEIR PROTECTION FROM OXYDATION, BY COATING OR COVERING THEM WITH ZINC.

In our last volume (p. 122), we published a general description of the invention of M. Sorel, of a means of preservation of metals from oxydation, by taking advantage of the galvanic or electric action which results from the contact of two metals, relatively negative and positive; also reports of various eminent chemists upon the efficiency of the preservative means. The process was originally patented in France by the inventor, and was brought into public notice by a joint-stock company, the shares of which were in great demand, and attained to an enormous premium in the market. In consequence of the favourable reception the invention met with, monopolies were secured in almost every country in Europe, and in America. The British patents were obtained in the first instance by communication of the invention to Captain Craufurd, and of the subsequent improvements of M. Sorel, to M. Lecomte de Fontainemoreau. The " British Galvanization of Metals Company " is, we understand, preparing to apply the invention to practice on a most extensive scale; Messrs. Crawshay, and some of the most extensive iron-masters in the kingdom, being interested in the matter.

In a matter which bids fair to be one of the most important of the age, in connexion with metallurgy, we are happy to be able to lay before our readers full and authentic particulars of the various processes employed—including the substance of all the patents which have been obtained, and the most recent improvements in the process.

First ;—the mode of preparing the metals, or articles of metal, to be coated or covered with zinc.—The plates, or other articles of metal, are to be cleansed of rust, or other extraneous matters which may be adhering to them, in the following manner: They are to be immersed in a bath of water, acidulated with sulphuric acid, which acidulated water, if used hot, should be used in a leaden vessel, or if used cold, in a wooden vessel. Plates or sheets of metal should be placed vertically in the bath. The plates of metal, or articles of metal, are allowed to remain in the acidulated bath till the rust, or other extraneous matter, becomes loose, or easy of being detached; they are then to be taken out and thrown into cold water, from whence each piece or article is taken separately, and scoured with fine sand and a piece of cork, or by any other convenient and suitable means; rubbing them occasionally with a brush as the scouring proceeds, after which they are again thrown into water to remove the loosened particles of rust or dirt. Small articles, such as nails, hooks, and the like, which cannot be conveniently scoured with sand, need only be washed or rinsed when taken out of the acidulated bath. Another method of cleansing, or preparing the metals, or articles of metal, is to wash them separately (except where they are so small that this cannot be done conveniently) either in a solution of sal-ammoniac, or in a bath composed of about equal quantities of water and muriatic acid. In cases where the sulphuric acid bath is used, the metals or articles may remain therein without injury for some time; but when the muriatic acid bath is used, the metals or articles of metal should be dried and coated or covered with the zinc as quickly as possible, otherwise the metals or articles of metal are speedily rusted or oxydized, by the muriatic acid adhering thereto. Where, however, it is not convenient to coat or cover the metals, or articles of metal, immediately after they are taken out of the bath, they may be preserved, temporarily, from rust or oxydation, by depositing them in lime-water, or some alkaline solution. In all cases the metals or articles should, previously to being coated or covered with zinc, be dried, by being held or placed near a fire or over a reverberatory furnace.

Second ;—as to the crucibles or vessels in which the zinc is to be melted.—The zinc is to be melted in a crucible of earthenware, or in a crucible of iron, or iron vessel similar to those generally used in tinning sheet iron, lined internally with bricks, connected with potter's clay or other earthen substances so as to prevent any contact between the zinc and the iron, which would not only produce an alloy of zinc and iron, but cause the crucible to be destroyed, besides preventing the zinc from adhering to the metals, or articles of metal, desired to be coated or covered. The method which has been found to answer most efficiently in practice is, to melt a considerable quantity of lead in an iron crucible, and to drop upon the melted lead an iron hoop of such diameter as just to fit within the interior of the crucible, so as that a part of the hoop shall be immersed in the lead, and part project above the surface. The zinc is added to the melted lead, and when it is fused, being lighter than lead, and not combining to any material extent therewith, always remains at the surface, and only comes in contact with the inner rim of the iron hoop plunged in the lead in the crucible. Any opening which there may be between the hoop and the crucible, is stopped up with clay, to prevent any zinc getting between the hoop and the crucible. By these means the zinc, instead of spoiling or destroying the crucible, only spoils or destroys the hoop, which can easily be replaced by another, or others, from time to time. Instead of an iron hoop, an earthen one may be used in the same manner as already described of the iron hoop. The best method of heating the crucibles is, to surround them with either coke or charcoal.

Third ;—as to plates of metal, or articles of metal of a large size.—The zinc being melted in manner before described, must be skimmed carefully, and its surface covered with sal-ammoniac, rosin, borate of soda, black or white, or other flux.

The operations of covering, enveloping, or plating, should be performed as soon as may be after the zinc, or alloy of zinc, is in a state of fusion, as at a high temperature it would rapidly volatilize the sal-ammoniac, or other material employed as just described, to cover its surface. The plates or articles of metal having been cleansed, or prepared as before described, are to be taken hold of by means of a pair of tongs or forceps, with very narrow-pointed nippers, in order that as small a mark as possible may be left upon the plates, or articles of metal; they are then inserted in the fused zinc, moved about gently, and drawn out slowly, so as not to take up too much of the zinc, and yet so as that the zinc shall not have time to set, or become hard upon the surface of the plates, or articles of metal, covered. The plates, or articles of metal must then be thrown into clean water, and rubbed therein with a sponge or brush, and afterwards dried by passing them through bran, or saw-dust: without this washing and rubbing, the plates, or articles of metal, would be defaced by black spots. Additional clearness and whiteness may be given to the plates or articles of metal, by dipping them rapidly in water acidulated with sulphuric acid, previously to their being thrown into the clean water. Large plates, or articles of metal, must be introduced very slowly into the fused zinc, because when suddenly plunged in, explosions are apt to take place. Where plates of metal warp or bend in consequence of the heat of the zinc into which they have been dipped, or otherwise, they must be passed between rollers, being first powdered or sprinkled with rosin, sand, or some powder of a nature to prevent slipping, as they pass between the rollers. Plates of about the size of ordinary tin plates may be zinced several at a time, by being placed in a grated case with handles, care being taken to keep the plates apart from one another in the grated case. When chain cables or other large chains are zinced, upon being withdrawn from the crucible they are to be shaken, in order to prevent the links from becoming joined or soldered one to another. Cannon balls and other large articles should be heated in a reveberatory furnace, or otherwise, previous to their being zinced. Where articles have screws, or other parts which do not require to be protected, the said screws and other parts may be covered with a thin coating of clay. Vacant, or cut-out places in articles, may be stopped up with wood.

Fourth ;—in certain cases it may be found expedient to apply a second coating of tin.—For example, in cases of large sheets or pieces of metal which may be placed in contact with substances which might corrode the zinc, or in cases of vessels intended to receive acids, or to be used in the preparation of food, where it is necessary that the zinc coating should be covered with tin, pure tin, or tin mixed with two thirds of lead, is in such cases to be melted in an iron crucible, and covered with a layer of fat or tallow, about two inches thick; the articles having been rubbed with a sponge or brush, moistened with a solution of sal ammoniac or muriatic acid, are to be quickly plunged into the melted tin and drawn out slowly, that the tin may properly cover the zinc. The melted tin, or melted tin and lead, should be kept at such a heat as almost to inflame the tallow upon the top thereof.

Fifth ;—process of covering with zinc small pieces of metal, such as nails, small chains, and the like.—Articles of this description being immersed in the acidulated bath before described, must be moved about so that the acid may act as equally as may be on every part, and so that by the friction of the pieces one against another, the oxide or dirt may be removed. They are then dipped in muriatic acid and dried in a reverberatory furnace, and the coating of zinc given in either of the following modes: the zinc (which should be very pure) being fused in a crucible, (small if of iron, to avoid spoiling any large quantity of zinc), and covered with sal ammoniac or other suitable material, the articles to be coated are thrown therein, and allowed to remain for about one minute. They are then taken out slowly with an iron skimmer, and by small portions at a time, allowing as much as possible of the superfluous zinc to drop off. The articles so taken out of the crucible are, upon cooling, necessarily soldered or attached together by the zinc, of which there is besides a considerable excess attached to them; to free them from all that zinc which is not wanted for the purpose of protection, the articles are put into a reverberatory furnace and covered with charcoal: a red heat is then maintained for about a quarter of an hour, and the combined mass of charcoal and zinced articles is mixed, stirred, and shaken by means of an iron rod or poker, until the excess of zinc has been separated from the articles; being then drawn to the front of the furnace by an iron rake or other means, they are kept in agitation until the zinc or alloy is set. The remainder of the process is the same as in the case of large articles before described. But a still more convenient and efficient process of coating nails and other small articles is that which follows:— the articles are placed in a wire basket, which is plunged in melted zinc, covered with a layer of sal ammoniac or other suitable material; the basket is then drawn out of the fused zinc, and carefully shaken, so that any excess of zinc adhering may fall off; after which it is thrown into water or other fluid, not of a nature to attack the sal ammoniac or other covering material used. It is of importance to seize the proper moment for throwing the articles into the water, but the tact necessary in this respect is probably to be acquired by practice only. A perforated cylinder, or any machine of a similar kind, revolving on its axis, may also be used for finishing the nails and other like articles. As soon as taken out of the melted zinc they should be thrown into the machine then revolving on its axis; and this machine should be kept at a degree of heat sufficient to prevent any superfluous zinc stopping up the holes.

Sixth;—to zinc wire.—The following is the method adopted; the wire having been previously prepared in manner before prescribed, is wound upon a drum or cylinder, and one end of the wire being introduced into the fused zinc, covered with sal ammonia or other suitable material, it must be made to pass horizontally through it, and maintained in a proper direction, and with proper degree of motion by any suitable contrivance, and as it emerges from the zinc bath, it must be rolled upon a second drum or cylinder, and afterwards cleaned and finished in the same manner as recommended for other articles.

Seventh:—process for zincing fire-arms

and polished and filed steel articles.— The zinc employed for this description of articles must be very pure, and if melted in earthenware crucibles, these crucibles must, where the articles are large, be placed inside other crucibles of iron or cast iron, and the empty space between the earthen and cast iron crucibles being filled up with fine sand or lead; the melted zinc is then covered over with sal ammoniac or other suitable material, as in the other methods; but to prevent the sal ammoniac or other suitable material from staining the articles by adhering to them when they are drawn from the melted zinc, some earthy substances, such as gravel, sand, lime, or chalk, in pieces or powder are mixed therewith. After the articles to be zinced are well covered over with the zincing, they are polished. First, all unevenness in the zinc coating is made to disappear, which is effected by means of files or scrapers; afterwards, pumice stone, sand stone, or emery or glass paper is used, and then the polishing is terminated as in the case of any other metals. It is important not to use hard bodies to make the coating smooth and give the polish; cork, leather, rags, or other similar things, should be employed. Without this precaution the zinc might be rubbed off in some places. It is to be observed, that to take the polish well, the articles should not have been polished previously to the zincing.

Eighth;—process of making zinc powder.—The zinc is put into a reverberatory furnace, every opening that would admit the atmosphere being carefully closed, and the temperature of the melted zinc raised to a degree approaching red-heat. The door of the furnace is then opened or lifted, and the zinc skimmed; after which, one-tenth of its weight of wrought (not cast) iron filings moistened with muriatic acid (to which some sal ammoniac may be added) is thrown into the fused metal, it being stirred all the while. After the introduction of the filings, the surface of the zinc must be covered over with fine charcoal powder, and the temperature carried to about the cherry-red heat, at which temperature the zinc must be maintained for about one hour, stirring it occasionally with an iron rod or poker. The metal is then conveyed into a brick, earthen, or cast iron trough, and covered over with a lid of cast iron to exclude all contact with the atmosphere, and is stirred by means of an iron rod passing through a hole in the cast-iron door or lid, until, by cooling, it becomes so solid, that it can be stirred no longer. When quite cool it is ground into powder, and the more friable it is found to be the better the operation has succeeded. The powder thus obtained will preserve from oxydation copper and iron, polished and filed steel, such as clock work, articles of hard ware, ironmongary and the like, by merely covering the articles with the powder, and this, even though they should be exposed to wet or damp.

Ninth;—process of making and applying zinc paint and paste.—The zinc may be also applied through the medium of a paint or paste, in the following manner: The powder of zinc is mixed with any of the unctuous substances generally used in paints and varnishes. The substances which have been found to answer best, are those which partake of the nature of the zinc, that is, are conductors of galvanic fluid, and hence this paint has been commonly called galvanic paints. Very good paint is to be made with oil distilled from the refuse tar of gas manufactories, adding one-third of spirit of turpentine. Oil varnish may also be used, but it is objectionable on account of its high price. Linseed oil may be employed, as it is in common paint, but it is not quite so favourable to the galvanic effect. White lead, or ceruse, may be added to the paint to give it more consistence. The proportions of the materials of the paint depend on the substances with which the zinc powder is mixed, and on the uses to which the paint is to be applied. With the same powder a paste may be made by rubbing copper, steel, and filed or polished iron, with which, they will be effectually protected. This paste is made with melted wax, into which is put ten times its weight of powder of zinc, and about 1-50th of tallow or oil.

Tenth;—process of making a zinced paper.—For protecting small polished metal articles, a zinced paper, or wrapper, is employed, manufactured by mixing powder of zinc, ground very fine with the pulp of the paper while making, or by powdering common paper pre-

viously covered with some adhesive substance, such as gum, or flour paste, taking care always to exclude the use of animal glue, which has a tendency to cause the iron to rust.

Eleventh and lastly;—process of plating copper, iron, and other metals with zinc.—Solid plates of zinc may be combined with other metals for their protection from oxydation, by plating in the following manner:—The iron, copper, or other metal, being previously cleansed by the acidulated bath, in the manner before described, it is covered with thin sheet zinc, well powdered with sal ammoniac, and the two sheets of metals are passed through heated rollers, from which they are received into water in a state of perfect adherence.

DR. LARDNER'S INSTRUMENTS FOR EXPERIMENTING UPON RAILWAYS AND THE MOTION OF RAILWAY CARRIAGES.

[From Dr. Lardner's article on the Great Western Railway Inquiry, in the *Monthly Chronicle.*]

Instrument for detecting vertical deflection.—To test the formation and stability of the road, it was determined to observe the effects which the rails and their supports suffered by the action of the wheels in passing over them. Mr. Wood contrived and constructed instruments for this purpose, consisting of a simple lever, the shorter arm of which was placed either under the lip of the rail itself, or under a staple attached to the rail, so that when the rail would sink the arm of the lever would be depressed, and if the rail would rise, the arm of the lever would rise also by the superior weight of the longer arm. Thus every motion of the rail upwards and downwards would produce a contrary motion in the opposite end of the lever, and as the arms of this lever were unequal in the proportion of about six to one, the actual vertical deflexion of the rail was exhibited on a proportionally magnified scale by the motion of the longer arm. In order to register these deflexions, which usually were produced with great rapidity and in considerable number by the wheels of a train successively passing over the rail to which the instrument was attached, Mr. Wood adopted the same method as was previously used in several other self-registering machines. A narrow strip of paper of considerable length, being rolled upon a small cylinder, was gradually unrolled from it to another cylinder, and as it passed from the one to the other it was drawn over a disc against which a pencil was pressed, which was carried by the longer end of the above-mentioned lever. The motion of this pencil upwards and downwards produced by the deflexion of the rail would, if the paper were quiescent, merely draw a vertical line upon it; but by the motion of the paper under the pencil every separate motion of the pencil upwards and downwards produced a waving line, the summits of each wave exhibiting the magnitude of each deflexion. Three of these instruments were constructed by Mr. Wood, with a view to expedite the taking of the observations, so that being applied to different parts of the rail, three sets of deflexions would at the same time be taken by one passage of a train.

Instruments for measuring lateral and horizontal deflexions.—It will be perceived that the effect of the last instrument was only to measure the deflexion of the rail downwards or upwards. After Dr. Lardner had been some time engaged in experimenting with these, he succeeded in constructing another set of instruments, capable of measuring similar effects in the lateral or horizontal direction. These instruments consisted of a compound lever by which any motion of the shorter arm was magnified fifty times, so that when the shorter arm was drawn back or drawn forward in the horizontal direction through the fiftieth part of an inch, the end of the longer arm was moved upwards or downwards according to the direction of the motion of the shorter arm through the space of an inch. The shorter arm of this lever bore by a hardened steel point upon a flat circular disc of steel constructed on the end of a short rod or cylinder moving horizontally in guides. The other end of this cylinder was presented to the side of the rail to which was attached a hardened steel point which bore upon the disc; so that the cylinder thus moving in guides was placed between the two steel points, one attached to the rail, and the other to the short arm of the lever of the indicating instrument. The longer or indicating arm was furnished with a pencil, which registered its indications on paper in the same manner as in the instruments contrived by Mr. Wood for registering the vertical deflexions. The two sets of instruments combined rendered the means of observation of the effects of carriages upon the rails complete. It is evident that the rail could not suffer any effect, which would not be felt, measured, and registered by one or both of these instruments. To the experiments made with these instruments, at least one third of the whole period of this inquiry was devoted, and many hundred diagrams were taken, exhibiting the effects produced not only on the rails themselves, but on the chairs by which they are sup-

ported on the timbers, where timbers are used, and on the stone blocks on which other railways are supported.

Instrument for testing the laying of rails, &c.—In addition to these tests of the effects produced upon the rails by the traffic over them, Dr. Lardner proposed to apply another which would show the state of perfection with which the rails were laid, or their state after the lapse of any length of time. It is evident that on a straight line of railway the two rails on which the wheels of the same carriage rest ought to be at the same level, so that the carriage may stand in a truly horizontal position. A newly-constructed road ought to be laid with sufficient precision to effect this; but after being worked for any length of time, it cannot be expected to preserve it. One rail will subside more than the other, owing to the different degree of firmness of its supports, and of the ballasting beneath them; in fact the rails will lose the correctness of their relative level, and the carriage, when resting on them, will not be as truly vertical in its position, as it would be on a well and newly made railway. An instrument was contrived and constructed, which being rolled slowly along the rails, wrote upon paper as it went with considerable precision the extent to which the rails of the same line departed from a common level. The operation of this instrument may be easily explained. An iron tube of about an inch in diameter is formed of a length equal to the gauge of the line, or the width of the rails; at each end of this are two shorter legs at right angles to it, open at their ends; thus, when the intermediate tube is placed in the horizontal position, the two short legs may be brought to the vertical position; and if the horizontal tube be extended between the lines of rails, the vertical tubes will be immediately over the centre of each rail. Now let us suppose this instrument fixed to a vertical frame, and placed on wheels or rollers, which shall rest upon the rails; let mercury be introduced into it until the horizontal tube and about half of each of the vertical tubes are filled. If the rollers which support the instrument be now made to rest upon the rails, the short tubes being in an upright position, the two surfaces of the mercury in the short tubes must, by the laws of fluids, be at the same level. If the rails be not at the same level, then the mercury will stand higher in the tube which is over the lower rail, than in that which is over the higher one. If the instrument be reversed, the mercury will also reverse its position relatively to the instrument, and will still stand higher in the tube which is over the lower rail,

When the instrument is adjusted, which it may easily be by this process, so that when the rails are truly level, the height of the mercury in one of the tubes is accurately known, then every change which that column of mercury undergoes, while the instrument is rolled over the rails, will indicate a corresponding departure in the rails from the common level, that departure being twice as great as the rise or fall of the mercury.

In order to make this instrument register its own indications, Dr. Lardner placed on the column of mercury in the tube a float, the rod of which, resting above the tube, moved in guides, so as to rise and fall regularly on the surface of the mercury on which it rested, rose and fell; to this rod was attached a pencil, under which, paper being moved in the usual way, a curve was described, whose height above a datum line was always equal to half the departure of the rails from a common level.

Among the several instruments, the invention and construction of which have arisen out of this important inquiry, there is not one which has equal general utility with this self-registering level, and it is only to be regretted that its construction was completed at so late a period, that it has not been applied as extensively to the different lines as might have been wished. Its use, however, will not be confined to this investigation. The advantages which it will offer as a test of the condition of a newly-made line, or of the manner in which the contractor will preserve one in operation, is obvious. It will be a check whose indications cannot be disputed, and they are indications which involve the best qualities of a well-made line. It is evident that its usefulness in practice may be extended by adding to it two other instruments on the same principle, to be rolled each along the same rail. The object of these would be to register every change of level of each rail, independently of the other, in addition to the register preserved by the present instrument of the departure of the two rails from a common level.

Instruments for Measuring the Vibration of Carriages.—An iron tube is extended across the floor of the carriage from door to door, from which rises two perpendicular legs at each door to the height of about twelve inches. The horizontal part of this tube extending along the floor is filled with mercury, which likewise fills the legs to the height of some inches from the angle of the tube, being similar in all respects to the tube used in the instrument for recording the relative levels of the rails. The principal irregularity of motion to which railway carriages are liable, being a lateral swinging to the right and to the left between the rails, this motion immediately affects the horizontal column of mercury which fills the tube extending along the floor, and the inertia of this column causes the column in the vertical tubes to oscillate in proportion to the lateral vibration of the carriage. A float is placed on the mercury in one of the vertical tubes, which bears a pencil similar to that described in the self-registering level, which pencil inscribes on paper each particular oscillation of the mercury, and its exact extent.

This, however, is only one of several irregular motions to which the carriages are liable. Another of these is a rocking motion arising partly from the former lateral vibration, and partly from the irregularity of the level of the rails, either side of the carriage alternately sinking and rising, either as the relative levels of the rails change, or as the conical tires of the wheels mount upon them and descend by the lateral vibration. This rocking motion would cause a body placed at either side of the carriage alternately to ascend and descend in the vertical direction through a corresponding space, and at similar intervals. This motion was measured in the apparatus in the following manner:—A siphon barometer, formed of an iron tube of nearly an inch in bore, was placed at the side of the carriage near one of the doors. This barometer would be raised and lowered as the side of the carriage itself was elevated and depressed by the irregularity of the motion; and this alternate vertical motion being imparted to the mercury in the barometer, the latter, in virtue of its inertia, would receive a corresponding oscillation upon the same principle as the horizontal column in the tube was affected by the lateral motion. A float was placed in the shorter leg of the barometric siphon, which was made to inscribe the vibrations at paper in the same manner as the other instruments.

Besides this rocking motion, railway carriages, like others, are liable to more or less alternate vertical shake common to the whole body of the carriage; and although it was manifest that this was the smallest in amount of all the irregularities of motion, it was deemed right to ascertain it. This was accomplished by a small self-registering siphon barometer placed in the centre of the carriage. All these three instruments were probably mounted upon the same frame, and their three pencils were made to act upon as many discs over which the paper was moved. The rolls of paper were all moved by the same winch, which acted upon a worm and a system of wheels driven by a common band, so that all the papers moved on the respective discs at the same rate, and received upon them the inscriptions corresponding to the different motions. In front of each disc was provided a stamp, bearing upon it the letter indicating the kind of motion recorded on the paper. Thus to the disc at which the horizontal motion was written the stamp H was printed; to that on which the vertical motion was inscribed the stamp V was printed; and that on which the rocking motion was recorded was exposed to the stamp R. All these punches were attached to a common rod, and moved together by the lever provided for that purpose. A person stationed at the window of the carriage at the moment of passing each quarter of a mile, struck the lever with his hand, and punched a letter on the paper which moved over each disc. These letters divided the paper into spaces corresponding to each quarter of a mile, and vertical lines were subsequently drawn upon it, which resolved the diagrams thus formed into portions corresponding to each particular quarter of a mile of the road traversed.

In this manner the number of jolts of the carriage, and the nature and amount of each jolt which took place in each quarter of a mile, were registered.

So satisfactory have been the indications of this instrument, that by inspecting the diagrams the general state of the road can be with great certainty pronounced. In passing along a newly made line, for example, it is at once rendered manifest when the train passes from a cutting to an embankment, the latter being in a state of settlement, and therefore presenting more irregularity of surface.

———

TURF FUEL FOR MANUFACTURING IRON, AND FOR STEAM-ENGINE FURNACES.

The application of turf or peat, prepared in various ways, with and without admixtures of pitch or bituminous matters as fuel, has lately been the subject of considerable discussion and experiment. From the *Dublin Advertiser* we extract the following account of the modes in which the article is used in various parts of Ireland and England, and proposals for the extension of its utility, particularly with regard to the manufacture of iron.

" Ireland has a valuable resource for its reviving industry in its abundant supply of turf. It has been stated that turf could be had on some of the great inland waters of Ireland at 1s. a ton; but let us suppose that good mountain turf could be had for 2s. a ton, two tons of such may make about one ton of charcoal. Turf is charred in Ireland by two methods. The horse-shoers, in those parts of Ireland where coal is much enhanced by carriage, make turf charcoal in small quantities as they want it : a cone of dry turf is built on hard ground, covered partly with dust—it is then inflamed, closed up with dust, and extinguished by water. But on the Mourne mountains and in Roscommon, a chamber is dug in the bog, field with dry turf, which, when sufficiently inflamed, is smothered by the wet stuff thrown out to form the chamber in which the charcoal was formed. This process produces a greater quantity of charcoal than the former method, and more cheaply. Two tons of turf, which will make one ton of charcoal, may be had in some places in Ireland for 4s., whereas a ton of wood charcoal, at the great iron-works in Gloucestershire for instance, will cost £4. To compare the prices of these fuels in the British market, we must add the expense of charring and freight to the first cost of that from the Irish turf; still the Irish charcoal will in most places be much lower than the British wood charcoal. But is it as good? Not for the use of the high furnaces which go from forty-five to sixty-five feet of elevation. The pressure of the contents of the high furnace acting on the copious ashes of turf or turf charcoal, may obstruct the draught; but in the processes of refining bar-iron, and in the making of steel, I suppose it to be as good. But charcoal of any kind is not used in many of the high furnaces of Great Britain. In the four great processes by which iron is brought from the ore to finished bar-iron, namely, fusing, refining, puddling, and reheating, one hundred tons of finished bars requires a thousand tons of coals, for four hundred and thirty of which, turf or turf charcoal may be substituted. Under the boilers of the steam-engines that supply power either for blowing the fires, shaping the metal, or for the production of other necessary forces, the use of coals or coke seems unobjectionable; but wherever the fuel and metal are in contact, charcoal gives a better metal than coal or coke, being free from sulphur and other contaminations. The processes for making refined iron and steel enhance the value so much, that refined iron is double the price of cast iron, and the best cast steel has its value increased in a far greater proportion. In all these processes, charcoal must be preferred to coke, and this, next to the quality of the ore, is a chief cause of the superiority of the Swedish over the British iron. Some of the iron-masters of Britain are well aware of the value of turf fuel. The proprietor of a large work in Gloucestershire bought an estate containing turf bog to supply his works ; the plan, however, was laid aside, from the difficulty of transport. We have five millions of acres of waste land. Let us suppose half of these are covered with turf bog from one to twenty feet deep, having a specific gravity a little less than water, and containing, probably, four to six parts water, to one of inflammable matter. Of what immense value, then, is the turf we possess and neglect! It could not be all converted into charcoal in a few years ; no, and so much the better. But much more turf might be made than can be on the present system, if a plan mentioned in an early number of the *Journal des Mines* was adopted. Turf partially dried was thrown, without order, into a house which was adapted for the communication of heat without flame, and the escape of vapour ; in which it was exposed to the heat of 200 degrees Fahrenheit, and rapidly dried ; thus the preparation of turf might become a continual employment, independent of season or weather. We therefore command a vast supply of a fuel nearly as valuable as the best used in the British iron-works, and in some situations cheaper than the worst. I have heard that there are extensive iron-works in England where the coals cost more at the mouth of the furnace than 12s. per ton.

" In the year 1827, iron was made in Great Britain in 284 furnaces, to the extent of 690,000 tons, which being converted partly into castings, and partly into bars, rods, and plates of malleable iron, was worth £6,290,000. It has greatly increased since. 240 vessels freighted with iron articles sailed from England to America in less than twelve months after the great fire in New York. Here is abundance of fuel in Ireland, and demand in England ; how are they to meet? By cheap railroads from the turf mountains to water carriage. The Americans make railroads for £1000 a mile ; we have labour and iron cheaper than they, but it is better to imitate them than the British, who can afford to spend £70,000 per mile on railroads. Where should they first be made ? From Dublin to near Lough Bray, where, before the military road was begun, a ten-foot rod was driven into the bog with little resistance. The turf might be conveyed to Dublin by a railroad, and thence to Liverpool ; the waggons may descend by their own weight to Rathfarnham, and draw up lime, tools, and industrious hands to cultivate one hundred square miles of neglected valuable mountain, adjoining a city where many thousand good labourers are perishing for want of employment.

"There will be also a demand for turf, charred or not, for the steam-boats. By this fuel the heat can be expeditiously raised. In London, charcoal is sold for lighting fires. In conservatories, turf charcoal has been found by the Dutch preferable to wood charcoal, when used in open braziers, so as to warm the house without the expense of flues. It was said to answer particularly with orange trees. From a series of four experiments which were made, as detailed in the *Annales des Mines*, and also from experiments carefully made in this country by Dr. Stokes, it was found that the calorific power of charcoal made from peat was equal to that of wood charcoal. The experiment by Dr. Stokes was the evaporation of a certain quantity of water, by the two descriptions of fuel, in the same apparatus, and under the same circumstances."

At a late meeting of the Institute of Civil Engineers the use of peat in the manufacture of fuel was the subject of discussion.

It had been remarked at a previous meeting that the iron made with peat fuel was more malleable than Swedish, and that the tools were of a superior quality. It was doubted whether peat fuel had been recently employed, or, indeed, whether it could be used at all in the puddling furnace, though it might in the refining or smelting furnace, but with a diminished produce. The working of iron by peat fuel was known to improve its quality in some respects, and the welds especially thus made were superior to those made with coal. The Dartmoor peat was frequently used for this purpose, and found exceedingly good. The improvement of iron by the use of a particular fuel seemed a very difficult question. The weld made with ligneous carbon, owing to the absence of sulphur and pyrites, must be better than that made with a fuel containing these impurities. The analysis of peats is very various. They all contain 5 per cent., and some 20 per cent., of earthy matter. Some kinds of peats were stated to produce three times as much gas as coal. Peat was said to contain no sulphur, but the experience of several gas works, in which peat was employed, proved that some peats contain large quantities of sulphur, as the purifiers become rapidly filled with sulphuretted hydrogen. All coal, however pure to the eye, contains pyrites and sulphur, so that sulphur must be considered as one of the elements of coal. Much is to be attributed to long practice in the use of fuels ; the smiths of Cornwall can use peat, and the smiths of Pembrokeshire anthracite, for all purposes of working iron ; both would, however, use pit coal could it be conveniently procured.

At the same meeting Mr. C. W. Williams presented specimens of peat, from the first state, as taken from the bog, to the last, when compressed and converted into a hard coke; and of his new resin fuel, or artificial coal, which is composed of resin and turf coke. This resin fuel is found of the greatest use in long voyages, when used with a proper proportion of coal, as it enables the fireman to maintain the requisite pressure of steam with great regularity, and also to raise steam more rapidly on any emergency. It is not adapted for use as a fuel by itself, but when about 2½ cwt. of this fuel is used with 20 cwt. of coal, by throwing it in front of the fire with each charge of fresh coal, a much better combustion of the coal takes place, and the effect is equal to that which would be produced by 27 cwt. of coal. Thus, 2½ cwt. of this fuel so employed is equivalent to 7 cwt. of coal. The cost is from 35s. to 40s. per ton. The transatlantic steamers carried from 40 to 60 tons of it, and besides the advantages attending its use, there was a saving in room, which was applicable to the stowage of cargo.

A long discussion took place amongst the members on the important facts which the application of this fuel had elicited.

These appeared in some measure contradictory to the results, which could not be doubted that 9 lbs. of coke will do as much in any department of the arts as 12 lbs. of coal; for on adding to coal a peat and a hydro-carbon far more inflammable than coal, the result is equivalent to that which is produced by all the carbon, hydrogen, and oxygen, in many times the quantity of coal. It was remarked, that the circumstances under which fuel was employed ought to be considered, as the consumption of fuel under steam-boilers could hardly be compared with the consumption for simply heating and keeping hot a large mass of matter as in a glass-house. It could not be believed, that the absolute quantity of heat from the coke of a ton of coals is the same as of the ton of coals, for in that case all the heat of a coke oven would go for nothing, and there were instances of this being beneficially employed.

Lord Willoughby's machine for the compression of peat we have already described, as also published various accounts of its working. We have lately heard further statements of the success of its operation.

METHODS OF CASTING IN PLASTER, &c.

[Delivered in a Lecture, at Liverpool, by Mons. Bally, the Travelling Assistant of the late Dr. Spurzheim.

To Cast in Plaster.—Obtain some fine plaster of good colour, and pass it through a muslin sieve to remove any coarser particles which may be present. By mixing gum arabic with the water intended to be used in the plaster, not only will the plaster be rendered very hard when it sets, but a beautiful gloss will be given to the surface. Care must be taken to drop the plaster powder gradually into the water, and to permit the bubbles to rise before the mixture is stirred, otherwise it will become lumpy. The plaster should be of the consistence of the yolk of an egg, and of course used immediately. If the medal intended to be copied, is a valuable one, with a smooth surface, it will be advisable not to oil it, as in cleaning the oil off, the polish may be injured; but if the surface be rough, there will be no remedy, and the oil must afterwards be removed by dabbing the surface of the medal gently with a soft cloth. A rim of thin lead, brass, copper, or even oiled paper, is then tied round the medal, and some liquid plaster, in the first place, stippled over its surface with a soft brush, to prevent the formation of air bubbles, as well as to insure its insertion into the most minute crevices; after which the plaster is poured upon the surface to the thickness of half an inch, or an inch if a large medal. To cause the separation of the mould from the medal all we have to do is to immerse it in water, when it is readily removed, otherwise the mould is sure to be broken. To obtain a plaster cast from this mould, we must oil it with warm boiled linseed oil, and allow it several days to dry. Whenever the mould is used, it must be well oiled, otherwise the surface of the casting will be destroyed. The best olive oil must be used, or the colour of the plaster will be injured.

To Cast in Wax.—The mould is first made in plaster, but before being used is placed in warm water, of which it is allowed to absorb as much as it will take,—oil not being used in this process. The surface must then be allowed to dry, or the wax would not adhere closely. Pure wax is too greasy for the purpose, and bladder flake white is therefore mixed with it. The quantity cannot be stated, but the addition of too much gives wax the appearance of plaster, by taking away its richness. The oftener the wax is remelted its colour is injured. In order to obtain a gray marble colour, a marble powder, procurable of any statuary, is mixed with the wax, which not only gives a beautiful appearance to it, but renders it more durable. The wax is poured into the mould, and allowed to flow over its surface, and by moistening the plaster mould in water when the wax has become hard, the cast is easily removed. Wax models may be fastened by means of boiled linseed oil and flake white, and also by a combination of bees' wax and rosin.

To Cast in Sulphur.—This is a very permanent mode, but as a mould it can only be used for plaster—for hot wax or sulphur would injure its surface. When sulphur is heated to the temperature suitable for forming casts it becomes nearly black, and has therefore to be colourd with vermillion in the proportion of one ounce of vermillion to three of sulphur. The surface of the mould, however, need only be coated with this expensive mixture, and common sulphur added in any quantity. You must use wood to stir the sulphur, as iron will take away its colour. The sulphur will take fire in melting unless it is properly stirred, and at first will become thick and viscid, but by continuing the application of heat it will again assume a perfectly liquid form.

To Cast in Glue.—If a medal is so much sunk and engraved that you cannot get a plaster cast off, a mould may be obtained by pouring glue upon it. In this manner a bunch of grapes can be taken in the natural state, and by cutting the glue down the centre the grapes can be extracted, and the mould used to produce a representation of the original in plaster. Isinglass may be similarly used, but it is first mixed with flake white in the state of powder.

To Cast in Bread Paste.—Take the inside of a penny roll and work it up well with vermilion, the longer the better, until it becomes viscid and tough; it is then to be worked well into the mould. After having obtained the mould, it must be fastened down upon a piece of wood by wetting it, so as to prevent it from warping as it dries. After it has been thoroughly dried, you may oil it, and then obtain as many casts as you please from it in plaster, wax, or sulphur. By means of bread paste a traveller may always take a mould of any small object of interest he meets with on his journey, and thus a proper knowledge of its mode of use becomes invaluable.—*Liverpool Mercury.*

MODES OF MAKING METALLIC AND COLOURED INKS.

Gold ink is made by grinding upon a porphyry slab, with a muller, gold leaves along with white honey, till they be reduced to the finest possible division. The paste is then collected upon the edge of a knife or spatula, put into a large glass, and diffused through water. The gold by gravity soon falls to the bottom, while the honey dissolves in the water, which must be decanted off. The sediment is to be repeatedly washed till entirely freed from the honey. The powder, when dried, is very brilliant, and when to be used as an ink, may be mixed up with a little gum water. After the writing becomes dry, it should be burnished with a wolf's tooth.

Silver ink is prepared in the same manner.

Red ink.—This ink may be made by infusing, for three or four days, in weak vinegar, Brazil wood chipped into small pieces; the infusion may be then boiled upon the wood for an hour, strained, and thickened slightly with gum arabic and sugar. A little alum improves the colour. A decoction of cochineal with a little water of ammonia, forms a more beautiful red ink, but it is fugitive. An extemporaneous red ink of the same kind may be made by dissolving carmine in weak water of ammonia, and adding a little mucilage.

Green ink.—According to Klaproth, a fine ink of this colour may be prepared by boiling a mixture of two parts of verdigris in eight parts of water, with one of cream of tartar, till the total bulk be reduced one half. The solution must be then passed through a cloth, cooled, and bottled for use.

Yellow ink is made by dissolving three parts of alum in 100 of water, adding 25 parts of Persian or Avignon berries bruised, boiling the mixture for an hour, straining the liquor, and dissolving in it four parts of gum arabic. A solution of gamboge in water forms a convenient yellow ink.

By examining the different dye-stuffs, and considering the processes used in dyeing with them, a variety of coloured inks may be made.

Sympathetic ink.—The best is a solution of muriate of cobalt.

REPORTS OF MEETINGS AT THE ROYAL INSTITUTION.[*]

February 8th.—*The Conversazione.*—Very few articles entitled to notice were laid on the table this evening. The principal curiosity was a print sent by Mr. Brockedon, taken in three different sizes from *the same* engraved plate. The singular artifice by which this is effected has already been the subject of remark and discussion in the *Mechanics Magazine.*' The largest impres-

[*] We are promised a continuation of these reports—Ed. M. M.

-sion was, as may be expected, very faint in comparison with the smallest. Notwithstanding the many conjectures which have been made by Mr. Babbage and others, the real process appears to be yet undiscovered.

The Lecture. — The Lecturer was Mr. Parsey, and the subject, as may be supposed, was "Perspective rectified." This is a very unfavourable theme for a popular lecturer of which character the Friday Evening Lectures are intended to be ; but Mr. Parsey had recourse to models more than to diagrams, and by which he very happily illustrated, and more than amply explained, his peculiar opinions on the subject. The publicity which is given, even to erroneous notions, is not in the end unfavourable to the promulgation of truth, for opposition is sure to be aroused ; and what previously, though correct, were feeble mental perceptions, are strengthened and rendered clearer to the mind by the self-prompting which is induced, of the true ground and reason of its belief. Such, to all appearance, was the effect produced on the audience ; they seemed to be only the more fully convinced that the Lecturer was wrong, or rather that he had taken a one-sided, and therefore a very imperfect view of his subject.

Mr. Parsey contends, that the measure of the angle under which an object (lineal) is seen, is the true quantity of its perspective delineation ; that therefore the base of an isosceles triangle, whose vertex is at the eye, and whose sides are the extreme rays proceeding from the object, is the only true line on which it can be represented ; that this triangle being taken vertically as well as horizontally, as it necessarily ought to be, it forms a cone, the base of which is the only true perspective plane on which objects can be delineated ; and that this plane is thus consistently made perpendicular to the axis of vision ; for objects, whether taken in the direction of their height or their width ; it being by its adoption in the latter particular, admitted to be the correct method. The conclusion is, that the usual mode of delineating objects being on a supposed intervening transparent vertical plane, which plane being in general *oblique* to the visual axis, is in all such cases *mathematically* incorrect, and that their true perspective can be represented on such supposed plane only when placed *perpendicular* to the said axis.

Mr. Parsey would have us believe that he is led to this conclusion by a mathematical necessity inherent in the very nature of things, and therefore he names his new system, "Natural perspective," as well as "Perspective rectified." There lies the

fallacy. Geometry prescribes the rules of perspective generally, and therein is able to teach the correct delineation of objects on any proposed perspective plane, but here science ends and art begins ; for as to the proper choice of the plane, it is to be determined by a variety of considerations, which appeal only to artistical skill and judgment. This is entirely a practical point, and to decide it on mathematical principles, is to mistake the true ends of science, and to make it an imperious master where it ought to take the place of an assisting servant. If Mr. Parsey will endeavour to show that on the principles of *art*, the perspective plan which he advocates is to be preferred in all, or in many cases, to that which is usually adopted, he will command attention ; but in no other way will he succeed or will he deserve success. To show that there is no geometrical or necessary reason for the adoption of either the oblique or perpendicular perspective plane, it is sufficient to observe, that the same object viewed from the same point, will appear in its delineation on either of these planes (supposed to be transparent and to invervene) to be exactly coincident with *both*, for both projections strike the eye by the same rays as does the object itself ; and this was clearly shown by one of Mr. Parsey's own diagramic models. In fact, the delineation on the one plan was seen to be the foreshortening of that on the other. So also any perspective drawing of the same object being delineated on either plane, or indeed on any plane, will present precisely the same appearance, when viewed from those respective points of position and distance, where alone they are to be properly appreciated ; but as in that extreme case of a very oblique plane called an "anamorphosis," the objects appear as monstrous when seen from the more usual points of view, so it may become a point for discussion, whether, in regard to drawings, as generally and irregularly seen, they appear to less disadvantage, by being delineated on the one or on the other of the perspective plans in question. Whether Mr. Parsey can *on this ground*, or on any other connected with *art*, establish superior claims, to attention in behalf of that perspective plan which he patronises, and in opposition to the uniform practice of all artists, is quite another question, but this task is what he ought to have attempted instead of hunting out mathematical mares nests. This is, not the first crotchet of the kind which Mr. Parsey has entertained, for if we recollect right, he professed a few years since to have discovered the solution of the celebrated problem of squaring the circle.

CONSTRUCTION OF LIGHTHOUSES ON SAND, BY MEANS OF MITCHELL'S PATENT SCREW MOORINGS.

We now lay before our readers one of the most important experiments of the present day, which promises to give to the engineer a foundation as secure in the sea as he has hitherto enjoyed on the surface of the earth. The success of this attempt will give us resources of battle with an antagonist, before which all our mechanical strength has too often proved defective, while to the maritime interests of the country it will afford new and further protection. We can appreciate the difficulty which Smeaton encountered in planting the Edystone on the firm rock ; but we have now the means offered to us of security even upon the shifting sand.

When we gave* a description of Mitchell's patent screw moorings, we did not then anticipate the application which they have since received. It having been brought under the notice of the Corporation of the Trinity House, that this instrument might be advantageously applied in establishing lighthouses on sands, their attention was immediately given to the subject, and accordingly directed an experiment to be made to ascertain its practicability, under the superintendance of their engineer, Mr. James Walker.

The spot selected is on the verge of the Maplin sand, lying at the mouth of the Thames, about twenty miles below the Nore, forming the northern side of the Swin, or King's Channel, which, on account of its depth, is much frequented by large ships, as also by colliers and other vessels from the North Sea, and where a floating light is now maintained. This spot is a shifting sand, and is dry at low-water spring-tides. The plan is to erect a fixed lighthouse of timber framing, with a lantern, and residence for the attendants. For this purpose, in August last, operations were commenced to form the base of an octagon, 40 feet diameter, with Mitchell's mooring screws, one of which was fixed at each angle, and another in the centre ; each of these are 4 feet 6 inches diameter, attached to a shaft of wrought iron about 25 feet long, and 5 inches diamer, and, consequently presenting an immense horizontal resisting surface. For the purpose, a stage for fixing the screws, a raft of timber 30 feet square, was floating over the spot, with a capstan in the centre, which was made to fit on the top of the iron shaft, and firmly keyed to it ; a power of about 30 men was employed for driving the screws ; their united labours were continued

until the whole force of the 30 men could scarcely ⸺ the capstan : the shafts were left standing about 5 ⸺ above the surface of the sands. The fixing of the ⸺ screws including the setting out the foundation and adjusting the raft, which had to be replaced every tide, did not occupy more than nine or ten days.

This is the portion of the work hitherto effected, and its continuation will be proceeded in when the proper season comes in the ensuing spring. Upon this foundation the superstructure of timber is to be constructed, consisting of a principal post, strongly braced and secured, with angle-posts made to converge until they form a diameter of about 16 feet at the top, giving the superstructure the appearance of the frustrum of an octangular pyramid, the feet of the angular posts and braces are well secured and keyed down to the tops of the iron shafts, and the whole is connected at top and bottom with strong horizontal ties of wood and iron. The entire height of the superstructure will be 30 feet above the top of the iron shafts ; up to a point about 12 feet above high water-mark spring-tides the work will be open ; the part above will be enclosed as a residence for the attendants ; in the centre and above this will be erected a room or lantern of about 10 feet diameter, from which the lights are to be exhibited.

The interval that has elapsed since the screws were fixed has fully proved the security of them, which, although driven into sand, seem as if fixed into clay, and in this state they have remained since the summer. The whole process confers the greatest credit both on the engineers and Mr. Mitchell, the patentee of the screws who superintended the work (assisted by his son), and we feel happy to hear that his ingenious invention daily obtains a greater extension.

The importance of this experiment certainly calls for a trial, and it was with due liberality that the Trinity Board sanctioned the expense. To them it involves the question of a better security of the light, and a less expense in its maintenance, both objects justifying the experiment, and counter-balancing the expense of prime cost in such construction. The insecurity of floating-lights has been too manifestly productive of disastrous consequences not to call for a remedy, and it will be fortunate if by this means it be obtained. Within the last month the Nore light was blown from her mooring ; and the breaking away of the North-west Light of the Mersey is supposed to have led to the lamentable shipwrecks at Liverpool.

We can perceive only one objection which can be started, and that is rather to be de-

* See Mech. Mag., vol. 28, p. 289.

termined by experience than conjecture, that is—how far the edifice is liable to be washed away by storms as one of the Edystone buildings was ; but this in our opinion, will be mainly provided against by the unity of construction and the breadth of base, well secured to the shafts by the screws.

The progress of this work will naturally be watched with interest, for it is one which in its influence is not limited to this individual case. It is of much more importance than chain piers, as it will enable us to obtain a foundation in positions where they cannot at present be used. It must be remembered that the screw can be employed where the pile is of no avail, and that it possesses a much stronger hold, and has greater durability.

We shall thus, therefore, be able to construct piers and breakwaters in localities inaccessible, and be enabled to render important service to the interests of commerce. We think, too, that the screw itself would be of great utility in securing the end chains of suspension-bridges, as its powers of resistance can be extended to any necessary degree by an increase in size. The greater employment of the screws, which would arise from their successful application, will have a further beneficial effect in enabling the patentee to supply them at a diminished expense, which under their present limited sale, is necessarily high.

The carpenter's work of the superstructure is about to be contracted for, which is intended to be erected and put together at the wharf at Blackwall, to save time of fitting, &c., at the spot.—*C. E. and Arch. Journ.*

RECENT AMERICAN PATENTS.

(Selected from the *Franklin Journal* for October and November, 1838.)

IMPROVEMENTS IN THE CONSTRUCTION OF THE SAW CYLINDER FOR COTTON GINS ; *Jacob Idler, Philadelphia.* — This patent is taken for the employment of metallic rings between each of the saws, in order to keep them apart. These rings are to be cast with arms, in the manner of ordinary cast wheels ; the spaces between the wheels will, it is said, allow room for the buckle in any of the saws to be relieved without the edges being affected thereby. An iron shaft is to pass through the centres of these discs, and of the saws, which are to be wedged or screwed up in the ordinary way.

The claim is to " the forming the cylinder of the said cotton gin with hollow cylindrical metallic sections ; with projections or bearings on the sides of the arms and ends of the sockets of the said sections, for sustaining, in conjunction with the sides of the rings or sections, the circular saw plate, in a firm and true position on the shaft, and parallel with each other ; the spaces formed between the arms and bearings, allowing room for the swellings or bucklings of the saws, so that the part of each saw plate outside of the cylinder, which runs between the ribs or combs of the gin, shall always be true and even."

We believe that gin saws are not now usually made of round plates of steel, like circular saws, but of two flat semi-hoops, the inner edges of which are confined between the discs by which the saws are divided ; these are cheaper than whole plates, and are less liable to buckle.

IMPROVEMENTS IN THE MACHINE FOR STEAMING AND MASHING APPLES ; *John Dimm, Pennsylvania.*—This apparatus is to prepare apples for fermentation by crushing and steaming them, so as to reduce them to a pulp or pommage. A rectangular box is made, across which there are two fluted rollers for crushing the apples ; this box is surmounted by a hopper, having a close lid to retain the steam which is to be introduced into it. Above the crushing rollers there is a horizontal sliding shutter, fitting the box, and supporting the apples in the hopper. This shutter is made of two boards placed one above the other, and having a space between them into which the steam is to be introduced through a suitable tube.

The upper board is perforated, so as to allow the steam to pass through, and to act upon the apples in the hopper ; when they are sufficiently steamed, the sliding shutter is withdrawn, and the apples then rest on the fluted cylinders. Motion being given to these by a winch, the apples are reduced to pulp, and fall upon an inclined trough below, which conducts the pulp away, prepared for the fermenting tub.

MACHINE FOR SAWING THROUGH TREES, &c.; *Samuel H. Hamilton, New York.*— This is an ingeniously contrived machine, intended principally for the purpose of sawing down trees, but which may also be adapted to the cross-cutting them after they have been felled. The machine is to be drawn from place to place upon two wheels. The saw is made stiff, as it has no straining frame, but stands out from the machine like a key-hole saw from its handle. The gearing by which it is driven is operated upon by means of a vertical wheel, which is turned by a winch. This acts also as a fly wheel, and is at the same time bevil-geared into a horizontal wheel, which gives motion to the crank that operates on the saw.

The frame of the machine is sustained, in front, upon a roller resting on a curved way, and this frame turns on a vertical shaft in the centre of the curvature of the way, allowing the frame, with the projecting saw, to move round, so that in the operation of cutting it may be made to bear continually against the tree; a weight hanging over a pulley serves to keep it up to its bearing. The claims need not be given, as they refer to the arrangement of the respective parts, as described.

Under particular circumstances, trees may no doubt be advantageously cut by such a saw; but for felling in the woods we do not look for any thing superior or equal to a good American axe and axeman. Such a machine as that described, cannot be moved about in the woods in a new country; and in most situations whilst it was being prepared and anchored, the axeman would be able, unaided, to complete the work. There are other serious objections to sawing machines for felling. The saw will bind in the kerf in spite of all precautions; even the wind alone, were there no other cause, would frequently produce this effect. The axe-man can always determine one of two ways in which his tree shall fall, but not so with the manager of the saw; and woe to his machine when a tree falls upon it, an event which is not only possible, but very probable.

ARTIFICIAL STONE: *Joseph Woodhull, New York.*—The materials to be employed are plaster of Paris, quick lime ground to fine powder but not slaked, oxides of iron, and of mnaganese, calcined at a full red heat, and the kind of iron ore denominated poor ore, or whites, calcined and reduced to fine powder. The proportions to be used are five parts of the plaster, two of lime, one of manganese, one of whites, and four of sand, all estimated by measure. These materials are to be well mixed in the dry state, 'the whole is then to have sufficient water added to adapt it for being put into suitable moulds; when this is done, stones broken small are to be added, until the mould is full. In from twenty to forty minutes the moulded material may be removed. Variegated marble is to be imitated by using suitable colours, and proper manipulation. For in-door purposes the matallic oxides may be omitted; for outside work the material is to be saturated with linseed oil. The claim is to "the combining together plaster of Paris, lime, the oxide of manganese, the description of iron ore denominated whites, and sand, all previously calcined, and the whole prepared and managed substantially in the manner set forth."

There will thus be made a very inferior kind of scagliola; an article when well compounded and managed, of great utility and beauty; which is what we dare not predicate of the compound above described. The validity of this patent admits of much doubt, as the combination can scarcely be denominated new, in any respect; at all events, the right extends only to the particular composition as given, and whilst there are many others which are better, equally cheap, and not covered by any special claim, it is apprehended that the right will be of little value to the owner, whilst it will not stand in the way of any one.

MACHINE FOR CLEARING BURRS FROM WOOL, IN THE SKIN; *Erastus Tracey, Duchess County.*—The skins with the wool on, as imported from South America, are filled with burrs, from which it has hitherto been found difficult and expensive to remove them, but with this machine it is said to be readily and perfectly effected. A cylinder is made about two feet in length, and one foot in breadth, and upon this are placed several rows of steel teeth, about an inch long, and an eighth of an inch apart. This cylinder is made to revolve with great rapidity, and the skins are fed up to it by means of a feeding apron passing around rollers, one of which is nearly in contact with the revolving cylinder, and has above it a pressing roller, to hold the skin firmly. The endless apron, with its appurtenances, is on a separate frame, which is made to approach or recede from the revolving cylinder by means of a rack and pinion. The skin is laid on the revolving apron, and carried forward so that one-half of it hangs down from beneath the pressing roller; it is then moved up so as to be acted upon by the revolving cylinder, and drawn back on the revolving apron, until the first half has been cleaned; the skin is then turned end for end, and the operation completed. The claim is to "a machine having a revolving cylinder furnished with teeth to which the skin, placed upon an endless apron, is fed up and operated upon, as herein set forth."

SAFETY RAIL ROAD CAR; *William Kinkead, Maryland.*—The claims are to "a method of sustaining the car in case of the breaking of the wheels, axles, &c. in the manner described. And to the manner of constructing the safety hooks with joint or hinges, screw rod and nut, so that said nut can be engaged with, or disengaged from, the centre guide bar, or safety rail, with great facility and despatch, in case of any trivial accident which might arrest the progress of the cars, or in case it might be desired to turn out of the main track."

This is an *improvement* in railroads, such as is frequently devised in the parlour, but which never finds its way on to the track, and which, were it placed there, would prove to be but a transient lodger. A third rail is to be laid in the centre between the ordinary rails, and is to extend up nearly to the bottom of the cars*; there is to be a plate on this rail, forming it into a T rail, and from the bottom of the car hooked pieces are to descend and clip under the top of the said T plate. Rods reaching into the car act upon and withdraw these hooks when necessary. Should a wheel, &c. break, the car is to rest on the centre rail, and should the car tend to leave the track, the hooks and safety rail are to prevent it.

STRENGTHENING THE AXLES OF RAILROAD CARS AND LOCOMOTIVES ; *Ziba Durkee, Philadelphia.*—Upon the middle of each axle there is placed an iron cylinder, having flanches rising from each of its ends, giving to it the general form of a common bobbin. These flanches are each to have holes through them to receive the ends of rods of iron, the other ends of which pass through iron plates placed on the face of each wheel, and surrounding the hubs; there may be four, or any preferred number, of such rods, which are to be drawn tight by nuts and screws. This, the patentee says, will "greatly strengthen the axle, and prevent it from breaking, and support the wheels should an axle break. With this attachment a smaller axle than those now in use will answer every purpose."

ATTACHING SPRINGS TO CARRIAGES ; *David A. Morton, New York.*—To the underside of the bottom of a carriage, are to be fixed cases, or tubes, running lengthwise, and near to each side. Within these cases there are to be spiral springs, wound so as to leave a space between each coil, to allow of their being drawn together. Rods passing through these springs have straps attached to them, which straps extend out at either end and wind round a cylinder fixed upon a shaft for that purpose. To the jacks by which the body is to be suspended are also attached straps which in like manner wind around cylinders upon the same shaft with the straps first named, but in a reverse direction. The straps which are attached to the springs pass over rollers at the ends of the spring case, to conduct them to the cylinders upon the shaft. Different modes of modifying these springs, straps, cylinders, and their appendages, are pointed out by the patentee ; there may, for example, be but one spring case along the centre of the

carriage body, or the spiral springs may be affixed without their being enclosed in a case.

The patentee says, "it will appear obvious that when the body is thus suspended and the carriage is in operation, a portion of all the straps are alternately unrolling and rolling up on the cylinders. When the body settles, a portion of the straps attached to the springs are at the same time taken up, which will consequently contract the springs. When the body rises a portion of the straps attached to the jacks are taken up, and a portion of those attached to the springs are at the same time thrown off, which will necessarily elongate the springs."

"What I claim as my invention and desire to secure by letters patent, is the combination of the above described rollers, the straps, and the irons by which the springs are contracted, with horizontal spiral springs attached to the under side of the carriage body as herein described."

There is certainly much ingenuity in the above described arrangement, and, so far as we know, it is essentially new ; and we think also that these springs will operate pleasantly ; the great objection to them is their liability to fracture, an accident of frequent occurrence in spiral springs similarly acted upon.

LOCOMOTIVES AND RAILROAD CARS ; *Jonas P. Fairlamb and L. C. Judson, Philadelphia.*—"What we claim as our invention," says the patentees, "and desire to secure by letters patent is the construction and application of the cylinder bunts, to graduate the concussion of two bodies coming in contact. The construction and application of the safety guard clamps to wheels, to avoid danger when an axle breaks. The application of small truck wheels to guard against danger when a large wheel breaks. The spring lever, and mode and manner of guiding cars from one track of railway to another without the use of switches; and the construction and application of the double universal joint, each part as specified and described."

What the patentees call cylinder bunts is a new form of buffing apparatus, to ease off the blow when two cars, &c., come into collision. It is proposed to have strong metallic cylinders bored out in the manner of cannon, and into these to fit pistons, with rods projecting out, and covered with some elastic material, to receive the blow ; within the cylinders are to be placed spiral springs; but the principal dependance is upon the elasticity of the air, re-acting upon the piston. "It is believed that with four such cylinders of twelve-inch calibre, eight feet in length, two inches thick, of sound cast iron,

---

* So far this plan seems to resemble that patented by Mr. Kollman, and described p. 304 of our present volume.

well hooped with first rate wrought iron upon each, two locomotives may meet when going at the rate of thirty miles an hour without sustaining any damage." *Credat Judaeus.*

The safety guard clamps are to extend over each wheel, and come nearly in contact with the hub on each side, so that in case of the breaking of an axle, they shall form bearings by being brought into actual contact.

The spring-lever, which is to be a substitute for switches, is to be attached to the frame of each set of wheels of an eight-wheel car in such a manner as that by pressing upon them they shall direct the cars; how this is to be effected is by no means clearly made known.

The double universal joint is to connect cars together, and to be used in lieu of the devices now employed.

MISCELLANEOUS OBSERVATIONS UPON GUNPOWDER.

In Bengal, mixing is performed by shutting up the ingredients in barrels, which are turned either by hand or machinery; each containing 50 lbs. weight, or more, of small brass balls. They have ledges on the inside, which occasion the balls and composition to tumble about and mingle together, so that the intermixture of the ingredients, after the process has been gone through, cannot fail to be complete. The operation is continued two or three hours; and I think it would be an improvement in Her Majesty's system of manufacture if this method of mixing were adopted.

In England two or three pints of water are used for a 42 lb. charge: but the quantity is variable; both the temperature and the humidity of the atmosphere influence it. Bramah's hydrostatic press, or a very strong wooden press working with a powerful screw, lever, and windlass, constitutes the description of mechanism by which density is imparted to gunpowder. The incorporated or mill-cake powder is laid on the bed or follower of the press, and separated, at equal distances, by sheets of copper, so that when the operation is over, it comes out in large thin solid cakes, or strata, distinguished by the term press-cake. The mill-cake powder at Walham Abbey, is submitted to a mean theoretic pressure of 70 to 75 tons per superficial foot.

Gunpowder should be thoroughly dried, but not by too high a degree of heat; that of 140° or 150° of Fahrenheit's thermometer is sufficient. It appears to be of no consequence whether it be dried by solar heat; by radiation from red-hot iron, as in the gloom stove; or by a temperature raised by means of steam. Her Majesty's gunpowder is dried by the last two methods. The grain should not be suddenly exposed to the highest degree of heat, but gradually.

The method of trial best adapted to show the real inherent strength and goodness of gunpowder, appears to be an eight or ten-inch iron or brass mortar, with a truly spherical solid shot, having not more than one-tenth of an inch windage, and fired with a low charge. The eight-inch mortar, fired with two ounces of powder, is one of the established methods of proof at Her Majesty's works. Gunpowders that range equally in this mode of trial, may be depended on as being equally strong.

Another proof is by four drachms of powder laid in a small neat heap, on a clean, polished, copper plate; which heap is fired at the apex, by a red-hot iron. The explosion should be sharp and quick; not tardy, nor lingering; it should produce a sudden concussion in the air, and the force and power of that concussion ought to be judged of by comparison with that produced by powder of known good quality. No sparks should fly off, nor should beads, or globules of alkaline residuum, be left on the copper. If the copper be left clean, i. e. without gross foulness, and no lights, i. e. sparks, be seen, the ingredients may be considered to have been carefully prepared, and the powder to have been well manipulated, particularly if pressed and glazed; but if the contrary be the result, there has been a want of skill or of carefulness manifested in the manufacture.

"Gunpowder," says Captain Bishop, "explodes exactly at the 600° of heat by Fahrenheit's thermometer; when gunpowder is exposed to 500° it alters its nature altogether; not only the whole of the moisture is driven off, but the saltpetre and sulphur are actually reduced to fusion, both of which liquify under the above degree. The powder on cooling, is found to have changed its colour from a gray to a deep black; the grain has become extremely indurated, and by exposure even to very moist air, it then suffers no alteration by imbibing moisture." *Dr. Ure's Dictionary of Arts and Manufactures.*

LIST OF ENGLISH PATENTS GRANTED BE-
TWEEN THE 25th OF JANUARY, AND
THE 22nd OF FEBRUARY, 1839.

Pierre Jean Isidore Verdare, of the Sablonere Hotel, Leicester-square, gentleman, for improvements in the manufacture of starch, and in machinery for preparing and in employing of the refuse matters obtained in such manufacture. January 25 ; six months to specify.

John Howard Kyan, of Cheltenham, Gloucester, Esq., and William Hyatt, of Lower Fountain-place, City-road, Middlesex, engineer, for improvements in steam engines. January 29; six months.

John Hillard, of Bread-street, Cheapside, merchant, for certain improvements in machinery, or apparatus for making or manufacturing screws. January 29; six months.

William Lukyn, of Lower Cowley House, Oxford, dentist, for certain improvements in applying, and attaching artificial and natural teeth. January 29; six months.

Thomas Collette, of Aylesbury, for improvements in children's cots. January 29 ; two months.

Charles James Blasins Williams, of Half-moon-street, Piccadilly, Esq., M.D., for certain improvements in two-wheeled carriages. January 29; six months.

Robert Carey, of Breadgar, near Sittingbourne, Kent, gentleman, for certain improvements in paving or covering streets, roads, or other ways. January 29; six months.

Frank Hilis, of Deptford, manufacturing chemist, for certain improvements in the construction of steam boilers, and of locomotive engines. January 29; six months.

Thomas Barnabas Daft, of Regent-street, gentleman, for certain improvements in inkstands, and in materials and apparatus for fastening and sealing letters or other documents. Feb. 2; six months.

Moses Poole, of Lincoln's Inn, gentleman, for improvements in the means of conveying and transporting persons and goods from one place to another; a communication from a foreigner. February 4; six months.

John Evans, of Birmingham, paper manufacturer for improvements in the manufacture of paper. February 4; six months.

Thomas Robinson, of Wilmington-square, gentlemen, for improvements in the process of rectifying or preparing spirituous liquors in the making of Brandy. February 7 ; six months.

Christopher Binks, of Newington, Edinburgh, manufacturing chemist, for certain improvements in obtaining or manufacturing, and in rendering useful chlorine, the chlorides of lime and soda, and other compounds of chlorine applicable in bleaching. February 8; six months.

Charles Gabriel Baron de Suaree, of Red Lion square, Colonel in the French service, and William Pontifex, of Shoe-lane, London, coppersmith, for a new mode of obtaining dyes, colors, tannin, and acids from vegetable substances. February 11; six months.

George Henry Manton, of Dover-street, Piccadilly, gun maker, for certain improvements in fowling pieces and other fire arms. February 11 ; six months.

Edward Pearson Tee, of Barnsley, dyer, for improvements in weaving linen and other fabrics. February 11 ; six months.

John Thomas Betts, of Smithfield-lars, rectifier, for improvements in the process of preparing spirituous liquors, in the making of brandy. February 11.

Frederick Cayley Worsley, of Hollywel-street, Westminster, Esq., for certain improvements in locomotive engines and carriages. February 14; six months.

Richard Prosser, of Birmingham, C. E. for certain improvements in apparatus, for generating steam, consuming smoke, and heating apartments. February 19 ; six months.

Moses Poole, of Lincoln's Inn, gentleman, for improvements in epauletts, and ornamental metallic wire fringes, and other ornamental articles, or fabrics of wire, a communication from a foreigner. February 21 ; six months.

Johann Andreas Stumpff, of Great Portland-street, musical instrument maker, for improvements in grand and other piano-fortes. February 21; six months.

Matthew Uzielli, of Fenchurch-street, merchant, for improvements in locks or fastenings. February 21 ; six months.

Herbert Reid Williams, of Gloucester, surgeon, for improvements in trusses and surgical bandages. February 21; six months.

Thomas Hall, of Leeds, brass founder, for a new combination or arrangement of parts, forming an improved furnace for consuming smoke, and economising fuel applicable to steam engine boilers, and other furnaces. February 21; six months.

William Nash, of Budge-row, merchant, for certain improvements in the constructions of bridges, viaducts, roofs, and other parts of buildings. February 21; six months.

John Sylvester, of West Bromwich, Stafford, whitesmith, for certain improvements in the arrangement and construction of apparatus for hanging and closing doors. February 21; six months.

William Joynson, of Saint Mary Cray, paper mills. Kent, paper maker, for a certain improvement, or certain improvements in the manufacture of paper. February 21; six months.

William Nash of Budge-row, merchant, for improvements in machinery, for winding, spinning, doubling, and throwing silk and other fibrous materials. February 23; six months.

———

LIST OF IRISH PATENTS GRANTED IN
JANUARY, 1839.

Thomas Sweetapple for an improvement or improvements in the machinery for making paper. January 4.

Richard Smith, for improvements in the means of connecting metallic plates for the construction of boilers and other purposes. Jan. 5.

Charles Felton and George Collier, for improvements in power-looms. Jan. 7.

James Timmons, for improvements in the manufacture of glass. Jan. 9.

Horace Cory, for improvements in manufacture of white lead. Jan. 9.

Thos. T. Berney, for improvements in cartridges. Jan. 24.

W. J. Curtis, for improvements in machinery and apparatus for facilitating travelling, or transport on railways, parts of which are also applicable to other purposes. Jan. 28.

———

NOTES AND NOTICES.

Sir James Anderson's Steam Carriage.—We have received a note from Mr. Shaw the Secretary of the " Steam Carriage and Waggon Company " stating that " Sir James Anderson's steam drag is finished and will be in Dublin in a few days." From *Saunder's News Letter* we extract the following description of one of the Company's passenger carriages, built by Mr. Dawson, of Dublin. " It presents a peculiarly safe and commodious appearance, and, from its construction, may be pronounced impossible to be over-set. The front body, which is entered at the side in the usual way, contains more than ample space for six passengers, each having an arm-chair, and as convenient, if not better and more comfortable accommodation, than the first-class railway carriages. The back body, which is entered at the rear, is intended for ten passengers, although affording sufficient room for twelve. It is so ample in its dimensions that one may walk perfectly erect, from end to end, without incommoding the passengers at either side; it is admirably ventilated and lighted, and is to be furnished with a peculiarly constructed table, supplied with the newspapers of the day. The outside passengers sit round the roof, fourteen in number, having footboards, like outside cars, and supported at the back by a railing: the carriage altogether containing 30 passengers. The front boot contains a cistern for water, and a space for coke or fuel for a stage of from ten to twenty miles; and there is room at different parts of the carriage for stowage of about 1½ tons of luggage if necessary. We were particularly pleased by the light and elegant appearance of a machine intended for so many passengers, and cannot but feel gratified that Dublin can compete with England in furnishing these carriages for a company exclusively English. But Mr. Dawson has already supplied a number for the English railways. We understand the drag and carriage will be exhibited in Dublin previously to their embarkation; and we look with much interest to an adventure of so much national advantage.

New American Flax Machinery.—Under the head Machinery I might have mentioned an invention, as it is considered here, of apparent importance: by some, indeed, it is compared, for *northern* interests, to the cotton-gin which has wrought such wonders in the *south.* I refer to machinery for the preparation of flax for spinning, after the manner in which cotton is spun. A large company, in Delaware, is now engaged in the manufacture of the "short staple" produced by this new invention. The advantages alleged are:—1. That there is no loss of fibre, as no tow is to be taken out, all the lint being used up; whereas, by the old plan of hackling, finger-spinning, &c., there was a loss of perhaps half the original weight. 2. That the expense of labour on the whole process of cloth-making is reduced to *one-tenth* of what it was. 3. That the expense of bleaching in the flax, as now, is much less than in the old plan, and the process less injurious to the texture. These statements may be a little sanguine, but they are, probably, not without foundation in fact. This invention is considered the more important, because much of the soil, of the middle states particularly, is well adapted to the culture of flax.—*American Correspondent of the Athenæum.*

American Steam Navigation.—Speaking of steam, the secretary of the navy at Washington has just issued an interesting document of statistics on this subject, from which we may gather material for settling a question mooted some time since in your columns, about the comparative steam-boat force of your country and ours. In 1836 you had 600, at home and abroad. Doubtless the number has increased since then. Our number is rated at 800 now, of which 600 belong to the western waters,—where, in 1834, there were but 254; so that you may get a glimpse of what *will* be a few years hence. About 140 belong to the state of New York. In tonnage we are more in advance of you, the total here being estimated at 155,000, to about 68,000 on your side, two years ago, at the same time, we have no boats equal in size to some of yours; our largest runs between New York and Natchez, and is of 860 tons; the next in size are on Lake Erie, and along the New England coast. The average is 200 tons.—*Ibid.*

American Locomotives for English Railways.—One of our Philadelphia workmen has lately received an order for the manufacture of *eleven locomotives for the Gloucester and Birmingham Railroad,* at the price of about 85,000 dollars, and seven more have been ordered of the sam . . . from *Austria.*—*Ibid.*

A Novelty of the Olden Time.—Bewick, the reviver of wood engraving in England, receives on all hands the credit of being the inventor of the method, now so much practised, of giving a softened effect by means of *sinking* the surface of the block in certain situations. Unluckily, however, there are yet extant some of the original cuts of Albert Durer, the first inventor of the art, in which the very same expedient is resorted to, for producing the very same effect! Some of these curious blocks are now in England, and display very striking specimens of old Albert's style of calling this resource of his great invention into play.

Government School of Design.—The extravagant terms of the "Government School of Design" have been recently reduced to a sum more within the means of the working mechanic. The admission to both Day and Evening School is now one shilling, instead of four, and to the evening school alone six pence per week. Besides lessons in drawing, instruction is now given in the art of figure weaving, as practised at Lyons, a loom having been fitted up for that purpose at Somerset House, and a *patriotic* French weaver having been engaged to explain in detail the method by which his fellow-countrymen have been hitherto enabled to preserve the superiority in the market for fancy goods.— Mr. Papworth has resigned his situation of superintendent of the "school," which is now filled by a Mr. Dyce.

Indian Steam Navigation.—The East Indian Mails by way of the Red Sea have for some time reached England very irregularly, in consequence of the vessels assigned to that service on the Indian side having been taken off the station at short notice, in order to co-operate in the measures taken in consequence of the apprehended rupture with Persia. The delay thus occasioned has naturally given rise to great dissatisfaction, and has had the effect of calling into existence a new joint-stock company, having for its object the keeping-up of a regular communication by steam with India, without the chance of interruption by the vessels employed being engaged in any service foreign to the plan. Sir Robert Wilmot Horton, late Governor of Ceylon, is at the head of the association, which seems to display more vigour than any of its predecessors, and will, it is to be hoped, succeed in the object in view, either directly, or by forcing the East India Company to give a distinct pledge that the Red Sea steamers shall not for the future be called off on any account from their special duty. Sir Wilmot Horton is also chairman of another company, not so promising, for promoting inland navigation in India, by means of steam-boats plying on the various rivers. Here also the East India Company are already in the field, and will not perhaps be easily beaten out of it.

Portuguese steam boats.—Although Portugal is in most respects at least a century behind England in the arts of life, it is singular that one of our most recent novelties is already in practice there. A small steam-boat plies on the Tagus, of similar construction to the miniature steamers which now swarm on the Thames for carrying passengers from one part of town to another. The Portuguese vessel crosses the river from Lisbon, and returns several times a day, at a fare about equal to threepence English, and it is so well patronized that the speculation is understood to answer extremely well.

LONDON: Printed and Published for the Proprietor, by W. A. Robertson, at the Mechanics' Magazine Office, No. 6, Peterborough-court, Fleet-street.—Sold by A. & W. Galignani, Rue Vivienne, Paris.

Mechanics' Magazine,

MUSEUM, REGISTER, JOURNAL, AND GAZETTE.

No. 812.] SATURDAY, MARCH 2, 1839. [Price 3d.

Printed and Published for the Proprietor, by W. A. Robertson, No. 6, Peterborough-court, Fleet-street.

HOLEBROOK'S PATENT SHIFTING AND REEFING PADDLE-WHEEL.

Fig. 1.

Fig. 2.

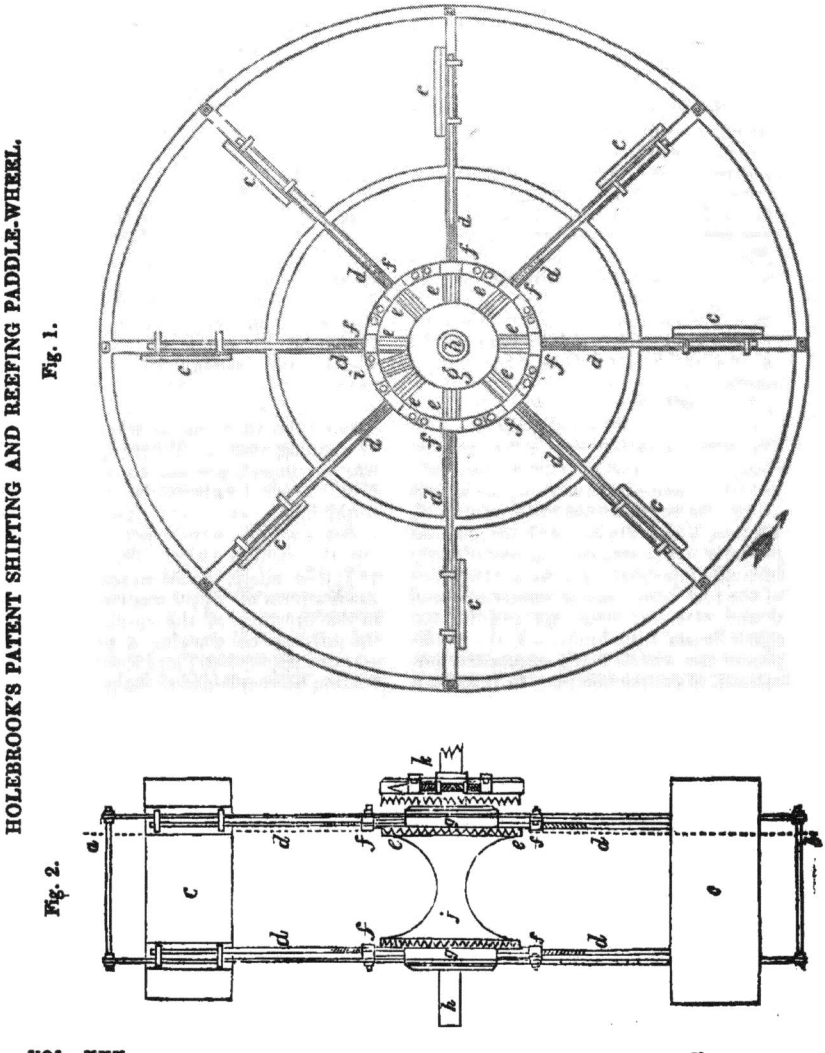

HOLEBROOK's PATENT SHIFTING AND REEFING PADDLE WHEEL.

One of the most obvious disadvantages under which the common paddle wheel labours, has long been considered to be that resulting from the impracticability of always apportioning the quantity of surface of paddle board, by a ready method, to the power of the engines.

It is admitted that there is a certain number of strokes which the engines may make in a minute, from which the best reaction can be obtained from the water in which the paddles move; and it does not admit of contradiction that, as the fuel carried in a vessel is lessened in quantity, the immersion of the vessel is diminished, and with that, the immersion of the paddle wheels. The wheels becoming too little immersed, the resistance of the water is lessened, and the engines make more than the proper number of strokes, and their power is not well employed; while if the wheels of a vessel are too much immersed, the engines make fewer than their proper number of strokes, and their power is also not well used.

Under the usual conditions of steam navigation it is impossible that a vessel should not be always more or less subject to a waste of power from the wheels being too much or too little immersed, because, if she sets out with her paddles properly immersed, they almost directly become improperly so, in consequence of the fuel being partly consumed; and if she sets out with her paddles too much or too little immersed, the condition of her wheels being improperly immersed, of course obtains. In practice it would seem, under the common circumstances of steam navigation, desirable to immerse the wheels to such a depth, that when half the fuel necessary for a voyage shall have been consumed, the wheels may be properly immersed : in such a case a vessel would of course set out with her wheels too much immersed, and arrive at her destination with them too little so; and, unless the paddles of the wheels of such a vessel were moved nearer or further from the centres of the wheels, by being shifted, a mis-employment of the power of the engine must always, to a greater or less extent be continually occurring. But it does not appear to be advantageous in conse-

quence of the time required to shift the paddles, (considering that time, in voyages by steam vessels, is even of more importance than cost of fuel,) to resort to this mode of obviating the objection arising from an improper immersion of the paddle wheels; and thus it will be perceived, that under the usual circumstances of steam vessels, either the engine must be improperly tasked, or time wasted; and the former will probably be the alternative adopted.

Under a conviction of this dilemma, in which steam navigation with common paddle wheels is involved, and with a view to obviate its consequences, the paddle wheel about to be described was produced; by which it will be seen that the engineer of a steam vessel has, at all times, a perfect control over the immersion of the paddles of the wheels. In the specification of the patent for this and another paddle wheel, which has been lately described in the *Mech. Mag.* (No. 800) a form of wheel was delineated and explained, by which the paddles of a wheel could be removed entirely from the water to near the centre of the wheel, and the vessel be rendered, except as regards the skeletons of wheels, similar to a sailing vessel. Where speed alone was an object, a wheel so constructed could never be desirable, for all that could be necessary under these circumstances, would be the capability of moving the paddles within a limited range, and this range, could never be great. Such a form of wheel was only adopted in the drawings of the specification for the purpose of showing a principle in extreme application; and another modification was explained in the specification itself, such as that which is the subject of the following observations.

Figs. 1 and 2 represent an improved paddle wheel, the paddles of which are placed as in the common paddle wheel, but in which they are capable of being moved, within a certain range, from or towards the centre of the wheel, so that any desired degree of immersion may be produced, and, by a variation in the construction, the paddles can also be reefed occasionally.

Rods having their interior ends screwed, are fastened to the bolts of the paddle boards, and the screwed ends of these rods are inserted in nuts, whose outsides are furnished with teeth in the manner of pinions; the teeth of which nuts take into the teeth of a reel having the teeth of a crown wheel at each end; this reel moving easily upon that part of the paddle wheel shaft, which is between the two circles of which the paddle wheel is composed. Between the side of the vessel and the circle of the paddle wheel nearest the vessel is a crown wheel, so placed upon the paddle wheel shaft that whenever the motion of the wheel is stopped, the free motion of this paddle wheel shaft is not impeded; and between this crown wheel and that end of the reel before-mentioned, which is nearest to the vessel's side, there is a small toothed wheel, whose bearings are fixed to the inner circle of the paddle wheel, which small toothed wheel takes both into the reel and the crown wheel, maintaining a connection between both of these parts, while it is so placed on the paddle wheel that it does not take into the teeth of the nuts before-mentioned. In connection with the crown wheel is a clutching apparatus, to grasp at times the circumference of this wheel whenever the paddles are required to have their places changed; the operation of which clutch will hereafter be described.

Fig. 1 represents a sectional side view of the paddle-wheel, taken at the end of the reel before mentioned, which is nearest the vessel's side; or, in plainer terms, of a section made by a plane in the direction of the dotted line $a b$ of fig. 2. In fig. 1 let the parts c represent the paddles in section, these being furnished with bolts by which they are enabled to slide upon the arms of the wheel, and by which they are also connected with rods, marked d, having screwed ends; let the parts e represent the nuts in which the screwed ends of the rods d work; these nuts turning by their outer ends in sockets, f, and by their inner ends in holes, made through the block, marked g, which steadies the circle of the wheel on the paddle-wheel shaft which shaft is represented in section by the part marked h; the holes in the block g serve, also, for the passage of parts of the rods d, when the paddles are drawn as near the centre of the wheel as the combination will permit. Besides the parts of fig. 1 already described, there is still another part to notice, namely, the small-toothed wheel,

i, which maintains the connexion between the reel and the crown-wheel before mentioned, this wheel having one end inserted in one of the sockets, f, and its other end in a hole in the block g. This figure, it has been said, represents the interior side of that circle of the paddle-wheel which is nearest the vessel's side, and this circle exactly resembles the other circle in structure, except as regards the small-toothed wheel i, which appertains to the inner circle exclusively; and except also, the threading of the screws thereof, the screwed parts of the two circles having their threads in opposite directions.

Fig. 2 shows an edge view of the paddle-wheel, representing the paddle-wheel shaft, the reel, the crown-wheel, the clutching apparatus, and the uppermost and lowermost paddles, with their rods and pinion nuts. In this figure, h represents the paddle-wheel shaft, j the before-mentioned reel, k the crown-wheel and clutching apparatus, the parts marked f the outer sockets of the nuts, and those marked g, the steadying blocks of the paddle-wheel; the parts marked e represent the nuts, those marked d the rods of the paddles, and the parts c the uppermost and lowermost paddles themselves. The small-toothed wheel i of fig. 1 is not shown in this figure (fig. 2), in order to avoid confusion in the drawing, but its proper position, from what has been before stated, may be very readily imagined. It may be specially noticed of this figure, that the screws of one circle of the wheel are differently threaded from those of the other; the reason of this must be obvious, when it is considered that the reel, by putting the pinion nuts into opposite motion, would, unless a difference of threading was made, cause the rod at one end of a paddle to advance towards, while the rod at the other end of the same paddle would recede from, the centre of the paddle-wheel. The clutching apparatus which, together with the crown-wheel which it is intended to clutch, is marked k, consists of two pieces of iron, playing by one end on a strong pin, which in this figures is supposed to be hidden by part of the paddle-wheel shaft, while the other ends have two screwed holes to receive a screwed spindle, which spindle, be it observed, is screwed half-way with a right-handed thread, and the other half-way with a left-handed thread, the object of these different threads being to make both parts of the clutch approach and depart equally from each other at the same time; this spindle works in two sockets, and has a tapering squared end to receive a key, by which it may be turned; and it may be further observed, that as the parts of the clutching apparatus, which are in contact with this spindle, do not move exactly in a straight line, it will be necessary to have the screwed holes, in these parts, somewhat larger than the screwed part of the spindle, in order that, when these parts are put in motion, neither the spindle nor they should be strained.

The parts of the wheel having now been described, it will be proper to show the action of the combination, when the paddles are required to be removed nearer to the periphery of the wheel than they are in figs. 1 and 2. The wheel being at rest, it will be necessary that a key be applied to the tapering end of the spindle of the clutching apparatus, and this spindle turned until the crown-wheel is firmly grasped by the clutching irons; when this has been done, the wheel must be set in motion by the engine in the direction of the arrow against fig. 1, upon which, as the paddle-wheel carries with it the small-toothed wheel i, of fig. 1, and as this wheel i cannot move round with the paddle-wheel with its teeth locked in the crown-wheel which is at rest, without also turning on its centre; and as this wheel i cannot turn round on its own centre without also turning round the reel j in fig. 2, it follows that the position of the reel upon the paddle-wheel shaft, will always be changing, while the wheel i is in motion on its own centre. The reel being thus in motion, will, by taking into all the pinion nuts, turn such of these nuts as are on one side in one direction, and such as are on the other side in a contrary direction; the effect of which motion of the nuts, upon the screwed ends of the rods d, will be, that parts of these rods will be continually emerging from within the nuts, and, as the nuts are prevented from moving towards the centre of the paddle-wheel, the screwed ends of the rods d, and with these, the other parts of the rods and the paddles fastened to them, will move towards the periphery of the wheel; and these parts may

so move until the paddles come in contact with the exterior ring of the wheel. To move the paddles nearer the centre of the wheel, it will only be necessary to reverse the direction of the motion of the wheel, and continue it until the paddles touch that ring of the wheel which is next to the external one, at which time it will be found, that the screwed ends of the rods d, will be within the pinion nuts and the blocks g, having their inner extremities in contact with the paddle-wheel shaft.

By the manner of moving the paddles which has just been described, it would seem requisite that a vessel should be stopped in order to have the places of the paddles upon its wheels changed, but this is not necessary, though, in practice, it may generally be found advisable; for, supposing fig. 1 to represent part of a larboard wheel, and the vessel with which it might be imagined to be connected to be in motion on its voyage, it will be perceived, upon consideration, that, if it were desired to place the paddles nearer the periphery of the wheel, the only operation which would be necessary to be performed would be to bring the clutch into operation; the motion of the engine would do all the rest: it would not even be necessary to firmly grasp the crown wheel, but only to prevent it from revolving as fast as the paddle-wheel, because a difference in speed between the crown-wheel and the paddle-wheel would effect, only more slowly, all that could be attained by totally arresting the progress of the crown-wheel, while the paddle-wheel was in motion. By what has been stated, it will appear, that, until the clutching apparatus is brought into action, the paddles, and the nuts, and the reel, and the wheel i, and the crown-wheel k, all revolve as the paddle-wheel revolves—in fact, that there is not the slightest action between any of the parts of which the paddle-wheel is composed: but, when the clutch is brought into operation, then only, is there any motion among the parts of the paddle-wheel. When the paddles are desired to be pressed more towards the periphery of the wheel, all that would be necessary to effect such a purpose, would be to bring into slight contact the wheel k and the clutching irons, and to keep this contact until the paddles arrived at the position desired.

From figs. 1 and 2, it appears that the paddles would not be reefed, but only moved from their places; but, by the same means that the whole of a paddle is moved, a part of one could also be moved; and, if a moveable part of a paddle were brought under cover of another fixed part, such a paddle would really be reefed. It follows, therefore, that the modification shown in figs. 1 and 2, in order to allow of its paddles being reefed, only requires to have permanently fixed paddles, and small moveable ones pushed out beyond the fixed ones; and then, upon moving the smaller ones under cover of the fixed ones, the paddles of the wheel would in reality be reefed.

Besides the advantages resulting from a proper apportionment of the surface of paddle board to the power of the engines, by partly withdrawing the paddles from the water and by reefing them, afforded by the employment of such a wheel as that which has now been described, there are others, arising from the power of differently tasking the two wheels of a vessel. In many situations the desired course of a vessel can only be maintained by means of a wasteful application of the rudder, or by an inconvenient use of sails. In cases of towing vessels lashed to the sides of the towed ships, it must be evident, that, the load being all on one side, if the wheels propel equally, the towed vessel and its tower could only be made to keep their proper course by a free use of the rudder; under which circumstances, it is plain, that power is consuming only to be destroyed by the action of the rudder. By the employment of wheels, such as that under consideration, any continued inequality in the load upon each wheel might be obviated at once, by pressing outwards or drawing inwards the paddles of each wheel until the power of each was properly proportioned to the load upon it. Other reasons there are than those already stated why this wheel is advantageous, in affording the power of properly immersing the paddles of a wheel; but into a statement of these it would not be advisable now to enter.

The diagrams accompanying this description show a wheel which admits of its diameter being diminished in the proportion of 11 to 9, an extent which will probably be as large as any that will generally be necessary. The means of putting the paddles in motion from or towards the centre of the wheel may be much simplified by furnishing the wheel i, of fig. 1, with a square tapering end, and applying manual force to turn a key fitted upon such an end; but, by this means, it will always be necessary to stop the progress of the vessel, and the paddles cannot be moved from their places by the power of the engines, though, on the other hand, it is true that the clutching apparatus will not be necessary.

It may be objected, at first sight, to this wheel that the wear of it must be great, but, it should be observed, that the apparatus for moving the paddles, which would be entirely inadmissible if required to be continually used, is of such a nature, that though it should be at least daily used, yet would not be required to be in action more than a minute in each day, and the extent to which the paddles would be required to be moved during a voyage to New York would only be 2-11ths of the whole diameter of the wheel; and before a vessel started on its return voyage they would require to be removed to the same extent. Thus, it will be seen, that as far as wear from friction is concerned, any objection on that account is idle; but, there is a wear from the corrosive action of the sea water, which is more destructive than that from friction; and yet the portion of the surface of the iron necessarily exposed to this action in this wheel, as now drawn, is so small as to reduce this wear to almost nothing, and even any corrosive action may by various contrivances be actually entirely prevented. It will be seen that no rubbing surfaces are exposed to the action of the sea water, except the interior parts of the bolts and the portions of the arms of the wheel upon which these slide, the other rubbing surfaces are only exposed to the action of the spray, which cannot be materially destructive; besides, it should be recollected that every ship has a rudder, parts of the iron-work of which must necessarily rub each other and become exposed to the most destructive action of the sea water, and yet this action does not appear, from complaints made of it, to be of much consequence.

Carefully considering what has been

before stated, and the ideas thereby induced, it must be admitted, unless the loss by improper immersion of the paddles be chimerical, (an idea which an attentive examination of the subject must at once dissipate,) that a combination affording a ready means of obviating that loss is a desideratum not lightly to be disregarded. Bearing in mind that the wheel which has been described does afford such a means, and that without any deduction on the score of strength, (for it can be made equally strong with the common paddle wheel, and, also, that every common paddle wheel can be converted into one like it,) it seems but fair to conclude that this combination is worthy of such a trial as would effectually determine whether there exist such practical objections to its employment as will outweigh the advantages which are derivable from it: particularly, as every wheel, having a similar object in view with this modification, must, in the very nature of things, be more or less subject to the objections which may be urged against this one; and as, also, unless these objections are, to a certain extent, overlooked, it would appear likely that any prospect of avoiding the waste of power which has been mentioned, must be shut out by the almost necessarily continued existence of such objections.

Z.

PHENOMENA PRODUCED BY WATER UPON HIGHLY HEATED METALS— THEORY OF THE GENERATION OF STEAM—CAUSES OF THE EXPLOSION OF THE "VICTORIA" BOILERS.

Sir,—I have read with much interest and pleasure in the 703rd number of your valuable journal, the article under the head of "Phenomena produced by water upon highly heated metals," furnished by Mr. C. Tomlinson; another article in your 799th number under the head of "Dr. Schafhaeutl's experiments on the conversion of water into steam;" and another in your 802nd number under the head of "On the generation of steam—Dr. Schafhaeutl's experiments," and request the favour of an insertion of a few observations thereon, in your miscellany, as a supplementary article to my former communication inserted in your 795th number, under the head of "On the causes of steam-boiler explosions."

Those communications furnish the details of several experiments, conducted with a view to the discovery of the cause of drops of water being repelled from the surface of highly heated metals, the rotatory motion to which they are subject, and the much greater period of their dissipation by evaporation, than such as come in close contact with metal of a lower temperature; but as the several experimentalists furnish opinions only as to the cause, and as there is some discrepancy in their conclusions, it must be evident that after all, we are but furnished with the practical facts, and that we have yet to obtain an indubitable proof of the cause.

Without attempting a trespass on the patience of your readers, or my own time, by a review of those articles, or presuming to endeavour to disprove the correctness of the theories advanced, I will, with your permission, venture to furnish a theory for the cause, differing in some material points from the rest, and leave your readers to judge of the probabilities of either.

In the communication furnished by me in your 795th number, under the head of "Heat," the prominent feature of the theory endeavoured to be inculcated relative thereto, is, that heat is a fluid composed of spherical, imponderable, and indestructible atoms, of less magnitude than the ponderable atoms which enter into the composition of all material bodies in conjunction with the imponderable; and that heat is subject to two laws—equal diffusion as the primary —and recession from the centre of the earth, or gravity toward the centre of the sun, as the secondary. The secondary feature of the theory is, that metallic and other bodies, are composed of spherical, ponderable, and imponderable atoms, and consequently that the structure of composition, must furnish interstices resulting from the close contact of those spherical atoms in solid bodies, and that the ponderable atoms being of greater magnitude than the imponderable, and the difference in magnitude being sufficiently great to cause the interstices presented by their union to be greater than the imponderable atoms, the latter are able to permeate, or pass through those interstices without separating the ponderable atoms from each other; and that, consequently, in the generation of steam, the atoms of heat evolved in the combustion of fuel, are subject to transit through the interstices presented by the structure of composition of the vessel, to the lower surface of the water contained in that vessel, as induced by the law of equal diffusion; and that having entered into the water, its passage from the lower to the upper surface is induced by the law of recession.

The third feature of that theory is, that the transition of heat from one body to another, is with a celerity proportionate to the difference in the temperature of the imparting body and the receptive.

It is the principles of this theory then, which I call to aid in furnishing an explanation of the cause of the non-contact of the drop or mass of water with highly-heated metal, and the rotatory motion it assumes, as pointed out by your correspondents.

It appears, by their representations, that to produce such an effect, the vessel must acquire a temperature far exceeding the temperature of ebullition of the several liquids experimented upon, and that the lower the temperature of ebullition of the fluid, the greater is the effect of recession and rotation.

This declaration appears to me to be confirmatory of the correctness of the third feature of my theory, proving that the greater the difference between the attainable amount of temperature of each fluid, the greater is the difference in the celerity of transition of heat from the vessel to the several fluids, the rapidity of transition being greatest to the fluid of the lowest temperature of ebullition, and consequently that the recession and rotation of water must exceed that of mercury, &c.

The following extract from the communication of Mr. Tomlinson, taken from page 308 of your 703 Number, will serve to describe the general results of the experiments tried, to suit my purpose of explanation:—

"From a dropping tube I filled the crucible with water about one-fourth, drop by drop. The water fell in with scarcely any audible sound; the drops collected into one large button, which was very convex, and resembled mercury in form; it moved with great rapidity, with the results before noticed (recession from the surface of the vessel, and rapid rotation). Its specific gravity must have been small, for minute portions of charcoal which fell in, generally sunk through the water to the bottom of the vessel. The water remained in this state seven minutes, during which time it gradually decreased in bulk until the whole was gone. Several times, when a fragment of charcoal did not sink, it partook of the motion of the water, and described an equilateral triangle within the globule, touching its sides three times during one revolution. When the crucible was removed from the furnace, and allowed to cool, a burst of vapour ensued in about one minute and a quarter from the time of its removal, and more than one-fourth of the water was dissipated. This occasioned so great a loss of heat, that in a very few minutes the crucible and the remaining water could be handled."

From the details of this experiment we learn that the platinum crucible was intensely heated, and that water was inserted, drop by drop, to the amount of one-fourth of its content; that such drops amalgamated, and together formed a spherical mass, and that without touching the bottom or sides of the crucible, it revolved with rapidity.

The phenomena of this experiment then consists in the water not falling to the bottom of the vessel in obedience to the law of gravity, or not remaining motionless (as a mass) at the bottom of the vessel, and being subject to a most rapid evaporation.

The cause of an effect so contrary to what might be expected from the natural deductions of usually received theories relative to the action and transmission of heat (and the correctness of which theories appear to me to invest the results with the dignified name of phenomena), I think may be traced to the circumstance of the water possessing but little heat, being placed by dropping into the interior space of the vessel containing much heat, both as relates to its atmospheric contents and the amount present in its metallic substance.

The transmission of heat from the major to the minor possessing body, occurring by the law of equal diffusion to which heat is subject, with a celerity proportionate to the difference in the amount of their respective possession, the very great difference therefore between the temperature of the crucible and its aerial contents, and the water inserted, will necessarily cause a rapid rush, to an extensive amount, of heat from the bottom and the sides of the crucible to the body of falling water; and the transition of heat from the bottom of the crucible to the water, as induced both by the law of equal diffusion and recession, being of greater force than the power of gravity to which the water is subject, its contact with the bottom of the crucible is prevented so long as the difference in temperature is kept up to an amount sufficient to cause a transition of heat from the bottom of the crucible to the water, of greater force than the power of gravity, which would otherwise impel the water to the bottom of the crucible.

An illustration of this effect is well exemplified by a child's pastime, when engaged in causing a currant berry, with a pin through its centre, to dance round the orifice of a piece of tobacco pipe, which he effects by a continuous and unintermitting ejection of air from his lungs through the pipe, upon the surface of the berry.

To account for the rotatory motion of the water, I should ascribe it either to the insertion of the water, by dropping into the crucible, not being accurately centrical; or if so, that it may be subsequently forced from the centre by the irregular radiation of heat from the bottom of the crucible; when in such case the radiation of heat from the side of the crucible to which the water was the nearest situated, as induced by the law of equal diffusion, would necessarily impel it towards the opposite side, when the radiation of heat from the latter side, would impel it back to the former, and under such alternate impulsion the water would necessarily assume a rotatory motion.

Having thus endeavoured to account for the suspension of the water at a distance from the bottom of the crucible, and the cause of its rotation, I will next endeavour to explain what appears to me to be the cause of the slow evaporation of the water.

The generation of steam consists in causing the imponderable atoms of heat to combine with the ponderable atoms of water, and the imponderable and ponderable atoms thus united being of less specific gravity than an equal number of atoms of which the atmosphere is composed, those which may be termed the aqueous atoms evolute from their associate atoms, and ascend among those which may be termed atmospheric atoms; and their ascension is induced by the law of recession to which heat is subject, although they, as ponderable atoms, are subject to the law of gravity, because they are in combination with imponderable atoms, subject to the law of recession, the balance of power being in favour of the latter, resulting from their majority in number.

By the usual mode of generating steam, the water is in close contact with the material of which the containing vessel is composed. Heat is impelled through the material composing the bottom of the vessel, both by the law of equal diffusion and of recession. From the material of the vessel it is transmitted to the lower surface of the water, and promptly ascends through the body to its upper surface.

Heat is also impelled to the material composing the sides of the vessel by the law of equal diffusion only, and from such material it is transmitted to the water within the vessel, through the body of which it quickly ascends, as impelled by the law of recession.

Heat being accumulated at the upper surface of the water to a sufficient amount to combine with the ponderable atoms of water, so as to create a conjunction of atoms of less specific gravity than an equal number of atoms composing the superincumbent atmosphere, a transition of such heat, and the ponderable atoms of water, constituting together a fluid called steam, evolates from the surface of the water in the vessel, and by a continuation of the transmission of heat, the generation of steam, or the evaporation of the water is continued to be effected.

In the customary prudential mode of generating steam, the heat is not attempted to be transmitted to the water faster than it is capable of receiving it; and its receptive capabilities, after it has attained to the temperature of ebullition are proportionable to the amount of abstraction of heat from its upper surface, in the generation of steam; and the capabilities of generating steam are proportionate to the rapid abstraction of the amount previously generated, to the extent of the upper surface of the water in the vessel, and to the amount and rapidity of the transmission of heat to its upper surface. But if the transmission of heat to the lower surface or bottom of the vessel or its sides, is greater and more rapid than the receptive capabilities of the water, and to an amount exercising a greater force than the power of gravity to which the water within the vessel is subject, then will the water be repelled from the bottom and sides of the vessel, and an accumulation of heat in the material of the vessel instead of at the upper surface of the water will occur, and thus would be established a relative position of the water to the vessel, as the suspended water was to the crucible, with this difference in the causation of the suspension, that in the case of the crucible, the impartation to its material was previous to the insertion of the water, whereas to the vessel it was subsequent.

Having endeavoured to point out the customary mode of prudentially generating steam, and to show the analogy between an intensely heated crucible previous to the insertion of water, and a vessel or steam-boiler intensely heated after the insertion of water and the analogous results, let us now come to close quarters with the subject of the cause of the slow evaporation of the mass of water in the crucible referred to.

The position of the water in such crucible we find to be a state of suspension in a highly heated atmosphere, repelled from its bottom and sides, by a rapid current of heat, induced by the law of equal diffusion; such heat impinging upon the external surface of the water instead of passing through its bulk, receding from the chacing heat, the impelling force being superior to the gravity of the floating and the flying mass; and the circumambient atmosphere, of such extreme rarity and small specific gravity, as to prevent the generation of steam to a great amount, because of the diminished receptive capacity of the atmosphere, and from the extraordinary impartation of heat thereto from the intensely heated crucible; and herein do we find, I conceive, the several causes of the slow evaporation of the revolving mass of water.

I will next endeavour to investigate the probable cause of the sudden burst of vapour from the mass of water in a short period after the removal of the crucible from the furnace, and the consequent rapid diminution of the bulk of the water.

As soon as the crucible was removed

the impartation of heat thereto ceased, the rarefied air which it contained ascended and was replaced by air of less temperature, which, in its turn, being rarefied by an abstraction of heat from the crucible, ascended also and gave place to another supply of air, and so on continually; while at the same time an abstraction of heat from the external surface of the crucible was continually occurring effected by the same medium, and the water to a small amount participated in effecting the abstraction of heat from the crucible. This conjoint action continuing until the temperature of the crucible was diminished, when the force of the radiation of heat from the crucible to the water was no longer superior to the force of gravity of the water, the latter fell to the bottom of the former, and was thus placed in a position the most favourable for the generation of vapour, and the air in the crucible being then of a much lower temperature, its effective capacity for heat or vapour was much increased; and consequently the generative and receptive power being both increased, the production of the effect followed as a matter of course; and so large an amount of heat being abstracted from the crucible by these several means, a reduction of its temperature to a state of handling soon occurred.

Before I proceed to an endeavour to accomplish the ultimate object of this communication, I will call the attention of your readers to one most important feature in the details of the crucible experiment,—that the water inserted was to an amount of about one-fourth of its content. And although the motive for not inserting more is not stated, yet, I think it may be fairly inferred, that it was from an apprehension that a greater amount of water would not have been productive of the required effect. And why not? Because its specific gravity would have exceeded the repelling force of the current of heat proceeding from the crucible.

And what an important lesson do we learn from this experiment! We find that a vessel intensely heated will cause a repulsion of water from its surface, to an amount equal to a fourth part of its content.

'The ultimate purpose of this letter is to point out the great similarity in the results of this crucible experiment, and the several circumstances connected with the bursting of the steam boiler of the *Victoria*, as furnished by the evidence contained in the report of inquest in the 785th Number of your Journal, and to deduce therefrom the probable cause of the explosion. And in order the better to accomplish this endeavour, I will place before your readers the leading features of that evidence, as follows:—that a diminution of the draught of the furnace was experienced, supposed to have been effected by an alteration of the fire bars, and considerable difficulty was found in generating the necessary quantity of steam; and that in order to obviate such difficulty, the fires were urged to an unusual point of intensity: that the vessel ran foul of a collier, and a considerable concussion was the result: that the explosion occurred from three to five minutes after the collision: that immediately after the explosion, the boiler which burst, and the adjacent boilers, were found to be red hot: that immediately after the collision the feed-cocks were shut: that the river water fermented more than the sea water: that the water in the boilers always appeared to be in a state of ferment: that a constant bodily fear was entertained: that the boiler-flue collapsed, and its rupture ensued: that the thickness of the plates of the boiler were a quarter of an-inch, instead of half an-inch, as usual: that the smallest space for the water between the flue or furnace cylinder, and the boiler cylinder, was but two and a-half inches, or two inches, and three inches in the largest: that the fire-place of the boiler was very large, and the water-space not proportionate to it.

Having furnished the material part of the evidence adduced, I will next endeavour to point out that which appears to me to be the several causes of the collapsion and rupture of the boiler-flue, and the intimate connexion between those causes and the resulting effects, with such as have been detailed relative to the crucible experiment.

Although the length of the furnace, from the door to the bridge, is not given, yet, from the longitudinal section furnished (fig. 1 in your work) it appears to be about a fourth of the length of the boiler, and with a properly regulated draught, the heat emitted from the burn-

ing fuel would be subject to perpendicular ascension, as induced by the law of recession, and if the receptive capacity of the metallic plates immediately above the fuel, was equal to the quantity of heat evolved in combustion, a transmission of such heat from the plates to the water in contact would occur, and the diffusion of such heat throughout the whole mass of water in the boiler, would be effected by the law of equal diffusion, subsequent to its transition to the water situated immediately over the furnace; but if the receptive capacity of such plates was not equal to the amount of heat evolved, then would the surplus quantity of heat above the receptive capacity of those plates, be carried forward beyond the bridge, and in its progress be transmitted through the more distant plates; and by such a proportionate length of the furnace to the boiler, the impartation of the greatest portion of heat evolved, to the water in the boiler, is economically effected, and the generation of steam more rapidly accomplished. But it appears, that in this ill-fated voyage, the draught was insufficient, and, consequently, the passage of air between the furnace-bars was not sufficiently rapid: therefore the combustion of the fuel was languid, the heat evolved insufficient, its progress beyond the bridge too tardy, its transmission to the water inefficient, and the production of steam of too small an amount. To furnish a remedy for this evil, it appears that the engineers and stoker endeavoured to cause, by personal exertions, that combustion which would have been more properly performed by a sufficient and well-regulated draught, and the result of their exertions caused more than a fair proportion of the heat evolved to be received by those plates immediately above the furnace, causing them eventually to become red hot without the generation of steam to an amount proportionate to the quantity of heat evolved for the purpose; and herein do we find, I conceive, a resemblance of that part of the furnace-flue thus improperly heated, to the intensely heated crucible.

Finding that with all their exertions to urge the fire on to an intense degree, they could not raise a sufficient amount of steam, they tried the effect of a diminution of water, whereby its amount in the boilers was reduced below such a gravity as was equivalent to the force of the radiating heat from the surface of the furnace-flue; and consequently its repulsion from such surface, and the resulting motion, (termed fermenting by one witness, and unusual ebullition by another) ensued. And herein will again appear a similarity between the effect produced by the red hot boiler on the water contained, and the intensely heated crucible upon the water which had been inserted to about one-fourth of its content.

I now come to that which appears to me to have been the immediate cause of the bursting of the boiler, the collision of the vessel with the collier, the concussion, the consequent addition of momentum to the gravity of the water, its resulting contact with the highly-heated metal, the generation thereby of a sudden and immoderate amount of steam, and the extraordinary pressure upon the metal, injudiciously weakened by excessive heat, and which metal was never sufficiently thick to incur such fearful liabilities.

Should the addition of momentum resulting from the concussion to the gravity of the water, not have been sufficient to cause an impact of the water to the metallic surface, yet may we, I think, furnish an additional, and of itself a sufficient cause, in the partial diminution of the temperature of the boilers, resulting from the measures adopted by the engineers, &c., to check the speed of the vessel, causing the water to fall into contact with the surface of the boiler, as induced by its own gravity, becoming superior to the force of the radiation of heat from the metallic surface, the latter force being diminished by a reduction of temperature; and, in such case, we again find a similarity in the sudden and extraordinary generation of steam, causing the boiler to burst, and the sudden burst of vapour on the removal of the crucible a minute and a-quarter from the furnace.

Before I close this letter, I beg to be allowed a few remarks on the dangerous tendency of what I should term a too great refinement in the spirit of economy. From an examination of the engravings, and a perusal of the description of the boilers of the *Victoria*, as furnished in your Journal, and an attentive consideration of the evidence adduced, I think

that every reflective mind must be convinced, that in their principles of construction, economy, to the verge of rashness, appears to have been too much studied, and it is much to be regretted, that such a spirit should exist to tarnish one of the brightest discoveries of the age, by the immolation of human life upon its shrine.

Apologizing, Sir, to yourself, and the readers of your valuable pages, for the great length of this letter,

I am, your obedient servant,

G. A. WIGNEY.

Brighton, Feb. 15th, 1839.

———

ON LOCOMOTION ON COMMON ROADS.
ARTICLE II.—REASONS OF PAST FAILURES.

The failure of most of the attempts to obtain locomotion by steam power on common roads, I should say, are mainly attributable to the way in which the projectors have attempted to combine the mechanical power with the carriage wheels, with a view of converting the power of the engine into a rectilinear motion of the carriage. This has generally been attempted, at least in all those public exhibitions which have borne any thing like the appearance of locomotion, by a crank on the axle of the driving wheels; the attempt being directed to make the *wheels go round*, and, in consequence of the wheels going round, to make the carriage go forward, these two motions having been observed to be concurrent in all draught carriages. But in so doing they are clearly inverting the order of nature, or rather of art, which has existed from the time that wheels were first invented. Hitherto the cause of a carriage going forward has been the muscular power of man or horse applied, so as to draw or push it; and the rotary motion of the wheels has been the consequence of such progressive motion of the carriage,—the wheels go round *because* the carriage goes forward. But in the application of power here contemplated, this order is inverted; the progressive motion of the carriage is the *consequence* of the rotatory motion of the wheels, effected by the action of the engine on the axle to which they are attached. The carriage goes forward *because* the wheels go round, thus clearly seeking the object in the reverse way to that in which we have hitherto been accustomed to obtain it; making that which is the cause in one instance, the consequence in the other, and seeking the same effect from a directly contrary application of means. One of the principal reasons which have induced the parties to seek locomotion in this way, appears to have been the successful application of it in the case of railways; and the attempt has been made without taking into due consideration the differences which exist between common roads and rail roads; which, on examination, will prove so many, so various, and so great, as almost to preclude the possibility of similar results arising from similar application of means. Amongst the peculiar properties of the rail road, the principal ones are smoothness, hardness, and a near approach to a level, or limited variation from it; all which properties are essential for the production of progressive motion from the rotatory motion of the driving wheels Not one of these properties exist in the common roads in any thing like a sufficient degree to warrant us in the expectation of success through the same means. The common roads, from their very nature, and the materials they are made of, must be in their best state, rough and uneven; and the constant repairs they are liable to, and which are indispensable, render them more so: and though Mc Adam has done much for their improvement by cutting away hills and filling up vallies, he has not reduced their slopes to any thing like the gradients in a railway, by confining their deviations from a level to a certain standard, in no case to be exceeded; nor by converting their undulations of surface and tortuous courses to any thing like the straight lines, the graduated rises and falls, and extended regular sweeps of the rail road.

But the greatest difficulty we have to contend with is this, when progressive motion is to be derived from the circular motion of the wheels, it is necessary we should have a firm, continuous, *biting* hold of the wheels upon the ground, that the friction between the two surfaces, the wheel and the ground, should be constant and unbroken, and sufficiently strong to throw the power of the engine into progressive motion,

instead of suffering it to run to waste by acting against the air, or by allowing the wheels to go round without the carriage advancing. This is the peculiar property of the rail road; the surfaces of the rails and wheels are made sufficiently smooth as to be in continual contact with each other; and, at the same time, the slight oxidation on the surface of the rails furnishes the wheels with a sufficient hold to prevent their slipping round without advancing; but which would even then be frequently the case were the gradients not confined to such a limit as to prevent its occurrence. Now neither of these properties exist in the common roads; there may be occasionally patches of ground to answer the purpose, but, in the long run, continual obstacles from hills and rough roads would bring us to a stand still.

Whoever has noticed the motion of a carriage wheel on common roads, at a velocity anything beyond a walking pace, must have perceived that its progressive motion consists of a *series of jumps* from one obstacle to another; this is distinctly visible in the progress of a hackney coach or break, through the streets of London, where the leaps of the wheels over the paving stones, for a foot or eighteen inches at a time, may be clearly perceived; the wheel, instead of being in continual contact with the ground, is for one-half its time suspended in the air, the instant it reaches the ground, another stone or pebble throws it off again; thus constituting what is understood by the rattle of the wheels, and which is neither more nor less than the wheels encountering with, surmounting, and leaping from the summits of the innumerable and minute obstacles which every where present themselves on common roads. It is evident here, that one of the principal requisites of the driving-wheel is wanting; there is not that continuous, biting contact between the wheels and the ground which is indispensable to give the wheels a proper and firm leverage. In all the successive springs of the wheel from the ground, the power of the engine is exerted against the air only, and serves but to make the wheels go round without at all tending to advance the carriage; and as these springs and jumps are incessant during the progress of the carriage, and cannot by any possibility be avoided, it follows that any application of mechanical force, where this indispensable contact cannot be effected, must be altogether inefficient and misdirected.

But, were it possible to get over this objection, or could the roads be made so smooth, or the driving wheels so large, as to afford sufficient contact, we should still have to contend against another, of equally serious consequence. In the railroad the variations from the level are tied town to a certain extent, the rises and falls are limited, and long tracts of level country are often made unlevel, in order to approach some distant obstacle at a graduated inclination : — this rise and fall in the London and Birmingham Railway is restricted to 16 feet in a mile, or 1 foot in 330, and even at this slight inclination the power of the engine is said to lose one half its efficiency, and it is an ascertained fact, that was the inclination much greater, the gravity of the engine and train would overcome the contact between the wheels and the rails, and progressive motion would cease; the power of the engine having no other effect than that of grinding the two surfaces together, without advancing,— hence the necessity of having stationary engines to *draw* the train up an inclination, when it is found necessary to exceed the limited departure from the level, and to which the locomotive engine is, confessedly, incompetent.

What then would be the case with the numerous hills which present themselves all over England, and which so indefinitely exceed in steepness any of the gradients allowed on a railway. What would be the case when we should have to encounter a patch of new made, or new mended road; with nothing but the loose gravel or stones for the driving-wheels to act against—and what would be the case when these two difficulties presented themselves in a combined form, which they occasionally must do,—viz., in a newly repaired hill ? To propose stationary engines in all these points of difficulty would be absurd, because, so far as rough roads are concerned, they would have continually to alter their position, as what was rough road yesterday, would be smooth to-morrow, and *vice versa*.—there appears to be but one way, and that liable to many objections, which is, by providing a facility in the

engine of detaching the power from the driving wheels, and applying it to an axis or capstan capable of winding up a rope or hawser carried out a-head, and fixed to a tree or some object in advance of the difficulty to be got over, and by these means to drag the carriage forward till it again becomes situated on passable ground, when the power might again be directed to its original channel, that of the driving wheels.

It must be evident from these observations that all attempts to procure locomotion on the common roads, by the same means, on the same principles, and through the same adaptation of machinery as it is obtained from on the railways, must be altogether futile while such a vast difference exists between the roads in question, in all the essentials requisite to its production ; and that, to be obtained effectually, it must be sought for in a quite different way : the application of the power must be directed through a new channel, and the continued and firm contact between the power and resistance, which is a *sine qua non* in the case, must be sought for in another direction, and by other mechanism than has, as yet, been resorted to for the purpose.

G. S.

IMPROVED BREAK-JOINT FOR PORTABLE FIRE-ESCAPE LADDERS.

Sir,—While describing the French portable fire-ladders, at page 250 of your present volume, I observed that, "the Parisians might with great advantage adopt the pair of small wheels on the top ladder, and also the escape-belt for lowering females, or infirm persons, &c., which are now considered an indispensable appendage to this elegant and convenient form of ladder." There is one other simple contrivance, that would be found exceedingly useful if generally adopted. I allude to the "break-joint," of which a sketch accompanies this communication. I have had a little apparatus of this kind in constant use for some time, and derive great convenience from its employment. It is an improvement upon the original suggestion of Mr. John Gregory, a description of which appeared in your 27th volume, page 137. My "break-joint" is of beech, it consists of two upright pieces *ab*, set out at the base to the width of the upper step of the ladder, and made concave on the

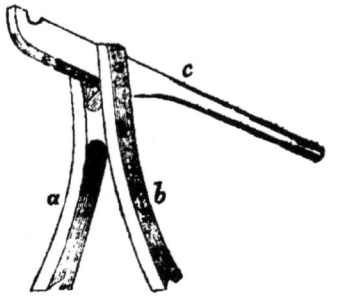

under side to prevent slipping; *c* is a long-handled lever with a notch at the reverse end for the round of the upper ladder to rest in. The legs of the apparatus *ab* being placed on the upper step of the lowest length of ladder, the end of the lever is brought under the last step of the length above, when by pulling down the handle *c*, the ladders are prized up and the bottom length may be removed.

There is no provision made in this instrument for tightening the joints, nor is it needful; if the ladders are tolerably well made their own weight will always be sufficient to effect this object; but by long continued exposure to water, or to damp weather, the wood is apt to swell, and the joints become liable to set fast. If several lengths are joined up under these circumstances, their separation becomes a matter of some difficulty, without a mechanical aid is at hand. The one I have employed and here depicted is as simple and as efficacious as can well be desired; if required to be made more portable and to go into a smaller compass, the fulcrum-pin may be so arranged as to take out easily, when the whole can be packed flat, occupying but little space. It may very conveniently be strapped to one of the lengths of ladder.

I remain, Sir,

Yours respectfully,

WM. BADDELEY.

London, Feb. 15, 1839.

ROYAL INSTITUTION. — MR. FARADAY'S LECTURE ON GURNEY'S OXY-OIL LAMP.

Feb. 15th.—*The Conversazione.*—Among the articles submitted to the notice of the members, that which attracted the principal attention, was a very large working model of Arnold's Chronometer Escapement, sent by Mr. Dent—so large, indeed, that the spectator perfectly unacquainted with these matters, could comprehend this beautiful, yet simply-contrived movement, almost at a glance. Mr. Dent had also on the table one of his time-pieces, with the pendulum of his invention.

The Lecture.—Mr. Faraday lectured on Mr. Gurney's oxy-oil lamp. The importance which is attached by the public, and the interest which attaches of itself to subjects of utility, was evinced by the crowded state of the theatre. The admirable and very attractive manner in which Mr. Faraday illustrates such subjects, must be allowed, however, to have had its influence. The oxy-oil lamp is simply a common argand, adapted for the burning of oil in the usual manner; but instead of being supplied wholly with atmospheric air, it is, in respect to the *interior* of the flame, fed with oxygen gas. This is effected by making a tube, proceeding from a gas-holder, rise up nearly to a level with the top of the wick, where, as it has the same area as the wick, its end is closed all but a very fine aperture, whence issues a jet of gas under the pressure of a few inches of water. The effect of the jet is to draw the oil flame to itself, and thus it has the appearance of a gas-light issuing from a single aperture, but possessing a greater brilliance. The illuminating power is thus augmented $2\frac{1}{2}$ times for the same consumption of oil.

The lecturer explained the *rationale* of this effect in his usual interesting manner. It is in bringing related and analogous facts, experimentally exhibited, to bear upon the subject in hand, that Mr. Faraday so eminently excels as a lecturer. If the facts are new, so much the better; but if they are old and thoroughly familiar, it is no matter, he presses them into his service, and by exhibiting them in an aspect illustrative of new views, or of new phenomena, he confers on them all the freshness and charm of novelty. With that sagacity which distinguishes the successful experimentalist, and which qualifies for, and is further inspired by, the pursuit of practical knowledge, his eyes are ever open to the observation of things in all the varying phases of their appearance, which new circumstances beget a capability of disclosing. Hence he is ever presenting us with old acquaintances in new faces. He leads us at first by a path which is strikingly elementary, both in point of information and experiments ; and yet his observations never appear as being common-place, for it is perceived that they are always pointing at—what in the end they never fail of conducting us—the most refined and discriminative investigations.

In explanation of the increased illuminating power of the lamp, Mr. Faraday exhibited the scintillating and highly luminous appearance of charcoal when burned in oxygen gas, and showed that the effect is due to the highly ignited or incandescent state of the particles of carbon, and not simply to their combustion. This fact he confirmed, by projecting into a gas flame magnesia powder, which being incombustible, the greater illuminating power of the flame, which was produced could not be caused by the combustion, but merely by the ignition of the particles of the powder. So in regard to the argand burner, the volatilized carbon of the oil is brought by means of the oxygen gas to an intense state of ignition, and thus a greater quantity of light is produced, without an increase in the consumption of oil. Other experiments were introduced in further elucidation of this fact, and such as showed the unignited condition of carbon in the body of flame ; but the proposed application of the invention, as being a matter of great importance, requires now to be noticed.

It appears that the system in which the French light-houses are established, of having lenses instead of reflectors, has its advantages ; but these cannot be properly developed without a most intense central light, but in which the height of the flame is not a matter of consequence, or rather, is undesirable. It also appears that it would be of vast importance to have the power of increasing this light according to circumstances. Fresnel's lamp of concentric wick does not fulfil these conditions, besides which, the expenditure of oil is very great in proportion to the light produced. But the oxy-oil lamp fully answers these purposes. At the same instant that the flame becomes more brilliant by the admission of oxygen gas, it shortens in its length, whilst within certain limits its quantity and intensity can be varied at pleasure. The arrangement proposed for a light-house is, to have a number of hollow or argand wicks, of about half an inch in diameter, placed in a circle of about six inches in diameter. each wick may thus be trimmed separately without interfering with the general light, and, of course, only a portion of these need be burned, if thought sufficient for the occasion. Such a light was introduced in the lecture, and its power of illumination was

indeed astonishing. We could have wished to have seen it in juxta-position with an oxy-hydrogen light, and under circumstances suitable for instituting a comparison. We understand that there is some probability of this invention being applied to the lighting of the House of Commons.

In regard to economy, it remains to be observed that, as the ordinary light is augmented two and a half times for a given quantity of oil, one and a half times that quantity is saved, the cost of which at 6s. 8d. per gallon, will nearly cover the expense of the oxygen gas. Allowing for the wear and tear of the apparatus, the interest of capital, and other items, the cost of the gas may be stated as being twice that of the oil. It appears that one pint of oil will require ten cubic feet of oxygen gas, and give a light in one hour equal to thirty-seven argands, the cost of which being

		s.	d.
Ten pence for the oil and twenty pence for the gas, will be..............		2	6
In the ordinary mode of burning oil it would take two pints and a half to produce the same effect.........		2	1
By Fresnel's lamp of concentric wicks there would be required four pints and a half of oil................		3	9

The light, therefore, is but little more expensive than ordinary oil-light ; but cost, in the case of light-houses, is but a minor consideration when put in comparison with increased efficiency in answering their important purposes.

Mr. Faraday, in the course of his lecture, made some very just remarks—elicited by the originality of this invention being questioned as due to Mr. Gurney—on the little estimation in which mere suggestions, although claiming priority, should be held in comparison with the talent and energy required to overcome the difficulties which are ever consequent on all attempts to carry them into practice.

If we may subjoin a remark of our own on the utility of this invention, we would observe that the superiority of effect appears to us to be attainable only by means that are too elaborate, and involving too much manipulation. The additional supply of oxygen may be obtained more directly and more readily from the atmosphere by an extension of the ordinary means by which combustion produces a draught or current of air ; and even should it be necessary to resort to mechanical means to procure the strength of current required, as in Mr. Beale's patent lamps, still it would be a simpler method than the chemical process, and one less liable to failures. But then oxygen supplied in this form is diluted with so large a proportion of nitrogen, that the additional aid to combustion, or rather in the case of oil, to a higher state of ignition, is completely neutralized by the *cooling* effect of the current, so that the argand oil-lamps already afford, by means of the ordinary draught of air, the maximum of illuminative power. This, however, is not the case with the more volatile oils, for here the mass of free volatilized carbon is so dense, that on being more abundantly supplied with oxygen, the combustion and heat become so intense as not to be *quelled* by the accompanying nitrogen, and thus the particles of carbon are raised to a more luminous state of incandescence. Hence it is that the naptha lamps require, and will bear so strong a draught, and that they produce so white and brilliant a flame, equal, in our opinion, to the oxy-oil flame in intensity, though not in height. Hence also it is, that in the combustion of still coarser materials Mr. Beale employs a yet stronger current of air (mechanically produced) with so excellent an effect. The combustion of the volatile oils may, in the case of light houses, be very easily aided, if need be, by a chimney; but whether this mode would be equal in efficiency to the oxy-oil light; besides being, as it undeniably would be, superior to it in point of cheapness and simplicity, is a question that can be determined only by experiment.

LIST OF SCOTCH PATENTS GRANTED BETWEEN THE 22nd JANUARY, AND THE 22nd FEBRUARY, 1839.

Edward Cooper, of Piccadilly, Middlesex, stationer, in consequence of a communication made to him by a certain foreigner residing abroad, for improvements in the manufacture of paper. Sealed Jan. 23, 1839; four months to specify.

Peter Taylor, of Birchen Bower, within Chadderton, Lancaster, rope maker, and slate merchant, for improvements in machinery for propelling vessels, carriages and machinery, parts of which improvements are applicable to the raising of water. Jan. 23.

Frederick Cayley Worsley, of Holywell-street, Westminster, Middlesex, Esq., for certain improvements in locomotive engines and carriages. Jan. 24.

Thomas Walker, of Birmingham, Warwick, clockmaker, for improvements in steam-engines, which improvements are also applicable to the raising or forcing fluids. Jan. 24.

Thomas Sweetapple, of Catteshall Mill, Godalming, Surrey, paper-maker, for an improvement or improvements in the machinery for making paper. Jan. 2d.

John Wilson, of Liverpool, Lancaster, Lecturer on Chemistry, for certain improvements in the process of manufacturing alkali from common salt. Jan. 30.

Sally Thompson, of North-place, Gray's Inn Road, Middlesex, for certain additions to locks or fastenings for doors of buildings, and of cabinets, and for drawers, chests, and other receptacles for the purpose of affording greater security against intrusion, by means of keys improperly obtained. Jan. 31.

Job Cutler, of Lady-Pool-lane, Birmingham, Warwick, and Thomas Gregory Hancock, of Princess-street, machinist, for an improved method of condensing the steam in steam-engines, and supplying their boilers with the water thereby formed. Jan. 31.

Horace Corey, of Narrow-street, Limehouse, Middlesex, batchelor of medicine, for improvements in the manufacture of white lead. Feb. 7.

Edward Tamwell, of Liverpool, merchant, for improvements in the manufacture of soda. Feb. 7.

Timothy Burstall, of Leith, Scotland, engineer, for certain improvements in the steam-engine, and in apparatus to be used therewith, or with any other construction of the steam-engine, or any other motive power for the more smooth and easy conveyance of goods and passengers on land and water, part of which will be applicable to water power. Feb. 11.

Charles Gabriel Baron de Suaree, of Red Lion-square, Middlesex, Colonel in the French service, and William Pontifex, of Shoe-lane, London, coppersmith, for a new mode of obtaining vegetable extracts. Feb. 12.

Morton Balmanno, of Queen-street, Cheapside, London, merchant, for a new and improved method of making and manufacturing paper, pasteboards, felt, and tissues, communicated by a foreigner residing abroad. Feb. 14.

Joseph Birch, of Bankside, Blackfriars, Surrey, calico printer and designer, for certain improvements in printing cotton, woollen paper, and other fabrics and materials. Feb. 19.

Harrison Grey Dyer, of Cavendish-square, gent., and John Hemming, gentleman, of Edward-street, Cavendish-square, Middlesex, for improvements in the manufacture of carbonate of soda. Feb. 19.

Edward Rearson Tee, of Barnsley, York, dyer and linen manufacturer, for improvements in weaving linen and other fabrics. Feb. 20.

Joseph Bunnett, of Deptford, Kent, for improvements in steam-engines. Feb. 20.

NOTES AND NOTICES.

Patent Inflated Saddles.—Sir,—I have been informed of a paragraph being inserted in your Magazine, No. 809, for Feb. 9, stating that a new invented, air-seat saddle, by a Mr. Collinson of Burneston, was about to be made known to the public. Being the patentee of the inflated air-seat *saddles*, I consider Mr. Collinson's to be an *infringement*; my *claim* and *legal right* being to all saddles of that kind as the first *inventor*. The patent inflated saddle has been in use and on sale for three years past. Thomas Taylor, Banbury.—Feb. 23, 1839.

Improvement in Woolcombing.—A very important invention in the woollen manufacture has lately been patented in Great Britain—it consists in heating the carding engines and combs by steam, which has the beneficial effect of allowing the wool to be stretched or extended as it is operated on by the teeth of the cards or combs, without breaking the fibre. By the use of the process, out of every 100 pounds of undressed wool, 95 of the best wool may be obtained; whilst by the methods now in use 65 pounds is the best result. The invention is of French origin, and so great are the advantages considered to be, both in point of quantity and quality, that until machinery can be got ready here, engagements have been made to send wool over to France to be combed, and then to be returned to England; and which it has been calculated will render a good profit, notwithstanding a duty of 40 per cent. upon its importation.

LONDON: Printed and Published for the Proprietor, by W. A. Robertson, at the Mechanics' Magazine Office, No. 6, Peterborough-court, Fleet-street.—Sold by A. & W. Galignani, Rue Vivienne, Paris.

Mechanics' Magazine,

MUSEUM, REGISTER, JOURNAL, AND GAZETTE.

No. 813.]   **SATURDAY, MARCH 9, 1839.**   [Price 3*d.*

Printed and Published for the Proprietor, by W. A. Robertson, No. 6, Peterborough-court, Fleet-street.

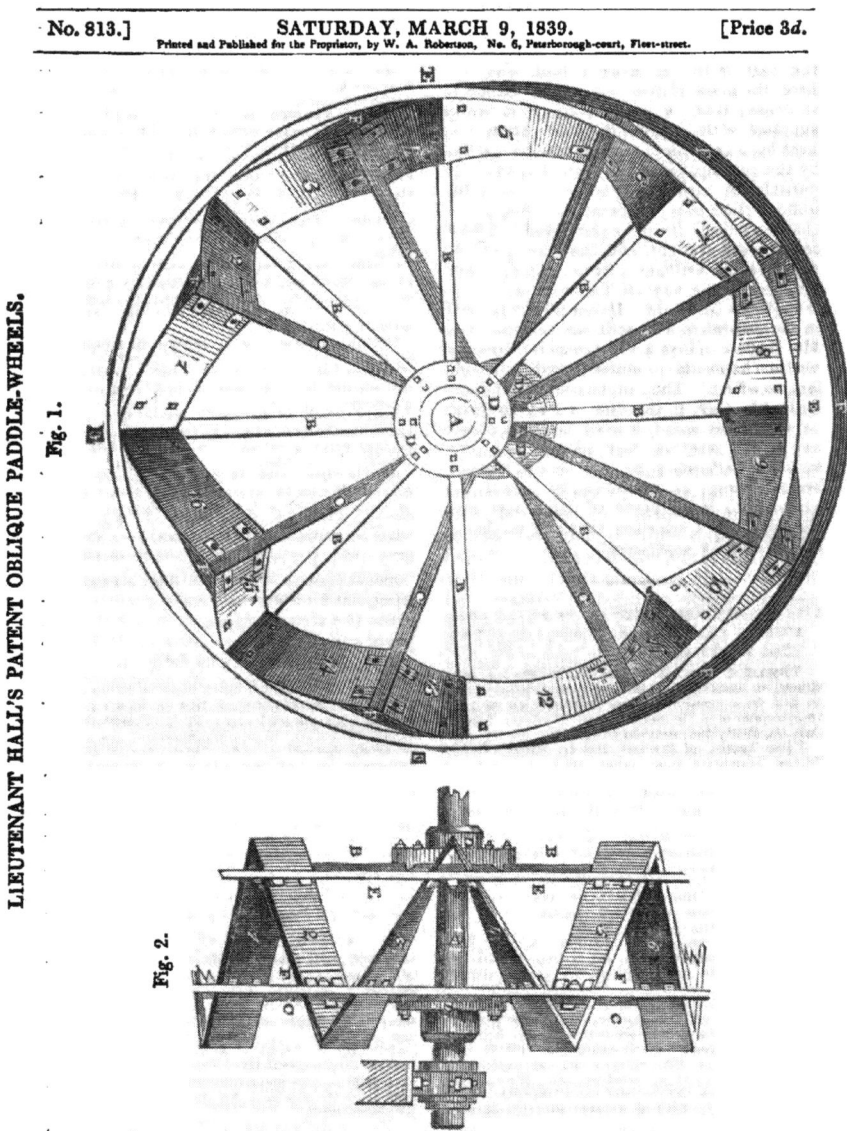

LIEUTENANT HALL'S PATENT OBLIQUE PADDLE-WHEELS.

Fig. 1.

Fig. 2.

LIEUTENANT HALL'S PATENT OBLIQUE PADDLE-WHEELS.

The objects of this invention of Lieut. Hall, R.N., are those usually proposed in improved paddle-wheels, viz:—The removal of the distressing and injurious tremor in steam-vessels, occasioned by the stroke of the paddle-boards upon the water; the avoidance of the lift of back-water; and the employment of the powers of the engine to the greatest possible advantage. In Lieutenant Hall's wheel the arms or spokes diverging from each extremity of the shaft, are not opposite and parallel to each other respectively, as in the ordinary wheel, but those at one end are placed alternately with respect to those at the other end of the shaft. The paddle-boards uniting these arms are consequently at an angle with the axis of the wheel. They are also joined together so as to form angles with each other throughout their entire breadth, and salient and re-entering angles with the side of the vessel. The paddle-boards are made to assume the requisite form by being slightly twisted, from right to left and left to right alternately, which is readily effected by previously steaming them; and they are sufficiently rounded to preserve an equal dip in the water in all dispositions of the wheel. There is thus obtained a continuous surface representing a single paddle-board, carried in alternate directions from arm to arm round the wheel until the extremities meet.

The paddles are affixed to each wheel, so that the salient angles of the one wheel shall enter the water at the same instant with the salient angles of the other, and, as necessarily follows, the re-entering angles of each wheel also enter simultaneously. The resistance is then identical with that of oars when rowing, with the advantage of being continuous. By this coincidence of the entry and exit of the corresponding boards on both paddles, that lateral motion to which the paddle-shaft would be liable in its bearings, were this not attended to, is prevented.

In action, the paddle-boards thus arranged enter the water in an endless series, and increment by increment, without noise or any concussion upon the water, and present to it, throughout the entire revolution of the wheel, an equal and constant resistance; while the action

upon the water is at right angles with the shaft or line of motion.

The results of this construction are stated by the inventor to be the accomplishment of the before stated objects:—perfect freedom from all vibration communicated to the vessel by the paddles—absence of any disagreeable noise or flapping of the paddles upon the water—no lift of back-water by the emerging paddles—the greatest regularity and smoothness in the action of the engine—increased speed imparted to the vessel, (beyond that hitherto obtained with equal power,) by the avoidance of the lift of back-water and the application of a continuous propelling power in place of the alternating or reciprocating one heretofore employed.

When the vessel is laden beyond her ordinary trim, or where it may be deemed desirable to employ deeply immersed wheels, the advantages derivable from this construction are proportionally augmented.

Although the expression, " paddle-boards," has been exclusively used in the above description, iron or other metal may be substituted for wood. The construction partakes of the properties of consecutive arches resting alternately upon each other, and consequently presents the *strongest* form of which divided parts are susceptible. Simplicity is also a prominent characteristic of these wheels, and, as regards expense, they do not exceed that of the most ordinary paddle-wheels in present use.

Description of the Figures.—Fig. 1 is a front view of the wheel, its axis being a little below the level of the eye; and fig. 2 an edge view of the same wheel. The same letters indicate similar parts in each figure wherever they occur. A A, the shaft or axis; B B, the arms nearest the eye; C C, the arms diverging from the distant end of the shaft; D D, the bosses; E E, the exterior ring nearest the eye; F F, the exterior distant ring. The figures on the float-boards indicate the manner in which they follow each other in consecutive order, without any opening or space between them—each being supported by that which precedes and follows it throughout the entire circumference of the wheel.

Fig. 3 differs from Figs. 1 and 2, in having two sets of float-boards, but they are set on in precisely the same manner

Fig. 3.

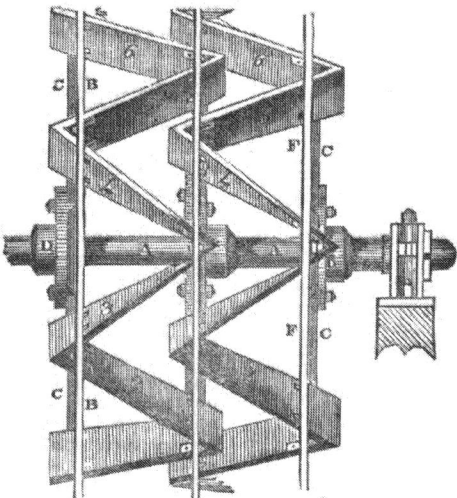

as those in Figs. 1 and 2, and are marked with the same letters of reference. This arrangement is to be preferred when the breadth of the wheel is such as to render three arms requisite to support each paddle on the ordinary construction.

Models of these wheels having been submitted to the Government authorities, the Lords of the Admiralty were pleased to order a trial to be made of their merits by their Chief Engineer and Inspector of Machinery. Paddles of the required form were consequently constructed at Woolwich Dock-yard, and temporarily attached to the previously-existing frame-work of the *Dasher*, a new steamer just fitted out for the packet service. The official Report made to their Lordships, of which they were pleased to inform the inventor by Sir John Barrow's letters of the 5th of March and 4th of May last, was, that the results of the trial were *favourable* and *satisfactory*.

These wheels have been since permanently fitted to a steam-boat of fifty-horse power at Liverpool, and have been in constant daily service between Woodside and Liverpool, or in towing heavy vessels out of the harbour, for upwards of three months, and during unusually severe weather.

From amongst a number of most satisfactory testimonials which we have seen, we select the following by Captain Denham, as being the most comprehensive, and coming from the best authority.

"*Marine Surveyor's Office, Liverpool,*
November 22, 1838.

"My Dear Sir,—According to my promise, when your diagonal floats were fitting at this port to the *Helensburgh* steamer, I have taken the opportunity of trying their comparative effects with those of the ordinary water entrance and exit, and I am satisfied that you have completely obviated that disagreeable and bolt-loosening tremor common to steamers, and promoted the impetus in proportion as you avoid the usual lift of back-water. These advantages resulting from *no additional expense* must carry an adoption wherever known. In the *Helensburgh's* cabin I experienced no more vibration than in a sailing vessel; nor on the paddle-box could a drop of water be detached by agitation from the convex surface of a brimful tumbler; whilst at night you could steal along an enemy's shore, or into his fleet, without imparting any of the telltale flapping of the usual paddle-float. Most heartily wishing it the early *general* adoption it deserves.

"I remain, yours, very truly,

"(Signed)   "H. M. DENHAM.

"Commander R.N., Resident Marine Surveyor,
Port of Liverpool."

It should be observed that, in both the instances of the *Dasher* and the *Helens-*

z

burgh, the previously-constructed frames of the wheels were employed, which were of the ordinary form. Greater advantages would accrue, if the wheels were made throughout upon the principle before explained, although, in most cases, the existing proportions of paddle-wheels admit of an easy adaptation to this construction, at a very moderate cost. Z

MR. BADDELEY'S REPORT ON LONDON FIRES.

Sir,—In his last annual Fire Report, your valuable correspondent, Mr. Wm. Baddeley, takes occasion to observe, that, " under the protection of the present highly-efficient Fire Establishment, the comparative security of lives and property is so much increased, as to be a subject of general remark." Any where else, this observation might pass without particular notice, backed as it is by nothing better than a quotation from "a humorous ballad published a few weeks, since,"—but, unluckily, several of the columns of the *Mechanics' Magazine* preceding that in which it appears, are filled to such repletion with appalling narratives of the ravages of " the destructive element," that few readers will be inclined to give in their adhesion to its correctness. Let us examine a little those facts in this very report, which have the *appearance,* at least, of contradicting most completely the allegations of that great authority, the doggrel " lament of the penny-a-liners," and take the liberty of testing Mr. Baddeley's assertion as to the *increased comparative security of lives and property,* by a few instances in point supplied us by no other than Mr. Baddeley himself. We need not be long in arriving at a result perfectly satisfactory.

First, as to security of property : the number of fires has vastly increased since the Fire Brigade commenced operations; in 1832, the last year of the old system, they amounted to 209; during the past year the number was 568. But this is not all :—the firemen certainly cannot prevent fires breaking out, and, according to Mr. Baddeley, the Brigade men are called to a far greater number of merely frivolous cases than the old firemen, which may account for much

of the apparent increase;—the testing point is, what number are prevented from spreading to a serious extent *after they are discovered?* Mr. Baddeley will inform us :—

1832. " Consumed," and "partly consumed" 56
1838. "Totally destroyed," and " very seriously damaged"......................... 185

So much for " increased comparative security of property." Why, without allowing anything for the great number of slight fires which the brigade attend, it appears that the absolute proportion of serious fires to the total number has amazingly augmented under their *highly-efficient* management! The old firemen-watermen, turning out only when "something like a fire" required attention kept down the serious losses to one in four, while the Brigade, always on the *qui vive* for any conflagrations however small, have one " total destruction," or " very serious damage" to every *three;*—nay, to less than every three! Under the protection of the *highly-efficient* Fire-Brigade, the serious fires have become much more in number than *treble* what they were when our fire-engine system was unimproved! And yet, in the face of such facts as these, Mr. Baddeley,—by whom those facts, in both cases, are supplied,—would wish us to believe that the " comparative security" under the new system has " become the subject of general remark." And this, also, in the very same number of a work which contains a long and circumstantial account of an immense number of large conflagrations, at which the fire generally seems to have had matters entirely its own way (the Royal Exchange to wit) —and in which the alarming details of total destruction are wound up by the quiet paragraph—" This fire completes the list of total losses, and *such a list as has hardly ever before occurred!*"—It is scarcely credible, but nevertheless true, that this paragraph, and the one alluding to " increased security," both proceed from one and the same pen, the former at page 312, and the latter no further off than page 319!*

It has not always, indeed, been held by Mr. Baddeley that " increased protection of property" is a state of things at all creditable to the firemen, but ra-

---

* Vide, for confirmation of this fact, the *Mechanics' Magazine,* No. 809, February 6, 1839.

ther the reverse. Thus, when Mr. Simeon asserted the superiority of the *Sapeurs Pompiers* of Paris to the Fire Brigade of London, how did Mr. Baddeley defend the latter? Not by enumerating the large fires that they *did*, but the large fires that they did *not* put out! He first gives a long catalogue of " fires of very serious magnitude" occurring in London within the last five years, commencing with the Houses of Lords and Commons, and closing with the Royal Exchange, and then triumphantly observes that, *against this list,* " the Parisians can set nothing beyond the Theatre Italia* in January last." Now, considering that the question was, which party had been most successful in *suppressing* fires, and not in allowing them to rage unchecked, one would suppose that a champion of the London Brigade would hardly endeavour to support their cause by adducing the instances in which they had totally failed in their object, and twitting their adversaries with having *extinguished all serious fires but one*, during a period in which the Brigade could point to a large number of first-rate extent and importance, within their district, where the devouring element had reigned uncontrolled. It would sometimes seem, to be sure, that Mr. Baddeley imagined the duty of the Fire-Brigade consisted in securing to the public the spectacle of " a good fire," whenever an opportunity presented itself. Thus, in the same reply to Mr. Simeon, he observes, " whenever a comparison has been made, in my hearing, between the promptness, skill, and intrepidity of the London and Parisian firemen, it has always been to the advantage of the former; a glance at the foregoing elements will show the fairness of the conclusion :"—the said elements, including the very *conclusive* list of " fires of serious magnitude" alluded to, so that we are to infer the " promptness, skill, and intrepidity" of the Brigade

from the extensive ravages made at the fires they have attended! At this rate, Mr. Baddeley had better suggest to the Parisian firemen the use of spirits of turpentine, instead of water, in their engines: an improvement which would soon enable them to turn the tables on their London rivals, and to produce in due time a catalogue of calamity surpassing, perhaps, that furnished by Mr. Baddeley. At present, with only one " total loss" of any consequence to refer to in a space of five years, they must be allowed to cut a truly contemptible figure!

On the subject of the " security of *lives*," the facts furnished by Mr. Baddeley enable us to come reluctantly to the same melancholy conclusion, as with respect to the "security of property." Mr. B. was some time since exceedingly angry with a gentleman who supplied to the British Association a statistical account of the fires of London, from 1833 to 1837, for asserting that " the number of fatal fires had greatly increased" Yet the gentleman in question had deduced this opinion solely from the facts furnished in Mr. Baddeley's own reports; and how, from a perusal of them, he could draw any other inference, it is not easy to imagine. The number of persons burnt to death in 1832, the last year of the old system, Mr. Baddeley stated to be *nine;* the number similarly reported *by him* in 1838, is *twenty-one;* the year before, which was that immediately preceding the drawing-up of Mr. Rawson's paper, it was *nineteen.* But Mr. Baddeley explains (vol. xxix, page 457) that the " apparent increase" arises from the firemen being now called in when wearing-apparel only is ignited. From his last report, however, it appears that *two* only out of the twenty-one cases reported belonged to this class. Besides, the number of fatal fires *of all descriptions* has greatly increased, if we may judge from Mr. Baddeley's statements, that during the last year *one hundred and thirty* lives were lost from this cause, while in 1831 the whole number was only *thirty-five.*

Seldom, indeed, have the readers of the *Mechanics' Magazine* had their feelings harrowed up by such horrible details of human suffering as crowd several of the pages devoted to the last An-

---

* So called several times over in the letter referred to; the " Théâtre Italien," is, of course, intended. Mr. Baddeley elsewhere makes frequent mention of an article which he says is called the " Ital*ien* ladder :"—By whom? The " *Italien*" seems meant for French, but it is needless to say there is no such word as " ladder" in that language; besides, the real French word *(échelle)* being feminine, would require the adjective to be spelt " *Italienne*". Why not stick to plain English, and say " Ital*ian* ladder" at once.

nual Fire Report. The *firemen* do not appear to be culpable: from Mr. Baddeley's account it would appear that they have never arrived on the fatal spot until after the mischief has been done. What is this owing to? Can it be to the reduced scale of the Fire Establishment of the last few years, during which both "serious" and "fatal" fires have increased in at least as great a ratio as the expenses of the insurance offices have diminished? At all events, it need excite no wonder that the firemen now seldom arrive in time to be of service, if the fact be (as stated by Mr. Baddeley, page 41) that, while it requires *eighty* firemen to efficiently protect the city of Edinburgh,—a city of not one-tenth the population of the metropolis, — "the Fire Engine Establishment of London musters *not quite one hundred!*"

And I remain, Sir,

Very respectfully yours,

AQUARIUS.

London, Feb. 20, 1839.

PREVENTING THE OXYDATION OF, AND COLOURING METALS.—MESSRS. ELKINGTON AND BARRATT'S PROCESSES.

In our 811th number we published a full account of the processes of M. Sorel, for the coating of iron and other metals, with zinc, to prevent their oxydation. We now lay before our readers descriptions of some further processes for the same purpose, and for giving iron and steel a brass colour, invented and patented by Messrs. Elkington and Barratt, of Birmingham, which, although they have not been found so effectually to answer the desired purpose, are nevertheless very ingenious, and worthy of the attention of those interested in the matter. We may observe, however, that M. Fontainemoreau, in his patent, claims the application of amalgams of zinc, as also alloys of zinc with lead, tin, or bismuth; how far the peculiar processes of Messrs. Elkington and Barratt will enable them to uphold their patent, we will not take upon ourselves to say.

Messrs. Elkington and Barratt's invention consists of certain modes of coating metals with zinc, and zinc and mercury; and a mode of colouring iron and steel.

In order to coat copper and brass with zinc, there are mixed in an earthen vessel seven parts of muriatic acid, (specific gravity about 116,) and one hundred parts of water, both by weight; and to these are added four parts of zinc, in the state of powder, or pieces. These articles are allowed to remain twenty-four hours, or until the acid and zinc cease to act upon each other, and the solution thus obtained is poured into a convenient vessel for boiling it, adding a quantity of zinc in powder, or in thin pieces. While boiling, the articles to be acted upon are immersed therein, bringing them in contact with the metallic zinc, and they will speedily become coated therewith. They are then removed, and washed with water and dried. In using this solution of zinc, if the articles are of iron or steel, they are previously coated with copper; and this is effected as follows: The articles are first cleansed, or pickled in dilute sulphuric acid, composed of one part concentrated acid to sixteen parts of water; and having prepared a solution of sulphate of copper, commonly called blue vitriol, the iron is immersed therein while cold, for few seconds, and speedily removed and washed. This is repeated one, two, or three times, or until it is found that the iron is perfectly coated; care must be taken not to allow it to remain too long in the solution of copper, or the copper precipitated on the surfaces becomes loose. If a strong coating of zinc be required, the processes of coppering and zincing are repeated, and it has also been found, that if the articles, when of copper, or if of iron, after they have been coppered, are introduced into a dilute solution of nitrate of mercury, and then again boiled in the solution of zinc, that the same object is obtained. The nitrate, or any other convenient solution of copper, may be substituted for the sulphate.

Another process is as follows: Take dilute muriatic acid in about the proportion of one part acid, (the specific gravity 116,) and thirty parts of water, into this introduce a quantity of zinc, in powder or in small pieces. The articles of iron are then to be placed in the acid, and kept in contact with the zinc during the process, which will require from two to five minutes, or until they are evidently coated with zinc; then remove, wash and dry them, as before.

Various metals, as iron, steel, copper, brass, &c., may be coated with the amalgam of zinc, and although this may be effected by using the two metals in almost any proportions, it has been found that six or seven parts of zinc, with one part of mercury, will answer best; these are amalgamated by heat, or by agitating the two metals in contact with dilute muriatic, or other convenient acid; the zinc being previously granulated, or reduced to small pieces. To this amalgamated zinc, when effected by heat, add dilute acid, as before, and then introduce the articles, which may require to be kept occasionally stirred. Instead of using the muriatic acid, some salts are employed, as the muriate, or sulphate of ammoniac, in the proportion of one of salt to thirty ounces of water, or thereabout; and other acids than the muriatic may be employed, as ascetic sulphuric &c., and which require no other directions than to employ them of about the same strength as directed for the muriatic acid. It is preferable to employ these solutions in a hot or boiling state, as the effect is thereby obtained in a shorter period; but the processes where the free acid is used, may be successfully performed in a cold state, the acid and water to be added occasionally, as the solution becomes reduced in strength or quantity, by the boiling or action upon the metals to be coated.

An amalgam of zinc may be employed in a melted state for some articles, in which case, and particularly if they are of iron, they require to be well pickled or cleaned and also to be immersed in a solution of muriate of ammonia, to induce the perfect adhesion of the amalgam, which amalgam may be varied in almost any proportion of the two metals, but the proportions now given are considered best.

The oxides of the metals may also be used in the same manner, instead of the metals, or in conjunction with them; as, for example, a solution of zinc may be made with the oxide of that metal, instead of using the metal itself, and so also with the oxides of mercury. Messrs. Elkington and Barratt's processes may be applied in connection with the means patented by Messrs. Craufurd and Fontainemoreau, and detailed by us last week.

The process for colouring metals is as follows:—To colour iron and steel to imitate brass; first wet the iron or steel by means of a solution of copper as already described, and having afterwards boiled it in the saturated solution of zinc, having excess of zinc therein, until perfectly covered, remove it and dry it in saw dust and then submit it to heat in a closed oven until the required colour is obtained, and which is easily observed by looking occasionally at the articles during the process. They are afterwards to be pickled in a dilute acid and washed and dried.

A process called "Similoring," from the words "*simile l'or*," has been before practised for colouring copper and brass, and which consists in obtaining on the surface a thin coating of zinc, and submitting the articles so prepared to the action of heat till a colour approaching to that of gold is obtained. The object of the present invention, so far as it relates to coating copper and brass, is to obtain a good and sufficient coating of zinc on the surfaces, in order to prevent or retard oxidation. The use of heat is omitted, which would be prejudicial to the coating of zinc.

MODE OF LIGHTING GAS LAMPS BY ELECTRICITY.

Sir,—The following is a brief description of a plan by which, in my opinion, electricity may be applied to the important use of lighting the gas lamps in the public streets. I propose to run an insulated copper wire parallel with the gas pipes, terminated at each jet by two small balls, as in fig. 3, and communicating with an electrical machine stationed at the gas works, a few turns of which will cause an electric spark to pass across the unconnected parts of the wire at the summits of the jets, and inflame the gas issuing at the apertures, thus lighting up in an instant all the lamps so provided. To prevent loss of gas, a ball-valve has been contrived, the pressure of which need not exceed two or three ounces, and which will retain the gas in the pipes when the main is turned off at the gas-works.

Without further detail I will endeavour to develope my scheme by an explanation of the figures.

Fig. 1 represents a section of jet and valve. A, jet; B, holes for the gas to escape at the orifice H; C, fastening for spiral spring; D, spiral spring attached to—E, brass cup to cover—F, glass sphere to fit—G, glass socket.

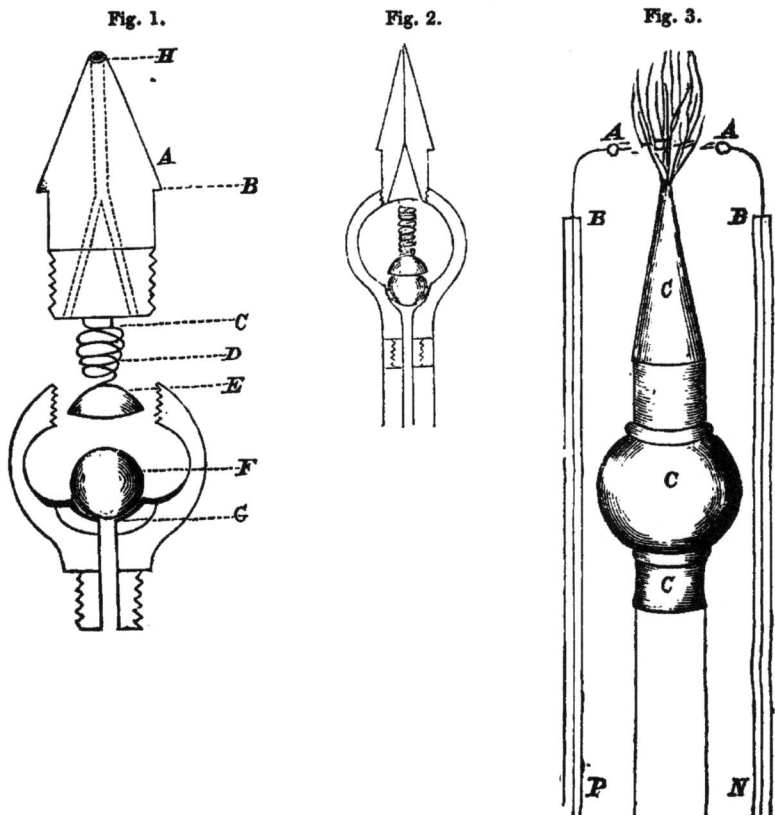

Fig. 1.   Fig. 2.   Fig. 3.

Fig. 2, section of fig. 1, joined together.

Fig. 3; A A, brass balls attached to wires insulated in—B B, glass insulating tubes; C C C, jet and valve complete; P, positive wire; N, negative wire.

My plan might be easily adopted in public buildings, as a small electrical machine might be kept in any part of them for the purpose.

The advantages of this plan, if carried into execution, are obvious. Besides superseding the expensive and tedious process of turning the gas on and off at each lamp separately, a considerable sav- ing would also be effected by the gas being retained in the pipes instead of partly wasted as under the present system; and the facility it would afford in cases of emergency, such as dense fogs coming on early in the evening, of lighting up the metropolis in an instant, may also be mentioned. With respect to public buildings one other recommendation may be added—it would very materially diminish the chances of accident by fire.

JOSEPH BECK.

12, Thomas-street, Lambeth,
Jan. 21, 1839.

WHITELAW'S IMPROVED EXPANSION VALVES.

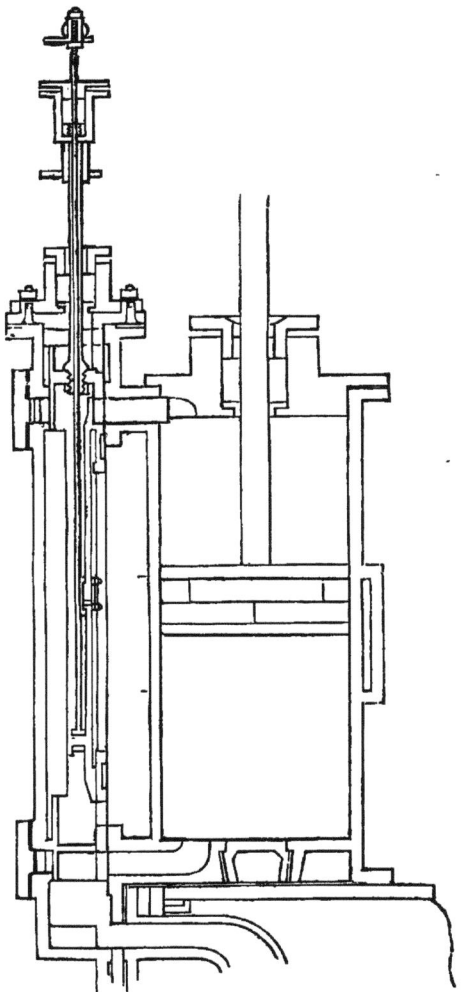

Sir,—The accompanying is a drawing of the cylinder and nozles belonging to one of a pair of steam-engines intended to be put up here with my expansion gear to them. The large slide is of the D construction, and cast in one piece; its top and bottom parts are connected by two ribs, one on each side, in order to get the cut-off valve into its place; it is made in two pieces, and coupled in the manner shown in the figure.

As the large slide will wear so as to bring its rod always nearer to the cylinder, I have made the stuffing-box on the cover of the nozles in a way that it may be shifted closer to the cylinder at any time. From the descriptions now and formerly given, your readers will easily understand the other parts of the figure.

My expansion valve might be made to work without having the additional part cast on the steam side of each face of the large slide, by allowing the expansion valve to slide upon the face of the nozles in the same way as the large slide; and when the steam was cut off, one end of the expansion valve would then fit upon the steam end of one of the faces of the large slide. A small variation in the length of one of the levers which work the cut-off valve, and making its cross head, or some other of the parts slightly flexible, will be the only other alterations necessary to be made in this case.

I expected that the Cornwall Polytechnic Society would have made some remarks upon my former communication to your Magazine, on this subject, but as they have not done so, I will, the first time I have nothing else to do, (it may be long enough till then, and it may be early,) give your readers my ideas respecting them, for I would not like that the *Royal Cornwall Polytechnic Society, for the encouragement of the Sciences, &c.* should have it again in their power to insult any other of your readers in the way they did me.

I feel obliged to Mr. Trevelyan for the notice which he has taken of my feeding apparatus. I am of opinion that he is right in thinking that it will answer well; but I also believe that if he had read my letters in answer to Mr. Baddeley's remarks, he would not have written the last part of his notice. It appears to me that some of your correspondents have rather too much of the disposition to find fault; now, I do not see that any one should object to a thing, even if it is no more than passable—unless he is prepared to give something better.

J. WHITELAW.

Glasgow, 26th Jan., 1839.

LOCOMOTION ON COMMON ROADS.
ARTICLE III.—PROPER MECHANISM THE DESIDERATUM.

It is an observation no less common than true, that "Nature has done nothing in vain;" that the means she takes to effect her ends are always the most direct, and the best calculated for her purposes: and, without referring to sacred writing for authority, we have only to look around us and examine the workmanship of the creation, the more minutely the better, to convince us that when the Great Mechanic turned it out of hand, He had reason to pronounce that "All was good," that every animated being was furnished with sufficient powers, not only for its existence and progress to perfection, but also for its protection, enjoyment, and happiness, according to the wants and propensities He had been pleased to gift it with.

These wants and propensities in the animal creation are as numerous and various as the forms impressed upon them; and the first impression which strikes us, when taking a survey of the multitudinous collection which the study of natural history opens to our view, is the three grand divisions of matter or spheres of action, destined for the location in which this mass has to exist and perform its functions—the air, the water, and the earth; each of them furnished with the means of sustaining animal life, but peculiar and distinct from each other.

Amongst this immense mass of creation there is one want or propensity pervading every species, and common to the whole; that is to say, Locomotion, or the power of transporting the person from one place to another; and whether the immediate object we are contemplating may be an inhabitant of either of the elements above mentioned, or any two or more of them, we find every species furnished with means specially adapted for its station, and calculated, according to its sphere of action, to gratify and carry out at pleasure, this universal propensity, of ranging abroad, of changing its place or situation.

The method which Nature has universally taken for effecting this object is muscular power, exhibited through machinery of body, proportioned to meet the densities, resistance, and other properties of the element in which it is to be exhibited; thus, in the case of marine animals, the muscular power is lodged chiefly in the tail, which striking at right angles to the direction of the body, against the water with which it is surrounded, gives the necessary impulse forward; the steerage and direction being effected by the fins, and by contortions of body the animal has always at its command. In the case of birds the muscular power is lodged in the wings, the steering or guiding power in the tail; and so on through the whole of creation; varied indeed in amount and proportion to the wants and propensities before mentioned, to which each species is subject.

Now, if we look abroad and take a survey of the different human inventions bearing on the subject under consideration, viz. locomotion, we shall find that those have always been most successful, that have copied nature closest, that have adopted her principles, and followed the beaten track she has marked out for them; for instance, what are the hulls of our ships but a *fac-simile* of the bodies of aquatic birds? What are our masts, sails, and rigging, but a close imitation of the wings of the feathered creation, with this difference, that instead of motion being produced by their striking against the air, it is produced by placing them in such a position as that the air should strike against them. What are our keels and rudders but a contrivance copied from the fins, &c. of fishes, or the tails of birds? And what, after all, are the paddles of our steam boats but the identical action of the fishes' tail in the direction before mentioned, or the web-foot of the water-fowl similarly exerted? We have here experience to convince us that Nature having done all that is necessary for her own purposes, has given man the cue in what points he may reasonably hope, by cultivating, lengthening, strengthening, or multiplying her processes, to improve upon or carry out to a greater extent and perfection the rough material of motion she has placed at his command.

Let us now apply this doctrine to the subject immediately under our consideration, that of locomotion by means of artificial power, on the common roads; and having made up our minds as to the object we are in pursuit of, let us as a preliminary step, look into Nature and observe what means she has taken to effect a similar object; and here we most certainly shall not meet with *wheels* as a motive power; no animal that we are acquainted with in creation, being furnished with them. As a contrivance of human ingenuity, for the purpose of lessening friction or sur-

mounting small obstacles, they are all very well; but still, nothing more than the passive medium of a power generated elsewhere. As an active medium, Nature has, and Art must, repudiate them.

We have seen in our ships, and steam boats, the success attendant upon following up the processes of nature, on the water, and in the air. What then are the means of motion she has adopted in the third grand division, on the land? Why, *legs—legs*—invariably *legs*, whenever she has intended celerity of motion; and whatever may be our attempts to arrive at the object of our desires, by trying to supersede her laws, or to strike out a new, and independent course of action for ourselves, "to this complexion we must come at last," if we would reasonably entertain the expectation of arriving at the object of our search, in a natural way.

Having observed this universal law of nature, which constitutes, in the great bulk of the animal creation, biped, or quadruped, legs as the source of locomotion on land, we have the line of action clearly marked out, in which we should direct our experiments; and, instead of obstinately persisting in the attempt to attain our object through driving wheels, we must go back to the simplicity of nature; and by accurately examining, watching, and comparing her motions, endeavour to find out, if we can, the means of improving upon her; and keeping her productions in our eye, as the pattern piece of workmanship, direct our attention in the way before observed, as to whether we cannot, by lengthening, strengthening or multiplying her processes, produce a mechanism, acting precisely in the same way as the legs of horses; but with increased strength— greater stretch—quicker stroke, or improved general effect.

It is a curious observation, but true, that all the early attempts at locomotion had a tendency in this direction, but were abandoned without hitting their object, in consequence of not keeping the model here recommended in view,— the propellers of the day were, in their principles, and action, anything but that which they were intended to supersede. The major part existed only upon paper, while those which we actually tried failed, from want of being directed in the course here pointed out. In the mean time, the success of driving wheels on rail roads, adapted expressly for them, drew the current of experiment into a new, and false channel; and the adage, that "circumstances alter cases" being lost sight of, so far as regarded the roads in question, the natural results have been miscarriage, and disappointment.

The object of this letter is to point out the way in which this discovery should be sought,—the line of direction our inquiries and experiments ought to take :—and as the improvements which have been made since the time mentioned, in the management, and processes of generating steam, present a complete substitute for the muscular power of the horse, all that remains for us to do, is to find out the mechanism, through which this power is to be conducted, so as to lead us effectually to the result we are aiming at—which is neither more, nor less, than a close imitation of the action of heels.

　　　　　　　　　　　　　G. S.

10, Poultry.

———

THE CHRONOMETER PRIZES.—REPLY OF MR. CARTER TO PHILO-VERITAS.

Sir, — The *Mechanics' Magazine* of 23rd of February, contains an article impugning the validity of a certificate given in favour of my chronometers, several of which have been honoured with the prizes bestowed by the Government for the best instruments.

Had the writer of the communication confined his attacks to myself, I should not have deigned to notice the mere invention of an anonymous accuser, but when his malevolence is directed against a gentleman so estimable for scientific attainments and moral worth as Professor Airy, I feel myself called on to repel with scorn and contempt, allegations so utterly unfounded and untrue.

The letter is signed " Philo-Veritas;" but I shall attempt to prove, in continuation, that this lover of truth is a " Liar of the first magnitude." I am also honoured with the name of Mr. Caster, an appellation which does not belong to me ; a mere trifle in itself, but worthy of notice, since it serves to show the

carelessness and falsehood which characterizes the whole production*.

In the year 1828 I find, upon reference to the official public documents relative to chronometers which have obtained Government rewards, that the prize was awarded to a Mr. Guy of Radnor-street, for his chronometer, No. 1410. Surely this could not be either Mr. Parkinson or Mr. Frodsham? In 1830 the first prize was granted to Mr. Baker of Pentonville, the second to Mr. Carter of Tooley-street, and the third to Mr. Murray ; here again there must be some mistake. In 1831 the name of Cottrell, Oxford-street, is first on the list; then Mr. Frodsham appears for the *first* and *only time* in the annals of successful competition as entitled to the second prize ; Mr. Webster third. What then becomes of the fact, that the whole of the prizes in this year were obtained by Messrs. Parkinson and Frodsham? In 1832 the first prize was awarded to Mr. Molyneux, the second to Mr. Young, and the third to Mr. Webster; here again I submit that it is impossible to find the name of Parkinson & Co. In 1833 the second premium was granted to Mr. Appleton of Burton-street, Burton-crescent, and the third to Mr. Molyneux ; no chronometer during this year performing sufficiently well to entitle it to the first prize. In 1834 the third and only premium was granted to John Carter, 207, Tooley-street ; and in 1835, although in competition with fifty-five chronometers, on the termination of the trial, February 1836, only one chronometer remained, named John Carter, Tooley-street, No. 160, which, as the subjoined letter will testify, obtained the third and only prize.

So much then, Mr. Editor, for the *impartial* truths of "Philo-Veritas !"

I cannot take leave of Philo Veritas without exemplifying another proof of his amiable veracity ; he unblushingly asserts that the certificate of Professor Airy is erroneous, because, he declares, "Mr. G. B. Airy did not succeed to the post of Astronomer Royal, until long after the public trials alluded to, had ceased." The gross and unpardonable falsehood of this assertion is immediately detected and exposed by the following

———

* This error was the fault of the printer.—ED. M.M.

official letter received by me from the Astronomer Royal, dated—

"Royal Observatory, Greenwich,
March 9th, 1836.

"Sir,—I beg to inform you that at the conclusion of the trial (at the end of last month) for prizes offered by the Lords Commissioners of the Admiralty, for chronometers of certain degrees of merit, the trial number of your chronometer, Carter, No. 160, has appeared to be 4ˢ·53, exceeding by a very small quantity, the number prescribed by their Lordships for the third prize.

"In consideration of the very near approach to the prize number, and the severity of the weather in the past winter, I am directed by their Lordships to state that they have awarded to the chronometer, 'Carter, No. 160,' the third prize.

"(Signed)   "G. B. AIRY.
"To Mr. John Carter."

I have now to apologise, Mr. Editor, for trespassing so much on your valuable pages, and, in conclusion, allow me to repeat that though I was not surprised at the effusion of professional envy, falsehood and meanness directed against myself, yet I do most deeply regret, that the station of Astronomer Royal, combined with the splendid talents and honourable bearing of the distinguished philosopher who fills that exalted post, were insufficient to protect him from the malevolence and insult of an anonymous assailant.

Feeling assured that after this exposition your columns will not again be made the vehicle of such unworthy attacks,

I beg to remain, Sir,
Your obedient servant,
JOHN CARTER.

61, Cornhill, and 207, Tooley-street, London.

OMNIBUS-STOPPING SIGNALS.

Sir,—Any person who is in the habit of travelling in omnibuses must be aware how difficult a matter it is for those seated at the inner end to communicate with the conductor : if they do not calculate their distance well they are sure to overshoot their point by some 50 or 100 yards. Experience of this has led me to trouble you with the following plan.

Let a cord be stretched close under the roof, along the centre, from end to end.

Let it be made fast at the inner end and supported by three or four rings at equal distances; the end nearest the door being attached to the lever of a clapper, placed to strike a large bell fixed upon the roof. A spring bearing upon a lever from the axis of the clapper would return it to its place. Thus a passenger, wherever seated, by merely pressing upon the cord would make a single stroke upon the bell, thereby instantly gaining the conductor's attention. If the bell were sufficiently deep-toned, the effect would not be unpleasant.

In another way the signal might be made still more musical. Let a horizontal wheel (say 6 inches diameter) be suspended under the roof of the omnibus, close to the door, with a pulley fixed on its axis, having the end of the cord wound twice or thrice round it; the wheel being kept in position by a spiral spring. The edge of the wheel should just appear outside the omnibus through an opening made for the purpose. In the same plane with this wheel let two or three steel springs of different lengths (like those used in musical boxes, only much larger) be fixed to the outside of the omnibus, pointing forwards, to be struck by studs or pins on the edge of the wheel. All this might be packed in a very small compass, the wheel, &c., being protected by a casing, and the musical springs by a projecting ledge or roof. A slight pressure upon any part of the cord would thus produce a chime of two or three notes.

For Savory's clock, it appears to me that Mr. Gray's explanation, given in your 796th number, is the true one. My opinion would not be worth mentioning had it not been formed independently, without being aware, at the moment, how exactly it agreed with that gentleman's. Besides the *appearance*, there is not I think room for *glass* plates of sufficient strength, nor is the plan of "Nautilus," however ingenious, applicable to the present case, as the boss in the centre of the hand is only about a quarter of an inch in diameter.

I remain, Sir,
Your obedient servant,
JOHN RIVINGTON, jun.

Sydenham, Feb. 22, 1839.

MR. TALBOT'S PROCESSES OF PHOTOGENIC DRAWING.

[Read at a late meeting of the Royal Society.]

The subject naturally divides itself into two heads—the preparation of the paper, and the means of fixing the design. In order to make what may be called ordinary photogenic paper, the author selects, in the first place, paper of a good firm quality, and smooth surface; and thinks, that none answers better than superfine writing paper. He dips it into a *weak* solution of common salt, and wipes it dry, by which the salt, is uniformly distributed throughout its substance. He then spreads a solution of nitrate of silver on one surface only, and dries it at the fire. The solution should not be saturated, but six or eight times diluted with water. When dry, the paper is fit for use. He has found, by experiment, that there is a certain proportion between the quantity of salt and that of the solution of silver which answers best, and gives the maximum effect. If the strength of the salt is augmented beyond this point, the effect diminishes, and, in certain cases, becomes exceedingly small. This paper, if properly made, is very useful for all ordinary photogenic purposes. For example, nothing can be more perfect than the images it gives of leaves and flowers, especially with a summer sun. The light passing through the leaves delineates every ramification of their nerves. If a sheet of paper, thus prepared, be taken and washed with a *saturated* solution of salt, and then dried, it will be found (especially if the paper has been kept some weeks before the trial is made), that its sensibility is greatly diminished, and, in some cases, seems quite extinct. But if it be again washed with a liberal quantity of the solution of silver, it becomes again sensible to light, and even more so than it was at first. In this way, by alternately washing the paper with salt and silver, and drying it between times, Mr. Talbot has succeeded in increasing its sensibility to the degree that is requisite for receiving the images of the camera obscura. In conducting this operation, it will be found, that the results are sometimes more, and sometimes less satisfactory, in consequence of small and accidental variations in the proportions employed. It happens sometimes that the chloride of silver is disposed to darken of itself, without any exposure to the light—this shows, that the attempt to give it sensibility has been carried too far. The object is, to *approach* to this condition as near as possible, without *reaching* it; so that the substance may be in a state ready to yield to the slightest extraneous force, such as the feeble impact of the violet rays when much attenuated. Having, therefore, prepared a number of sheets of paper, slightly different from one another in the composition, let a piece be cut from each, and, having been duly marked or numbered, let them be placed side by side in a very weak diffused light, for about a quarter of an hour; then, if any one of them, as frequently happens, exhibits a marked advantage over its competitors, Mr. Talbot selects the paper which bears the corresponding number to be placed in the camera obscura.

With regard to the second object—that of fixing the images—Mr. Talbot observed, that, after having tried *ammonia*, and several other re-agents, with very imperfect success, the first which gave him a successful result, was the iodide of potassium, much diluted with water. If a photogenic picture is washed over with this liquid, an *iodide of silver* is formed, which is absolutely unalterable by sunshine. This process requires precaution; for, if the solution is too strong, it attacks the dark parts of the picture. It is requisite, therefore, to find, by trial, the proper proportions. The fixation of the pictures in this way, with proper management, is very beautiful and lasting. The specimen of *lace*, which Mr. Talbot exhibited to the Society, and which was made five years ago, was preserved in this manner. But his usual method of fixing is different from this, and somewhat simpler—or, at least,. requiring less nicety. It consists in immersing the picture in a strong solution of *common salt*, and then wiping off the superfluous moisture, and drying it. It is sufficiently singular that the same substance which is so useful in *giving* sensibility to the paper, should also be capable, under other circumstances, of *destroying* it; but such is, nevertheless, the fact. Now, if the picture which has been thus washed and dried, is placed in the sun, the white parts colour themselves of a pale lilac tint, after which they become insensible. Numerous experiments have shown the author that the depth of this lilac tint varies according to the quantity of salt used, relatively to the quantity of silver; but by properly adjusting these, the images may, if desired, be retained of an absolute whiteness. He mentions, also, that those preserved by *iodine* are always of a very pale primrose yellow, which has the extraordinary and very remarkable property of turning to a full gaudy yellow, whenever it is exposed to the heat of a fire, and recovering its former colour again, when it is cold.—*Athenæum.*

ROYAL INSTITUTION. — MR. JOHNSTONE'S LECTURE ON THE DISTINCTIONS IN THE INVESTIGATION OF MENTAL AND MATERIAL PHENOMENA.

Feb. 22nd.—*The Converzatione.*—There were some interesting specimens on the table of the rock salt of Cheshire, and of the several varieties of refined salt as prepared for the different markets, sent by Mr. Hemming. Mr. Palmer's pneumatic filter was exhibited. It will be unnecessary to notice it here, as it was described in a recent number of the *Mechanics' Magazine.* We will only observe, that though in most cases it will form a useful apparatus, still there is such a thing as making more haste than good speed. Good filtration cannot always be effected through the medium of pressure as a substitute for surface, but where it can be used with advantage, this is a ready and elegant mode of applying it.

The Lecture. — Mr. Johnstone lectured " on the leading distinctions in the investigation of mental and material phenomena." This is a subject which, to have justice done to it, would have required a mind of the highest order, and a larger allowance of time than could be afforded to a single lecture. As thus treated, it would not have been inappropriate either to the occasion or to the character of the audience ; but as treated by the lecturer it was an agreeable essay on the advantages and on some of the characteristics of metaphysical science, interspersed with pleasant remarks and quotations, and having just such a sprinkling of allusion to physical facts as would afford a colouring of justification of the title given to the lecture. The title was a misnomer also, in a yet more important particular, for the few " leading distinctions" to which the lecturer adverted, referred to what is incidental to the subjects themselves, and not to discriminations in the mental process implied in their " *investigation.*" In justice to the lecturer it should be observed, that besides being compelled by the limitation of his time (the rule, which grants only an hour, being always strictly enforced) to give only fragmentary notices, as it were, of his subject, he was almost obliged to introduce physical illustrations, if merely to relieve the tedium of his theme, and even though it should require a most indulgent latitude of allowance to deem them as being at all relevant to it, in the manner at least, in which he treated it.

An abstract of the lecture would exceed our limits, and a complete syllabus would be uninteresting ; we shall therefore give in the latter form only two of its heads, the first, on account of the justness of the observations, and the second, to substantiate our criticism.

1. Physical science depends greatly as to its progress and its truthful character, on the mind's procedure in the investigation of its objects ; hence the value of mental science, the utility of its culture, the futility of the ancient philosophy, the efficiency of the Baconian method, and the surprising advances which have followed its adoption. Ancient, was not less vigorous than modern intellect, but it took the wrong path.

2. The evidence of facts in mental investigations is internal and cogitative, whilst in those which are physical it is external and experimental—illustrated by chemical experiments, giving evidence of the facts that there is fire in water, and that in a clear saline solution (Epsom salts) there exists an insoluble material—illustrated by the mathematical theorem, that the square of the hypotenuse of a right-angled triangle, is equal, &c.

It was in this part of his subject—for no where else had it any bearing on physical phenomena—that the lecturer appeared to think, that by adducing such obvious differences in the *form* of investigation, as are denoted by the terms. experiment and cogitation, he was pointing out "distinctions" in the *nature* of investigation. The form is determined by the purport or the peculiar subject of inquiry, and may be experimental or cogitative, or both, either in mental or physical science, but the process of investigation, in itself, is purely a mental procedure, whichever the form, or whichever the science, and to it belong such mental distinctions only, as are indicated by the terms deduction and induction. Indeed to speak still more correctly, experiment, like observation, is one of the conditions of investigation, rather than one of its forms, but cogitation is essential to it, as being the plastic spirit which uses the materials that the other, as the instrument, obtains, and moulds and fashions them into coherent forms. In the order or method of this mental proceeding, two "leading distinctions" may be observed, according to which the " investigation of phenomena" may be characterised as either deductive or inductive. To have had those distinctions pointed out, and the advantages of their corresponding methods elucidated, as being equally subservient in mental as in material philosophy, would indeed have been a treat.

In his observations on the uncertainty of the evidence afforded through the medium of the senses, the lecturer quoted Lord Brougham in support of his own opinion, that we are not so clearly cognizant of the existence of matter as of mind, and illustrated the deceptive nature of sensible evidence, by the exhibition of specimens of machine medallion engraving, which conveys the illusive idea of relief, and of a specimen on a large scale of the philosophical toy, the invention we believe of Dr. Roget, which conveys to the eye the impression of a whole and perfect image, although depicted in moieties on the opposite sides of a revolving tablet. The lecturer, in conclusion, alluding to the vast difference which exists in different individuals in respect to their mental faculties, instanced the cases of Shakespeare and Newton, and eulogised them in glowing terms.

NOTES AND NOTICES.

Stephenson Memorial.—We observe, with pleasure, that a well-deserved tribute of admiration and respect is about to be paid to an individual, to whose genius and untiring energy, his country is deeply indebted for one of her grandest modern improvements—the formation of railways, and the application of locomotive power—we allude, of course, to George Stephenson, Esq. A committee has been formed, embracing many of the first names connected with the iron trade, to consider the subject of a "Stephenson Memorial," and we hear that a colossal statue is spoken of, to be erected in such part of the kingdom as may hereafter be determined upon, and formed of that most appropriate material—cast-iron.—*Mining Journal.*

The Iron Sailing Ship.—The first sailing vessel ever built of iron was, it will be recollected, constructed in Liverpool, and was, very appropriately named the *Ironsides.* She sailed for Pernambuco, which she reached after a passage of forty-seven days. Much interest, indeed, anxiety was felt to know whether the iron would influence the needle. We are happy to state, that the compass was correct, throughout the whole passage; and that, therefore, no fear need be entertained as to its general correctness on board of iron-built ocean-going vessels. The *Ironsides* may be shortly expected on her return passage.

Roberts's Self-acting Mule.—The petition of Mr. Roberts, of the firm of Sharp, Roberts, & Co., of Manchester, praying for a prolongation for seven years, of the several patents granted to him in the year 1825, for the United Kingdom of Great Britain and Ireland, for that important invention, was heard lately before the judicial committee of her Majesty's privy council:—present, the Marquis of Lansdowne, lord president; Lord Lyndhurst, Lord Brougham, and Sir Herbert Jenner.—Counsel for the petitioner, Sir Frederick Pollock and Mr. Teed. No caveat having been entered to oppose the application, the Attorney-General briefly addressed their lordships, stating that on the part of the crown he had no objection to offer to the prayer of the petitioner being granted. A model of the machine was then exhibited, and the nature and objects of the invention were explained to their lordships. Evidence was next given in support of the allegations contained in the petition; and after a short consultation, the Marquis of Lansdowne, on behalf of the judicial committee, granted a prolongation of the several patents for the term prayed for, on account of the great ingenuity and merit of the invention, and the obstacles which had from time to time been opposed to prevent the patentee from deriving a fair remuneration for so important an invention during the original term of the patents.

New Patterns and Inventions Bills.—We have received copies of Messrs. Mackinnon and Baines' "Bill for the better encouragement of arts and manufactures, and securing to individuals the benefit of their inventions for a limited time;" and Messrs. Thomson and Labouchere's "Bill to secure to proprietors of design for articles of manufacture, the copyright of such designs for a limited time." The former bill is similar to Mr. Mackinnon's of last year;—the subject of fourteen years patents, we grieve to see, remains unattended to. We shall offer a few remarks on the subject in our next.

Beale's Patent Air Light.—Modes of obtaining artificial light seem now to be the scientific mania of the day. We have almost every week occasion to notice some new luminary with greater or less pretensions to superiority and economy. Mr. Beale some years ago patented a method of burning coal tar and other cheap oleaginous substances, by forcing a current of air through the burner at about the point of combustion; but in consequence of the unpleasant smell which the tar or other cheap oil produced, its use was confined to manufactories and places where this was defect of little consequence. Mr. Beale, some time ago, at the instigation of Mr. De Mourier, entered deeper into the theory of his invention, and has succeeded in so improving his apparatus as to render it capable of burning oil of various kinds, or rather generating gas therefrom and burning it mixed with a stream of atmospheric air, obtaining a most clear and beautiful light, and without smell or smoke. We have seen five lamps in action at Mr. De Mourier's private offices in Lombard-street, in which the merits of the invention are clearly evidenced. The stream of air is supplied to the lamps by means of a double bellows, similar to the bellows of an organ, kept in motion by a weight. Instead of the weight a spring may be used. There are, of course, numerous ways of supplying the air, but the method just described appears well adapted for private and public buildings. The proprietors of the patent, we understand, intend to contract to supply parties at half the price of gas. As we intend shortly to publish a detailed description of the whole apparatus, we shall not therefore now enter more particularly into its construction.

New Dibbling Machine.—Mr. James Hitchins, of this city, is about to introduce to the notice of the farmer and the public, a machine for dibbling wheat, turnips, or any kind of grain and seed. By a simple arrangement, seed and bone-dust, in small or large portions, are deposited together in holes made at exact intervals. The whole apparatus can be worked at the full speed of a horse in a walking pace. It is well known that bone-dust in drill-rows is of small value except in the first stage of vegetation, and when not in actual contact with the seed, it is rather detrimental than beneficial. By this dibbling machine, 75 per cent. will be saved in the amount of bone-dust required, and nearly all the expense of thinning the plant and keeping them clean in future, will be economised.—*Lincoln Gazette.*

Salt Manufactory.—Manufactories are about to be established in various parts of the agricultural districts, for the purpose of supplying the farmers with marine salt, soda, and lime, for manure, at a reduced rate—and this is to be effected by establishing these works in such stations as are most easy of general access.—*Mark-lane Express.*

LONDON: Printed and Published for the Proprietor, by W. A. Robertson, at the Mechanics' Magazine Office, No. 6, Peterborough-court, Fleet-street.—Sold by A. & W. Galignani, Rue Vivienne, Paris.

Mechanics' Magazine,

MUSEUM, REGISTER, JOURNAL, AND GAZETTE.

No. 814.]   **SATURDAY, MARCH 16, 1839.**   [Price 3*d.*

Printed and Published for the Proprietor, by W. A. Robertson, No. 6, Peterborough-court, Fleet-street.

MR. JEFFREYS'S PATENT PNEUMATIC GRATE.

Fig. 4.

W.C.W. DU.K.G.

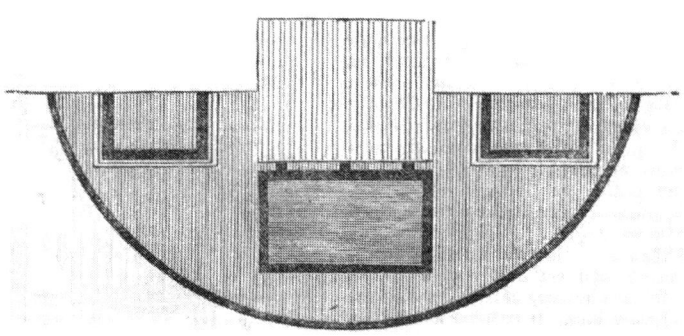

Fig. 5.

THE PATENT FRESH-AIR PNEUMATIC GRATE.

Mr. Jeffreys's new warming and ventilating apparatus, "the fresh-air pneumatic grate," is so named on account of the peculiarity in its structure, which causes the point of greatest pressure of the several currents of air entering the chimney to be on a vertical plane, instead of a horizontal one, as in common fire-places, where the pressure is in the chimney throat.

It has been the desire of all judicious improvers of grates and fire-places to bring the grate as far forwards as possible; but so long as the ascent of the smoke is in the usual course this can only be effected by lowering the front or breast-work. If the fire is brought quite out of the chimney the front has to be closed down to, or near to, the upper bar, forming almost a close stove. If such a fire has the bars exposed the consumption of fuel is very large, and its warming power is small. There is a great rush of air through the fire, burning away the fuel, and the upper surface being shut in its heating influence is lost. Such a fire, though preferable to a close stove, is less cheerful than a common fire. In some particular cases an open fire can be set out in front of a chimney, as in some places in the north, but then a very large draught of air is required to force the current of smoke backwards into the hole leading to the chimney, and puffs of smoke are at all times liable to escape into the room.

- The elevating of the front or breast-work of a fire-place so much improves its appearance and effect, that many persons are ready to sacrifice the advantage of a forward position of the grate, in order to command it. By putting the grate far in, as was common in former days, a very lofty and stately appearance may be given to the open front, but then the warming effect is lessened.

The grate we have now to describe commands both the advantages in their fullest degree. It is more open in front than any other grate, and is, nevertheless, so well set out, that a person quite in the chimney corner, has the fire right before him. It radiates as directly on both sides as it does forwards. Its range of radiation is in fact doubled.

Upon first seeing it the impression of most persons is that such a fire must smoke; but so far is this from being the

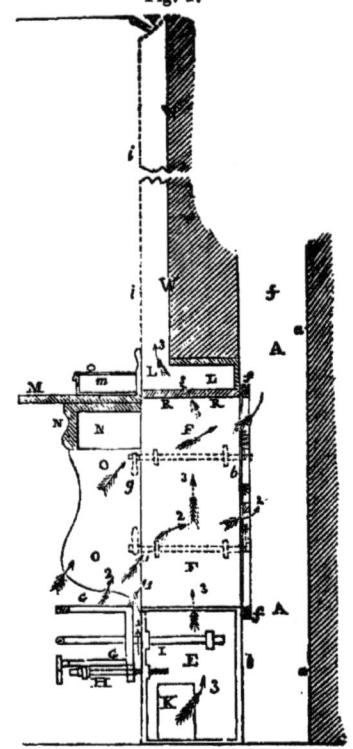

Fig. 1.

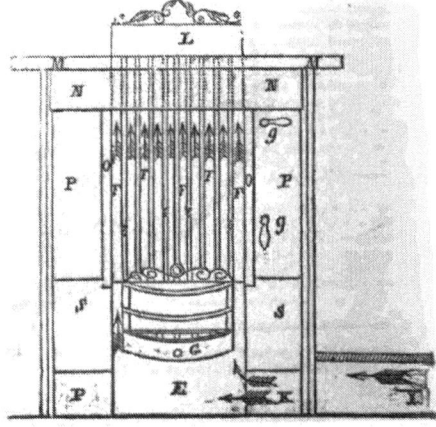

Fig. 3.

case, that it might be guaranteed to cure all ordinary kinds of smoking, except that occasioned by gusts of wind, where a higher stack of chim-

Fig. 2.

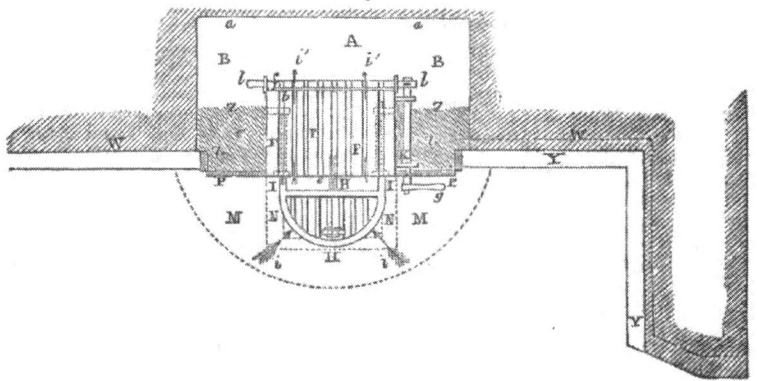

Fig. 7.

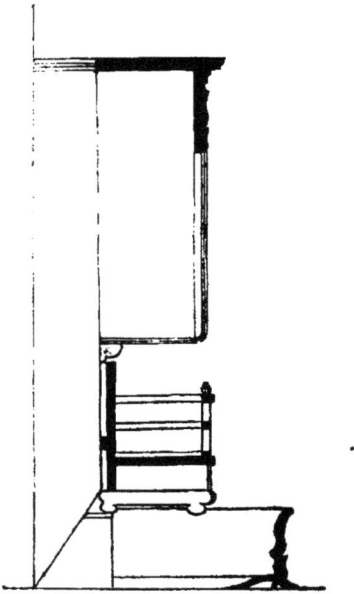

Fig. 6.

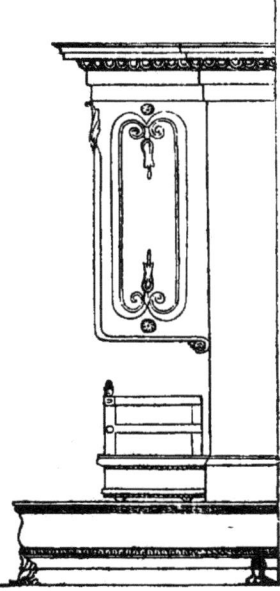

Fig. 8.

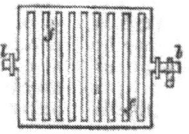

A A 2

neys, or a wall out-tops the chimney it is connected with. In this case no fire-place can be proof against puffs, though the pneumatic grate is said to be much freer from this influence even than others.

An apparatus of this description is in operation at a house of business belonging to its inventor, Mr. Jeffreys, No. 148, Regent-street, being one of the depots for the well-known instrument, the "Respirator," also that gentleman's invention. It has been visited by the most eminent men of science in London, and has received their unqualified approval, as also that of men of taste in design, who consider the plan to have great capability for tasteful decorations, and elegance of form.

We have ourselves seen it, and examined it attentively, and as far as we can judge from a single inspection, the foregoing statements are fully borne out.

Description of Mr. Jeffrey's Fresh-air Pneumatic Grate.—Fig. 1, is a vertical section; fig. 2, (p. 419) a horizontal section or plan, and fig. 3, a front view. On the front page fig. 4 is a front elevation, and fig. 5 a plan. Fig. 6 (p. 419) is a side elevation, and fig. 7 another side section. A is a common house chimney. B B is the capacious space in the chimney in which common grates are set. C C, the large front opening partially closed by the brick work C C, into which is set the apparatus E E, F F, which consists of a box below E E nearly square, having flat tubes F F F, placed parallel, very near to each other, and upright, or nearly so, and passing through a partition of metal R R, above, so that the current which is making its exit at L L from the tubes denoted by the vertical arrows, fig. 1, is kept distinct from the current passing between the tubes into the chimney denoted by the oblique arrows. This apparatus is set in the chimney only so far, that the back of it, *b b*, is many inches in advance of the back of the chimney *a a*, figs. 1 and 2, the greater the distance, the better will be the effect. G G, figs. 1, 2 and 3, is a grate of any suitable form placed in front of the above apparatus supported by the iron rods I I, which slide in tubes or collars fixed in the box E E, so that the grate may be moved backwards or forwards by means of the screw H, turned in front by a winch. The use of this construction is

to allow air to pass up behind the grate between it, and the apparatus E E, F F, shown by the arrows 1, 1, 1; which air keeps the flaming current (arrows 2, 2, 2) from entering between the tubes too low down; the current *above the summit of the flame* is thus only allowed to enter between, and play against the tubes; and much more heat is thrown forwards into the room. O O, figs. 1 and 3 are two cheeks of marble, common stone, or metal, which prevent the ascending current from the fire from being blown sideways; and N N, figs. 1 and 3 is a frieze of similar materials forming a screen for turning any smoke that may rise so high into the fissures between the tubes F, F, F. Thus there is no upright passage into the chimney, and it will be perceived that the very back of the fire is several inches in front of the wall of the room, W W, figs. 1 and 2; and the space from the upper bar of the grate G to the frieze N, is even much more than in ordinary grates, causing the opening in front to be very capacious, while the grate also faces the room at the sides as much as in front. A grate placed in this position, unaccompanied with any particular provision, would send most of its smoke into the room, although the frieze N should be brought to within half the distance, at which it is here placed above the grate; and it has been supposed by persons not familiar with the movements of heated currents, that by placing in its way the large apparatus, occupying most of the passage towards the chimney, the tendency of the smoke to enter the room would be increased; whereas, by thus obstructing the greater part of the passage the opposite effect is produced, the whole of the smoke enters the chimney with a steadiness unequalled by the draught of any common fire if the apertures are vertical, and their summit R R, is above the level of the frieze N. Hence, by this construction most smokey chimneys may be effectually cured. Nearly the same effect would be produced by plates of metal placed parallel with the edges forward between the fire and the chimney; but by employing the tubular apparatus F F, in addition to the above important effect of enabling the fire to stand forward into the room, the tubes may be made to receive air from the box below E E, and to discharge it into the room at L L L L fig. 1, which is a space fronted by any neat plate L fig. 3. This air (arrows 3, 3, 3)while passing up the tubes is exposed over a great surface to the action of the smoke-current passing between the tubes, and it draws from this current a large part of its warmth. By increasing the draught up the tubes by a hollow half-column *l l* fig. 3, receiving the air from L L, and discharging it at the top of the room; and by making the area of the tubes much freer than that of the fissures between them, as much or more air may be made to circulate up them than passes up the chimney itself, and thus, nearly half of the warmth of smoke may be saved. The air enters the box E E at the opening K, figs. 1 and 3, and is brought thither by a passage made behind the skirting board to an orifice below the nearest window, where the wall is thin and the opening easily made. At K there is a two-way slot, or door, by opening or closing which the air of the room, or that from without doors, may be made to enter the box E E, at pleasure. By substituting sheet metal for the skirting, and clearing away the plaster, and widening the cornice, or moulding a little, a passage of from 3 to 5 inches deep, and from 6 to 10 inches high, may be obtained; or a pipe may be conveyed under the floor between the joists; or where the back of the chimney is towards the open air, the easiest and best passage will be found in that direction. This bringing in and warming of fresh air in great abundance, is a point of the greatest importance to health, and it puts an end to all cold draughts from the windows and doors. P P are marble jambs on each side of the grate. On one side the lower portion of the jamb is made to slide out, leaving a passage Z Z Z Z, fig. 2, below the brick work, for the entry of a person into the chimney. S S, fig. 3, are polished reflectors of steel. Fig. 8 is one of two frames of cast-iron, placed behind the tubing; one frame behind the upper, the other behind the lower half of the tubes. By raising or depressing the handle *g*, the frame is shifted so as to leave open or to close the fissures between the tubes, and thus to regulate the draught. The arrows *i i* and *i′ i′*, fig. 2, show the manner in which oblique currents of air which would disturb the course of the smoke are changed in the passages into parallel'

currents, and as such in the chimney they have no tendency to compress or obstruct the course of the smokey current between them.

This apparatus admits of a much cheaper form than that shown in the frontispiece. It may be wholly of cast-iron, and made for a cost trifling, when compared with the expected economy of fuel.

Some persons not prepared with the knowledge of pneumatics which renders manifest the peculiarity of the construction of this stove, have looked to the point in which it is alone similar to inventions of former dates, namely, the provision for bringing in fresh air from without doors by a passage or pipe carrying it to the back of the fire. But in all other cases the air-tubes are set in the fire, and even when of earthenware must become overheated, and render the air passing through them unwholesome. Whereas, in the pneumatic grate the tubes cannot be overheated, and instead of robbing the fire, they derive their heat entirely from the waste smoke. They are, in fact, the pneumatic apparatus itself, which in a very happy manner is made to perform the double duty of ensuring a steady and certain in-draught of the smoke from a well projected fire, and of warming thoroughly, yet moderately, a very large body of fresh air, without robbing the fire itself of any heat for the purpose.

MR. S. CROSLEY'S PNEUMATIC TELEGRAPH.

Sir,—It is a matter of surprise and regret that in a commercial country like England no improvement has yet taken place in telegraphs so as to render them available by night as well as by day, and also in foggy weather. Two projects, viz.: the hydraulic and electro-magnetic have already been laid before the public —a third, the pneumatic, is now offered, and it is earnestly hoped the very great importance of the subject may attract the attention of government with a view of examining into their respective merits, in order, if practicable, that the most eligible project may be adopted.

The following is a description of the pneumatic telegraph recently proposed by Mr. S. Crosley :—

1. Atmospheric air is the conducting agent employed in the operation of the pneumatic telegraph.

2. The air is isolated by a tube extending from one station to another; one extremity of the tube is connected with a gas-holder or collapsing vessel, as a reservoir to compensate for any diminution or increase of volume arising from compression or from changes in the temperature of the air in the tube, and for supplying any casual loss by leakage. The other extremity of the tube terminates with a pressure index.

3. It will be evident to every one acquainted with the physical properties of atmospheric air, that if any certain degree of compression be produced and maintained in the reservoir, at one station, the same degree of compression will speedily extend to the opposite station, where it will become visible to an observer by means of the index.

4. Thus, with ten weights producing ten different degrees of compression, distinguished from each other numerically, and having the index at the opposite station, marked by corresponding figures, any telegraphic numbers may be transmitted, referring in the usual way to a code of signals, which may be adapted to various purposes and to any language. The only manipulation is that of placing a weight of the required figure upon the collapsing vessel at one station, and the same figure will be represented by the index at the opposite station.

5. In establishments where the telegraphic communications do not require the constant attendance of a person to observe them, and where periodical attendance is sufficient, the signals may be correctly registered on paper, by connecting with the air tube an instrument called a *pressure register*, also invented by Mr. Crosley, which has been successfully employed in large gas-light establishments upwards of fourteen years, for registering the variations of the pressure of gas in street mains. The same instrument produces also an increased range of the index scale, by which means the chance of errors from minute divisions is obviated.

6. There being now three different projects for improvements in telegraphic communications, viz.: the Electro-Magnetic, the Hydraulic, and the Pneumatic Telegraphs,—and assuming that such improvements are of importance to the state, as well as to railway proprietors and the community at large, it seems desirable that their merits should be thoroughly investigated by competent engineers, and that the aid of government should be solicited, for the purpose of establishing, on a practical scale, the most eligible project.

7. It may be observed, that the introduction of railways, has not only created an additional use for telegraphic communications, but the important difficulty which previously existed in the expense of providing a proper line and safe foundation is, at once, removed by the side of the railway itself, possessing as it does, by its police, the most ample security against injury, either to the tubes or electric wires.

8. The prominent questions for consideration seem to be—the certainty and accuracy of the communications, the first cost, the expense of repair and superintendence, also the time required for transmitting intelligence.

9. On the question of time, it is quite clear that neither the hydraulic nor the pneumatic can compete with the electro-magnetic telegraph in rapidity. No doubt, on investigation, each project will be found to possess its peculiar advantages. Thus, in considering the advantage one may have in point of time, another may possess a greater degree of certainty or accuracy in the communications; sufficient to outweigh the difference of time, for instance, between 1 second and 1 minute, or even between 1 second and 5 or 10 minutes.

10. The projector of the Pneumatic Telegraph is not in possession of any experimental results on a practical scale by the electro-magnetic or by the hydraulic telegraphs, employed at any considerably extended distances, or of their continued operation for any long period of time; nor can he offer much decisive information, of a practical nature, analogous to the operation of the pneumatic telegraph on these points; the following circumstances may, however, be referred to :—

11. There has been upwards of twenty years' experience in the transmission of gas for illumination through conduit pipes of various dimensions. In several instances, the gas has been supplied at the distances of five to eight miles by low degrees of pressure. As one proof of great rapidity of motion, it has been observed, that when any sudden interruption in the supply has occurred at the works, the extinction of all the lights,

over large districts, has been nearly simultaneous. Another instance of the great susceptibility of motion which frequently happens, is the flickering motion of the lights at great distances when water has accumulated in the pipes.

12. The only experience in the transmission of atmospheric air through conduit tubes, which applies more particularly to this subject, may be referred to at three railway establishments ; viz., Edinburgh, Liverpool, and Euston-square, London. In these establishments, air-tubes, from 1¼ to 2 miles in length, have been employed for the purpose of giving notice when a train of carriages is ready to be drawn up the inclined plane by the stationary engine at the summit, so that it may without delay be put in motion. This notice is communicated by blowing a current of air through the tube at the foot of the inclined plane, and sounding an organ-pipe, a whistle, or an alarm-bell at the stationary engine. It will be satisfactory to know, that this operation has been regularly performed from two to four years without one single failure or disappointment.

13. It may further be noticed, that a trial was made with a tube of one inch in diameter; very nearly two miles in length, returning upon itself, so that both ends of the tube were brought to one place :— the compression applied at one end, was equal to a column of seven inches of water; and the effect on the index at the other end, appeared in fifteen seconds of time.

14. Laws have been propounded by eminent men on the expenditure of aeriform fluids through conduit pipes, and of the resistance of the pipes; but these are not strictly applicable to the present question. Under all circumstances, it seems desirable that experiments on a practical scale, at extended distances, should be resorted to, as the most satisfactory guide for carrying into effect telegraphic communications of this kind.

A model of the pneumatic telegraph is placed at the Polytechnic Institution, where its operation may be seen daily at half-past 12 and half-past two o'clock.

S.

London, March 2, 1839.

ON THE SUPPLY OF WATER TO THE METROPOLIS.

Sir,—I have on two previous occasions adverted to the benefits that would be derived by rendering the supply of gas *uniform* and *uninterrupted ;* I now beg leave to allude to the more pressing necessity that exists for, and the greater advantages that would result from, the uninterrupted supply of another fluid, of far more vital importance, viz. : *water.*

Much has been said and written, and sometimes justly, complaining of the limited and irregular manner in which this necessary of life is doled out by some of the London water-companies. Well-grounded complaints are continually being made either of inferiority of quality, or deficiency of quantity, or of both ; the latter cause of complaint, upon some recent occasions of fire, has occasioned the most disastrous consequences. At the same time, every candid person who is conversant with these matters will at once admit, that great improvements have of late years taken place, both in the quantity and the quality of the water generally supplied. On the other hand, I imagine the most zealous friends and advocates of the water-companies must acknowledge there still remains room for considerable improvement.

I apprehend it is unfortunate that among our many mechanical inventions, we have no such thing as a *water-meter :* at least nothing that approximates sufficiently near to the character and offices of a *gas-meter ;* no practical method of registering the quantity of water supplied to different parties.

The introduction of a machine, not too expensive in its construction nor easily deranged, that would accurately register the quantity of water supplied, and thereby enable the water-companies to stipulate for payment for the quantity used, at a given rate per hundred or thousand feet, would create a new era in these matters. It is well known that wilful waste and frauds of various kinds are continually practised upon the several water-companies ; many people seem to imagine they can never get enough for their money. Upon the occasion of a serious deficiency of water occurring at some fires in the city a few years since, an inquiry into this matter was instituted in the Court of Common Council, when Mr. Mylne stated, that the New River Company had discovered that immense waste took place, and that the water was appropriated in many extraordinary ways. He mentioned one instance where a *smoke-jack* having been found inefficient, a *water-wheel* was formed in the chimney, and the meat roasted by its hydraulic power! My own observations lead me to believe, that in those districts where the supply of water is tolerably abundant, the quantity positively and absolutely wasted, far exceeds the quantity that is fairly used. In the districts of some of the southern companies, the period of supply is limited to one hour three times per week, and in these districts nearly one-third of the consumers are without the ordinary means for shutting off the water when their butts, &c. are filled—consequently the superfluous water runs to waste, to the great annoyance of those, whose cisterns being on a higher level seldom receive more than a very scanty supply. Persons sometimes quarrel with their water-rate, and dispense with the public supply. They sink a well on their premises—erect a pump—and obtain any quantity of water they please (as they say) *for nothing*. Taking the cost of the pump, the expense of sinking the well, and to these add the charge for labour in pumping (which is in exact proportion to the quantity raised) and I question whether there is in any case much saving. There is some little convenience in being able to obtain upon any particular occasion an extra quantity of water—but against this must be set the risk of being left at some seasons of the year, without any water at all.

Calculated as the present water-charges are, to cover the enormous waste of the extravagant, they may perhaps seem high, although in many places, a like supply on similar terms, would be hailed with the utmost delight.

I believe whenever it shall become feasible for the water-companies to charge for the quantity actually consumed, the cost will be found extremely moderate, and quite as low, if not cheaper, than the expense incurred by individuals for private supplies.

Many persons would gladly pay for an increased supply of water, who, under the present arrangements cannot possibly obtain it; while those whose consumption was small, would reap the full benefit of their economy, and find their

charges proportionally diminished. Under such a system, the mains and services would always remain charged, and every cistern remain filled. The permanent and plentiful supply of water thus afforded, in the event of fire, would prove of incalculable advantage. It was well observed by Mr. Braidwood upon the investigation before alluded to, that " had the supply of water for domestic purposes been more liberal, the want of it would not have been so severely felt in case of fire." Whether we regard the health, the morals, or the security of the community, an abundant supply of pure water must be considered a public and a private blessing.

Commending the subject of a *water-meter*, to the consideration of the ingenious,

I remain, Sir, yours, respectfully,
WM. BADDELEY.

London, March 7, 1839.

TAKING IMPRESSIONS OF DIFFERENT SIZES FROM THE SAME PLATE.

Sir,—In the report of the meeting of the Royal Institution, February 8th, in No. 811, there is a notice of a print having been laid on the table by Mr. Brockedon, remarkable for having impressions of three different sizes taken from the same plate ; and it is stated that the process is as yet undiscovered. Allow me to suggest the following as the method by which it is, or at least by which it may, be effected; and I communicate it with the more confidence, from Mr. Brockedon having told me that he believes it to be the plan which is really adopted, and that it has already been suggested by two or three individuals, among whom I believe may be reckoned Mr. Brockedon himself, and Professor Wheatstone.

I imagine then, that an impression is taken from an engraved copper plate in some soft metal, or probably in some more fusible yet ductile alloy, by the French method of stamping, called, *En cliché;* this impression is used to obtain its reverse in a similar manner, and that these two, thus fitting exactly into each other, are then passed together through the rolling mill until the desired extension is obtained. Of course care is taken that the plates are extended in both their dimensions alike, so as to preserve the original proportions of their sides, and it is probable they may require cautious annealing during the process. Thus plates of various sizes, and renewable from time to time as they become impaired by use, may be easily obtained. It cannot be correctly said that the impressions are *printed* from the same plate, but it may be said with truth that they are taken from the same *engraved* plate. The invention is French.

I am, Sir, yours, &c.,
BEN. CHEVERTON.

RAILWAY IMPROVEMENT SOCIETY.

A private meeting very numerously attended by deputations from most of the leading railway companies, was held on Saturday last at the chambers of Messrs. Burke and Venables, in Parliament-street, for the purpose of considering the propriety of forming a Society for promoting and advancing the scientific improvement of railways throughout the kingdom, and for protecting, generally, the interests of railway proprietors.

Mr. George Carr Glyn, the chairman of the London and Birmingham, and North Midland Railway Companies, was called to the chair, and opened the proceedings by adverting to the great and manifest importance of the proposed society, as affording a means of bringing the united experience and influence of the principal persons connected with railways, to bear upon all questions which may arise respecting them.

The honourable chairman further alluded to the very great ignorance which exists among many, even at this day, on the subject of railways, and the consequent prejudices which prevail against them, and pointed out the great advisability of having some regularly organised association, which would be looked up to as an authority on all subjects in which their interests were involved.

The meeting was subsequently addressed by several other gentlemen present, who all concurred in the importance of the proposed association, and dwelt on the advisability of forming, at its outset, a collection of maps, reports, models, and other scientific and statistical details relating to railways, which should be accessible to the several members of the society, and which would, in time, become a most valuable and interesting museum of reference on matters connected with railways.

Some discussion took place as to the amount of the subscriptions, and the name to be given to the proposed association:—viz., whether it should be called the " Railway Society" or the " Railway Institute," but eventually this, with all other matters of detail, was left to a committee of management, formed of some of the Directors of the principal railway companies present, who were empowered to add to their number if they should see fit.

Resolutions, embodying the substance of the foregoing remarks, were unanimously passed, and the several persons present having enrolled their names as the first members of the society, the meeting separated.

ON CAOUTCHOUC—BY ANDREW URE, M.D.,
F.R.S., &c. &c.
[From the *London Journal of Arts.*]

Sir,—Since writing the article " Caoutchouc" for my *Dictionary of Arts, Manufactures, and Mines*, now in course of publication, I have received, from several quarters, some valuable information; and have also made a series of experiments upon the subject, the results of which I have now the pleasure of transmitting to you for insertion in your journal.

Hitherto the greater part of the caoutchouc has been imported into Europe from South America, and the best from Para; but of late years a considerable quantity has been brought from Java, Penang, Sincapore, and Assam. About twelve months ago, Mr. William Griffith published an interesting report upon the *Ficus-elastica*, the caoutchouc tree of Assam, which he drew up at the request of Captain Jenkins, agent in that country to the Governor-General of India. This remarkable species of fig tree is either solitary, or in twofold or threefold groups. It is larger and more umbrageous than any of the other trees in the extensive forest where it abounds, and may be distinguished from the other trees, at a distance of several miles, by the picturesque appearance produced by its dense, huge, and lofty crown. The main trunk of one was carefully measured, and was found to have a circumference of no less than 74 feet; while the girth of the main trunk, along with the supports immediately round it, was 120 feet. The area covered by the expanded branches had a circumference of 610 feet. The height of the central tree was 100 feet.

It has been estimated, after an accurate survey, that there are 43,240 such noble trees within a length of 30 miles, and breadth of 8 miles of forest near Ferozepoor, in the district of Chárdwár, in Assam.

Lieutenant Veitch has since discovered that the *Ficus-elastica* is equally abundant in the district of Naudwar. Its geographical range in Assam seems to be between 25 deg. 10 min., and 27 deg. 20 min. of north latitude, and between 90 deg. 40 min., and 95 deg. 30 min. of east longitude. It occurs on the slopes of the hills, up to an elevation of probably 22,500 feet. This tree is of the banyan tribe, famed for "its pillared shade, where daughters grow about the mother tree," which has furnished the motto "*tot rami, quot arbores*," to the Royal Asiatic Society. Species of this genus afford grateful shade, however, in the tropical regions of America, as well as Asia.

Many species of other trees yield a milky tenacious juice, of which birdlime has been frequently made; as *Artocarpus integrifolia*, and *Lakoocha*, *Ficus indica* and *religiosa*, also *F. Tsiela, Roxburghii, glomerata*, and *oppositifolia*. From some of these an inferior kind of caoutchouc has been obtained.

The juice of the *Ficus-elastica* of Chárdwár is better when drawn from the old than from the young trees; and richer in the cold season than in the hot. It is extracted by making incisions a foot apart, across the bark down to the wood, all round the trunk, and also the large branches, up to the very top of the tree; the quantity which exudes increasing with the height of the incision. The bleeding may be safely repeated once every fortnight. The fluid, as fresh drawn, is nearly of the consistence of cream, and pure white. Somewhat more than half a *maund* (42 lbs.) is reckoned to be the average produce of each bleeding of one tree; or 20,000 trees will yield about 12,000 maunds of juice; which is composed in 10 parts, of from 4 to 6 parts of water, and, of course, from 6 to 4 parts of caoutchouc. The bleeding should be confined to the cold months, so as not to interfere with, or obstruct the vigorous vegetation of the tree in the hot months.

Mr. Griffith says, that the richest juice is obtained from transverse incisions made into the wood of the larger reflex roots, which are half exposed above ground, and that it proceeds from the bark alone. Beneath the line of incision, the natives of Assam scoop out a hole in the earth, in which they place a leaf of the *Phrynium capitatum*, Lin., rudely folded up into the shape of a cap. He observes that the various species of *Tetranthera*, upon which the *Moonga* silk-worm feeds, as also the castor oil plant, which is the chief food of the Eria silkworm, do not afford a milky caoutchouc juice. Hence it would appear that Dr. Royle's notion of caoutchouc forming a necessary ingredient in the food of silkworms, and being " in some way employed in giving tenacity to their silk," seems to be unfounded. If Botany discountenances this idea, Chemistry would seem to scout it altogether, for silk contains 11.33 per cent. of azote, and caoutchouc contains none at all;[*] being simply a solid hydro-carburet, and, therefore, widely dissimilar in constitution to silk, which consists of oxygen 34.04, azote 11.33, carbon 50.69, and hydrogen 3.34 in 100 parts.

This hydro-carburet emulsion is of common occurrence in the orders *Euphorbiacia* and *Tulicea*, which may be looked on as the main sources of caoutchouc. The American caoutchouc is said to be furnished by the *Siphonia elastica*, or the *Hevea guianensis* of Aublet, a tree which grows in Brazil, and also in Surinam.

Dr. Royle sent models of cylinders, of $1\frac{1}{2}$ to $2\frac{1}{4}$ inches in diameter, and 4 or 5 inches in length, to both the Asiatic and Agricultural Societies of Bengal, to serve as patterns for the natives to mould their caoutchouc by. Mr. Griffith says that this plan of forming the caoutchouc into tumblers or bottles, as recommended by the committee of the London Joint-stock Caoutchouc Company, is, in his opinion, the worst that can possibly be offered; being tedious, laborious, causing the caoutchouc to be blackened in the drying, and not obviating the viscidity of the juice when it is exposed to the sun. He recommends, as a far better mode of treating the juice, to work it up with the hands, to blanch it in water, and then subject it to pressure. I shall presently describe a still better method which has recently occurred to me, in experimenting upon the caoutchouc juice. This fluid, with certain precautions, chiefly exclusion from air, and much warmth, may be kept in the state of a creamy emulsion for a very long time.

New Experimental Researches on Caoutchouc.

The specific gravity of the best compact Para caoutchouc, taken in dilute alcohol, is 0.941567
The specific gravity of the best Assam, is . 0.942972
   "   "   Sincapore . 0.936650
   "   "   Penang . 0.919178

In the process of making the *elastic tissues*,[†] the threads of caoutchouc are first of all deprived of their elasticity, to prepare

---

[*] See my paper on the ultimate analysis of vegetable and animal substances, in the Phil. Trans. for 1822.

[†] See *Dictionary of Arts, Manufactures, and Mines.*

them for receiving a sheath upon the braiding machine. For this purpose they are stretched by hand, in the act of winding upon the reel, to 7 or 8 times their natural length, and left two or three weeks in that state of tension upon the reels. Thread thus *inelasticated* has a specific gravity of no less than 0.948732; but when it has its elasticity restored, and its length reduced to its pristine state, by rubbing between the warm palms of the hands, the specific gravity of the same piece of thread is reduced to 0.925939. This phenomenon is akin to that exhibited in the process of wire-drawing, where the iron or brass gets condensed, hard, and brittle; while it disengages much heat : which the caoutchouc thread also does in a degree intolerable to unpractised fingers, as I have experienced.

Having been favoured by Mr. Sievier, managing director of the Joint-stock Caoutchouc Company, and by Mr. Beale, engineer, with two different samples of caoutchouc juice, I have subjected each to chemical examination.

That of Mr. Sievier is greyish brown, that of Mr. Beale is of a milky grey colour; the deviation from whiteness in each case being due to the presence of aloetic matter, which accompanies the caoutchouc in the secretion by the tree. The former is of the consistence of thin cream, has a specific gravity of 1.04125, and yields, by exposure upon a porcelain capsule, in a thin layer, for a few days, or by boiling for a few minutes, with a little water, 20 per cent. of solid caoutchouc. The latter, though it has the consistence of pretty rich cream, has a specific gravity of only 1.0175. It yields no less than 37 per cent. of white, solid, and very elastic caoutchouc.

It is interesting to observe how readily and compactly the separate little clots or threads of caoutchouc coalesce into one spongy mass in the progress of the ebullition, particularly if the emulsive mixture be stirred; but the addition of water is necessary to prevent the coagulated caoutchouc from sticking to the sides or bottom of the vessel and becoming burnt. In order to convert the spongy mass thus formed into good caoutchouc, nothing more is requisite than to expose it to moderate pressure between the folds of a towel. By this process the whole of the aloetic extract, and other vegetable matters, which concrete into the substance of the balls and junks of caoutchouc prepared in Assam and Java, and contaminate it, are entirely separated, and an article nearly white and inodorous is obtained. Some of the cakes of American caoutchouc exhale, when cut, the foetor of rotten cheese; a smell which adheres to the threads made of it, after every process of purification.

In the interior of many of the balls which come from both the Brazils and East Indies, spots are frequently found of a viscid tarry-looking matter, which, when exposed to the air, act in some manner as a ferment, and decompose the whole mass into a soft substance, which is good for nothing. Were the plan of boiling the fresh juice along with its own bulk of water, or a little more, adopted, a much purer article would be obtained, and with incomparably less trouble and delay, than has been hitherto brought into the market.

I find that neither of the above two samples of caoutchouc juice affords any appearance of coagulum when mixed in any proportions with alcohol of 0.825 specific gravity; and, therefore, I infer that albumen is not a necessary constituent of the juice, as Mr. Faraday inferred from his experiments published in the 21st vol. of the *Journal of the Royal Institution*.

The odour of Mr. Sievier's sample is slightly acescent, that of Mr. Beale's, which is by far the richer and purer, has no disagreeable smell whatever. The taste of the latter is at first bland and very slight, but eventually very bitter, from the aloetic impression upon the tongue. The taste of the former is bitter from the first, in consequence of the great excess of aloes which it contains. When the brown solution which remains in the capsule, after the caoutchouc has been separated in a spongy state by ebullition, from 100 grains of the richer juice is passed through a filter and evaporated, it leaves 4 grains of concrete aloes.

Both of these emulsive juices mix readily with water, alcohol, and pyroxilic spirit, though they do not become at all clearer; they will not mix with *caoutchoucine* (the distilled spirit of caoutchouc), or with petroleum-naptha, but remain at the bottom of these liquids as distinct as mercury does from water. Soda caustic lye does not dissolve the juice; nitric acid (double aquafortis) converts it into a red curdy magma. The filtered aloetic liquid is not affected by the nitrates of baryta and silver; it affords with oxalate of ammonia minute traces of lime.

In a continuation of this paper I shall lay before your readers, next month, several interesting facts concerning the manufacture of caoutchouc on the great scale, supplementary to the account given in my *Dictionary of Arts*, &c.

13, Charlotte-street, Bedford-square,
February 18, 1839.

THE INVENTION OF PHOTOGENIC DRAWING OR SOLAR DELINEATION.

Sir,—I was somewhat amused and still more astonished on perusing, in your Magazine of the present month, a letter addressed to you by Mr. Thomas Oxley, teacher, on the subject of M. Daguerre's method of obtaining fac-similes of objects by the action of the solar light. Before I proceed to rectify the strange mistake which the writer has committed with respect to myself, I shall transcribe a portion of his letter. Mr. Oxley's object is to prove that he and some other Englishmen had conceived the idea of obtaining fac-similes of different objects by the action of the rays of the sun long before M. Daguerre had made those experiments which have lately been announced in the French journals. He proceeds as follows :—

" This honour, I believe, is due to England and to Englishmen, and I claim it for myself and two other gentlemen, namely, Mr. John Turneau, (Turmeau,) a very excellent artist and eminent portrait painter, the inventor and patentee of the Liverpool lamps, so much celebrated before the introduction of gas lights,—the other Egerton Smith, Esq., a proprietor and editor of the *Liverpool Mercury,*" &c.

Mr. Oxley, after some personal compliments which it is unnecessary here to repeat, states that about the year 1823, having then a moderate knowledge of chemistry, he was aware that certain chemical preparations, especially those of silver, were subject to great changes of colour from the effect of the rays of light, and he communicated his ideas on the subject to the gentlemen just named, &c. Mr. Smith, and Mr. Turneau (Turmeau) both told him that they had long " entertained ideas of the possibility of accomplishing such a thing." Mr. Oxley adds that he suggested to Mr. Turneau (Turmeau) and Mr. Smith that the camera obscura might be advantageously used to facilitate the process.

Now, Sir, when I inform you that although I have not been an inattentive observer of what has been going forwards in the scientific world, I never had the most remote idea of the phenomenon in question until I met with a notice of the experiments of M. Daguerre in the French journals,—you may judge of my astonishment at finding my name so conspicuously blazoned forth in your pages; and as I do not wish to " strut in borrowed plumes," I lose no time in disclaiming all right to the compliment so unexpectedly bestowed upon me. Mr. Oxley has mistaken me for Mr. Charles Seward, who was a partner with Mr. John Turmeau, of this town, in the Liverpool Patent Lamp.

These gentlemen, it seems, in the prosecution of their chemical investigations, had made some experiments with a preparation of silver, and had succeeded in obtaining very faint outlines of different objects from the action of the sun's rays, admitted through an aperture in a window shutter; but finding these solar delineations very evanescent, and not being able to fix them, they pursued their experiments no further. Mr. Turmeau has assured me, however, that he never conceived the idea of using the camera obscura in the process, and that he has no recollection of any hint of such appropriation having been suggested by Mr. Oxley.

As far as the mere knowledge that certain preparations of silver spread out on a surface of paper, metal, &c., were susceptible of receiving impressions from the solar light, your respectable and intelligent correspondent, Sir Anthony Carlisle, has stated in your last publication, that he and the late Mr. Wedgwood were acquainted with the fact forty years ago.

The discovery of this property of light would be of little value, unless permanence can be given to the outlines thus obtained; and it seems that M. Daguerre and Mr. Talbot have both succeeded in supplying the desideratum. As for the claim to priority of discovery, we must leave that point to be decided by those gentlemen themselves.*

The insertion of this explanation in your valuable Magazine will oblige,

Sir, yours respectfully,

EGERTON SMITH.

Liverpool, March 6, 1839.

---

* It appears, from an article in the last *Literary Gazette,* that the first hint of this discovery was communicated to M. Daguerre by M. Neipce, who died a few years ago. M. Daguerre intimates, however, that the suggestion was so vague that it cost him long and persevering labour to bring it to its present perfection. The correspondent of the *Gazette,* who appears to have clearly established the claim of priority for *M. Neipce,* observes, " I do not think that M. Neipce could have given such a very imperfect idea fifteen years ago, as the specimens M. Neipce brought and exhibited in 1827, in England, (and some of them are still in my possession,) are quite as perfect as those productions of M. Daguerre described in the French newspapers of 1839." As far as we can gather, from the rather scanty information now before the public on this subject, there seems to be considerable difference, both in the process and effect, between the methods adopted by M. Daguerre and Mr. Talbot. The following paragraph on the subject is derived, we believe, from some of the French journals :—" The scientific folk of Paris are busied in endeavouring to find out the composition of the plate by which M. Daguerre is enabled to obtain an exact representation of any object or scene. This plate, placed in a camera obscura, receives from the impingement of light certain impressions, varying according to the intensity; so that, in about a quarter of an hour, the cathedral of Notre Dame, for example, engraves itself perfectly on the plate. It was, at

INVENTION OF PHOTOGENIC DRAWING.
[From the *Athenæum*.]

During the discussions which took place in Paris respecting the priority of the discovery of M. Daguerre and Mr. Talbot, the name of M. Niepce was incidentally mentioned as the person to whom the former was indebted for the first idea of fixing the images represented in a camera obscura. Subsequently, M. Niepce's claim to honour has been more fully admitted; and this has been singularly confirmed by Mr. Bauer, in a letter published in the *Literary Gazette*. Mr. Bauer therein states, that, in 1827, he became acquainted with M. Niepce, then on a visit to his brother at Kew; that M. Niepce made known to him, and others, that he had discovered a means of " fixing, permanently, the image of any object by the spontaneous action of light," and exhibited several specimens. That, by the advice of Mr. Bauer, he, M. Niepce, drew up a memoir on the subject, dated 8th December, 1827, which he forwarded to the Royal Society, but which was subsequently returned, because it is contrary to the rules of the Society to read a paper referring to a process which is not disclosed. That shortly after, and when about to return to France, M. Niepce presented Mr. Bauer with specimens of the newly-discovered art, which are now in his possession. Thus, then, the question of priority, as between England and France, is settled beyond all dispute* at the same time, we must observe, that the processes of M. Daguerre and Mr. Talbot are manifestly different. As to the relative merit of M. Niepce and M. Daguerre, there is no doubt, in our opinion, that, though the first idea was suggested, and the earlier specimens produced by M. Niepce, yet that he was long and zealously assisted by M. Daguerre, who had been for many years engaged in similar pursuits; and there is legal proof that, so early as 1829, they entered into an agreement, by which they declare themselves " *associés pour exploiter le pro-*

cédé à *l'invention duquel ils avaient concurru l'un et l'autre.*" Mr. Bauer is, indeed, in error, when he states that the specimens presented to him, in 1827, by M. Niepce, are quite as perfect as those produced by M. Daguerre, and described in the French papers in 1839. The specimens in the possession of Mr. Bauer, and others, given at the time to Mr. Cussels, of Richmond, have been obligingly submitted to our examination. They may be divided into—pictures copied from engravings, and pictures copied from nature. The best specimens of a copy from an engraving belongs to Mr. Cussels; and, though somewhat different in its style and general effects, it is not, considering that it has been exposed for more than twelve years to all the casualties of dust and damp, much inferior to similar copies shown to us, when lately in Paris, by M. Daguerre. Mr. Bauer possesses the only copy of a picture taken from nature; but this, so far from being equal to the specimens produced by M. Daguerre in 1839, is even more shadowy and indistinct than any of the earlier specimens of the art which we saw in Paris, and immeasurably inferior to the latter works. That the early process of M. Niepce, and the present one of M. Daguerre, are essentially the same, though greatly improved, we cannot doubt, As M. Daguerre has good and sufficient reason for not making his secret known for the present, the pictures exhibited by him are covered to the very edge with paper; notwithstanding which, we came to the conclusion that the material was either pewter highly polished, or washed with silver; and all the specimens in the possession of Mr. Bauer and Mr. Cussels are on pewter, apparently covered with a very thin coating of transparent varnish; but whether this varnish was applied before receiving the impressions, or subsequently, to fix them, is not obvious : we incline to the latter opinion. The most curious fact, in relation to this discovery, yet remains to be told. It would appear, considering the character of the pictures, all but impossible that impressions from them could be multiplied after the manner of an engraving; M. Daguerre, indeed, stated to us that it was impossible, and it is but reasonable to believe that he is as fully informed of the nature and extent of the discoveries as M. Niepce himself. Yet, in 1827, M. Niepce not only declared that it was possible, but produced specimens of such multiplied copies : and Mr. Bauer has now in his possession, not only copies of engravings, fixed permanently by the action of light, not only scenes from nature, *but metallic plates engraved, and engravings copied from them :* and he understood and

---

first, supposed to be the chlorure of silver, known to be susceptible of change from the effect of light; but on this substance light produces shade, and *vice versa*, nor is the effect permanent. On M. Daguerre's composition, on the contrary, dark spots produce corresponding shade, and that in every gradation of tint. The moon's ray had no effect on the chlorure of silver; it has on M. Daguerre's composition, and reproduces its own image perfectly."—E. S.

* Not so. The Editor of the *Athenæum* and these other claimants to the invention of photogenic drawing, appear to have overlooked the claim made by Sir Anthony Carlisle for himself and Mr. Wedgwood, as having performed experiments upon the subject 40 years ago, many years prior to the dates of the evidence of the earliest of these new claimants. We refer our readers to Sir Anthony's letter, No. 809, p. 329.—ED. M. M.

believes that no engraving tool was used, but that *the drawings were fixed by the action of light, and the plates subsequently engraved by a chemical process, discovered by M. Niepce.* If so, the greatest secret of all remains yet to be made public, and is, we believe, as unknown to M. Daguerre as to others.

ERICSSON'S STEAM-BOAT PROPELLER.

The experimental iron steam-boat *Robert F. Stockton*, constructed for testing Capt. Ericsson's propeller, which we noticed some time since, being on the eve of departure for the United States, at the request of a number of scientific gentlemen who were desirous of witnessing her performance, the proprietor consented to another trial being made, and on Saturday last a large party was invited for this purpose. Among those present were Major-General Sir John Burgoyne, Chairman of the Board of Public Works, and Commissioner for Steam Navigation, &c., in Ireland; Major Robe, of the Royal Engineers; Mr. James Terry, of Dublin, largely concerned in canal navigation; Messrs. Vignoles, Delafield, Reid, Napier, and Thomas; several distinguished Swedish naval officers; Captain Stockton, of the United States navy; Mr. Ogden, Consul of the United States at Liverpool; Mr. Young, an American civil engineer, &c. Some 30 gentlemen were present, and the result of the trial gave universal satisfaction.

One of our correspondents having before described the construction of the new propeller, we will now more particularly direct attention to the effect produced during the trial, which appeared quite conclusive as to the success of this important improvement in steam navigation. The distance from the West India South-dock to a point opposite Woolwich church and back, measuring 37,000 feet, was passed in 45 minutes precisely (21 minutes with, and 24 against the tide), the boat towing at the time a heavy city barge on the one side, a large wherry on the other, and another wherry astern. The speed of the engine being repeatedly timed by one of the gentlemen present, Mr. Young, an intelligent American engineer, it was found to average 66 revolutions per minute, or 2,970 during the 45 minutes. The inventor demonstrated, by accurate working drawings, that the spiral planes of the propeller are set at such an angle, that had the resistance of the water been perfect, the progress of the boat could

only have been 132 feet at each revolution, or 39,204 feet during the time, instead of 37,000 actually performed, thus showing a loss of less than 6 per cent. Respecting the engines for working the propeller, it was observed, that they may be made much stronger and more compact than ordinary marine engines, in consequence of the power being applied directly to the shaft which works very near the bottom; this, for sea-going vessels, will be very important, and their original cost must be considerably reduced, as all the paraphernalia of shafts, wheels, wheel-guards, &c., will be dispensed with. We were struck with the great regularity of the motion, not the slightest jar being perceptible. The engines consist of two cylinders 16 inches in diameter, with 18 inches stroke, and are worked by steam of a pressure varying from 35 lb. to 55 lb. to the square inch; their construction is extremely simple, and evinces a knowledge of steam machinery in the inventor which is calculated to give additional confidence in the success of his propeller in all the varieties of its application for canal, river, or ocean navigation.—*Times.*

ROYAL INSTITUTION.—MR. BRANDE'S LECTURE ON STEEL.

March 1st.—*The Conversazione.*—Mr. Palmer had on the table his pneumatic filter applied to the making of coffee, (as suggested by a friend of ours,) and exhibited filtered infusions both cold and hot. The former possessed in a much greater degree the *fragrance* of the berry, but as to its other qualities we say nothing. Mr. Macdonald sent some specimens—which are now said to be scarce, though we know not for what reason—of those interesting deposits from the water of the baths of San Filippo, in Tuscany, which being received in suitable moulds, take any form at pleasure. In the present instance it was that of portrait medallions; the appearance is more that of wax than of marble.

The Lecture.—Mr. Brande, on Steel. The lecturer gave a slight notice of the history of the metal; described the process of its manufacture, explained the difference between blister sheer and cast steel; exhibited specimens of each, and of the crucibles employed in the latter operation; pointed out the chemical and mechanical distinctions between cast iron and steel; and enlarged on those pecularities of steel, as distinguished from other metals, which confers on it so many valuable properties, available in arts, manufactures, and science, illustrating the whole with appropriate experiments.

The great body of observations filling up the above outline, would, in point of information, be too familiar to the readers of the *Mechanics' Magazine* to require notice; we therefore select only those which it is thought may be interesting. In the lecturer's opinion, steel was known to the Egyptians, for he could not conceive the possibility of executing their stupendous monuments, without a knowledge of its use; he supposes that they derived it from India. He drew attention to the rather singular fact, that there is at present imported into this country, from the East Indies, a very pure iron, and of which the Vauxhall steel company avail themselves in making steel. ' He observed, in regard to the process of cementation, that

it may be conceived, either that the particles of carbon pass by transference through the substance of the iron, or else, that the carbon penetrates it in the form of gas, and thus effects the union constituting steel. In adverting to the doubts which have been entertained respecting the chemical nature of steel—alluding, we believe, to a remark of Dr. Dalton, that he had never been able to detect any carbon in it—Mr. Brande said, that it was his firm belief it could not be made without carbon. He admitted, however, that phosphorous is an essential ingredient in *good* steel, and to its existence in animal charcoal, he attributed the superior efficacy in the process of case-hardening, of that kind of charcoal to the common sort.

We subjoin tables of the constitution of cast-iron and steel, as exhibited by Mr. Brande:—

	White Cast-iron.		*Grey Cast-iron.*	
Carbon........	2.33	.... 2.64	.... 2.45	.... 1.66
Silicium	0.84	.... 0.26	.... 1.62	.... 3.00
Phosphorous ..	0.71	.... 0.28	.... 0.78	.... 0.49
Manganese	a trace	.... 2.14	.... a trace	.... a trace
Iron	96.12	.... 94.68	.... 95.15	.... 94.85
	100.00	100.00	100.00	100.00

Charcoal absorbed.

$\frac{1}{170}$ soft cast steel.
$\frac{1}{100}$ common steel.
$\frac{1}{80}$ the same but harder.
$\frac{1}{70}$ ditto but too hard.
$\frac{1}{73}$ white cast iron.
$\frac{1}{70}$ mottled.
$\frac{1}{70}$ black.

Taking the liberty to append a few observations of our own, we would remark that in regard to the use of steel by the ancient Egyptians, it is not probable that either they or the Hindoos were acquainted with the art of making it by cementation, but it is not unlikely that they used a kind of native steel, or *wootz*, or that they adopted that primitive practice, which exists even at the present time, particularly in Styria, of de-carbonising cast-iron, where, in consequence of the process falling short, by one or two stages of complete refinement, it sells even at a lower price than the pure iron obtained at the same time.

With respect to the two distinct conceptions of the mode in which cementation is effected, we do not perceive that any material difference is involved in those distinctions, unless we further suppose, that the outer portion of metal is overcharged with carbon, and parts with its excess to the next, and so on, until the middle is reached; but in this case we should expect to find

that by the time the interior has become steel, the exterior would be converted into cast-iron, or that at least a greater difference of stratification would be produced than is observable. Should, however, this be the real effect, and we are not certain that it may not be so in an extreme case, still it would not be conclusive in favour of the mode by transference, for a greater absorption of carbon would take place near the surface, by the mode of gaseous penetration. On the other hand, though it is certain that the gaseous form of carbon is found sufficient for the formation of steel, yet the inference from this fact in favour of the gaseous mode is also not conclusive, for the carbon may still be imbibed only at the surface. May it not, however, be asked, in behalf of the latter mode, whether the mere density of iron could possibly prevent its fermentation by gaseous carbon? Whether there would not be required a repulsive force? Whether it is not to this force that we are indebted for the retention of the gases in vessels? and, whether the gas of carbon would not penetrate freely enough, when this force is nullified by that which is connected with chemical attraction? But there is no end to conjecture on the subject of molecular action, and it will never be otherwise until new mediums of research are brought within our reach.

No one will question the great ability of Mr. Brande as a lecturer, but we could almost indulge the wicked wish of being able to temper with, and manœuvre a little, the progress of the clock in order to see the good effect for once of surprising him out of the extreme deliberateness of his delivery; but we question whether he would not rather forego half of his lecture, than quicken his paces in the least. Unfortunately, also, he does not think it necessary to raise his voice higher than in common conversation, but though his enunciation is so beautifully distinct, that with *listening* attention, he may be heard in most parts of the theatre, yet the continual effort to give such attention, especially at this season of the year, when every body thinks it to be his duty to cough, soon becomes very fatiguing, and at last so painful as quite to mar the pleasure of hearing him lecture.

NOTES AND NOTICES.

Chronometer Prizes.—Sir,—A recent number of your Magazine (No. 811,) contains an article signed " Philo-veritas," and entitled " *Chronometer Premiums—Abuse of official influence,*" in which a statement regarding the number of our chronometers which have obtained prizes at Greenwich is brought forward, in apparent contradiction of a certificate granted by the Astronomer Royal to Mr. Carter, of Cornhill. It has come to our knowledge that copies of the number of the *Mechanics' Magazine* referred to have been sent to different persons, amongst whom we may name the Astronomer Royal himself, and the Hydrographer of the Admiralty, labelled " *with Parkinson and Frodsham's compliments.*" As we had no concern whatever, directly or indirectly, in writing, communicating, or publishing the article in question, and as we have no means of knowing whether the deception attempted on Mr. Airy and Captain Beaufort may not have been practised on other gentlemen, we trust you will not refuse, in justice to us, to publish this protest against so unjustifiable a use of our name. We remain, Sir, your obedient servants,
Parkinson and Frodsham.
Change Alley, March 12, 1839.

Great Western Steam Ship Company.—A half-yearly general meeting of the proprietors of the *Great Western* steam-ship was held in Princes-street, Bristol, last week. Mr. Maze took the chair. Mr. Claxton read the report, which stated that the company's first ship had disproved all unfavourable auguries, and promptly rewarded the enterprise of the projectors. It was impossible to speak too highly of the qualities of the *Great Western* steam-ship; after having run 35,000 nautical miles, and encountered 36 days of heavy gales, her seams required no caulking, and when she was docked she did not show a wrinkle in her copper. The average of her passages out was 15½ days, and home 13 days; the shortest passage out was 14½ days, and the shortest home 12¼. About 1,000 passengers had gone in the ship. After alluding to the great expense necessary to combine speed, security, and enjoyment, it expressed a hope that through the liberality of the American Congress the duty of 2d.

per bushel on coals would be given up, and thus a saving of nearly 1,000*l.* a year would be effected. The company have decided on constructing their next vessel of iron, for which the preparations are far advanced. It appeared from the statement of accounts, that after paying 2,000*l.* for additions to the ship, and insurance to October next, 1,500*l.* for goods damaged in the hurricanes in October last, and upwards of 2,000*l.* being set apart for a reserve fund, there remained from the profits sufficient for a dividend of 5 per cent., making with the former one of 4 per cent., 9 per cent. for the year. The report was unanimously adopted.

Iron.—Every person knows the manifold uses of this truly precious metal; it is capable of being cast in moulds of any form; of being drawn out into wires of any desired strength or fineness; of being extended into plates or sheets; of being bent in every direction; of being sharpened, hardened, and softened at pleasure. Iron accommodates itself to all our wants, our desires, and even our caprices; it is equally serviceable to the arts, the sciences, to agriculture, and war; the same ore furnishes the sword, the ploughshare, the sythe, the pruning-hook, the needle, the graver, the spring of a watch or of a carriage, the chisel, the chain, the anchor, the compass, the cannon, and the bomb. It is a medicine of much virtue, and the only metal friendly to the human frame. The ores of iron are scattered over the crust of the globe with a beneficent profusion, proportioned to the utility of the metal; they are found under every latitude, and every zone; in every mineral formation, and are desseminated in every soil.—*Dr. Ure's Dictionary of Manufactures.*

British Association.—The period for holding the ensuing meeting, which was left for the decision of the council, has now been fixed. The first meeting of the general committee will be held at Birmingham on Saturday, August 25, and the various sections for scientific business will meet on the 27th, and through the week. The local council and committees are already busily engaged in making the preliminary arrangements.

Ericsson's Propeller Company.—A company has been formed for building a ship of 1,000 tons burthen, to run between England and the United States, fitted with Captain Ericsson's patent propeller (a very favourable experiment with which is detailed in another part of our number.) The working plans of the vessel and engine have been prepared, and everything is in a state of forwardness for the commencement of operations. We observe that Mr. Ogden, the American consul at Liverpool, well known for his scientific attainments and enterprise, is one of the directors of this undertaking, which is called the " Atlas Steam Navigation Company."

Light-drawn-pictures. — Our vivacious Parisian neighbours have certainly the faculty of making the most, in words, of anything they chance to lay hold of. " M. Daguerre's ingenious discovery," says the Paris correspondent of the *Post,* " which has assumed the name of ' Daguerrotype,' continues to excite very great curiosity and admiration. It is affirmed that the Emperor of Russia has offered 500,000f. for his secret, and that he has declined the munificent reward. It is not likely that his friend, Mr. Arago, will succeed in obtaining a larger national one from the Chambers." Mr. Talbot's communications to the Royal Society have made the processes public property here, had they not been so before. Sir John Herschel has turned his attention to the subject and has already obtained pictures from the light of Daniell's great galvanic battery; Sir David Brewster too, has commenced an investigation into the matter.

LONDON: Printed and Published for the Proprietor, by W. A. Robertson, at the Mechanics' Magazine Office, No. 6, Peterborough-court, Fleet-street.—Sold by A. & W. Galignani, Rue Vivienne, Paris.

𝔐𝔢𝔠𝔥𝔞𝔫𝔦𝔠𝔰' 𝔐𝔞𝔤𝔞𝔷𝔦𝔫𝔢,

MUSEUM, REGISTER, JOURNAL, AND GAZETTE.

No. 815.] SATURDAY, MARCH 23, 1839. [Price 3d.

Printed and Published for the Proprietor, by W. A. Robertson, No. 6, Peterborough-court, Fleet-street.

GLADSTONE'S PATENT WINDLASS.
Fig. 1.

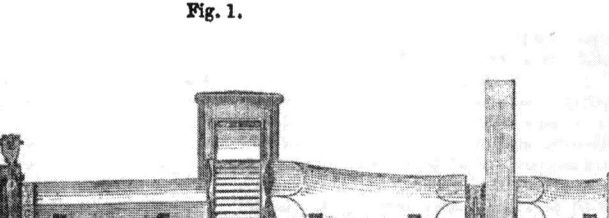

3.

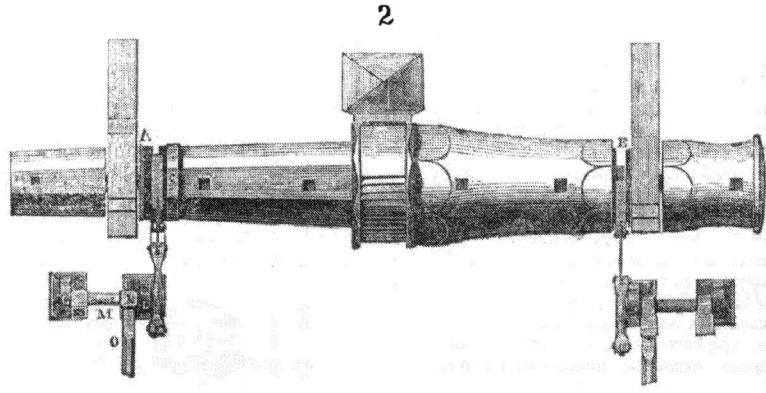

2

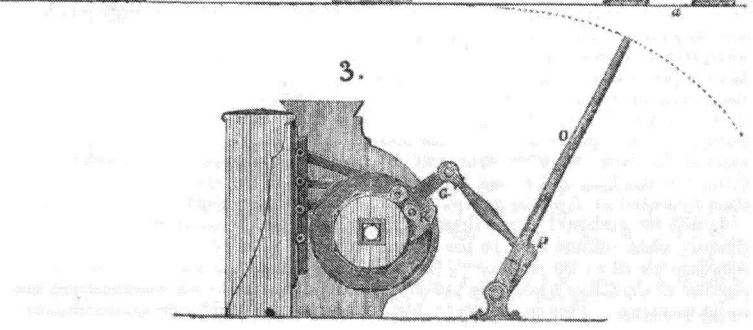

GLADSTONE'S PATENT WINDLASS.

In nautical mechanics, as well as in all other applications of machinery, the object is, to obtain the greatest effect with the least power, and in the shortest time; but there is an additional requisite, not to be wholly overlooked in applying machinery on ship board, that is the prejudices and ordinary modes of action of the seamen must be humoured as much as possible. Where so much so often depends upon the men working "with the will," and where bare and tardy fulfilment of duty would be ruin, it is necessary that the predispositions, likings, and prejudices, to which sailors are so proverbially prone, should be combatted for the purpose of being overcome, with gentleness and discretion. Equal in importance to the rudder, as regards the safety of the ship, is the anchor, and, in connection with the anchor, the windlass. Many patents have been obtained for improvements in this important apparatus. Some, otherwise good, have failed of success in consequence of the dislike of the seamen to change their usual way of working. The turning of a winch handle, or the up-and-down pump action, appears to have been considered by Jack as *infra dig.*, and nothing but the handspike has been more than tolerated in his iron grasp.

It will be gathered from these introductory observations that in the patent windlass about to be described, the old routine of the sailor's action is preserved in its working. This is, however, but a minor advantage, in comparison with the others, gained by this very ingenious invention of Mr. Gladstone, of the firm of Gladstone, Eddowes, and Co., of Liverpool.

In Mr. Gladstone's patent windlass, or, as he more properly terms it, his "windlass propeller," the windlass being the same as is ordinarily used, the mode of producing rotation only being changed. The motive force is applied, not directly to the windlass, but to a system of levers of the first order, distinct from, yet connected therewith, and by which the following advantages are obtained:—1st, a longer leverage is obtained than by the common handspike, because the length of the handspike, which is the primary lever, must be limited to a convenient height for a man to work, and from this length must be deducted the height of the ordinary windlass above the deck, whereas Mr. Gladstone's handspike works from a socket fixed immediately on the deck. By this longer leverage the weighing of the same anchor would be easier work for the usual number of men, or it could be done quicker, or with fewer men. 2nd, any fleeting or shifting of the handspike is rendered unnecessary after it is once placed in its socket, whereby a great deal of time is saved, especially at night, when the seamen cannot, or only with difficulty, see the holes for the handspike in the barrel; even in daylight it requires considerable practice to fleet the handspike quickly and regularly. 3rd, By making the two sets of men, one each end of the windlass, work with alternate strokes, a continuous rotation may be given to the windlass, and the anchor weighed in considerably less time than by the intermitting action, in little more, in fact, than half. 4th, the additional parts add very much to the strength of the windlass, so that its upsetting is next to impossible.

We shall now proceed to describe the invention. Fig. 1 is an elevation of a ship's windlass, as seen when viewed from behind, or looking forward, with Mr. Gladstone's improvements applied thereto; the starboard half of the windlass being shown with the casing on, and the larboard with the casing off. Fig. 2 is a ground plan of fig. 1; and fig. 3 is a side elevation of part of the windlass and its accompaniments detached and laid open, in order to show more clearly the mode in which the windlass works. (Figs. 1, 2, and 3 are on our front page.) The windlass is provided as usual with a ratchet wheel and palls in the centre, but close to the side bits (X X), or at any other convenient points. Upon the windlass spindle are fixed two smaller supplementary ratchet wheels. The construction of the wheels A and B, which are precisely alike, is shown separately in figs. 4, 5, 6, 7, and 8. Fig. 4 is a plan of the inner side of the wheel; fig. 5, a side view of fig. 4;

Figs. 8. 7.   Fig. 4   Figs. 5   6

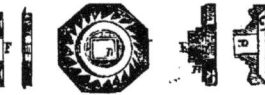

fig. 6, a section of fig. 5; fig. 7, a view

of the outer side of the wheel; and fig. 8, a section of fig. 7. The wheel, it will be observed, is made in its outer circumference of an octagonal form, in order that it may fit upon the body of the windlass, as shewn at C. D is a square hole, within which the windlass spindle is wedged, and E a square piece, in which the said square hole is made, that fits into the opening F, in the outer part of the wheel (figs. 7 and 8).

The means by which the ratchet wheel last described, is acted upon, are shown as in combination, in figs. 1 and 2; and more or less detached, in figs. 3, 9, 10, 11, and 12. G, of which fig 9 is a separate view, is a first lever, which revolves on the journal H, of fig. 5. I is a drag pall, attached to the lever G, and shown sideways in fig. 10, and the plan in fig. 11. K is a push pall, also attached to the first lever. Fig. 12 shows the first lever G complete, with its drag and push palls, as applied to the ratchet wheels.

Fig. 12.   Figs. 10.   9.

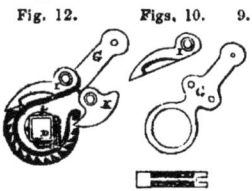

Fig. 11.

L L, fig. 1, are two journals, secured by bolts or screws through the deck *a a*. M is a shaft, which passes through the eyeholes of the journals L L, and rotates freely therein. N N, two square sockets, for the reception of handspikes, as shown by O O, figs. 1, 2, and 3. P is a crank, attached to the end of the inner shaft M. Q, a rod which connects the crank P with the first lever G, which acts immediately on the ratchet wheel A or B. The handspike O when worked, describes a quarter of a circle. When at its highest point of elevation, the system of levers, of which it forms a part, present the form shown in Fig. 1, on the larboard side. When pulled a little way over they take the position represented in fig. 3; and when the quarter revolution is completed, that exhibited on the starboard side of fig. 1. Should it be desired to increase the power of the leverage, that may be effected by unforelocking or unscrewing the connecting rod, and attaching it to the crank P, at a point lower down than

at first, for which purpose there are holes made on the crank P, as shown at *b b b*, fig. 3.

The two ratchet wheels A and B may, of course, be either worked simultaneously by the means hereinbefore described; or, if a continuous rotation is desired, this may be effected by working them alternately.

The number of handspikes applied may also be increased from one on each side, to three or four or any other convenient number, by making a corresponding addition to the sockets with which the working shaft M, is provided.

So generally have the merits of these inventions been appreciated by the shipping interest that though it has not been more than a twelvemonth patented, it is already in very extensive use. The following testimonial which we select from a great number equally satisfactory, with an inspection of which we have been favoured, is from a party who had full opportunity of ascertaining the working properties of Mr. Gladstone's windlass, both in America and in the river Mersey, during the heavy gales of last winter :—

Copy of a letter from Capt. Higgins of the ship Leila, of Baltimore, U. S., 650 tons.

Messrs. T. M. Gladstone and Co.,
        Liverpool.

Dear Sirs,—Your Patent Windlass Propeller came duly to hand, and I put it on board my new ship *Leila*, and with its operation I am highly pleased.

I have built ships for more than 20 years, and have used nearly all the purchases that have been invented; and have no hesitation in saying that yours is the greatest combination of *quickness, strength, and power* that has come within the reach of my knowledge, and still it is perfectly simple.

My ship's anchors weigh upwards of 26 cwt. with 1¾ inch chains, and *four or five men can heave them up with ease and dispatch.* It is my opinion that *my ship's company* could heave up an anchor *four or five tons* weight with it.

        I am, Dear Sirs,

    Your most obedient humble servant,

                ASA HIGGINS.
January 15, 1839.

ON THE NATURE AND PURPOSE OF A CAVEAT, IN RELATION TO PATENTS FOR INVENTIONS.

We have found, in the course of our extensive intercourse with inventors, that a most erroneous idea generally prevails with regard to the nature and object of a legal proceeding known by the term of "lodging caveats." It is thought, very generally, that a caveat gives the inventor a right, for a limited time, to his invention, or that it secures to him a priority over any after applicant for a patent; protecting him from being forestalled in obtaining a patent; and enabling him to make experiments in safety. By many it is thought, in fact, to be a kind of preliminary patent.

To remove these incorrect impressions, as far as this can be affected by the publicity which the circulation of the *Mechanics' Magazine* will give to correct information upon the subject, is the purpose of the present article. We are certain that it will save our inventive friends and ourselves the expense of many postages, and much letter-writing.

In obtaining a patent for an invention, a variety of formalities have to be complied with. It has to run the gauntlet of a dozen or so of government offices, at each of which certain documents have to be prepared, signed, recorded, and fees paid. This is the business performed by the patent agent, and the fees paid at these offices constitute nearly the entire expense of the patent.

The custom is to grant all patents petitioned for, as a matter of course, unless an opposition be entered to the petition by some interested party; but as the proceedings in obtaining a patent are not in any way published or made known to the community, no one could have an opportunity of preventing a post-inventor, or the fraudulent possessor of an invention, from obtaining a patent, to the damage of the first and rightful inventor. The lists which are published in the *Mechanics' Magazine* and other journals, are of patents which *have been* granted. It is true, that even when a patent is granted, it will stand only on its originality and the true inventorship of the patentee, and may be repealed on writ of *scire facias*; but the expense of this legal proceeding, the difficulty of obtaining the necessary evidence to support it, and the almost penal consequences which follow a failure in the prosecution, viz., payment of double costs, renders it almost a dead letter. Moreover, even when successful, it is little or no satisfaction to the true inventor, as by the patent being repealed, the invention becomes public property, and he has then no more interest in it than a stranger, or the party from whom he has wrested the wrongful monopoly.

Again; there are many inventions which a patent will not protect, and also some which inventors may think it advisable, and find it perhaps more profitable, to work in secret. Now, it has been decided, that the secret working an invention by one inventor does not invalidate a patent obtained by another inventor for the same invention. For inventors, or parties interested in the secret working of an invention, therefore, to protect themselves, it is necessary that they should have notice of applications being made for patents for inventions likely to interfere with theirs. To obtain such notice is the purpose of the caveat, as we shall proceed to explain.

A petition for a patent having been presented to the Queen, she refers it to either of her law officers, the Attorney or Solicitor General, to take the petition into his consideration, and to report whether or not the prayer may be granted.

At the Attorney and Solicitor-Generals' offices it is that caveats are entered. The caveat is to the purport that you desire that no patent be allowed to pass for (say, for instance,) *any improvements in steam-boilers*, without notice being given to the party entering the caveat.

Suppose, therefore, an inventor petitions for a patent for an "improved method of generating steam";—on the petition being referred to the Attorney or Solicitor-General, the clerks of these officers would examine their registries of caveats, and finding the caveat above instanced therein, would give notice to the caveator. The information contained in this notice is, that a certain petitioner, whose name and address is given, is applying for a patent for an "improved method of generating steam," (no further description of the invention is given) and that if the caveator thinks the invention will interfere with his, he must enter an opposition, within seven days, to the patent being granted, or that otherwise it will be allowed to pass.

In the caveat the exact invention desired to be protected is never described —all that is stated is some generic title, such as we have before instanced, and upon this title the caveator will have notice of all applications for patents coming in any degree within the compass of its meaning.

Had no caveat been upon the registries the patent just instanced would have been allowed to pass as a matter of course;—that is, the Attorney or Solicitor-General would have reported in favour of the grant; and the same also would be the case should the caveator fail to enter an opposition within the seven days prescribed.

It will be seen, therefore, that unless followed up by oppositions to all applicants for patents which may be judged *from their title* to be likely to interfere with the invention caveated, the caveat affords no protection whatever, but is useful merely as a matter of information. In relation to such subjects as steam-engines, such a course of opposition would entail enormous expense, and would in fact be one which no one at all acquainted with the matter, would for a moment think of adopting. An Asphalte Pavement Company lately followed this course, opposing every patent applied for, in any way relating to pavements, asphalts, cements, or other like objects; but individuals collectively do not always act so discreetly as they do severally.

We will suppose now that an inventor has entered a caveat; that another inventor has applied for a patent for an invention upon the same subject; and that the caveator has received notice of this application. From the meagre information in the notice, he has to judge whether it would be adviseable to enter an opposition. The following are the only grounds upon which a judgment can be formed :—1st. It will be well to enter an opposition where the subject matter of the caveator's, and of the petitioner's invention is of an uncommon or peculiar nature, and where variety of improvement is unlikely:—2nd. Where the invention is revealed in the title, or words giving a clue to it, are used:— 3rd. Where the petitioner happens to be one whom the caveator may have trusted with the particulars of his invention in confidence; one whom he may have employed to make apparatus, or experiments, or to assist him in perfecting his invention; or any artificer, workman, or

other person whom he may know, or suspect of having means of obtaining a knowledge of his invention :—4th. Where the address of the petitioner is in a neighbourhood where any experiments of the caveators may have excited interest, or set inquiry astir, so giving an impetus to the inventive faculties of others.

Judging from any of these circumstances that an opposition is adviseable, and the same being entered through the Patent Agent, within the time prescribed, the progress of the patent applied for will be stayed. In some instances, on an opposition being entered, the petitioner never proceeds to have the impediment removed, either from fear of meeting his opponent, or from the delay occasioned thereby, giving him time to think better of his project, and perhaps, to discover some hitherto unobserved and insurmountable objection. In this case the caveator benefits the petitioner by preventing him expending money in patenting an invention which is useless. When the petitioner is ready to meet the opposition, the Attorney General appoints a time to hear the case of the petitioner, and of the opposer; of which time the opposer must have seven days' notice. At the time appointed, the petitioner appears himself, or by his agent, or in the company of his agent, and privately describes and explains to the Attorney-General the invention, for which a patent is desired. The petitioner having withdrawn, the opposer then also privately describes the invention, upon the ground of which he has entered his opposition. The Attorney-General being now in possession of both inventions, is able to decide whether the one invention interferes with the other. If they be different, he reports in favour of the petitioner. If they be alike, he refuses the patent to the petitioner. Supposing the opposer to be a patentee of the invention, upon existence of which the patent is refused, the refusal is of course absolute; but if the invention be new, unpatented, and confined to the petitioner and opposer, if they can come to terms, the Attorney-General will grant a patent to them jointly.—Or if the petitioner can prove that the opposer has obtained a knowledge of the invention fraudulently, or surreptitiously from him (the petitioner,) then of course the patent would be allowed to proceed.

We have now, we think, put our readers fully in possession of the nature, object, and operation of general caveats, as regards the protection they afford to inventors. There are others, that is to say, *special* caveats, which a party may, under extraordinary circumstances, avail himself of, at a stage of the proceedings of the patent, ulterior to the Attorney-General's reporting thereupon, as against some particular patent, the name of the petitioner and title of the patent being set forth in the caveat. These are some-

times useful where an opposer obtains additional evidence against the petitioner subsequent to the hearing upon the report. Under special caveats, the opposer has to pay the expenses of both sides in the matter, and as a security for this, has to deposit 30*l.* at the Attorney-General's bill office. In oppositions under the general caveat, each party has to pay his own expenses, which usually amount to less than five pounds on each side, supposing, as is usually the case, the matter be decided at one hearing.

REPORT OF THE COMMITTEE OF THE FRANKLIN INSTITUTE ON THE EXPLOSIONS OF STEAM-BOILERS —PART II. REPORT OF THE SUB-COMMITTEE ON THE STRENGTH OF MATERIALS EMPLOYED IN THE CONSTRUCTION OF STEAM-BOILERS.

[From the *Franklin Journal.*]

(Continued from Page 125, No. 798.)

Specific heat of Iron.—In determining the specific heat of the standard piece of the steam pyrometer, it was deemed advisable to employ processes somewhat independent of each other, and especially to vary the circumstances of the experiments, so that if possible the discordances on this subject might be reconciled, or at least referred to their probable origin. The first step was to ascertain the specific heat of the standard piece from 212° to the ordinary temperature of the atmosphere.

Bath for heating the standard piece.—The apparatus for this purpose is represented in Plates VI. and VII. In Plate VI. is seen a vertical section of the apparatus in which the iron standard piece S was heated W is a cast iron vessel in the form of a frustum of a cone, 12 inches high, 7 inches in diameter at top, and 5 at the bottom ; the upper edge being furnished with a flanch to match a corresponding flanch on a cover D, which was turned to fit it, and to form with it a steam-tight joint as at ff. In the central part of this cover is the mouth of another vessel or tube M, $2\frac{1}{2}$ inches in diameter, the bottom of which is about half an inch above the bottom of W. The space between the outside of the tube M and the inside of W, is nearly filled with water, as seen at $w\,w$, introduced by an aperture ordinarily closed by the iron stopper P. The interior of M is filled to about the level $m\;m$, with mercury.

In the cover D on the side opposite to P, is another aperture into which is fitted by grinding, an iron tube G, to receive and convey away to a water vessel I, the excess of vapour generated in W. The vessel thus constructed is placed in a circular opening in a sheet-iron plate, B B, 14 inches square, which rests on two guide rods R R, of the frame of the machine. (Plate I., see p. 17)

F is the furnace already described, as applied to the purpose of heating the bars of iron. X is a sheet-iron cylindrical case, one-eighth of an inch thick, open at both ends, and having along one side, a slit $s\,s$, one-sixteenth of an inch wide, left for the purpose of permitting the wire n, by which the standard piece S is suspended, to be introduced or withdrawn laterally, while the loop at its top is held by one hand of the operator ; and by the other, the handle Y. The rod which carries this handle is fastened by rivetting to the inside of the case K, and below the rivets the point is turned inward toward the centre of the case, preventing the iron cylinder S from rising above that point. This gives the operator entire command of the latter, notwithstanding the buoyancy of the mercury.

The thermometer T, graduated above the boiling point of water, was placed with its bulb on a level with the centre of S.

By raising or lowering the furnace F, the ebullition was maintained at a nearly uniform rate during the time of several consecutive experiments.

When the mercury in T was found to be stationary at 212°, the shield K was withdrawn from the mercurial bath ; the operator held Y in one hand, while the other supported the wire n; and in this manner, without allowing S to come in contact with the air, conveyed it to the mouth of the water vessel of the cooling apparatus, where K was held in a vertical position, and the wire quickly lowered till the standard piece was immersed, when the shield was immediately removed, the wire escaping through the slit $s\,s$.

The cooling apparatus.—The arrangement of parts in the cooling apparatus, is seen in Plate VII., (p. 440) where A is the cylindrical containing vessel, filled with water to such a height as to be completely full, when the standard piece S, and the bulb of thermometer t are immersed.

B is a cylindrical vessel of tinned iron, 14½ inches high, and 9 inches in diameter, to which is adapted the cover D of the same material, having in the middle a circular aperture $a\,a$, 3 inches in diameter, for receiving the thermometer t, and the standard piece S; allowing likewise sufficient space to move the suspending wire and thermometer. Another aperture through D, near

its circumference, admits the lower part of the thermometer i; and an opening near the bottom of B, receives the bulb of the thermometer l.

The thermometer O was suspended just without the vessel B, to mark the temperature of the room. C is a support for the water vessel A, formed of a cylindrical block of charcoal, 4 inches high.

The method of adjusting the weight of water in A, in this series of experiments, was the same as that subsequently described

Plate VI.

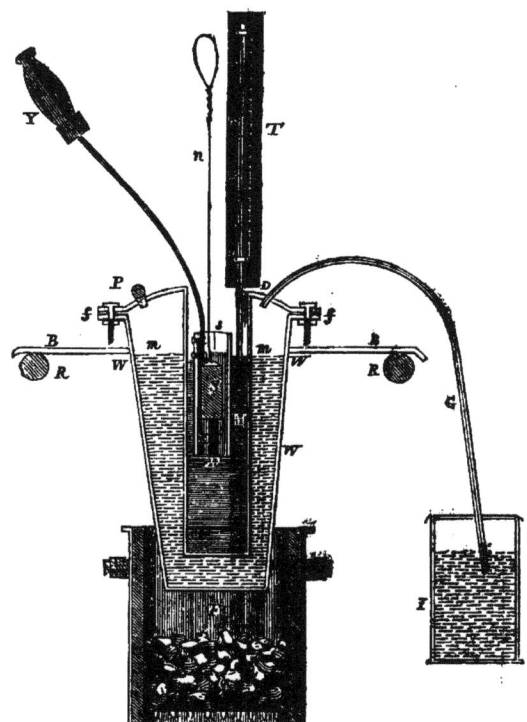

in the experiments on the latent heat of vapour, except, that in the present case, no process of weighing was required, after the heating had been performed.

The exact adjustment of the quantity of water to be used in every repetition of a given series, was made by means of a tube of small dimensions, open at both ends; by the aid of which it was easy to add or to

remove minute quantities of liquid to render the balance true.

Thermometer in the Water Vessel.—The most important of the thermometers used in this part of the investigation was that marked t, but which was in fact the thermometer A, elsewhere referred to, the bulb of which was 7½ inches long, intended to reach from the top to the bottom of the liquid, and which it actually did in some of the containing vessels with which it was used. Its diameter was about three-quarters of an inch. The glass constituting the

bulb of this thermometer weighed 433.85 grains, as ascertained by actual weighing, after the instrument had been broken. It was filled after the elongated bulb had been

Plate VII.

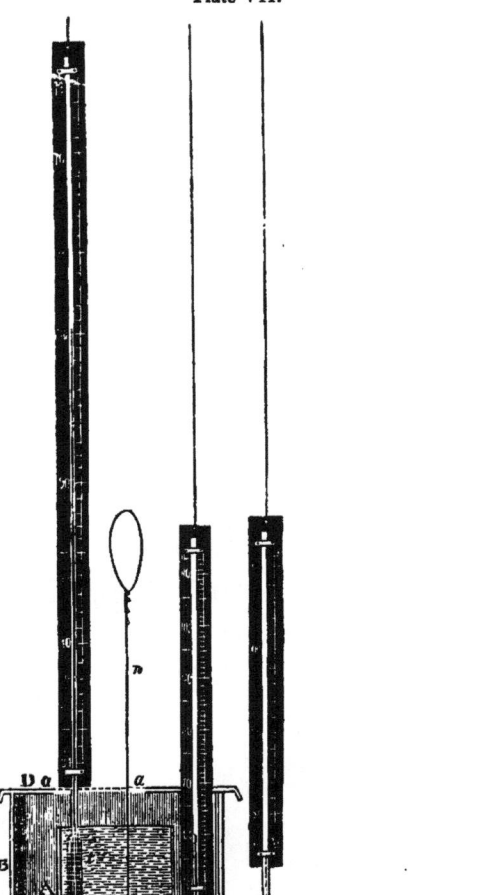

joined to its stem, and the separate weight of these ascertained. The mercury precisely filled the bulb alone at 32°, and the weight of

mercury required for this purpose was found to be 4682 grains. As most of the experiments with this instrument were made with an ini, tial temperature about 62°, it has not been deemed necessary to make more than one correction for the quantity of mercury expelled from the bulb, or excluded from the influence of the water vessel. At 62° the bulb must have held, by calculation,* 4670 grains of mercury, which, by a mean of several determinations already published, possesses a specific heat of .0327, and gives an equivalent, in grains of water, of 152.7.

When this thermometer was used in connection with a containing vessel of glass, the weight of its bulb was added to that of the vessel; and in other cases, the specific heat attributed to it, was that obtained by means of several trials of it with glasses of different thicknesses, instituted with a view to determine the effect of that material, towards cooling the heated body or standard piece. The specific heat thus found was .10036, according to which the bulb would be equivalent to 43.45 grains of water. The length of the scale of this thermometer was 37 inches, and the graduation extended from 34° to 74°; so that each degree was nearly $\frac{4}{5}$ of an inch in length, and the degrees were divided each into 50 parts, each part being of such magnitude that the eye could, when necessary, easily subdivide and read them into hundredths.

The graduation of this thermometer was obtained by direct comparison with a well-tried standard instrument, the degrees of which were about one-quarter of an inch in length. For this purpose, the bulbs of both were immersed in a large quantity of water, contained in a Hessian crucible, surrounded by another of black lead; thus affording a combined mass which changed its temperature with extreme slowness, and enabled us to mark, with deliberation and accuracy, every degree on the long scale, after having for some time agitated the two in contact with each other, and tempered the water to the point required.

The general mode of operating with the apparatus, Plate VII., was, after giving the water vessel and thermometer t, a temperature a few degrees below that of the surrounding air, to take simultaneous observations of all the thermometers which were recorded by one assistant, while another person, bringing the hot standard piece, sur-

rounded by its shield, from a distance of about 4 feet out of the heating apparatus (Plate VII), immersed and held it suspended, as already described. The moment of immersion was observed by a second assistant, on a time-keeper marking seconds. The manipulator, continually moving the thermometer t, and the standard piece about in the water, read off the degrees and parts as successively attained by the mercury, while the second assistant noted, and the first recorded them, together with the time of each observation.

The method just described, afforded the means of determining approximately, the proper temperature below that of the air, at which the water ought to be, when the standard-piece was immersed, in order that the heating and the cooling power of the atmosphere should be equal to each other. Table XIV will be found to contain a synopsis of the experiments conducted in this manner with reference to different containing vessels. Some trials were made to ascertain, with different vessels, the rate at which the air alone would, under given circumstances, produce certain elevations or depressions of temperature. But the interference of extraneous causes, such as the presence or absence of a stove in the apartment, the heat derived from the person of the observer, and others near the scene of the experiment, made it evident that a good defence against the influence of the air would be a better guarantee against error, from that source, than any table of corrections which could be constructed amidst so many modifying causes.

The results of a series of trials made in part without employing the vessel B to defend the water from the air and from radiation, are exhibited in Table IV. Some attempts were made, as above referred to, towards the correction of the irregularities therein observed; but the uncertainty attending the process, induced the committee to prefer, when practicable, the *prevention* to the correction of these anomalies. As the vessel B held 924½ cubic inches, the whole quantity of air it could contain did not exceed 277½ grains; which, supposing the specific heat of air to be .26, would not be equivalent to more than 72 grains of water, but as a considerable portion, amounting to at least $\frac{1}{4}$ of the whole, was occupied by the water vessel and its support, the remaining air could in no instance have been equivalent to more than 62 grains of water, and as the greatest change which occurred in the temperature of this portion of air during any experiment, was but 1.7°, and as the mean of all the changes of this kind observed during the progress of the investiga-

---

* Petit and Dulong found the expansion of *mercury in glass* between 32 deg. and 212 deg., to be $\frac{1}{63.8}$ consequently its expansion for 1 deg. $\frac{1}{180+63.8}$ $=\frac{1}{11484}$, and for (62 deg.—32 deg.)=30 deg., it will be $\frac{30}{11484}$ of its bulk at 32 deg. But 4682+ $\frac{30}{11484}$=12.2 grains and 4682—12=4670 as above stated.

tion, was a grain of 0.325°, while the correspondent mean gain of temperature in the water was 7.26°, and the mean weight of the latter 13.100 grains, it is evident that the relative influence of the air and of the water will be represented by 62 grains × 0.325° = 20.15 and 13100 grains × 7.26° = 95106, or the former is $\frac{1}{4710}$ part of the latter, from which it appears that from this cause the expression for the specific heat could not have been affected under the fifth place of decimals.

In a series of 13 trials in which the water, amounting to about 40,000 grains, was contained in copper vessels (Table VIII.), and the rise of temperature in the same was at a mean about 2.5° the air gained about .292 of a degree, which would indicate that the cooling power of the air confined in the vessel B, compared with that of the water in A was but as 1 to 5550, a result which would still less affect the general correctness of the determination.

[The limits to which the size of our Magazine confines us, prevents our republishing the elaborate tables of experiments, now and hereafter referred to, which would of themselves occupy a goodly octavo volume. The results of these experiments here stated, will, however, we have no doubt, be sufficient to satisfy a great majority of our readers. We refer those who desire to enter more deeply into the subject and to attain a more intimate knowledge of the experiments (we should particularly recommend them to the attention of Mr. Parkes and Captain Pringle, the engineers appointed by the British government to investigate the subject of Steam Boiler Explosions; also to some of our respected correspondents, among others, Mr. Tomlinson, Mr. Wigney, and Dr. Schafhaeutl) to the *Journal of the Franklin Institute*, from which we extract these reports, vol. xix, pp. 94 to 105, for tables IV. to IX; the pages of the Journal where the subsequent tables are to be found, we shall point out in their proper places as they occur in the report.]

After the preliminary series already given (Table IV.), two other sets of experiments were made, one in each of two glass vessels similar to that in which the preceding trials had taken place,—equal to each other in liquid capacity, but of different thicknesses; the one being more than four times as heavy as the other. Table V. contains the experiments with the thicker, and table VI. those with the thinner of these vessels. The particular object of these trials was to determine, if possible, the effect of the containing vessel on the general result of the experiment; in other words, to decide its specific heat, by observing the difference which would arise from a mere change of thickness in the containing vessel, while all other circumstances of the trial were the same in both cases. A comparison of several experiments in each table, with corresponding ones in the other, will show that when the water at the commencement was from 60° to 63.5°, the actual difference in the rise of temperature, due to a difference in the weight of the containing vessels of (12272—2996) = 9276 grains of glass, was about three-tenths of a degree; and from the comparison of nine experiments in the first of these tables, with the same number in the second, it will be seen that we obtain for the specific heat of glass .111063[*]. A part of the trials in these and the subsequent series were made by means of the spirit thermometer C, the equivalent of which was only approximately found, on account of not having taken the precaution to weigh the bulb and tube separately before filling the instrument. It is also, like all other spirit thermometers, liable to some uncertainty in its indications owing to the different quantities of the liquid which may at different times be taken up in wetting the tube, an uncertainty, which is the greater, the more sudden are the changes to which we submit the instrument.

The equivalent value assigned to it by finding the weight and capacity of an equal length of the same tube is 117.4 grains of water, as hereafter mentioned.

The next apparatus used in this part of the investigation consisted of two glass jars, smaller than those above described, both of the same capacity, but differing from each

---

[*] The principle of calculation applied to all these comparisons is embraced in the formula $x = \dfrac{(T't - Tt').(w+e.)}{T't'g - T'\,tg}$, where x is the specific heat of the container; T' is the gain of temperature by the water when the thinner glass is used; T the gain when the thicker vessel is employed; t', is the loss of temperature by the iron when the thinner, and t, that when the thicker is employed; w, is the weight of water in both cases, and e the equivalent in grains of water, of the liquid in the thermometer; g is the weight of the thicker jar; g' that of the thinner. Thus comparing the two identical experiments 7 and 8, table V., with experiment 6, table VI., in which the initial temperature of the water, and other circumstances, coincided with the former, we have $T' = 5$ deg. 76; $t = 143$ deg. 1; $T = 5$ deg. .4; $t' = 142$ deg. .74; $w = 16494.5$ grs.; $e = 152.7$ grs.; $g = 12.706$ grs., and $g' = 34.30$ grs. Hence $T\,t' = 824.256$; $T\,t' = 770.796$; $T't - Tt' = 54.46$; $w + e = 16.647.2$; $T't g = 9796733.976$; $T'tg' = 2827198.08$; from which $x = .130156$ the specific heat of the glass by this comparison.

other in weight, being nearly in the proportion of 3 to 1. The experiments in these two vessels were made in two sets of 6 each, three of each set being commenced in the thicker vessels, at a temperature of 60°, and three at 60.5°; and the same number at the same two points in the thinner. The results are contained in table VII., where it will be perceived that, from five comparisons between the trials in these two jars, the influence of the glass is such as to indicate a mean specific heat of .103086, which taken with the above result of the nine comparisons, between Tables V. and VI. gives a mean specific heat of flint glass of .107074.

As we are now only referring to the *apparatus* employed, we shall reserve our remarks on the results presented by these tables, respecting the specific heat of *iron*, until we have described the other methods of verifying their correctness.

The fourth set of apparatus for this purpose, consisted of two cylindrical copper vessels, of the same height as the glass ones already described : but of such diameter as to contain about 38600 grains, or a little over 5½ pounds avoirdupoise of water, and so differing in thickness, that the one weighed nearly four times as much as the other. The mode of conducting experiments in these two vessels, and the principle of calculation applicable to them, is entirely similar to that already given for the two pairs of glass jars,—except that the equivalent of the glass in the thermometer, was now separately computed.

The results will be found in table VIII., in which it will be perceived that the number of comparisons furnishing data for determining the specific heat of copper, is but two, and of these only one can be considered entirely unexceptionable.

From this it should seem that the specific heat of copper is .10431, whereas the four determinations of Wilkie, Crawford, Dalton and Petit and Dulong give .10750 for the specific heat of that metal.

A fifth mode of determining the specific heat of iron was by employing as water vessels two cylindrical sheet iron jars of the same capacity, but of thicknesses differing from each other in about the proportion of 3 to 1. As in the preceding sets, the specific heat of the container may here be found by comparing together experiments made at the same temperature, in the two vessels; and this ought to give their variation, if any exist, from the specific heat of the standard piece itself. Another method is to assume that the specific heat of the standard piece and of the sheet iron con-

tainers is the same.* The use of the two containers in this latter case serves only to verify each others results, since each furnishes a separate and independent calculation.

The results of experiments in the two iron vessels will be found in Table IX. A comparison furnished by two experiments in each vessel, gives by calculation on the principle used in the case of the glass containers the specific heat of the Russian sheet iron, of which they are composed = .101714.

Results of Experiments on Specific Heats.—When it is considered that numerous causes interfere with the operations on specific heats, it cannot be expected that one, or a few trials, should be deemed sufficient to settle so difficult and intricate a question. For this reason the committee preferred the method of multiplying and varying the trials, and making a deduction from the mean results, in order to verify the general efficacy of the standard piece, in producing vapour.

1. The first part of the *preliminary series* (table IV.) indicates the effect of radiation from surrounding objects in the apartment to the water-vessel. The 13 experiments constituting this part of the table, exhibit a mean result of .123004 as the specific heat of iron.

2. The second part of the same series in which the cylinder B. was employed, indicates a decided effect from that precaution, and gives as a mean result .11294, for the specific heat.

3. Experiments No. 16, 20, 23, 25 and 26, the greatest number of comparable results in this part of the series, (differing only in the fourth place of decimals,) gives a mean of .11346.

4. In table V., where the thicker of the two glass cylinders of the same capacity was used, we have the mean result of the whole 14 experiments .11288.

5. The four experiments No. 2, 5, 6, and 9, which are the greatest number that conform to the third place, give a mean result of .11349.

6. The 10 experiments in table VI., made in the thin cylinder of the same capacity as the foregoing, give the specific heat = .11308.

---

* The formula in this case gives the specific heat of iron $z = \dfrac{T(w+e)}{it-i'T}$ where T is the temperature gained by the water, w = the weight of water in grains, e = the equivalent of the thermometer in grains of water; t = the temperature lost by the standard piece; i = the weight of the standard piece in grains, and i' = the weight of the sheet iron containing vessel. See *Am. Jour. of Sci.* vol. 27, p. 277.

7. Experiments No. 2, 5, 6 and 7, the greatest number of those which may be regarded as conformable to the third place of decimals, give a mean of .11361.

8. Table VII. contains 3 experiments made in each of the two vessels used in that series, which were performed under a cone of tinned iron to defend the water-vessel from radiated heat, but as it was set loosely on the table which supported the container, it did not prevent the motion of air around the latter, and as the experiments made in this manner terminated from three to four degrees above the temperature of the room, there is reason to suppose that the results of those 6 experiments are all below the truth. Taking then the other six of this table, which were made with the same precautions as those in table V. and VI., we have as the mean result in the thicker glass .112952; and that in the thinner .113631.

9. The two experiments which conform entirely with each other, for the thicker vessel, give the specific heat .113498, and the two for the thinner .113489.

10. The mean of all the results, including both those obtained with the cone, and those with the cylinder of tinned iron, to defend the water vessel, give a mean result of .112350, and the six rejected experiments taken by themselves .111511.

11. In the thinner copper vessel, the trials as recorded in table VIII, exhibit the mean of seven results equal to .115752.

12. Rejecting those which began and ended too low, and hence gained heat from the air, as well as from the iron, we have in the thinner vessel .112577 as the mean of four experiments which are considered comparable.

13. In the same table eleven experiments in the thicker vessel indicate a mean of .114900.

14. With the same jar, seven experiments which are considered comparable, give a result equal to .113261.

15. In the thick sheet iron cylinder, weighing 5167 grains, we find by table IX., that the mean of five trials gave a result = .113253.

16. In the same vessel, three experiments which differ only in the 4th place of decimals give a mean specific heat = .113622.

17. In the thinner sheet iron jar weighing 1733 grains, nine trials gave a mean result of .112972.

18. Three experiments in this vessel which differ only in the fourth place of decimals, give a mean of .113365.

The following table embraces a synoptical view of the experiments on specific heat thus far detailed.

TABLE X.

No. of the comparison.	No. of the table referred to.	Kind of containing vessel used.	Weight of the vessel in grs.	No. of experiments compared for the general mean.	Mean of specific heats from a comparison of all the trials.	No. of comparable experiments differing in the fourth place of decimals.	Mean specific heat by the comparable results.	No. of rejected experiments.	Mean specific heat by the rejected experiments.
1	IV.	Thin glass.	3325.	13	.112940	5	.113460	13	.123004
2	V.	Thick glass.	12272	14	.112880	4	.113490		
3	VI.	Thin, containing same as preceding.	2996.	10	.113080	4	.113610		
4	VII.	Thick small glass.	6923.	3	.112952	2	.113498	6	.111511
5	VII.	Thin small glass.	2465.	3	.112931	2	.113489		
6	VIII.	Thick copper.	19738	11	.114990	7	.113261		
7	VIII.	Thin copper.	5178.	7	.115752	4	.112577		
8	IX.	Thick iron.	5167.	5	.113953	3	.113622		
9	IX.	Thin iron.	1733.	9	.113972	3	.113365		
				75	.113716	34	.113374	19	.117257

It hence appears that by a mean of 34 out of seventy-five experiments in nine different vessels, with five different liquid capacities, and composed of three different kinds of materials, we obtain a result not sensibly varying from .1134, as the specific heat of the iron standard piece between ordinary temperatures and 212° Fahrenheit.

Hence it appears that by a mean of 34 out of seventy-five experiments in nine different vessels, with five different liquid capacities, and composed of three different kinds of materials, we obtain a result not sensibly varying from .1134, as the specific heat of the iron standard piece between ordinary temperatures and 212° Fahrenheit.

Plate VIII.

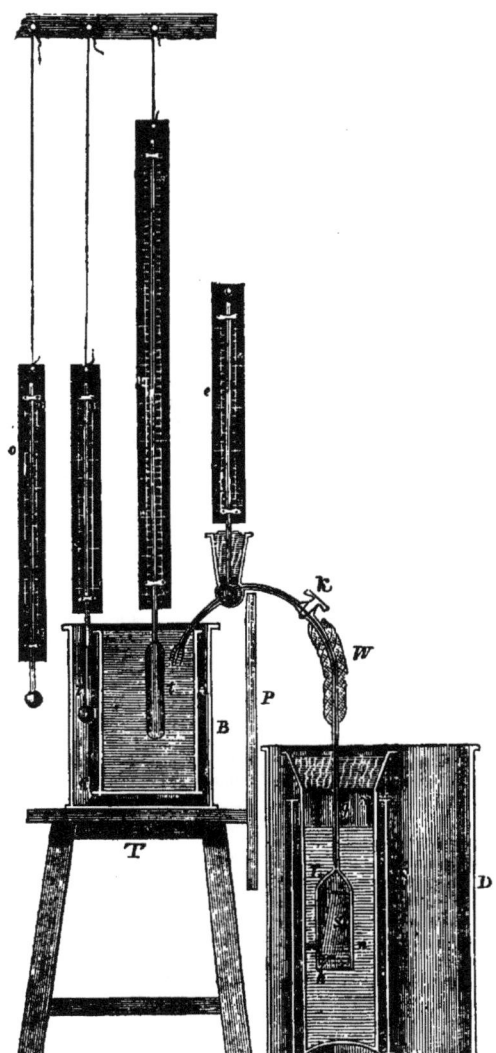

The next step in this investigation again required the use of the copper cylinders, but instead of the *heating apparatus* W being filled with water, it was made to contain mercury, which allowed a higher temperature to be given to the vessel M. This latter vessel was now filled with melted tin, instead of mercury, as well to avoid the inconvenience from mercurial fumes, as to obtain the specific heat of iron at a second fixed temperature, the melting point of tin. It needs scarcely be mentioned, that the same degree of exactness in the accordance of experiments at temperatures above 400°, as at 212°, is hardly to be expected. Table XI. (*Frank. Jour.*, vol. 19, p. 158) exhibits a number of trials made in the manner just pointed out.* A certain amount of error may possibly have been introduced into these experiments by the want of uniformity throughout the mass of melted tin; for, after withdrawing the standard piece and lowering the thermometer to the bottom of M, it was found that a difference of a few degrees, was a possible occurrence; but as the bulb of the mercurial thermometer, which marked the temperature of the melted tin, was generally kept at the same level with the centre of the standard piece, any difference between the two, must be trifling in amount.

The last arrangement for demonstrating the specific heat of iron, was in the nature of a verification of the methods already detailed, by means of a direct application of the standard piece to the purpose for which it is ordinarily employed—that of generating vapour instead of heating water.

It was, for this purpose, heated as before described, in the bath of melted tin to such temperatures, above 212°, as were deemed necessary, and immediately plunged into boiling water. The *effect* produced, was now ascertained by multiplying together the weight of vapour generated, and the latent heat of steam; while the *cause* was found in the weight of iron, its specific heat and the temperature which it expended. The shield to defend the iron in transitu was employed, and the other precautions to avoid error were still persevered in. The results will be found in table XIII. (*Frank, Jour.*, vol. 19, p. 165.)

Before proceeding, however, with the detail of those trials, it is necessary to state the mode of ascertaining the latent heat of the vapour of water, which enters as an essential element into the calculations of that table. It will be perceived that the prin-

ciple of the method is similar to that of Count Rumford.

Apparatus for the latent heat of vapour.—The apparatus by which the latent heat of vapour was examined, is represented in Plate 8, in which A is a cylindrical vessel to contain water; B a larger cylinder formed of pasteboard, higher than the preceding, surrounding and defending it from the air; C is a stand of charcoal on which the vessel rests; D is a vessel of tinned iron, 14 inches high by 9 in diameter, to prevent the vessel F (the same which has already been described as the boiler of the steam pyrometer,) from affecting by radiation the temperature of A. P is a sheet of tinned iron, attached in a vertical position to the edge of the table T, serving still further to defend A from the influence of radiation from the boiler or steam pipe. S is a cylindrical piece of cast iron, having round its lower base a ridge r, adapted to retain a hold of the small hook h, within the copper case L, intended to receive it when hot. From the upper conical part of this case rises a pipe g, quarter of an inch in diameter, curved into a semi-circle at the top to dip under water. Within the curved part is a stopcock K, adapted to regulate, or entirely to prevent when required, the flow of steam from L. At q is an enlargement of the pipe g, with a funnel-shaped tube to receive the bulb of a thermometer e, sustained and made tight by packing around the lower part of its stem. The purpose of this thermometer is to mark the temperature of the effluent steam. w is a handle, formed by a number of folds of flannel made fast to the pipe. x is a thick roll of cloth surrounding g, and preventing the escape of vapour exterior to the tube. The thermometer o marked the temperature of the apartment in the immediate vicinity of the apparatus. i gave that within the pasteboard case, while t gave the temperature of the water in A.

(To be continued.)

ROYAL INSTITUTION. — MR. BRAYLEY'S LECTURE ON THE EQUILIBRIUM OF THE ATMOSPHERE.

March 8. — *The Conversazione.*—Mr. Read exhibited what we think to be a very important application of his patent instrument, commonly called the " stomach pump," namely, to the restoration of suspended animation. Besides extracting noxious gases in other cases, and the naturally vitiated state of the air in the lungs of drowned persons, the apparatus is competent to keep up an artificial breathing, the supply of air being both warm and pure, and rendered more exciting if thought proper

---

* It will be evident on a comparison of this table with those which have preceded, that the general law observed by Petit and Dulong (*Ann. de Chim. et de Phys.*, vol. VII.) of an increase of specific heat by increase of temperature, when the method of heating water is employed, is confirmed by these results; but experiments on the production of vapour hereafter given, exhibit a very striking conformity, in regard to specific heat, with those made below 212 deg.

by a mixture of oxygen gas with it. The former particular we esteem to be of great importance, for cold being perhaps the most powerful agent in destroying life, it cannot be less necessary to communicate warmth to the interior, than by baths and other means, to the surface of the body. The Royal Humane Society will assuredly hasten to add this to their other means of recovery. We would advise Mr. Read to furnish with the apparatus, air tight bags, which being applied to the extremities, and the air exhausted by the same instrument, would tend to relieve congestion of blood in the head and other parts, and facilitate its circulation.

Among other interesting articles on the table, was a section of a nettle, discovered by Mr. Cunningham, the Australian traveller, at Morton Bay, north of Sydney, growing 100 feet high, and having leaves (some of which were exhibited) at least 8 inches by 9 inches. It will create, therefore, the less surprise, when it is stated that the section was not less, by our estimation, than 18 inches in diameter. It was sent by Mr. Lambert. Mr. Hemming forwarded some photogenic drawings, with the engravings from which they were copied; but neither these specimens, nor those exhibited on a former occassion by Mr. Talbot, and which Sir Anthony Carlisle has by an oversight attributed. in No. 809, to M. Daguerre, are not worth for a moment the attention of the artist, for, besides other defects, the lights and shades are reversed. Still the subject is curious, and to a trifling extent may be useful even at the present stage of its progress, where however it will not stop. The French, indeed, can show something more deserving of notice than these specimens, and what we conceive is totally different, both in the effect produced and in the process itself, except so far as the light of the sun is concerned.

The Lecture.—Mr. Brayley's subject was '' The equilibrium of the atmosphere, as dependant on the united action of gravity and temperature.'' The purport of this lecture was the introduction of some speculations concerning the mechanical constitution of the atmosphere. The Lecturer first imparted the usual information respecting its weight, its elasticity, the law of its pressure and density, its sensible height, as determined by different methods, and the opinions which have been held at various times as to its limits, until they pretty generally accorded with that inferred from the researches of Dr. Wollaston, of its being impossible that the rarity of the air can exceed that point at which the repulsive force between its particles becomes less than the force of gravitation. He then adverted to

M. Poisson's hypothesis, according to which, the limit of the atmosphere, instead of thus being one of almost insensible gradation, is abrupt and well-defined, through a process in the upper regions of the air, no less singular than that of its conversion by cold into a *liquid* or even a *solid*. Without taking this extreme view of the subject, nor yet controverting it, the Lecturer contended that, admitting that extra mundane space to be colder than the mean temperature of the air, it must follow that, at a certain point in the rarity, or in the height of the atmosphere, an inverse order of its density commences; so that beyond the region of greatest rarity (greatest under the supposed conditions), another region exists, in which the density *increases* in some ratio with the altitude, but whether it terminates in the liquidity of the air, it was no part of his inquiry to determine.

He attributed this effect to the extreme cold which is supposed to reign in space, and advanced in confirmation of his views, the singular vaccillating appearance which is observed in the progress of the oscultation of fixed stars by the moon; that is, supposing the moon to have an atmosphere, and that it is similarly constituted. But what the Lecturer depended on principally for argument and illustration, is the remarkable deficiency of luminosity in a middle region of the atmosphere of comets, and which he had the authority of Sir John Herschell to attribute to the dissipation of its vapours by heat, and the luminous appearance of the outer region or coma, to the condensation of those vapours in an exterior sphere of greater cold. This he imagined to be a case in point, or analogous at least to his supposed constitution of the earth's atmosphere, the different nature of the aerial fluids being matter of no consequence. The only experimental illustration of his theory that he advanced, was the condensation of the attenuated vapour in Dr. Wollaston's cryophorous by the application of cold. This he gave as an instance of gradation in density, of a column of vapour as it approached the source of cold and became converted into water; and adduced the fact of liquefaction of the vapour, as a proof, as we understood him, of the high degree of density of the immediately previous and proximate state of the vapour.

Remarks.—M. Poisson, like a true mathematician, supports his hypothesis of the liquid state of the air at the extreme limit of the atmosphere, by a mere analytic investigation, resting on assumed atomic and other data, particularly that of the extreme cold existing in universal space—but which by the way is opposed to the deductions of M. Fourrier, also mathematically derived—

and not only does he thus support it, without any reference to physical considerations and probabilities, and the recognised methods of legitimate induction, but in direct opposition to phenomena, particularly those which belong to optics. But Mr. Brayley, in behalf of his modification of, or rather addition to, Poisson's theory, argued the point on better grounds, and we will add in a more philosophical manner, by taking the principles and appearances appertaining to natural philosophy for his guide, though we greatly fear that they will by no means bear him out in his views. We have, however just given him credit, in one particular, for more than he really performed; he having strangely omitted the argument, and advanced only the evidence he was able to collect for the inference from it, that cold produces an increasing density in the upper regions of the atmosphere. We must also say, that of one portion of this evidence relating to the occultation of stars, it cannot be allowed him, if what astronomers inform us be true, that the singular appearance attending it, is not common to all the stars. And in regard to the supposed constitution of the atmospheres of comets, it scarcely amounts to analogical evidence, for their vapours are imagined to be condensed in the sense of incipient liquefaction by the operation of cold, and not brought into a state of greater density as an elastic fluid. The nature of the phenomenon, if it be such as Sir John Herschell supposes, is corroborative rather of Poisson's than of Mr. Brayley's theory, allowing only for that immense distance of separation between the particles of condensed vapour, which its extreme tenuity denotes, and which prevents the formation of anything more approaching to a liquid than a fog. But Mr. Brayley may say that a state of greater density in the elastic fluid must necessarily precede its condensation into liquescent or visible vapour. However this may be, the appearance alluded to as belonging to comets, gives evidence only in favour of the former fact, although it must have been in reference to the latter that Mr. Brayley could be supposed to have cited it. But what proof is there of the necessity just referred to? Does Mr. Brayley think the fact to be so established by the experiment with the cryophorous as to allow him to dispense with argument? It is scarcely necessary to say, and certainly it is not necessary to prove, that his statement of a difference in the density of the vapour in this instrument, being produced in proportion to its proximity to the source of cold, is by no means correct. The formation of water, to

which he pointed in proof of his assertion, was an inapposite fact, for the vapour was already of the greatest density consistent with its temperature, and on the application of cold, it collapsed at once into the state of water, without passing through any intermediate stages of density. Of course variations can be effected in the density of aqueous vapour, as well as in gases, and simply by the operation of difference of temperature, but it must be under other circumstances than those which were given. Cold, although spoken of in positive terms that are too apt to mislead, is not a positive but a negative cause, and cannot of itself produce an increase of density in air or vapour. It is an abstraction, or it is a diminution of repulsive force, but there the matter ends and the particles will remain as distinctly separate as before, unless there is an approximating force present, or elicited, to take advantage of the circumstance. But our limits are exceeded—still we must find room to say, that the lecture was very well got up.

NOTES AND NOTICES.

Launch of the Steamer Nicholai.—The ceremony of launching a splendid steam-vessel, named the *Nicholai*, took place on Saturday, from Gordon's Dock-yard, Deptford. She is the largest steamer belonging to Russia, and is intended to ply as a packet between Lubeck and St. Petersburgh. The *Nicholai* was built by Mr. Taylor, in the incredible short period of four months from the time of laying the keel, after the designs of Mr. Carr, of the firm of Ritherden and Carr, surveyors to the Hon. the East India Company. After being launched the vessel was immediately towed across the river to the establishment of Messrs. Seaward and Capel, of the Canal Iron Works, Limehouse, the firm to whom, by the special order of the Emperor, is entrusted the execution of the steam machinery. This machinery is to be precisely similar to that adopted on board her Majesty's steam-frigate the Gorgon, which has been so much approved of; and with a speed proportioned to that used in the construction of the wood work, the whole will be fitted up and the vessel entirely ready for sea in one month from the date of the launch. The engines will be of 240-horses' power, and the peculiar merits of the mode of construction adopted by the eminent engineers referred to are such that the machinery is greatly simplified, and the space which it occupies proportionably diminished. Thus the length of the engine-room in the *Nicholai* is only 45 feet, whereas on the ordinary plan it would exceed 62 feet. The advantage of having so much space in the most superior part of the vessel cannot be too highly estimated. This will be immediately evident form the fact that the *Nicholai*, of 800 tons burden, will carry 150 passengers—a number equal to that of the Great Western, of 1,400 tons. The success of the Gorgon, just returned from a six months' cruise on the coast of Spain, has been so pre-eminent that the Lords of the Admiralty have given orders for five more pair of engines on the same principle, to be fitted into five new frigates. One of these (the Cyclops) will be of 1,300 tons burden, with engines of 320 horses power. She will carry 20 guns, and be the largest man-of-war steamer in the world.

LONDON: Printed and Published for the Proprietor, by W. A. Robertson, at the Mechanics' Magazine Office, No. 6, Peterborough-court, Fleet-street.—Sold by A. & W. Galignani, Rue Vivienne, Paris.

Mechanics' Magazine,

MUSEUM, REGISTER, JOURNAL, AND GAZETTE.

No. 816.]   SATURDAY, MARCH 30, 1839.   [Price 6*d.*

Printed and Published for the Proprietor, by W. A. Robertson, No. 6, Peterborough-court, Fleet-street.

SPIRAL-SCREW WATER MILL ON THE
MISSISSIPPI, NORTH AMERICA.

Sir,—I have been kindly favoured with
the annexed description by a gentleman
who possesses the rare qualities of a fondness for mechanism, and a determination
of keeping his eyes open when travelling.
The combination of ingenuity with economy, is a striking feature in this machine, and in the American character generally. With respect to its power, that
is still to be learnt, and we in this country are very sceptical with respect to the
efficiency of these screws as a moving
power used for propelling by steam, for
in no other shape have we seen it. In
this application of it, however, there appears scarcely any limits to its power
by adding to the number of the spirals
as far as the river will admit. The application is so novel that you will, no
doubt, readily give it publicity The
following is an extract from the notes
accompanying it:—

"'On passing the small town of Cape Girardeau (about 45 miles above the junction
of the river Ohio), two grist mills attracted
my attention as being of an entirely different
construction from what I had ever seen before. A part of the mill is erected upon logs
over the water, in which a cog-wheel at an
angle of 45°, which works the mill, is connected by several long and stout poles, fastened together by joints, that stretch down
for some distance into the current of the
river, where they are attached to a sort of
long spiral wheel upon the principle of a
patent cork-screw. This simple piece of
mechanism is made of oak, and floats upon
the surface of the water, about half immersed in it, or rather more, and the force
of the current between the spirals gives it a
rotary motion, and that motion, by means
of the connecting rods, sets to work the mill
above. It is quite a new invention (12th
August, 1833,) and has a very singular appearance in passing down the river; but I
understand that it grinds rather slowly,
though this, of course, must depend much
upon the strength of the current on which
it is erected, as well as upon the depth of
the water, to allow of a considerable diameter to the spiral wheel. The method,
however, is simple and cheap, and its unusual appearance interested me much."

The same sort of machine is used
for towing or warping vessels out
of the Mauritius harbour against the
trade winds—sometimes by the force of
the current, and sometimes worked by
men with a windlass or handle in the
boat.

The town of Cape Girardeau is, it appears, a French settlement, and therefore, from a similar machine used in the
Mauritius, perhaps the hint was borrowed.

I am yours respectfully,

H. A. M.

Bristol, Feb. 25, 1839.

THE LONDON FIREMEN DEFENDED
FROM THE UNJUST ASPERSIONS OF
"AQUARIUS."

"The Fire King sat on his throne of flame,
  And fever'd he looked, and felt ;
And his colour went, and his colour came,
  As his Courtiers before him knelt.

"His passion got up two hundred degrees,
  As his sword through the air he sawed ;
And those who with fear were all of a freeze,
  With perspiration thawed.

"'Ye shivering dogs !' cried the King of Fire,
  And he thrust his sword in its sheath—
"'Who is this *Braidwood* that stirs my ire ?
  Now answer, on pain of death !'

"'I'll do for that Braidwood !' then said he,
  'And the whole of the Brigadier brood ;
I'll make them respect my dignity,
  As every fire-man should !'"

Sir,—Your testy correspondent, who
endeavours to conceal the fiery fierce-
ness of his wrath under the cool and
watery signature of "Aquarius," has at
page 404, vented his annual discharge
of spleen, upon my "Report of London
Fires for 1838," published in your 809th
number. Like the implacable hatred of
Hannibal to the Romans, is that of
"Aquarius" to the London Fire-brigade,
and the slightest allusion to any of their
meritorious exertions throws him into
strong hysterics; in one of his recent
paroxysms he has evidently penned the
communication, which graces page 404
of your 813th number.

At the beginning of last year, "Aquarius" endeavoured, by twisting of some
facts—by colouring of others, and by
misrepresenting of many more—to make
out a charge of inconsistency against me,
and of misconduct, or at least, mishap,
against the Fire-brigade. I considered
that letter altogether beneath notice, nor
is the present more deserving a reply,
but that some few of your readers not
conversant with these matters, may perchance, imagine there is some slight

foundation for the view which "Aquarius" is pleased to take of the subject.

To prevent wrong impressions, therefore, I think it right to state, in the first place, that the comparisons made by "Aquarius," and upon which the whole of his arguments are founded, viz., between the last year of the old system, (that is of separate bodies of fire-men acting independently and regardless of each other,) and the results of the *improved system*, characterized as it is by concert and harmonious co-operation; these comparisons, I say, are unjust. "Comparisons are odious," but especially so when founded on false premises.

It may, perhaps, be in the recollection of most of your readers, that I commenced my annual report of London fires some time antecedent to the formation of the present Fire-Establishment; and that I have more than once explained that my source of information was then limited to the experience of *one* fire-office only. Now it must be pretty well known, that under the old *regime*, no one fire-office in London was in possession of information respecting *all* the fires that occurred, nor of any thing like it; in fact, out of my own observations and attendance, I was frequently enabled to add to the list. Now, if we just compare this state of things, with the present very complete arrangements, by means of which every alarm of fire throughout the whole extent of this vast metropolis, be the result what it may, is reported twice a day to the head-quarters of the establishment, this alone would account for a vast discrepancy in point of numbers. I should like to know what sort of a fire-report I could have drawn up for 1838, had my sources of information been confined to the *practical experience* of the *County Fire-Office*, which continues to jog on in the *unimproved* state? I can vouch that it would have looked somewhat more companionable with its like of 1832, than did the voluminous records afforded by the comprehensive knowledge of the better plan. In the next place, the classification of the damages is now somewhat different to the former practice, making an apparent increase of *serious fires*, when the fact is, the per centage of really serious fires is very considerably reduced. Again, under the old system, fatal fires were seldom or ever reported. Under these circum-

stances, therefore, the disparaging comparisons instituted by "Aquarius," and so boastingly brought forward continually, are founded (as he well knows) on erroneous data, and are altogether false in the inferences attempted to be drawn from them.

It must be palpably evident, that the fire-men cannot prevent fires taking place, and that the number of these accidents may be expected annually to increase, with the spread of population and of buildings. Neither is it in the power of fire-men under all circumstances, to prevent the occurrence of serious fires, because it continually happens, that fires have really become what is *now* called *serious*, before they are discovered; in fact, I have already mentioned several instances in which *total losses* have actually taken place before the fire-men have been called out; and because the results are honestly reported, are these men to be censured by such hypercritical observers as "Aquarius"? This writer says, "the testing point of the fire-men's success is the number of fires that are prevented from spreading to a serious extent *after they are discovered*." This list, then, in strictness, would include *all* the *slightly damaged*, and very many of the *seriously damaged*, with a tolerable share of the *total losses!*"

It is not, however, by a dry enumeration of number alone, that any correct idea can be formed of the good or ill success of the fire-men's exertions. The ever varying circumstances of each particular case, require to be known and duly weighed, before any satisfactory conclusions can be drawn from them.

In the truly appalling list of last year's serious fires, I have entered into the details of most of them as far as was consistent with the limits of your work; and I would ask any intelligent and impartial reader, if there was a single instance in which blame would seem to attach to the fire-men. Take for example, the destruction of the Royal Exchange; there the flames had attained a most serious ascendancy before the fire was discovered—great difficulty arose in obtaining access to the building—water was with difficulty obtained, from the intense frost which prevailed—the building was filled throughout with timber galleries and passages most intimately connected—and yet some portions of the building were preserved, and the fire most effectually prevented from extending its ravages to the adjoining buildings.* Look again at the burning of Messrs. Edgington's tarpaulin manufactory in the Kent Road; an extensive range of wooden buildings filled with the very daintiest food for fire; the first intelligence given to the fire-men was by the illuminated horizon, the flames bursting forth all at once with terrific fury—on reaching the spot, no water was obtained for upwards of an hour, in fact, not until the whole of the premises were burned to the ground.

At the Temple fire again, intelligence was most reprehensibly with-held from the fire-men till the flames had gained an ascendancy that for a time defied suppression, and only a scanty supply of water at hand for the purpose. In this way I could go on, had I no respect for the patience of your readers, through the whole of the serious fires in my obnoxious list, but these instances will sufficiently explain my reasoning.

So long as the public continue to manifest a worse than apathetic indifference to the *prevention* of fires, conflagrations must and will increase; and so long as the public persist in disregarding all precautionary measures, and continue to set at defiance, all wholesome provisions, and legislative enactments for limiting the damages of fire, conflagrations may be expected to be more and more extensive in their ravages.

Look at Fenning's wharf and warehouses, the site of one of the most tremendous fires of modern date, the destruction of which was solely attributable to the illegal dimensions of the buildings; and yet, in defiance of law, and in spite of the dictates of reason and of common sense, a new pile has risen, like a phœnix from the ashes, as unlawful in its undivided dimensions as the former. In the event of another fire breaking out, the same consequences as before seem inevitable—the entire destruction of the whole!

The vast warehouses of St. Katharin'e Docks, present another immense pile, the preservation of which, if once well on fire, seems scarcely within the possibility of man, and yet if any of these consequences should follow, what a precious outcry would be raised by "Aquarius" against the fire-brigade, forsooth!

It was an intimate practical acquaintance with these and similar co-operating circumstances that led me to assert the greater liability to serious and extensive fires of *London*, as compared with *Paris;* this argument "Aquarius" has attempted to hold up to ridicule, by a comparison as invidious as it is unjust. A greater piece of injustice can hardly be conceived, than to take isolated statements of facts, brought forward upon different occasions, to illustrate different arguments, and to place them in juxta-position, without the explanation required to reconcile their seeming contradiction. This, however, is the unworthy artifice resorted to by "Aquarius," and in this way only, does he attempt to support his jaundiced view of these matters.

In conclusion, with respect to *fatal fires,* I most expressly stated, that during last year, *no life had been lost after the arrival of the fire-men;* a thing of too frequent occurrence under what "Aquarius" himself designates the *unimproved system.*

The fire-brigade, therefore, are wholly free from blame—provided their attendance has been sufficiently prompt after being called. Now, Sir, I assert without fear of contradiction, for it does not admit of contradiction, that the attendance of the fire-men was not under any circumstances near so quick formerly as it is now. The location of the men at their respective engine-stations—a sufficient force constantly on duty day and night—with the admirable system of calls, ensures a rapid attendance of a larger force than was ever witnessed prior to the formation of the London fire-establishment. The constant effect of this plan has been in most cases such a timely attendance of men and engines as to confine to the list of "slightly damaged" accidents, which, under the old system, would inevitably have turned out "serious fires," or even "total losses."

In addition to the benefits thus conferred upon the community, I consider we are under considerable obligations to the Committee of Management of the London Fire-establishment for the handsome and liberal manner in which the valuable information by them collected,

* It is right to observe, that a portion of the fire-brigade were at the same time occupied with a serious fire at *Chelsea.*

is always made accessible to persons in any way connected with the public press, or who take an interest in these questions. The readiness with which this is afforded, is, I am aware, extremely annoying to certain narrow-minded persons, who therefore take every possible opportunity of abusing, and underrating the advantages of this excellent establishment.

Although I have endeavoured *briefly* to reply to the remarks which provoked this communication, I find I have greatly exceeded due limits, I must therefore conclude, and remain,

Sir, yours respectfully,
Wm. Baddeley.
London, March 15, 1839.

PREVENTION OF RAILWAY ACCIDENTS

Sir,—When we take into consideration the vast sacrifice of human life from accidents on the various railroads, it is amazing, that in a country so prolific of improvements of every kind, no one has undertaken to produce plans for obviating these calamities. Surely means might be adopted, for at least *lessening the chances* of such accidents. It is a disgrace to an enlightened nation like Great Britain, that men of capital should seem careful for nothing beyond acquiring the best returns on money laid out on speculation. A man should unquestionably make the most of his employed capital; but, he should not at the same time shut out those philanthropic feelings which we should all exhibit towards our fellow men. Let, therefore, the various companies set about devising the most efficient means for averting the horribly destructive accidents that almost daily occur on the various railroads.*

In the humble hope of drawing more efficient pens to the subject, I now offer the following hints :—

1. When a person is in danger of being run over by a train, if he had the presence of mind to get into *either of the outside spaces it would be safer* than in *the middle one* (a presumed reason for which will be seen in the 5th article.)

2. Whether the individual be in *the middle space*, or in *either of the lines*, he should immediately *fall flat on the ground;* it would, in the former case prevent the commotion of air, consequent on the rapid motion of the train, from drawing any portion of his dress towards the carriages; and in the latter case, the entire train would roll over him without doing him the slightest injury, as was exemplified in the case of a Pole, an officer on the Great Western Railroad, who, some time ago, escaped unhurt, with the exception of a hot cinder falling from the furnace and slightly scorching his face: it is consequently *preferable to lie on the face, the hat* should also be *thrown off*, as there might otherwise be a chance of it coming in contact with some projecting point of the train.

3. It is therefore *safest*, when the individual cannot get to either of the *outside spaces* (which is undoubtedly the best) to throw himself *flat on the ground* in the *middle space*, or in *either of the lines*.

4. It is consequently an obvious duty which the various companies owe to humanity, to have their carriages so constructed, that there would be sufficient space from the bottom of each (including cross beam or iron work) to the bottom or bed of the lines that a man of the largest dimensions might lie there unhurt.

5. The *middle space is particularly unsafe* for any one to *stand* on, when there are two trains going in contrary directions and passing each other at the same instant : in proof of which, a poor man about a fortnight ago, going to his daily labour, and having to cross the railroad at Kinton near Harrow, whilst a train was approaching from the Euston Square Terminus, instantly ran to *the middle space*, thinking no doubt, that *there* he would be perfectly safe—but another train at almost the same instant of time coming up in a contrary direction, caused such a *commotion of air*, first from being agitated by the one train from east to west, and next this agitated air being met by the other train going from west to east, that the poor man must have been, as it were, in the midst of a powerful *whirlwind* and entirely under its im-

---

* These introductory remarks of our correspondent might well have been spared. It has been proved by incontestable evidence, that travelling by railway is incomparably safer than by any other means of conveyance. Take any number of passengers travelling a given number of miles by railway—and a like number of passengers travelling a like number of miles by horse-coach—and it will be found that the proportion of accidents is fifty to one in favour of the former.—Ed. M. M.

pulse: we may, therefore, without hesitation, come to the conclusion, that his dress must have been blown about in every direction, and consequently come in contact with one of the carriages—thus drawing him towards inevitable destruction. The *commotion of air* here hinted at, may be supposed *hypothetical;* but let a reflecting mind pause before it comes to this conclusion; let a rational being ask himself what effect a body, of the weight and magnitude of an ordinary train rushing through the air, at the rate, let us say, of 30 miles an hour, or 14⅔ yards in one second: and another similar body passing by in an opposite direction at the same velocity, through a fluid so subtile as atmospheric air is known to be, and he cannot but be convinced that the agitation must indeed be terrific: hence the poor fellow lost his life. I would say then—*avoid the middle space by all means,* or, *if you prefer it—fall flat upon the ground with the face downwards.*

6. Might not some simple contrivance, say of the form of an arc of a circle, or that of the fin of a fish with a spring attached to it, be so placed on either side of *the tender or first carriage,* as to throw off to the right or left any body that might accidentally come in contact with the same?

7. There ought to be the greatest possible vigilance enforced on those officers whose duty it is to see that no impediments be permitted to the free ingress and egress of the trains; and upon no account whatever should their attention be directed to any object unconnected with their duty, more especially when there is a train either on the point of starting, or when near any of the places where they stop. The following occurrence shows the listlessness—I might say heartlessness of one of those men. "Lord Lichfield and three of his friends were nearly killed about a month ago: *the hour* of the Manchester train *was changed without any notice,* which is a very common occurrence; and when his Lordship got to Birmingham it was gone. His Lordship felt very much annoyed, as his Royal Highness the Duke of Sussex and a large party were coming to dine at his seat, Ranton Abbey; he asked whether he could have an engine for himself and party, which was immediately provided.

Away they went at great speed, but, owing to the *negligence of a policeman* in not turning a plate they were carried with great force off *the line* and upset into a pit. They were all much injured."

8. In tunnel transits I would recommend fire works (say a small Catherine wheel) to be placed on the first carriage, to be lighted at the moment of entry, and so constructed as to burn during the entire transit; that the lights should likewise be of one colour for the left-hand line, and of a different colour for the right-hand one: this would to a certainty be the means of preventing such serious accidents as that which happened at the tunnel running from Chalk Farm to the vicinity of Kilburn, when Pickford's train and another came so furiously in contact as to demolish some of the carriages, besides seriously injuring many and alarming all of the passengers.

9. In conclusion. If methods such as I have now stated, or others that may be more efficient be not adopted, the public will, or at least ought, to demand, I say *emphatically, demand,* that *the speed be lessened.* Let the companies look to this; it would certainly not, in a pecuniary point of view, be to their interest to lessen the motion—let them therefore apply other remedies.

I am, Sir,
Yours very truly,
WILLIAM RUSSELL.
Edgeware Road, March, 11, 1839.

IMPROVEMENTS IN PHOTOGENIC DRAWING—THE CAMERA OBSCURA.

Sir,—In order to obtain real images from the sun's beams, concentrated by refraction through a convex lens, into a dark chamber or camera obscura, we must receive them on a paper prepared for the purpose, whose colour is dark, and whose material is of a nature to suffer absorption from the rays of light, in a just proportion to their intensity. Now all the methods of producing pictures, hitherto promulgated in England, are the reverse of this, as nitrate of silver, or the lunar caustic dissolved, gives shadows for lights; we must, therefore, seek some medium of a different nature, such as stains from vegetable bodies, or coloured pigments ground up with volatile essences, the shadows from which, after the lights have been created by absorption, may be fixed by re-agents; such a sensitive paper would secure a permanent design from nature by means of the sun's reflected rays; for we know their power in extracting stains of smoke or dirt from ancient engravings when placed under water about a quarter of an inch, so as to be acted on at noon by refraction; hence, I conclude, that when our sympathetic paper receives the rays of light in a moist state, the dark colouring matter will be more readily discharged. Observe, also, that the camera must be without a mirror, and the lens a compound one, to correct distortion. My camera, or box, is constructed as follows:—

The compound lens is about one inch and a half in diameter, composed of two plane convex lenses nearly touching each other by their convex sides; its focus is two inches and a quarter, and it is inserted into a box four inches square, whose opposite end is open, and cut off exactly at the focal point of the image; to the open end of the box there is a cover, like that of a canister, whose sides are half an inch deep; to the bottom of this cover I affix the sensitive paper, so that when it is put on, it exactly comes up to the focal point of the lens, which also has an outside cover, only to be taken off when presented to the object in sunshine intended to be represented: in order to examine the progress made by the influx of light, the cover of the larger end may be removed from time to time without altering the position of the camera, and quickly replaced, until satisfied with the effect; and thus, having your sympathetic papers all of the size of the inside of the cover, you may change them as often as you please, in trying your experiments; finally, carry them home in the closed camera, with both ends shut, whose interior, I need scarcely add, should be carefully blackened. But, after all, until we shall have discovered a dark pigment that can *quickly* be absorbed by light, concentrated from the reflection of natural bodies, we shall not have arrived at that perfection now sought after, in consequence of the reports of M. Daguerre's extraordinary success.

The liberal disclosures of Mr. Talbot have set many people to work here, in copying lace, prints, and leaves, by means of the direct rays of the sun, through the glass of windows, and his

sensitive paper; this may be serviceable in mechanic arts, and is very amusing, for they are so speedily effected that the earth's motion can never disturb them. But, before we can get real photogenic pictures from reflection, we shall have numerous disappointments to encounter, which time and patience alone can conquer, and, therefore, the more hands are employed the better, and such portable machinery as I here recommend.

I am, &c.,

G. CUMBERLAND, Sen.

Bristol, March 15, 1839.

MR. OXLEY'S PROCESSES FOR PREPARING PHOTOGENIC DRAWING PAPER, &c.

Sir,—In my communication of the 2d February, 1839, which appeared in No. 809 of your valuable publication, I wrote in haste and from memory only, and therein very briefly pointed out how the paper might be rendered susceptible of being acted upon by the rays of light, and how the impressions produced thereby might be permanently fixed thereon. I will now, by your permission, transcribe the results of some of my experiments, which will enable any persons desirous of photogenic drawings to prepare their own paper by such methods as will answer their wishes in the most satisfactory manner.

First Way.—Moisten the paper first with a solution of muriate of soda in proportion of one of the muriate to ten or twelve times its weight of water; then dry it, and when dry moisten it equally over with a solution of the nitrate of silver, and then dry it again before the fire; let this be done by candle light, for day light would immediately act upon it; this should be done five times, alternately with the solutions of muriate of soda and of silver, and dried each time, th last time being with nitrate of silver.

Second Way.—Moisten the paper with muriatic acid diluted with water to the strength of strong vinegar, and then dry it as aforesaid after each time of moistening; do this four times, and each time after being dried moisten it with the solution of the nitrate of silver, drying it also each time after the nitrate has been spread over its surface.

Third Method.—Moisten the paper with the solution of the nitrate of silver, and dry it before the fire; do this three or four times, and being dried the last time after silvering, then moisten it well with the diluted muriatic acid, or chlorine, as the French call it, and then dry the paper. I have several other methods, but these will be found satisfactorily to answer the purpose.

Proportion of nitrate of silver; one dram, or sixty grains to an ounce of soft water, and I would not advise using less than 45 grains of the nitrate of silver to the aforesaid quantity of water.

These papers began to change colour in a few seconds after being exposed to the solar light, and in fifteen minutes became of a slate colour, or nearly of an indigo blue, and in two hours were nearly black.

Remarks.—Frequent moistening in the manner aforesaid, though it wonderfully increases the susceptibility to change colour, yet the often wiping of the paper, whether with a sponge or otherwise, destroys the smoothness by opening the pores thereof. The best way would be to dip the paper in the solutions, and to absorb the superfluous moisture by pressing the moistened paper between two sheets of *white* blotting paper. I have no doubt that Bath drawing board would answer, or still more solid substances, such as the metallic plates which I mentioned in my communication sent to you six weeks ago, but not yet published;* for it is evident that porous substances, such as paper, are not the best adapted to receive the very minute impressions which are required to give faithful delineations of natural objects.

For fixing the objects when drawn by the camera, the proportion of muriate of soda or common salt may be one part to about six or eight times its weight of water, in which the photogenic drawings may be immersed and then the superfluous moisture wiped off, or absorbed by blotting paper as aforesaid; or if left to soak for half an hour or an hour, the effect will be still better, and then dried as aforesaid.

I had intended to have forwarded this, or a similar communication, within a few days after I had sent my last to you, but the pressure of business has prevented me. I have seen nothing of Mr. Talbot's communications nor any others, except what has appeared in your journal. The first piece of Mr. Talbot's appeared in No. 810 of your Magazine, but did not give the process, which more fully determined me to do this myself, as I had expected that Mr. Talbot had intended to keep the processes secret. And now, Mr. Editor, that you have seen what Mr. Talbot's processes are both for preparing and for fixing the images thereon after they had been delineated by the camera obscura, you will find, on the perusal of my paper which appeared in No. 809 of your journal, that that communication contains in a few words the sum and substance of Mr. Talbot's processes, and that four or five weeks previous to my seeing any thing of that gentleman's method, which appeared, and where I saw it for the first time, in No. 813 of your Magazine. In conclusion it may be well to remark that this kind of drawing paper will be found very expensive, as a dram of nitrate of silver, which costs a shilling, will not be more than sufficient for the preparation of a moderate sized sheet of drawing paper.

I remain, sir,
Yours very respectfully,
THOMAS OXLEY.

No. 3, Elizabeth-place, Westminster Road, 14th March, 1839.

[We have received another communication from Mr. Oxley in reply to the letter of Mr. Egerton Smith, which appeared in our 814th number; we have only room for the following extract, but which, we think, is quite suf-

---

* The passage which Mr. Oxley refers to, follows it was overlooked inadvertently, in consequence of being appended to a communication upon the subject of "Mechanical Flying."—"It is well known that some substances are powerfully acted upon by the rays of light, whereby some are precipitated from the menstruum that holds them in solution, some are oxygenized or corroded, and others are, to a certain extent, de-oxygenized, by the action of the rays of light; seeing this is the case I would not despair, had I but time to devote to a series of experiments, in going so far as to find out a menstruum, or composition, which being spread over the surface of a plate of steel, zinc, or copper, and then having the pictures of the objects thrown upon it by the camera obscura, or other optical apparatus, the plate would be etched or corroded by the composition according to the intensity of the light and shade, whereby chemistry and optics may be made to supersede, and far surpass, the laborious process of engraving, as hitherto practised. I can see nothing here to be doubted of, it will be certainly brought to pass by somebody, for we know that machines have already been made which perform what has always been considered as mental operations, and why not succeed in making nature operate on herself, and become her own artist?"

ficient for the vindication of Mr. Oxley.]
—Ed. M. M.

"I hope you will permit me to reconcile the apparent discrepances of that communication. I cannot boast myself of having the most tenacious memory in the world, but at all events my memory is as likely to serve me as faithfully as those of Mr. Smith or Mr. Turmeau can serve them, they being my senior by some 15 or 20 years each; and it is by no means unlikely that these gentlemen may, in the space of 16 years, have forgotten what I then communicated to them. But in regard to Mr. Smith, no person could scarcely be quicker at conceiving ingenious inventions, and I believe it possible that within the space of 30 years he may have conceived as many notable projects, but from the great multiplicity of business pressing continually on his attention, (although he might have thought much of them for a time), they have, one after another, sunk into oblivion, and have long since been obliterated from the tablet of his memory.

"It may be, that Mr. Smith has even forgotten that I was a frequent visitor at his house for some years, and that during that time I wrote different pieces which appeared in Mr. Smith's *Literary Kaleidoscope*, and also in the *Liverpool Mercury*; and Mr. S. may have forgotten that he has more than once complimented me, by saying, that he wished that all his correspondents wrote and composed with the same neatness and care that I did, for he never found it necessary to alter a single word or letter in any of my communications; this fact also may have *slipt his memory*, but not mine. And as respects my former friend, Mr. Turmeau, I appeal to your judgment, Mr. Editor, if you can think it at all likely that I should have referred to Mr. Turmeau, if I had not firmly believed that he would have both recollected and confirmed all the circumstances I had therein stated? But if Mr. Turmeau has forgotten that I suggested to him the use of the camera obscura in photogenic drawings, I do hereby most positively assure him, and the whole world, that *I did so;* and that, without having ever heard of anything of the kind from any person whatever; and whether my memory be better or worse than that of Mr. T. I will not take upon me to say; but I consider I had more reason to remember these circumstances than Mr. Turmeau had."

MESSRS. HERAPATH AND COX'S NEW PROCESS OF TANNING.

Sir,—For some time past I have been working in my tannery upon Messrs. Herapath and Cox's new method of tanning, under license from the patentees, and have much pleasure in bearing testimony to the value of the process, and in corroborating the remarks of your correspondent, Mr. E. Wilkins, tanner and currier, of Southampton, whose letter, inserted in your valuable Magazine, No. 804, I have read.

I should premise that I am a tanner of butts for sole leather from South American hides; that my accustomed time for the tanning process by the old mode has been twelve months; that my tanning ingredients are English bark and Smyrna valonia, in about equal proportions; and that during the time in which I have been tanning and selling nearly fifteen hundred butts upon the new plan, it has been personally superintended by myself. I do not rely, therefore, upon the dictum of others.

The following is a comparative statement of actual results produced by the new method upon five different lots, amounting to 497, and by the old method upon four lots, amounting to 425, salted Buenos Ayres and Monte Video hides :—(See table next page.)

By this statement it will be seen that the *new process* has produced *in three months* 3¼ *lbs. more leather from a* 58 *lb. hide*, than is produced by the *old method in twelve months from a* 59½ *lb. hide !*

These hides were bought in Liverpool by the same agents, were of similar quality, were tanned in the same yard, in the same year, with the same ingredients, *and no stronger liquor was administered to those operated upon by the patent process* than to the others. I have selected the above quantities out of many others, because, both in numbers as whole lots and in their averages as hides in their raw state, they approximate nearer, and of course give a more correct comparison, than any others I could take. The time taken to tan those upon the patent mode has varied; some

	No. of Hides.	Nett payable weights of Hides.	Payable average of Hides.		No. of butts, shoulders, & bellies.	Nett weights of butts, shoulders, and bellies.	Nett average of butts, shoulders, and bellies.	Total weight of leather.	Total weight of leather per Hide.
		cwt. qr lb.	lb.			cwt. qr lb.	lb.	cwt. qr lb.	lb.
New method	497	257 1 15	58	Butts	497	137 0 12	30½		
				Shoulders	497	11 2 27	2½		
				Bellies	994	40 2 25	{ per pair 9½	189 2 8	42½
Old method	425	224 2 13	59¼	Butts	425	103 0 9	27½		
				Shoulders	425	12 1 10	3¼		
				Bellies	850	33 2 25	{ per pair 8⅔	149 0 16	39⅔

have had *three months tanning*, whilst some have been brought *from the bundled state of salted hide to the market in two months and twenty-one days!*

With respect to colour, I produce what I please, from a dark brown to a straw; but the chief recommendation of sole leather tanned by the patent process is its *extraordinary impermeable, elastic, and durable qualities;* so that its advantages to the consumer are even greater than to the tanner. I could bring proof upon proof to substantiate this assertion from my numerous customers, for the use of this leather in coal mines, engine manufactories, iron foundries, and public works, as well as from public and private individuals who have long used and tested the leather.

This simple statement *of facts* I leave for the consideration of your readers, and in conclusion I beg to observe that I shall be happy to answer any inquiries which may be made of me through the medium of your Magazine in a straightforward and open manner, and as strictly correct as the close attention I have given to the process will enable me.

I am, sir, yours most respectfully,
STEPHEN F. COX.

Patent Leather Tannery, Nailsea, near
Bristol, March 21, 1889.

[The following are descriptions of samples which we have received from Mr. Cox, of leather tanned upon Herapath and Cox's patent process, and which we shall be happy to show any parties interested who may call at our office. ED. M. M.]

No. 1, a piece of shoulder cut from a well-grown Buenos Ayres hide, tanned in three months.

No. 2, a piece of very coarse hide cut from the belly inwards, tanned in ten weeks.

No. 3, a piece of Monte Video hide cut from near the back, of a milder tannage and softer finish and lighter colour than the two former, tanned in eight weeks.

No. 4, a firm, prime, and bright coloured piece of Buenos Ayres salted hide cut near the ribs, limed, tanned, *dried, and sent to market, in two months and twenty-one days!*

No. 5, a milder tanned piece of bright coloured leather cut from flank and belly inwards, tanned in seven weeks.

The most impermeable pieces are Nos. 1 and 4. No. 1, although a shoulder, and consequently a loose part, will bear soaking for twenty-four hours without saturation.

OMNIBUS STOPPING SIGNALS.

Sir,—By a curious coincidence a friend suggested to us, about six months back, the introduction of the trifle explained in the first part of the letter from Mr. John Rivington, jun., page 413 of your interesting journal. Our friend has described it in almost the same words; he tried a bell on Walter's Brixton coach about six years ago, and he fixed one of the bells, about Christmas last, on Martin's Brixton coach; he also exhibited it at the Royal Institution, January 18th.

If your columns are not better employed you will oblige us by inserting the following description of it, it was printed with the instructions for the management of Mr. Cowper's parlour printing press, a very ingenious and effective apparatus for the use of young persons and amateurs. We manufacture both articles and shall be happy to show them to any of your readers.

We remain, Sir,
Your obedient servants,
HOLTZAPFFEL & Co.

64, Charing-cross, and 127, Long-acre, London, March 16, 1839.

The Coach Bell, for calling the attention of the Conductor or Driver of a Coach or Omnibus.—No person can ride in an omnibus without witnessing the difficulty there is in attracting the attention of the conductor. One person shouts as loud as he can, another begs his neighbour to call out; another stretches his stick or umbrella, or perhaps himself, across the faces of his fellow travellers, and pokes or thumps the conductor; and every body knows that, notwithstanding all these expedients, the passenger is frequently carried beyond his destination; the same inconvenience occurs in stage coaches, with this variety, that the passenger has to put his head out of the window, now and then losing his hat. To prevent these inconveniences, and to furnish a ready and quiet way of calling the attention of the conductor, is the object of this simple invention, called the coach bell.

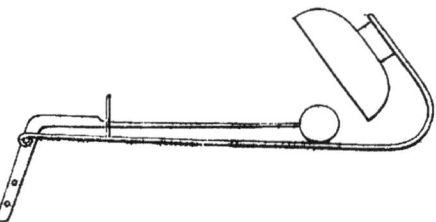

The bell is placed on the outside of the roof of the omnibus, just over the door, and, consequently, close to the conductor. It is struck by a hammer; the hammer is a crooked lever, and is formed of elastic steel; a string is fastened to the short end of the lever, and passes through a hole just over the door, and then through brass rings, or eyes, along the inside of the roof of the omnibus; six small tassels are attached to the string, and on pulling any one of these, the hammer is jerked upwards, and the bell struck once, which is quite sufficient to call the attention of the conductor.

No shaking of the omnibus, nor the roughest road, can possibly make the bell ring. Mr. Martin, the coach master of Brixton, has been the first to adopt the "coach bell;" and as it can be had for a trifling expense, there is little doubt of its being generally adopted.

E. C.

REPORT OF THE COMMITTEE OF THE FRANKLIN INSTITUTE ON THE EXPLOSIONS OF STEAM-BOILERS —PART II.—REPORT OF THE SUB-COMMITTEE ON THE STRENGTH OF MATERIALS EMPLOYED IN THE CONSTRUCTION OF STEAM-BOILERS.

[From the *Franklin Journal*.]
(Continued from Page 446, No. 815.)

No fire was kept in the apartment where the experiments were performed. The container A, when quite dry, was first accurately counterpoised in a scale pan, and then taken out and weights substituted to avoid any possible error in the scale beam. The vessel being next returned to the scale pan instead of the weights, was filled so nearly with water, that the vapour to be condensed, would make it quite full; after which a counterpoise was once more effected, and the vessel being taken to its place within B, the weighing by substitution was repeated, giving the sum of the weights of the water and of its container. The counterpoise being once adjusted for a series of experiments, it was only necessary in repeating the trials with new portions of water, to replace the vessel A in the scale pans, and pour in a fresh charge of water till the equilibrium was obtained. The several thermometers were next adjusted in place, that which was to indicate the temperature of the water being made with special reference to these trials, and the weight of mercury which it contained, as well as that of the glass immersed in the water exactly ascertained.

While the operations of weighing and arranging, as just described, were performed by one assistant, another having heated to a low red heat the cylinder S, inserted it in the copper case L, where being retained by the hook h, it was plunged into the boiling water contained in F, and the latter placed within D ; the packing x was adjusted, and stop-cock K opened, to allow the air to be driven from the pipe g, and the whole apparatus, including the thermometer e, to be raised to the temperature of 212°. This being done, the apparatus was conveyed to the room where the water vessel was placed, and after turning the stop-cock for an instant, to expel any water which might have been condensed during the transit, the mouth of the pipe was brought briskly round and immediately plunged under the water. Here it was continued until the vessel A was perceived to be full, when it was withdrawn and the stop-cock closed, but not until the mouth was quite above the water. During this manipulation the thermometer t was was kept constantly moving, to equalise the temperature of the water. This being done, the increase of weight was ascertained by again counterpoising the vessel and its contents. It is apparent that heat imparted to the water, by the condensed vapour, must be employed for heating at least three different bodies, the water, the container, and the thermometer t.

It was therefore necessary to know the weight and specific heat of each, in order to determine the relation of the heating power of a given weight of steam, compared with the cooling power of a quantity of water equivalent to the sum of these three bodies. As the quantity of air which was included in the box B was small in amount, as the specific heat of air is, weight for weight, but about one-fourth that of water, and as the experiments were generally performed in such a manner as to allow the air to operate partly in favour, and partly against the heating power of the steam, it was not deemed important to take into the calculation the minute quantity due to the cooling power of this mass of air. In a few instances, however, its effects are noted in the column of remarks.

The results of experiments on this subject are found in table XII. (Frank. Jour., vol. xix, pp. 162,3.

As the trials on specific heats below 212° had preceded those now under consideration, it was found convenient, to employ as containers some of the same cylindrical vessels which had been used in that investigation. The materials of each are specified in the table. The calculations are very simple when we have obtained an expression for the equivalent quantity of water equal to the three terms above specified.*

Results of Experiments on Latent Heat. —The accompanying table presents the determination, in the manner already described, of the latent heat of the vapour of water. The trials were made in four different cylindrical vessels, one of copper, two of glass, and one of sheet iron.

The quantities of water varied from about 13000 to upwards of 39000 grains.

The equivalents of the thermometers were either approximately estimated by knowing the size and thickness of the bulbs, or were actually determined by weighing before and after filling, and in every instance the calculation for the thermometrical equivalent, was made only on the part of the instrument actually immersed.

It will be perceived that three out of the four vessels, give mean results which differ from each other by not more than 3 degrees. The third set, or that made in thin glass, and which differs widely from all the rest, ought probably to be rejected. If this be done, the other three sets give a mean result equal to 1037 degrees ; which is 3.8 degrees less than that obtained by Count Rumford. Including the third set, the mean result will be 1026.83. As the steam rising up in the case L, necessarily came in contact with the hot iron S, it became, to a certain extent surcharged with heat ; but as the thermometer indicated its temperature at the moment of escape, an allowance is easily made for the surcharge. The rapidity of flow being duly regulated by the stop-cock k, the steam was prevented from carrying over any water in an unvaporized state. As the amount of surcharge seldom exceeded 3 degrees, it was not considered necessary to calculate for the difference between the

---

* The latent heat of steam was calculated by the following formula.

Putting w = the weight of water in the vessel.
    n = that of the vessel itself.
    g = that of the glass in the immersed part of the thermometer.
    m = that of the liquid in the bulb.
    s = the specific heat of container.
    y =   do.   of glass.
    z =   do.   of the thermometric liquid.
    v = the weight of vapour condensed.
    T = the temperature gained by the water.
    t = the distance of the final temperature of the water below that at which the steam enters it,
and =l the latent heat of vapour at the boiling point.

Then the heating effect is represented by $T \times (w + ns + gy + mz)$, and the cooling effect by $v \times (l + t)$; whence $v \times (l + t) = T \times (w + ns + gy + mz)$ and consequently $l = \dfrac{T (w + ns + gy + mz)}{v} - t$.

specific heat of vapour and that of water. By the experiments of Delaroche and Bérard, the specific heat of vapour, compared with that of water, is .847 to 1.000. Admitting this to be true, the result must, in any case which has occurred to the committee, be but little affected by allowing for the difference.[*]

Specific Heat by Vaporization.—Having determined the latent heat of vapour, it is not difficult to verify our preceding determinations of the specific heat, by operating in precisely the same manner as we do to obtain the temperature of a body, except that the temperature of the bath of melted metal is now first ascertained by the mercurial thermometer; and the actual temperature of the standard piece being then known, is compared with the weight of vapour which it produces, by cooling in boiling water from its initial temperature down to 212°. These experiments were made both before and after the screw beam and counterpoise were changed. The weight of the standard piece is, in both cases, taken in degrees of the pyrometer scale as existing at the time.

It will be seen, that, assuming as correct the determination of latent heat, made by the committee (1037°), the experiments given in the accompanying table (No. XIII., *Frank. Jour.*, p. 165) afford results for the specific heat of iron as follows:—

1. Taking the mean of 29 experiments, it is 11325
2. Taking only those made before the screw beam was changed (9 experiments), we obtain.... 11340
3. Taking together the last 20 experiments of the table, we have.. 11324
4. Experiments Nos. 7 and 8 with the first screw and counterpoise, differing only in the 5th place of decimals, give........ 11336
5. Six out of the last 20, differing only in the fourth place, give.. 11356

6. The mean result of these 5 comparisons, is.................. 11336

---

[*] The experiments hitherto published, had left some doubt as to the true latent heat of vapour. Black first obtained the number 810 deg.; Watt afterwards gave it 950 deg.; Southern produced 945; Lavoisier made it rather more than 1000 deg.; Rumford 1040.8; Despretz 955.8; Ure 1000; Thompson "more than 1000 deg." Watt and Clement have both established the position that the latent heat of steam, added to the sensible heat above 32 deg., is nearly a constant quantity. As, however, the point 32 deg. is entirely arbitrary, and as no temperature is now *known*, at which vapour does not rise from water or ice, there is reason to suppose that in strictness, the constant—if there be one—is different from that which these experimenters have derived. If not, the latent heat of vapour must diminish below 32 deg. as the temperature diminishes.

As a mean of the nine sets of experiments in different vessels, the specific heat below 212°, determined by heating water, as above detailed, was found = .113374. As the calculations just detailed are carried only to the 5th place, the two results may be considered as differing from each other only by $\frac{1}{111377}$th part of the total value.

Of those experiments which differ considerably from the general result, about the same number was found above, as below the *mean*, showing that if these discrepancies be due to errors of observation, they are as we ought to expect, liable to be either in excess or defect; and that they counterbalance each other.

The eight experiments, of which the results differ only in the 4th place of decimals, were made at temperatures varying from 392 to 595, without indicating any decided difference in the specific heat of the metal within those limits.

Of the extreme results in the table, the highest was obtained at 480° and the lowest at 488°;—the next to the highest, at 500°, and the next to the lowest, at 292°.

From the exact conformity of the general results of the method of evaporation, and that of heating water, in trials below 212°, it appears that if the specific heat of the standard piece be determined by the latter method, and its weight be duly regulated to conform to the length of the threads of the screw beam, and to the weight of the revolving counterpoise, its indications of temperature will be such as to connect themselves immediately with those of the mercurial thermometer.

Heating and cooling of Liquids.—In determining the specific heat of the iron standard-piece, it became evident that the influence of the air and other extraneous objects upon the temperature of the vessel of water could not be omitted, at least while the experiments were conducted, without enclosing the container in some other vessel which might shield it from the radiating and conducting power of surrounding bodies.

But in order to neutralize, as far as practicable, the disturbing influence of the causes just mentioned, it was evident that with a given state of the air and of other bodies, the water-vessel must be made to receive during an experiment, as much heat from surrounding objects as it imparted to them. This could be effected only by commencing each experiment, so much below the temperature of the air, that, during the cooling of the iron in water, the temperature of the latter should, in rising, pass through the temperature of the air, and not only rise above it, but so divide the duration of the experiment that the cooling effect of the air in the latter portion of time should precisely equal its heating influence in the former.

It therefore became necessary to, discriminate between the respective influences of hot iron and of the air, in order that the temperature of the water might be adjusted to that of the apartment before commencing the experiment.

By an examination of table XLV. (*Frank. Journ.*, vol. xix, p. 167 to 170) it will be perceived that in twenty-three different experiments the times of rising through different stages of temperature are given, together with the initial and final temperatures of the air and of the water. It will not fail to be observed, that in comparisons of this nature, the materials and construction of the thermometer are elements of quite as much importance as the quantity of liquid heated, or the materials and other circumstances of the container.

Thus, it will be seen, that by a mean of 8 sets of observations in which the mercurial thermometer A (calculated to be equivalent to 152.7 + 43.45) = 196.15 grains of water, was employed, the time required to obtain the full effect of 6000 grains of iron heated to 212°, and cooled in water at 60 or 64 degrees, was 148 seconds.

The quantity of water was varied from 12,000 to 18,000 grains.

With the mercurial thermometer B, estimated at about 43 grains of water, the time by 7 sets of observations was found to be 126 seconds; the quantity of water from 13,000 to 20,000 grains, and the vessel either of glass or sheet iron; the two latter circumstances serving to produce comparatively little effect on the time required to bring the temperature to a stationary condition.

With the spirit thermometer C, 8 sets of observations gave a mean duration of 295 seconds, the weights of water varying from 13,000 to 18,000 grains and the container being either iron, weighing 1,733 grains, or glass, weighing from 2,900 to upwards of 12,200 grains. This last thermometer has a bulb 6 1/10 inches in length, and .5 inch in diameter, weighing 264 grains; and contained about 142 grains of alcohol, which, by the mean of 8 different determinations[*], has a specific heat of .641, and consequently is equivalent to 91 grains of water, and the glass to, 26.4 grains, whence the whole portion immersed was equivalent to 117.4 grains of water, whence the whole portion immersed was equivalent to 117.4 grains of water.

A circumstance which deserves attention in examining this table, is, that a few hundredths of a degree in rise of temperature, often required, at the commencement of an experiment, a much longer time than in the periods immediately following. In fact, it was sometimes observed, that the plunging of the hot iron into the water was accompanied by an instantaneous minute depression of the liquid in the thermometer; subsequent to which, a stationary period occurred, and then a rapid rise—as indicated by the observations in the table. This phenomenon is to be ascribed to the sudden expansion of the glass composing the bulb of the instrument, by the first impression of the heat, affording an enlarged cavity for the liquid, before the latter begins to feel the same influence, and consequently to expand. This effect is the more striking, the greater is the difference of temperature to which the instrument is suddenly exposed. It needs hardly be mentioned that the opposite effect of a rise in the liquid, accompanies the sudden immersion of the thermometer in a mass of fluid colder than itself.

It is also worthy of remark, that the time required by the thermometer to attain the same final temperature as the liquid in which it is plunged, is greater than that employed by the iron in giving up its excess of heat to the same liquid. This will generally require some deduction from the total observed time, in order to arrive at the true time of cooling of the standard piece.

The deduction will be less the more sensible is the thermometer. Experiments with thermometer B, require less correction than those made with A, and the latter less than those with C.

Owing to the fact just stated, it is not always easy to determine the precise moment when observations ought to cease; consequently the last rise noted may, for our present purpose, often be rejected, when the amount observed does not exceed five or six hundredths of a degree, and the remaining time taken as the true duration of the cooling. By the aid of these observations, we shall be enabled to determine, very nearly, the relation between the respective *augmentations* of temperature in the water, and the *times* in which they severally occur. The more exact termination would equisade that the standard p ece and the thermometer should be either both rapidly moving, or both at rest in the same relative positions, for every experiment. It was easily perceived that no slight influence might, in the earlier parts of the process, be ascribed to these circumstances.

An inspection of the table shows that the general relation to which we have referred, is such, that *two-thirds of the change of temperature in the water occurs during the first-third of the entire period of observation.* This supposes the proper *correction* to have been applied to the latter as above pointed out.

Thus in experiment 3, table VIII., thermometer C gave a change of temperature 6.98°, two-thirds of which is 4.66°. During the time of the 10th observation, the rise of temperature came to 4.66°, and the time then elapsed was 84″ from the beginning, the *whole time* being 251″. Difference .33″.

In table VI., experiment 3, with the same thermometer, we have a total rise of 5.78° in 249″. Two-thirds of 5.78° is 3.85°, and one-third of 249″ is 83″. It appears that a rise of 3.85° had been attained during the sixth observation, and that at the moment when this took place, the time elapsed was 78.7″. Difference 4.3″.

Again, in table V., experiment 2, with the same thermometer, the total time was 290″; but the last observation gave a change of only $\frac{9}{100}$ of a degree in 37″. This being omitted, we have the time 253″, and the change 5.61°, two-thirds of which is 3.74°. One-third of the time is 84.3″. The observations prove that a rise of 3.74° took place during the 5th observation, when the total time elapsed was 79.58″. Difference 4.72″.

When thermometer A was used, in experiment 5, table VI., a gain of 5.71° took place in 154″. The last $\frac{2}{100}$ of a degree required 11″; this being omitted, we have 5.69° in 143″. Two-thirds of 5.69 is 3.78; which by observation was attained in 44″, whereas the calculation would give 47.6″. Difference 3.6″.

With the same thermometer used in experiment 10, table VII., it appears that the total time, exclusive of 29″ taken up in rising through the last $\frac{6}{100}$ of a degree, was 154″, one-third of which is 51.3″. The total rise in this time was 8.5°, two-thirds of which is 5.67°, which by observation was attained in 46″. Difference 5.3″.

The thermometer B, of which the action was more prompt than that of either of the others, gives results more nearly agreeing with the law above stated. Thus in table IV., experiment 24, we find a rise of 4.6° in 94″. Two-thirds of 4.6° is 3.06° and one-third of 94″ is 31.3″. By observation 3.06° had been attained in a trifle less than 30″. Difference 1.3″.

Again, in table VIII., experiment 2, a rise of 7.4° took place in 109″, one-third of which is 36.3″. The observation shows that a rise of 4.72°. Difference 1.3″.

Again, in table VIII., experiment 2, a rise of 7.4° took place in 109″, one-third of which is 36.3″. The observation shows that a rise of 4.72° had been attained in 37″. Difference .7″.

If in table VII., experiment 4, thermometer A, we omit the time of the last observation, we have a gain of 8.36° in 103″. Two-thirds of 8.36° is 5.57°, this rise of temperature had occurred at the end of 35″ by observation—whereas by calculation we should have 34.3″. Difference .7″.

In table VII. experiment 5, we obtained a gain of 8.25° of temperature in 141″. Two-thirds of 8.25° is 5.5°. This last number of degrees had been gained by the water about the middle of the 6th observation, or when the time from the commencement was 34″. As the last $\frac{2}{100}$ of a degree required 52″, we may safely attribute to the sluggishness of the thermometer the same retardation as in the preceding experiment; in which case we should have the total time 107″, one-third of which gives the calculated time for a rise through two-thirds the range equal to 35.6″, and the difference between the observed and the calculated times =1.6″.

In table VII., experiment 6, two-thirds of the gain of temperature was observed to have taken place at the end of 37″. The total time during which observations were made, was 125″, and as this time is much less than either of the two preceding, we may suppose that a less allowance is required for the tardiness of the thermometer, in consequence, perhaps, of more rapid agitation in the liquid while the latter received its augmentations of temperature. Hence, if we deduct 15″ we have remaining 110, one-third of which is 36.6″ for the calculated time of attaining two-thirds of the gain of temperature. Difference .4″.

In table IV., experiment 25, we found that a gain of 4.7° was effected in 145″, the last 18 of which were taken up in raising the thermometer B. $\frac{8}{100}$ of a degree. Omitting this period, we have a remainder of 127″, one-third of which is 42.3″. Two-thirds of 4.7° is 3.14°, which, on inspecting the column of *rise of temperature*, we find was produced in 38.8″ from the time of beginning. Hence the calculated exceeds the observed time by 3.5″.

Of these eleven comparisons it will be observed that eight give the time by observation for two-thirds rise of temperature less, by a small amount, than one-third of the total time, while the others give the former greater than the latter quantity. The mean result, however, is a difference of only 1.5″. The results might probably be found to conform more exactly to the law, if the liquid were indefinite in quantity, and its rise indefinitely small, compared with the number of degrees through which the iron cooled.

Heating by Contact of Air.—The result just obtained, combined with another on the rate of heating of the vessels of liquid exposed to the action of air, will show on which of the experiments the greatest reliance is to be placed, as exhibiting the true specific heat of iron, without requiring a deduction for the influence of air.

The manner of performing these experiments, has been already adverted to. It consisted merely in filling the cylinders with water of a low temperature, and inserting in them, the same thermometers which had been used in experiments on specific and latent heat ; placing other thermometers outside of the cylinders, to mark the temperature of the air.

The time of ariving at, and of leaving each mark on the scale, was then noted ; and the mean taken as the point of time for attaining each degree.

Table XV. (*Frank. Jour.*, vol. 19, p. 174.) contains the result of these observations. The first 9 are, perhaps, from the particular attention directed to them, deserving of the most confidence, and from these it appears that *the rate of heating or of cooling, of a mass of liquid acted on by the air at a higher or a lower temperature, is directly and simply propoportional to the difference of temperature between the liquid and the air.**

This is no more than a verification of the Newtonian law which is well known to be sensibly true only for very moderate differences, such as those observed by the committee which never exceeded 20°. The same law is also well known to fail entirely, when carried to very great differences. Assuming then the correctness of our result it enables us to determine, that while the iron in experiments on specific heats, was imparting its excess of heat to the water, *the air gave to the liquid as much heat as it received from it, whenever the initial temperature of the water was twice as much below that of the room as the final temperature was above it.*

Strength of rolled Copper.—Tables numbered from XVI. to XXIII. inclusive, (*Frank. Jour.*, vol. 19, pp. 176 to 191,) present the results of experiments on the strength of boiler copper, both at ordinary and at elevated temperatures. From these tables it appears that at temperatures vary-

ing from 62 to 82 degrees Fah., the strength of rolled copper is by a mean obtained from 66 experiments on 8 different specimens within those limits, equal to 32826 pounds to the square inch. The *irregularities* of strength vary in the different specimens from $1\frac{9}{10}$ to $4\frac{6}{10}$ per cent. of the mean tenacity of the specimen in which they occur, and the mean value for the 8 bars is $3\frac{5}{10}$ per cent. The strips of copper as received from the manufacturers were of 4 different thicknesses, two of each thickness, and they were reduced by filing to a nearly uniform size throughout the whole length. By an attentive observation, it will be seen that the thicker specimens give in general the higher results.

Thus, No. 1, of which the original thickness was two-tenths of an inch, (called by the manufacturers three-sixteenths,) broke at eight trials, with an average force of 30704 lbs. per square inch.

No. 2, with the same thickness, broke with 31468 lbs. as the average weight, at seven different trials. Hence the mean strength of these two bars is 31086 lbs. per square inch.

Nos. 3 and 4, the thickness of which was a " *scant quarter*" of an inch, broke, the former at ten trials, with 33428 lbs., and the latter at six trials, with 33243 lbs. giving a mean of 33335.

Nos. 6 and 7, having a thickness differing but little from the two preceding, but rather greater, gave, the one at seven trials, 33411, and the other at nine, 33005 lbs. per square inch; or, as a mean of the two specimens, 33205.

Nos. 5 and 8, with a thickness before filing of not less than .27 of an inch, exhibited a tenacity of 33771 and 33780 lbs., the former being the mean of eleven and the latter of eight successive trials, showing a mean of 33775.

The manufacturers have not, in their note accompanying the specimens, referred to any difference either in the kind of pig metal, the melting and refining which took place previous to rolling, or in any other circumstance attending the manufacture of the different bars, which could lead the committee to assign a probable cause for the difference in point of cohesion between the respective pairs.

That difference between Nos. 5 and 8, and 1 and 2, is no less than 3071 lbs. per square inch, or 9.3 per cent. of 32836, which we have found to be the average strength of eight specimens.

But, as already stated, the irregularities observed in any one specimen, did not exceed $4\frac{6}{10}$ per cent. of its mean strength. It seems therefore probable, that in reducing the lighter specimens to their final thick-

---

* This results from a mean of 14 comparisons between the differences of temperature, and the corresponding times of heating through a given indefinitely small range of temperature, as one-tenth of a degree, by the formula $Dz : dz : : t : T$. Where D and d are observed differences of temperature, between the water and the air, t and T the corresponding numbers of seconds required to raise the temperature 0.1 deg.; and z the power of the difference of temperature according to which the times vary. These fourteen comparisons give a mean value of $z=1.002$.

ness, the operation was extended so far as to reduce below a proper point the temperature of the copper, and thus to injure its texture. It will be seen that the highest results obtained by the committee, are almost identical with that given by Mr. Rennie.

In every calculation of the strength of materials for a steam boiler, the *least strength* known to be possessed by any part of the sheet, is that which alone can be relied on for fixing the pressure to which it may be subjected.

For *copper*, at ordinary temperatures, the lowest result obtained by the committee was 30406 lbs, per square inch, and the mean minimum for the 8 bars 32146 pounds. To other temperatures subsequent developments apply.

Effect of increased temperature on Copper. —The effect of temperature on tenacity, has been hitherto but slightly examined, either for theoretical or practical purposes. The general truth that heat diminishes, and eventually overcomes cohesion, is too well established by daily observation to admit of question.

The temperature of *no tenacity*, is generally supposed to be that at which the fusing point of the given substance is placed, and the point of maximum tenacity ought, upon general principles, to be found at the point where least heat prevails, that is, at the natural zero, or point of *absolute cold*, if such a point exist in nature. Between these two extremes, it might be supposed that the tenacities of different substances, particularly such as are capable of passing immediately from the solid to the liquid state, would be found to obey certain laws. As the total cohesion at the maximum would present to a mechanical agent tending to overcome it, the whole of its resistance, and as, at more elevated temperatures, a part of that tenacity would be overcome by heat, and the rest must be destroyed by the mechanical force, it is evidently a question of experiment, to decide what relation the two forces have to each other at the several temperatures between the two extremes to which we have just alluded. To decide the theoretical question, or in other words, to deduce, from the experiments, a law which might be expressed in an abstract form corresponding to all the possible phenomena, would require a state of the materials different from that usually found in commerce or employed in the arts. It would also, as we have seen, require a knowledge of that, about which philosophers no less than practical men, are far from being agreed;— namely, the point of absolute cold. As the purposes of this committee did not lead

them to investigate the problem in all its *possible* bearings, but only in view of the limits which practice assigns, and with the conditions commonly given to the materials, it will not perhaps, be easy, to construct from the tables a formula in all respects unexceptionable.

The general course of experiments involved the necessity of operating, at the different temperatures, on different bars of copper, and as all the bars are not found to give, even at ordinary temperatures, the same strength, for equal areas of section, it became necessary to deduce from experiments on each bar, at some assumed low temperature, a standard tenacity with which to compare its strength at every other point. The part of this *standard tenacity* which was taken away by the heat at the higher temperatures, becoming known by the experiment, a comparison was furnished for deciding approximately the relation between the temperature given, and the portion of tenacity which it had overcome. It will be found on an inspection of table XXIV. containing the comparison of these experiments, that on the eight different bars, the whole number of trials which furnished standards of comparison, at ordinary temperatures, was sixty-six, and consequently on an average about eight trials to each bar; while at the elevated temperatures there were made thirty-nine different experiments at nineteen different points on the scale, the greater number of points, however, having but one experiment each.

An inspection of plate IX., where these experiments are represented, will show that at nearly all parts of the scale, within which the trials were made, the strength diminishes more rapidly than the temperature increases, but some of the higher experiments indicate that the conditions of the law are such as to be represented by a curve, having a point of inflection. It will also be noticed that the three experiments which appear anomalous, and which in the plate are marked with queries, are all found in trials of the same bar of copper, (No. 7,) and that all these might be referred to a curve, varying but little in form from that which we have traced. It is not however *necessary* to suppose that these experiments belong to a different curve, for upon recurring to the table of bar, No. 7, (table XXII,) it will be found that one of the anomalies is satisfactorily accounted for by a delay in taking the temperature after the fracture had occurred, and that one of the others and probably both, were cases of weakening by a slight alloy of the copper by the melted metal through which it passed, in consequence of not having been defended by oxide. The other bars tried at high temperatures were treated with dilute nitric acid, creating a thin film of oxide, which effectually defended the surface, without sensibly diminishing even the smoothness of the bar.

It will be observed that the difference of tenacity, at the *lower temperatures*, for a difference of from 60 to 90 degrees, is scarcely greater than the actual *irregularities* of structure in the metal at common tempera-

Plate IX.

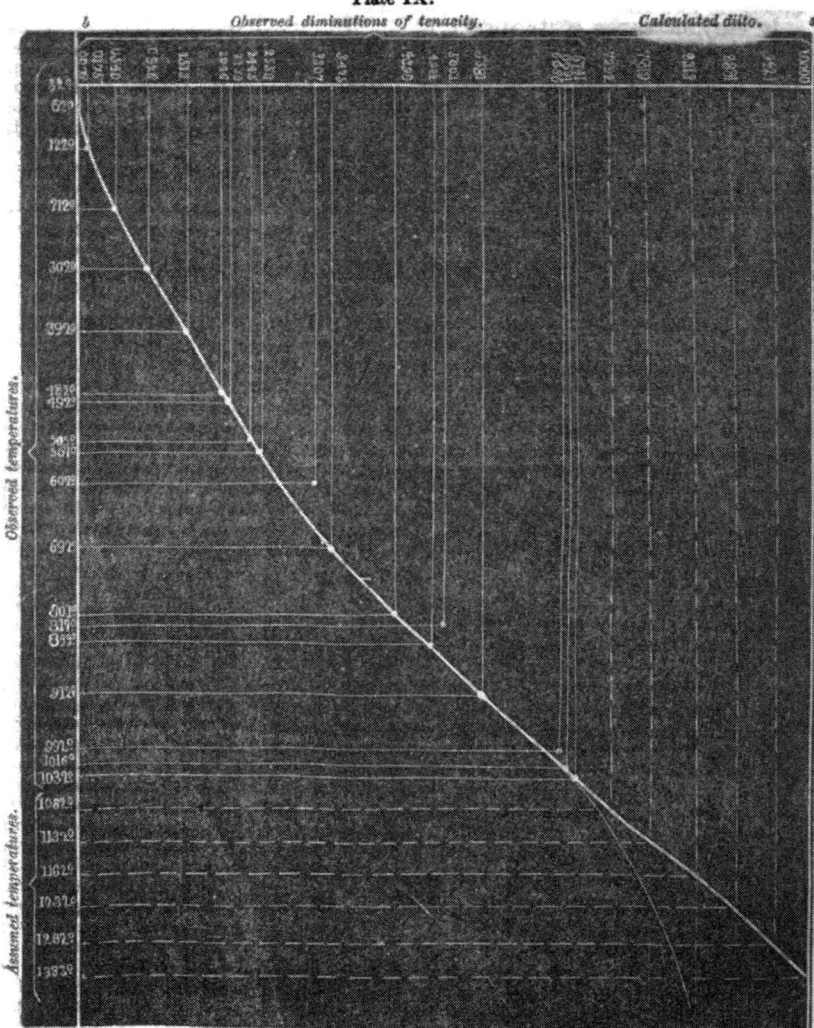

tures, and consequently, it was not practicable from these experiments alone to deduce a law which should express the tenacities at all points between the maximum above referred to, and the melting point of the metal.

Nor would much confidence probably have been reposed in results thus obtained.

In laying down the results in plate IX, the line *a b* is made to represent the total tenacity of copper at 32°. The horizontal dotted lines express the observed temperatures above 32°, and the vertical ones, the diminutions of tenacity at the respective points.

In examining the eleventh and twelfth columns of Table XXIV. (Frank. Journ., vol. xix, p. 242), with a view to a relation which may afford a practical rule for calculating the strength of copper at any given temperature, it will be found, that with the exception of the three anomalous cases, Nos. 11, 14, and 17 of the table, they may be referred to a species of parabola, of which the ordinates representing the temperatures above 32°, have to the abscissas representing the diminutions of tenacity, a relation expressed by *the cube roots of the squares* of the latter quantities; or, in other words, that the *squares of the diminutions are as the cubes of the temperatures*[*].

By applying the law above stated[†], and assuming the greatest diminution observed, or that obtained at 1,000° above the freezing point, as a true standard of comparison, we get the calculated results contained in the 13th column of the table, and a comparison of that with the twelfth, furnishes the *differences* in column 14th. This last, compared with the 9th, shows that the greatest deviations, even of the anomalous experiments already noticed, do not amount to so much as the actual irregularities sometimes found in the metal at common temperatures, for while the highest numbers in the 14th column are less than four and one-tenth per cent. of the total strength, several of those in the 9th amount to more than four and a half per cent. of the same sum.

The curve traced (Plate IX.), represents the column of calculated results, and is continued to the opposite side of the figure to show to what point this law would lead as the temperature of no tenacity[*]. This is seen to be 1,332 degrees Fah. which is 663 degrees lower than any determination of the melting point of copper hitherto made. It is well known that metals in general pass, in coming to the state of fusion, through a condition, in which though disintegration is nearly or quite complete, fluidity is not

perature above 32 deg., at which each experiment was made, and the diminution of strength corresponding, agreeably to the preceding note.

TABLE XXV.

No. of the comparison.	Temperature above 32 degrees.	Diminution from the ascertained strength at 32 degrees.	Mean value of x, by 13 comparisons at each point.	Difference of each from the mean value of all the trials.
1	90.	.0175	1.586	+.036
2	180.	.0540	1.462	—.068
3	270.	.6926	1.518	+.018
4	360.	.1613	1.444	—.036
5	450.	.2046	1.489	—.011
6	460.	.2133	1.474	—.026
7	513.	.2448	1.447	—.053
8	529.	.2359	1.466	—.034
9	660.	.3496	1.474	—.026
10	769.	.4398	1.570	+.070
11	812.	.4944	1.565	+.065
12	880.	.5581	1.542	+.042
13	984.	.6691	1.557	+.057
14	1000.	.6741	1.458	—.042
Mean of 14 means = 1.500 = x				

Hence $t.1.5 : t'1.5 :: d : d'$, or $t3 : t'3 :: d2 : d'2$, which is the practical rule above given.

[*] The application of the law deduced from the research in the preceding note, to the purpose of getting the column of calculated diminutions, as well as to that of extending the curve to the limit of tenacity, requires but a transformation of the proportion $t3 : t'3 :: d2 : d'2$ into the equation

$$\overline{\frac{t'}{t}}\Big|3 = \frac{d'}{d}\Big|2, \text{ whence } \overline{\frac{t'}{t}}\Big|\frac{3}{2} = \frac{d'}{d}, \text{ and } d' = d \times \overline{\frac{t'}{t}}\Big|\frac{3}{2}$$

or $(Log\ t' - Log\ t) \times Log\ d = Log\ d'$. Thus, to obtain the strength of copper at 1232° Fah., we have $1232° - 32° = 1200° \quad \therefore \quad Log\ 1200 = 6.0791821$
$$Log\ 1000 = 3.$$

$$\begin{array}{r} .0791821 \\ 3 \\ \hline 2)2375463 \\ \hline .1187731 \end{array}$$

Add Log .6741 = —1.8287243

Gives Log .8861 = —1.9474964

Hence 1.000 — .8861 = .1139. is the remaining strength, or $11\frac{4}{10}$ per cent. of the strength at 32°, is all that remains at 1232°, which is a visibly red heat in day light.

[*] To determine whether any, and if any, what, single function of the temperature will at any point express the diminution of strength, as compared with that observed at other points, it was not deemed expedient to rely on a single comparison. The following method was therefore employed to obtain an expression corresponding with each of fourteen different points, compared with thirteen others. Putting t = any observed temperature above 32 deg.; t' = any other temperature above the same; d = the diminution of tenacity by the former temperature, and d' = that by the latter: also x = the power of the temperature, according to which the diminution of tenacity varies, we have

$t'x : t'x :: d : d'$ whence $\frac{t'x}{t'x} = \frac{d'}{d}$ from which we get $x = \frac{Log\ d' - Log\ d}{Log\ t' - Log\ t}$. Thus at a temperature of 984 deg. Fah., the tenacity was found by experiment to have been diminished .6691, its amount at 32 deg. being 1.0000; and its diminution at 492 deg. was .2138; hence by the above formula,

$\frac{Log .6691 - Log .2138}{Log (984-32) - Log (492-32)} = 1.508.$

[†] The following table exhibits the mean results of the several sets of comparisons with the tem-

fully established; and in this granular state they, in some cases, continue through a considerable range of temperature. The melting points are those at which fluidity is clearly established. But, notwithstanding this fact, and the very close accordance of the law above mentioned, with the observed diminutions of tenacity, we do not venture to assert that the theoretical law which might be derived from operating on copper absolutely pure, and of uniform tenacity throughout the specimen, would not give a form so varied as to change the parabolic curve into one possessing a *point of inflection*. An inspection of the figure, as well as a reference to the table in the preceding note, will be found to favour the supposition that the rate of increase in temperature corresponding to a given decrease of tenacity, does in fact pass through a minimum near the point where one-half of the absolute tenacity is overcome. The right hand branch of the curve indicates the probable course *after inflection*.

Extensibility of Copper.—In producing the rupture of bars of copper it became evident that this metal undergoes during the mechanical strain to which it is subjected, a degree of elongation, dependent in some measure on the temperature to which it is raised. The mode of ascertaining this point consisted in measuring, after the trial of each bar had been completed, the united lengths of all its fragments. In reconstructing the bars for this purpose, care was taken to bring the corresponding portions into as close a contact as possible, and also to allow by estimation for any imperfection in the same from roughness of the fracture. A second mode was, to select from among the fragments of each bar one or more which retained the original inch-marks, and which had at the same time been apparently strained to the full extent of its resistance without actually parting. By this latter method of trial it was ascertained that the extensibility of all the 8 bars, with the exception of Nos. 6 and 7, was nearly uniform, varying only between 40 and 44 per cent. of the original length. A section measured on No. 6, gave the length between two-inch marks only 1.25 inch, and on No. 7, 1.28. The trials on both these sections had been made at ordinary temperatures. When comparing the total lengths after fracture, with the original length of each bar, we obtained as a general result, very nearly the same extension as when employing the several inch-marks as just stated. The mean elongation of the whole after 116 fractures, was 43.5 per cent. of the original length. Other things being equal, the bars of least area appeared to

have been most extensible. No. 2 was stretched, by 18 fractures, from 30 to 46¼ inches. No. 8, by 14 fractures, from 30 to 43 inches. But the circumstance of most importance is the temperature of the bar at the moment of trial. Thus, on bar No. 7, (Table XXII.) the first fracture was made at 912° and the area of section afterwards was $.744 \times .244 = .181536$ square inch, and the diminution from its original size only .002571, while at the thirteenth fracture, when the temperature was 81.5°, the area, after trial, was $.550 \times .174 = .095700$, a diminution of .088316, or 34 times as much elongation as before.

Strength of Boiler-Iron at ordinary temperatures.—The results of experiments on boiler-iron, at ordinary temperatures, will be found included in thirty-two tables, from XXVI. to LVII., inclusive.—(*Frank. Jour.* pp. 246 to 275, and 326 to 360.) On some few of the specimens, the strength of which is exhibited in these tables, all the experiments were made with a particular view to the irregularities of the metal, and at, or near the same temperature, while on other bars much diversity in the objects of the experiments prevailed, and consequently of these, only a few trials can be selected which may be considered entirely appropriate to the present topic. When making comparisons with a view to the mean strength of sheet iron, even from the same plate, it is necessary to consider that the question may be answered differently according to the direction in which the specimen was cut off; to the condition in which it was submitted to trial, whether rough from the shears, filed to a uniform size and smooth surface, or filed away in notches to overcome the influence of the shears; or, finally, according to the previous treatment of the specimens, whether subjected or not to annealing or other influences of heat after leaving the rolls. The tables furnish, under appropriate heads, the information necessary to answer, separately, the several questions arising out of these different aspects of the subject. With regard to the method of preparing the specimens, by reducing them to an uniform size throughout the whole extent of the bar, it may be remarked, that on the 41 bars of iron, which in the course of this report are described as having undergone that preparation, there were measured 1049 points, or sections, and there were made 517 fractures, showing on an average but little more than two inches between two adjacent points of fracture. It also appears that on only two of those bars (Nos. 220 A. and 224 B.) did the mean area, of all the points measured, correspond exactly with that of all the sections of fracture. On 22 bars the mean area of the fractured parts is *less* than that of the measured sections, by an average of .000340 of a square inch, and on 17 bars the mean section of fracture is *greater* than the mean measured sections by an average amount of .000187. This proves what might indeed have been anticipated, that the fractures would, in general, take place at the smaller sections, and as the mean area was about .175 sq. inch, it appears that the difference between the measured and the fractured sections, due solely to irregularities of filing, that is, between the condition of our specimens and that of others which should be absolutely uniform in size, amounts to not more than $\frac{125}{175000}$ or $\frac{1}{1400}$ part of the total strength. This portion is less than the irregularities in the structure of rolled iron, as may be shown by referring to tables LV. and LVII.

Methods of manufacturing Boiler Iron.— For the information of the general reader, who may not be familiar with the several processes in the manufacture of iron. referred to in some of the foregoing, and in several of the subsequent tables, it may be proper here to make a few remarks explanatory of the methods pursued in the United States for producing wrought iron of the descriptions embraced in this part of the report.

It has already been stated that the iron furnished to the committee was, with a single exception, manufactured by the aid of charcoal. This remark applies, of course, to the first process, that of *smelting* it from the ore, which is, for the most part, performed in the usual blast furnaces, from 30 to 40 feet in height, and about 8 feet in their greatest interior diameter, producing the different varieties of *pig metal*.

It has been mentioned to us, that in Missouri this process is sometimes dispensed with, especially when working the ore of the "iron mountain," a rich, heavy, magnetic oxide, of a bluish or iron-grey colour, and of the extraordinary specific gravity of 5.36. The ore is there put into open forge-fires, resembling the Catalan forges of the south of Europe, and by a similar treatment to that which there prevails, brought at once to the condition of malleable iron, without passing through the state of *cast* or *pig* metal.

The process of manufacturing blooms, or as they are, when intended for boiler plate, technically termed, "blocks," is to subject the pig metal of the blast furnace to the combined action of heat and air in an open forge-fire of charcoal, drawing off the melted cinder, or "slag," by a suitable opening ; and after stirring and compacting the iron as it begins to agglutinate, or "come

round to nature," to carry the ball to the heavy forge-hammer, and form it into a prismatic mass, from 15 to 20 inches in length, and from 5 to 9 inches in diameter, according to the weight of the plate intended to be obtained from it. These *blocks* when taken to the rolling mill are heated in an air furnace, supplied generally with bituminous coal as a fuel, and at the first heat are reduced by a heavy hammer into slabs, two or three inches thick, and of a length nearly corresponding with that of the blocks. This operation discharges much of the remaining cinder, and other impurities left in the block by the *bloomery* treatment. At the second process they go to the *rolls* where they are placed first, with the length of the slab corresponding in direction with that of their axis ; secondly, with the length of slab *across* the diameter of the rolls, until it has been increased to the required breadth of the finished sheet ; and, finally, by placing the original length of slab once more parallel to the axis, and extending the plate till it has been reduced to the requisite thickness.

Sheet iron by the process of *puddling*, is, for some purposes, manufactured from pig metal into malleable iron, without the intervention of any other process of refining, than that which takes place in the puddling furnace itself. But for the boiler plate, it is believed to be customary, first to subject it to the action of the "run-out" refinery fire, in an open charcoal, or coke furnace urged by a powerful blast. As, in this fire, a large mass of metal is melted down at a time, and the cinder drawn off separately, the earthy impurities which in simple puddling would be retained in the balls, are at once removed, and by the aid of a small stream of water which is occasionally made to accompany the blast, a partial decarbonization of the metal is probably effected. When in full fusion, the metal is drawn off or "run out" into an oblong bed, and while still hot, is broken up into blocks of a few pounds weight each, to be conveyed to the puddling surface. In the latter it undergoes a second fusion, and the usual operation till agglutinated into "balls." Bituminous coal is the fuel here employed, and the furnace is of the reverberatory form. The balls pass from this furnace first to the large hammer, by which they are moderately compacted ; and immediately after to the rolls by which they are reduced into broad bars or slabs. The latter are reheated and at once rolled into plate, the former cut up into lengths of about 15 or 18 inches and piled, three high, to be reheated and welded into slabs of sufficient magnitude for plates of boiler iron.

Filed iron, when manufactured into blooms, does not, generally, it is believed, undergo increased hammering after being received at the rolling-mill. At the first heat it is reduced to bars an inch or more in thickness, when it is cut up, piled as before mentioned, and rolled into plate.

The practice of *piling* appears to be followed in some instances, from a supposition that a greater security from flaws and other blemishes, must result from combining the strength of three distinct laminæ, and fortifying the weak points of one by the strong parts of the two others, than could probably be derived from the simple unlaminated sheet, in which any imperfection would, it is supposed, extend through the entire thickness.—But the uncertainty of uniform welding between the members of a pile is sufficient to warrant some hesitation in approving this method. Table XXXIII., at experiments 12 and 13, affords evidence that the structure of this description of boiler iron is sometimes exceedingly imperfect, owing to a want of complete welding between the laminæ of which it is composed.

In a subsequent part of this report, will be found discussions on the relative influence of the different processes above described, and also on the *repetitions of piling* upon tenacity. It will there be seen that the practice of piling or *fagotting* may not in all cases prove detrimental to the iron, but will depend in some measure upon the degree of refining which it has previously undergone,—and its consequent freedom from earthy or other impurities which might interfere with accurate welding, as well as upon the temperature employed for that process. If, after reducing either *refinery blooms* or *puddled balls* to thin bars, fit for piling, there be made a perfect union of surfaces during this operation, the latter has evidently the advantage of affording to the impurities a more ready escape from interior portions of the metal, than would otherwise be obtained.

Strength of Iron made by other processes than rolling into Plates.—The tables numbered from LIX to LXXVIII inclusive, (*Frank. Jour.*, vol. 19, p. 410 to 449,) will be found to contain the results of experiments on various specimens of iron manufactured by other processes than rolling into boiler-plate, particularly those of hammering into bars, slitting into rods, rolling into bolts and drawing into wires.

In the number of specimens here tried the committee have included a few of foreign iron, Russian, Swedish and English, as well with a view to compare the results of their method of trial with those of former experimenters, as to show how far the processes generally adopted in manufacturing the article in this country may admit of improvement.

A few experiments on boiler-iron, made upon original or on filed sections, will be found in table LIX, and a small number of trials on cast iron, which does not, however, appear to have been of a very favourable character. Table LX also contains accounts of a miscellaneous collection of specimens obtained from different quarters. The remaining tables, in this series relate to bars which had been reduced to approximate uniformity of size throughout their whole length, and not a few of them were tried at elevated as well as ordinary temperatures; but of the former we shall speak more at length in a subsequent part of this report.

Results of experiments on wrought iron not rolled into Plate.—Among the facts disclosed in this range of results, those repeating English cable-bolt iron, are given in tables LXI, LXII and LXIII. But much of the matter in the first two of these tables, refers to the influence of high temperatures. They, however furnish a mean result, on bar 214, for the strength in the cold state, of 57987 pounds to the square inch. On bar 212, an experiment, on a deeply filed section, gave 59975, while two trials on 213 gave a mean of 59351 pounds. Hence the mean of these three results, viz.: 59105 pounds, represents the strength of the best English cable-bolt iron under ordinary circumstances. Table LXIII presents the results on two portions of the above iron, one cut from bar 213, the other from 214, and on these portions, the effect of *hammer-hardening* was tried. The bars when drawn out under the hammer, previous to being filed down, were hammered until nearly, or quite cold. It will be seen that the lowest result on these two specimens was 65718, the highest 75045, and the mean of eight trials, 71000 pounds. From this statement, it is apparent that the process applied augments, very sensibly, the tenacity of the material; for the lowest result in this table, is 5743 pounds, or 9.5 per cent. above the highest of the three just detailed, as given by the metal in its ordinary state; while the mean of the hammer-hardened specimens, is 11282 pounds or 19.2 per cent. above the mean strength of those which had only hammered out in the ordinary way, and left to cool off from a red heat without the simultaneous application of any mechanical action.

Table LXIV contains the experiments on a specimen of wire, about one-third of an inch in diameter. The maximum strength at 50°, is 88354 pounds per square inch, the minimum, 72325, (the latter being on a part annealed before trial,) and the mean of all the trials 81387.

Experiments 3, 4, 5, and 6, on this wire, gave results so nearly identical, that we may perhaps more properly assume their mean as its true average strength, equal to 84186 pounds per square inch, at from 60 to 66 degrees Fahrenheit. From this mean the diminution by annealing, is 14 per cent.

By the first five experiments, in table LXXV, it appears that the strength of Russian bar iron, at ordinary temperatures, is 76069 pounds per square inch. It will be perceived that the specific gravity of this specimen is considerably higher than that of most other samples of metal, which we have examined. Its superiority in point of tenacity is, probably, attributable, in a great degree, to the refining process to which it had been subjected. The fracture was of a peculiarly fine, fibrous appearance, and had generally a tolerably regular bevel or chisel edge, across the thickness of the bar.

In the tables numbered from LXV to LXIX, will be found an account of experiments on five bars of iron, manufactured in Missouri; and in those numbered from LXX to LXXIV, are recorded the operations on the same number of bars made near Nashville in Tennessee. In the case of these, as well as other bars on which some of the the trials were marked as at elevated temperatures, the fractures often took place at points so remote from the source of heat, that the results really belong to "ordinary temperatures." Including fractures made under the circumstances just alluded to, the number of results obtained at those temperatures, on the Missouri iron, is 22, and the mean strength 47909 pounds per square

inch. On the Tennessee iron, were made, under similar circumstances, 21 experiments, giving a mean strength of 52099 pounds. The Missouri bars appeared to possess a course fibrous structure, and were judged to have undergone but little refining in bringing the metal to a malleable state.

Table LXXVIII. exhibits the tenacity of iron manufactured by Messrs. Grubb, of Lancaster county, Pennsylvania, as 58661 pounds per square inch.

While on this subject, we may refer to some following tables of experiments on iron from Salisbury, Connecticut, manufactured from different sorts of pig-metal, reserving, however, the particular discussion of those tables to a subsequent section of this report. It will be found on inspecting table LXXIX. that forty experiments, at comparable temperatures, were made on the materials from that quarter, the mean result of which is a strength of 58009 pounds per square inch.

In connexion with the present topic, may also be mentioned the result of experiments on specimens of iron manufactured in Centre County, Pennsylvania, an account of which will be found in table XCVII. The mean strength of three bars, as given by 15 experiments, is 58400 pounds. Table CII. includes, among others, 10 experiments, at ordinary temperatures, on Phillipsburg wire of smaller sizes than that already mentioned. Of these, the larger—.19 inch in diameter—will be found to have exhibited, at five trials, a mean strength of 73880 pounds; and that which had a diameter of .156 inch, a strength of 89162 pounds per square inch.

Collecting together the foregoing details, we have for the strength—
Of Missouri bar iron, at ordinary temperatures, by.... 22 exp. 47909 pounds.

Slit rods, (Nos. 180 and 182.)	2	50000
Tennessee bar	21	52099
Salisbury, Conn.	40	58009
Swedish bar	4	58184
Centre Co., Pa.	15	58400
Lancaster Co., Pa.	2	58661
English Cable iron	5	59105
Do. hammer hardened	8	71000
Russian bar	5	76069
Phillipsburg wire, diam. ⎰ .333	13	84186
.190	5	73880
⎱ .156	5	89162
Cast steel, (Table LIX.) specimen	1	130681

Strength of Iron made from different sorts of Pig-metal.—The experiments to determine the effect of different kinds of pig-metal, either separate or in mixture, when converted into wrought iron, by the same refining process, were performed on bars furnished by the Salisbury Iron Company, of Salisbury, Connecticut, of which one specimen, from which were formed the two bars, 208 A., 218 B., was produced from lively gray pig; 219 A., and 219 B., were from a

specimen formed from lively gray pig; 220 A., and 220 B., from mottled pig; 221 A., and 221 B., from white pig; and 222 A., and 222 B., from a mixture of all these kinds together. By a reference to the tables (from LXXX. to LXXXVIII. inclusive, Frank. Jour., vol. x. pp. 3 to 23) containing details of the trials upon these bars, it will be seen that on all of them, some experiments were made at high temperatures, and of course that the purpose of the present comparison can be properly accomplished, only by referring to those, which were made at, or near the same temperature. The experiments at ordinary temperatures embracing the mean strength, as well as the irregularities of structure, are preferred as most satisfactory in reference to this point. In presenting these results, care has been taken to exclude all those trials in which the effect of heat would be appreciable, either during or subsequent to the time of trial.

It will be seen that if we take into view all the bars of this iron, on which experiments were made after they were reduced to a uniform size, and exclude only 219 B., on which the sections were all deeply filed, the advantage will appear to be in favour of the metal manufactured from *white* pig; next to which, is that produced from *lively gray*, giving 98¼ per cent. of the strength of the first. Next, in the order of strength, will be found the iron from *dead gray pig*, inferior to the first by 1 2-3 per cent. ; next, that from the mixture of the four kinds of pig, which appears to have been weaker than the same by 4 4-10 per cent. ; and, finally, that from mottled pig, in which the inferiority extended to 5 per cent. The following table (LXXIX., *Frank. Jour.*, vol. xx., p. 23) exhibits, at a view, the comparative strength, and the respective degrees of uniformity of the several bars, with the strength of some of them at high temperatures.

At elevated temperatures, the results, except that on No. 219 B, are much nearer to each other, than those at the points selected for our general comparison. On that bar, the trials were upon filed sections. The experiment at 573°, giving a strength of 66620, exceeded those at corresponding temperatures on the other bars, by an average of about 6222 lbs., or 10¼ per cent. ; while the two experiments which were made upon it at low temperatures, as will be seen by table LXXXII., gave results, the mean of which being 66724 lbs., surpasses that of the other nine bars by 9275 lbs., or by 16 1-10 per cent. Hence we are compelled to believe that this specimen, as it came to hand, had undergone the process of hammer-hardening,—a process which the direct experiments of the committee have proved to be capable of essentially modifying the tenacity of the metal.

From the above, it appears, that the greatest difference of strength which under ordinary circumstances, can be attributed to differences in the pig-metal[*] from which

wrought iron is produced, is about 5 per cent., and that under every mode of trial, the article formed from a mixture of different kinds of pig, is inferior in tenacity and uniformity to those derived from either of the ingredients, unless we except that from *mottled gray*. And even this latter will, on a comparison of all the experiments made upon it, under every circumstance, be found superior to the bars from *mixed* castings.

If we take into the amount 219 B., the order of values, beginning with the highest, will be *lively gray, white, dead gray, mixed pigs, mottled;* and if we arrange them in the order of their values, as deduced from a comparison of *all the experiments*, on each kind of iron, with the number of trials made on each, we have :—1. *lively gray* 15 experiments.—2. *white* 27 experiments.—3. *mottled gray* 36 experiments.—4. *dead gray* 21 experiments.—5. *mixed metals* 31 experiments.

So far as these experiments may be considered decisive of the question, they favour the lighter complexion of the cast metal, in preference to the darker and mottled varieties, and they place the mixture of different sorts, among the worst modifications of the materials to be used, where the object is mere tenacity.

Effect of high temperature on Iron.—The experiments on bars of iron at high temperatures, were made either on sections deeply filed, or on those specimens which had been reduced by filing to a uniform size.

The trials below 600° were chiefly conducted in a bath of oil, arranged round the bar as already represented in Plates III. and IV., and the temperatures marked by the mercurial thermometer. For temperatures above that point the bath of tin and lead was substituted, and, when necessary, the steam pyrometer took the place of the common thermometer.

The view already presented of the influence of heat on copper, indicated partly by each of these two instruments, has enabled us to observe that they connect themselves in their indications in a manner to prove that no serious errors can be anticipated in the temperatures assigned in the higher parts of the scale when operating on iron.

If, however, in examining the effect of temperature on copper we meet with some difficulties in consequence of the irregularities of structure in the material, of want of conformity in different bars, and of the occasional weakening effects of alloying, on the total tenacity as we approach a red heat, the obstacles there encountered are comparatively trifling, when contrasted with those which are to be surmounted in the investi-

---

[*] It will be understood that this remark does not apply to pig-iron, contaminated with sulphur, phosphorus, copper, or other similar impurities ; but only to such as contain different proportions of the *ordinary* ingredients.

gation of the effects of heat upon the tenacity of iron. Here we have, not only the variations due to the original composition of the metal; the differences resulting from the variety of pig-metal used in its manufacture, and the defects of the mechanical structure, owing to the want of uniformity in welding, or of regularity in the temperature of working the bars; but we have superadded to all these, a singular anomaly in the effect of heat itself on the tenacity of this material, which is believed never to have been before made the object of special inquiry.

Notwithstanding these impediments, the committee have not felt authorized to leave so important a point of inquiry, without a faithful attempt to unravel its intricacies. It would have been easy to devise a set of experiments, which, for a theoretical purpose, might have afforded the analyst some interesting problems, and probably served to clear the subject of heat from certain difficulties with which its investigation is encumbered. Such, however, was not the purpose in view of the committee.

When we attempt to form a scale of the weakening effects of elevated temperatures, founded, as in the case of copper, on trials at ordinary temperatures, or even at the freezing point, we shall find that many of the first numbers in the scale will be negative, instead of positive, and this will continue to different points of temperature, according to the nature or condition of the iron on which the experiments are made. In fact, some of the very first experiments at high temperatures rendered this manifest, by showing that on a bar of uniform size, the fracture would not take place within the heating bath; and even that *much* filing of the part in the oil or melted metal, was necessary in order to prevent the fracture from taking place at unfiled sections *out of the hot bath* rather than at the filed one in it. This circumstance was noted at 212°, 392°, and 572°, rising by steps of 180° each from 32°, at which last point some trials had been made in melting ice. At the highest of these points, however, it was perceived that some specimens of the metal exhibited but little, if any, superiority of strength over that which they had possessed when cold, while others allowed of being heated nearly to the boiling point of mercury before they manifested any decided indication of a weakening effect from increase of temperature.

It hence became apparent that any law, taking for a basis the strength of iron in its ordinary condition, and at common temperatures, must be liable to great uncertainty, in regard to its application to different specimens of the metal. It was evident that the anomaly above referred to, must be only apparent, and that the tenacity actually exhibited at 572°, as well as that which prevails while the iron is in the state in which it was left by forging, or rolling, must be below its maximum tenacity. To determine what ratio exists between the ordinary strength of a bar and its maximum strength when in the most favourable condition for resting a longitudinal strain, experiments were made on several bars by heating them to 572°, and then applying weight enough to cause a fracture, either within or without the heated part. The bar was then taken out and allowed to cool, when the strength which was obtained on parts influenced by the heat became a standard of comparison for experiments at more elevated temperatures. A mean of thirty-five comparisons, conducted in the manner just described, afforded a standard 16.2 per cent. greater than the ordinary strength of the metal: but the standard most relied on for furnishing the basis of calculations, and for determining a law of diminution of tenacity, was derived from the five varieties of iron, manufactured by the Salisbury Iron Company, which, being of a tolerably uniform texture, were considered rather more suitable than others for supplying the ground work of a law for calculating the effect of temperature on this metal generally. An examination of the trials on those bars will be found to furnish a standard of maximum tenacity 15.17 per cent. greater than their mean strength when tried cold. When, however, an unexceptionable standard was given by any bar after trial at 572° and subsequent cooling off, its own standard for increased strength was used in computing the true effect of heat at other high temperatures.

Thus, at a temperature of 1317°, the bar No. 226, which had possessed, when cold, a strength of 54758 lbs., gave a remaining strength of only 18913 lbs. Now, 54758 lbs. increased 15.17 of itself, gives 63065, and from this deducting 18913 we have 44152 lbs. for the *diminution* of its absolute tenacity by the temperature just mentioned, or .7001 of the maximum strength.

On the same bar, (No. 226,) were made at different points, two other experiments with the same weight each time in the scale. The first of these sections gave way when the temperature had reached 1237°. The strength per square inch given in this case was 21298, and comparing this with the maximum strength, 63065, we obtain 41767 as the diminution, equal to .6622 of that maximum.

The second trial on a larger area of section required a higher temperature to cause the fracture to take place under the given weight, viz.: 1245°, given at this temperature a tenacity of 20709lbs., and by the same computation showing a diminution from the maximum 63065 of .6715. Both of these trials having been made with the precaution of raising and lowering the suspended furnace, to regulate the heat, it is believed that no essential error in regard to temperature can have existed. The first was conjectured to be, if anything, a trifle in excess.

If we take the mean of these wo results, viz. .6668 for the diminution of tenacity at 1241° the mean temperature, it cannot vary far from the true effect. On bar 227 an experiment was made at 1187°, giving a tenacity of 21913lbs. per square inch. Within two and a half inches of the same point a cold fracture gave a strength of 52186 lbs., from which the calculated maximum is 60102, and the diminution is 60102—21913, or 38189 ; which is .6352 of the same maximum tenacity.

On No. 229 was made an experiment at 1159°, which exhibited a tenacity of 25620 lbs. Three experiments on the same bar when cold, gave a mean strength of 55774lbs. Hence 55774 × .1517 = 8460 ; and (55774 + 8460)—25620 = 38614, which is .6011 of the maximum tenacity.

On No. 227 we have an experiment at 1155°, giving a tenacity of 21967, and the four cold experiments nearest to the same point give a mean of 47749, from which we obtain the maximum 54992, and the diminution = .6000.

On No. 226 was made an experiment at 1142°, but as the iron at the part in which the fracture took place was defective from flaws, and had probably been impaired by the previous straining of the bar, it was not considered necessary to attempt to reduce its apparent tenacity to the standard, being entirely anomalous.

At 1111° the bar No. 227 had a strength of 27502, and another trial on the same at 1097°, 27502.

The weight in the scale was the same in both cases, and the temperatures would probably have been the same, had not the standard piece in the latter case accidentally risen above the melted lead a short distance just before the fracture. Taking the mean of these two results 27603, for the strength at 1111°, and the mean of six trials on this bar near the two points where these fractures occurred, viz. ; 53426, we obtain the maximum tenacity at those points 61531, and the diminution by heat .5614.

On bar 152 an experiment at 1037° gave a tenacity of 37764, and on No. 214 an experiment at 1022° gave 37410. The mean cold strength of these two bars was 59105, from which we deduce the maximum 68071; and the diminution for the mean temperature 1030°, equal to .4478 of the maximum.

At 947° an experiment on bar No. 232 gave a strength of 42401, the mean of the two experiments subsequently made nearest to this point gives the experimental maximum strength 66193 from which the diminution is .3593.

At 932° bar No. 214 had a tenacity of 45531, while its cold strength was 59319, and its maximum 68202, hence the diminution is .3324.

An experiment was made on bar 149 at a temperature marked 825°, but as the furnace was not lowered during the performance of it, and as the time during which the bar continued to stretch after the strength had been fairly overcome, was considerable, the temperature is in all probability too high ; and the experiment is not considered comparable with the rest of the series.

In bar No. 214, at the temperature of 824°, the remaining strength was 55892, the original strength, 60850 and the maximum by calculation, 70080, whence the diminution is .2010.

On bar 149 was an experiment at 770°, giving a tenacity of 54781.; while the original strength was 56825, and the diminution from the calculated maximum .1627.

On No. 16 we find an experiment at 766°, giving 54819. Two subsequent experiments yielded maxima, the mean of which is 65174; whence the diminution is .1586.

In bar No. 150, a temperature of 734°, left a strength of 57903. The first experiment on the bar afforded 59397, from which we calculate the maximum 68407, which proves the diminution at this temperature to be .1535.

On No. 14 we obtained a strength of 53378, at 732°, and the mean of three experimental maxima, is 62736, hence the diminution by heat is .1491.

On No. 152 we had at 722° a tenacity of 54442, and three experiments gave a cold strength of 55990, from which a calculated maximum of 64483 is obtained, and consequently a diminution of .1547. But an experimental maximum of 62799 was obtained on this bar, which on account of the remoteness of the point where it occurred, from the point on which the hot fracture was made, is believed to be rather too low. Calculating, however, from this maximum, we find the diminution 1316.

If we take the mean of the two results, .1557 and .1316, we have the probable diminution from the true maximum, .1436.

On No. 150 we find an experiment at 662°, giving a tenacity of 58182. On the same bar an experimental maximum was found of 65785, from which we get the diminution equal to .1155.

On No. 16 was made a trial at 636°, yielding a result of 50039, a result far lower than

that given afterwards on the same bar at
700°; we are therefore compelled to believe
that this experiment was made on a defective
part of the bar.

On No. 219 A was a trial at 636°, which
exhibited a tenacity of 66010. An experi-

ment subsequently made within 1½ inches of
the same point, gave a tenacity of 57633,
and consequently the diminution is .1045.

On No. 90 an experiment at 606° gave a
tenacity of 56938, and three experiments on
the same bar, when cold, gave a mean of

Plate X.

Observed diminutions of Tenacity.

54715, from which the calculated maximum
is 63015, and the diminution .0964.

On the same bar (No. 90) another trial
took place at 596°, giving a strength of
57682 lbs., and if we assume the original
strength of this section equal to that given
by the third experiment on the same bar,

55037, we shall have the maximum by calculation 63386, and the diminution .0899.

By a mean of 5 sets of experimental
maxima derived from 65 trials on the 5
varieties of Salisbury iron we have a standard
of 66146. The six trials at the mean temperature of 570° referred to in our remarks

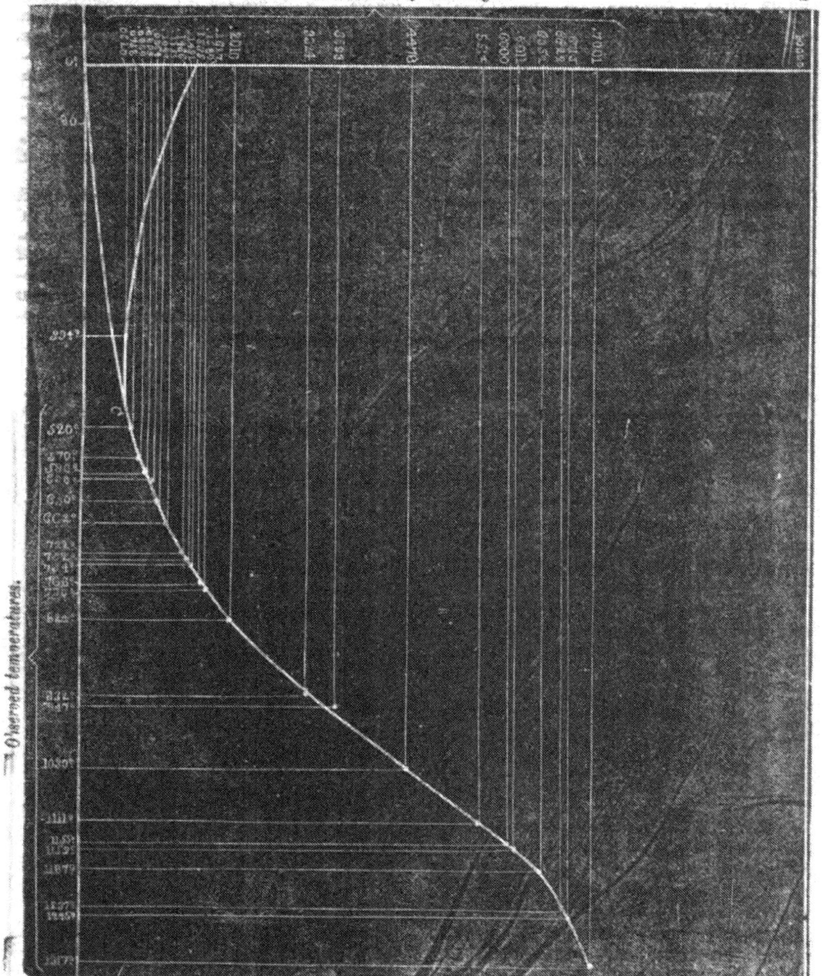

in Table LXXIX. of the effect of employing different kinds of pig-metal, show that at a mean temperature of 570° those trials gave a strength of 60398 lbs., whence the diminution is .0869.

Of 224 B, at 520° the tenacity was 58451. On the same bar, four cold experiments gave a mean strength of 54934, which by calculation gives a maximum of 63267, and a consequent diminution of .0761.

On a survey of the preceding discussions it will be seen that in determining the maximum belonging to each point of fracture, it has been necessary to resort sometimes to experimental, and sometimes to calculated results, but that in several cases the two operate as checks upon each other.

On attempting to extend the principle to trials made below the temperatures already cited, we are liable to encounter an ambiguity in the results, owing to the fact that the maximum tenacity is not generally to be obtained without having carried the previous temperatures to about 550° or 600°, and the tension to nearly or quite that of the original strength of the metal when cold.

In projecting into a curve as in Plate X. the data furnished by the experiments above described, and of which a synopsis is given in the following table, it becomes at once apparent that what was *conjectured* with respect to copper, in regard to a point of inflection, is here presented in a manner to admit of no uncertainty. Indeed it could hardly be otherwise, when we consider that the melting point of wrought iron, at which all tenacity must be overcome, is doubtless situated above 3000°; and by the experiments of Clement and Desormes, is as high as 3945°. Now it appears that at a temperature no higher than about 1050° one-half of the strength is destroyed; at 1240°, two-thirds; and at 1317°, seven-tenths of the maximum tenacity is overcome.

The following table exhibits the observed temperatures, and corresponding tenacity of the metal with the calculated, or experimental maximum of strength,—the ratio of the observed diminution to the maximum tenacity, and the irregularity of the metal in parts of the original strength at ordinary temperatures.

TABLE XC.

No. of the comparison.	Marks of the bar.	Temperature observed.	Tenacity observed.	Maximum tenacity at the point of fracture.	Manner in which the maximum was obtained.	Diminution by heat in parts of the maximum tenacity.	Irregularity of the metal in parts of the original strength.
1	224 B.	520°	58451	63275	Experiment	.0738	.0992
2	Salisb. iron	570	60398	60398	do.	.0869	.1125
3	90	596	57682	57682	Calculation	.0899	.2401
4	90	600	56938	63086	do.	.0964	.2401
5	219 A.	630	60010	67033	Experiment	.1047	.1440
6	150	662	58182	65785	do.	.1155	.0644
7	152	722	54442	64483	Calculation	.1436	.0507
8	14	732	53378	62736	Experiment	.1491	.1310
9	150	734	57903	68407	Calculation	.1535	.0644
10	16	766	54819	65176	Experiment	.1589	.1563
11	149	770	54781	65445	Calculation	.1627	.0234
12	214	824	55892	70080	do.	.2010	.0413
13	214	932	45531	68202	do.	.3324	.0413
14	232	947	42401	66193	Experiment	.3593	.0446
15	{ 214 152 }	1030	37587	68071	Calculation	.4478	.0460
16	227	1111	27603	61531	do.	.5514	.0330
17	227	1155	21967	54992	do.	.6000	.0330
18	229	1159	25620	64234	do.	.6011	.1102
19	227	1187	21913	60102	do.	.6352	.0330
20	226	1237	21298	63065	do.	.6622	.1147
21	226	1245	20703	63065	do.	.6715	.1147
22	226	1317	18913	63065	do.	.7001	.1147

From the eighth column of the preceding table, it appears that of these 15 different specimens of iron, the mean irregularity of structure is 10 per cent. of the mean strength when tried cold.

For the purpose of ascertaining, approximately, the law of decrease in strength by temperature, an investigation was made similar to that adopted for copper, embracing, however, only 12 of the points contained in the preceding table.

(To be continued.)

PROGRESS OF THE THAMES TUNNEL.

A meeting of the Proprietors of the Thames Tunnel took place on the 5th instant, at which the Directors " reported progress," both as regards works and pecuniary matters. Upon the latter point we cannot trust ourselves to say a word; and upon the former we must allow the " Report" to speak for itself. After a few general prefatory observations, like in substance to many of those which have preceded it, lamenting difficulties, and expressing certainty of their being overcome, citing the approbation of the Duke of Wellington, Earl Grey, and other members of successive Governments in the scheme, they proceed :—

" It had been found by experience that the attempt to continue the work of excavation when the soil was peculiarly loose and disturbed was disadvantageous, inasmuch as the cost of the work was enhanced without a proportionate progress, and that an occasional suspension of this portion of the work was even favourable to its ultimate progress, inasmuch as during that time the ground became naturally consolidated, and more easily and safely excavated. In order however to prevent any loss to the Company, whenever it was necessary to stop the advance of the shield, Mr. Brunel proposed to commence the work on the north side of the river, with a view to ensure employment for the men above, when their labours were suspended below ground. Thus, though the work was to be suspended in one part, it was intended to be progressing in another, and the whole establishment was to be kept in constant employment, to the manifest advantage of the public and the Proprietors.

" This plan was, however, abandoned with regret, upon its being considered to be inconsistent with the condition under which the public money was granted, viz., that " the most hazardous part of the work" should be first finished before any other portion was commenced.

" Your Directors, therefore, after the most careful and anxious deliberation, determined to adopt the second plan proposed by Mr. Brunel, and which was in partial operation at the period of the last General Meeting. It was not in perfect and general operation for some short time afterwards, and previous to which the river once more forced its way into the works. The delay, however, on this as on former occasions was inconsiderable, and the week after the irruption the work proceeded.

" This plan consisted of three principal features, viz :—

" 1st, To divert the navigation from that part of the river immediately over the mining operations.

" 2ndly, To gain the command of that part of the river, without interruption, and to be thus enabled to load and cover its bed, both over the works in progress and in advance of them ; and to compress this artificial bed, directly over the shield, by grounding upon it, at every fall of the tide, a vessel when ballasted, of about 900 tons burthen. And,

" 3dly, To make alterations in the auxiliary parts of the shield, still further to add to its security and power.

" Your Directors are now enabled to report the decided success which has attended this plan throughout the past year, and which has indeed exceeded their most sanguine anticipations.

" The brickwork of the tunnel has been advanced, since the last meeting, 90 feet, and is now within 60[*] feet of low water mark ; and if the same rate of progress continues, which there is every reason to expect, low water mark will be reached in the course of the autumn.

" It will be clear to those who are best acquainted with the work, that when this is accomplished the most hazardous portion of the tunnel will be completed ; and that however novel, and even bold, the work which then remains to be done, in order to realise the original design, yet its completion becomes comparatively safe and easy, and calculable within a reasonable time.

" Your Directors, in their report of last year, acknowledged the great and liberal assistance afforded by the Board of Admiralty on very many occasions ; and the generous and cordial assistance they had at all times received from the Corporation of the City of London, the Navigation Committee, and Harbour Masters, and which the Directors continue to receive. These authorities enabled the engineer to, fully and entirely to

---

* Since this Report was read to the General Meeting, five feet have been excavated ; the distance to low water is therefore only 55 feet.

his satisfaction, to perfect and enter upon
the plan he was constrained to follow, and
which was in strict accordance with the
Treasury Minute.

" Thus the Directors have now the gratification to meet the Proprietors, after so
many years of anxiety and delay, with the
most reasonable hope, of having but one
more Annual Meeting intervene between the
speedy termination of all the peculiar
hazards and difficulties of the undertaking ;
and such an advance of the works as shall
bring them near to the time when their
anxious trust will be discharged, by the final
completion of this most arduous enter-
prize."

LIST OF ENGLISH PATENTS GRANTED BE-
TWEEN THE 23rd OF FEBRUARY AND
THE 27th OF MARCH, 1839.

George Augustus Kollmann, of the Friary, St.
James's Palace, professor of music, for certain improvements in the mechanism and general construction of pianofortes, being an extension of
former letters patent for the term of seven years.
February 23.

Charles Louis Stanislas Baron Heurteloupe, of
Queen Ann-street, for certain improvements in fire-
arms, and in the balls to be used therewith. February 23 ; six months to specify.

Thomas Pratt, of South Hylton, Durham, mechanic, for an improved capstan and winch for purchasing or raising ships' anchors, without the application of a messenger, in which there is no fleeting or surging, and for drawing or working of coals
and other articles and things out of coal and other
mines ; and also for the drawing and working en
railroads. by drawing pulleys, with flat or round
ropes. February 23.

James Russell, of Headsworth, Stafford, gas-tube
manufacturer, for certain improvements in manufacturing tubes for gas and other purposes, being an
extension, for the term of six years, of former letters patent, granted to Cornelius Whitehouse, and
assigned by him to the said James Russell. February 26.

Moses Poole, of Lincoln's Inn, gentleman, for
improvements in constructing and applying boxes
to wheels. February 28 ; six months.

Moses Poole, of Lincoln's Inn, gentleman, for
certain improvements in tanning. February 28 ;
six months.

John Leigh, of Manchester, surgeon, for an improved mode of obtaining carbonate of lead, commonly called white lead. February 28 ; six
months.

Richard Whytock, of Edinburgh, manufacturer,
and George Clink, of the same place, colour maker,
for further improvements in the process and apparatus for the production of regular figures or patterns in carpets and other fabrics, in relation to
which a patent was granted to the said Richard
Whytock, on Sept. 8, 1832, and generally in the
mode of producing party colours on yarns or threads
of worsted, cotton, silk, and other fibrous substances. March 1 ; six months.

Morris Platow, of 19 Poland-street, Oxford street,
engineer, for improvements in pumps or engines for
raising or forcing liquids. March 6 ; six months.

John Dickson, of Brook-street, Holborn, engineer, for certain improvements in rotatory steam
engines. March 6 ; six months.

Auguste Victor Joseph Baron d'Asda, of Millman-
street, Bedford-row, for improvements in pro-

ducing or affording light, which be denominates
solar light. March 6 ; six months.

Walter Hancock, of Stratford, in the county of
Essex, engineer, for certain improvements in steam
boilers and condensers. March 6 ; six months.

George Robert d'Harcourt, of Howland-street,
Fitzroy-square, gentleman, for improved artificial granite, stone, marble, or concrete, in which
said invention neither asphaltic nor bituminous
substances are used. March 6 ; six months.

William Vickers, of Birshill, Sheffield, merchant,
for a mode of obtaining tractive power from carriage
wheels under certain circumstances. March 6 ; six
months.

John Clark, of Upper Thames-street, London, engineer, for a new or improved form or construction
of a leg and foot for propelling carriages on rail or
common roads, and a new combination or arrangement of machinery, for locomotive carriages, by
means whereof the weight of the load to be carried
is rendered applicable as a part of the power for
moving or propelling the carriage on which it is
supported or rests. March 6 ; six months.

Charles Schafhautl, of Cornhill, London, gentleman, for an improved method of smelting copper
ore. March 6 ; six months.

Orlando Jones, of Rotherfield-street, Islington,
accountant, for improvements in the manufacture of
starch, and the converting of the refuse arising in
or from such manufacture, to divers useful purposes.
March 6 ; six months.

George Holworthy Palmer, of Surrey-square, Old
Kent-road, C. E., and George Bertie Paterson, of
Hoxton, engineer, for certain improvements in gas
meters. March 6 ; six months.

Thomas Horton, of Princes-end, Stafford, boiler-
maker, and Thomas Smith, of Horseley-heath, in
the same county, mine agent, for certain improvements in the making or constructing of chains for
pits, shafts, mines, or other purposes. March 6 ; six
months.

Edward Ford, of Liverpool, builder, for certain
improvements in conducting the manufacture of
salt cake or sulphate of soda, and hydrochloric or
other acids and alkalies, or other chemical processes
wherein deleterious vapours are given off, and in the
erection of furnaces and works connected therewith.
March 8 ; six months.

Joshua Christopher Gamble, of Saint Helens,
Lancaster, manufacturing chemist, for improvements in apparatus for the manufacture of sulphate
of soda, muriatic acid, chlorine, and chlorides.
March 14 ; four months.

Elisha Haydon Collier, late of Boston, but now
of Globe Dock Factory, Rotherhithe, C. E., for improved machinery for manufacturing nails. March
14 ; six months.

Christopher Nickels, of York-road, Lambeth, for
improvements in the modes of manufacturing of
fabrics from linen, woollen, silk, and other fibrous
materials. March 15 ; six months.

Richard Lamb, of Davids-street, Southwark,
gentleman, for improvements in apparatus for
supplying atmospheric air in the production of
light and heat. March 15 ; six months.

Alexander Francis Campbell, of Gross Stanstead,
Norfolk, Esq., and Charles White, of Norwich, mechanic, for certain improvements in ploughs. March
15 ; six months.

Thomas Henry Ryland, of Birmingham, screw
manufacturer, for an improved manufacture of
screws for wood, in iron, brass, or any mixed metals,
commonly known as "wood screws." March 15 ;
six months.

John Ruthven and Morris West Ruthven, of
Edinburgh, civil engineers, for improvements in
boilers for generating steam, economising fuel, and
propelling vessels by steam or other power, and
ventilating vessels, and which may be applied to
mines and buildings. March 20 ; six months.

Edward Law, of Downton Hook, Kingsland,
gentleman, for certain improvements in various

ing-sea water and other fluids, and in the manufacture of salt: March 20; six months.

Joseph Amesbury, of Burton Crescent, surgeon, for a certain apparatus for the support of the human body. March 20; six months.

Andrew Smith, of Princes-street, Leicester-square, for certain improvements in the manufacture of ropes, for cables and other purposes to which ropes are applicable. March 20; six months.

George Nelson, of Milverton, Warwick, chemist, for a new or improved method, or new or improved methods of preparing gelatine, which has the properties of or resembles glue. March 23; six months.

Thomas Fisher Salter, of Great Hallingbury, Essex, farmer, for an improved machine for winnowing and dressing corn and other grain. March 23; six months.

Richard Roberts, of Manchester, for an improvement or certain improvements of, in, or applicable to the mule billy jenny stretching frame, or any other machine or machines, however designated or named, used in spinning cotton wool or other fibrous substances, and in which either the spindles recede from and approach the rollers or other deliverers of the said fibrous substances, or in which such rollers or deliverers recede from and approach the spindles; being an extension for the term of seven years of former letters patent. March 26.

Henry Montagu Grover, of Boveney, Buckingham, clerk, for improvements in brewing by the use of a material not hitherto so used. March 26; six months.

Joseph Lees, jun., of Manchester, calico printer, for certain improvements in the art of printing calicoes, muslins, and other woven fabrics, and in certain processes connected therewith. March 26; six months.

Edmund Butler Rowley, of Manchester, surgeon, for an improved steam engine applicable to locomotive, marine, and stationary purposes. March 26; six months.

Elisha Hale, of Leadenhall-street, London, for improvements in umbrellas and parasols. March 27; six months.

William Newton, of Chancery-lane, for certain improved machinery for cutting and removing earth, which machinery is applicable to the digging of canals and the levelling of ground for railroads or ordinary roads and similar earth works. March 27; six months.

LIST OF SCOTCH PATENTS GRANTED BETWEEN THE 22nd APRIL AND THE 22nd MAY, 1839.

Alexander Borland, of Paisley, in Renfrew, in Scotland, for a machine for measuring water and other liquids, and registering the quantity thereof. Sealed 23rd February, 1839; four months to specify.

Sir James Caleb Anderson, of Buttevant Castle, Cork, baronet, for certain improvements in locomotive engines, which are partly applicable to other purposes. Feb. 26.

Orlando Jones, of Rotherfield-street, Islington, Middlesex, accountant, for improvements in the manufacture of starch, and the converting of the refuse arising in or from such manufacture to divers useful purposes. Feb. 27.

Frederick le Mesurier, of New-street, Saint Peter's Port, Guernsey, gent., for a certain improvement or certain improvements in the construction of pumps for raising water or other fluids. Feb. 28.

Richard Whytock, of Edinburgh, manufacturer, and George Clink, of the same place, colour-maker, for further improvements in the process and apparatus for the production of regular figures or patterns in carpets, and other fabrics in relation to which a patent was granted to the said Richard

Whytock, on the 8th of September, 1832, and generally in the mode of producing party colours on yarns of worsted, cotton, silk, and other fibrous substances. March 6.

Pierre Armand Lecomte de Fontainemoreau, of Charles-street, City-road, Middlesex, for certain new and improved metallic alloys to be used in various cases as substitutes for zinc, cast iron, copper, and other metals, being a communication from a foreigner residing abroad. March 8.

Benjamin Goodfellow, of Hyde, Chester, mechanic, for certain improvements in metallic pistons. March 8.

John Hawkshaw, of Manchester, Lancaster, C.E., for certain improvements in mechanism or apparatus applicable to railways, and also to carriages to be used thereon. March 8.

John Muir, jun., merchant, Glasgow, for certain improvements in the apparatus connected with the discharging press for conducting, distributing, and applying the discharging liquors and the dyeing liquors. March 11.

Thomas Vaux, of Woodford Bridge, Essex, land surveyor, for improvements in tilling and fertilising land. March 13.

Alexander Croll, of Greenwich, Kent, chemist, for improvements in the manufacture of gas for the purpose of affording light. March 13.

Moses Poole, of Lincoln's Inn, Middlesex, gent., in consequence of a communication from a foreigner residing abroad, certain improvements in tanning. March 13.

Henry Ross, of Leicester, worsted manufacturer, for improvements in machinery for combing and drawing wool, and certain description of hair. March 13.

James Walton, of Sowerby-bridge, York, cloth dresser and frizer, for certain improvements in machinery for making wire cards. March 13.

Henry Huntley Mohun, of Regent's Park, Middlesex, M.D., for improvements in the composition and manufacture of fuel, and in furnaces for the consumption of such and other kinds of fuel. March 13.

Josias Christopher Gamble, of Saint Helens, Lancaster, manufacturer, for improvements in apparatus for the manufacture of sulphate of soda, muriatic acid, chlorine and chlorides. March 13.

James Russell, of Handsworth, Stafford, gastube manufacturer, assignee of Cornelius Whitehouse, of Wednesbury, for an extension of six years from May 26, 1839, of a patent granted to the said Cornelius Whitehouse for an invention of certain improvements in manufacturing tubes for gas and other purposes. March 15.

Joseph Rayner and Joseph Whitehead Rayner, late of Birmingham, Warwick, but now of the city of Coventry, C.E., and Henry Samuel Rayner, of Ripley, Derby, C.E., for divers new and important improvements in machinery for roving, spinning, and twisting cotton, flax, silk, wool, and other fibrous materials. March 15.

John Leigh, of Manchester, Lancaster, surgeon, for an improved mode of obtaining carbonate of lead, commonly called white lead. March 18.

Samuel Clegg, of Sidmouth-street, Gray's Inn Road, Middlesex, engineer, for a new improvement in valves and in the combination of them with machinery. March 18.

Joseph Bennet, of Tumlee, near Glossop, Derby, cotton spinner, for certain improvements in the machinery for carding, drawing, slubbing, roving, and spinning silk, wool, cotton, worsted, flax and other fibrous substances, which improvements are also applicable to other useful purposes. March 20.

John Robinson, of North Shields, Northumberland, engineer, for an invention of a nipping lever for causing the rotation of wheels, shafts, or cylinders, under certain circumstances. March 22.

LIST OF IRISH PATENTS GRANTED IN FEBRUARY, 1839.

Thomas R. Williams, for certain improvements in machinery for spinning, twisting, or curling and weaving horse hair, and other hairs, as small as various fibrous substances. Feb. 1.

William Brindley, for improved arrangements in the construction of screw presses. Feb. 16.

Robert Logan, for a new cloth or cloths constructed from cocoa-nut fibre, and certain improvements in preparing such fibrous materials for the same and other purposes. Feb. 22.

W. Thorp, and Thomas Meakin, for certain improvements in looms for weaving, and also a new description of fabric to be produced or woven therein. Feb. 22.

NOTES AND NOTICES.

Soldering of Lead.—A new method of soldering lead together has been invented by a M. Desbassyers de Richemont, consisting of a portable apparatus, which he calls aerhydric pipes. From these issue the most brilliant and intense flames, which rapidly melt the lead, and strike each part at the same time; the liquid lead may be pushed away with the flame if it should run too far, and portions of fresh lead may be applied to effect the soldering, in case it should be impossible to borrow any from the neighbouring parts.

Submarine Explosions by Electricity.—An experiment was made with complete success with one of Daniel's galvanic batteries, under the superintendence of Colonel Pasley, of the Royal Engineers, at half-past two o'clock last Saturday, off the gun wharf, Chatham. 35 lbs. of powder were exploded in about 10 fathoms of water, the length of the wire conveying the electric fluid being 500 feet; it caused a most tremendous explosion. Three smaller ones were afterwards tried, but only one succeeded; there was a numerous assemblage of spectators. The *Royal George*, at Portsmouth, we understand is to be blown up in a similar manner, and this experiment was preparatory to the attempt.—*Maidstone Journal.*

A Wooden Country.—Wood (in America) is not only used with prodigality for all the purposes to which it is necessarily applied, but is substituted in numberless instances for substances which, under other circumstances, would have been more suitable. Not to speak of wooden houses, bridges, and roads, of wood for fuel and fencing; we find it adopted in the west for purposes more anomalous, where wooden pins are substituted for nails, and wells are curbed with hollow logs, where the cabin door swinging on wooden hinges, is fastened with a wooden latch, and the smoke escapes through a wooden chimney. Engineers have proposed to substitute wood-work for masonry in the construction of railways and canal locks; and it is said, that an eminent lawyer in Missouri had a very convenient office, made of a single section sawed from a hollow sycamore. Well may ours be called *a wooden country.*—*Hall's Notes.*

Wood.—The experiments of M. Payen have led him to the conclusion, that the ligneous body so universally existing in phanerogamous vegetables, is not an immediate principle of vegetation, but that it is composed of two parts, chemically distinct. Having obtained the cellular tissue in its earliest state, from various ovula, and the radicles, or radicle fibrils of several plants, he only found in it various combinations of carbon, hydrogen, and oxygen, and consequently it is not truly ligneous, but that the thickening substance in the interior of the fibrous cells, is operated on by agents which have no effect on the elementary tissue, such as soda, potash, and azotic acid. Remarkable differences take place in the composition of woods, according to their species, and the same species according to climates. Hence the proportion of carbon relative to that of hydrogen and oxygen, the predominance of hydrogen over oxygen in the strongest woods. In combustion, an excess of hydrogen tends to the production of heat, and offers a reason for preferring what are called heavy woods, with the exception of the birch, which owes its superiority to a principle named betuline.—*Athenæum.*

Blood Painting.—Sir John Robison, of Edinburgh, has communicated to the Académie des Sciences a curious fact, which he considers may possibly have some relation to the discovery of M. Daguerre. A medical man, having occasion to bleed a patient, caught the blood in a porcelain basin, at the bottom of which was painted a bouquet of flowers. Some time after, the blood having become coagulated, he was about to throw it away, when, on removing the clot, he observed on the surface which had been in contact with the bottom of the basin, a perfectly distinct design in bright red of such parts of the bouquet as consisted of green colour, the leaves, &c. Sir John has since repeated the experiment several times, and always with the same result.—*Athenæum.*

Mr. Crosse's Electrical Experiments.—Tuesday week Mr. Andrew Crosse delivered a lecture on atmospheric electricity, at Taunton, illustrated by a number of beautiful experiments. He illuminated 400 feet of iron chain, hung in festoons about the room, the whole extent being brilliantly lighted at the same instant by the passage through it from the spark from the battery, and melted several feet of wire. Mr. Crosse afterwards detailed the results of many experiments on thunder clouds and mists. By means of a wire apparatus suspended in his park, he had discovered that a driving fog sweeps in masses, alternately, negatively and positively electrified; and once the accumulation of the electric fluid in a fog was so great, that there was an incessant stream from his conductor of sparks, each one of which would have struck an elephant dead in an instant.—*Times.*

New Railway Carriage.—The other day we inspected a new railway carriage, of what is usually called the third class, having neither roof nor enclosed sides, which is destined for the Manchester and Leeds Railway. It is 17 feet 10½ inches in length, and 7 feet 11 inches and a half in width. The form of the carriage is not so square as those hitherto used, but more nearly resembling the form of a long boat, with the stem and stern cut off square. A bench seat extends the whole length of the carriage on each side, and down the middle is another broad bench, 27 inches and a half in width, divided in the middle by an open rail, or back of wood, rising to a height of 14 inches from the seat, so as to form two benches, on which the passengers sit back to back. Allowing 14 inches to each passenger, the carriage would seat about sixty persons. Ascent is had by two broad iron foot plates, at each corner of the carriage, so that there are four doors, affording ready ingress and egress. An iron rail extends along the sides of the carriage, so as to prevent any thing from falling over. The exterior of the carriage is painted an olive green, and is formed into pannels. The whole has a neat appearance, and is capable of accommodating a greater number of passengers than any carriage we have seen of equal dimensions.—*Manchester Guardian.*

END OF THE THIRTIETH VOLUME.

Lightning Source UK Ltd.
Milton Keynes UK
UKHW022023081121
393637UK00003B/196